KWÄDAY DÄN TS'ÌNCHI

TEACHINGS FROM LONG AGO PERSON FOUND

KWÄDĄY DÄN TS'ÌNCHÌ

TEACHINGS FROM LONG AGO PERSON FOUND

Edited by RICHARD J. HEBDA, SHEILA GREER AND ALEXANDER P. MACKIE

ROYAL **BC** MUSEUM
VICTORIA, CANADA

Champagne and Aishihik First Nations

Published by the Royal BC Museum, 675 Belleville Street, Victoria, British Columbia, V8W 9W2, Canada.

Pages and cover designed and produced by Lime Design Inc.

Typeset in Bunday Sans and Chaparral Pro.

Printed in Canada by Friesens.

MIX
Paper from responsible sources
FSC® C016245
www.fsc.org

Cover image: Canadian Museum of History, Photograph NO. 98065. See page 168.

Back cover image: Sarah Gaunt (CAFN) photograph. See page 72.

Library and Archives Canada Cataloguing in Publication

Kwäday Dän Ts'ìnchi : teachings from Long Ago Person Found / edited by Richard J. Hebda, Sheila Greer, Alexander Mackie.

Includes bibliographical references and index.

ISBN 978-0-7726-6699-4 (softcover)

1. Ice mummies—British Columbia—Tatshenshini-Alsek National Park. 2. Human remains (Archaeology)—British Columbia—Tatshenshini-Alsek National Park. 3. Excavations (Archaeology)—British Columbia—Tatshenshini-Alsek National Park. Tatshenshini-Alsek National Park (B.C.)—Antiquities. I. Greer, Sheila, editor II. Hebda, Richard J. (Richard Joseph), 1950–, editor III. Mackie, Alexander (Alexander P.), editor IV. Royal BC Museum, issuing body

E78.B9K83 2017   971.1'00497   C2017-905947-5

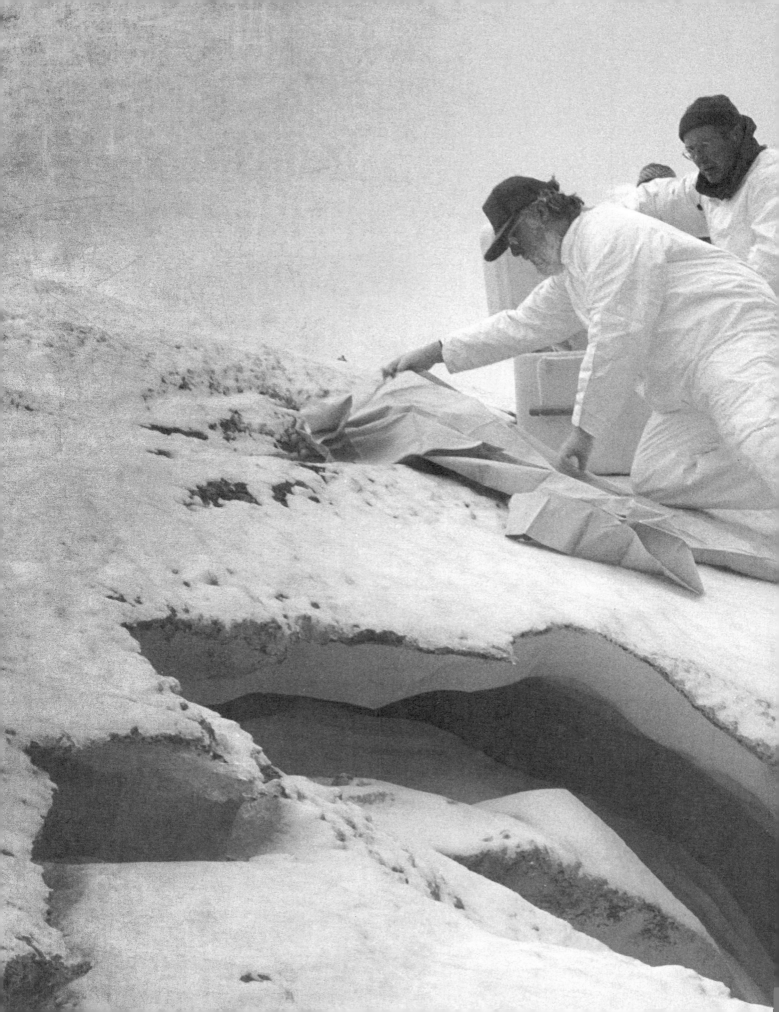

Dwarf birch, alder and willow thickets, Kelsall Lake, Haines Highway.

Richard Hebda photograph.

Acknowledgements

AS YOU TURN THE PAGES OF THIS VOLUME you will quickly come to appreciate its size and complexity. Bringing it together was an enormous and challenging task involving many and diverse people. Individuals who helped with the numerous facets of the management and operation of the Kwädąy Dän Ts'ìnchį discovery, as well as those who contributed to individual studies or sub-projects, are mentioned in the Acknowledgements sections of the relevant chapters. Here we would like to thank those who helped this volume come to be.

We acknowledge the vital role the leadership of the Champagne and Aishihik First Nations (CAFN) has played since the discovery was first made. This includes councils under the leadership of Chief Bob Charlie (1998–2002), Chief James Allen (2002–2006, 2010–2014), Chief Diane Strand (2006–2010), and Chief Steve Smith (2014–present). Though it hasn't always been easy, CAFN leadership has stuck with us on this long journey. Their support has allowed not only for the generation of the CAFN chapters but also for senior staff to offer a steady guiding hand on the project as a whole and on the process of bringing this book into production. Those senior staff include Lawrence Joe (former director of Heritage, Lands and Resources) and, when she wasn't being chief, Diane Strand (former director of Language, Culture and Heritage, one-time heritage officer, presently director of Community Wellness). Through the various iterations of this publication, the support of these individuals was always certain.

We thank the contributing authors for their patience and especially their willingness to make changes to their work to meet the needs for consistency throughout the volume, in particular the quirky standards of scientific names and the critical spellings of Indigenous languages. We thank you for replying in a timely manner to our requests and hope that you are pleased with the result.

We acknowledge too the work of the scholarly reviewers who ensured that the quality of the contributions met a high standard. Kjerstin Mackie of the Royal BC Museum ensured that the manuscripts went out and returned from the reviewers in a timely manner.

Thank you to Julie Cruikshank and Madonna Moss. You recognized the importance of this volume in your forewords and confirmed that we had indeed produced something special and different: a work that might stand as an example of a progressive approach to a subject that encompassed both the technology of science and traditional knowledge.

Special thanks are due to the Royal BC Museum publishing staff, past and present, and several copy editors who took on the challenge of a very difficult series of texts. Gerry Truscott managed publication of the volume during several years of its development until his retirement. During this time the text and illustrations were compiled and organized by Margery Hadley and copy edited by Jesse Holth and Amy Reiswig. Michelle van der Merwe replaced Gerry Truscott as publisher and piloted the volume to its final publication, with the assistance of in-house editor Annie Mayse. We also thank Lara Minja for producing the final design and Stephen Ullstrom for compiling an incredibly complex index.

We thank the Royal BC Museum for providing the financial support to publish the volume so that these ideas and stories could be shared with so many others who showed such strong interest. Grant Hughes, the museum's former director of curatorial services, provided the impetus in the early stages to publish this volume. We thank Professor Jack Lohman, chief executive officer of the Royal BC Museum, for his support and foreword to the book. We thank also Dr. Duncan McLaren and the University of Victoria planning committee of the 2008 Northwest Anthropology Conference who hosted initial drafts of these chapters and contributed surplus conference funds to the production of this book. And many thanks to the staff of our respective institutions, as well as our families—we can finally say it's done and thank you for your patience and support too.

To everyone, we acknowledge this book has involved a long journey; thank you all for sticking with us.

Richard J. Hebda
Sheila Greer
Alexander P. Mackie

CONTENTS

1

Respecting the Discovery

2

People, History and Honouring

CONTENTS

5

Journeys

6

Connections

Professor Jack Lohman CBE

CEO, Royal British Columbia Museum and Archives

Foreword

MORE THAN 17 YEARS HAVE PASSED since the Kwädąy Dän Ts'ìnchį man was found on a glacier in the traditional territory of the Champagne and Aishihik First Nations. In that time, a remarkable amount has been learned about his life, his death and the landscape that is so inextricably linked with this story of one young man. No one could have predicted the true depth of collaboration and learning that would result from his discovery.

The backdrop to this time period has been that of mounting global recognition of Indigenous rights and the urgent need for environmental protection. I am pleased readers will have access to a work that is both collaborative and holistic in approach and that works to bridge divides between scientific institutions and Indigenous peoples. This publication tells the story of the Kwädąy Dän Ts'ìnchį discovery. It is an extraordinary exploration of the life of one individual who lived far in the past, and of the present-day First Nations, researchers and institutions that came together to learn from him.

We cannot speak with the young man who set out across the glacier so long ago. What led him to take the dangerous path across the ice? Was he going to visit family or friends, to trade or to learn? How many languages did this young man speak? What rich social networks did he engage with during his short years?

And what about the objects found with him? How do the belongings we carry every day and the things we surround ourselves with tell the story of who we are? Some of the most captivating clues about his life can be glimpsed in the construction of the Kwädąy Dän Ts'ìnchį man's remarkably preserved belongings. A fur robe, carefully stitched together from the skins of tiny ground squirrels, was found to have been repaired with the sinew of a Blue Whale—the largest animal that has ever lived! These small details that capture our imagination also ground us in the humanity of this young man, transporting us to the time he lived and provoking us to reflect on the episodes that made up his life.

The precious belongings that the Kwädąy Dän Ts'ìnchį man carried with him can reveal so much about the world in which he lived. A hat masterfully woven from spruce root kept him sheltered from snow, rain and sun, but where did he get it? Who dug the roots, split the fibres and wove their artistry into this precious item? Was it a favourite auntie who brushed slugs, spiders and damp earth from her hands as she searched for the perfect roots? Or perhaps a dear friend? Some things we can never know about his life. What we can say as a certainty is that something as simple as a hat provides a direct link between the Long Ago Person Found and people alive today. The same style of weaving is still practised by First Nations to create hats just like the one this man wore as he crossed the glacier all those years ago. But there is no way to reduce the life of an individual to the items that they owned. Kwädąy Dän Ts'ìnchį's story is not simply the story of a hat, or a robe, but of how these objects tie him to communities of people alive today.

The expansive scope and detail of this inquiry into the Kwädąy Dän Ts'ìnchį remains were only possible because of a partnership launched between First Nations, the Government of British Columbia and their partners. We have been able to learn much, thanks to the generosity of the Champagne and Aishihik First Nations and their neighbours. That sometimes differing priorities, worldviews and modern-day cultural divides could be bridged with a project centring on the life of a young man is something that should be celebrated.

It is my hope that this publication will serve as an example for those seeking to undertake meaningful and collaborative research with First Nations. Projects that incorporate diverse cultural perspectives and respect the values and priorities of Indigenous peoples can create far deeper and impactful learning experiences than those carried out in isolation. This publication is proof of the great success that can be achieved by supporting scientific enquiry alongside the traditional knowledge and oral histories of Indigenous peoples.

Julie Cruikshank

University of British Columbia

Foreword

IT IS AN HONOUR to be invited by Champagne and Aishihik First Nations to write some introductory words for this important volume. I feel fortunate to have known members of this community over the years, and I have followed this project from a distance.

Almost 40 years ago, several First Nations women elders from the southern Yukon Territory made the choice to document their life histories for younger generations. Three in particular—Mrs. Annie Ned, Mrs. Kitty Smith and Mrs. Angela Sidney, all born at the end of the 19th century—insisted that glacier stories be included in their accounts. This seemed perplexing at first. All of these women lived far from glaciers by the time they told their stories. Yet their determination, even urgency, to document their own experiences of glaciers, as well as glacier narratives they had heard from parents and grandparents, was clear. Learning, they would all agree, comes first and foremost from direct primary experience—from actively living on the land, as each of them had. Their concern was that younger people did not have this opportunity. So they drew a strategy that had long served to pass on knowledge in the absence of direct experience: their oral tradition.

A series of events in the 20th century had dramatic consequences for the First Nations who had formerly considered the Tatshenshini River basin their home. The establishment of a provincial boundary dividing the Yukon Territory from British Columbia bisected their territories in 1900. The construction of the Alaska Highway in 1943 resulted in the establishment of "protected areas"—a game sanctuary, later a national park, and then a World Heritage Site—where they were no longer permitted to live. Prohibitions against hunting, trapping or fishing in these areas had the effect of forcibly relocating them some distance northeast of the Tatshenshini. The women who chose to present accounts of lives lived near glaciers concluded that without their intervention, young people who had been denied the experience of living in their traditional territories might never know them. They believed that their descendants would need those stories someday. And they were right.

They chronicled travels between the Gulf of Alaska and the Yukon plateau made by coastal and inland peoples who traded, travelled and intermarried, crossing glaciers on foot or travelling in dugout canoes under the glacier bridges that periodically spanned the lower Alsek River during the 1800s. They recorded narratives about glaciers that dammed rivers and formed lakes that eventually erupted catastrophically. They explained hazards of glacier travel and spoke about individuals who had survived falls into crevasses. They would be pleased that their stories about travels in the Tatshenshini basin and on the surrounding glaciers have been revived in the course of this project.

The loss of a young man in the prime of life is a tragic story at any time. Yet he might have been gratified to know about the important role his life would play in communities some centuries later. The unexpected discovery of his remains in August 1999 occurred at a crucial moment in the history of Champagne and Aishihik First Nations (CAFN). Land claims and self-governance agreements had been settled with federal and territorial governments a few years earlier, and CAFN members were actively involved in park co-management. The man was found melting from a glacier in the traditional territories of Champagne and Aishihik First Nations—territories also now encompassed by Tatshenshini-Alsek Park. His death was probably accidental. His woven fibre hat, his fur robe, some tools and a piece of fish he was carrying were preserved with him. The robe carried traces of spruce and pine pollen and some fish scales. As we learn in these papers, the physical evidence of his last days has offered remarkable insight about his times for present-day community members, for regional historians and for scientists. Local elders named him Kwädąy Dän Ts'ìnchį—Dákwanjè for "Long Ago Person Found".

From the outset, there was close cooperation between archaeologists and members of Champagne and Aishihik First Nations, without which the scientific research could not have proceeded. The Native American Graves and Repatriation Act structures such relationships in the United States, but in northern Canada, partnerships are being carefully negotiated as part of implementing recent land claims and self-governance agreements. Notably, this collaboration coincided with the period when Champagne and Aishihik First Nations were regaining rights for both use and management of the lands within these provincial and national parks. Members of the CAFN were interested in learning where he might have travelled from, where he was going and how his travels overlap with their own oral histories. Women, for instance, have sometimes speculated about which coastal grandmother might have made his exquisite fibre hat, or which auntie might have made his soft fur robe. Scientists are especially interested in how this rare discovery, which preserved flesh and hair as well as bones, might contribute to our understandings of health, nutrition and disease; they have also considered what his perfectly preserved hat, tools and fragments of robe may reveal about everyday life from that time. The communities in whose territories he was found agreed to allow scientific investigations that included First Nations representatives on the management team and to let the materials travel for scientific analysis. As agreed, his remains returned to the community within a specified time frame. First Nations hosted a funeral potlatch for him on July 21, 2001.

In the processes documented in these pages, new knowledge about human history has been jointly created in a remarkable collaboration among First Nations, the BC Archaeology Branch, the Royal British Columbia Museum, the Yukon Heritage Branch and national and international scientists. Champagne and Aishihik First Nations have taken on a stewardship role, regularly inviting other coastal and interior First Nations to participate in the projects described here. Indeed, this collaboration has brought to light and strengthened connections among coastal and interior communities, an outcome that elders hoped might flow from their stories. Champagne and Aishihik First Nations' Heritage Department has made the Kwädąy Dän Ts'ìnchį man a major focus of their research since 1999, bringing community members together to discuss what might be known about his life. The gopher robe (*säl ts'ät*) project, for instance, brought together senior women consultants with younger women eager to learn their skills, reactivating older traditions and practices. The community DNA project has begun to unravel his kinship connections with contemporary families. Ties between the young man who lost his life so long ago and living people are building relationships across communities.

Throughout the life of this project, Elder Paddy Jim has continued to take a guiding role, providing advice on many subjects, particularly the ceremonies conducted along the way, and making sure that everything was done correctly. In his words:

> *For all our history, they haven't got any story like that, you know,*
> *find a body. We're talking about Native history. That's why it's so*
> *important to us, you see.*

Elders who told their stories in previous generations would be proud to see that ten of the papers in this volume—almost a third of the total number—are authored or co-authored by Champagne and Aishihik First Nation members: Lawrence Joe (former director of Heritage, Lands and Resources); Diane Strand (former chief of Champagne and Aishihik First Nations and also deeply involved in heritage research); Frances Oles, artist and coordinator of the *säl ts'ät* gopher robe project; Ron Chambers, a well-known guide and outdoorsman with a lifelong interest in archaeology; Sheila Joe-Quock and John Fingland, both committed long-time staff members—all participated in writing these chapters. They did this together with consulting elders Moose Jackson, Wilfred Charlie, Jimmy G. Smith and John Adamson, all of whom have now passed on. Former chief James Allen and the late Gerald Brown have both provided ongoing counsel and guidance. They accomplished this with the invaluable assistance of Sheila Greer, whose role in this project has been crucial, and with remembered contributions from their friend and colleague, Sarah Gaunt.

———

Madonna L. Moss

University of Oregon

Foreword

AS YOU WILL DISCOVER in reading this extraordinary book, the Kwädąy Dän Ts'inchį man is an ambassador from the past. This young man lived during the Little Ice Age and died while travelling over a glacier in what is now Tatshenshini-Alsek Park, in northern British Columbia. Miraculously, portions of his body and some of his belongings were preserved in the ice and protected from the glacier's movement. In 1999, while out hunting Dall's Sheep, three modern travellers found the man's remains emerging from the edge of a melting glacier. They contacted Yukon Heritage Branch archaeologists, who in turn notified the authorities responsible for this remote area of Tatshenshini-Alsek: the Champagne and Aishihik First Nations, BC Parks and the BC Archeology Branch. The condition and context of the finds would require that all parties show great expertise, cooperation, flexibility and diplomacy in the recovery and study of the Kwädąy Dän Ts'inchį man and his belongings and the reburial of his body. As you read these pages, you will marvel at the relevance of oral history, the meticulousness of conservators, the exactitude of scientists, the technological skill involved in making a gopher-skin robe and the generosity of Champagne and Aishihik First Nations in sharing this knowledge with the world.

The Kwädąy Dän Ts'inchį man has challenged conventional cultural categories and crossed various political divides. Although he was found in northern BC, he was very close to the borders of Yukon and Alaska. The fact that his homeland now encompasses parts of two separate countries might surprise this young man, and he might also find it odd that maps of his homeland display a complicated mosaic of tribal boundaries. He was found within the bounds of what was once a "wilderness park," but the discovery of his remains and nearby artifacts indicate this has clearly been a humanized landscape for centuries.

The Kwädąy Dän Ts'ìnchį person lived both on the coast and in the interior. He and his people travelled inland from the coast and back, and they were probably as adept at handling seaworthy canoes as they were at travelling across glaciers. Although we archaeologists might classify groups based on their marine or terrestrial lifestyles, economies or diets, the Kwädąy Dän Ts'ìnchį project shows this man relied on marine foods for much of his life, but in the several months prior to his death, his diet was more terrestrially based.

Although his remains and belongings have been radiocarbon dated (he is estimated to have lived between 1720 and 1850 AD), we cannot say definitively whether he lived prior to European contact or during the "protohistoric" period. While the presence of metal with an archaeological find often indicates a post-contact date, the bead found among the man's belongings was made from native copper using Indigenous technologies. His knife had an iron blade, but the iron might have originated in a shipwreck prior to direct contact between his people and Europeans. The discovery of an apparently European strain of *Mycobacterium tuberculosis* DNA in his organs also raises questions. Was this latent infection from pre-contact or introduced tuberculosis? Could this disease have found its way to the Northwest Coast from distant contact with European colonists, such as the Spanish to the south or the Russians farther north and west? Despite all our investigation, the Kwädąy Dän Ts'ìnchį man has kept some secrets to himself, even as he has challenged scientists to re-evaluate what we think we know about his culture and times.

The studies presented here provide a remarkably rich cultural and environmental biography of the Kwädąy Dän Ts'ìnchį man. Yet we will never know precisely where he was headed at the time of his death or exactly why he died. It appears he died of exposure, perhaps during an early snowstorm. A young man heading up into the high country on a purposeful journey must have been both skilled and brave. He reminds me of my son, and many of my students, who love to climb mountains and explore the alpine. I think of the piece of Sockeye salmon the Kwädąy Dän Ts'ìnchį man carried, and I imagine him chewing it slowly, enjoying its crispy skin, knowing that it would sustain him for many more hours of walking. I cannot help but imagine that his coastal-style hat was made by his mother, whom he had perhaps recently visited and with whom he'd enjoyed a feast of beach food, including Dungeness crab and beach asparagus. Perhaps he had especially relished those foods since his move to the interior, where they relied more on mountain sheep, beaver and freshwater fish. Perhaps his wife made the luxurious gopher robe he wore, or perhaps it was made by her mother and aunties working together. Perhaps he was headed to meet his family at their fishing camp along the middle reaches of the Tatshenshini River. He may have intended to live with his wife's family for several years, helping them fish, hunt and trap and making seasonal trips to the coast to trade. Perhaps the couple would have returned to the coast later in their lives, had he not died on that glacier. These thoughts are of course speculative, and the Kwädąy Dän Ts'ìnchį man may have been a trader, a messenger or a traveller on a personal quest. Yet I am certain that when he did not return from his trip, his relatives across the region ached with longing to know what had befallen him.

First Nations stories tell of individuals who lost their lives during glacier travel, but they also record successful long-distance migrations and inspiring rescues. Many decades later, his kin have finally put the Kwädąy Dän Ts'ìnchį man's body to rest with an appropriate funeral service. They have also allowed us to share in getting to know this young man, his people and his culture.

In this book you will learn about the efforts of an exceptional group of world-renowned scholars and scientists who brought their skills, inventiveness and creativity to the Kwädąy Dän Ts'ìnchį project. I salute them all, but especially those who facilitated the cross-cultural communication and interdisciplinary exchanges at every stage in the process. You have demonstrated great patience, stamina and grace. You have shown the world how powerful genuine collaboration can be and you have set a new standard for the many disciplines represented herein. *Gunalchéesh*, Kwädąy Dän Ts'ìnchį and your relatives. *Gunalchéesh*, Al, Sheila and Richard.

———

1

RESPECTING THE DISCOVERY

INTRODUCTION

Richard J. Hebda

Exceptional discoveries often occur in exceptional places and under exceptional circumstances. Part 1 of this book explores key aspects of the Kwädąy Dän Ts'ìnchį discovery and the remarkable place where it was made. It begins with a description of the discovery itself and its unique circumstances, followed by an exploration of the magnificent landscape—especially its glaciers. Descriptions of the native flora and challenging fieldwork conditions provide more context, and part 1 concludes with a report on the age of the human remains and associated materials.

DURING ONE OF THE WARMEST YEARS ON RECORD, three sheep hunters encountered human remains on a small icefield on the north side of an unnamed mountain in Tatshenshini-Alsek Park. In their own words, the hunters describe the story of their find and begin to relay the first lessons from a young man who lost his life centuries ago on a glacier in what is today British Columbia.

Fortunately, the hunters were aware of the potential significance of the discovery and recognized their responsibilities. They quickly contacted the appropriate people, and the work set out in a respectful manner and in full collaboration with First Nations, scientists and government institutions in a way that made the project unique.

Although this story is about a person and people, the natural landscape played a profound role in its unfolding. The Tatshenshini-Alsek landscape is stupendous and varied, from verdant oceanside forests to bleak, craggy mountains and massive glaciers. Visitors from outside may view it as a harsh land with strong coastal-to-inland contrasts, but it is rich in plant and animal resources that were well known to the Aboriginal people of the region.

Glaciers are among the most formidable and dynamic elements of the Tatshenshini-Alsek landscape, and glacial ice played a central role in the Kwädąy Dän Ts'ìnchį discovery and evolving story. Special circumstances in the character and behaviour of the entombing glacier led to the exceptional preservation of the human remains and artifacts, and extraordinary circumstances also led to the melt that exposed those materials.

The site's isolated location and fickle weather imposed tight constraints on work with and recovery of the remains. These conditions permitted only a few short visits to the site, yet a remarkable amount was accomplished. Field activities were respectful of the human remains and held to an extremely high standard, ensuring clean, high-quality samples. The end result was the recovery of study material that could undergo the most rigorous analysis.

Many of the artifacts and much of the biological material that was investigated came from plants. Until the time of the discovery, the study area's flora was poorly known, especially elements of the moss and algal life. Botanical collections from the study site and adjacent area added much to our understanding of the region's natural environment. Most importantly, improved documentation of the flora has provided key insights to help unravel the travels and activities of the Long Ago Person Found man before he perished on the glacier.

The age of ancient human remains and artifacts is central in decoding the human story. Obtaining a reliable age for the discovery site posed several challenges. Eventually, radiocarbon dates on various items and the body itself demonstrated that the site represented several centuries of material. Repeated strategic dating revealed that the Long Ago Person Found man had perished about 200 years ago, during that most interesting time when Europeans were first making contact with this part of North America but had yet to visit the discovery region.

——

1

THE DISCOVERY OF KWÄDĄY DÄN TS'ÌNCHĮ

William H. Hanlon

"**Hey, Mike, look—a stick.**" The stick appeared odd in such a barren landscape of rock and ice. I picked up the stick, examined it and showed it to Mike. "Looks like a part of a walking stick, doesn't it? It looks like we're not the first people to hunt here."

THE DATE WAS AUGUST 14, 1999. Mike Roch, Warren Ward and I were hunting Dall's Sheep deep in the spectacularly wild Tatshenshini-Alsek Park. For the first time in six years of applying we had drawn permits to hunt one of the most limited, micromanaged big-game species in British Columbia. Dall's Sheep inhabit only a small, rugged area in the northwestern corner of the province, and most of the sheep population lives in the park. Tatshenshini-Alsek Park combines with the adjacent Kluane National Park in the Yukon and the Wrangell-St. Elias and Glacier Bay national parks in Alaska to create more than eight million hectares of remote wilderness. It is the largest continuous piece of protected wilderness on the planet and a UNESCO World Heritage Site. It's just the kind of place where Mike, Warren and I go to immerse ourselves in the arduous pursuit—the "doctrine of the strenuous life," as Theodore Roosevelt put it. He said, "Skill and patience, and the capacity to endure fatigue and exposure, must

be shown by the successful hunter." This is the kind of experience that comes from wilderness hunting.

Our adventure had begun five days earlier, after a 3,000-kilometre drive that started in the very southeastern corner of the province and took us across parts of Alberta, British Columbia and Yukon, then back into BC to the banks of the Tatshenshini River. We shouldered our packs (fig. 1), carrying a 10-day supply of provisions and hunting gear, and anxiously waded into the swollen and muddy river. Freeze-dried food, down-filled GOR-TEX sleeping bags, quality optics and rifles were just a few of our hunting necessities. The objective of our first day was to hike as far as we could away from the highway and into good sheep habitat. We managed to travel 15 kilometres the first day—not a record by any means, but a good distance considering we were hiking in awe of the beautiful scenery, wild rivers, jagged peaks and rams!

Figure 1. Mike Roch (*left*) and Warren Ward beside the Tatshenshini River. Bill Hanlon photograph. Colour version on page 579.

Figure 2. Bill Hanlon, Warren Ward and Mike Roch with Dall's Sheep ram horns, just west of Fault Creek. Bill Hanlon photograph.

Figure 3. First ice patch at the head of the west fork of the Fault Creek Glacier with Bill Hanlon (*left*) and Warren Ward at the base. Mike Roch photograph. Colour version on page 579.

By the fourth day of what we considered to be our hunt of a lifetime, Mike and I had already harvested a full-curl ram each (fig. 2) and were now concentrating our efforts on a ram for Warren. We had drawn tags in what is called a group-hunt application, so we planned to hunt until we each shot a ram in our allotted 10 days.

On the morning of the sixth day, we decided to hunt near a high-hanging basin at the headwaters of a glacial stream flowing into the O'Connor River, where we had spotted a distant band of rams the previous day. A lot of the decisions we make while out on the land are based on weather, air currents, game movements, habitat and topography. We approached the basin from above, because we wouldn't arrive until mid afternoon and in the heat of the day. By that time, the afternoon heat would force the air in the basin to rise and we would be upwind from the rams. If we approached the rams from below, the wind would carry our scents to the sheep long before we ever entered the basin.

We left our base camp at 8:00 am and climbed a pass that allowed us to descend into a rocky valley at the base of a glacial moraine. We crossed a small glacier-fed stream at the valley bottom in the cool of the early morning. Rock-hopping with light packs across a braided stream bed proved to be a relatively easy and dry affair. Traversing these streams in the heat of the day makes for a more challenging event; the melting glaciers this time of year cause freshets, turning streams into muddy glacial torrents. The year 1999 was the hottest year on record in British Columbia, and the north was no exception. The concept of climate change and global warming was not yet a household issue, but we witnessed first-hand the ice melting at an alarming rate.

We climbed out of the valley bottom and crossed the first of many ice patches we would encounter (fig. 3). The terrain seemed like it had been mined.

Figure 4. Bill Hanlon looking southeast down the the so-called Empty Valley from the ridge above the discovery site. The Samuel Glacier is visible in the background. Mike Roch photograph. Colour version on page 579.

Freshly thawed and broken glacial moraine was calving from the base of the receding ice patch.

Footing was extremely treacherous in this newly thawed earth. We managed to scramble our way out of a maze of mud, rock and ice, onto the edge of an azure tarn, where we refilled our water bottles. We continued to climb to the ridge that would give us the first distant view of the basin. The sky was grey, and the air was relatively warm, with no haze, fog, smoke or heat waves to limit the distance we could see. We soon reached the ridge and witnessed the most awe-inspiring landscape we had ever seen—not just beauty, but sheer and utter grandeur. The main body of the Samuel Glacier, flanked by granite spires, not long exposed to the elements, took our breath away (fig. 4). The Samuel Glacier is the source of three major river systems in Tatshenshini-Alsek Park: the Tatshenshini, Parton and O'Connor rivers. To be among such enormity and wildness doesn't just make you feel alive; it makes you bristle with vitality! All this while

hiking great distances, carrying heavy packs, eating minimally and sleeping even less—yet we were feeling absolutely fantastic! A great place to stop for lunch and discuss hunting strategy.

While digging out our lunch of beef jerky, trail mix and energy bars, I said to Mike and Warren, "Let's just skate across this ice patch in between us and the basin and cut our distance."

Mike replied, "Nope, Warren and I promised our wives we wouldn't climb on any glaciers."

"We're not climbing," I replied. "We're just sliding down a gradual slope."

Mike and Warren were concerned about crevasses and that the sheep might be able to see us, or at least that was their excuse. After more discussion we decided to hike along the edge of the ice patch to keep out of sight.

Our route paralleled a 10-metre wall of turquoise ice undercut by a meltwater channel. Mike jokingly said to Warren, "Why don't you crawl under the ice and I'll take your picture?"

Figure 5. Bill Hanlon on Fault Creek Glacier standing next to the robe on August 14, 1999. Warren Ward photograph.

As usual, Warren ignored Mike and we kept hiking along a recently exposed gravel bar. We were walking in single file, picking our way along the ice edge when I said, "Hey, Mike, look—a stick!"

We stopped and examined the wooden object. "Looks like part of a walking stick, doesn't it?"

"We're not the first people to hunt here."

I put the stick back on the ground and kept walking. A few seconds later Mike asked, "Hey, Bill, what did you do with that stick you found?"

"I put it back where I found it. Why?"

"Because I think I found the other half of it."

I ran back and picked up the stick. When I returned, we put the two similar-sized ends together, and they fit. A few metres further on, Mike picked up another interesting stick and stopped to examine it. This stick was about a metre long and curved the entire length, with a blackened end and a carved end.

Mike exclaimed to me, "These aren't sticks, they're artifacts!" All the while he waved the stick over his head and tried to come up with the right word to describe his discovery. "It begins with 'A' and they used to throw

spears with these things," he explained. He was trying to remember the word "atlatl".

While all this artifact finding was going on, Warren had continued to wander down the edge of the ice. He stopped, put his binoculars to his eyes, looked up about 10 metres at the top ridge of the ice and said in a matter-of-fact way, "I think I just found the poor fellow who lost all this stuff."

The air was suddenly electric and the hair on the backs of our necks stood on end. Since Mike and Warren weren't allowed to climb on any glaciers I took one for the team and climbed up an icy chute to get up on the edge. I carefully walked over to the smudge on the ice and stood above what looked like the hide of an animal (fig. 5). Mike called me to look up for a photo opportunity, and then I began looking more closely at the "hide".

It was evident that whatever we were looking at had just emerged from the ice and had not been exposed for long. The first thing I noticed was a small wooden-handled object sitting on top of the hide. I picked up the object and slid it out of its sheath. The tool was unlike anything I had ever seen before, yet it somehow looked familiar. Wound around the worn wooden handle was a leather lace, which lashed a small, corroded tip of some kind. By then Mike and Warren had violated their wives' orders and were standing beside me.

Mike took a few more photographs, and then we looked closer at the hide. We noticed tiny X-shaped stitches along the seams, giving the hide a patchwork appearance. We also noticed a small piece of leather with cut fringes. There was nothing modern or contemporary about what we were looking at, and we soon realized that this was someone's clothing or pack. Scattered around the object were small fragments of bone and hair. I picked up a small bone fragment, examined it, found it to be rather spongy, and put it back down where I found it.

It was then I noticed something else protruding from the ice about two metres away from us. Upon closer examination, we identified it as a human pelvic bone. We could see the attached legs disappearing into the ice below the pelvis. We took one photo of the pelvis and a few more of the clothing and carefully walked away from the site. Our discussion by this time was bordering on hysterics because we had far more questions than answers, but we realized that it was human remains and had the common sense and respect not to disturb the area. We put a few artifacts in small plastic bags to take with us so we could prove we had actually found something old and maybe of some significance—the small piece of leather with the fringes, the small tool in its sheath, two of the walking stick pieces and the atlatl-shaped stick, all carefully stored in our packs.

We continued on our way, but our minds were heavy with questions about what we had just found. Could it be very old? Who might it be? Was it a missing person? There was one thing that threw a wrench into our belief that we were the first to discover these remains. Lying right on top of the remains, like it was melted there, was a survey stake with an orange ribbon and some fine string. Was this site already identified? Had it already been investigated? Why was survey material right there? Despite this concern, we were very excited and now very distracted from the original purpose of our trip.

Regardless, we continued to hunt further down the ice and came across the remains of a bull Moose, also just emerging from the same ice patch. We knew about this Moose because a friend of mine had hunted this basin 10 years earlier and had shown me a photograph of the Moose completely encased in ice except for the antlers. We took a photo of the Moose antlers and continued on. After thoroughly glassing the basin and not seeing any rams, we decided to begin the long hike back to our base camp. The streams were swollen with the day's melt, and it took a while to find safe places to cross them.

We awoke early on the morning of the seventh day and decided to begin our long hike back to the truck and hunt our way out. We knew it would take us two full days. Mike and I were both carrying packs weighing over 100 pounds, including our gear, the butchered meat, salted cape, and the skull and horns from each of our rams (fig. 6). We were still very excited from the day before and really wanted to talk to someone who could tell us about what we had found. This gave us something to keep our minds off the weight on our backs while we trudged along.

Two days and 40 kilometres later we reached the truck. Now, where should we report our discovery? Whitehorse, Fort Nelson, Fort St. John? On the way to Whitehorse, Warren found a tourist brochure on the back seat and pointed out an advertisement for the Beringia Museum and Interpretive Centre. "This looks like the place we should stop," he said. "Maybe they might be able to answer some questions."

We all agreed, and three hours later we found ourselves knocking on a locked door a minute after 8:00 pm. A summer student was vacuuming the carpet after a long day but opened the door to three unshaven, desperate-looking characters with their faces pressed against the glass. We bombarded the poor fellow with a million questions and gestures, but he managed to understand our story. He knew of a person who might be able to help us out and led us to a phone with the contact number. "His name is Greg Hare and he is an archaeologist with the Yukon government. He works closely with the centre."

Figure 6. Mike Roch (*left*) and Bill Hanlon with full packs, just west of Fault Creek. Warren Ward photograph. Colour version on page 579.

I managed to get hold of Greg Hare immediately and began blathering our story about the artifacts we had found. He asked a few pertinent questions, didn't get excited, and suggested we find a hotel, hot shower and dinner, and he would meet us at the centre at 8:00 the next morning. Then I added, "Oh, by the way, Greg, there is a body in the ice along with the artifacts I described to you." That got him excited.

The next morning, August 17, we met Greg Hare and Ruth Gotthardt at the Beringia Centre at 8:00 am sharp. Here were two people who would finally answer a few of our hundred questions. After brief introductions we showed them the artifacts we had packed out from the site, which had been kept overnight in the centre's freezer. Although they were quite excited about the interesting artifacts, they were both very careful not to jump to conclusions or speculate on age or origin. It wasn't until we began describing the human remains that they began to answer some of our questions.

According to Greg, the site had never been identified in the past, nor had any ancient preserved human remains ever been found in the north. He would know, because he had been project coordinator for an ice-patch research project for the past 10 years in the Yukon. The project team had been studying the receding ice and finding numerous artifacts as they became exposed. Both Greg and Ruth concurred that the chances of someone being caught in a glacier long ago was one in a million, and the chance of someone finding a body emerging from a glacier today, among hundreds of thousands of glaciers, was even slimmer. Combine these odds with the short time that any organic remains would survive intact once exposed to the elements, predators, and so on, and this could truly be a once-in-a-lifetime find.

Once again, both Greg and Ruth were careful not to jump to conclusions, but their enthusiasm and excitement was highly evident and infectious.

It wasn't long before Greg had made the appropriate contacts—RCMP, First Nations, BC Parks, Helicopter Service, etc.—to deal with any issues that might affect our next steps. Greg asked if one of us would like to return to the site with them, but with the number of people he had to transport and the distance we had to drive home, we declined the invitation. We gave Greg the location on a topographical map and the instructions to look for the Moose antlers sticking out of the ice. We agreed to contact Greg at his home in the evening once he returned from the site.

It would be a long drive while we waited for Greg and Ruth's full professional opinions on their preliminary visit of the site and the remains. What Greg and Ruth were about to confirm would make headlines around the world.

———

2

THE NATURAL SETTING OF THE DISCOVERY REGION

Landscapes, Ecosystems and Species

Richard J. Hebda

The story of a person is very much the story of the places familiar to and experienced by that person. Places are sources of food and shelter, clothing, medicines and more. They are sources of knowledge. They underlie belief systems. The places and the people met there provide the stimulus for actions, identity and creativity. Without place there is no person, no personality and no story.

LANDSCAPES HOLD PLACES and are part of the canvas upon which the detailed story of a life—and related lives—is painted. Landscapes connect the bits and pieces of a person's life. Understanding the landscape, the biological and physical framework in which individual studies are set helps not only in the interpretation of those studies but also in drawing many life stories together.

Biological and physical spaces have a powerful role in the story of the Kwädąy Dän Ts'ìnchį man. As hard as we may try, it is difficult for a person living in today's world, surrounded by places of human making and design, to appreciate the intensity of the natural landscape around the discovery site. The region of Alaska, BC and Yukon is more than just untamed

nature—it is a region of spectacular power and extreme natural variation. There are few places on earth where the variety in climate, landform and ecology is so great over such short distances. Powerful rivers, bountiful seas, massive and rugged mountains, deep forests, dense alpine thickets, open rocky meadows, mild ocean shorelines and the extreme climates of the continental interior all come together.

This chapter provides the physical and biological background for the chapters that follow and the context for the individual stories in them. It is also intended to give you a sense of the complexity and intensity of the environment that the Kwädąy Dän Ts'ìnchį person inhabited, witnessed and depended upon. The landscape is as much a part of the story as the person in this case.

The contrasts exposed in this chapter will become a recurring theme in the story of the Kwädąy Dän Ts'ínchį man.

The account that follows is based on information from a wide range of technical literature, as well as my experiences in the region over many days and the experiences of others who have visited the area and the discovery site. But it cannot capture the impressions, knowledge and deep understanding of those who have lived there, or those who have inherited the stories of their elders.

Location

THE STUDY REGION lies in northwestern North America, spanning the province of British Columbia and the Yukon Territory in Canada and the state of Alaska in the United States of America (fig. 1, page 580). The major communities in the region include the towns of Haines, Alaska, and Haines Junction, Yukon, more than 200 kilometres apart at either end of the Haines Highway, a paved road. Several important settlements, such as Klukwan and Klukshu, occur on or near the highway. The nearest settlement west of the discovery site is Yakutat, Alaska, 160 kilometres away on the Pacific Ocean. The discovery site itself is in British Columbia within the boundaries of Tatshenshini-Alsek Park. This park together with adjacent national parks in Canada and the United States—Wrangell-St. Elias National Park and Preserve, Glacier Bay National Park and Preserve and Kluane National Park and Reserve—makes up the largest protected natural area in the world, a UNESCO World Heritage Site.

Modern political boundaries may appear important on maps, but they are of little significance in the natural world, for this region's real features are its mountains, glaciers, rivers, inlets and ecosystems. Human boundaries can be stepped over or crossed by vehicle. Natural features must be contemplated, understood and experienced. They are real in a way that border crossings are not, and they have always been real.

Physiography

FIRST AND MOST OBVIOUS TO A VISITOR, visible even from space, is the physical complexity of the landscape—a complexity matched by few places on earth. Extending inland from the Pacific Ocean, long, steep-sided inlets penetrate into massive, angular mountains. These stupendous mountain ranges are cut by great river valleys, and separated by high plateaus. Farther inland, they yield to gentle mountain slopes in the transition to major continental basins, such as that of the Yukon River. The dramatic physical changes span scarcely 150 kilometres, about the distance from Vancouver to Hope in southwestern British Columbia. Many mountain valleys are further choked by grinding glaciers or flooded by ice-dotted glacial lakes.

Technically, the region is part of the Outer Mountain area of the Western System in the physiographic classification of Holland (1976). The main upland features of the Outer Mountain area are the Alsek Ranges (figs. 1 and 2, page 580), the Icefield Ranges and the Fairweather Ranges, and the Boundary Ranges to the east. Just north of the area, in southern Yukon, is the Teslin Plateau (Mathews 1986). The discovery site is located in the Alsek Ranges of the Saint Elias Mountains unit of Holland (1976) (figs. 1 and 2). Mathews (1986) includes adjacent Alaska and Yukon in his classification, and subdivides the region to include two physiographic depressions: the Glacier Bay Depression and the Chatham Trench. Both are important to the story that unfolds in this volume, and of course extremely meaningful to the people who have inhabited and still inhabit the region.

Mountain Landscapes

THE ST. ELIAS MOUNTAINS are familiar to many people, illustrated in numerous images of Alaska as a great wall of white peaks rising sharply above the sea, with great glaciers flowing down from them. In the western portion, the Fairweather Ranges include BC's highest peak, Mount Fairweather, at 4,663 metres, (15,300 feet) on the BC–Alaska border, and farther north, Mount St. Elias and Mount Logan. At the eastern limit of this massif run the Tatshenshini, Kelsall and Chilkat rivers.

The westernmost part of the St. Elias Mountains comprises the southern portion of the Fairweather Ranges and the northern portion of the Icefield Ranges, with the highest peaks and the greatest glacier cover and sources of much of the ice that fills the Tatshenshini-Alsek River depression. The Alsek River valley is the only lowland elevation feature that breaks through the great mountain front to the sea.

The Alsek Ranges, which include the Kwädąy Dän Ts'ìnchį discovery site (fig. 2), lie immediately southeast of the Alsek River and east of the Fairweather Ranges, separated by Melbern Glacier and its deep trench. The Alsek Ranges is an imposing mountain mass by any standards but is generally lower than the Fairweather Ranges, with peaks reaching from 1,980 to 2,590 metres (6,500–8,500 feet). Being inland in the rainshadow of the Fairweather Ranges, it carries less snow and ice cover than its neighbours to the west (J.S. Peepre and Associates 1992).

Glaciers and icefields dominate the west and south portions of the discovery region adjacent to the ocean and inlets. In the west, most of the St. Elias Mountains are buried under ice marked by protruding peaks. In Alaska, there are the Nunatak, Novitak and Yakutat glaciers and their feeder fields. In BC, the Vern Ritchie, Tweedsmuir, Melbern and Grand Pacific glaciers grind to the inland valley bottoms. To the south, Muir, Carroll and Casement are large ice systems. Smaller ice fields and upper valley glaciers occur widely in many of the mountain masses, among them the Samuel Glacier complex at the discovery site (fig. 2). In the Alsek Ranges, almost every high-elevation valley contains a small glacier tongue that delivers cold melt water and sediment into the streams that flow to the valley bottoms. Closer to the Yukon border the ranges are ice-free.

Valleys are strongly shaped by past streams of glacial ice and are choked with masses of debris delivered from their icefields and spread out by rivers into networks of separating and reconnecting, braided channels on active valley flats. These braided flood plains are often more than 1.5 kilometres wide. As Holland (1976, 29) describes it, they "give the appearance of a land just emerging from its glacial cloak". Some rivers and streams, such as the Kelsall and O'Connor rivers, cut deeply into the surface and have relatively narrow valley floors at the base of steep sides.

Features not well expressed by broad physiographic descriptions are the extensive areas of flat rolling terrain between mountain masses occurring at high elevations (fig. 3). The most obvious place of relatively gentle upland terrain occurs along the Haines Highway between Klukshu and Pleasant Camp at the Canada–US border to the south (fig. 2). This is not a landscape of extreme slopes but rather a high-elevation rolling terrain. It is covered in extensive, shrubby alpine plant communities, medium and small lakes (Kelsall Lake, Mineral Lakes) and wetlands. Patches of rolling and modestly sloping terrain occur even at high elevations adjacent to icefields, including one immediately west of the discovery site. This gentle topography extends gradually inland to connect with the plateaus and valleys in the Yukon.

Valleys, Rivers and Inlets

LOWLAND LANDSCAPES figure prominently in the chapters that follow, despite the importance of mountains and glaciers in the immediate neighbourhood of the discovery site. Leaving Three Guardsman Pass at just over 900 metres the Haines Highway begins a gradual and long descent to the ocean, at the heads of Chilkat and Chilkoot inlets. The descent follows a tributary of the Klehini River to the river itself at Rainy Hollow. In contrast to the steep upper-mountain-valley slopes, the valley-bottom slopes are relatively gentle. The valley floor is a wide, flat flood plain, much of it covered in braided networking channels (fig. 4). Several small, eastern tributaries of the Klehini flow down gentle gradients compared to the

Figure 3. A rolling alpine landscape in the Chilkat Pass, Haines Highway, Alaska. Richard Hebda photograph. Colour version on page 581.

Figure 4. A braided channel flood plain on the Chilkat River, Haines Highway, Alaska. Richard Hebda photograph. Colour version on page 581.

deeply incised main stem of the river to the west—an important point when considering access to the gentle rolling terrain above. The Klehini joins the Chilkat River at Klukwan, and the braided flood plain widens and extends southeastward to large tidal flats at the head of Chilkat Inlet near Haines, Alaska. At flood times, large parts of the Chilkat River flood plain, between the confluence of the Klehini and the tidal flats, are under water. Along the Klehini River, glacier snouts are well above the valley bottom. But on the lower Chilkat River and along Lynn Canal, glacial snouts occur at low elevations and contribute to great fan deltas on the river flood plain or along the ocean shore. Except for these features, the valley walls of the Lynn Canal of the upper Chatham Depression rise steeply to icefields or peaks well over 1,000 metres, some nearing 2,000 metres, above sea level in a distance of only five to ten kilometres.

Climate

THE LATITUDE AND HIGH RELIEF of the region result in a remarkably diverse climatic pattern. Two general features are central to understanding the climatic variation and sharp changes over very short distances within the region. First, the Kwädąy Dän Ts'ìnchį discovery area is located in the transition zone from coast to continental interior. Cold, dry arctic air masses compete with cool, moist ocean air masses along a steep and varying gradient. Indeed, the region may exhibit one of the steepest ocean–inland climate gradients on earth. Second, the region's high relief results in sharp climatic changes, with moist marine climates scarcely 10 kilometres from high-elevation alpine climates.

These regional contrasts can be appreciated from climate stations nearest to the discovery site: Haines Junction, Yukon, compared to Haines and Yakutat, Alaska (table 1). But these records do not capture the reality on the ground, especially at high elevations.

Table 1. Climatic characteristics of three stations near the discovery area. US data converted to metric values.

Station location	Elevation	Mean annual temperature (°C) (Mean temperatures in January and July)	Mean annual precipitation (mm)	Mean annual snowfall (cm)
Haines Junction, Yukon[1]	600 metres	-2.9 (-21.4, 12.6)	305.8	159.8
Haines, Alaska[2]	Sea level	8.9 (-4.3, 14.7)	1214.1	311.7
Yakutat, Alaska[3]	Sea level	3.9 (-3.8, 12.0)	3843.0	492.8

[1] "Canadian Climate Normals," Government of Canada, accessed October 2009.
[2] "Recent Climate in the West," Desert Research Institute, accessed October 2009.
[3] "Yakutat," ClimateZone.com, accessed October 2009.

Haines Junction, Yukon, has a climate typical of a continental regime. The mean January temperature is only -21.4°C, but the mean July temperature is 12.6°C. With just 305.8 millimetres of precipitation, the climate is very dry, but the precipitation, whether snow or rain, is more-or-less evenly distributed throughout the year.

About 200 kilometres to the south at Haines, Alaska, the climate is much milder, with the January mean just below freezing, at -4.3°C, but the July mean of 14.7°C only slightly warmer than at Haines Junction. Annual precipitation is much greater, at 1,214.1 millimetres, falling heavily in the autumn after a relatively dry summer. Much snow falls in late autumn and winter.

At Yakutat on the open coast (180 kilometres southwest of Haines Junction, 225 kilometres west of Haines), mean daily January and July temperatures are similar to those at Haines, but the precipitation is three times greater, and almost five metres of snow falls each year. Considering the rate of cooling with elevation, it is easy to appreciate why the glaciers are so extensive near the coast, even at low elevations. Inland at Haines Junction there is only one-tenth the precipitation and less than one-third the amount of snow—consequently there are few glaciers.

Snowfall occurs at Haines Junction 10 out of 12 months of the year, though only in trace amounts during August and September. At Haines, Alaska, no snow falls from June through August, and only trace amounts occur in May and September. Considering the much higher elevation of the discovery site than Haines Junction (1,600 vs. 600 metres above sea level) and its inland position compared to Haines, snowstorms and snowfall should be expected in August and September. For example, Al Mackie reported that snow fell on August 22 and 23, 1999, while he visited the discovery site (see chapter 4). J.S. Peepre and Associates (1992) describe a climate more representative of the discovery site, from short-term observations near Tats Lake at 750 metres above sea level and 50 kilometres to the northwest. Mean daily temperatures were 10.5°C, 12.4°C and 10.4°C for June, July and August, respectively. They estimated the mean January temperature to be -23.4°C and the mean

July temperature to be 3.5°C, much lower than at the main stations in the region. Clearly even summer temperatures can be cold.

Geology

THE COMPLEX REGIONAL GEOLOGY is well illustrated in the online geological compilation map of British Columbia (Cui and Erdmer 2009), which incorporates mapping by Campbell and Dodds (1983) and Mihalynuk et al. (1993). The region spans the Duke-Denali fault system, which includes slices of non-North American crust belonging to the Wrangellia and Alexander terranes (Coney et al. 1980). Much of the area between the Haines Highway and the discovery site, including most of the upper Tatshenshini drainage, was sliced and shuffled by these faults. This great series of faults compares in scale and character to the San Andreas system. One of the western fault traces in the series occurs immediately east of the discovery site in the north-northwest trending valley into which the Samuel Glacier flows. This same fault marks the eastern limit of contiguous Alexander crustal terrane and of the Lower Paleozoic sedimentary rocks (shale and sandstone, limestone and dolomite) that form much of that part of the crust. The rocks in the immediate area of the discovery site are mainly Ordovician to Silurian limestone and silty limestone with minor volcanic rocks deposited at a tropical continental margin about 400 million years ago.

Rocks of the Tats geological group, exposed near Tats Creek to the west of the discovery site, are of particular interest to the story of the Kwädąy Dän Ts'ìnchį man. About 220 million years ago, in the Late Triassic period, a rift formed in the crust and was filled by fine-grained sediments and volcanic rocks together with a massive deposit of copper and cobalt-rich sulphide (MacIntyre 1984; Mihalynuk et al. 1993). Chapter 24 by Cooper and others describes a copper artifact recovered with the Kwädąy Dän Ts'ìnchį man.

Younger rocks, such as Cretaceous and Tertiary volcanic rocks (about 140–50 million years old), occur as remnants of once widespread surface rocks now eroded away and preserved in down folds or down faulted basins within the ancient Alexander terrane basement. Widespread blankets and packages of unconsolidated Ice Age deposits (Quaternary Period, less than 2.6 million years old) have been modified by rivers within broad valleys of the Tatshenshini and major tributaries such as the O'Connor River.

Vegetation

THE REGION'S VEGETATION reflects the marked variation in climate and landforms, changing accordingly up slope and inland. Beginning along the marine coastal lowland, climatically mild sites support rich and productive coastal rainforest dominated by great coniferous trees (fig. 5). These give way over a short distance inland and up slope to cool, damp forests dominated by Mountain Hemlock (*Tsuga mertensiana*; fig. 6). Not too much further inland and further up slope the Mountain Hemlock forest changes to parkland with scattered trees and intervening low-growing shrubs and herbaceous vegetation (fig. 7). Further inland and up slope dense, shrubby thickets of alder and willow blanket the landscape (fig. 8). At high elevations, alpine tundra consisting of scattered patches of herbaceous species replaces the shrubs (fig. 9). On the highest places, patches of alpine turf alternate with lichen-covered rock fields (see chapter 5, fig. 4), and in some places the conditions are too harsh to support any plant growth at all. The increasingly cold and dry climate inland sees the coastal conifer forest of hemlock and spruce replaced by forests of White Spruce (*Picea glauca*) and Lodgepole Pine (*Pinus contorta*), here and there with stands of Trembling Aspen (*Populus tremuloides*) (fig. 10).

Figure 5. Western Hemlock–Sitka Spruce coast forest. Richard Hebda photograph.

Biogeoclimatic Vegetation Types of the Region

THE GENERAL PATTERN OF VEGETATION described in the preceding paragraph is more precisely described through the biogeoclimatic zone classification system used to represent ecosystems in British Columbia (Meidinger and Pojar 1991). These zones and their subunits permit a detailed description of the landscape's ecosystems according to the key species, and the vegetation subtypes reflect the nature of the environment as experienced by travellers in the region.

Figure 6. Mountain Hemlock forest below Three Guardsmen Pass. Richard Hebda photograph.

Figure 7. Mountain Hemlock parkland in a sea of alder thickets. Richard Hebda photograph. Colour version on page 582.

Figure 8. Alder and willow thickets with meadow patches in the foreground at Three Guardsmen Pass. Richard Hebda photograph.

Figure 9. Alpine tundra and rolling terrain looking west to the discovery site and Samuel Glacier. Richard Hebda photograph.
Colour version on page 582.

Figure 10. Boreal forest, north approach to Chilkat Pass, Haines Highway. Richard Hebda photograph. Colour version on page 583.

There are six biogeoclimatic zones in the study region in British Columbia, a high diversity for this relatively small area (fig. 11 , page 583). On the coast, lowland sites are occupied by the Coastal Western Hemlock zone and immediately above it the Mountain Hemlock zone. The Coastal Mountain Alpine zone occurs at the highest elevations, which are strongly influenced by damp and relatively mild coastal air masses. Under more continental inland climates, such as in the pass along the Haines Road, mid elevations equivalent to the Mountain Hemlock zone support the Spruce-Willow-Birch zone and above it the Boreal Altai Fescue Alpine zone, a recently recognized division of the Alpine Tundra zone. Forested areas inland are occupied by the Boreal White and Black Spruce zone (fig. 12). These large ecological units are not uniform throughout BC, and this is reflected by variations called subzones and variants specific to a region. Although these zone classifications are not used in adjacent Alaska and Yukon, the ecosystems are the same or similar and the

descriptions are suitable for these parts of the study region, too.

The particular kind of Coastal Western Hemlock forest in the region is the wet maritime (wm) subzone, which occurs along the marine shoreline of Lynn Canal and its off-shoots. According to Banner et al. (1993), it can be found from sea level up to 300–500 metres in elevation. In general the climate is rainy and snowy, and the summers are cool and moist.

The dense, closed forests consist mainly of Western Hemlock (*Tsuga heterophylla*) and Sitka Spruce (*Picea stichensis*), with Black Cottonwood (*Populus balsamifera* ssp. *trichocarpa*) forming great stands on the floodplains of the valley bottoms such as along the lower reaches of the Klehini and Chilkat rivers. Red Alder (*Alnus rubra*), Paper Birch (*Betula papyrifera*) and Subalpine Fir (*Abies lasiocarpa*) occur in some forest stands. Under the trees, common shrubs include several species of blueberry (*Vaccinium* spp.), False Azalea (*Menziesia ferruginea*) and Devil's Club (*Oplopanax horridus*). There is a diversity

Figure 12. Alpine above spruce, willow and birch scrub at north approach to Chilkat Pass. Richard Hebda photograph. Colour version on page 583.

of understory herb species such as Bunchberry (*Cornus canadensis*), Five-leaved Bramble (*Rubus pedatus*) and Spiny Wood Fern (*Dryopteris expansa*), and a rich moss cover.

Only a small patch of the Mountain Hemlock zone occurs in the region, and is encountered by visitors descending the Haines Highway southward from Chilkat pass. A forest or parkland of Mountain Hemlock and Subalpine Fir is characteristic (fig. 7). At lower elevations Mountain and Western hemlock occur together. Shrubs of the heather family grow under the trees and in openings. Blueberries and huckleberries (*Vaccinium* spp.) predominate in the more forested sites, whereas various species of mountain heather (*Cassiope* and *Phyllodoce* spp.) predominate in open conditions. Herbaceous plants are relatively uncommon.

The Coastal Mountain Alpine zone is restricted to high elevations in the western part of the region immediately above the Mountain Hemlock zone. It is typically dominated by extensive mats of Pink and

White mountain heathers (*Phyllodoce empetriformis* and *Cassiope tetragona*). The widespread occurrence of glaciers and snowfields in the Coast Mountains limits the area of occurrence of this ecosystem compared to the inland alpine tundra known as the Boreal Altai Fescue Alpine zone.

Ecosystems largely isolated from the strong influence of mild, moist coastal air predominate in the study region and present a very different collection of species from those of the coastal zones. Low-elevation forests of the dry cool and very wet cool subzones of the Boreal White and Black Spruce zone occupy the major inland valley bottoms and adjacent slopes such as along the Tatshenshini and Alsek rivers and the northern approaches to Chilkat Pass (see fig. 11). Typical stands of the dry cool forests are composed mainly of White Spruce, but in places there can be abundant Trembling Aspen and stands of Lodgepole Pine. Common shrubs include Soopolallie (*Shepherdia canadensis*), Highbush-cranberry (*Viburnum edule*), Labrador Tea (*Rhododendron groenlandicum*), Twinflower (*Linnaea borealis*) and Bunchberry.

The very wet cool subzone is unique to the Tatshenshini and Alsek valley bottoms west of the discovery site, in the transition from the interior to the coastal climate (Pojar 1993). Heavy snowpack and strong winds shape the vegetation here, and fire is infrequent. White Spruce stands typical of the Boreal White and Black Spruce zone are uncommon; instead, wide-ranging forests of Black Cottonwood form the canopy on the valley floor and the outwash terraces. Scattered Paper Birch and Trembling Aspen also occur. Common forest shrubs include Sitka Alder (*Alnus viridis* ssp. *sinuata*) and Highbush-cranberry. Willows and Sitka Alder also form shrub thickets. Gravel sites support mats of Mountain Avens (*Dryas drummondii*) with lichens.

Open areas occur widely in this spruce-forest zone, including wetlands and grasslands. Wetlands, mainly bogs and fens, appear on valley bottoms and poorly drained flat terrain. Bogs occur where the water is stagnant, and are often associated with the accummulation of cold air off the mountain slopes, restricting tree growth. Scrub Birch (*Betula nana*) and Labrador Tea grow in the shrub layer, Crowberry (*Empetrum nigrum*), Cloudberry (*Rubus chamaemorus*) and sedges dominate the herb layer, and sphagnum forms the moss cover. Along with sedges and relatives, willows, Mountain Alder and Scrub Birch predominate. South-facing slopes on coarse-textured soils support grassland scrub patches. Under such conditions grow stunted Trembling Aspen, Lodgepole Pine, Prickly Rose (*Rosa acicularis*), Saskatoon (*Amelanchier alnifolia*), junipers, Kinnikinnick (*Arctostaphylos uva-ursi*) and Pasture Sage (*Artemisia frigida*). Several grass species occur in these settings but rarely do they dominate.

The most widespread ecosystem type is the Spruce-Willow-Birch zone, which is characteristic of mid-elevation inland landscapes occurring above the Boreal White and Black Spruce zone. The Spruce-Willow-Birch zone occupies what might be thought of as the subalpine zone (600–1,400 metres in the study region), where the winter conditions are particularly harsh and the growing season is short and cool. Based on limited climatic data, this zone appears to be relatively dry, averaging about 700 millimetres of precipitation annually. The lower part of the zone is more-or-less forested with Subalpine Fir and White Spruce, but the upper portion is often a mass of shrubby thickets of willow, alder and birch (fig. 12).

The general pattern in the lower part of the zone consists of closed to partly open forests of spruce and Subalpine Fir. Trembling Aspen and Lodgepole Pine may also be present. The higher up you go, the more Subalpine Fir there is. Below the canopy, shrubs are abundant, predominantly Grey-leaved Willow (*Salix glauca*) and Scrub Birch, but also including widespread shrubs such as Soopolallie, Shrubby Cinquefoil (*Pentaphylloides floribunda*) and willow species. Many herbaceous and ground-level woody species occur too, such as Crowberry, Lingonberry (*Vaccinmium vitis-idaea*), Dwarf Blueberry (*Vaccinium caespitosum*), Twinflower and Arctic Lupine (*Lupinus arcticus*), as well as the grasses Altai Fescue (*Festuca altaica*) and Fireweed (*Epilobium angustifolium*). Mosses and reindeer lichens carpet the ground.

The extensive shrubby landscapes of the upper Spruce-Willow-Birch zone are one of the remarkable features of the region. Seemingly endless tracts of deciduous willows and Scrub Birch, 50–100 centimetres tall, blanket valley bottoms and lower slopes at mid and higher elevations (fig. 13). Several species of willow—such as Grey-leaved (*S. glauca*), Barclay's (*S. barclayi*), Tea-leaved (*S. planifolia*), Barratt's (*S. barrattiana*) and Woolly (*S. lanata*)—combine into dense masses of upright and creeping stems that are nearly impenetrable. Wet and dry openings occur widely. For example, valley bottoms subject to cold air ponding are a patchwork of thickets with fens of cotton grasses (*Eriophorum angustifolium* and *E. chamissonis*) and Altai Fescue grasslands, according to soil drainage. At the upper edge of the zone, the thickets intersperse with herb-dominated alpine communities rich in wildflowers.

In the Tatshenshini-Alsek portion of the region, which has relatively abundant moisture and a heavy snowpack, Black Cottonwood occurs abundantly, even forming the treeline (Banner et al. 1993). Sitka Alders form

Figure 13. Dwarf birch, alder and willow thickets, Kelsall Lake, Haines Highway. Richard Hebda photograph.

dense, impenetrable thickets several metres tall along with Barclay's Willow and Alaska Willow (*Salix alaxensis*). Lush meadow communities occur widely, characterized by Fireweed, Red Raspberry (*Rubus idaeus*) and Cow-parsnip (*Heracleum maximum*). Dwarf shrub patches include Arctic Willow (*S. arctica*), Crowberry, and Bog Blueberry (*Vaccinium uliginosum*) (Pojar 1993).

Alpine tundra can be seen from almost any place in the region, covering many of the mountain peaks (fig. 14). Above the Mountain Hemlock zone and adjacent to coastal mountain glaciers, there are relatively small patches of the Coastal Mountain Alpine zone starting at about 950 metres in elevation. Inland, the Boreal Altai Fescue Alpine zone has generally less snow and harsher conditions; the zone begins at 1,000–1,600 metres above sea level.

Tree species grow only as low shrubs, and Subalpine Fir is the most common, except on lands adjacent to the Mountain Hemlock zone, where Mountain Hemlock also occurs. In wet coastal-mountain alpine areas, dwarf members of the heather family, including pink and white mountain heathers (*Phyllodoce empetriformis*; *Cassiope* spp.) and Partridgefoot (*Luetkea pectinata*), occur widely. Rich meadows full of colourful wildflowers such as Cow-parsnip and lupine are widespread. Arrow-leaf Ragwort (*Senecio triangularis*) and Sierra Larkspur (*Delphinium glaucum*) occupy sites with moist and relatively deep soils.

The open landscape of the Boreal Altai Fescue Alpine zone is one of the remarkable features of the region. Tussocks of fescue and patches of lichen dominate a largely herbaceous tundra populated by scattered wildflowers, sedges and Mountain Sagewort (*Artemisia*

Figure 14. Alpine tundra in Chilkat Pass, Haines Highway, Alaska. Richard Hebda photograph. Colour version on page 584.

norvegica). Where fescue is less common and soils less stable because of freeze-thaw and wind disturbance or coarse substrates, an alpine turf of sedges, mat-forming dwarf willows (e.g., *Salix reticulata, S. polaris*) and low-growing herbs forms a firm surface. At the highest elevations, the landscape is often dominated by patches and even fields of angular gravel and boulders up to a metre or more across. Plants grow sparsely on these surfaces and often only crust-forming lichens occur (see fig. 4 in chapter 5). Cushion-forming species such as Moss Campion (*Silene acaulis*) provide striking colour in midsummer. Travel on foot in the Boreal Altai Fescue Alpine zone is relatively straightforward compared to parts of the Spruce-Willow-Birch zone, where dense thickets frustrate progress.

At about 1,600 metres above sea level, the Kwädąy Dän Ts'ìnchį discovery site is located just inside the distribution of the Boreal Altai Fescue Alpine zone. The description of the flora (chapter 5) and photographs of the site surroundings give a good impression of the environment. Vegetation is relatively sparse over the abundant coarse rocky debris, but where it occurs it is dominated by alpine turf species, such as sedges and low-growing herbs typical of the upper alpine tundra. More lush plant cover occurs at slightly lower elevations and on gentle slopes.

Environmental History

THE RECENT AGE OF THE DISCOVERY indicates that vegetation and plant communities have changed little to the present time. But studies of pollen and spore remains from lakes and wetlands in northern BC reveal that wide-ranging ecological and climatic changes have occurred over the past 10,000 years, resulting in the ecosystem pattern visible today (Hebda 1995). No investigations have been carried out close to the discovery site or even along the Haines Highway, but two investigations—one along the White Pass–Yukon corridor (Spear and Cwynar 1997) and one on the Chilkat Peninsula in Alaska (Cwynar 1990)—provide insight into the region's history.

Waterdevil Lake is located in the northern part of the White Pass basin at 875 metres elevation in woodlands of Lodgepole Pine and Subalpine Fir, with abundant Sitka Alder. White Spruce occurs abundantly a few kilometres to the north. Before 8,500 radiocarbon years ago, shrub tundra occurred with alder on moist sites and dwarf birches on dry sites. There were likely a few groves of poplar trees (presumably Balsam Poplar [*Populus balsamifera*]). In the next interval, 8,500–5,200 years ago, warming and drying is revealed by the abundance of Common Juniper (*Juniperus communis*) pollen and the occurrence of juniper needles. The landscape was still covered in shrub tundra but it was apparently drier than in the previous interval. White Spruce arrived near the lake just before 6,000 years ago. Between 5,200 and 4,000 years ago the landscape was covered by forest-tundra in which spruces (probably White Spruce) were scattered among thickets of alders and dwarf birches. The modern pine and Subalpine Fir woodland arose during the last 4,000 years, partly in response to a cooler and moister climate than in the preceding interval but also because pine may have only at this time migrated into the region.

At a slightly higher elevation of 910 metres, in the White Pass–Yukon corridor, just below today's alpine tundra zone (probably equivalent to the upper Spruce-Willow-Birch biogeoclimatic zone), the cover of heaths and dwarf birches has changed little over the past 10,000 years (Spear and Cwynar 1997). The area began and has remained as some form of "alpine shrub tundra", in Spear and Cwynar's (1997) words, until the present. Scrubby Mountain Hemlocks and Subalpine Firs may have persisted through most of the record.

Much of the inland landscape around the Kwädąy Dän Ts'ìnchį discovery is at even higher elevations than White Pass and farther along the migration route of pines and spruces. Consequently, it is likely that the high-elevation land near the discovery site has been covered by shrub tundra for all of the past 10,000 years and true alpine tundra at the highest elevations. Trees, except perhaps for scrubby Subalpine Firs, are likely only recent immigrants to the region, presumably arriving mostly from the Yukon to the north (MacDonald and Cwynar 1985, Ritchie and MacDonald 1986).

The history of the milder coastal zone is very different from that of the interior uplands and valleys (Cwynar 1990). As already mentioned, conifer forests of the Coastal Western Hemlock biogeoclimatic and Mountain Hemlock zones (or their Alaska equivalents) occur here in a narrow band adjacent to the marine shore. The earliest vegetation began with woodland of Lodgepole Pine (presumably the coastal form), a widespread coastal vegetation type at the time (Hebda and Whitlock 1997). The woodland at first included abundant ferns and herbs such as sagewort (*Artemisia* spp.) and later a lot of shrubby alder. Sitka Spruce, which grows there today, and a species of poplar became elements of the vegetation as pine declined 9,500–8,000 years ago. Western Hemlock, now a dominant forest species, likely arrived 8,000 years ago and was well established as a forest component by 6,700 years ago. From 6,700 to 2,900 years ago, essentially modern shoreline forests became established, consisting of a mix of Sitka Spruce, Western Hemlock and some Mountain Hemlock. Apparently tree birch also joined the forest canopy. Fully modern closed forests were established at about 2,900 years ago with a decline in Mountain Hemlock and alder.

Wildlife

- -

THE ECOLOGICAL DIVERSITY, ruggedness and relative inaccessibility of the Tatshenshini-Alsek and adjacent territory result in it being one of the great wildlife regions of the continent (J.S. Peepre and Associates 1992). The search for wildlife brought the hunters to the Samuel Glacier area and led to the discovery of the human remains (see chapter 3 on the discovery). The southern Tutchone and Tlingit people know the wildlife well and have depended on it for millennia (chapter 9). The Kwädąy Dän Ts'ìnchį person was surrounded by artifacts fashioned from wildlife on the glacier before he died (chapter 26).

The region's wildlife can be broadly described using the biogeoclimatic subzones and variants (Banner et al. 1993), with key values highlighted from the Management Direction Statement for Tatshenshini-Alsek Park (BC Parks, 2001).

The terrestrial wildlife of the Coastal Western Hemlock zone's wet maritime subzone is not well documented, but its diversity is thought to be relatively low because of the variant's northern position and scarcity of estuaries (Banner et al. 1993). Nevertheless, American Black (*Ursus americanus*) and Grizzly (*Ursus arctos*) bears are known to be abundant. Northern and interior species such as the Wood Frog (*Lithobates sylvaticus*) and Northern Red-backed Vole (*Myodes rutilus*) are noted, too.

The wildlife of the Spruce-Willow-Birch zone, which predominates at mid to high elevations, is not well known but is recognized as being diverse. Grizzly and American Black bears occur widely, especially in summer. Dall's Sheep (*Ovis dalli*), Moose (*Alces americanus*) and Caribou (*Rangifer tarandus*) also inhabit this zone in summer. Wolverine (*Gulo gulo*), Canada Lynx (*Lynx canadensis*) and Grey Wolf (*Canis lupus*) are characteristic medium-sized species. Other notable mammals include the American Marten (*Martes americana*), Snowshoe Hare (*Lepus americanus*) and Arctic Ground Squirrels (*Urocitellus parryii*), as well as several vole species. The list of bird species is long, including in forest habitats, Spruce Grouse (*Falcipennis canadensis*),

Common Raven (*Corvus corax*), Boreal Chickadee (*Poecile hudsonicus*), Grey Jay (*Perisoreus canadensis*), Red-breasted Nuthatch (*Sitta canadensis*) and Ruby-crowned Kinglet (*Regulus calendula*). The widespread shrub habitat is favoured by Willow Ptarmigan (*Lagopus lagopus*), Gyrfalcon (*Falco rusticolus*) and Wilson's Warbler (*Cardellina pusilla*). Wetland lake habitats are used for breeding by a large number of shoreline and water birds; for example, I observed at least two species of gulls at the outlet of Kelsall Lake in the summer of 2008.

The relatively dry Boreal Altai Fescue Alpine zone hosts notable populations of Caribou, Dall's Sheep and Mountain Goat (*Oreamnos americanus*). Other characteristic mammal species are Grizzly Bear, Grey Wolf, Red Fox (*Vulpes vulpes*), Wolverine, Hoary Marmot (*Marmota caligata*) and Arctic Ground Squirrel. Dall's Sheep and Collared Pika (*Ochotona collaris*) are the only mammals that live at high elevations in the region in BC. The list of characteristic birds is also diverse in this zone, ranging from the Rosy Finch (*Leucosticte tephrocotis*) and Snow Bunting (*Plectrophenax nivalis*) to Gyrfalcon and Golden Eagle (*Aquila chrysaetos*). Ptarmigan—both Rock (*Lagopus muta*) and White-tailed (*Lagopus leucura*)—are widespread and often encountered in the summer.

In general, the Boreal White and Black Spruce zone in northern valleys in the region has relatively high wildlife species diversity (Banner et al. 1993). Boreal forest habitats are especially important for overwintering species. Fire disturbances in these forests provide a diversity of habitats and vegetation structure. Common mammal species noted by Banner et al. include Grey Wolf, American Black Bear, Canada Lynx, Moose, Caribou, Mule Deer (*Odocoileus hemionus*), Ermine (*Mustela erminea*), Snowshoe Hare, Red Squirrel (*Tamiasciurus hudsonicus*) and Deer Mouse (*Peromyscus maniculatus*). Beavers (*Castor canadensis*) are active in wetlands and on small streams (Champagne and Aishihik First Nations n.d.). The forest habitat supports abundant and diverse bird populations. Year-round resident species include Northern Goshawk (*Accipiter gentilis*), Great Horned Owl (*Bubo virginianus*), Ruffed Grouse (*Bonasa umbellus*), Common Raven, Grey Jay,

Downy Woodpecker (*Picoides pubescens*) and Black-capped Chickadee (*Poecile atricapillus*). There are also numerous small summer-resident birds such as Dark-eyed Junco (*Junco hyemalis*), Yellow-rumped Warbler (*Setophaga coronata*) and Purple Finch (*Haemorhous purpureus*). Many water birds breed in the zone, especially Northern Pintail (*Anas acuta*), scaups (*Aythya* sp.) and Green-winged Teal (*Anas crecca*).

The Management Direction Statement for Tatshenshini-Alsek Park (BC Parks, 2001) notes the occurrence of the rare Glacier Bear (*Ursus americanus emmonsi*), a subspecies of the American Black Bear. It also notes populations of the sensitive or vulnerable Great Blue Heron (*Ardea herodias*), Bald Eagle (*Haliaeetus leucocephalus*), Arctic Tern (*Sterna paradisaea*) and Northern Shrike (*Lanius excubitor*) in the park area.

The region is modestly rich in fish species (McPhail 2007; J.S. Peepre and Associates 1992), including several species of salmon important to First Nations. McPhail (2007) notes that this fisheries subregion is of biogeographic interest because the rivers support Pacific coastal species in their lower reaches and Berginian (northern) species such as Arctic Grayling (*Thymallus arcticus*) and Northern Pike (*Esox lucius*) in their headwaters portions.

Five species of salmon—Pink (*Oncorhynchus gorbuscha*), Coho (*Oncorhynchus kisutch*), Sockeye (*Oncorhynchus nerka*), Chinook (*Oncorhynchus tshawytscha*) and Chum (*Oncorhynchus keta*)—occur in the river systems, with Chinook, Coho and Sockeye in the inland parts of the Alsek and Tatshenshini rivers and tributaries (J.S. Peepre and Associates 1992; BC Parks 2001). The Tatshenshini contributes by far the most to the commercial fishery (J.S. Peepre and Associates 1992; chapter 20 this volume). The salmon are important food for other wildlife, especially Bald Eagles and bears, and are extremely important to the human inhabitants of the region (chapter 20). Other commonly occurring fish include Steelhead and Rainbow Trout (*Oncorhynchus mykiss*), Dolly Varden (*Salvelinus malma*), Lake trout (*Salvelinus namaycush*), Arctic Grayling, Round Whitefish (*Prosopium cylindraceum*) and Slimy Sculpin (*Cottus cognatus*) (J.S. Peepre and Associates 1992).

The mid to upper portions of the Tatshenshini and its tributaries and associated lakes (such as the outlet of the O'Connor River) are exceptionally important rearing habitat for Chinook and Sockeye salmon. Chinook spawning runs begin in May, with the peak run in the Tashenshini in late June to mid-July; Sockeyes migrate from late July to early September, and Coho migration peaks in mid-September to early October (J.S. Peepre and Associates 1992).

This region is one of diverse and contrasting climates and landscapes ranging from a wet and relatively mild coast to towering ice-covered mountains to cold, dry interior plateaus. This physical complexity is reflected in a diversity of ecosystems, from lush coastal conifer forests to cold, dry, sparse alpine rock fields. The fauna and flora include many common widespread species, as well as notable species restricted to either coastal or inland realms. The wildlife species that abound today have been well known by the people of the region for hundreds or thousands of years.

ACKNOWLEDGEMENTS

Alexander Mackie (BC Archaeology Branch), Jim Haggart (Geological Survey of Canada), Mitch Mihalynuk (BC Geological Survey), Will MacKenzie and Allen Banner (BC Ministry of Forests) and Jim Pojar helped obtain information for this chapter. Alexander Mackie and Mitch Mihalynuk and Allen Banner reviewed all or parts of drafts of the chapter.

3

GLACIAL SETTING AND SITE SURVEY OF THE KWÄDĄY DÄN TS'ÌNCHĮ DISCOVERY

Erik W. Blake

Since the end of the Little Ice Age (LIA), approximately 150 years ago, glaciers in northwestern British Columbia have been losing mass (Reyes et al. 2006). This loss has accelerated in recent decades, with ice cover decreasing by approximately eight per cent between 1985 and 2005 (Bolch et al. 2010). A serendipitous combination of geographical setting, climate trends and a visit by modern-day hunters led to the preservation and discovery of the Kwädąy Dän Ts'ìnchį man. These human remains stayed surprisingly intact for 167–297 years (Richards et al. 2007; Richards et al., chapter 6) within a small, actively flowing, unnamed glacier in the Datlasaka Range of northern BC. For the purposes of this chapter, I will refer to this glacier as the Fault Creek Glacier.

IN THIS CHAPTER I explore the importance of the glacial environment to the discovery and present a brief glacial history of the area, with a focus on the glaciers near the discovery site. I discuss aspects of travel in glaciated terrain as it might have related to the Kwädąy Dän Ts'ìnchį man's route choices, and I also present survey results from the 1999 and 2003 site visits showing the original layout of the discovery and how it changed over this short interval.

GLACIERS AND THEIR MECHANICS

THE CHARACTER OF THE GLACIER entombing the Kwädąy Dän Ts'ìnchį man strongly influenced the physical context of the discovery site and the nature of the discovery itself. Understanding the preservation of the remains requires an understanding of what glaciers are and how they work.

Glaciers form wherever the rate of snow accumulation exceeds melt (ablation) over many years. Under these conditions, there is a net accumulation of snow, which eventually compresses under its own weight to form ice. Accumulation occurs primarily in the upper reaches of a glacier and ablation in the lower reaches. The *ablation zone* (the lower part of a glacier, where the rate of snow melt exceeds that of accumulation when averaged over a year) is replenished continuously by down-slope flow from the *accumulation zone* (the upper part of a glacier, where accumulation exceeds melting). In the accumulation zone, snow is compressed and transformed into a porous, sintered mass called *firn* (sintering is a sublimation process that causes snow crystals to grow together over time). Further compression seals off the air passages between ice crystals, forming bubble-filled, non-porous ice. Ice can also form when surface melt percolates down into snow and freezes, thereby filling and sealing the voids. This is called regelation (or refrozen) ice.

The retreat of a glacier appears to be an uphill motion of the glacial terminus, but the ice never stops moving downhill. Retreat occurs when ice melts faster in the ablation zone than it is replenished by flow from the accumulation zone. A retreating glacier is said to have *negative mass balance*.

Equilibrium Line

AN IMAGINARY LINE called the equilibrium line separates the accumulation and ablation zones. This is the line where the rate of annual accumulation equals annual melt (Østrem and Brugman 1991). The equilibrium line altitude (ELA) changes in response to long-term shifts in temperature and precipitation. The line separating snow from ice on the glacier surface at the end of the melt season approximates the location of the equilibrium line (Østrem and Brugman 1991, 65). One can see this approximate equilibrium line on several glaciers in the aerial photograph figure 1, taken in late August 1979 (see point *f*), where the snow appears white and the ice appears darker (in part because a thin layer of windblown debris, accumulated over successive summers, covers it).

No systematic studies of historical ELA exist for northern British Columbia. Studies of ELA changes in the European Alps from the mid 19th century to the present show an upward shift of 100 metres in the ELA corresponding to a temperature rise of 0.5°C (Haeberli 1994), and similar studies in Norway yielded comparable changes in ELA (Torsnes et al. 1993). Studies on Canada's Baffin Island give larger increases in ELA since the end of the LIA (Williams 1978). As these authors note, ELA response to climate change exhibits regional variation, so these results cannot be applied quantitatively to glaciers in northwestern BC.

Objects on the glacier surface above the ELA will become buried in snow, firn and ice, whereas objects below the ELA will melt out of the surface. Relevant to the discovery site is that the ELA can change with time.

42

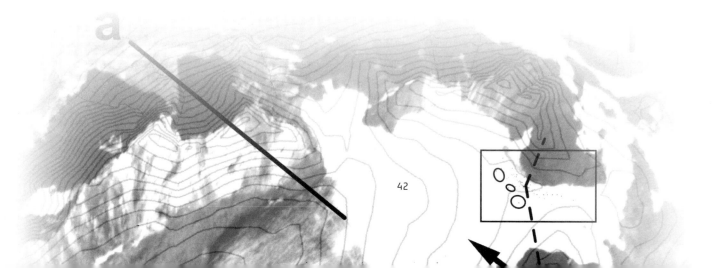

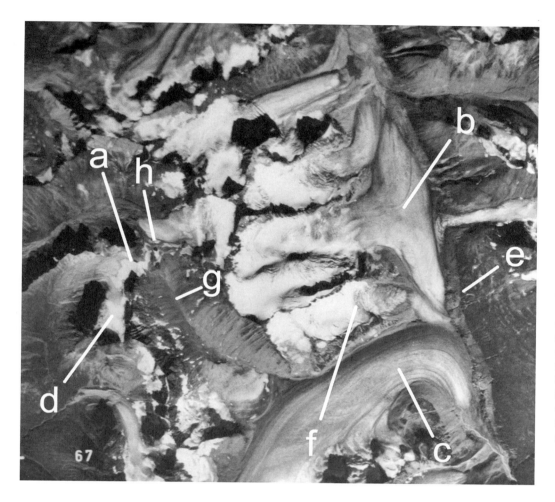

Figure 1. Aerial photograph from 1979 showing A, The Kwädąy Dän Ts'ìnchį discovery site; B, Samuel Glacier, which initially flows eastward and then splits to flow north and south; C, Samuel II Glacier; D, Fault Creek Glacier; E, Little Ice Age maximum trim line for the Samuel Glacier; F, approximate equilibrium line on a Samuel Glacier tributary; G, unglaciated valley east of the discovery site; H, marginal moraine headwall to the G valley. Photo 67, Series A25292, August 23, 1979. Courtesy of Natural Resources Canada, 2017. Aerial photograph A25292-67.

Ice Flow and Deformation

THOUGH WE NORMALLY THINK of ice as a rigid solid, it can flow like a viscous liquid. As ice thickens, the increasing pressure on the ice below changes its mechanical properties and makes it easier for the ice to flow. The flow law for ice is non-linear and depends on depth, slope angle, temperature, crystal size and contaminant (debris) content (Glen 1958, Weertman 1973). A simplified form of the flow law is

$$\dot{\varepsilon}_{xy} = A_0 e^{-Q/RT} \tau_{xy}^n$$

where $\dot{\varepsilon}_{xy}$ is the shear strain rate tensor and τ_{xy} is the deviatoric shear stress tensor. T is absolute temperature, R is the gas constant and Q is the activation energy for creep. A_o, in turn, depends on the hydrostatic pressure (imposed by the ice above), the size and shape of the ice crystals, ice crystal orientation and the concentrations of air bubbles and impurities in the ice. Some of these parameters (e.g., crystal size and orientation) depend on the strain history of the ice, so the interaction between the variables is complicated. There is some debate about the value of n, but a reasonable value is $n = 3$ (Cuffey and Paterson 2010). Therefore, if stress is doubled, strain (deformation) increases eightfold.

Some of the implications of this formula are as follows:

- Ice flows faster when it is warm.
- Ice in a bowl (e.g., a volcanic caldera) will not flow because there is little or no shear stress (this is also a rather intuitive conclusion).
- Glaciers flow by internal deformation even if the base is frozen to bedrock.
- Ice near the bottom of a glacier deforms far faster than ice at the surface because of the third-power functional relation between stress (pressure and shear) and strain (deformation).

This last point is particularly interesting in relation to objects buried in a glacier. If objects are near the surface or in stagnant (non-flowing) ice, then they will tend to be preserved intact. An excellent archaeological example of this preservation is found in southern Yukon, where artifacts of wood, sinew, feathers and bone have been preserved in largely stagnant ice for as long as 9,400 years (Hare et al. 2004; Dove et al. 2005). Conversely, if an object is buried deeper in a glacier (for example, by falling into a deep crevasse), it will be subjected to significant shear and compression and will sustain damage by being stretched and torn apart.

Temperature in a Glacier

THE BOUNDARY CONDITIONS imposed by the environment determine ice temperature within a glacier. Because of annual fluctuations in air temperature, snow accumulation, meltwater percolation and solar warming, there is a surface layer (about 15 metres thick) within which temperature changes annually. Below this surface layer, the cumulative effect of these processes sets the ice temperature. Geothermal heat flux (heat escaping from the Earth's interior), friction and meltwater warm the glacier base. Cuffey and Paterson (2010) provide a thorough discussion of temperature in glaciers; below is a brief overview.

A glacier's surface cools in winter, which creates a wave (or zone) of below-average temperature that moves down into the glacier. This cold wave continues downward even as the surface begins to warm in spring. The amplitude of the annual cold wave decreases with depth until it vanishes at a depth of about 15 metres. With some exceptions, the temperature of the ice below this surface layer will experience no annual variation.

Some glaciers are below the melting point at the bed. Such a glacier is frozen to its bed, and the entire glacier is below the melting point. These so-called polar glaciers are found at high altitudes and in polar regions. They are characterized by slow flow (entirely by deformation of cold ice) and generally small size.

An intriguing phenomenon occurs when glaciers experience significant surface melt in the summer. Meltwater percolating down through snow, firn and cracks warms the ice below. Eventually, the glacier's entire thickness (with the exception of the surface layer) rises to the melting point. These glaciers are termed temperate glaciers. At first glance, one might expect that a temperate glacier cannot be stable since the ice is at the melting point—it seems that it should rapidly melt away. Yet this does not happen because no sufficient source of heat exists to melt all the ice.

Because the bed of a temperate glacier is at the melting point, water is present (owing to geothermal, frictional and strain heating, as well as introduction of surface melt), and this can result in sliding. The glacier's base may lie on solid rock or on a layer of deforming sediment. Temperate glaciers move faster than polar glaciers because they move by both sliding and deformation mechanisms (fig. 2A). Indeed, because the ice is warmer, they also deform faster, according to the flow law discussed earlier. Both types of flow contribute to the movement observed at the surface. Basal water lubricates glacier sliding, and the flow velocity of temperate glaciers varies in response to the amount of meltwater reaching the bed (Fountain and Walder 1998).

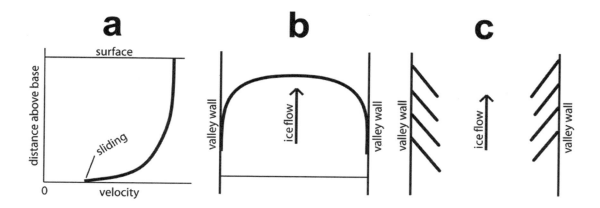

Figure 2. *A*, graph of depth vs velocity profile in glacial ice. Note that ice near the surface moves almost as a solid block. As one nears the bed, the ice becomes progressively more plastic, and therefore the velocity decreases exponentially. This diagram also shows a velocity contribution from sliding. *B*, A schematic aerial view showing the velocity profile across the width of a valley glacier. Maximum velocity occurs at the centre of the valley. Frictional drag at the margins decreases velocity. *C*, Aerial view of glacier showing marginal crevasses (diagonal lines). Tension from the lateral velocity gradient (*B*) creates these crevasses.

Many glaciers in an alpine environment display characteristics of polar glaciers high in the accumulation area and transform into temperate glaciers (except for a layer of cold surficial ice) as the ice flows into the ablation zone. These glaciers are termed subpolar or polythermal (multiple temperature) glaciers (Clarke and Blake 1991).

The Fault Creek Glacier is almost certainly a polythermal glacier. Studies show that Tats Glacier, 50 kilometres west of the study site, is a polythermal glacier, with sub-freezing temperatures below the surface layer yet significant surface melt in summer (J.P. Schmok, pers. comm.).

Crevasses

CREVASSES are cracks in a glacier that can extend downward for tens of metres. They form when the strain imposed by glacier flow cannot be absorbed by ice deformation. For example, if a glacier is travelling over a basal bump, the surface ice experiences tension (stress) as the glacier passes over the obstacle. If the ice flow is slow enough, then the ice can accommodate this by stretching (straining) according to the flow law. As

we have seen, ice at a glacier's surface is generally colder (a lower value of T) and has less pressure from overlying ice (resulting in lower values of τ_{xy}), so the strain rate ($\dot{\varepsilon}_{xy}$) is lower. Simply put, if the glacier stretches too fast, it cracks. This process causes the dramatic crevasse fields and ice falls in glaciers as they move down irregular mountainsides. Nevertheless, crevasses rarely cut completely through a glacier, because the glacier's ability to deform in response to stress, rather than crack, increases considerably with depth.

In a valley glacier, frictional drag at the margins (valley walls) causes the flow velocity to decrease (fig. 2*B*). This results in strain rates high enough to crack the ice and form marginal crevasses. These crevasses angle up-glacier from the margins (fig. 2*C*) and make travel on glacier margins quite dangerous.

Cryoconites (Melt Pockets)

CRYOCONITES are shallow melt holes found on the surface of glaciers. An object on an ice surface is invariably darker than the ice (it has a lower albedo) and therefore absorbs more solar energy. As a result, the object will warm preferentially and sink into a puddle of its own making. Cryoconites appear wherever ice temperatures reach the melting point. For example, Boon et al. (2003) discuss cryoconites forming up to 20 centimetres deep on John Evans Glacier in the Canadian high Arctic.

Cryoconites can continue to grow even if the glacier surface has frozen over. This is because light penetrates the surface and can warm a dark object buried in the ice. As long as the ice around the object is at the melting point, any solar heat absorbed will cause the object to sink deeper into the ice. Cryoconites support ecosystems consisting of bacteria, algae and diatoms (e.g., Mueller and Pollard 2004). If there is a drain for meltwater around an embedded object, then an air gap between the object and the ice can develop. The gap would be above and to the sides of the object (see chapter 4).

Glacial Landforms

WHEN A GLACIER RETREATS, it leaves behind a U-shaped valley quite distinct from the V-shaped valley carved by liquid water. Typically, a series of terminal moraines arc across the valley floor, creating a series of dammed lakes and tarns. These moraines are made up of basal material being ploughed forward or carried by the glacier.

Marginal moraines form when material is excavated by the glacier margin or when material falls onto the glacier surface from the valley wall. Where two glaciers meet, the marginal moraines merge to form a medial moraine. Medial moraines appear as longitudinal dark stripes on the glacier surface. In the ablation zone, a sufficiently thick medial moraine will insulate the underlying ice, causing the medial moraine to form a ridge as the surrounding ice surface melts downward.

Marginal moraines often contain a core of orphaned (remnant) glacial ice. As this insulated ice melts over a period of many years, it keeps the surface of the marginal moraine wet and unstable.

Other landforms are also created by glacial retreat. Water reaching the glacier bed accumulates into subglacial (under-ice) drainage channels and eventually exits at the terminus. Subglacial channels may fill up with size-sorted material, and when the glacier melts away, these water-channel "casts" form sinuous ridges, called eskers, that meander across the valley floor. As well, buried or partly buried blocks of ice may be left behind as a glacier retreats. When these blocks melt, they leave behind small, deep lakes, called kettle lakes (see Flint and Skinner 1974 for a basic overview).

Regional Glacial History

A WIDESPREAD COLD PERIOD between about 1280 and 1860 AD (Holzhauser and Zumbühl 1999) known as the Little Ice Age (a term coined by Matthes 1939) was only the last in a series of "cold snaps" that occurred after the end of the Wisconsin Glaciation, more than 10,000 years ago. For example, two outlet glaciers from the Juneau Icefields (about 150 kilometres southeast of the Fault Creek Glacier) experienced four or five periods of growth in the past 2,000 years (Clague et al. 2010). In addition to the LIA maximum, there was a glacial advance between 1,700 and 1,400 years ago (Reyes et al. 2006). Unfortunately, the complete history is hard to decipher because glacial advances tend to erase traces of earlier advances. In this paper, LIA refers to the last large and significant glacial advance, which ended at the close of the 19th century. The LIA and earlier Holocene glacial advances are distinct climatic events separate from the Wisconsin Glaciation maximum, when ice completely covered this region (Clague 1989).

The Kwädąy Dän Ts'ìnchį man died between 1720 and 1850, during the LIA (Richards et al., chapter 6). Dendrochronological (tree-ring dating) studies at Kaskawulsh Glacier, about 150 kilometres northwest of the study site, suggest that the maximum ice extent occurred around 1750 AD (Reyes et al. 2006), but that retreat did not start until the early or mid 19th century (Borns and Goldthwait 1966). Lichenometric studies at the Wheaton Glacier, located 84 kilometres east-northeast of the study area, suggest the LIA maximum occurred in the late 19th century (Church and Clague 2009). The LIA maximum observed at southwest Yukon ice patches (Farnell et al. 2004), approximately 100 kilometres north of the site, was sustained until the end of the 19th century, based on lichenometry (V. Bowyer, pers. comm.).

Today the LIA maximum extent of ice is often visible on valley walls as a distinct change in colour tone. Rocks above the high mark are often darker, as they have a greater cover of lichens and other vegetation (fig. 1*E*). At lower elevations one can sometimes observe an inverted treeline, where trees are present *above* a certain elevation on the valley wall, but not below, where they were removed by the ice.

Since the LIA maximum, glaciers in the Yukon and northern British Columbia have been shrinking. Glaciers in the St. Elias Range (100–200 kilometres northwest of the study site) have lost 22 per cent of their surface area (Barrand and Sharp 2010) since the 1957–58 International Geophysical Year. Bolch et al. (2010) note a 7.7 per cent decrease in area for glaciers in the northern Coast Mountains of British Columbia between 1985 and 2005, and Church and Clague (2009) observed a 50 per cent loss in ice cover since 1907, with the majority since 1944. The melting has been extensive and dramatic. Such decreases in area would necessarily be associated with a rise in elevation of the equilibrium line for these glaciers.

TRAVEL IN GLACIATED TERRAIN

ALTHOUGH EVERY GLACIER IS DIFFERENT, some general comments can be made about travelling near and on glaciers. First-hand experience travelling in glaciated terrain has no substitute, but it is my hope that these comments might help the casual reader to visualize travel in a glaciated environment. One can then postulate where travel routes might be established, how those routes might change seasonally and how the Kwäday Dän Ts'ìnchį man came to the Fault Creek Glacier.

Travelling in a valley occupied by a glacier—particularly when the valley is quite full of ice, as it would have been when the Kwäday Dän Ts'ìnchį man died—can be both unpleasant and unsafe. The hazards of glacier travel today are the same as in the Kwäday Dän Ts'ìnchį man's

time. The human tendency to seek the safest and easiest route is presumably similar, too, although these two objectives can often be at odds when choosing a travel route.

Route choices through a glaciated valley can be divided into four zones: on the glacier, beside the glacier, on valley slopes and on adjacent ridges. Travel on the ridges, while providing a spectacular view, is not practical if one is travelling a long distance carrying supplies. Loose talus slopes, rock ridges and precipitous drops make for tricky and slow travel. Regardless of one's fitness level, climbing ridges is the choice of the tourist, not the traveller.

Movement along valley slopes, in addition to the dangers noted above, is strenuous because of the constant traversing. A lucky traveller may find an animal track to follow that makes for more level-footed walking, but sheep and Caribou tend to angle up- and down-slope, not walk straight along the valley. Nevertheless, it is conceivable that a well-travelled trade route would have an established side-valley track, probably shared with sheep and Caribou (that would not refuse the easier passage).

Glacier margins do not make pleasant or safe walking. Marginal moraines are often ice-cored (i.e., melting ice is just beneath a surface of loose, wet till), making for unstable, slippery footing. As noted previously, the margins of the glacier itself are rife with crevasses that may be covered with snow bridges for much of the year. In 2008, I was unsuccessful in descending into the Samuel Glacier valley, in the region of figure 1*E*, although others, including Darcy Mathews, have found a route down in later years (D. Mathews, pers. comm.). The descent today is about 150 metres (far more than it would have been in the Kwäday Dän Ts'ìnchį man's time), but the terrain is steep and unstable. The danger from falling rock is high—small rock avalanches are ongoing all along the east and west valley walls.

Older marginal moraines will stabilize as their ice cores melt and their surfaces become vegetated. Once stabilized, a marginal moraine may provide a comfortable sidewall route. Such a marginal moraine can be seen in figure 3*H*. Today there is an established

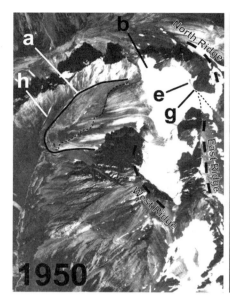
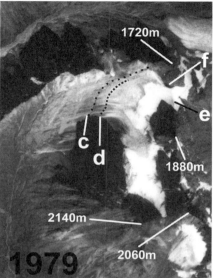

Figure 3. Comparative aerial photographs of the Fault Creek Glacier. Images are not corrected for perspective distortion. The elevations of some peaks are marked. *A*, The solid line marks Little Ice Age limit, and the dashed line follows the 1950 glacial terminus; *B*, a tongue of ice extending into the valley to the north, missing in 1979; *C*, 1979 terminus; *D*, approximate 1987 terminus; *E*, shadow of snow ridge feature, which in 1950 divides into two in the east (dashed lines—the shadow is faint on original air photographs) and is also visible in 1979; *F*, a marginal lake (level with the approximate ELA) beside the discovery site; *G*, a smaller marginal lake, also beside discovery site; *H*, terminal moraines. Courtesy of Natural Resources Canada, 2017. Aerial photographs A12820-374, A25292-67.

game trail along the crest of this moraine. During the LIA, this trail would have led onto the glacier surface near the point marked by figure 3*A*.

The safest (and easiest) place to travel might be straight up the middle of the glacier, and there is historical evidence of such routes in the area (Cruikshank 2001). Nevertheless, an on-glacier route carries risks, especially in the snow-covered accumulation zone. If the glacier flows over a subsurface rock bench (step) or change in slope, crevasses may form on the surface. As noted earlier, the Kwädạy Dän Ts'ìnchị man is thought to have been travelling sometime in late July or August, the time of maximum melt. Snow bridges across crevasses are water-saturated and weak, making travel especially dangerous. The threat of weakened snow bridges is one of the leading reasons that modern climbers in the St. Elias Mountains mount their glacier-crossing expeditions in April and May.

When getting onto or off of a glacier in the ablation zone, crevasses are generally easy to identify and avoid, particularly if one is familiar with the angled trend of marginal crevasses and the appearance of crevasses where the glacier flows over bumps. It would be reasonable to expect that a seasoned traveller and observer of his environment would be aware of these hazards.

FAULT CREEK GLACIER

THE KWÄDẠY DÄN TS'ÌNCHỊ MAN was found on the north margin of a small, unnamed glacier, which we are calling the Fault Creek Glacier (Beattie et al. 2000). The Fault Creek Glacier is located 5.5 kilometres east of the Samuel Glacier (see all locations marked on fig. 1). Glaciers in this area have been experiencing negative mass balance (melting) since the end of the LIA approximately 150 years ago. Evidence of glacial retreat is found in historical air photographs of the area. For example, the LIA trim line (ice limit) is clearly visible in the Samuel Glacier valley wall (fig. 1*E*). This trim line represents a decline in glacier surface elevation of approximately 120 metres (up to the date of the photograph in 1979).

The Fault Creek Glacier begins high on a col, 1,960 metres above sea level (MASL), south of the Kwädạy Dän Ts'ìnchị site. It flows northward over a modest ice fall, makes a sharp left-hand turn and then flows westward a short distance down the valley before terminating. A small spur of glacier to the east feeds into the westward flow at the corner. In figure 1, a medial moraine can be seen where the two flows join. The Fault Creek Glacier ranges from 400 to 600 metres wide.

A series of distinct terminal moraines lie beyond the terminus of the Fault Creek Glacier (fig. 3H), marking the maximum extent of ice during the LIA. These terminal moraines form dams for a series of small lakes, which remained present in 1999 and 2003. At the LIA maximum, the terminus was at 1,430 MASL. By 1950 the terminus had receded approximately 700 metres up-slope, although there was a longer remnant tongue of ice in the shadow of the West Ridge. The terminus of the glacier in 1979 was at approximately 1,450 metres elevation (fig. 3C), and a tongue of ice branching to the north was gone (fig. 3B). In 1987 air photographs (table 1), the terminus (fig. 3D) was farther up the valley, at approximately 1,520 metres elevation, having retreated about 1.2 kilometres from the LIA maximum (fig. 3A).

Snow Ridge Feature

AERIAL IMAGERY from 1950 shows a distinct ridge-like snow feature along the north edge of the Fault Creek Glacier (fig. 3E). This feature splits into two arms up-glacier (as shown by the dashed line in fig. 3E) and is also visible in the 1979 air photographs, although less distinctly. The 1987 imagery (not shown) is too over-exposed to discern whether the ridge is present. In 1999 the feature, now transformed into regulation ice at the discovery site, is very prominent (figs. 4D, 5D and 5E). This ridge appears to be caused by blowing snow accumulating in a fixed pattern. In years of melt, the snow transforms into regulation ice and, in years of low melt, the snow accumulates (see chapter 4). Historical

Table 1. Aerial photographs available through the Canadian National Air Photo Library and the BC government (1974 and 1987). Snow cover varies among these photograph sets, but they are usable for study of the discovery site.

Date	Series	Frame pair	Parameters	Comment
1950-08-09	A12820	373/374	20,000' ASL, 6" lens	Some snow cover at discovery site.
1955-06-23	A15093	27/28	35,000' ASL, 6" lens	Considerable snow cover. Difficult to discern glacier margins and surface features.
1974-09-07	15BC5625	31	19,000' ASL	Excellent image quality, but flight line stops short of discovery site. Fault Creek Glacier off SW corner of image.
1979-08-23	A25292	66/67	42,000' ASL, 153.37 mm	Discovery site free of snow. Small scale.
1987-08-25	15BC87076	193/194	42,700' ASL	Poor exposure – bleached out snow.

records derived from journals kept at three Hudson's Bay Company posts (Francis Lake, Fort Selkirk and Pelly Banks) for the period 1842–52 mention a predominance of east and southeast winds in winter (Tompkins 2006). But these sites are over 300 kilometres inland from the Fault Creek Glacier, so the wind patterns may not apply. The modern prevailing wind direction in Whitehorse, Yukon (150 kilometres northeast), is southeast in both summer and winter (Klock et al. 2001), a pattern driven by semi-permanent regional circulation patterns, including the Aleutian Low and Pacific High (Klock and Mullock 2001).

A south or southeast wind blowing up the valley to the east (fig. 1G) would curl around the East Ridge (see fig. 3), depositing snow on the glacier surface. Over time, with the rise in equilibrium line altitude, this snow has turned into regelation ice as seen in figure 5D.

Unglaciated Valley

THE VALLEY TO THE EAST (figs. 1G and 4D) of the Kwädąy Dän Ts'ìnchį find, referred to elsewhere in this volume as the "Empty Valley", likely remained free of glacial ice during the LIA. The valley's side walls are slopes of uninterrupted *talus* (loose cobbles and boulders). This suggests that it has been some time since a glacier occupied the valley. There is also no evidence of lateral or terminal moraines left behind by a retreating glacier. The valley could have been kept free of ice during the LIA because of its southern exposure and lack of feeder glaciers. The Fault Creek Glacier valley is quite lush below the LIA terminal moraines (fig. 3A), but the unglaciated valley is bare. In part, this may be because the unglaciated valley has steeper, unstable walls and has no lakes or flat areas to trap water.

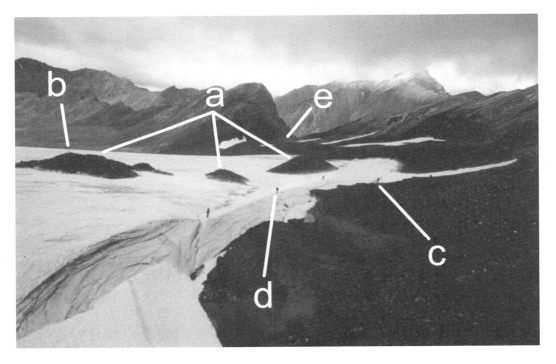

Figure 4. Oblique view of the Kwädąy Dän Ts'inchį discovery site on August 22, 1999. The view is looking west from approximately the location of figure 3F. The North Ridge (see fig. 3) is to the right. *A,* The three nunataks to the south of the discovery site; *B,* the survey reference position; *C,* the survey instrument position; *D,* the remains of Kwädąy Dän Ts'inchį man, with people walking along the ice ridge; *E,* The col that was filled with ice in 1950 (fig. 3B). Erik Blake photograph.

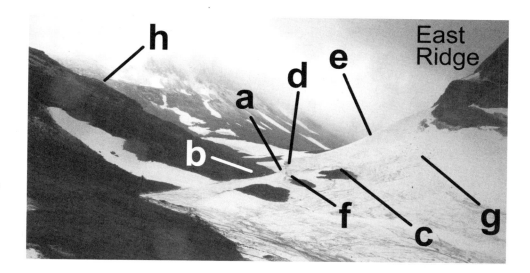

Figure 5: Oblique aerial photograph looking southeast in August 1999. *A,* Discovery site; *B,* survey instrument position; *C,* reference position; *D,* regelation ice ridge upon which Kwädąy Dän Ts'ìnchį man was found; *E,* extension of snow/ice ridge to the East Ridge (compare with fig. 3*E*); *F,* shallow trough to the south of the regelation ice ridge; *G,* rock fall on glacier surface; *H,* 30° slope on North Ridge above discovery site. KDT image 99-08-138.

If this valley was free of ice during the LIA, it may have been a route choice that brought the Kwädąy Dän Ts'ìnchį man to the Fault Creek Glacier. Greer et al. (chapter 23) discuss artifacts found in the region that date from other times during the LIA, among them a birch bark scroll found at the south end of this valley. The Kwädąy Dän Ts'ìnchį man is thought to have been travelling inland (Dickson et al. 2004; chapter 28), and if he was travelling up this valley and had reached the top, he may have chosen to turn west into the Fault Creek Glacier valley rather than attempt to climb over the lateral moraine and glacier blocking the headwall of the valley (fig. 1*H*). Nevertheless, accessing the Fault Creek Glacier valley would have entailed a steep 100-metre climb on loose material.

DEATH, BURIAL, PRESERVATION AND EXPOSURE

A FEATURE OF THE FAULT CREEK GLACIER important to understanding the preservation of the Kwädąy Dän Ts'ìnchį man's remains is the series of three bedrock bumps arranged like a bulwark to the south and west of the discovery site (fig. 5*A*). In 1999 these bumps appear as a series of nunataks (rocky outcrops emerging through the ice), though they are covered with snow or ice and not visible in aerial photography from 1950 through 1987.

Figure 6 shows that the topography around the discovery site (rectangle) is relatively flat compared to the rest of the glacier. This figure also shows the location of the nunataks, as visible in 1999, in the context of the glacier overview. It is clear the nunataks separate the discovery site from the main bulk of the glacier as it makes its westward turn. Figure 7 shows a sketch of the discovery site, based on surveyed coordinates, superimposed on the 1950 air photograph. The nunataks prevent the flow of the Fault Creek Glacier from reaching the discovery site and, aided by the lower surface slope, they also prevent the ice at the discovery site from flowing downhill.

Given that the wind-generated ridge is prominent in 1950, it may have been present when the Kwädąy Dän Ts'ìnchį man came walking. Indeed, he may have chosen the wind-packed snow ridge as a preferred route. Had I been travelling along this glacier margin, I would have chosen to travel on this ridge, where a few metres of extra climb afford a better view and the walking is fine.

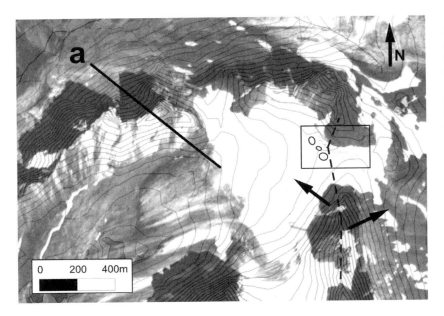

Figure 6. Aerial photograph from 1950 superimposed on digital elevation model derived from 1987 photogrammetry. The contour interval is 20 metres. The photograph has not been corrected for perspective, so registration is not precise. The rectangle identifies the Kwädąy Dän Ts'ìnchį discovery site shown in figure 7. The three ovals mark the approximate location of the nunataks as revealed in 1999. The dashed line indicates the elevation profile in figure 8. The arrows indicate possible avalanche paths from the East Ridge, both missing the discovery area. The approximate 1950 equilibrium line (*A*) is well below the discovery site, so in 1950, the Kwädąy Dän Ts'ìnchį man would have continued to be buried in more snow. Courtesy of Natural Resources Canada, 2017. Aerial photograph A12820-374.

Avalanche Danger

FIGURE 8 shows a topographic profile across the Fault Creek Glacier valley, following the dashed line in figure 6. The slope to the north of the discovery site (North Ridge in fig. 3) has an average slope of 30 degrees. In general, avalanches occur on slopes steeper than 25 degrees (McClung and Schaerer 2006), so this slope is steep enough to cause avalanches. Most avalanches, though, initiate on steeper slopes than this. For an avalanche triggering at the top of the North Ridge, the vertical drop of 100 metres would produce a Size 2 avalanche (a Canadian avalanche size classification, McClung 2003), which could bury, injure or kill a person. Not shown in figure 8 (because of the 20-metre contour interval in the Terrain Resource Information Management data) is a series of steps on the slope that would have a tendency to slow or arrest an avalanche. Some of these steps can be discerned in figure 5*H*.

Snow avalanches are a mixture of snow and entrained air, and they can travel a considerable distance. As a rule, a line drawn from the initiation point of an avalanche to the end of the avalanche run-out forms an angle of 10 degrees to the horizontal (McClung and Schaerer 2006). For an avalanche off the North Ridge, this is more than enough to cover the discovery site (dashed line in fig. 8). It is difficult to predict whether this south-facing slope would have supported an unstable snow-pack in late summer. This is generally not the time of year that avalanches occur, because unstable winter accumulations have already avalanched and sustained warmer temperatures tend to stabilize remaining snow. Nevertheless, a late summer storm could have deposited fresh snow on this slope. Conversely, to the south, the East Ridge looms high and steep, but the avalanche faces are oriented away from the discovery site (arrows on fig. 6).

No evidence of snow avalanches (off the North Ridge or East Ridge) was found in any historical air photographs or during the 1999 and 2003 site visits. But these photographs and visits were made late each summer, so it is possible that such evidence had already melted away.

There is, however, evidence for avalanching of rocks from the East Ridge. The glacier surface below the mountain is strewn with boulders up to 50 centimetres in diameter (fig. 5*G*). These boulders are broken from the mountain by freeze/thaw cycles and are released as the ice holding them in place melts. Unlike a snow avalanche on its cushion of air, these rocks are not carried far, and none was found near the discovery site.

Crevasse Danger

THOUGH CREVASSES are often found at the margins of glaciers, a prerequisite is that there be significant glacial flow (fast enough to cause the ice to crack). The discovery site is very close to the col between the Fault Creek Glacier valley and the valley to the east, and the surface slope is low (about 3 degrees). Had there been enough ice to cause rapid flow in this geographical setting, the ice in the vicinity would have become a spreading centre with ice spilling over into the valley to the east as well as to the west, but there is no evidence of ice flow to the east (no moraines).

Thick ice could also have caused flow over the then-buried nunataks, which might have caused tension crevasses in the surface. However, the Kwäday Dän Ts'ìnchi man was discovered upstream from, and to the north of, the nunataks.

Because there is no evidence of ice flow to the east, it is unlikely that any crevasses (either marginal or tension crevasses) were present when the Kwäday Dän Ts'ìnchi man died.

Burial and Preservation

IN 1950 the ELA was below the discovery site (fig. 6A). Given the retreat of the Fault Creek Glacier since the LIA maximum, it is safe to assume that the ELA at the time the Kwäday Dän Ts'ìnchi man died was even further below the elevation of his remains (1,587 metres). He died in the accumulation zone and was buried by subsequent snowfall. I speculate that the Kwäday Dän Ts'ìnchi man did not get buried beneath more than 10–15 metres of snow, but there is no way to verify this estimate.

The Kwäday Dän Ts'ìnchi man would have been kept at or below 0°C. As long as the remains were within 15 metres of the surface—in the surface layer—they would have experienced annual temperature variations. But the average annual temperature would have been below freezing for the entire period of burial, even when the remains were in the surface layer.

Figure 7. Overview of the Kwäday Dän Ts'ìnchi discovery site superimposed on a 1950 aerial photograph. The survey instrument and reference position for both 1999 and 2003 are marked. In 1999 the three nunataks were separate. The circled dots and reference position mark their peaks. In 2003 the upper and middle nunataks were joined owing to a lowered glacier surface (dashed line), whereas the lower nunataks had joined with the terrain beside the glacier (dashed 2003 snow margin). The 1999 ridge where the Kwäday Dän Ts'ìnchi man was found is marked, as is the trough in the glacier surface located south of the ridge. Remains from 1999 and 2003 were found where marked. Though the glacier surface had melted down two to three metres by 2003, there was a wide apron of snow beside the glacier shown by the spread between the 2003 ice and snow margins. The location of the stick, discovered in 1999, is shown.

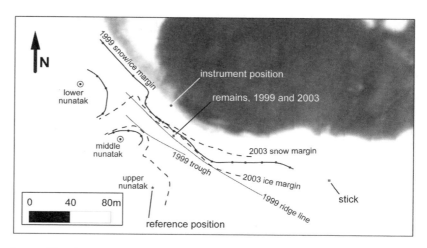

Courtesy of Natural Resources Canada, 2017. Aerial photograph A12820-374.

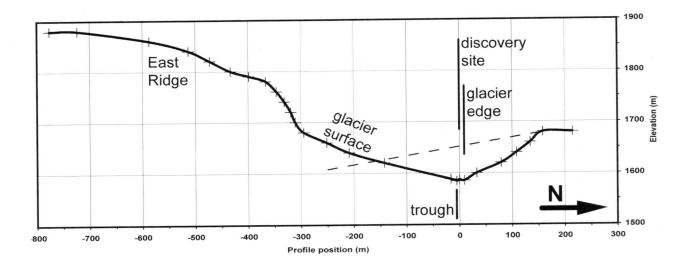

Figure 8. Topographic (surface position) profile along the dashed line in fig. 6 derived from 1999 and 2003 survey data, and the 1987 British Columbia Terrain Resource Information Management (TRIM) digital elevation model. Each data point is designated with +. The discovery site is at the origin (0) of the profile.

Given that the Kwädąy Dän Ts'ìnchį man was buried in stagnant snow and ice, the principal effect of the overlying snow would have been compression, not shearing. But the man was not discovered intact. Figure 4 shows a series of dark bands crisscrossing the edge of the glacier. These bands are past and present cracks in the ice. The cracks are unlike crevasses (which are near-vertical) and are probably caused by a combination of thermal and mechanical stress caused in turn by solar heating and undercutting of the glacier margin by meltwater streams. This cracking process may be responsible for breaking apart the human remains. Figure 9 shows a detail of the glacier margin during the recovery of the Kwädąy Dän Ts'ìnchį man's leg in 1999. Cracks and collapse features are clearly visible, despite the fresh dusting of snow, and water was observed draining from a crack in the hollow containing the Kwädąy Dän Ts'ìnchį man's torso (see chapter 4).

Melting and Exposure of the Remains

WARMING CLIMATE since the end of the LIA caused an up-slope rise in the ELA and melting of the glacier surface at the discovery site. The 1979 air photograph shows the equilibrium line near the discovery site (fig. 1F), so it is likely that ice above the Kwädąy Dän Ts'ìnchį man began melting around this time.

When the remains were discovered in 1999, the ridge crest stood 1.5–2.5 metres above the surface of the glacier immediately to the south (figs. 4 and 5). To the north, the ridge sloped directly down to the glacier margin at an angle of 25–30 degrees, and to the south there was a shallow trough in the glacier surface (fig. 5F). In 1999 the glacier's margin was undercut by a meltwater stream. Since the gap between the rock and the ice was only a few tens of centimetres, it was not possible to measure how far under the ice this gap extended. Nevertheless, the undercut was a distance of at least several metres, and water could be heard flowing in this opening. Sizable undrained melt lakes at the edge of the ice late in the summers of 1950 and 1979 (figs. 3F and 3G) demonstrate that the glacier margin blocked water flow and was likely frozen to the bed (i.e., not temperate). The conclusion is that the margin of Fault Creek Glacier has only recently warmed to the melting point.

In 2003 the ice ridge had completely melted away and the surface of the glacier had dropped by 2.5–3 metres (fig. 10). In 2004, the ice was completely gone from the discovery site (fig. 5A).

SITE SURVEY

DURING THE 1999 AND 2003 visits to the Kwädạy Dän Ts'ìnchị site, the positions of glacial features, artifacts and human remains were established using a site survey. These measurements inform the context of the discovery and help us draw conclusions about how the Kwädạy Dän Ts'ìnchị man came to be in this location, how he was preserved through time and how he came to be discovered.

In 1999 an instrument position (IP) and reference position (RP) were established off the glacier margin and on the upper nunataks (figs. 4B, 4C, 5B and 5C). These same stations were occupied again during the 2003 visit. The IP was on a small bedrock spur on the north margin of the glacier. This spur has a clear view up- and down-glacier, as well as to the south overlooking the discovery site and glacier margin. Only locations on the south side of the three nunataks were not visible (masked by the nunataks), but these locations did not reveal any finds of interest. The RP was placed on the upper (easternmost) nunatak. The RP is near the highest point on this nunatak and is located on bedrock.

Survey Baseline Coordinates

THE DATUM of 1:50,000 National Topographic System of Canada map sheet for the site (Nadahini Creek, 114P/10 114P076/114P066) is NAD27. This datum varies significantly from the current WGS84 datum used for our surveys. In the region of the discovery site, the discrepancy between NAD27 and WGS84 is approximately 200 metres.

The coordinates of the IP, as measured in August 2003 using a GPS receiver averaging for 110 minutes,

Figure 9. A view looking southwest from beside the glacier during the 1999 recovery effort. The remains being retrieved are off the crest of the ice ridge, where most were found. Notice the cracks and collapse features in the glacial ice. Cracks such as these might have contributed to the breakup of the Kwädạy Dän Ts'ìnchị man's remains. Running water could be heard in the open cracks. Erik Blake photograph.

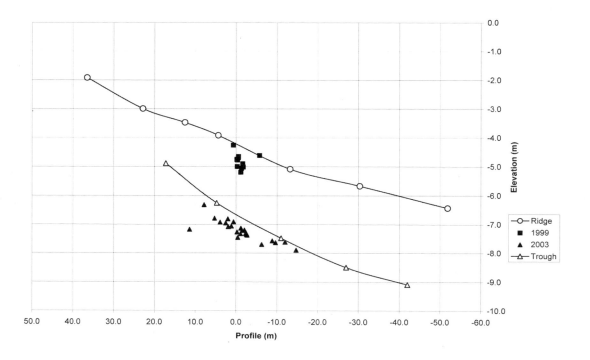

Figure 10. A vertically exaggerated profile of all surveyed finds (remains and artifacts, except the stick) from 1999 and 2003 projected onto a profile line that runs along the 1999 regelation ice ridge crest. The trough is the 1999 glacier surface immediately south of the ridge (see figs. 5 and 7). Elevation is relative to the survey instrument position, and positive profile position is up-glacier. The cross-section in figure 8 cuts these profiles at an approximate right angle at the origin, which is the discovery site.

were UTM 08 394126E 6619422N, elevation 1,592 metres (datum WGS84). Expressed as latitude and longitude, these coordinates are 59°41.957 N, 136°52.876 W. This GPS location is believed accurate to four metres (no differential correction was performed). The coordinates of the RP are UTM 08 394112E 6619339N, elevation 1,596 metres (datum WGS84, averaged for 10 minutes). The GPS-derived true-north bearing from the IP to the RP is thus 189.6 degrees. The corresponding magnetic bearing is 163.5 ± 0.5 degrees (measured with a Silva mirror sighting compass). After applying published declination and drift corrections, the true-north bearing is 191.8 degrees. Given that the error in the GPS-derived bearing is over 3 degrees (owing to the horizontal position error), the corrected magnetic bearing was used to translate survey locations from this IP/RP survey baseline into easting and northing coordinates.

Survey Procedure

IN 1999 AND 2003 a tripod with a retro-reflector was placed over the RP and a theodolite with a built-in laser rangefinder (a total station) was placed over the IP. A second retro-reflector on a pole was carried by various members of the recovery teams to mark items of interest. By measuring the horizontal angle between a target item of interest and the RP, the vertical angle to the target, the distance to the target and the heights of the IP and target pole, it is a modest trigonometric exercise to derive the position of the target relative to the IP.

Survey Results

TABLE 2 (page 58) shows the positions of various features and finds relative to the IP. Positions are in metres and are accurate to ± 1 centimetre. For the sake of brevity, repeated measurements of the RP position, all of which repeat to within 1 centimetre for both 1999 and 2003, are omitted.

Figure 10 is a vertically exaggerated side view showing the relative location of the remains found in 1999 and new remains found in 2003. It is evident that the remains had dropped in elevation as the glacier surface ablated. Some horizontal spreading is seen, perhaps caused by meltwater flow carrying material across the glacier surface. Many of the remains found in 2003 could have been hidden just below the surface in 1999, having been buried in active or refrozen cryoconites (melt pockets). The light snowfall during the visit in 1999 may also have hidden smaller features on the surface.

Exposure of Kwädąy Dän Ts'ìnchį man's remains may have begun many years before 1999. Processes of reburial, cryoconite formation, water flow, regelation and freeze/thaw cycles in the surface layer would have contributed to the dispersal of the remains, both vertically and horizontally.

CONCLUSIONS

MELTING ICE in Yukon, Alaska, Northwest Territories and Scandinavia has in recent years revealed a treasure trove of archaeological artifacts and faunal remains, dating back thousands of years (Dixon et al. 2014). The burial, preservation and discovery of the Kwädąy Dän Ts'ìnchį man required a series of specific events—omission of any one of them would have left him lost to the First Nations of today and to science. First, he died in the accumulation zone of a glacier and was therefore buried and frozen. His location on the snow ridge (presuming

it was present at that time) was an area of enhanced accumulation because of blowing snow. He would have been buried faster, and possibly deeper, than had he died elsewhere. Second, the glacier where he died was stagnant. The low surface slope and the obstacle presented by the then-buried nunataks shielded him from the destructive forces of glacial flow. Finally, in a period of warming and melt, modern-day hunters discovered his remains. These hunters happened by at the right time of year and at the right time in the melting of the Fault Creek Glacier. A few years earlier and the remains might have stayed hidden; a few years later and the body may have been completely exposed and dispersed.

We will never know for sure why the Kwädąy Dän Ts'ìnchį man was hiking one of the highest passes in the area. The sidewalls of the Samuel Glacier valley to the east are almost 300 metres lower than the Fault Creek Glacier (even above the LIA maximum), and the valley to the northwest that bypasses this group of mountains entirely is also much lower in altitude.

Of the hazards listed above, a snow avalanche would be the most likely to have caused injury. But my own feeling, standing at the discovery site, is that it is quite safe. The fact that the Kwädąy Dän Ts'ìnchį man was found on such a convenient walking surface suggests to me that he was not struck and swept away by an avalanche. I consider it more likely that he died of exposure.

If he came from the east, he would have just completed an exhausting 100-metre climb up a talus slope. Perhaps he was disoriented by fog or a summer storm that caused whiteout conditions, and he wandered off his intended path.

What we do know is that a combination of fate, geography, climate change and a serendipitous discovery—all acting together on a remote, previously unstudied glacier—resulted in an unparalleled opportunity to learn about life in northwest British Columbia in the time of the Little Ice Age.

ACKNOWLEDGEMENTS

I would like to thank the Champagne and Aishihik First Nations for the honour and privilege of my twice visiting the Kwädąy Dän Ts'ínchį site on their traditional territory. Al Mackie and the BC Archaeology Branch gave me the opportunity to make those visits. Many thanks go to Mackie and to Richard Hebda for their constructive editorial comments and great patience, as well as to Weston Blake Jr. for reviewing the paper. I would also like to thank my coworkers at Icefield Instruments Inc. for their forbearance during those times when I abandoned them to work on this project.

Table 2. Positions of various features and finds relative to the instrument position (IP). Targets are listed in the order they were measured. For clarity, all but one reference reading have been omitted. UTM coordinates of the IP are UTM 08 394126E 6619422N, elevation 1592 metres (NAD83/WGS84 as determined by uncorrected GPS).

Date	Comment	Easting	Northing	Elevation
	Reference station (RP)	-17.31	-82.84	0.82
1999-08-22	ice edge starting at low edge	-67.38	62.59	-7.24
1999-08-22	ice edge starting at low edge	-37.16	29.67	-6.85
1999-08-22	ice edge starting at low edge	-23.80	8.32	-6.84
1999-08-22	ice edge starting at low edge	-24.90	1.76	-6.10
1999-08-22	ice edge starting at low edge	-14.24	-11.79	-6.04
1999-08-22	ice edge starting at low edge	-10.84	-15.28	-5.96
1999-08-22	snow below body	-7.68	-16.84	-7.21
1999-08-22	snow below body	2.98	-25.94	-6.48
1999-08-22	snow below body	14.20	-33.32	-7.55
1999-08-22	ice edge starting at low edge	24.55	-44.89	-6.28
1999-08-22	ice edge starting at low edge	40.37	-55.95	-6.83
1999-08-22	ice edge starting at low edge	60.35	-56.59	-6.38

Date	Comment	Easting	Northing	Elevation
1999-08-22	ice edge starting at low edge	73.05	-56.73	-4.41
1999-08-22	ice edge starting at low edge	89.27	-56.71	-2.21
1999-08-22	ice edge starting at low edge	103.13	-56.21	-1.17
1999-08-22	ice edge starting at low edge	116.75	-61.45	0.47
1999-08-22	stick	155.19	-74.40	4.06
1999-08-22	top of ridge	86.04	-88.33	6.35
1999-08-22	photo 17	62.38	-75.65	4.51
1999-08-22	photo 18	33.22	-57.99	-0.53
1999-08-22	start of ridge	28.42	-53.00	-1.92
1999-08-22	start of ridge	17.57	-44.70	-3.00
1999-08-22	start of ridge	9.48	-38.23	-3.47
1999-08-22	start of ridge	2.70	-33.52	-3.92
1999-08-22	neck	0.64	-30.28	-4.26
1999-08-22	head? (false identification)	-1.38	-31.25	-4.75
1999-08-22	ridge	-10.26	-21.58	-5.08
1999-08-22	ridge	-22.35	-9.48	-5.67
1999-08-22	ridge petering out	-37.28	6.11	-6.43
1999-08-22	lower nunatak	-66.81	-8.62	-10.41
1999-08-22	lower nunatak	-63.96	17.04	-8.60
1999-08-22	lower nunatak	-72.57	28.88	-8.31
1999-08-22	top of lower nunatak	-88.99	20.90	-4.17
1999-08-22	middle nunatak	-56.16	-27.68	-10.25
1999-08-22	middle nunatak	-41.46	-25.86	-9.00
1999-08-22	middle nunatak	-28.90	-30.96	-7.93
1999-08-22	top of middle nunatak	-49.35	-34.44	-7.05
1999-08-22	trough behind body	-41.22	-13.06	-9.09
1999-08-22	trough behind body	-29.89	-22.91	-8.50
1999-08-22	trough behind body	-16.80	-32.23	-7.48

Date	Comment	Easting	Northing	Elevation
1999-08-22	trough behind body	-2.67	-40.05	-6.25
1999-08-22	trough behind body	8.47	-46.35	-4.88
1999-08-22	hair #2	-2.15	-31.99	-5.00
1999-08-22	hair #3	-2.92	-31.53	-5.19
1999-08-22	hair #1	-1.57	-30.88	-4.65
1999-08-22	hair #4	-2.82	-30.73	-4.90
1999-08-22	bag 14	-4.04	-25.91	-4.61
1999-08-22	bag 2	-1.66	-31.03	-4.69
1999-08-22	bag 12	-1.67	-31.37	-4.77
1999-08-22	bag 30	-2.91	-31.26	-5.06
1999-08-22	pole 2	-7.25	-22.64	-4.88
1999-08-22	bag 24	-3.00	-30.74	-5.01
1999-08-22	trench corner 1	-0.66	-30.72	-4.49
1999-08-22	trench corner 2	0.08	-30.10	-4.50
1999-08-22	trench corner 3	0.60	-30.69	-4.49
1999-08-22	trench corner 4	-0.15	-31.41	-4.53
1999-08-22	legs end 1	2.78	-29.64	-4.97
1999-08-22	legs end 2	3.48	-29.88	-5.04
1999-08-22	pole 1	12.53	-59.68	-2.64
2003-08-28	facial bones	-2.62	-29.03	-7.34
2003-08-28	parietal	-2.80	-29.27	-7.35
2003-08-28	occipital	-2.87	-29.54	-7.29
2003-08-28	loose teeth	-2.11	-28.90	-7.32
2003-08-28	right temporal	-2.05	-30.09	-7.21
2003-08-28	left temporal	-1.12	-30.89	-7.26
2003-08-28	right scapula	-0.84	-30.32	-7.46
2003-08-28	robe	2.69	-30.78	-6.95
2003-08-28	P01	5.15	-31.81	-6.84

Date	Comment	Easting	Northing	Elevation
2003-08-28	phalanx	-1.40	-29.93	-7.33
2003-08-28	metacarpal #A	-5.24	-26.46	-7.70
2003-08-28	down-slope metacarpal #B	-7.69	-25.32	-7.57
2003-08-28	hair sampled	1.28	-31.43	-7.08
2003-08-28	phalange B with tissue	-2.73	-29.98	-7.19
2003-08-28	small artifact	-2.72	-31.25	-7.13
2003-08-28	small artifact	5.06	-32.26	-6.79
2003-08-28	small artifact	5.07	-30.18	-6.92
2003-08-28	small artifact	1.82	-29.70	-7.06
2003-08-28	small artifact	-2.24	-28.48	-7.38
2003-08-28	small artifact	-7.64	-24.09	-7.63
2003-08-28	small artifact	-11.84	-21.14	-7.89
2003-08-28	small artifact	-15.14	-28.87	-7.62
2003-08-28	small artifact	-4.13	-35.51	-6.91
2003-08-28	small artifact	3.72	-37.62	-6.32
2003-08-28	video camera (2003)	4.65	-34.90	-6.23
2003-08-28	OP (on snow)	5.62	-26.72	-6.81
2003-08-28	pole 2 top	-7.53	-22.98	-7.44
2003-08-28	pole 2 bottom	-6.99	-24.13	-7.34
2003-08-28	pole 1 top	12.02	-58.58	-3.82
2003-08-28	pole 1 bottom	12.76	-60.27	-3.64
2003-08-28	nunatak	-3.38	-92.29	-2.44
2003-08-28	nunatak	-2.10	-72.00	-3.44
2003-08-28	nunatak	-13.42	-61.35	-5.15
2003-08-28	nunatak	-20.89	-53.61	-6.39
2003-08-28	nunatak	-27.48	-49.93	-7.40
2003-08-28	nunatak	-21.41	-39.18	-7.57
2003-08-28	nunatak	-24.41	-29.88	-8.20

Date	Comment	Easting	Northing	Elevation
2003-08-28	nunatak	-37.26	-20.51	-9.62
2003-08-28	nunatak	-56.89	-26.35	-11.16
2003-08-28	ice edge	-66.93	-12.18	-12.29
2003-08-28	ice edge	-54.23	-4.77	-11.28
2003-08-28	ice edge	-49.13	-0.96	-10.92
2003-08-28	ice edge	-38.89	8.44	-9.46
2003-08-28	ice edge	-32.83	9.80	-8.92
2003-08-28	ice edge	-30.03	2.49	-8.83
2003-08-28	ice edge	-17.49	-14.05	-8.44
2003-08-28	ice edge	-12.81	-17.48	-8.11
2003-08-28	snow edge	5.78	-22.61	-7.26
2003-08-28	distal phalange	19.97	-24.86	-7.18
2003-08-28	snow edge	32.09	-39.46	-5.77
2003-08-28	snow edge	47.50	-43.59	-6.06
2003-08-28	snow edge	70.82	-50.49	-3.51
2003-08-28	ice/snow	63.26	-69.85	0.31
2003-08-28	ice/snow	47.84	-64.62	-1.05
2003-08-28	ice/snow	32.34	-53.95	-3.28
2003-08-28	ice/snow	14.71	-36.97	-6.00
2003-08-28	ice/snow	10.82	-32.94	-6.51
2003-08-28	ice/snow	3.01	-25.44	-6.94
2003-08-28	ice/snow	-6.09	-17.58	-7.58

4

ARCHAEOLOGICAL FIELDWORK AT THE KWÄDĄY DÄN TS'ÌNCHĮ SITE AND SURROUNDING AREA

Alexander P. Mackie and Sheila Greer

Between 1999 and 2005, 12 visits were made to the Kwädąy Dän Ts'ìnchį site to collect and document archaeological materials and human remains. These visits involved 81 person days spent in the area, though some were very short days and some involved activities not related to data collection or recording, such as returning the cremated remains to the mountain. In total, about 300 hours were spent conducting various forms of research in the field.

IN THIS CHAPTER we describe the methods used for identification, collection and recording of the human remains, the man's belongings and the other artifacts collected in the larger site area. While further discussion, description and analysis of the artifacts, specimens and samples collected during these visits is reported in various chapters of this book, this chapter describes the work undertaken and the conditions prevailing during each site visit. Descriptions are based on field notes and related manuscripts that will be transferred from the BC Archaeology Branch to the Royal BC Museum, in Victoria, BC, on project completion and those kept at Champagne and Aishihik First Nations (CAFN) offices in Haines Junction, Yukon.

Table 1 lists all the site visits and the organizations involved, and table 2 lists each person that visited the site in any capacity related to the Kwädąy Dän Ts'ìnchį discovery.

The northern latitude and high altitude of the discovery area limit opportunities for fieldwork. During the life of the project, this annual window for studying the site has averaged from four to six weeks, sometimes as little as two to three weeks, and occurs during the last week or two in July and much of August. The most important access condition is for sufficient snow to have melted from the glacier surface and adjacent terrain to expose the site. The ideal timing for fieldwork occurs if and when the previous season's snowfall has melted away, but before snow from the new season has begun to accumulate. The 2000 and 2001 site visits, for example, both encountered a blanket of snow on the glacier and adjacent landforms that prevented fieldwork in the immediate vicinity of the site (figs. 10 and 11). As these conditions were also observed in the bordering southwestern Yukon, we believe it is unlikely that the site was exposed in either of those two years.

Access to the site during two hunting trips in 1999 and 2003 was on foot. For the 10 other visits, access to the site was by helicopter—always with flying services provided by Capital Helicopters, piloted by company owner and CAFN citizen Delmar Washington. In addition to the site visits discussed in this chapter, two other trips were made into the area on foot to explore possible travel routes that the ancient man may have taken. Neither resulted in an actual visit to the site or

involved documentation or collection, and so they are not included in the summary tables. An account of one of these efforts is presented in detail in chapter 32.

Safety concerns also hinder investigations at the site. Weather can change quickly and dramatically, and field workers risk being stranded in dangerous conditions. Weather conditions affected some fieldwork days by reducing access and limiting the amount of equipment that could be transported in the helicopter. On some days it also meant being ready to leave the site on short notice.

Funding for the site visits discussed in this chapter, including staff time, was provided variously by CAFN, BC Archaeology Branch, the Royal BC Museum, Yukon Heritage Branch, BC Parks and Dr. J. Dickson's research grants. Much volunteer time was also contributed during these visits, and the discoverers of the site paid for their own two visits to the area.

Table 1. Site visits and organizations, agencies and researchers involved.

Site visit	CAFN	BC Archaeology Branch	BC Parks[1]	Yukon Heritage Branch	Glaciologist	Scientific advisor	Other researchers
1999							
August 14, Hunters discover site	+	+	+	+	+	+	
August 17, Site assessment visit	+		+	+			
August 22, Site excavation and collection, day 1	+	+	+	+	+	+	
August 23, Site excavation and collection, day 2	+	+	+		+		

Site visit	CAFN	BC Archaeology Branch	BC Parks[1]	Yukon Heritage Branch	Glaciologist	Scientific advisor	Other researchers
2000 August 10, Monitoring site visit, snow blankets site, aerial inspection only	+		+				
2001 July 21, Snow blankets site, artifact and environmental sample collection incidental to reburial ceremony	+	+	+	+			
2002 August 9, Dickson and Mudie field trip, one partial day of reconnaissance and artifact collection	+		+				+
2003 August 16, hunters discover more human remains	+	+			+		+
August 28, Recovery and burial of cranium and other bones, collection of artifacts	+	+	+		+		+
2004 August 20, Monitoring visit, recovery and burial of more human remains, reconnaissance and artifact collection	+		+				
2005 August 12, Dickson and Mudie field trip, reconnaissance and artifact collection in site area	+		+				+
August 13, Dickson and Mudie field trip, reconnaissance and artifact collection	+		+				+

Site visit	CAFN	BC Archaeology Branch	BC Parks[1]	Yukon Heritage Branch	Glaciologist	Scientific advisor	Other researchers
August 14, Dickson and Mudie field trip, transit site on way to trail, observation of artifact too big to pack out	+		+				+
September 22, Return trip to collect artifact left behind on August 14	+		+				
Notes: 1 – BC Parks column includes visits where Park Management Board members and/or staff were present							

Table 2. Site visits and personnel participating.

Participant	Affiliation	August 14, 1999	August 17, 1999	August 22, 1999	August 23, 1999	August 10, 2000	July 21, 2001	August 9, 2002	August 16, 2003	August 28, 2003	August 20, 2004	August 12, 2005	August 13, 2005	August 14, 2005	September 22, 2005	Total Visits to Site
Bill Hanlon	Hunter	+							+			+	+	+		5
Mike Roch	Hunter	+							+							2
Warren Ward	Hunter	+														1
Gord MacRae	BC Parks		+	+	+											3
Lawrence Joe	CAFN and Park Board		+								+					2

Participant	Affiliation	August 14, 1999	August 17, 1999	August 22, 1999	August 23, 1999	August 10, 2000	July 21, 2001	August 9, 2002	August 16, 2003	August 28, 2003	August 20, 2004	August 12, 2005	August 13, 2005	August 14, 2005	September 22, 2005	Total Visits to Site
Diane Strand	CAFN	·	+								+					2
Micheal Jim	CAFN and BC Parks		+												+	2
Delmar Washington	CAFN and Capital Helicopters		+	+	+	+	+	+		+	+	+	+	+	+	12
Sarah Gaunt	CAFN and Park Board		+	+		+	+									4
Dr. Ruth Gotthardt	Yukon Heritage		+													1
Greg Hare	Yukon Heritage		+	+			+									3
Al Mackie	BC Archaeology			+	+		+			+						4
Ron Chambers	CAFN and Park Board			+	+		+			+						4
Erik Blake	Icefield Instruments			+	+					+						3
Dr. Owen Beattie	University of Alberta			+												1
John Fingland	CAFN				+											1
Ty Heffner	University of Alberta				+											1
Sheila Greer	CAFN					+		+		+	+	+				5

Participant	Affiliation	August 14, 1999	August 17, 1999	August 22, 1999	August 23, 1999	August 10, 2000	July 21, 2001	August 9, 2002	August 16, 2003	August 28, 2003	August 20, 2004	August 12, 2005	August 13, 2005	August 14, 2005	September 22, 2005	Total Visits to Site
James Allen	CAFN						+									1
John Adamson	CAFN Elder						+									1
Calvin Lindstrom	Carcross-Tagish First Nation						+									1
Gord Joe	CAFN						+									1
Dr. Petra Mudie	Geological Survey of Canada							+				+	+	+		4
Dr. James Dickson	University of Glasgow							+				+	+	+		4
Dr. Dan Straathof	Vancouver General Hospital									+						1
Dr. Bryce Larke	Yukon Medical Services									+						1
Frances Oles	CAFN										+	+	+	+		4
Greg Eikland	CAFN											+	+	+	+	4
Al Harvey	Photo-grapher for Dickson											+	+	+		3
Total personnel per visit		**3**	**8**	**8**	**7**	**3**	**9**	**4**	**2**	**7**	**5**	**8**	**7**	**7**	**3**	**81**

August 14, 1999: Site Discovery

SHEEP HUNTERS Bill Hanlon, Warren Ward and Mike Roch—referred to as "the hunters" in this chapter—noticed some artifacts and human remains at the edge of a glacier while traversing the site area near the end of their multi-day hunt (see chapter 1, this volume). There was clear, high cloud and good visibility during their visit. They took a few photographs of the site location, including some of the artifacts in place.

The hunters carefully collected five small artifacts to ensure that their observations were taken seriously. These include the hand tool (now referred to as the man's knife) and its sheath; these two items would later be catalogued as artifact IkVf-1:102 and IkVf-1:101, respectively. They also picked up a fringed piece of leather (fig. 1) and a small fragment of hide thought to be from the knife sheath. They packed these items in empty Ziploc bags that had been used for food storage during their hunting trip. They also brought out from the site two larger items: a carved stick (catalogued as IkVf-1:104 and later recognized as part of the gaff pole listed in chapter 23) and the wooden piece referred to as the hooked stick (IkVf-1:103).

The hunters recovered the artifacts from the ice and adjacent hillside at midday on August 14, 1999. They carried these pieces out to the Haines Highway, reaching their vehicle on August 16, and handed the items over to CAFN and Yukon Heritage Branch in the afternoon of that same day. By this date, the artifacts were completely thawed and probably had thawed shortly after their removal from the glacier. The artifacts were examined in Whitehorse and then placed in the freezer late afternoon on August 16. The original contents of the plastic bags used by the hunters to carry out the artifacts were also documented in case there was residue on the artifacts that might contaminate subsequent analyses.

The following year, artifact locations were discussed with some of the hunters while they were on a visit to Victoria. The locale where each artifact was seen or collected was documented on air photos during this session and during a subsequent meeting with Ward and again with Hanlon when he was in Whitehorse in 2005.

Figure 1. The robe mass. Arrow indicates fringed leather piece that was collected. August 14, 1999. Bill Hanlon photograph.

During later study, these various objects collected by the hunters would be subsampled more than a dozen times in order to identify their constituents, date them and recover materials such as pollen for analysis.

August 17, 1999: Preliminary Site Assessment

THE DAY AFTER THE HUNTERS ARRIVED in Whitehorse, Champagne and Aishihik First Nations organized an assessment visit to the site to confirm the discovery and determine the next steps. Participants in the initial assessment effort on August 17 included CAFN representatives Lawrence Joe, Diane Strand and Sarah Gaunt; Micheal Jim and Gord MacRae on behalf of BC Parks; and Yukon government archaeologists Ruth Gotthardt and Greg Hare. Weather conditions were poor at the time of the visit, with low cloud and rain, so less than two hours was spent at the site. During this time, the site was documented with still photographs, and a brief search was made of adjacent slopes and the ice surface. The locations of all items collected on this date

Figure 2. Aerial view looking east towards the site, with the Fault Creek Glacier and the discovery site in middle distance. The site is marked by the black arrow and is beyond the three nunataks barely visible along the ice. Along the valley wall to the left (*white arrow*) is a visible ridge from a lateral moraine that predates the Little Ice Age. The unvegetated areas indicate the maximum level of ice during the LIA. August 17, 1999. Sarah Gaunt (CAFN) photograph.

were not precisely documented, but GPS locations were taken for some pieces to establish the site location and boundaries.

On this visit, the man's pelvis and legs were observed lying parallel to the surface of the glacier, partially melted free, with the lower half or so still encased in ice. The man's upper body was not observed, as it was still below a thin layer of ice that had formed beneath the robe. The layer of ice had presumably formed after the man's head had melted from the glacier in a previous year.

Twenty-one artifacts or bags with assorted material were collected during this site visit. These pieces included the main mass of the robe, miscellaneous hide and robe fragments, a bead on a thong, the hat and various stick artifacts. In the studies that would eventually be carried out, the robe mass was sub-sampled nearly 250 times, and more than 80 additional subsamples were taken from other items collected at the site on this day.

During the first helicopter approach on this visit an object was observed to blow away from the site and land on the rocks a short distance off, in the same location where the hat (catalogued as IkVf-1:113) was subsequently found. We consider it most likely that the hat melted free of the ice and snow after the hunters left the area and had been resting near the human remains until the arrival of the group visiting on August 17.

Two carved wooden sticks were found on and in the snow immediately below the man's body. Both were at the snow's upper edge, where it meets the ice. One (believed to be a piece of IkVf-1:106) was on the surface, and the longer piece (IkVf-1:116) was partially buried in the snow. A third long, carved stick piece (IkVf-1:115) was found at 8–10 metres down-glacier from the body (fig. 3). This piece was partially frozen into the lower edge of the snow and recent ice, where the lens of snow pinches out at the edge of the glacier, and was situated within two or three metres of a robe fragment that was found on the adjacent ice ridge. These three carved stick pieces are assumed to have melted from the ice in an earlier season and become incorporated into snow that had remained from a previous year. These same fragments, along with a fourth—the notched stick fragment catalogued as IkVf-1:104, collected by the hunters on August 14—would later be fitted together in the lab in Whitehorse. This refitted artifact is identified as a gaff pole/walking stick (see chapter 23 on wooden tools). It is possible, though not certain, that this artifact is associated with the man. Other

Figure 3. Greg Hare *(left)* and Lawrence Joe recovering the longest of the carved stick pieces (the 115 portion of refitted artifact number IkVf-1:104) from snow and ice at the edge of the glacier, August 17, 1999. Sarah Gaunt (CAFN) photograph.

materials collected on August 17 were wooden artifacts IkVf-1:107 and IkVf-1:108. These two pieces had first been picked up by the hunters further up-slope and to the east of the site, only to be set down near the main find area. Later, in the lab, these two pieces were refitted together into a second, incomplete gaff pole. Wooden artifacts IkVf-1:117 and IkVf-1:118, which were refitted together into a slender pointed stick with carved end, were also collected on this date, as was IkVf-1:109. The latter catalogue number refers to small wood fragments previously suggested to be parts of a possible arrow (Beattie et al. 2002) but subsequently identified merely as "fragments of a small sapling" (chapter 23).

All objects recovered on August 17 were in an unfrozen state by the time they arrived in Whitehorse, if not already thawed (or very nearly so) when collected on the mountain. With the exception of the hat, the collected artifacts were placed in a freezer on arrival in Whitehorse. The hat was placed in a cold water bath in a fridge.

August 22, 1999: Site Excavation, Survey and Recovery of Human Remains

ON THIS VISIT to the site the weather was variable, with intermittent low cloud cover and the temperature slightly above freezing. Overall, conditions were good for the recording, excavation and collection activities that dominated the day (fig. 4). More than seven hours were spent in the area. Snow had fallen the night before, leaving a thin but welcome layer that prevented the glacier ice from being too slippery, but crampons were necessary when working on the steep edge of the glacier (see chapter 3, fig. 10).

Al Mackie (BC Archaeology), Owen Beattie (University of Alberta) and Erik Blake (Icefield Instruments) came on the first flight and scouted the area while waiting for the remainder of the crew, which included Gaunt (CAFN and Park Board), MacRae (BC Parks), Hare (Yukon Heritage) and Ron Chambers (CAFN and Park Board). Before work began in earnest, the group gathered together at the edge of the glacier

Figure 4. View south to site *(centre of image by individual with stadia rod)* during mapping of site; photo taken after removal of the human remains, August 22, 1999. Sarah Gaunt (CAFN) photograph.

and Chambers spoke a few quiet words of respect in honour of the person who had lost his life.

For safety reasons, Chambers set up the tent in case an overnight stay or storage became necessary. The rest of the day was concentrated on the recording and recovery of human remains and artifacts. The site and surrounding area were recorded with still photography and video, and an instrument survey was conducted of the glacier and key collection locations (fig. 4), as described by Blake (chapter 3).

The man's torso was discovered just visible at the surface of the ice, in the same location from which the robe mass had been removed on the previous visit. The torso was embedded in the ice from the clavicles down, which meant that an excavation about one metre deep was required to recover it (fig. 4). The chainsaw (fig. 6) used to cut the trench through the adjacent ice had been set up by Blake exclusively for his glacier research in order to avoid contamination of ice samples. The cutting of the trench proceeded by alternately sawing along the trench edges at a distance from the torso and then breaking out blocks of ice with ice axes. A narrow space

just a few millimetres wide had melted around the torso, which effectively left the torso frozen in an air pocket, suspended from only small points of attachment. This situation greatly simplified removing the body, as it was possible to chip and gently lever the ice away in blocks with no apparent damage to the man's skin from adhesion or flying chips of ice.

During the nine days since the site had been discovered, the man's pelvis and legs had melted completely free of ice, so that when Beattie reached to wrap up these remains, they shifted under his hands (see chapter 3, fig. 10). The recovery party felt it was likely that a strong wind or another day of melt would have seen this half of the body slide down-slope to the rocks below, where it would have decayed quickly. It is believed that the man's legs became detached in just this way, as only the foot bones were present on the rocks below the body and the lower part of the legs was largely reduced to bone. The legs and feet had likely melted out and started to decay in an earlier year. The photographs taken by the hunters show that only the pelvis was visible on August 14, and the legs were largely encased in ice.

A trench was dug around the man's body, and by mid afternoon the body was ready to be lifted from the ice. It was noted that the ice was harder near the bottom of the trench. The body was wrapped in sterile hospital wraps by Beattie and Hare, with Chambers providing backup assistance. The torso was placed in a purpose-made Coroplast box for transport, and the lower body was placed in a newly purchased, factory-clean cooler. Following recovery, this cooler was buried in the snow until the group departed the site, when the human remains and other samples were flown directly to Whitehorse. The trip to Whitehorse, including transfer times, took less than two hours.

On this day, 11 containers or bags of material were collected from the ice. Most were small plastic bags holding scattered pieces of robe and a few other artifact fragments, or a whole artifact in the case of the copper bead on a hide thong. Some bags included dozens of items retrieved from the same small, local area of the ice surface. A number of bags or containers of human remains were also collected in and on the ice, as well as on the rocks at the edge of the glacier. During analysis, these collections would eventually be subsampled 327 times to provide materials for further study.

Beneath the snow, and partially frozen into the firn on the ice surface, were several masses of the man's hair, which were excavated and collected on both days of this recovery visit (fig. 7). These were located slightly down-slope (39–94 cm lower in elevation) and between two and a half and four metres from the torso. We believe that during a previous thaw the man's head had likely melted out and moved down-slope a short distance to a small ridge of snow. Once exposed, the head would have absorbed sufficient radiant energy to melt the surface beneath it, and later the hair froze into the surface of the ice.

Upon arrival in Whitehorse, all collected materials were placed in freezers. At the Yukon Heritage Branch the human remains were placed in a new and sterilized freezer that had been bought earlier for treating frozen artifacts but had not yet been put into use. The cooler holding the lower body was also placed inside this freezer. Beattie established forensic-level protocols for access to the freezer. Forensic controls were required to prevent accidental contamination of the human remains, which might inhibit research such as DNA analysis. These controls were maintained as the freezer, with the remains inside it, was transferred to Victoria on September 2, 1999.

An identical replacement freezer was purchased for the Yukon Heritage Branch, and this empty freezer was used on a test run to see how to fit it into the airplane that would take it to Victoria. This proved an important exercise, as it turned out the plane required modification and the freezer had to be tilted on its side to get in the door before it could be turned upright inside the cabin. This knowledge allowed the freezer contents to be properly packed for transport while avoiding unnecessary delays on the tarmac.

The provenance of the robe mass, as observed during recovery, indicates that it was in a position around the man's neck prior to melting from the ice. It is not possible to tell how the robe was arranged when the man died, as it could have been moved by wind or gravity prior to being covered in snow, or it could have shifted during the ice movement that broke the body apart. Many small pieces of what is considered part of the robe were found scattered on the ice ridge surface for a few metres down-glacier from the main robe mass. Pieces of the robe were also found in the ice near the man's waist and next to the ankles, though, so we believe that the robe moved up to his neck area some time after his death. It seems most likely that at the time of death, the man was wrapped in or lying on his robe. Possibly, he was under the robe with some corners tucked under his body.

Information from the hunters indicates that the man's knife (artifact IkVf-1:102) in its sheath (IkVf-1:101) was found directly with the robe. The knife was suspended on a thong, which was broken when found; other long fragments of thong were found in and around the robe. The location where the knife and thong fragments were found suggests that the knife was suspended around the man's neck.

Two parts of a bag made of what was later determined to be beaver skin (IkVf-1:31) were found in the ice next to the man's torso. One part of the bag was next to the man's side, at about the lower rib cage, but separated by a thin layer of ice. The other part of the bag was found adjacent to the body near the right clavicle. During the autopsy some beaver hair was found adhering to the man's skin. Two other pieces of beaver hide (IkVf-1:Robe 28-5 and 28-70) were found with the main robe mass and may also be part of this same object. The bag appears to have been against the man's right side and possibly under his right arm. Though the right arm was never recovered, we know from the position of the muscles and bones in the shoulder area that this arm was raised at the time of death. This positioning, as well as the observation that many of the bones of his right hand were mixed up with a mass of his hair or found, in a later visit, in association with the cranial bones, suggests that the man's head had been resting on his right hand. He appears to have been lying on his right side when he died, possibly on the bag made of beaver skins. The latter might have provided insulation from the ice below.

August 23, 1999: Site Documentation and Systematic Surface Collection

DAY TWO OF THE RECOVERY EFFORT focused on systematic surface collection and site documentation, but involved a different field team. Mackie, Blake, MacRae and Chambers returned, but Beattie, Hare and Gaunt had other commitments and were replaced by Ty Heffner, an archaeologist, and John Fingland, who represented CAFN.

The first flight arrived at the site just before 10:00 am, and the group was able to stay between five and six hours. As there was a low ceiling and falling snow, pilot Washington warned the group that they should be ready to depart on very short notice due to uncertain weather conditions. As a result, the survey instruments were left at the helicopter, and locations were established by triangulating from points that had been recorded the previous day. The resulting map data, while less accurate than that of day one, were judged to

be sufficiently precise considering the field conditions and the apparent recent movement of most of the material due to thawing.

Much of this day was devoted to clearing surface snow over a wide area around the primary site, where the human remains had been recovered the previous day, in order to facilitate the search for additional materials. A trench was dug into the previous winter's snow (fig. 5A)—which was banked up at the foot of the glacier, below where the human remains were found—so that the ground surface could be checked for items that might have melted out in previous years. Nothing was found, but the trench provided access to the undersurface of the ice in the area where the torso was recovered. A piece of the robe, dripping with meltwater, was observed dangling down from the lower surface of the ice. For safety reasons, the fragment was not recovered on this day, though it is likely one of the pieces of robe picked up during later visits.

Slush and snow were cleared from the other slope of the adjacent ice ridge (fig. 5C) and along the meltwater course that crossed the ice surface parallel to the ridge (fig. 5B). One piece of fur hide, likely part of the robe, was found on the side of the meltwater channel opposite to the rest of the remains. Everything else collected on and in the ice, including a lot of material from beneath the fresh snow (fig. 7), originated from the ridge or a few metres down-slope toward the channel. A few items were also recovered this day on the adjacent hillside and on the rocks below the main find area.

A few hundred metres down the glacier from the site, the hunters had observed the remains of a Moose melting from the ice. This was revisited and bone samples were obtained for dating to see if they were possibly related to the man's death. Radiocarbon dating subsequently showed they were from the early 1960s.

Thirty-one bags of material were collected off the surface of the ice, including bits of hide, some fish, small plant remains and assorted other materials. The bare ground within about 500 metres of the site itself was surveyed for objects of interest, and a few wooden artifacts were picked up on the rocky slopes in the surrounding area. The bags of material collected that

Figure 5. Site area near end of day on August 23, 1999: *A*, trench in snowbank; *B*, portion of cleared meltwater channel; *C*, clearing of snow underway in centre of photograph. In the distance, Blake records data at one of the glacier monitoring stations. View to south. John Fingland (CAFN) photograph.

Figure 6. Commencing excavation of the trench to expose the torso and, in background, exposing and collecting hair masses. View to northwest, August 22, 1999. Sarah Gaunt (CAFN) photograph.

Figure 7. Excavation and documentation of materials from surface of ice and beneath recent layer of snow. View to northwest, August 22, 1999. Sarah Gaunt (CAFN) photograph.

Figure 8. Horizontal distribution of surveyed human remains and principal artifact collections for 1999 and 2003. The scale is in metres relative to a reference point. North is to the top. Colour version on page 585.

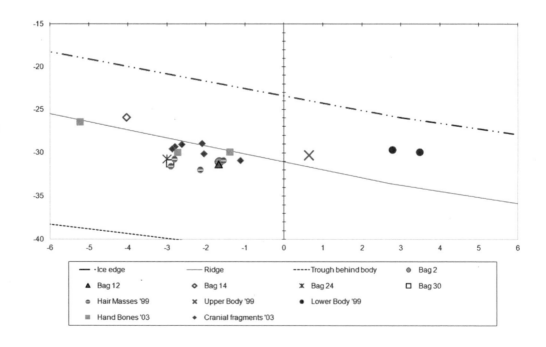

Kwädąy Dän Ts'ìnchį Glacier X-section
1999 and 2003

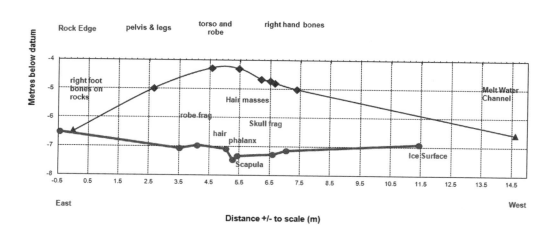

Rock Edge   pelvis & legs   torso and robe   right hand bones

Metres below datum

right foot bones on rocks

Hair masses

Melt Water Channel

robe frag

Skull frag

hair   phalanx

Scapula

Ice Surface

-4
-5
-6
-7
-8

-0.5 0.5 1.5 2.5 3.5 4.5 5.5 6.5 7.5 8.5 9.5 10.5 11.5 12.5 13.5 14.5

East   West

Distance +/- to scale (m)

Figure 9. Cross-sections of the glacier comparing ice levels in 1999 (*top*) and 2003 (*bottom*) using instrument survey data. This view is looking east up the glacier along the ridge.

day were later subdivided in the lab for a total of 395 samples, including ice, sediments and debris, artifacts, fish and the man's body.

The weather improved briefly mid afternoon, and as all that could safely be done had been completed, the group left the site. The collected materials were taken to the Yukon Heritage Branch in the late afternoon for freezing.

A set of Moose antlers, frozen in the glacier several hundred metres from the human remains, was also documented and photographed during the August 17, 22 and 23 site visits. The occurrence of these antlers is considered to be the result of natural death.

August 10, 2000: Site Monitoring

THE SITE was reached by helicopter on a bright sunny day. But with much of the previous winter's snow still present in the area (fig. 10), the crew felt that little could be accomplished on the ground and instead took aerial photographs.

Figure 10. Aerial view looking westward over site (*at arrow*), showing snow-covered glacier, August 10, 2000. Sarah Gaunt (CAFN) photograph.

July 18, 2001: Burial Ceremony, Environmental Sampling

IN 2001 a party of nine people visited the site to return the man's ashes to the area where his body was found (fig. 11). The group travelled by air to the site from the funeral service at Klukshu, Yukon, specifically for this purpose. It was culturally important to the First Nations responsible for managing the find that the ashes be carried back by people who had brought the body off the mountain in 1999, so archaeologists Mackie and Hare assisted with this task (for more on the decisions around cremation and cultural ceremonies, see chapter 8).

The main purpose of this trip was to place the ashes in a cairn on the mountain (fig. 12). Nevertheless some research and collection work was completed while the cairn was being built to receive the ashes, and people were being shuttled to and from the mountain. Chambers and Hare scouted nearby snow-free slopes and recovered one stick artifact several hundred metres from the main find area. This piece (IkVf-1:119) was

transferred to the Yukon Heritage Conservation lab at the end of the day. Mackie spent about one hour collecting 25 background environmental samples for future study. These included sediments from next to the glacier, snow from the surface of the glacier, insects and plant remains identified on the surface of the snow, and plants growing on rocky areas near the site. Photographs were taken of the ceremony preparations, site and sample locations.

The conditions for this visit were good, with a high, patchy cloud cover and no serious concerns about weather or flying conditions. The site was covered by snow (fig. 11), as expected at this time of year; the date of this visit was based on the scheduling of the funeral and not in consideration of suitability for fieldwork. As the site was not revisited later that same season, we do not know if it became snow-free at any time in 2001. In the adjacent southern Yukon, this year was noted as a low-melt one, so it is quite possible the site remained snow-covered throughout the summer season.

Figure 11. Funeral party members returning to helicopter after interment of cremated remains. The discovery site area is located in centre background. *Left to right*, James Allen, Gord Joe, Calvin Lindstrom, Ron Chambers and John Adamson, July 18, 2001. Sarah Gaunt (CAFN) photograph.

Figure 12. Sarah Gaunt placing flowers on the burial cairn on behalf of Champagne and Aishihik women with Elder John Adamson, July 18, 2001. Al Mackie (BC Archaeology Branch) photograph. Colour version on page 585.

Figure 13. Micheal Jim surveys the route during the attempted hike to the site, August 2001. The proposed route would have taken the party up the glacier shown above Jim's head. View to northwest. Sarah Gaunt (CAFN) photograph.

August 2001: Attempt to Follow Most Likely Travel Route

IN AUGUST 2001 a Champagne and Aishihik party attempted to hike the route that the Kwädąy Dän Ts'ìnchį individual was most likely to have taken to the site (see chapter 32). The party included Haines Junction residents Micheal Jim and Will Jones, as well as Sarah Gaunt of Whitehorse. They left the Haines Road on foot, heading west in the area between Mineral Lakes and Nadahini Mountain. Their planned route required crossing a lobe of a glacier located in the uppermost reaches of the northeastern fork of the O'Connor River (fig. 13). They found that this glacier had shrunk to such an extent that it left very steep and loose valley walls along their planned route. By the time they found a safe route down to the glacier, they no longer had enough supplies to make it to the site and back out again safely, so they abandoned their effort to reach the site on foot via a route from the southeast.

August 9, 2002: Botanical Research, Site Monitoring

THE PURPOSE OF THE 2002 VISIT was to collect plants, make botanical observations and monitor the site condition. The weather was variable, with intermittent low cloud overhead. The pilot put the party on notice that an early departure might be required if the weather deteriorated, and so work was carried out largely within viewing distance of the helicopter. While James Dickson (University of Glasgow) and Petra Mudie (Geological Survey of Canada, Dartmouth) undertook their botanical studies, Sheila Greer (CAFN) spent more than three hours doing reconnaissance. The area where the human remains were recovered was entirely snow-covered (fig. 14), with no signs of melt, and no additional finds were made. Southeast of the site area, overlooking the valley that is free of snow (also referred to as the Empty Valley), one stick artifact (IkVf-1:131) was found.

The site area was photographed, and GPS readings were obtained from various locations and topographic features. Two locations marked by flagging tape, which represented places where Blake had collected stick artifacts on August 23, 1999, were also documented. Rocks with recognizable copper inclusions were

Figure 14. Petra Mudie, James Dickson and Delmar Washington standing near the site. View to west, August 9, 2002. Sheila Greer (CAFN) photograph.

collected on this visit, although it was later established that these samples did not contain significant amounts of native copper. On the flight out from the site, the party stopped at Mineral Lakes, where Dickson and Mudie collected further botanical and mineralogical samples.

August 16, 2003: Discovery of Additional Human Remains

IN 2003 Hanlon and Roch returned to Tatshenshini-Alsek Park for a second sheep hunt, stopping by the site on their multi-day trip. At the time of their visit, the weather was good but visibility was poor, with patches of low cloud and fog. Their visit corresponded with a period of warm weather and high-melt conditions, and though snow was still present next to the glacier, the ice surface was free of snow (fig. 15). They also noted that the glacier had receded considerably in the area of the original discovery. Hanlon and Roch noticed a cranial

and other bones resting on the ground and remaining ice, as well as some fragments of hide in this same area. They took pictures and reported their observations in Whitehorse three days later.

CAFN had scheduled a site monitoring visit for a couple of weeks later in August and probably would have found these additional human remains. Nonetheless, as there is always a chance that any planned site visit might not happen due to poor flying conditions or unexpected snowfall, the project again owed a big debt of gratitude to the hunters for their excellent timing and conscientious reporting of discoveries.

August 28, 2003: Burial of Human Remains, Recovery of Artifacts

ON AUGUST 25 CAFN contacted the BC Archaeology Branch with news that the hunters had found additional human remains at the site. The First Nations had decided that these additional remains, likely from

Figure 15. Site situation, looking south over the site area, as workers document finds, August 28, 2003. Sheila Greer (CAFN) photograph.

the same man, should be recovered from the glacier and placed with those found in 1999, which had been cremated and returned to the mountain in 2001. The First Nations also agreed that these additional finds could be subject to scientific investigation in the field prior to burial.

The decision to allow study of these additional human remains on the mountain meant that appropriate expertise was needed for a recovery and reburial visit to the site. Beattie was not available, but fortunately Vancouver General Hospital pathologist Dan Straathof, who had conducted the autopsy of the remains recovered in 1999, was able to participate. Arrangements were quickly made, with August 28 selected for the site visit based on helicopter availability. Whitehorse-based Dr. Bryce Larke, the Yukon Medical Officer of Health, volunteered to assist Straathof with the assessment of the remains. Others involved in planning or executing the recovery and reburial effort included Yukon Heritage Branch conservator Valery Monahan and glaciologist Blake, as well as Chambers, Gree and Mackie.

While the site visit lasted less than three and a half hours, weather conditions were excellent, with sunny skies and reasonably warm temperatures (figs. 15 and 16). The glacier's ice surface was not slippery, as the

consistency had changed to that of coarse snow. But the glacier had shrunk considerably, and nothing remained of the snow or ice ridge where the remains had been recovered in 1999 (figs. 17 and 19). The ice surface now sloped gradually north and west to the bedrock below, and two of the previously observed three nunataks south of the find area now formed a single mass of exposed bedrock (fig. 16). In the ridge area where the remains had been recovered in 1999, about three vertical metres of ice had melted away (fig. 9). While the changes in glacier topography were disorienting, the survey instrument showed that the human remains exposed in 2003 were situated in the same horizontal location as those found in 1999 (fig. 8).

The exposed human remains were mapped with the theodolite, photographed in detail, measured and described. The horizontal distance between the skull bones mapped in 2003 and the location of the man's neck as mapped in 1999 ranged from 75 to 400 centimetres (fig. 8). The skull bones were 295 to 309 centimetres lower in elevation than the mapped position of the neck due to the wasting of the glacier in the intervening years. Swabs from the man's teeth were taken for pollen analysis. Six teeth were retained for possible DNA analysis as, at that time, the outcome of the DNA study of community members and of the man's tissues was uncertain. As the DNA analysis was successfully completed with other samples, these teeth were not needed for that purpose and together with sample residues from the studies of human tissues were returned to CAFN for reburial at the site. Various small artifact fragments present in the same area as the remains were also mapped and collected; most of the artifacts appeared to be fish and pieces of the robe. Environmental background sampling was also conducted. In all, 24 samples were collected, including ice, snow, sediment, fragments of fish (fig. 18), pieces of robe and a small rodent (a vole or shrew). Away from the immediate vicinity of the site area, Chambers and Greer also found four stick artifacts (IkVf-1:120/121, 122, 123 and 124).

While the documentation and research were underway, Chambers prepared the burial cairn to

Figure 16. Aerial view looking southeast down to the site area *(arrow)*, located to the left of the emerging knolls, August 28, 2003.
Sheila Greer (CAFN) photograph.

receive the human remains. When the field studies were concluded, he conducted the interment of the recovered remains, placing them beside the cremated remains from the 1999 find (see chapter 8).

Collected materials were immediately flown to Whitehorse and deposited with the Yukon Heritage Branch conservation lab. Wooden artifacts were placed in freezers, following recommended conservation protocol (see chapter 22), while the other materials were kept cool and prepared for transport south for further conservation and study at the Royal BC Museum. The latter pieces were transported to Victoria by Mackie on his return south.

Figure 17. The team setting up for documentation in the site area, looking east and up-glacier, August 28, 2003. Sheila Greer (CAFN) photograph.

Figure 18: Fragment of salmon showing skin and scales, approximately 10 cm long, August 28, 2003. Al Mackie (BC Archaeology Branch) photograph.

Figure 19. Looking northwest across site area as surface finds are being documented, August 28, 2003. Sheila Greer (CAFN) photograph.

Figure 20. The view west-northwest over site, all ice now melted from the find area, August 20, 2004. Sheila Greer (CAFN) photograph.

August 20, 2004: Site Monitoring

ON AUGUST 20, 2004, CAFN representatives Diane Strand, Lawrence Joe, Frances Oles and Sheila Greer travelled to the site for a monitoring visit. The site inspection lasted about two and a half hours, and conditions were very good, with high cloud and warm temperatures. The group found that the ice in the site area was melted completely away (fig. 20). In the same place where human remains had been found in previous years, they found a parietal bone—the only missing piece of the man's skull. It was found exposed on the rock and muddy ground surface. Less than three metres away was a wooden artifact (IkV-1:125, the carved and painted stick; fig. 21), and in between the parietal bone and the wood artifact were fragments of fur, hide lashing, possible pieces of salmon, miscellaneous pieces of tissue that might have been human or fish, plus two human teeth and some hair.

Figure 21. Wooden artifact IkVf-1:125, the carved and painted stick, on the ground surface in the area where human remains had been found, with the glacier now completely melted away. August 20, 2004. Sheila Greer (CAFN) photograph.

These remains were scattered over an area approximately three metres by three metres and located within three to four metres of the ice margin at that time (fig. 20). The orientation and arrangement of these materials suggested they had been dispersed by moving water, though only the parietal bone appears to have been transported any distance—a few metres at most.

As during the August 28, 2003, site visit, the entire area was carefully combed for artifacts and human remains, so that the site was cleared of all such materials. Both definite and possible human remains were documented and then interred with the other human remains. Artifacts and artifact fragments were collected in two bags, with the small, fragmentary materials ultimately transferred to the museum and the larger wooden artifacts turned in to the Yukon Heritage Branch conservation lab. In addition to the carved and painted stick found beside the skull bone, the wood pieces recovered included a second artifact (IkVf-1:126, an unmodified stick) found away from the main site area, overlooking the so-called Empty Valley.

The party visiting the site in 2004 concluded that the ice which had contained the Kwäday Dän Ts'ìnchį man had completely melted away, that all related materials had likely been collected and that there is little chance for further finds.

August 12–14, 2005: Biological Studies and Extended Reconnaissance

A BOTANICAL EXPEDITION led by Dickson and Mudie travelled to the site in mid August 2005. The main purpose was to collect botanical and environmental samples from water and sediments to help reconstruct the man's activities in the days before he died. CAFN also wanted to confirm that all human remains had been recovered from the site area.

The expedition flew in and established a camp down-slope from the site. This camp was used as a base for undertaking botanical observations and collecting samples in the general site area on August 12 and 13. After two nights at this location, the group hiked out to Haines Road via a route that took them first into the northernmost tributary of Fault Creek and then into the Parton River basin, camping three nights on the route out. Hanlon acted as a guide for the expedition, while Oles represented CAFN. The party also included BC Parks warden Greg Eikland, who is also a CAFN citizen, and professional photographer Al Harvey. Greer travelled to the site with the party on August 12, spending a few hours in the site area before leaving with the helicopter.

The weather was very good through the entire expedition, with sunny, blue skies. In the two days spent in and around the site area, the party was able to complete a wide-ranging examination of the ground surface and collect a variety of botanical samples. The area was documented photographically, and GPS coordinates were recorded for all finds made. Oles also left an offering of respect by the burial cairn that holds the cremated remains.

During her time at the site, Greer surveyed for additional human remains, examining the area where the earlier finds had been made, as well as downslope, but found nothing on the now ice-free surface.

Additional artifacts were, however, discovered in the surrounding area. All were found some distance away from the find site of the human remains, though some were located near previous wood artifact finds. The new finds included a stick fragment with a carved end

(IkVf-1:127) and badly fragmented wood pieces that may represent the base of a gaff pole/walking stick (IkVf-1:128). These two artifacts were taken to Whitehorse by Greer when she left.

Other artifacts found by members of the 2005 expedition include stick artifact IkVf-1:129, which is the distal end of the third gaff pole/walking stick (fig. 22). This artifact was considered too heavy to carry out, so it was left at the site and collected by Champagne and Aishihik BC Parks warden Jim during a site visit later that season, on September 22. Two additional small stick fragments found by Hanlon were also left at the site for later recovery. Unfortunately, these pieces—one segment about 30 centimetres long, the second approximately 9 centimetres; both thought to be part of the same minimally modified stick artifact—could not be found during the September site visit due to snow cover. These pieces were found near the IkVf-1:127 stick with a carved end and may be part of the same artifact. They remain at the site at this time.

Various faunal remains that had been observed in the site area during previous site visits were also collected in 2005. These included Dall's Sheep bones that appear to be from a natural death assemblage, likely a wolf kill, which were found scattered over the rock ground surface uphill and to the east of the site. Another single Dall's Sheep bone was recovered on the hill overlooking and to the north of the site. Canid bones (mandible and skull) were collected from the surface of the glacier, up-slope and to the west of where the human remains were found. All of these faunal remains were turned over to Yukon paleontologist Paul Matheus for study and identification. Matheus confirmed the identification of the sheep skeletal pieces and subsequently identified the canid bones as wolf, not dog, noting that a definitive identification based solely on physical characteristics was not possible, given the bones' distorted condition after their extended stay in a water-logged glacier environment (Matheus, pers. comm. to Greer).

A possible hunting blind structure found by the hunters during their 2003 visit to the site area was also investigated by Greer during her August 12, 2005,

visit. This feature consists of a rock wall beside an area where pavement rocks had been laid out, creating a flat surface. While man-made, the presence of a tent tie-down cord as well as a lack of lichen on the wall structure indicated that it was of recent vintage and likely represented a place where a small tent had been pitched. The structure, along with a tin can and a bundle of staking sticks also found by party members in the site vicinity, is likely attributable to mineral exploration work. Evidence of such activity, in the form of a pile of metal staking plates marked with the label of a mining company (Noranda) and a year (1981), was also seen beside the small lakes down-slope from the glacier. The tin can was taken out from the site as garbage; the staking sticks and plates were left where they were found. These may be associated with a claim stake found

Figure 22. Greg Eikland with a carved stick artifact IkVf-1:129, believed to be part of a gaff pole—one of several stick artifacts found in the area surrounding the site on August 12 and 13, 2005. Petra Mudie (Geological Survey of Canada) photograph.

in the glacier near the site in 1999, as well as a strand of hip-chain thread that was found running along the glacier ridge in the site area during 1999 fieldwork. These scant remains indicate some of the recent historical uses of the site area before it became protected.

In the bottom of the Empty Valley, southeast of the discovery site, Hanlon found a piece of cut and rolled birch bark in 2005. This object was recovered and later registered as part of a separate site (IkVf-2), since it is about one kilometre from the closest part of the Kwädąy Dän Ts'ìnchį site.

September 22, 2005: Artifact Recovery

IN ORDER TO COLLECT artifacts left behind during the August site visit, Jim and Eikland stopped at the site on September 22, 2005, when returning from fieldwork in Tatshenshini-Alsek Park. Significant accumulations of snow were present over much of the site, and weather conditions were poor, with blowing snow, so less than an hour was spent at the site. The party was, however, able to document the location of stick artifact IkVf-1:129, collected during the August visit, as well as the canid skull collected on the glacier surface. Owing to snow cover, they were not able to relocate the two stick fragments found by Hanlon.

Summary and Conclusions

BETWEEN 1999 AND 2005 the Kwädąy Dän Ts'ìnchį site was visited 12 times—twice by the site's discoverers travelling on foot, with the remaining visits by helicopter. The site's inaccessible location and challenging weather conditions constrained all visits to a short duration, limiting the investigations. The work focused on honouring both cultural and scientific protocols—that is, recognizing cultural needs while following accepted archaeological field recovery methods and meeting conservation and forensic standards.

In spite of difficult conditions, field crews were able to achieve their principal objectives. These included the assessment and recovery of the human remains in 1999 and, following the completion of the autopsy, the return and interment of the cremated remains in 2001. Human bones recently melted out of the glacier were recovered, documented and then interred with the cremated remains in 2003 and 2004, all without the remains leaving the mountain where the Long Ago Person Found lost his life.

In addition to recovery of human remains, a variety of field studies were conducted at the site and in the surrounding area in order to better understand who this individual was and what befell him. These studies included surveying for artifacts and mapping of finds, as well as collection of biological, geological and other environmental samples. The glacier morphology was mapped and its shrinkage documented. Remains of other animals—including a Moose and a canid that, like the man, had died and been trapped in the glacier—were also noted and sampled. These environmental samples, as well as the man's belongings and other artifacts, continue to inform and educate us about his life and times.

Recovered items considered to be the belongings of the Kwädąy Dän Ts'ìnchį individual include the robe and its many fragments, the beaver-skin bag, the hat, the bead (sample 24-37; identified in the lab in the man's hair), the knife and sheath, and a wooden artifact of uncertain function (IkVf-1:125, the carved and painted stick). Numerous fragments of fish (later determined to be salmon) were also collected in and around the area where the human remains were found. We believe that this fish may have been preserved dry fish the man carried for food. It is also possible that the four-metre gaff pole/walking stick artifact (IkVf-1:104) may be associated with the Long Ago Person Found, though this is less certain.

Excepting steep and dangerous slopes, an area of roughly one kilometre around the original site was surveyed for archaeological finds, with additional discoveries made. The materials recovered away from the location of the human remains consist solely of wooden artifacts—some shaped into recognizable objects, others not. A single piece of cut and rolled birch bark was found more than one kilometre from the site. These artifacts from further afield cannot be considered directly associated with the Long Ago Person Found. They nonetheless provide evidence for use of this area by others in times past.

The collection of artifacts recovered from this site is extremely unusual, as it includes preserved organic materials such as fur, hide, sinew, root and wood that rarely survive in archaeological sites. Proper recovery of such material adds complexity to fieldwork as well as to the subsequent care, handling and analysis efforts. The only comparable group of preserved organic artifact finds from a pre-contact Indigenous context in western Canada comes from ice patch archaeological discoveries made in neighbouring southern Yukon (Farnell et al. 2004; Hare et al. 2004; Hare 2011 et al.) and the Northwest Territories (Andrews et al. 2012). These ice patch finds have similarly been recovered from sites situated at higher elevations in melting snow and ice contexts.

By 2004 the portion of the glacier that had preserved the remains of the Kwädąy Dän Ts'ìnchį individual for centuries had completely melted away. On the last visit to the site in late August 2005, no additional human remains were found, and the team noted that the glacier had shrunk so much that the ground was exposed in the area where the original find had been made. It was apparent that the site associated with this man's death no longer existed. But additional artifacts found in the surrounding area that same year suggest that additional cultural finds may well occur.

During all fieldwork, as well as other work, every effort was made to treat the remains of the Long Ago Person Found with, at the very least, the same dignity and respect shown for the remains of other people who have died through accident. Considering the First Nations cultural context, the approach taken was comparable in many ways to what would have been done for any modern young hiker whose missing body had finally been located. This is an emerging standard for archaeologists to follow. In the past, archaeologists have been prone to treating ancient human remains more as specimens of scientific inquiry and less as the remains of persons warranting dignified respect. Yet these two approaches to the recovery of human remains, the retrieval for autopsy and archaeological excavation, have much in common, including the care taken to preserve information and to avoid contamination that might preclude further investigation. Our experience shows that in an archaeological context, many cultural values can be explicitly and sensitively addressed with little compromise to scientific investigation.

5

VASCULAR PLANTS, BRYOPHYTES, LICHENS AND ALGAE FROM THE KWÄDĄY DÄN TS'ÌNCHĮ DISCOVERY SITE IN NORTHWESTERN BRITISH COLUMBIA

James H. Dickson, Petra J. Mudie, Alexander P. Mackie, Brian Coppins, Roxanne Hastings, Richard J. Hebda and Kendrick L. Marr

This list of flora growing around the Kwädąy Dän Ts'inchį discovery site is the result of J.H. Dickson and P.J. Mudie's involvement in a study of plant remains found both within and outside the ancient body that melted from a glacier in northern British Columbia in 1999. The corpse was that of a young Aboriginal male who had died on the Samuel Glacier in Tatshenshini-Alsek Park, near the headwaters of Fault Creek (Beattie et al. 2000; Pringle 2002; Dickson et al. 2004; Mudie et al. 2005; Richards et al. 2007; Dickson and Mudie 2008).

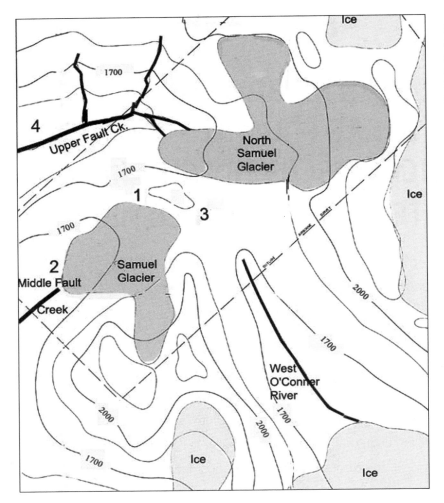

Figure 1. Topographic map of study area, showing the Samuel Glacier (also referred to as Fault Creek Glacier), main rivers and nearby mountain peaks. 1, 2 and 3 are the plant collection locations. Geographic coordinates are not given in order to protect the archaeological site.

AT ABOUT 1,600 METRES ABOVE SEA LEVEL, the discovery site (fig. 1) is in the heart of a very inaccessible region of the southern St. Elias Mountains. Although a mere 15 kilometres west of the Haines Highway on a map, the site is nonetheless a 60-kilometre walk distant and very difficult to reach on foot because of the mostly trackless, rugged, glaciated topography, with fast-flowing rivers and streams and densely thicketed subalpine scrub vegetation.

Camping near the site and walking around the area in the course of botanical collecting led to a general familiarity with the topography and ground vegetation that the Kwädąy Dän Ts'ìnchį man had travelled across

centuries ago. Concerning scientific results, there was one very important reason and one subsidiary reason for compiling a list of the flora in the immediate vicinity of the Kwädąy Dän Ts'ìnchį discovery site.

First, knowing what is growing there now greatly helps us identify and interpret the provenance of plant remains (pollen, spores, seeds, leaves, charcoal, mosses, lichens and algae) found with the body, on the associated clothing and within artifacts such as the beaver-skin bag. Second, the discovery site's remote location is one which had been previously unexplored botanically, and it lies in a vast area that has been little investigated for plant species, ecology and phytogeography. For example, the most relevant studies with regard to diatoms are those dealing with several lakes along the Haines Highway (Wilson and Gajewski 2002) and recent yet unpublished studies of the diatoms in the subalpine lakes Nesketahin Klukshu and Little Klukshu, and in Howard Lake (D. Selbie, pers. comm. 2008).

There have now been three plant-collecting visits to the discovery site. The first was very brief, by BC government archaeologist Al Mackie on July 21, 2001. The second was on August 9, 2002, by Dickson and Mudie—the first botanists to collect at the site (but only for a mere three hours, since bad weather forced the helicopter's early departure). Finally, a more protracted visit by Dickson and Mudie took place August 12–14, 2005, when they went in by helicopter and camped two days near the discovery site (then travelled out by foot, over the glacier and along an old mining trail to the Haines Highway August 14–17).

Above the discovery site on the southern Samuel Glacier are jagged peaks and razorback ridges that are part of a high-grade regional metamorphic complex consisting of tightly folded (NW-SE trending) metasedimentary rocks, including limestone/marble, basic and silicic igneous intrusions and a variety of metagreywackes—blanketed in places by glacial deposits derived from the above (J. MacDonald, Glasgow University, pers. comm. 2008). Below the retreating glacier there are erratics of various rock types, including saccharoidal limestone, and there are also scattered outcrops of mineralised copper (figs. 2, 3, and 4).

The discovery site area that we examined closely (fig. 1, locations 1, 2 and 3) ranged in altitude from *ca.* 1,425 metres to *ca.* 1,625 metres and lies within the Alpine Tundra Zone as defined by Pojar and Stewart (1991a and b). Here, the treeline—with scattered stunted White Spruce trees (*Picea glauca*) and Balsam Poplar (*Populus balsamifera*)—occurs at about 1,300 metres on south-facing slopes and about 150 metres lower on cold north slopes (Danby and Hik 2007). Location 3, at 1,450 metres, is at the uppermost occurrence of stunted shrubs, such as the Alaska Willow (*Salix alaxensis*), a few kilometres down-slope from the discovery site.

From below 1,400 metres to the treeline, the density of shrubs, particularly willows and alder, increases greatly. The timberline (i.e., closed canopy forest) begins much lower, at around 1,000 metres.

Location 1 is at about 1,600 metres above sea level (from GPS in the helicopter) and about 100 metres from the discovery site. This is a mainly unstable slope facing somewhat southeast, with outcrops of bedrock and much scree. For the algal collections, location 1 includes the rapidly melting snow on the glacier surface immediately south of this rocky slope, as well as a small, shallow, temporary meltwater pond (pond 1, three by two metres across and a half-metre deep) between the glacier and the slope.

Location 2 is a large rock outcrop some 500 metres up-slope from the site, forming a divide between the two glaciated valleys draining into the Upper and Middle Fault Creek rivers, respectively (fig. 1). Most of the botanical collecting was carried out close to the commemorative cairn at about 1,625 metres above sea level. The algal sampling was in a small kettle pond (about 20 by 15 metres) of unknown depth (greater than 1 metre), called pond 2, located on the col between the rock outcrop and the dry valley south of the Parton River headwaters (fig. 1). The elevation of pond 2 is about 1,615 metres.

Location 3, campsite 1, is 2 kilometres southwest of the discovery site. It comprises grassy slopes, an area of mires and flushes, blocky lateral and terminal moraines and a linear series of three small, shallow, turbid morainal lakes, located at about 1,425–1,450 metres above sea level. The algal samples were taken from three ponds: a scour pool (*ca.* 1 by 0.5 by 0.5 metres) within a small stream draining the mire (pond 3-1); from the lowest of the morainal lakes (about 50 by 40 metres, 1–2 metres deep), called pond 3-3; and from the very turbid morainal pond closest to the glacier (pond 3-2). The sample from pond 3-2 contained only glacial flour (fine silt and clay-sized particles), so there are no algal results for pond 3-2.

Figure 2. Location 2, mid-distance behind the helicopter. Location 1 is the slope up to the left, August 12, 2005. Al Harvey photograph.

Figure 3. This aerial photograph shows location 3, August 12, 2005. Al Harvey photograph.

Figure 4. Near location 1, showing the banded rock (semiperlite), whitish-grey lichen *Stereocaulon* (Cauliflower Foam), dark yellowish-green moss *Dicranowiesia crispula* (Mountain Pincushion) and the pink flower of *Saxifraga oppositifolia* (Purple Mountain Saxifrage). July 21, 2001. Al Mackie photograph. Colour version on page 585.

Methods

AT ALL THREE LOCATIONS, Mudie collected and identified the algae with help from Barbara Zeeb (Royal Military College of Canada) for chrysophyte stomatocysts and from Daniel Selbie (Queen's University) and Reinhard Peinitz (Carlton University) for chrysophytes and diatoms, respectively. In the field, pond and stream algae were obtained using a large (40 ml volume) plastic syringe with tubing attached

to allow sampling to a water depth of 1 metre. A volume of 120 millilitres was obtained at each aquatic site and strained through a 0.65 μm (micrometre) micropore filter in the syringe. The micropore filter with algal filtrate was then removed from the syringe and stored in a glass vial containing ethanol. Snow algae were strained in a similar way after melting a volume of 1 metre^2 by 10 cm depth of snow. The algal samples and environmental scanning electron microscope (ESEM) images are archived at the Geological Survey of Canada, Atlantic, in Dartmouth, Nova Scotia.

Using the standard methods of pressing and drying, Dickson, Mackie and Mudie gathered all other plants. Brian Coppins and Dickson identified the lichens; Dickson and Roxanne Hastings identified the bryophytes (mosses and liverworts); and Dickson, Richard Hebda and Ken Marr identified the vascular plants. The first set of bryophytes and lichens examined are housed in the herbarium of the Botany Department of the University of British Columbia in Vancouver, and the second set are at the herbarium of the Royal British Columbia Museum in Victoria, where all the vascular plant specimens are kept.

Results of Systematic Studies

WITHIN EACH MAJOR CATEGORY, the lichens and plants are listed alphabetically by genera and then species. The list of algae is accompanied by critical taxonomic comments because so little is known about this group in the discovery site region. Lichen, bryophyte and vascular plant lists are presented in tables 1, 2 and 3, including the locations where they were collected (see fig. 1).

Level of identification: Where the exact identity is not given, such as *Pertusaria* sp., *Lophozia* cf. *excisa* or *Erigeron* sp., the collected material was inadequate for definitive recognition or the record was made in the field without a voucher specimen for subsequent verification. For the algae, we use the primary classification of freshwater algae given by Wehr and Sheath (2003). Precise identification of chrysophyte cysts is often not possible because the taxonomy of the resting spores (statocysts) is in a state of flux (Duff et al. 1995, Wilkinson et al. 2001). The taxonomy of diatoms is also a rapidly changing field of study, with little consensus among "lumpers"—who want to retain large, well-established groupings of genera with similar morphology—and "splitters," who see differences in evolutionary trends, DNA of living cells and other fine details. Whenever possible, the systematic treatment of Antoniades et al. (2008) has been followed.

English names: Apart from the green algae (chlorophytes), with a few common names such as Red and Green Snow Algae, most algae below the taxonomic level of family do not have English names. The English names for the lichens are taken from the large volume on North American lichens by Brodo et al. (2001). Goward (1994, 1999) gives English names for British Columbian lichens. Similarly, there is no difficulty concerning vascular plants, for which the names come from Douglas et al.'s multi-volume work *Illustrated Flora of British Columbia* as updated in E-Flora BC. The bryophytes pose problems because there is no agreed-upon list of English names for North America. Consequently, *English Names for British Bryophytes* by Edwards (2003) has been used in part, as well as the names provided on E-Flora BC (http://www.geog.ubc.ca/biodiversity/eflora/).

Algae (figs. 5, 6, 7)

----

1. Green Algae (Chlorophytes)

1A. Coccoid and Colonial Green Algae

- *Botryococcus* sp., boghead coal alga. Common in pond 3-1.

- *Trochiscia nivalis* Lagerheim, Red Snow Alga. Location 1, glacier snow, kryal stream (i.e., glacier margin stream) near the discovery site and in pond 1.

- *Trochiscia cryophila* Chodat, Snow Alga. Location 1, kryal stream near the discovery site.

1B. Flagellated Green Algae: Chlamydomonas

- *Chlamydomonas nivalis* (Bauer) Wille 1903, Common Red Snow Alga hypnozygotes and cysts (fig. 7.11). Location 1, in a pink-tinged melting snow layer on the glacier surface, August 2002 and 2005, and pond 1.

1C. Conjugating Green Algae: Desmids (Desmidiales)

(Note that all the desmids in the discovery area were found only in pond 3-3.)

- *Cosmarium anceps* Lund var. *anceps*. Differs from *C. pokornyanum* in its narrower apices and broader girdle area (fig. 5.15). Rare species.

- *Cosmarium crenatum* Ralfs (fig. 5.12). Rare at discovery site. Closest known site recorded is at Thompson, Alaska (Croasdale 1962). It was not found in ponds near Anchorage (Hilliard 1965).

- *Cosmarium debaryi* Arch. (fig. 5.9). Uncommon species at the site. Surface is strongly and irregularly pitted.

- *Cosmarium debaryi* Arch. var. *inflatum* Klebs 1897 (fig. 5.10). Croasdale (1965) says this morphotype differs from *C. debaryi* var. *debaryi* only in its inflated sides. However, our specimens also lacked the conspicuous pits present in var. *debaryi*.

- *Cosmarium pokornyanum* (Grun.) West and West (fig. 5.4). Rare.

- *Cosmarium speciosum* Lund var. *biforme* Nordstrom (figs. 5.1, 5.2). Very common; also present at the second sampling site at the Mineral Lakes in August 2002 (Petra Mudie collection).

- *Cosmarium speciosum* Lund var. *rostafinskii* (Gutw.) West and West (figs. 5.3, 5.7). Less common than var. *biforme*.

- Cf. *Staurastrum pachyrhynchus* Nordstedt, punctate variety (fig. 5.8). Rare in the discovery area and differing from Croasdale's (1962) smooth-surfaced var. *pachyrhynchum* in its conspicuous punctate ornament.

- Cf. *Staurastrum hexangulare* Friv. Rumel. Only found in sample 14-1 from the intestine of the Kwädąy Dän Ts'ìnchį man. The identity of this unusual spiny spore is uncertain because only a few specimens were found, all of which appeared to have dehisced. The specimens most closely resemble resting spores of the desmid *Staurastrum hexangulare*.

- *Staurastrum alternans* var. *basichondrum* Schmidle 1898 (figs. 5.5, 5.11). Very common in the discovery area.

1D. Conjugating Green Algae: Filamentous Types (Zygnematales)

Zygnema sp. Abundant in pond 3-1.

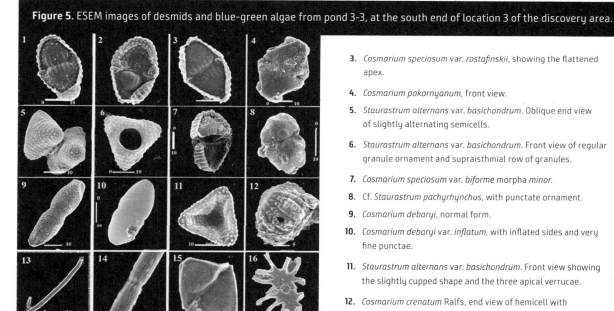

Figure 5. ESEM images of desmids and blue-green algae from pond 3-3, at the south end of location 3 of the discovery area.

3. *Cosmarium speciosum* var. *rostafinskii*, showing the flattened apex.

4. *Cosmarium pokornyanum*, front view.

5. *Staurastrum alternans* var. *basichondrum*. Oblique end view of slightly alternating semicells.

6. *Staurastrum alternans* var. *basichondrum*. Front view of regular granule ornament and supraisthmial row of granules.

7. *Cosmarium speciosum* var. *biforme* morpha *minor*.

8. Cf. *Staurastrum pachyrhynchus*, with punctate ornament.

9. *Cosmarium debaryi*, normal form.

10. *Cosmarium debaryi* var. *inflatum*, with inflated sides and very fine punctae.

11. *Staurastrum alternans* var. *basichondrum*. Front view showing the slightly cupped shape and the three apical verrucae.

12. *Cosmarium crenatum* Ralfs, end view of hemicell with 10 coarsely verrucate crenations.

13. *Phormidium* cf. *P. retzii*. Image 13. Entire strand *ca.* 100 micrometres long.

14. *Phormidium* cf. *P. retzii*. Detail of strand at start of vegetative division.

15. *Cosmarium anceps* var. *anceps*, front view.

16. Cf. *Staurastrum hexangulare* from KDT sample 14-1.

1. *Cosmarium speciosum* var. *biforme* morpha *minor*. Front view, showing paired marginal granules and six rows of vertical granules above and below the isthmus.

2. *Cosmarium speciosum* var. *biforme* morpha *minor*. Front view of basal tumor.

Figure 6. ESEM images of chrysophyte stomatocysts from pond 2, adjacent to location 2 of the discovery area.

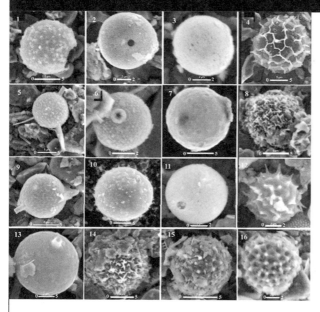

4. *Chrysidiastrum catenatum* Lauterborn stomatocyst showing large cyst with clearly marked continuous and discontinuous high ridges.

5. *Chryastrella* Chodat 1922 emend. Deflandre cyst-type. Cyst with spines about six micrometres long.

6. Cf. Stomatocyst 207 Duff and Smol 1994; apical view showing the verrucate pore collar.

7. Stomatocyst 3 of Duff and Smol 1988b.

8. Cf. *Cysta polygonata* Nygaard cyst-type of Gritten 1977. Debris lower right may include polygonal scales.

9. *Chryastrella* Chodat 1922 emend. Deflandre cyst-type. Cyst with apparently fused spine bases.

10. Cf. *Clericia frenquelli* 1925 emend. Deflandre cyst-type 1. Larger cyst, possibly with a very small obscure pore upper right.

11. Cyst 156 of Zeeb and Smol 1993. Oblique apical view.

12. Unknown stomatocyst 1, with long concave spinules.

13. *Mallomonas* cf. *M. fastigata* cyst-type of Sandgren 1983.

14. Cf. *Cysta polygonata* Nygaard cyst-type of Gritten 1977. Scales still attached upper left.

15. Stomatocyst 143 of Duff and Smol in Duff et al. 1992. Immature cyst, with scaly plates still attached.

16. Stomatocyst 143 of Duff and Smol in Duff et al. 1992. Apical view of more mature cyst, with obscure annulate pore in centre.

1. Cf. *Clericia frenquelli* 1925 emend. Deflandre cyst-type 1 Degraded cyst, with pore collar complex missing and a minutely pitted wall surface.

2. Cyst 15 of Duff and Smol 1988a emend. Zeeb and Smol 1993.

3. Cyst 156 of Zeeb and Smol 1993. Lateral view, with slightly raised pore rim upper right.

2. Golden-brown Chrysophyte Algae (Chrysophyta)

In our water samples, chrysophytes were observed only as silica-walled resting cysts called statospores; in the taxonomic literature, the statospores are commonly called cysts or cyst-types.

- Cyst 15 of Duff and Smol 1988a (emend. Zeeb and Smol 1993). Common in pond 1, rare in pond 3-3. A fairly generic smooth, round stomatocyst without

a thickened annulus or collar (fig. 6.2). It may be produced by a number of different chrysophyte species; similar stomatocysts produced by *Chrysosphaerella* are widespread in northern North America, being present from the Canadian Arctic (Zeeb and Smol 1993) to Ontario (Pick et al. 1984).

- Cyst 156 of Zeeb and Smol 1993 (figs. 6.3, 6.11). Common in pond 2 and rare in pond 3-1. Biological affinity unknown.

- Cf. *Clericia frenquelli* 1925 emend. Deflandre cyst-type, according to Cronberg (1986). A spherical cyst with short spines and granules (figs. 6.1, 6.10). Possibly a stomatocyst of *Dinobryon cylindricum* Krieger 1930, as illustrated by Sandgren (1983, fig. 8), but the collar was not seen in our samples.

- Cf. *Cysta polygonata* Nygaard cyst-type of Gritten (1977). A spherical stomatocyst with short spines, some of which are joined at the base. Differs from the stomatocyst illustrated by Gritten (1977) in that many spines terminate in short, stellate-branched structures, some of which support thin polygonal scales (figs. 6.8, 6.14). Occasional in pond 2.

- Cf. Stomatocyst 207 Duff and Smol 1994. A round stomatocyst with small, densely spaced conulae and a deep, conical pore ornamented with a fused ring of verrucae (fig. 6.6). The Samuel Glacier stomatocyst differs from the morphologically similar southern British Columbian Stomatocyst 207 Duff and Smol 1994 in its smaller size (four micrometres).

- *Chrysidiastrum catenatum* Lauterborn stomatocyst, according to Sandgren (1983) (fig. 6.4). Rare in pond 2.

- *Chryastrella* Chodat 1922 emend. Deflandre cyst-type, according to Cronberg (1986). A spherical cyst with simple elongate collar and three flattened spines; stomatocyst body is ornamented by lunate ridges (figs. 6.5, 6.9). Occasional in pond 2. The Samuel Glacier stomatocyst differs markedly from the collared spinose Stomatocyst 128 Duff and Smol (in Duff et al. 1992 emend. Duff and Smol 1994) from Ellesmere Island in its compound ornament of ridges and large spines, and the expanded width of the spine bases. This distinctive cyst-type does not appear to have been described previously.

- *Mallomonas* cf. *M. fastigata* cyst-type of Sandgren (1983). A smooth-walled, spherical cyst with simple elongate collar and several elongate antapical spines (fig. 6.13). Rare in pond 2.

- Stomatocyst 3 of Duff and Smol 1988b. A small, spherical cyst ornamented with shallow pits less than one micrometre in diameter, and with a low, apically flattened collar (fig. 6.7). Rare in ponds 2 and 3-3.

- Stomatocyst 143 of Duff and Smol in Duff et al. 1992. A spherical, reticulate stomatocyst with small, echinate spines (figs. 6.15, 6.16). This cyst is slightly smaller (six to seven micrometres) than those found on Ellesmere Island and elsewhere in North America, described by Duff et al. (1995).

- Unknown stomatocyst 1. A small, spherical cyst with a low collar and relatively long, acute spines having concave sides (fig. 6.12). This cyst is similar to the statocyst from the south Swedish highland illustrated by Cronberg (1986, figs. 20–27). Rare in pond 3.

- Unknown stomatocyst 2 (fig. 6.12). A relatively large cyst with scattered low ridges and a low annular collar, somewhat similar to Stomatocyst 334 of Wilkinson and Smol 1998, as illustrated by Wilkinson et al. 2001 (p 85, fig. 102). Occasional in melting snow near the discovery site.

3. Diatoms (Bacillariophytes)

- *Amphora inariensis* Krammer 1980 (fig. 7.16). Rare in pond 3-3.

- *Amphora* cf. *A. thumensis* (A. Meyer) Krieger 1929 (fig. 7.2). Common in pond 2. Variable in size, with small forms having fewer striae with larger pores than larger forms. Possibly includes some specimens of *Amphora pediculus* (Kützing) Grunow ex A. Schmidt 1875, which is present in Howard Lake near Klukshu (Bondy 2006).

- *Aulacoseira* sp. cf. *A. nivaloides* (Camburn) English and Potapova comb. et stat. nov. 2009 (fig. 7.1). A small, finely striate-centric diatom with a rounded valve surface. Rare in snow at location 1.

Figure 7. ESEM images of diatoms from ponds 2 (1–10) and 3-3 (13–16), and snow algae (11, 12) from the discovery area.

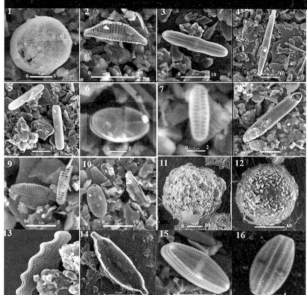

1. Cylindric diatom *Aulacoseira* sp. cf. *A. nivaloides*.
2. *Amphora thumensis*, a strongly asymmetrical pennate diatom.
3. cf. *Caloneis* sp. 1 of Antoniades et al. 2008.
4. *Fragilaria capucina*, a relatively large (30 μm) symmetrical monoraphid pinnate.
5. *Nitzschia perminuta*, with view of the rostrate-capitate apices (left) and rounded keel (right).
6. Very small oval pinnate *Fragilaria* sp.
7. Very small oval pinnate *Fragilaria* sp.
8. ?*Pinnularia glacialis*.
9. *Pinnularia intermedia*.
10. *Psammothidium kryophilum*.
11. *Chlamydomonas nivalis*.
12. Unknown stomatocyst 2.
13. *Luticola nivalis*.
14. *Cymbopleura anglica*.
15. *Luticola mutica*.
16. *Amphora inariensis*.

- Cf. *Caloneis* sp. 1 of Antoniades, Hamilton, Douglas and Smol 2008 (fig. 7.3). A small, pennate diatom common in pond 2; recorded as abundant in the eastern Canadian Arctic (Antoniades et al. 2008).

- Cf. *Eucocconeis flexella* (Kützing) Cleve 1895 sensu lato (fig. 7.6). Common in ponds 1 and 2. An ovate, pennate species with an axial mound (top image) within a central depressed region (bottom image), similar to *E. flexella* but much smaller than arctic forms in Antoniades et al. (2008).

- *Cymbopleura anglica* (Lagerstedt) Krammer 2003 (fig. 7.14). Common in pond 3-3.

- *Diadesmis* sp. (not illustrated). Common pennate diatom in pond 2, distinguished by strongly rounded apices, filiform raphe with teardrop-shaped end pores and very faint striae. A similar but smaller species is noted by Antoniades et al. (2008) as rare in the Canadian Arctic.

- *Fragilaria capucina* Desmazières 1825 sensu lato (fig. 7.4). Common in pond 2. A relatively large monoraphid widespread in the Arctic (Antoniades et al. 2008).

- *Fragilaria pinnata* Ehrenberg 1843 (no illustration). Common, small (six by two micrometres) pennate diatom in pond 2.

- *Hannaea arcus* (Ehrenberg) Patrick in Patrick and Reimer 1966 (fig. 7.7). Common in pond 1.

- *Luticola mutica* (Kützing) D.G. Mann in Round et al. 1990 (fig. 7.15). Occasional in pond 3-3; a cosmopolitan sub-aerial species (Antoniades et al. 2008).

- *Luticola nivalis* (Ehrenberg) D.G. Mann in Round et al. 1990 (also known as *Navicula nivalis*). Rare in pond 2 (fig. 7.13).

- *Meridion circulare* (Greville) Agardh 1831 (fig. 7.7). Rare in pond 2.

- *Navicula* spp. (not illustrated). Many small pinnate diatoms in pond 3-3 are probably species of *Navicula*, but they could not be identified with certainty in the study. Bondy (2006) reported occurrence of a low number of 34 *Navicula* species in Howard Lake near Klukshu.

- *Nitzschia commutata* Grunow in Cleve and Grunow 1880 (fig. 7.4). Common in pond 1.

- *Nitzschia perminuta* (Grunow) M. Peragallo 1903 (fig. 7.5). Most abundant monoraphid pennate diatom in pond 2.

- *Pinnularia intermedia* (Lagerstedt) Cleve 1895 (fig. 7.9). Common in pond 2; cosmopolitan in the northern hemisphere (Antoniades et al. 2008).

- *Pinnularia* sp. cf. *P. viridis* (Nitzsch) Ehrenberg (not illustrated). Rare in snow sample from location 1.

- *Pinnularia* sp. cf. *P. glacialis* C.G. Ehrenberg (fig. 7.8). Occasional in pond 2.

- *Psammothidium kryophilum* (Petersen) Reichardt 2004 (fig. 7.10). Pond 2; a common Arctic species.

4. Blue-green Algae: Filamentous Cyanobacteria

- *Phormidium* cf. *P. retzii* (Ag.) Gomont. Common in pond 3-1 (figs. 5.13, 5.14).

Lichens

Table 1. Scientific and common names and the observed location of lichens in the Kwädąy Dän Ts'ìnchį discovery site region. Locations are discussed on page 91.

LICHENS: Scientific name	Common name	Locations		
Alectoria nigricans (Ach.) Nyl.	Grey Witch's Hair	-	2	-
Alectoria ochroleuca (Hoffm.) A. Massal	Green Witch's Hair	-	2	-
Bryoria nitidula (Th.) Brodo & D. Hawksw.	Tundra Horsehair	-	2	-
Caloplaca tiroliensis Zahlbr.	Firedot Lichen	-	2	-
Candelariella terrigena Räsänen	Tundra Goldspeck Lichen	-	2	-

LICHENS: Scientific name	Common name	Locations		
Cetraria aculeata (Schreb.) Fr.	Spiny Heath Lichen	-	2	-
Cetraria ericetorum subsp. *reticulata* (Räsänen) Kämefelt	Iceland Lichen	-	-	3
Cetraria islandica subsp. *crispiformis* (Räsänen) Kämefelt	True Island Lichen		2	
Cetrariella delisei (Bory ex Schaer.) Kämefelt & Thell	Snow-bed Iceland Lichen	-	2	-
Cladina arbuscula subsp. *beringiana* (Ahti) N.S. Golubk.	Reindeer Lichen	1	2	3
Cladonia borealis S. Stenroos	Boreal Pixie-cup	-	2	-
Cladonia pocillum (Ach.) Grognot	Rosette Pixie-cup	-	-	3
Cladonia pyxidata (L.) Hoffm.	Pebbled Pixie-cup	-	2	-
Dactylina arctica subsp. *beringica* (C.D. Bird & J.W. Thomson) Kämefelt	Arctic Finger Lichen	-	2	3
Flavocetraria cucullata (Bellardi) Kämefelt & Thell	Curled Snow Lichen	1	-	-
Flavocetraria nivalis (L.) Kämefelt & Thell	Crinkled Snow Lichen	1	2	3
Micarea incrassata Hedl.	Dot Lichen	-	2	-
Ochrolechia sp.	Saucer Lichen	-	2	-
Peltigera leucophlebia (Nyl.) Gyeln.	Ruffled Freckle Pelt	-	-	3
Pertusaria sp.	Wart Lichen	-	2	-
Pertusaria gemmipara (Th. Fr.) C. Knight ex Brodo	Wart Lichen	-	2	-
Protopannaria pezizoides (Weber) P.M. Jorgensen & S. Ekman	Bottlecaps	-	2	-
Solorina crocea (L.) Ach.	Chocolate Chip Lichen	-	2	-
Stereocaulon sp.	Cauliflower Foam Lichen	-	2	-
Stereocaulon botryosum Ach.	Cauliflower Foam Lichen	-	2	-
Thamnolia vermicularis var. *subuliformis* (Ehrh.) Schaer.	Whiteworm Lichen	1	2	3
Umbilicaria hyperborea (Ach.) Hoffm.	Blistered Rock Tripe	1	-	-
Vulpicida tilesii (Ach.) J.-E. Mattsson & M.J. Lai	Gold Twist	-	-	3

Liverworts and Mosses (Bryophytes)

- -

Table 2. Scientific and common names and the observed location of liverworts and mosses in the Kwädąy Dän Ts'ìnchį discovery site region. Common names are from either E-Flora BC (Klinkenberg 2013) and or Edwards (2003).

LIVERWORTS: Scientific name	Common name	Locations		
Aneura pinguis (L.) Dum.	Greasewort	-	-	3
Anthelia juratzkana (Limpr.) Trevis	Scarce Silverwort	1	-	-
Asterella gracilis (F. Weber) Underw.	Star Liverwort	-	-	3
Barbilophozia floerkei (F. Weber and D. Mohr) Loeske	Mountain Leafy Liverwort Common Pawwort	-	-	3
Blepharostoma trichophyllum (L.) Dumort.	Finger-leaf Liverwort, Hairy Threadwort	-	-	3
Cephaloziella sp.	Threadwort	-	-	3
Gymnomitrion concinnatum (Lightf.) Corda	Braided Frostwort	-	2	-
Jungermannia exertifolia Steph.	Cordate Flapwort	-	-	3
Lophozia cf. *excisa* (Dicks.) Dumort.	Capitate Notchwort	-	2	-
Lophozia cf. *incisa* (Schrad.) Dumort.	Jagged Notchwort	-	-	3
Preissia quadrata (Scop.) Nees	Narrow Mushroom-headed Liverwort	1	-	3
Scapania sp.	Earwort	1	-	-

MOSSES: Scientific name	Common name	Locations		
Andreaea rupestris Hedw.	Black Rock-moss	-	2	3
Aulacomnium palustre (Hedw.) Schwägr.	Glow Moss, Bog Groove-moss	-	-	3

MOSSES: Scientific name	Common name	Locations		
Brachythecium sp.	Feather-moss	1	-	-
Brachythecium cf. *glaciale* Schimp.	Snow Feather-moss	1	-	-
Bryoerythrophyllum recurvirostrum (Hedw.) P.C. Chen	Red Beard-moss	-	-	3
Bryum pseudotriquetrum (Hedw.) P. Gaertn., B. Meyer & Scherb.	Common Green Bryum Moss, Marsh Bryum	-	-	3
Campylium stellatum (Hedw.) C.E.O. Jensen	Golden Star-moss, Yellow Starry Feather-moss	-	-	3
Catascopium nigritum (Hedw.) Brid.	Down-looking Moss	-	-	3
Ceratodon purpureus (Hedw.) Brid.	Fire Moss, Purple Fork-moss	1	-	-
Dicranella sp.	Forklet-moss	1	-	-
Dicranoweisia crispula (Hedw.) Milde	Mountain Pincushion	1	2	3
Dicranum acutifolium (Lindb. & Arnell) C.E.O. Jensen	Acuteleaf Dicranum Moss, Fringed Broom-moss	-	2	-
Dicranum elongatum Schleich. ex Schwägr.	Dense Heron's-bill Moss, Dense Broom-moss	-	-	3
Dicranum groenlandicum Brid.	Greenland Dicranum Moss, Greenland Broom-moss	-	-	3
Dicranum spadiceum J.E. Zetterst.	Dicranum moss, Confusing Broom-moss	-	-	3
Distichium capillaceum (Hedw.) Bruch & Schimp.	Distichium Moss, Fine Distichium	-	2	3
Ditrichum flexicaule (Schwägr.) Hampe	Bendy Cow-hair Moss, Bendy Ditrichum	-	-	3
Encalypta alpina Sm.	Alpine Candle Snuffer Moss, Elegant Extinguisher-moss	-	-	3
Eurhynchium pulchellum (Hedw.) Jenn.	Elegant Beaked-moss, Elegant Feather-moss	-	2	-

MOSSES: Scientific name	Common name	Locations		
Grimmia montana Bruch & Schimp.	Sun Grimmia	–	–	3
Hylocomium splendens (Hedw.) Schimp.	Step Moss, Glittering Wood-moss	–	2	3
Hypnum revolutum (Mitt.) Lindb.	Rusty Claw-moss, Revolute Plait-moss	–	–	3
Lescurea saxicola (Schimp.) Molendo	Rock Feather-moss	1	–	–
Mnium blyttii Bruch & Schimp.	Blytt's Calcareous Moss, Blue Star-moss	–	–	3
Orthothecium chryseum (Schwägr.) Schimp.	Orthothecium Moss, Yellow Lustre-moss	–	–	3
Palustriella commutata (Hedw.) Roth	Curled Hook-moss	–	–	3
Philonotis fontana (Hedw.) Brid.	Spring Moss, Fountain Fruit-moss	–	–	3
Pogonatum urnigerum (Hedw.) P. Beauv.	Grey Haircap Moss, Urn Haircap	1	–	–
Pohlia andalusica (Hohn.) Broth.	Roth's Thread-moss	1	–	–
Pohlia filum (Schimp.) Mårtensson	Fat-bud Thread-moss	–	–	3
Pohlia nutans (Hedw.) Lindb.	Nodding Thread-moss	–	–	3
Polytrichastrum alpinum (Hedw.) G.L. Sm.	Stiff-leaved Haircap Moss, Alpine Haircap	–	2	–
Polytrichastrum sexangulare (Brid.) G.L. Sm.	Polytrichum Moss, Northern Haircap	1	–	–
Polytrichum juniperinum Hedw.	Juniper Haircap	1	–	–
Racomitrium affine (Schleich. ex F. Weber & D. Mohr) Lindb.	Lesser Fringe-moss	–	2	–
Racomitrium canescens (Hedw.) Brid.	Grey Rock-moss, Hoary Fringe-moss	1	2	3
Racomitrium lanuginosum (Hedw.) Brid.	Hoary Rock-moss, Woolly Fringe-moss	–	2	–

MOSSES: Scientific name	Common name	Locations		
Sanionia uncinata (Hedw.) Warnst.	Sickle-moss, Sickle-leaved Hook-moss	1	-	3
Syntrichia norvegica (Web.) Lindb.	Norway Screw-moss	1	-	-
Tayloria lingulata (Dicks.) Lindb.	Tongue-leaved Gland-moss	-	-	3
Tomenthypnum nitens (Hedw.) Loeske	Golden Fuzzy Fen Moss, Woolly Feather-moss	-	-	3
Tortella fragilis (Hook. & Wilson) Limpr.	Brittle Crisp-moss	-	-	3

Vascular Plants

Table 3. Scientific and common names and the observation location of vascular plants in the Kwädąy Dän Ts'ìnchį discovery site region.

VASCULAR PLANTS (Pteridophytes, Horsetails, Club-mosses and Moonwort): Scientific name	Common name	Locations		
Botrychium crenulatum W.H. Wagner	Dainty Moonwort	-	-	3
Diphasiastrum alpinum (L.) Holub	Alpine Club-moss	-	-	3
Equisetum arvense L.	Common Horsetail	-	-	3
Equisetum scirpoides Michx.	Dwarf Scouring-rush	-	-	3
Equisetum variegatum Schleich.	Northern Scouring-rush	-	-	3
Huperzia selago (L.) Bernh.	Alpine Fir-moss	1	2	-
Selaginella sibirica L.	Northern Selaginella	1	2	

VASCULAR PLANTS (Flowering Plants): Scientific name	Common name	Locations		
Achillea millefolium L.	Yarrow	-	-	3
Anemone richardsonii Hook.	Yellow Anemone	-	-	3

VASCULAR PLANTS (Flowering Plants): Scientific name	Common name	Locations		
Antennaria monocephala DC.	One-headed Pussytoes	1	2	-
Arctostaphylos alpina (L.) Spreng.	Alpine Bearberry	-	-	3
Artemisia norvegica Fries	Mountain Sagewort	-	-	3
Astragalus alpinus L.	Alpine Milk-vetch	-	-	3
Campanula lasiocarpa Cham.	Mountain Harebell	-	2	3
Cardamine bellidifolia L.	Alpine Bittercress	1	2	-
Carex microchaeta Holm	Small-awned Sedge	-	2	-
Carex petricosa Dewey	Rock-dwelling Sedge	1	-	-
Carex podocarpa R. Br.	Graceful Mountain Sedge	-	2	-
Carex pyrenaica Wahlenb.	Pyrenean Sedge	1	-	-
Cassiope tetragona (L.) D. Don	Four-angled Mountain-heather	-	-	3
Castilleja sp.	Paintbrush	-	-	3
Crepis nana Rich.	Dwarf Hawksbeard	-	-	3
Deschampsia sp.	Hairgrass	-	-	3
Draba crassifolia Grah.	Rocky Mountain Draba	-	-	3
Dryas integrifolia M. Vahl	Entire-leaved Mountain-avens	-	2	3
Empetrum nigrum L.	Crowberry	-	-	3
Epilobium anagallidifolium Lam.	Alpine Willowherb	1	-	-
Epilobium angustifolium L.	Fireweed	-	-	3
Epilobium clavatum Trel.	Club-fruited Willowherb	-	2	-
Epilobium latifolium L.	Broad-leaved Willowherb	1	-	3
Erigeron sp.	Fleabane	-	-	3

VASCULAR PLANTS (Flowering Plants): Scientific name	Common name	Locations		
Gentianella sp.	Gentian	-	-	3
Hedysarum alpinum L.	Alpine Hedysarum	-	-	3
Juncus mertensianus Bong.	Merten's Rush	-	-	3
Juncus triglumis L.	Three-flowered Rush	-	-	3
Leptarrhena pyrolifolia (D. Don) Ser.	Leatherleaf Saxifrage	-	-	3
Luetkea pectinata (Pursh) Kuntze	Partridge Foot	-	-	3
Luzula arcuata (Wahlenb.) Sw.	Curved Woodrush	-	-	3
Luzula spicata (L.) DC.	Spiked Woodrush	1	-	3
Minuartia rubella (Wahlenb.) Graebn.	Boreal Sandwort	1	-	-
Myosotis alpestris Schm.	Mountain Forget-me-not	-	-	3
Orthilia secunda (L.) House	One-sided Wintergreen	-	2	-
Oxyria digyna (L.) J. Hill	Mountain Sorrel	1	-	3
Oxytropis huddelsonii A.E. Porsild	Huddelson's Locoweed	1	2	-
Parnassia fimbriata Konig	Fringed Grass-of-Parnassus	-	-	3
Parnassia kotzebuei Cham. & Schlecht.	Kotzebue's Grass-of-Parnassus	-	-	3
Pedicularis capitata Adams	Capitate Lousewort	-	2	-
Petasites frigidus (L.) Fries	Sweet Coltsfoot	1	-	3
Phyllodoce glanduliflora (Hook.) Cov.	Yellow Mountain-heather	-	-	3
Poa arctica R. Br.	Arctic Bluegrass	1	-	-
Poa paucispicula Scribn. & Merr.	Few-flowered Bluegrass	1	2	-
Polygonum viviparum L.	Alpine Bistort	-	-	3
Potentilla hyparctica Malte	Arctic Cinquefoil	1	2	-
Pentaphylloides floribunda (Pursh) A. Löve	Shrubby Cinquefoil	-	-	3

VASCULAR PLANTS (Flowering Plants): Scientific name	Common name	Locations		
Pyrola sp.	Wintergreen	-	-	3
Ranunculus eschscholtzii Schtdl.	Subalpine Buttercup	1	-	-
Rhodiola rosea L.	Roseroot	1	-	-
Sagina saginoides (L.) Karst.	Arctic Pearlwort	1	-	3
Salix alaxensis (Anderss.) Cov.	Alaska Willow	-	-	3
Salix arctica Pall.	Arctic Willow	1	-	3
Salix polaris Wahlenb.	Polar Willow	-	-	3
Salix pulchra Cham.	Diamond-leaved Willow	1	-	-
Salix reticulata L.	Net-veined Willow	1	-	3
Sanguisorba canadensis L.	Sitka Burnet	-	-	3
Saxifraga bronchialis L.	Spotted Saxifrage	1	-	-
Saxifraga cespitosa L.	Tufted Saxifrage	1	-	-
Saxifraga nelsoniana D. Don	Dotted Saxifrage	1	-	-
Saxifraga oppositifolia L.	Purple Mountain Saxifrage	1	2	3
Saxifraga razshivinii Zhmylev	Alaska Saxifrage	1	-	-
Saxifraga rivularis L.	Brook Saxifrage	1	-	-
Saxifraga tricuspidata Rottb.	Three-toothed Saxifrage	-	2	-
Sibbaldia procumbens L.	Sibbaldia	1	-	3
Silene acaulis L.	Moss Campion	1	2	3
Solidago multiradiata Ait.	Northern Goldenrod	-	-	3
Taraxacum lyratum (Ledeb.) DC.	Lyrate Dandelion	-	2	-
Tofieldia sp.	False Asphodel	-	-	3
Trisetum spicatum L.	Spike Trisetum	1	-	-

THE PRESENT FLORA (LICHENS, BRYOPHYTES AND VASCULAR PLANTS) AND ITS RELEVANCE TO THE DISCOVERY SITE

THE BOTANICAL RECORDING AND COLLECTING

at locations 1, 2 and 3 led to the preceding listing of *ca.* 50 algae, *ca.* 26 species of lichens, *ca.* 50 bryophytes and *ca.* 77 vascular plants. These are certainly underestimates in all four categories. The algae are discussed in the next section.

Only six species of microlichens have been collected, and the list of lichens is therefore very incomplete. The bryophytes received more skilled attention in the field, so the moss and liverwort flora (bryoflora) is more complete but still not exhaustive. The list of vascular plants is perhaps the most complete but is certainly also not comprehensive. Underestimated or not, these data provide a good basis for understanding the plant remains recovered from the Kwädąy Dän Ts'ìnchį discovery site. In relating the present flora to plants from the site, we considered samples from the robe, the surface of the body, the hair mass and the ice but not those from the intestines; the latter will be discussed in a separate paper.

With vascular plants, the microscopic remains (mainly pollen grains, spores and phytoliths) and macroscopic remains (mainly leaves, stems, wood and charcoal) are exceedingly familiar to paleoecologists and archaeobotanists. For many decades now, such remains from Holocene deposits, including archaeological samples from around the world, have been studied to great effect. In recent years, studies of the diatom and chrysophyte algae have also become important proxy-markers of climate change and water chemistry in the Canadian Arctic and subarctic regions (Lotter et al. 1999; Zeeb and Smol 2001).

In marked contrast, Holocene records of lichen remains are exceedingly sparse, as are those from the Pleistocene. Since 1959 the senior author (Dickson) has been extracting remains of mosses from numerous Holocene and Pleistocene layers, including many archaeological samples from Britain and elsewhere,

without having seen a single lichen. It seems that this whole group of organisms normally decays very quickly and so is not preserved in peat, mud or other types of deposits—with one striking exception. In the 1990s, while studying the many moss remains from the site of the Tyrolean iceman, Dickson saw some lichen fragments in a few samples (Dickson et al. 1996). This is likely to have resulted from instantaneous preservation by freezing. It is no great surprise, then, that lichens (albeit only a very few scraps) were found in the samples from the Kwädąy Dän Ts'ìnchį discovery site.

Lichens

ONE SMALL PIECE OF A LICHEN—two millimetres long with four short branches, reminiscent of the tips of a *Cladina*—was found in a sample from the Kwädąy Dän Ts'ìnchį robe (sample 24-14-1). The *Cladina* called *C. arbuscula* subsp. *beringiana* grows at location 2.

Two pieces of an as yet unidentified lichen, three millimetres long, came from the upper torso (sample 51-16). One "frilly" piece of lichen, possibly a *Flavocetraria*, was recovered from the Kwädąy Dän Ts'ìnchį man's skin (sample 51-5). *Flavocetraria* was found growing at all three locations.

Bryophytes

UNLIKE LICHENS, remains of mosses (leafy stems and detached leaves) can be abundant in Holocene, Pleistocene and archaeological contexts (Dickson 1973, 1986; Miller 1984). In contrast to mosses but like lichens, liverworts are markedly under-represented at archaeological sites and usually vanish before preservation unless the circumstances are exceptional. One leafy stem of a liverwort was found in a Kwädąy Dän Ts'ìnchį sample recovered from the upper torso (sample 51-16), but it was poorly preserved and not identified.

The following mosses were found growing at locations 1, 2 and/or 3, as well as recovered in association with the Kwädąy Dän Ts'ìnchį man's body.

These 10 mosses are all widespread in the temperate and cooler parts of the northern hemisphere. For example, every one of them has also been recovered from the ice and sediment samples taken at the Ötzi site, at 3,210 metres in the mountains bordering Austria and Italy (Dickson et al. 1996; Dickson et al. 2004).

- *Dicranoweisia crispula.* This species was found at all three locations, growing as the typical dense cushions on rock, but also occurs as taller, looser tufts. One leafy stem was found in a sample of the Kwädạy Dän Ts'ìnchị man's skin (sample 51-3-1).

- *Distichium* sp. Recovered as one stem from human skin (sample 51-3-1), and two others were recovered from the robe (samples 28-11 and 28-29). Specific identification cannot be established with certainty from such scraps, though it is likely to be *D. capillaceum*. That species was collected from locations 2 and 3, where it grew in rock crevices or on the ground.

- *Ditrichum flexicaule.* Only at location 3 was one small tuft found growing on the ground. Short, leafy stems were found in eight samples in all, though some were tentative identifications. Two of the more certain ones came from robe pieces (samples 18-5-1 and 20-7-1).

- *Encalypta/Tortula/Syntrichia.* Fragmentary, badly preserved leaves from the robe (sample 24-14), a piece of hide (sample 51-7) and human skin (sample 51-4) were assigned to this category. It is difficult to separate these genera based on such poor material. On the ground, *Encalypta alpina* grows at location 3, and *Syntrichia norvegica* occurs at location 1.

- *Hylocomium splendens.* Only two small patches were noted in sheltered positions on the ground, one at location 2 and the other at location 3. One stem, four millimetres long, was found in a sample from the robe (28-21).

- *Pogonatum urnigerum/P. capillare.* There was only one small patch of *P. urnigerum* growing on the ground at location 1, and one leaf was recovered from the upper torso (sample 51-16).

- *Polytrichastrum alpinum.* Only one patch was found growing in a sheltered position on the ground at location 2. The single leaf came from the human hair mass (sample 15-3-1).

- *Polytrichum/Polytrichastrum.* Poor material from samples 24-14-1 and 51-11, both from the robe.

- *Racomitrium canescens* sensu lato. Occurs at all three locations as patches on rock and on the ground. Leaves or leafy stems, mostly badly preserved, were found in many samples, including the human hair mass (sample 15-3-1).

- *Racomitrium lanuginosum.* Recorded only at location 2 but in large amount on rock at the south end of the outcrop. Only a single leaf from the upper torso (sample 51-16) was found.

If the occurrences of these 10 mosses a few hundred years ago were much the same as the present day, they could have become incorporated into the ice or onto the body and clothes from near the discovery site. But this does not exclude the possibility that they somehow got transported from further afield, even much further afield. Two particular cases can be taken as examples.

Growing above the timberline, but becoming much sparser and finally disappearing with increasing altitude, *Hylocomium splendens* is one of the most abundant mosses in the circumboreal coniferous forests. The small fragment, therefore, could well have adhered to the robe from lower altitudes, where it can be profuse in the coastal zone, such as at Glacier Bay and elsewhere. This is also true of *Racomitrium canescens*, which is the most frequently encountered moss in the Kwädạy Dän Ts'ìnchị samples. It is another very common moss in the Pacific Northwest, where it grows locally in abundance on a great variety of substrates, usually well drained, from sea level—as at Yakutat, Alaska—to high in the mountains (Schofield 1976, 1992).

The following three mosses were recovered from the Kwädąy Dän Ts'ìnchį site samples but were not found growing at locations 1, 2 or 3. While the first two could well have been overlooked during the fieldwork, the third is unlikely to have been overlooked in the field.

- *Bryum* sp. A small tuft was found in the ice around the upper torso (sample 51-16).

- *Pogonatum capillare*. A single, well-preserved leaf was recovered from a sample taken from mud and debris near the middle left thigh (sample P4).

- *Pleurozium schreberi*. Sample 10-3, a piece of leather or hide liable to have been part of the robe contained two short pieces of leafy stem. This species was not found in the vicinity of the discovery site and was not noticed elsewhere during the expedition between the discovery site and the Haines Highway. It is often associated with *Hylocomium splendens* and can be abundant.

Pollen

REGARDING POLLEN from the robe samples (chapter 28 this volume), all the pollen types given below could have come from plants growing at locations 1, 2 or 3. But because pollen is microscopic, there is the very real possibility of long-distance transport for the windborne pollen types such as *Artemisia* and Poaceae. Pollen found in the snow samples (see table 1, Dickson and Mudie 2008) largely records species transported long distances by wind, especially species in the pine and birch families.

- *Artemisia*. *Artemisia norvegica* grows at location 3.
- Asteraceae. Seven members of this family grow at locations 1, 2 and 3.
- Campanulaceae. *Campanula lasiocarpa* grows at locations 2 and 3.
- Caryophyllaceae. Three members of this family grow at locations 1, 2 and 3.

- Cyperaceae. Four members found at locations 1 and 2.
- Fabaceae. Three members grow at locations 1, 2 and 3.
- Liliaceae. *Tofieldia* sp. grows at location 3.
- Onagraceae. Three members grow at locations 1, 2 and 3.
- Poaceae. Four members found at locations 1, 2 and 3.
- Rosaceae. Four members found at locations 1, 2 and 3.
- *Sanguisorba*. *S. canadensis* grows at location 3.
- *Salix*. Leaf remains were also found. Five species of willow grow at locations 1 and 3.

Phytogeography of the Present Flora (Lichens, Bryophytes and Vascular Plants)

THE GREAT MAJORITY of the lichens, bryophytes and vascular plants listed here have often been recorded at high altitude and latitudes in North America, but several deserve brief comment due to their rarity.

Of the lichens, *Bryoria nitidula*, *Cetraria islandica* var. *crispiformis* and *Stereocaulon botryosum* are listed by Goward (1994, 1999) and Geiser et al. (1994) as more or less infrequent in British Columbia and southeastern Alaska. The latter can be used as an example, as Goward (1994, 287) shows a mere six localities in all of British Columbia, though with one in the Haines Triangle, not too far from the newly studied area. In his map for *Pertusaria gemminipara*, Thomson (1997) shows no occurrences close to the Haines Triangle.

However, lichens are a comparatively little-studied group with few scientists capable of identifying them. So in the case of these four lichens, the scarcities may reflect insufficient recording rather than a reliable indication of actual occurrence.

There is much more certainty about the genuine rarity of two mosses. The discovery location of the *Grimmia montana* so far north in British Columbia adds a significant range extension, since it is otherwise strongly southern in the province (Hastings and Greven 2007). Even further south in Colorado, it is "probably the most ubiquitous mat-forming moss of boulders and outcrops from the foothills to the alpine" (Weber and Wittman 2007).

Tayloria lingulata is a little-recorded species and rare according to Vitt et al. (1988) in their book *Mosses, Lichens and Ferns of Northwest North America*. It is also considered rare in the alpine zone according to Douglas and Vitt (1976), in their study of the mosses of the St. Elias–Kluane ranges, and it is very rare according to Weber's biogeographical study of the plants of Colorado (2003).

Two vascular plants also need special consideration. *Botrychium crenulatum* has only four widespread locations in British Columbia, as plotted in the *Illustrated Flora of British Columbia* (Douglas et al. 2000, 2002), with none in the Haines Triangle. This fifth locality is the northernmost, though the species' range extends south to Arizona and California. As well, the *Illustrated Flora of British Columbia* gives only one previous record for the amphiberingian *Saxifraga razshivinii* in the province. It had been found at Dollis Creek (previously known as Squaw Creek), some 35 kilometres north of the newly found locality, which is the southernmost in Canada. It is more abundant in the Yukon Territory (Cody 2000).

THE PRESENT ALGAL FLORA: ITS PHYTOGEOGRAPHY AND RELEVANCE TO THE DISCOVERY SITE

THE ALGAE from the snow samples and morainal ponds in the study area include at least 15 kinds of chlorophytes (mostly *Cosmarium* species of desmids), 12 chrysophyte stomatocysts, more than 20 diatoms and 1 filamentous blue-green alga (Zygnematales) that was noted in the field but not studied in detail. In the available time from fall 2005 to spring 2008 it was not possible to make detailed taxonomic studies of the algal flora from this previously unexplored arctic-alpine region; however, the most common taxa are listed here, and several are illustrated in figures 5, 6 and 7 in order to document the first report on algae from alpine morainal lakes and tundra fen land of the interior southern St. Elias mountains.

The snow alga flora (four taxa) at the discovery site is dominated by *Chlamydomonas nivalis*, a red snow alga. This species is widespread throughout alpine and arctic regions of the world, but Jones and Pomeroy (2000) state that it has not previously been reported for the Yukon and has only been noted for Mt. Seymour in southern British Columbia. Takeuchi (2001) reported seven snow algae species for a transect from a 1,270–1,770 metre elevation on the Gulkana Glacier in central Alaska, and he showed that only *C. nivalis* was abundant at elevations above about 1,600 metres. He did not report the presence of *Trochiscia*. Takeuchi et al. (2003) also found *C. nivalis* and two other genera of snow algae on coastal glaciers near Homer, central Alaska, and commented that there are fewer snow alga species in Alaska's maritime glaciers than inland.

The chrysophyte flora of the ponds in the discovery area represents only the species present near the shorelines during the few sampling days in mid August 2005, and it is therefore relatively low in species diversity (N = 12) compared to 37 taxa found in a pond near Anchorage, Alaska, when sampled weekly from May through December 1957 (Hilliard 1965). Zeeb and Smol (2001) also note that the golden-brown algae are most abundant in open lakewater offshore, whereas our sampling was limited to the pond edges. The low diversity may also reflect the fact that golden-brown algae usually favour acid water conditions, whereas the morainal ponds in the study area were mildly alkaline. Prescott (1953) reported only one genus (*Dinobryon*) from the neutral-to-alkaline tundra ponds around Barrow, northern Alaska, and Sheath et al. (1986) reported only one species, *Hydrurus foetidus,* associated with stream macro-algae in the Cook Inlet region of south-central Alaska.

Not enough is known about the relationship between the vegetative and cyst-production phases of the chrysophytes to say much about the phytogeographic affinity of the species in the discovery area, but it appears that Cyst 15 (found on the robe) is a cosmopolitan arctic species. Cyst 15 may also be present (but rare) in a saline pond near Whitehorse, sampled by Mudie in 2005; however, other associated distinctive cyst morphotypes (with three short, blunt, apical prongs) were abundant in this saline pond yet were not found in any of the samples from the Kwädąy Dän Ts'ìnchį man or his clothing (fig. 7 in Mudie et al., chapter 32, this volume).

Chrysastrella furcata and several species of *Dinobryon*, including *D. cylindricum*, have also been reported for a freshwater pond near Anchorage (Hilliard 1965). In contrast, the *Mallomonas* species reported for Kodiak Island and Cape Thompson, Alaska (Asmund and Hilliard 1961), did not include species with fimbriated scales like that found in pond 3-3. Duff et al. (1995) reported 110 stomatocyst morphotypes from freshwater to saline lakes in central British Columbia and commented that most of these were also present in lakes elsewhere in North America, thereby suggesting that most of the chrysophyte cysts are not very useful phytogeographic markers. It is notable, though, that no chrysophytes were found in the Alaskan salt marshes sampled during our fieldwork (see Hebda et al., chapter 28 this volume).

The occurrence of nine types of placoderm desmids (i.e., desmids with a central constriction) in the mildly alkaline morainal pond 3-3 was a most unexpected find because they are usually restricted to acidic environments, including temporary tundra pools and marshes in the Cape Thompson region of Alaska, where Croasdale (1962) found 157 taxa. Most of the species from pond 3-3 have a circumpolar arctic-subarctic distribution, although *Staurastrum alternans* var. *basichondrum* (one of the most common taxa) and *Cosmarium debaryi* may be restricted to Alaska, where they are found in flooded ground near Cape Thompson, at 167 metres above sea level. The punctate *Staurastrum pachyrhynchum* may be a new variety, as it is not reported for either the Cape Thompson or Devon Island arctic floras (Croasdale 1962, 1965). The most common desmid in pond 3-3, *Cosmarium speciosum* v. *biforme*, was also found in August 2002, in the subalpine zone at the second Mineral Lake sampling site. However, the only desmid found with the remains of the Kwädąy Dän Ts'ìnchį man was a *Staurastrum* species with about 12 large, branched spines, tentatively identified as *S.*

sexangulare (Bulnheim) P. Lundell (Dickson and Mudie 2008). This species has not previously been reported for the North American Arctic region, but elsewhere it has a cosmopolitan distribution from arctic to subtropical regions.

The filamentous algae found in the discovery area include the green alga *Zygnema* and the blue-green slimeweed *Phormidium*. Both genera were sampled only in the organic-rich pond 3-1, where *Zygnema* was producing abundant sparsely pitted zygospores. Both taxa are common in streams of the Cook Inlet drainage basin (Sheath et al. 1986), where *Phormidium retzii* is a dominant macro-alga in flowing streams.

The diatom diversity in the glacial ponds of the discovery area seems to be lower compared to that in 42 surface samples from subalpine lakes in the Kluane Lake area of the southwestern Yukon and along the Haines Highway (Wilson and Gajewski 2002). The species diversity of diatoms from the morainal ponds is also several times lower than in the subalpine lakes at Klukshu and Nesketahin, also in the Yukon (Selbie et al. 2009). It must be kept in mind, however, that our samples were from small volumes (120 ml) of suspended sediment removed from just above the mud surface, and they were not grab samples collected from the lake bottom. Therefore, our samples mainly record the population of the warm summer of 2005, not the integrated, multi-year populations that would be sampled by coring or grab sampling. It is likely that more detailed studies would reveal the presence of more diverse algal flora in the glacial lakes of the southern St. Elias Mountains. For example, the ice cave flora of the northern Yukon alone contains 97 diatom species in 29 genera (Lauriol et al. 2006).

Most of the diatoms that could be identified so far are also present in the subalpine lakes around Klukshu, Nesketahin and along the Haines Highway, from Pine Lake at Haines Junction, Yukon, to Three Guardsman Pass, BC. The exceptions are *Luticola nivalis* and cf. *Aulacoseira alpigena*. *L. nivalis* is found in Switzerland and is very rare in Ellef Ringnes and northern Ellesmere islands of the Canadian Arctic (Antoniades et al. 2008); it was also present in an ice cave of the Ogilvie

Mountains, northern Yukon (Lauriol et al. 2006). Reinhardt Peinitz (pers. comm. 2008) notes that it seems to be an "aerophilic" taxon, preferring wet moss and rock surfaces as habitats. *Aulacoseira alpigena* has a widespread distribution in arctic to boreal freshwater lakes of the northern hemisphere (English and Potapova 2009).

Although we have recovered relatively few diatom species from the discovery area, it is notable that this diatom flora is dominated by pennate types and that only one small cylindric species (cf. *Aulacoseira nivaloides*) was found (fig. 7.1). This low ratio of cylindric to pennate diatoms is a characteristic feature of subalpine and alpine freshwater environments, as is the overall small size of the diatom species (Lotter et al. 1999, Wilson and Gajewski 2002). In contrast, marine floras in both subarctic and arctic regions contain large populations of relatively large, chain-forming cylindric diatom species and large, thick-walled diatoms. This phytogeographical pattern has significant implications with regard to the diatom assemblage found by Mudie in four samples from the robe on top of the Kwäday Dän Ts'ìnchį man's torso (for details see Dickson and Mudie 2008, Hebda et al. chapter 28 this volume).

Conclusions

THE LISTING OF THE ALGAE, lichens, bryophytes and vascular plants growing around the Kwäday Dän Ts'ìnchį discovery site has helped considerably in understanding the plant remains found on and near the frozen body. It reveals which species could have originated in the immediate vicinity rather than far away during the man's last journey. It has also added considerably to the knowledge of a botanically little-known part of British Columbia.

ACKNOWLEDGEMENTS

The senior and second authors offer their thanks to the writer Jean Auel, whose subsidy helped greatly to make the 2005 expedition possible, as did subsidies from the Tlingit through help of Rosita Worl, director of the Sealaska Heritage Institute at Juneau. Financial assistance was also received from Judy Ramos, Dr. Elaine Abraham of the Yakutat Tlingit Tribe, and Robert Johnson of the Alaska Department of Fish and Game at Yakutat. We also thank the Champagne and Aishihik First Nations (CAFN) for their logistical support at all times, and Yukon government archaeologist Greg Hare, Whitehorse, for the loan of field equipment and for road transport in 2005. Helicopter transport was provided by Delmar Washington of Capital Helicopters in Whitehorse.

For his work on the Kwäday Dän Ts'ìnchį samples, the senior author is most grateful for financial backing from the Leverhulme Trust, the Carnegie Trust for the Universities of Scotland, the Royal Society of Edinburgh, the Royal Society of London, the British Council (Toronto) and the Bryological Fund of the Botany Department of the University of British Columbia. He is also very grateful to Wilfred Schofield and James MacDonald for helpful discussions on bryophytes and geology, respectively.

For field assistance, we thank the Tatshenshini-Alsek Park authorities, Greg Eikland and Frances Oles (CAFN), as well as Bill Hanlon (guide in 2005) and Al Harvey (photographer in 2005). For help in the laboratory, Mudie is indebted to Frank Thomas at the Geological Survey of Canada, Atlantic, for his skilful assistance with the ESEM studies. The authors would also like to acknowledge their fellow author Roxanne Hastings, who passed away in July 2017.

6

RADIOCARBON DATING OF KWÄDĄY DÄN TS'ÌNCHĮ MAN AND ASSOCIATED ARTIFACTS

Michael P. Richards, Sheila Greer, Owen Beattie,
Alexander P. Mackie and John Southon

We report here on a comprehensive radiocarbon dating program to date the age of the Kwädąy Dän Ts'ìnchį man (first reported in Richards et al. 2007). The individual was found in association with a number of artifacts, including a plant fibre hat, a robe-style fur garment and various wooden items. The first series of radiocarbon dates on this associated material ranged from 500 to 120 Before Present (Beattie et al. 2000), indicating that the area is a multi-phase site where various organic objects were deposited over at least a 400 radiocarbon-year time period. Of these, two initial conventional radiocarbon dates on clothing—a hat and a robe fragment—directly associated with the Kwädąy Dän Ts'ìnchį man (Beattie et al. 2000) indicated that the individual was at least 550 years old: Arctic Ground Squirrel fur from the robe, 450 ± 40 BP; and the hat's split-root plant fibre, 500 ± 30 BP (See table 1). To confirm the dating of this individual and better clarify the ages of the associated material culture, we undertook additional radiocarbon dating, including two separate dates directly on the

individual himself. We also re-dated the hat and robe fragments
reported in the Beattie et al. (2000) paper, obtaining dates that
were younger than initially reported but similar to the new dates
we obtained on associated artifacts and on the Kwädąy Dän Ts'ìnchį
man himself.

Direct Dating of the Kwädąy Dän Ts'ìnchį Remains

BONE COLLAGEN was extracted from the Kwädąy Dän
Ts'ìnchį individual following procedures detailed in
Richards and Hedges (1999) and Brown et al. (1988).
Two aliquots of bone were used, with one given
a chloroform-methanol pre-treatment to remove
possible lipid constituents. The two bone samples and
the single muscle sample were then demineralised in
0.5M HCl at 5°C for two days, and the resultant solid
was gelatinised in sealed tubes with ph3 HCl at 70°C
for 24 hours. The resulting gelatin was then filtered
through 30 kD ultrafilters and the >30 kD fraction was
lyophilised. Modern bone is approximately 20 per cent
collagen by mass. The two bone samples yielded 18.7
per cent and 19.2 per cent collagen-type material, and
the muscle sample yielded 3.7 per cent proteinaceous
material. The carbon and nitrogen contents and stable
isotope values of the solid proteinaceous material from
the two bone aliquots and single muscle sample were
measured using an elemental analyser and continuous
flow mass spectrometer at the Oxford Radiocarbon
Accelerator Unit (ORAU), University of Oxford. Those
data are also presented in chapter 19, this volume.
The C:N ratios of the extracted collagen-like material
from the bone samples are within the range observed
for well-preserved collagen (DeNiro 1985). There is no
statistical difference between the stable isotope values
and C and N contents for the two aliquots of bone.
The collagen extracted from the bone aliquot that did
not have the chloroform-methanol pre-treatment was
then radiocarbon dated using an accelerator mass

spectrometer (AMS) at the ORAU (table 1, page 119).
A separate fragment of bone was also sent to the
Rafter Laboratory, New Zealand, for pre-treatment
and subsequent AMS dating, for quality control. The
resultant radiocarbon ages are 952 ± 28 BP from Oxford
and 935 ± 75 BP from the Rafter Laboratory, which give a
combined age of 944 ± 80 BP.

As the isotopic analysis of the bone collagen
(see chapter 13) indicated that this individual had a
primarily marine-based diet, it is necessary to apply a
marine reservoir correction to these dates (Arneborg
et al. 1999; Richards and Sheridan 2000; Barrett et
al. 2000). Radiocarbon dates on marine organisms
require correction for the so-called marine reservoir
effect, due to the dilution of dissolved carbon dioxide in
surface waters—where photosynthesis takes place—by
deep-ocean dissolved, inorganic carbon depleted in ^{14}C
(Stuiver and Braziunas 1993). The global average marine
correction is taken to be approximately 400 years
and, in northern British Columbia and the Yukon, an
additional correction (the ΔR value) must be included
due to deep-ocean upwelling (Southon et al. 1990;
Southon and Fedje 2003). Unfortunately, there are no
published ΔR values from the coast near the discovery
site. The nearest available published ΔR value is from
Pavlov Harbour, on the Alaska Peninsula, Alaska (55.5°0'
N and 162°0' W), at 237 ± 50 BP (Robinson and Thompson
1981). Southon and Fedje (2003) have published shell
and wood pairs for Haida Gwaii (Queen Charlotte
Islands) and Prince Rupert. They found that for Haida
Gwaii, the offset is *ca*. 700 radiocarbon years for dates
older than 500 BP (and therefore a ΔR of 300), and the

average difference between two pairs of shell and wood from Prince Rupert is *ca.* 775 radiocarbon years (and therefore a ΔR of 375). Using these three ΔR values, we obtained three possible calibrated ages for the Kwädąy Dän Ts'ìnchį man (table 2, page 121). The full range of these possible calibrated ages is 1505–1853 cal AD.

Dating of Directly Associated Artifacts and Clothing

OF THE VARIOUS WOODEN ARTIFACTS dated in the earlier dating program, only the man's knife (IkVf1:1-2 wood, 150 ± 50 BP) was found directly in association with the Long Ago Person Found, as it was found—still in its sheath—on top of the robe covering the body.

We obtained one date on a repair patch of Arctic Ground Squirrel pelt (KDT-Robe-24-8-7, 183 ± 55 BP) from the robe that was found with the Kwädąy Dän Ts'ìnchį man when he was discovered (table 1, page 119). This new date, which calibrates to 1727–1812 cal AD (1sigma, using Calib, see table 1), is considerably younger than the date produced previously on fur from the robe: 450 ± 40 BP, which calibrates to 1422–1463 cal AD. We therefore obtained an additional date on animal sinew from a repair patch on the robe (24-34-9, 213 ± 55 BP), which was similar to the hide repair patch date but much younger than the earlier robe date, with a calibrated age range of 1778–1799 cal AD.

We also obtained two dates on a beaver-skin bag that was excavated from the ice directly in contact with the Kwädąy Dän Ts'ìnchį man's skin (fur, KDT-31-30a, and skin, KDT-31-30b). One date on fur from the beaver-skin bag (141 ± 65 BP) is similar in age to the repair patches from the robe described above, with a calibrated age range of 1720–1778 cal AD. A second date on skin from the beaver-skin bag (242 ± 60 BP) is older, with a calibrated age range of either 1630–1682 AD (36 per cent of the area of the probability distribution) or 1736–1804 cal AD (34 per cent of the area of the probability distribution). The fur on the beaver-skin bag appears to have been part of the bag's original construction, and there is no evidence that it was a repair or later addition.

The Age of the Kwädąy Dän Ts'ìnchį Man

DETERMINING THE AGE of the Kwädąy Dän Ts'ìnchį man has been challenging due to uncertainties about the correct marine offset to use for calibration, the varying ages for the artifacts and clothing, and the disparate ages of the robe. A further complication is the problem of dating artifacts made of old wood, which is wood taken from inner parts of long-lived trees that may already have been quite old when the artifact was made. Also, with perhaps a few exceptions, there are no indisputable direct associations between many of the artifacts and the human. The dating evidence and the abundance of artifacts recovered from the site (Greer 2005) indicate that this was probably a multiple-use site and that some of the artifacts could not have been directly associated with the Kwädąy Dän Ts'ìnchį man. Because of these uncertainties, and in order to determine an age for the Kwädąy Dän Ts'ìnchį man, we decided to only consider the dates of the human and the most recent age for a piece of the beaver-skin bag found in contact with his skin. If we consider the calibrated ages of the human and the most recent beaver-skin bag date together (tables 1 and 2), we can further narrow down a probable age for the Kwädąy Dän Ts'ìnchį man.

The range of calibrated ages for the human are from 1505–1853 cal AD, while the calibrated ages for the youngest sample from the beaver-skin bag (KDT-31-30a) range from 1720–1778 cal AD (31.3 per cent), with other probable age ranges later in time (table 1). Therefore, we conclude that the most likely age for the Kwädąy Dän Ts'ìnchį individual is between 1720 and 1853 cal AD. This is in agreement with the date on the man's knife obtained at the Lawrence Livermore National Laboratory (IkVf1:1-2 wood, 150 ± 50 BP). This tool was also found in close association with the body and has a calibrated date of between 1725 and 1780 cal AD (33.6 per cent, see table 1).

This age range, however, is at odds with the earlier date range based on two dates from hat and robe fragments originally reported in Beattie et al. (2000). In that paper, the date for the robe's Arctic

Ground Squirrel fur was 450 ± 40, which calibrates to between 1422 and 1463 cal AD (1 sigma). The date on the hat's split-root plant fibre was 500 ± 30 BP, which calibrates to between 1414 and 1437 cal AD (1 sigma). These calibrated ages are a few hundred years older than the new age we calculated using the direct dates on the Kwädąy Dän Ts'ìnchį person and new dates on associated artifacts. To further address this discrepancy, we undertook new radiocarbon dates on the robe and hat at the Oxford University Radiocarbon Laboratory. The new date on the robe was from an animal sinew repair patch and was discussed above: 213 ± 27 cal AD, which calibrates to 1778–1799 cal AD. The new date on the hat was from plant fibres and is 197 ± 23 BP, which calibrates to 1764–1800 cal AD. These new dates are statistically indistinguishable from each other and are approximately between 250 and 300 radiocarbon and calendar years younger than the dates produced previously on the hat and the robe. These two new dates are in clear agreement with the dates we produced previously on the associated artifacts and the human remains themselves, and they support our revised age range for the Kwädąy Dän Ts'ìnchį man.

Conclusions

THE RADIOCARBON DATES reported here on bone collagen and artifacts clearly associated with the human indicate the Kwädąy Dän Ts'ìnchį man is from the late pre-contact or early European contact period, *ca.* 1720–1850 cal AD.

ACKNOWLEDGEMENTS

We would very much like to thank the Champagne and Aishihik First Nations (CAFN) for allowing these dating studies on Kwädąy Dän Ts'ìnchį. The project is jointly managed by representatives of the province of British Columbia and CAFN, and we thank management committee group's Grant Hughes and Jim Cosgrove of the Royal British Columbia Museum and Lawrence Joe and Diane Strand from CAFN for their ongoing support of the project's scientific studies. We also thank the Heritage Program of the Yukon government and the Royal BC Museum for their support of the Kwädąy Dän Ts'ìnchį project, particularly the ongoing contributions of artifact conservators Valery Monahan (Yukon) and Kjerstin Mackie (RBCM). We also thank Robert Hedges (Oxford University) for running an AMS date on the bone collagen and Erle Nelson (Simon Fraser University) for his advice on this research.

Table 1. Uncalibrated radiocarbon dates of the Kwäday Dän Ts'ìnchị individual and uncalibrated and calibrated dates from associated artifacts found at the site (this study, Beattie et al. 2000, Southon unpublished on Cams-71937). Calibrated (1 sigma) age ranges were obtained using the Calib 6.0 program (Stuiver and Reimer 1993) using the 2009 IntCal09 atmospheric calibration curve (Reimer et al. 2009). The age ranges with the highest relative area under the probability distribution are given, with the relative area indicated as percentages. $\delta^{13}C$ values were measured relative to the vPDB standard.

Sample	Material	Sample number	Lab number	d13C	14C date uncalibrated BP	14C date calibrated AD
Human bone	Collagen	TORSO-1	OxA-10224	-13.8	952 ± 28	See table 2
Human bone	Collagen	TORSO-1	NZA-15675	-13.8	935 ± 75	See table 2
Hat (split-root)	Plant fibre	IkVf-1:R1 (sample from IkVf-1:113)	Beta-133765	-20.0	500 ± 30	1414– 1437 (100%)
Hat	Plant fibre	IkVf-1:113	OxA-16690	n.d.	197 ± 23	1764–1800 (54%)
Robe (Arctic Ground Squirrel)	Fur	IkVf-1:R2	Beta-133766	-23.9	450 ± 40	1422–1463 (100%)
Repair patch from robe	Animal sinew	28-34-9	OxA-16664	n.d.	213 ± 27	1778–1799 (46%)
Repair patch from robe (Arctic Ground Squirrel)	Skin	KDT-Robe-24-8-7	NZA-15674	-21.6	183 ± 55	1727–1812 (56%)
Beaver-skin bag found in direct contact with human skin, associated with KDT-31-30b	Fur	KDT-31-30a	NZA-15626	-21.0	141 ± 65	1720–1778 (31%)
Beaver-skin bag found in direct contact with human skin, associated with KDT-31-30a	Skin	KDT-31-30b	NZA-15673	-20.4	242 ± 60	1630–1682 (36%) 1736–1804 (34%)

Sample	Material	Sample number	Lab number	d13C	14C date uncalibrated BP	14C date calibrated AD
Throwing or snare stick (referred to as "throwing board or snaring implement" in Beattie et al. 2000)	Wood	Sample from IkVf-1:103	Beta-140633	-20.9	230 ± 40	1641–1680 (49%) 1764–1800 (40%)
Slender pointed stick with carved end (referred to as "throwing spear" in Beattie et al. 2000)	Wood	Sample from IkVf-1:117	Beta-140634	-27.0	360 ± 40	1464–1522 (53%) 1573–1628 (47%)
Gaff pole/ walking stick – distal portion (referred to as "walking or bear stick" in Beattie et al. 2000)	Wood	Sample from IkVf-1:106	Beta-140635	-25.5	140 ± 40	1722–1766 (28%) 1833–1879 (28%)
Gaff pole / walking stick – proximal end (referred to as "multi-notched stick" in Beattie et al. 2000)	Wood	Sample from IkVf-1:104	Beta-140636	-24.9	290 ± 40	1521–1578 (59%)
Arrow? shaft fragments (referred to as "dart shaft fragment" in Beattie et al. 2000)	Wood	Sample from IkVf-1:109	Beta-140637	-25.5	120 ± 40	1807–1891 (56%)

Sample	Material	Sample number	Lab number	d13C	14C date uncalibrated BP	14C date calibrated AD
Second gaff pole/ walking stick – distal end portion (referred to as "thick stick" in Beattie et al. 2000)	Wood	Sample from IkVf-1:107	Beta-140638	-24.0	500 ± 40	1409–1441 (100%)
Man's knife (handle portion)	Wood	IkVf 1:1-2 wood	Cams-71937	-25.0	150 ± 50	1725–1780 (34%)

Table 2. Calibrated ages of the Kwädąy Dän Ts'ìnchį individual using the combined age of 944 ± 80 BP and different marine reservoir offsets for three locations: Alaska (Robinson and Thompson 1981), Haida Gwaii and Prince Rupert (Southon and Fedje 2003). Data were obtained using the Calib 6.0 program (Stuiver and Reimer 1993) using the 2009 Marine09 marine calibration curve (Reimer et al. 2009). Calibrated ages were rounded to the nearest decade.

Region for offset	ΔR	Calibrated age (1 sigma)
Alaska	242 ± 50	1505–1682 cal AD
Haida Gwaii	300	1538–1724 cal AD
Prince Rupert	375	1663–1853 cal AD

2

PEOPLE, HISTORY AND HONOURING

INTRODUCTION

Richard J. Hebda

The lessons of this ancient teacher are best appreciated with an understanding of not just his physical environment but his cultural environment. This means learning who his people (in the broadest sense) were, being aware of how and where they lived and having a sense of their past and present worldviews. An appreciation of the contemporary social context is important because our understanding of the past is influenced by the present-day cultural and social environments.

THIS PART OF THE VOLUME begins with a personal view of the region's cultural landscape from Lawrence Joe, a member of the Aboriginal community most closely connected to the discovery and a staff member of the native government that manages the finds. Joe knows the land well, through hunting and fishing—using the land as his ancestors did. He provides modern-day insight into his culture, both past and present. He also explains how the Kwädąy Dän Ts'ìnchį discovery was received and accepted by the area's people, setting the community background to the story. Joe's account therefore provides important perspectives on the traditional Aboriginal relationship to the land and the working relationships among governments.

The account of the consultation and ceremonies (see chapter 8) explores the way in which the Champagne and Aishihik First Nations, a contemporary Indigenous government, acknowledged the significance of the discovery within and outside their community. They did not assume that the young man was one of their own, but rather understood him to be their responsibility because he was found in their territory. They recognized that their Indigenous neighbours had an interest in the find, as did the wider world. The discovery came with cultural responsibilities for the Champagne and Aishihik First Nations and generated a desire to do things right when dealing with the sensitive matter of human remains. The importance of respect for the human remains, the person they represent and his relatives was forefront in all actions taken. The people of the community met the challenges of the discovery through open consultation and meaningful ceremonies.

The complex history of the Tatshenshini-Alsek region comes to light using old maps, explorers' reports and ethnographic studies. An ethnohistorical account looking at recent centuries provides background for the contemporary social milieu. Trails and travel feature prominently in this story of changing settlement patterns and the 19th-century arrival of non-Aboriginal newcomers, with the Indigenous people's deep connection to the land emerging as a central theme.

Though the Kwädạy Dän Ts'ìnchị discovery region is recognized today by many as wilderness, several Aboriginal villages once existed along the Tatshenshini River, not far from the find site. Part 2 of the book concludes with an outline of what is known about these villages and where they might have been. While some of these places can be located with confidence, other locations have not yet been verified. These settlements may have been the home or destination of the man found frozen in the ice.

7

THE CONTEMPORARY CULTURAL LANDSCAPE OF THE KWÄDĄY DÄN TS'ÌNCHĮ DISCOVERY REGION

Lawrence Joe (*Ketäníä Tà*), Champagne and Aishihik First Nations

Dännch'e (How are you?). Let me introduce myself. I am Lawrence Joe; I belong to the Crow clan and my Tutchone name is *Ketäníä Tà*, which means "Walks in the Bush All the Time". It's an appropriate name as I do spend a lot of time in the bush, both on my own time and as part of my job as the director of Heritage, Lands and Resources for my First Nation. The Kwädąy Dän Ts'ìnchį discovery has been an interesting voyage of learning for me, a chance to learn from my elders, and to know more about their way of life and the land that was their home. In this paper, I will introduce you to the Tatshenshini-Alsek country, and to my people and our Aboriginal government, Champagne and Aishihik First Nations. I will also discuss some elements that I believe have been critical in making this project successful.

TATSHENSHINI-ALSEK COUNTRY is the southernmost part of Champagne and Aishihik First Nations traditional territory (figs. 1 and 2), with the main branch of the river, what is today known as the Alsek, draining into the Pacific Ocean near Dry Bay, Alaska. When my people talk about Tatshenshini-Alsek country, we say it is "all about the fish"—and by fish we mean salmon. The basin's rich salmon resource supported many generations of my nation's ancestors, allowing for a relatively high population density. The Tatshenshini-Alsek country is also an area where two cultures, two peoples, interacted. Our stories tell that, in times past, our people—the Tutchone or *Dän*, as we say in our language—moved down the river each season to meet the salmon as they migrated upstream from the coast, while the Tlingit of the Dry Bay area, around the mouth of the Alsek River, moved inland to a location with less rainfall so they would be more successful in drying their fish. Both cultures took advantage of this incredibly abundant resource. You will read about other aspects of Tlingit and Tutchone history in Tatshenshini-Alsek country in different chapters in this volume.

Our First Nation, Champagne and Aishihik, has close to 1,200 citizens. We are primarily of Dän (Southern Tutchone) cultural background, though some

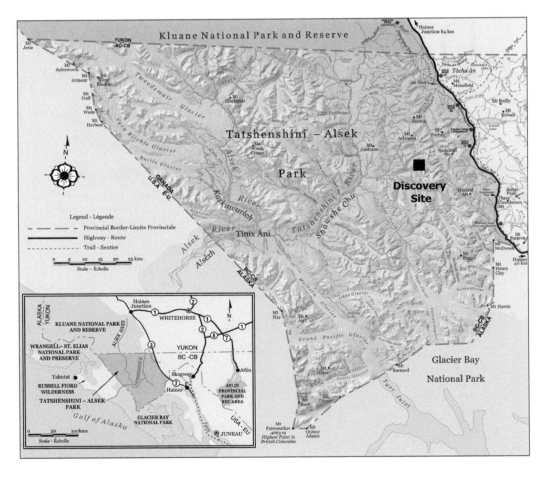

Figure 1. Map of Tatshenshini-Alsek Park, showing traditional villages and the location of the discovery site. Based on a BC Parks map. Colour version on page 586.

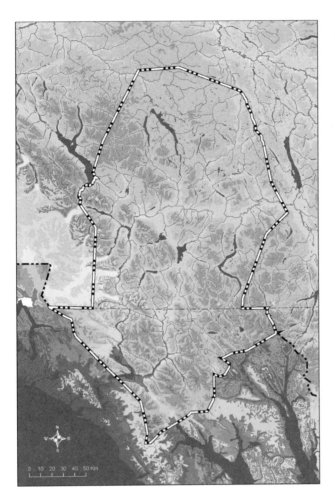

Figure 2. Champagne and Aishihik First Nations traditional territory (outlined), which includes lands in southern Yukon and northwestern British Columbia. Colour version on page 587.

many of our citizens reside. Others live in Whitehorse, the territorial capital, or outside the Yukon. Some of our citizens are residents of Haines or Klukwan in southeastern Alaska.

Our Traditional Lands

CHAMPAGNE AND AISHIHIK FIRST NATIONS traditional territory covers some 40,000 square kilometres, with about a third lying within the province of British Columbia and the rest in Yukon. The part of our land in British Columbia, Tatshenshini-Alsek country, is remote. It is an awe-inspiring landscape, dominated by high mountains and glaciers, with the Tatshenshini and Alsek rivers carving a route through to the Pacific. There is only one road through the area, the Haines Road, which connects Haines Junction, Yukon, with Haines, Alaska. Before the Haines Road was constructed, the Tatshenshini River was the main thoroughfare in this country. Our people once had many villages along this river, but the last of these ceased to be occupied on a regular basis in the last century.

Tatshenshini-Alsek country is difficult to travel through at all times of the year and challenging by most anyone's standards. Some parts of the basin, particularly in the northern areas, are easier to travel through when the ground is covered by snow. But these mountains generate their own weather, regardless of the time of year, and one can encounter blizzard conditions at any time. There is only a small window of opportunity, perhaps as little as 10 days over the course of the entire year, when you can safely get aircraft into some places. This small window occurs at the height of the summer melt—the time of year when the three hunters discovered the remains of the Long Ago Person Found. The area where the Kwäday Dän Ts'inchi individual lost his life is one of those narrow-window areas. Visiting these places by helicopter can be dangerous; if the weather changes, the pilot has to be ready to pull out with only a moment's notice.

It's a very dynamic landscape, constantly changing as a result of tectonic uplift, as mountains erode, and

of us have Tlingit ancestry. Our nation's name derives from two of our historic settlements: Champagne (traditional name Shadhäla), located on the Dezadeash River; and Aishihik, situated at the northern end of Aishihik Lake and the headwaters of the Alsek River drainage (fig. 1). Formerly, our population was spread throughout our traditional territory, living at least part of the year in these two villages or others such as Kloo Lake, Klukshu, Shäwshe and Hutchi. Today, our government's main administrative centre is in Haines Junction, Yukon (traditional name *Dakwakada*), where glaciers advance and retreat. This land is even changed by animals: as beavers dam streams, some of our old village sites along the Tatshenshini River have been flooded, and are now unsuitable camping places. This area is also being heavily affected by climate change. The warmer weather conditions have resulted in a severe spruce bark beetle infestation, which itself is changing the local hydrologic regime.

The two rivers, the Tatshenshini and the Alsek, have a strong cultural significance to our First Nation. One of our elders, the late John Adamson, used to tell a story about a giant that walked this land in long-ago times. As he travelled through the mountains, the giant carried a large spear with him. One time he dropped his spear, and as it fell, it opened up a pathway through the mountains. This pathway, located in the area of the O'Connor River, became a shortcut trail, a route where our people could pass through, allowing easier access to villages on the Tatshenshini River.

The Provincial-Territorial Boundary

IN 1900 the Government of Canada established the provincial-territorial boundary between Yukon and British Columbia at the sixtieth parallel. This action had severe consequences for our people, cutting us off from our lands in Tatshenshini-Alsek country. When I was growing up, we weren't allowed to hunt in the British Columbia part of our territory. Trapping was made difficult by the paperwork requirements of a government whose closest representative was located many hours away. We did not stop using the Tatshenshini-Alsek area entirely, however, and after the Haines Road was built through the British Columbia part of our territory in 1942, our people began to use it more, and it became a popular place for harvesting gophers (*Spermophilus* [now *Urocitellus*] *parryi*, or Arctic Ground Squirrels; *säl* in the Southern Tutchone language). Gophers may be small and ignored

Figure 3. View southwest to mouth of the O'Connor River with the Tatshenshini River winding its way to the Pacific in the distance. Sarah Gaunt (CAFN) photograph.

by game wardens, but in our culture, like salmon and Moose, they are highly valued and consumed in quantity.

While the enforcement of the provincial-territorial boundary brought about hardship for our people, the placement of this artificial boundary also illustrates our differing perspective on land use and land ownership. Around the turn of the last century, one of our chiefs encountered the survey crew that was cutting trees to mark the boundary line between the Yukon and British Columbia. His telling of the encounter with the boundary survey crew has been passed down orally in our community. Interpreting the boundary cut line as a trail, he was puzzled why anyone would build a trail where they were doing so. The provincial-territorial boundary runs in a straight line east-west, up and down the sides of mountains, without regard to the configuration of the land, the flow of the waterways or the habits of animals; it's a route that no person would ever travel. Our chief asked, "Why would they build a trail that goes this way, when we already have a trail that goes that way?" (that is, on a route that could be travelled). The European concept of land ownership was totally foreign to our people at that time, and our leader couldn't imagine the consequences that this strange, newly cut "trail" would have for his people.

1990s, Land Claims and Tatshenshini-Alsek Park

IN 1993 Champagne and Aishihik First Nations signed its final Land Claim Agreement, or modern-day treaty, with the governments of Canada and Yukon. We have been self-governing since 1995. Self-government is an important foundation that provides us—an Indigenous government—with the tools to operate our own programs, and to make decisions affecting our communities, our lands, our government and our people. It gives us rights, and allows us to pass our own legislation; our laws are paramount over parallel territorial or federal legislation. Self-government, it has also been said, allows us to make our own mistakes and, equally important, to learn from them. We undertook

our own archaeological survey to look for signs of our old villages on the Tatshenshini River, and since 1995 have operated our own heritage program that has done extensive interview work with our elders and compiled reports on the ethnohistory of Tatshenshini-Alsek country.

We do not have a treaty in respect of our traditional lands that lie in British Columbia, and it wasn't until 1990, with the *Sparrow* decision in the Supreme Court of Canada, that our Aboriginal Rights—our right to use and benefit from our traditional lands—were recognized. As a result of the *Sparrow* decision we were able to reclaim the use of our land in the province, and in the short period of time since the early 1990s, our use of Tatshenshini-Alsek country has been revitalized.

Although our claim to our lands in British Columbia is outstanding, this didn't stop Champagne and Aishihik First Nations from reaching an agreement with the province regarding the management of this area. In 1993, the province declared roughly a million hectares, nearly 80 per cent of our traditional territory that lies within the province, a provincial park. We strongly objected to the park's creation, since it was established without any consultation with us. We immediately entered into land-claim negotiations, submitting our claim on the first day that the BC Treaty Commission began operations that year. The following year, Tatshenshini-Alsek Wilderness Park, as it was known, was designated by the United Nations as part of a World Heritage Site, along with various neighbouring protected areas, including Yukon's Kluane National Park and Reserve, also in Champagne and Aishihik First Nations traditional territory, as well as in Alaska, Wrangell–St. Elias National Park and Preserve, Glacier Bay National Park and Preserve and Tongass National Forest (see chapter 29, fig. 2).

When our First Nation was not successful in concluding a treaty or an agreement regarding our lands in the province over the next few years, we took another approach. In 1996, we successfully concluded negotiations with the province on a co-management agreement for the park. We insisted that the word wilderness, with its connotation of a landscape devoid

Figure 4. Rafting down Tatshenshini River, Tatshenshini-Alsek Park. Sarah Gaunt (CAFN) photograph.

of people and history, be dropped from the park's name. Since that time, the park has been officially known as Tatshenshini-Alsek Park.

The park co-management agreement gives us shared decision-making in park management matters, with Champagne and Aishihik First Nations having exclusive responsibility for Aboriginal languages, Aboriginal language place names, the naming of our sites and interpretation of our culture and our history. If we wish to, we can take over the operations and maintenance of the Tatshenshini-Alsek Park. While Champagne and Aishihik First Nations have not yet taken this opportunity, the main on-the-ground staff in the park have been Champagne and Aishihik First Nations

people for a number of years. The park management agreement also gives us authority for the conservation, protection and management of Aboriginal heritage sites in this park, which would include the Kwädąy Dän Ts'ìnchį discovery site.

Working Together

OUR EXPERIENCE AS CO-MANAGERS of Tatshenshini-Alsek Park has been extremely valuable. Although our initial relationship with the BC government was strained, it has improved over time and become very positive. From experience, we know that relationships can be established formally at the negotiating table, but that they also develop outside of the board room. Senior staff and politicians of our respective governments periodically take the opportunity to work together in the field. Relationships build slowly, and this was also the case with the Kwädąy Dän Ts'ìnchį discovery. While the early details and days following the discovery are recounted in chapter 34, I will recount a few key details here to illustrate how this relationship took time to develop.

After the find was reported to Champagne and Aishihik First Nations, we flew out to the site to confirm what had actually been found. Government of Yukon archaeologists, with whom we had a prior working relationship, provided us with assistance on our first visit to the site. We returned from that site and went into a retreat with Champagne and Aishihik and Carcross-Tagish First Nation elders on Kusawa Lake. This pause provided us with an opportunity to talk about the discovery and to gain some advice as to what we should do. The elders we spoke with were clear that something had to be done, that the body should not be left out there, that it had to be removed from the glacier.

Coming out from our Kusawa Lake retreat, we received word that a representative of the BC government, archaeologist Al Mackie, accompanied by Professor Owen Beattie, a specialist in frozen human remains, had arrived in the Yukon. We had been advised that they had a ministerial order to recover the body. This concerned us, because we believed the government was overstepping its bounds, infringing on our authority and responsibility as specified in the park management agreement. Consequently, we made sure that they were fully aware of our concerns. We pointed out that this young man who met his end high up on a glacier in the mountains a long time ago was someone's father, son, brother or uncle. We conveyed that we had a cultural responsibility to him because he was found in our territory, and because our culture places strong direction upon us to do things right, especially when we are dealing with a situation where there has been a death. We had an obligation to ensure that our First Nation values and beliefs on these matters were respected. We could have chosen not to cooperate with the BC government and used legal means to stop

Figure 5. Champagne and Aishihik First Nations staff and citizen Micheal Jim on park-warden patrol duties in Tatshenshini-Alsek Park. Sarah Gaunt (CAFN) photograph.

the recovery. However, because we had already had the park management agreement in place, which ensured that our interests were protected and respected, it was apparent that another option was possible. Building upon our earlier experiences with the BC government, we opted for a shared decision-making structure as we negotiated an agreement with the BC Archaeology Branch regarding how this discovery would be managed. In these negotiations, respect for our community's wishes regarding any studies that may proceed and ensuring proper communication were both important considerations for us.

The Royal BC Museum joined the Archaeology Branch as a provincial partner as the project evolved over the years, fitting in well with the management framework. The project has also had the benefit of a silent partner of sorts in the Yukon government. Yukon Heritage staff (from the Department of Tourism and Culture) have provided expertise as requested at a number of key points in the history of the project, including the recovery of the remains, as well as ongoing assistance in the conservation of the artifacts and belongings that have remained in the north.

Our Government's Authority

CHAMPAGNE AND AISHIHIK FIRST NATIONS became involved in the Kwäd̲ay̲ Dän Ts'ìnchi discovery because our tradition of land stewardship forced us to take on a leadership role. Once engaged as a government, we moved forward with caution. We sought to bring our elders and our community members, and those of our neighbouring Tribes and First Nations, along with us. We've endured and experienced a wide range of opinions as to what may, could or should have been done. There was no clear consensus—indeed, some controversy—on the most appropriate course of action at different points in time; it has been challenging. Still, one thing is certain: if the Kwäd̲ay̲ Dän Ts'ìnchi discovery had occurred as little as four or five years earlier, the outcome would have been much different. If this discovery had been made before our

Figure 6. Lily Hume with salmon at the village of Klukshu, 1949. Catharine McClellan photograph. Courtesy of the Canadian Museum of History. Colour version on page 587.

Yukon Final Agreement and the Tatshenshini-Alsek Park Management Agreement had been successfully negotiated, we would not have had the standing as a government, nor the access to resources and capable staff that come with the latter, and I have no doubt that the discovery would have been handled much differently.

From the very beginning of our involvement, there has never been a question about our status as a government and our authority in these matters. This was evident from the first meeting with a British Columbia government representative in August 1999—a complete willingness to share decision-making with our First Nation. Over the course of this project, our relationships with other First Nations, as well as with the other governments and researchers, have been based on respect for us as a government, and structured around an understanding of each other's strengths, knowledge and expertise, as well as our respective interests and mandates.

Kwanischis (Thank you)

THE KWÄDĄY DÄN TS'ÌNCHĮ DISCOVERY has received considerable international attention, some because of its insight into matters related to climate change. While the discovery is certainly connected to glacial shrinkage as a result of climate change, I believe that the Kwädąy Dän Ts'ìnchį story is also of interest because of its potential to change attitudes about First Nations. Self-governing First Nations such as Champagne and Aishihik are playing a new role in contemporary Canada, as active participants in the management of lands, resources and even extraordinary finds such as the Kwädąy Dän Ts'ìnchį discovery. Although Champagne and Aishihik First Nations have only been self-governing for 20 years, our government has established a strong and very active heritage program that is involved in a number of exciting projects. We have developed a cultural centre, where people can learn about our culture and our past, where we share our story with visitors to our traditional territory.

The Kwädąy Dän Ts'ìnchį project has been successful because it has both provided answers and respected values. It has combined western science with First Nations traditional knowledge, and most importantly with traditional Aboriginal values. It has worked because of the commitment of the government researchers and administrators, community researchers and university-based scientists. All those involved wanted something good to come out of a long-ago tragedy, to learn something of the life and times of Long Ago Person Found.

I close by noting that the late anthropologist Catharine (Kitty) McClellan included a photo of an Indian woman harvesting Tatshenshini River salmon in her seminal book, *My Old People Say: An Ethnographic Survey of Southern Yukon Territory* (McClellan 1975). That photo, taken more than sixty years ago, shows my great-grandmother, Lily Hume, at the Tatshenshini River fishing village of Klukshu. At that time, my own people were not taking images of themselves, so I end by expressing my gratitude to people like Catharine McClellan, who had an interest in our culture, who spent time with our people, recording their stories and way of life. I am similarly grateful to all who have made this project the success it has been. As a First Nations person living in the north of Canada, I believe I am living at a special time in my peoples' history. Tremendous opportunities lie ahead for our growing population; today's youth may some day benefit from the many possibilities that a project such as this opens up.

8

CONSULTATION WITH OUR NEIGHBOURS AND CULTURAL CEREMONIES IN HONOUR OF THE LONG AGO PERSON FOUND

Diane Strand, Lawrence Joe and Sheila Greer (Champagne and Aishihik First Nations)

"The elders have indicated that we should use this situation, what appears to be an ancient tragedy, to learn more about this person, when he lived and how his clothes and tools were made and how he died." (Charlie 1999)

THE KWÄDĄY DÄN TS'ÌNCHỊ MANAGEMENT AGREEMENT (see chapter 34 on collaboration) established the Champagne and Aishihik First Nations as having the responsibility for consultation with other First Nations and Tribes on matters related to the discovery. Through the Management Agreement, Champagne and Aishihik First Nations was also recognized as having sole responsibility for matters related to cultural protocol, including cultural ceremonies such as a funeral. This chapter describes the varied consultation efforts that took place both with the citizens of our First Nation, as well as with neighbouring First Nations and Tribes located in adjacent southeastern Alaska, northwestern British Columbia and southern Yukon. Indigenous cultural ceremonies held to honour the Long Ago Person Found are also discussed. The actions taken to fulfil these two areas of responsibility provide an additional perspective on the contemporary

cultural context of the discovery and augment information presented in other chapters on the region's ethnohistory and the DNA study, including chapter 7 on the contemporary cultural landscape, chapter 9 on Tatsheshini-Alsek history, chapter 29 on stories of glacier travel, chapter 33 on the search for living relatives and chapter 37 on learning from the discovery.

In this chapter we use the Dákwanjè word for people, Dän, to refer to our people (Tlen 1993), as we provide some background information on Tlingit, Tagish and Dän ceremonies related to the deceased. These are all matrilineal clan societies, so there are many commonalities in our traditions, including with the funeral, where the clan opposite to that of the deceased assists the bereaved family at their time of loss. At the memorial potlatch, the deceased person's clan thanks the other side for the assistance they have received. There is variation in how these two main

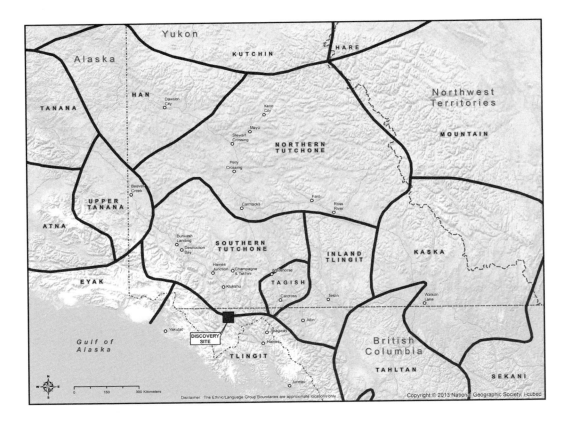

Figure 1. Location of the discovery site and local communities, showing ethnic/language group boundaries. After Catharine McClellan (1975).

ceremonies are practised across the southern Yukon, northwestern BC and southeastern Alaska, but the practices are familiar to Indigenous residents of all jurisdictions. An additional ceremony, the Forty Day Party, is restricted solely to the Alaskan Tlingit area.

The Indigenous groups of the greater area—the Tlingit, Tagish and Dän—share another common tradition: stewardship of the land. If a visitor died or became injured while travelling, the people responsible for the place where the mishap occurred would also be responsible for taking care of the person or dealing with the remains.[1] Because the Kwädąy Dän Ts'ìnchį discovery was made in the traditional territory of the Champagne and Aishihik First Nations, the Long Ago Person Found was our responsibility, and we took that responsibility seriously.

Consultation with Our Citizens and Neighbours

AS RECOUNTED IN CHAPTER 4, when Champagne and Aishihik First Nations first learned that this extraordinary find had been made in its traditional territory, we organized a party to visit the site. On August 17, 1999, a group consisting of Councillor Ron Chambers and staff Lawrence Joe, Diane Strand and Sarah Gaunt flew to the site and confirmed the nature of what had been found. Chief Bob Charlie and council were advised of the situation, and an emergency meeting of elders and citizens was immediately arranged. The emergency meeting was held in conjunction with a joint multi-day retreat that was taking place with the Carcross-Tagish First Nation out on the land at Kusawa Lake. At the meeting,

Champagne and Aishihik staff received direction to learn more about the person who had been found—in essence, to learn who he was or who his people were so that these people could make the necessary decisions on how to handle the situation. At a later meeting with CAFN elders in Haines Junction, the discovery was given the name Kwädąy Dän Ts'ìnchį, which means "Long Ago Person Found" in Dákwanjè.[2]

Once the decision had been made to take a leadership role in the matter, Champagne and Aishihik Heritage program staff began documenting information that might be relevant to understanding the discovery. At the same time, negotiations related to the management agreement were underway with the province of British Columbia. When the terms of agreement were finally set on August 31, 1999, Champagne and Aishihik's responsibility for consultation with neighbours and for cultural ceremony and cultural protocol matters was formally recognized. In the days immediately following the discovery, communication efforts focused on providing basic information to citizens of the area, as well as the wider public, about what had been found and the steps being taken.

Dialogue with our citizens on our government's response to and handling of the Kwädąy Dän Ts'ìnchį discovery took place through various community meetings, within the context of the First Nations' chief and council meetings and at several of the yearly summer general assemblies. The discovery was also discussed at a meeting hosted by the Southern Tutchone Tribal Council in 2000. Staff involved in the project had both formal and informal discussions with elders and others citizens, and in fall of 2000, Heritage program staff member Sarah Gaunt engaged in one-on-one consultations with citizens (Gaunt 2000).

Citizens also received information via a series of five community-oriented illustrated newsletters released from September 1999 to September 2001, and a colour newsletter produced for the 2008 Winter Event held in Haines Junction, reported below. Updates on project activities were also posted in the regular newsletter issued by the First Nation on a roughly bi-monthly basis

(at www.cafn.ca) and in special newsletters prepared for the general assembly we host each summer. We also provided information to our Klukwan contacts, for inclusion in editions of the periodic Klukwan Inc. shareholders newsletters.

Within two months of the discovery Champagne and Aishihik initiated the first in a series of consultation meetings with the neighbouring First Nations and Tribes. A Champagne and Aishihik First Nations delegation consisting of staff and political leadership visited several Alaskan communities—Chilkat Indian Tribe, Klukwan (CAFN 1999, 2000b); Chilkoot Tribal Council, Haines (CAFN 1999, 2000a); and Yakutat Tlingit Tribe, Yakutat (CAFN 2000c)—as well as Carcross, Yukon (CAFN 2000d). Meetings were held in 2000 with elders of and individuals representing the Little Salmon-Carmacks First Nation, Selkirk First Nation and Teslin Tlingit Council. At these various sessions, Champagne and Aishihik First Nations received support for having stepped forward to take a leadership role in managing the discovery.

The 2001 Regional Meeting of Tribes and First Nations

IN DECEMBER 2000 Dr. Owen Beattie from the University of Alberta—who was involved in his capacity as scientific advisor to the project—notified Champagne and Aishihik that the autopsy studies at the Royal BC Museum had been completed, which necessitated decisions on the final disposition of the remains. It was also apparent, however, that we were still a long way from knowing who this person was or who his family was. This was a concern, since the original intention was to have his relatives, if they could be identified, make the decision about these most sensitive matters. There was also a wide range of opinions within our community concerning our involvement with the find, some seeing it as a good thing, others not (cf. chapter 37).

Many decisions needed to be made, including on the final treatment of the remains. For instance, should they be buried or cremated, and should the artifacts be retained? We knew that cremation was once practised by both coastal Tlingit and interior (Dän) peoples, though the remains of Indian doctors were understood to have been handled differently (Greer 2000). While in-house subfloor burials were known in the coastal Tlingit area in the 19th century, "all sources indicate that cremation was the predominant mode of disposal of the dead" according to anthropologist Sergei Kan (1989, 38). Referring to the Tagish, Inland Tlingit and Southern Tutchone during the same period, McClellan (1975, 249) similarly noted that "all three practiced cremation".

Researchers such as McClellan (1975), Kan (1989) and de Laguna (1990) understood that all of these peoples gave up cremation in favour of inhumation under mission influence, with varying dates. A resident of the Champagne and Aishihik village of Shäwshe (see chapter 9) who drowned in the Alsek River in 1903 was cremated (Greer 2000), as were Dän who died as a result of the 1918 flu epidemic (Greer 2006). As de Laguna (1990) notes, cremation after 1880 was rare for the coastal Tlingit, though Kan (1989) learned that in some of the more isolated communities in southeastern Alaska cremation continued into the 1900s. While it was clear that cremation may have been common practice for both coastal and interior peoples in the 18th and 19th centuries, it was unknown if the practice prevailed during the Kwädąy Dän Ts'ìnchį man's time. In 2000–01, when these discussions on burial versus cremation were taking place, it was understood that the Kwädąy Dän Ts'ìnchį individual had lived approximately 500 years ago (Beattie et al. 2000); this date has since been revised to a few hundred years later (see chapter 6 on radiocarbon dating). There was also little or no information from an archaeological context, in either coastal or interior settings, to otherwise inform us on funerary practices in long-ago times (Greer 2000).

Decisions also had to be made about the man's belongings. According to McClellan (1987, 214), in the days when cremation was practised by First Nations in the southern Yukon, including Tlingit, Tagish and Dän, "goods that would be useful to the person in both the human and spirit worlds" would have been burned with the body. These might include personal items such as

bows and arrows, snowshoes, sewing kits, cooking baskets or horn spoons. Once burial was adopted, it was common practice for such items to be placed inside the small house or fence erected over the grave.[3] Thus there was debate and discussion within Champagne and Aishihik First Nations over whether the Kwädąy Dän Ts'ìnchį man's belongings should go with him, as per traditional practice, or if they should be retained, and if so, by whom.

With various matters needing to be sorted out, CAFN decided to hold a gathering of all the neighbouring First Nations in order to obtain guidance on choices that had to be made. We extended an invitation to First Nations throughout the greater region of southern Yukon, northwestern BC and southeastern Alaska to attend a meeting to discuss issues related to the Kwädąy Dän Ts'ìnchį discovery. Coming together in Haines Junction, Yukon, in May 2001, were representatives from the Yakutat Tlingit Tribe, Chilkoot Indian Association, Chilkat Indian Village and Sealaska Heritage Foundation, all from Alaska. Attending from the Yukon were representatives of Kwanlin Dün First Nation, Carcross-Tagish First Nation, Selkirk First Nation, Teslin Tlingit Council, Kluane First Nation, Little Salmon-Carmacks First Nations and Ta'an Kwächan Council. A Tahltan citizen was also in attendance. Champagne and Aishihik First Nations staff, council and citizens from the community attended, as did some resource people and scientists affiliated with the discovery and personnel from British Columbia Parks and Kluane National Park (CAFN 2001a, 2001b).

The 2001 regional meeting focused on the following questions: (1) whether the remains should be buried or cremated, (2) whether the man's belongings should go with him or be retained and (3) where the remains should be placed (i.e., their final resting place).[4] In-depth discussions over the course of the two-day meeting revealed a range of opinions, with varying interpretations of the find's significance and of why Long Ago Person Found came to be discovered at this point in time. Some wondered if the remains had been intentionally left or buried on the glacier. There was also discussion as to whether funerary ceremonies would have already been held for this person, even

Figure 2.
Regional meeting of representatives from Tribes and First Nations from southern Yukon, northwestern BC and southeastern Alaska, Haines Junction, May 2001. Sarah Gaunt (CAFN) photograph.

in the absence of his remains, and therefore whether it was appropriate to hold any type of services at the present time. Still, there was strong support for holding ceremonies for Long Ago Person Found, as there was no way of knowing if any had ever been held, particularly in light of the importance of such ceremonies for transitioning to the spirit world. The discovery was also viewed as a teaching opportunity, particularly for our youth. By the meeting's end, there was no clear consensus or firm conclusions on the questions posed to the group, but Champagne and Aishihik First Nations received general approval to continue as stewards of the discovery. The community DNA study, to see if living relatives of Long Ago Person Found could be identified, was also strongly endorsed at the regional meeting.

Despite receiving no clear direction from the regional meeting on the specific questions noted above, it was apparent that decisions had to be made, because timing was an issue. Clearly, it would take time to complete the study to identify living relatives of Long Ago Person Found. Yet many Champagne and Aishihik community members expressed strong concerns over

the delay in putting the Kwädąy Dän Ts'ìnchį individual to rest, viewing the time lag as culturally inappropriate or disrespectful.

It thus fell to Champagne and Aishihik First Nations Chief and Council to make a decision on the next steps. For a number of reasons—including cremation being a part of our collective history, but also because of concerns that the remains might be disturbed if buried—cremation was determined to be the best option. It was also decided that the cremated remains should be returned to the area in Tatshenshini-Alsek Park where the individual had lost his life. On the matter of the belongings, it was decided that for the time being, these should be retained, leaving decisions about what should happen with them for a future time. One item, however, was not retained. The Kwädąy Dän Ts'ìnchį man was wearing a small leather pouch around his neck; this item was understood to be his personal medicine pouch. As per Dän custom, such items are considered very personal in nature. The medicine pouch was never opened to examine the contents, and the pouch was cremated along with the human remains.

Making these decisions on next steps was very difficult for those tasked with this responsibility, especially when the clan of the deceased remained unknown at the time. As well, to alleviate concerns about the Long Ago Person Found being a person of power, plans were made to hold a ceremony to bless the artifacts to mitigate concerns about the objects retaining power or "bad medicine".

With direction received on the next steps, planning for a July 2001 funeral started, and arrangements were made for the remains to be cremated at a Victoria crematorium.

Ceremonies Overview

THE FUNERARY CEREMONIES of the Tlingit, Tagish and Dän of the greater study area of southern Yukon, northwestern BC and southeast Alaska include the funeral, the Forty Day Party and the Memorial Potlatch. The clan system plays a prominent role in funerary ceremonies, beginning with the funeral service that marks the passing of an individual (cf. Emmons 1991; McClellan 1975). With the death of a community member, the opposite side or moiety assists the clan of the deceased in their time of grief. That is, if an Eagle dies, Ravens will dress the deceased, dig the grave, prepare the funeral feast and so on. Even in contemporary times almost all of the work is done by community members.

The Forty Day Party, held approximately 40 days after the funeral, is only practised in the coastal Tlingit area in southeastern Alaska. This ceremony is held to help the spirit of the deceased to rest, and it brings the clans and the people together to begin planning the final Memorial Potlatch or Headstone Potlatch (cf. Central Council of the Tlingit and Haida Indian Tribes of Alaska 2008; Kan 2000). Both the funeral and the Memorial Potlatch involve payment of money, with clan and community members providing funds to the clan of the deceased to pay the workers and cover expenses. The Forty Day Party does not involve any payments.

The Memorial Potlatch is held at least one year after a person dies. In the Yukon this ceremony is also known as the Headstone Potlatch because it is when the grave marker is raised. In southeastern Alaska it has also been referred to as the final or "payoff" party for the deceased (Central Council of the Tlingit and Haida Indian Tribes of Alaska 2008; CAFN 2007; Kan 1989). At this feast or party, the clan of the deceased thanks the opposite clan/moiety for the support they received at their time of loss. The Memorial Potlatch represents the end of the mourning period; it brings closure for those involved and for the loss that has occurred. The Tlingit memorial party or ḵoo.éex' has been described as follows:

> The ḵoo.éex' is a ceremony in which the deceased and ancestors of the clan are remembered. It is a time for the surviving clan members to push away their sorrow after a year of mourning; to celebrate life; to reaffirm their social and kinships bonds; and to ceremoniously present their clan at.óow (clan regalia, objects, songs and stories). It is a time to honour the members of the opposite side [clan] (referred to as moiety in anthropological literature) who comforted the grieving clan and assisted with the funerary activities and who the host clan will now repay. (White and White 2000, 133)

The interior perspective on the tradition is similar, though a little bit different. Kluane First Nation member Mary Easterson writes:

> In earlier times when there were no written records, the potlatch functioned as "a way of publicly verifying history". It provided a record-keeping process of births, deaths, names and marriages. The principal purpose of the potlatch was maintenance and perpetuation of culture. (1992, 1)

In contrast to a funeral, which is a time of grief, the Memorial Potlatch is considered a happy event, with dancing (even competitive dancing), storytelling and gift giving. It is because gifts are given that the potlatch term, which is Chinook in origin and understood to mean "to give", became associated with the memorial event (McClellan et al. 1987, 215). Gift giving is a big part of the Memorial Potlatch:

> Both family and clan buy or makes [sic] gifts to give away to the opposite clan. You have to take care of opposite clan. An elder should be present to ensure everyone is treated equally. Make sure everybody from the opposite clan gets something, don't miss anybody.... Give generously to those who have come a long way. (CAFN 2007, 20)

Traditional names, which belong to the clans (see chapter 33), may also be given at a Memorial Potlatch but only to someone belonging to the same clan as the deceased. That is, if the memorial is in honour of a Wolf, a Wolf clan name—but not the name of the deceased—may be given to a Wolf individual.

Historically, potlatches may have also been hosted to commemorate a special event, such as the raising of a new clan house, or for an individual to honour a family member (e.g., a man honouring his mother). This is in addition to commemorating the passing of an individual (Central Council of the Tlingit and Haida Indian Tribes of Alaska 2008; McClellan 1975). Gifts were given to the opposite clan for witnessing the event and validating the passing of the recently deceased (Easterson 1992). Today, in the Yukon at least, "potlatch" may also be used to refer to the feast held at the time of a death—that is, a Funeral Potlatch (CAFN 2007).

2001 Ceremonies

AFTER THE REMAINS of the Kwädąy Dän Ts'ìnchị individual had been cremated in Victoria, the ashes were carried north by Champagne and Aishihik First Nations staff. This follows our cultural practice of staying with the remains of the deceased until they are put in their final resting place. For this same reason, a Champagne and Aishihik First Nations citizen had accompanied the remains when they were transported south to the Royal BC Museum for study in September 1999.

Planning of the 2001 funeral service was facilitated by a committee composed of representatives from both southeastern Alaska and the Yukon. Because the Kwädąy Dän Ts'ìnchị man's clan affiliation was unknown at the time, the committee included representatives of both the Wolf/Eagle and Crow/Raven clans. Committee members determined burial ceremony details, which typically would be decided by family members. These included identifying who would be the hosts for the funeral, the program order for the funeral service and the naming of individuals to serve as Dressers (honorary positions), Cooks and Servers. The committee also named Apprentices for key positions, highlighting that the funeral was also a learning opportunity for younger community members. Figure 3 presents the four-page funeral pamphlet distributed to attendees.

Ron Chambers and Frances Oles, both Champagne and Aishihik First Nations carvers, prepared a wooden box to hold the ashes (fig. 4). The cremation box was placed inside a metal container, which would afford better protection for the remains in the harsh environment of the final resting place.

The funeral service took place at Klukshu, Yukon, on July 21, 2001, with several hundred in attendance. The elders tasked with organizing the burial ceremony specified that, in order to restore balance, those who had been part of the original effort to recover the remains, including Yukon government archaeologist Greg Hare and BC government archaeologist Al Mackie, were asked to return the remains to the area where the individual has lost his life. Elder John Adamson, who had been a key advisor to Champagne and Aishihik staff responsible for the discovery, accompanied the group returning the remains to the burial site. Delmar Washington, the helicopter pilot who had provided transport during the various 1999 site visits, flew the remains and the group to the discovery site on the day of the funeral. All involved in transporting the remains to the site wore arm-bands made of red willow, as a form of protection, following traditional practice. The events at the site were brief; after a song by Calvin Lindstrom, and a few words of respect by others, the group laid the remains to rest beneath a cairn of rocks.

KWÄDĄY DÄN TS'ÌNCHĮ
Long ago person found

July 21, 2001

In August 1999, Kwädąy Dän Ts'ìnchį was discovered in a glacier in the Tatshenshini area within the traditional lands of the Champagne and Aishihik people. His appearance sparked world wide interest from those who wished to learn more about him and the time in which he lived, a time when the law of Crow, Raven, Wolf and Eagle dominated the lands of the coast and the interior.

After careful consideration, it was decided that if respectfully done, it would be appropriate to learn about this person - someone's lost son who perished away from home - so that something good may come out of a long ago tragedy.

Our Elders named him Kwädąy Dän Ts'ìnchį , meaning long ago person found, and have always been consulted in decisions which have often been difficult. Our ties with our neighbours have also been renewed in the decision making process about Kwädąy Dän Ts'ìnchį. People of the Interior and the Coast, from the Crow, Raven, Wolf and Eagle Clans, are involved in his funeral, since we don't know to what people or clan he belonged. As many have said: "*he belongs to all of us*".

A great deal of knowledge has been gained, and more will come forward in the years ahead. But now, the time has come to pay our respects to Kwädąy Dän Ts'ìnchį and to thank him for the opportunities he has afforded us. We will put him to rest where he was found, knowing that he has affected our lives and brought us together. He has given us many opportunities to learn about our culture, and our past, knowledge which we will use to educate our children in the future. For all these things we are grateful.

Let him now rest in peace.

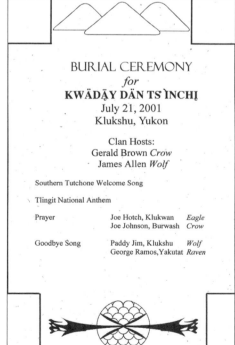

BURIAL CEREMONY
for
KWÄDĄY DÄN TS'ÌNCHĮ
July 21, 2001
Klukshu, Yukon

Clan Hosts:
Gerald Brown *Crow*
James Allen *Wolf*

Southern Tutchone Welcome Song

Tlingit National Anthem

Prayer	Joe Hotch, Klukwan	*Eagle*
	Joe Johnson, Burwash	*Crow*
Goodbye Song	Paddy Jim, Klukshu	*Wolf*
	George Ramos, Yakutat	*Raven*

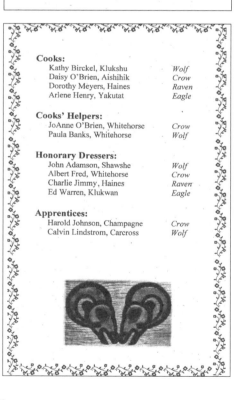

Cooks:
Kathy Birckel, Klukshu	*Wolf*
Daisy O'Brien, Aishihik	*Crow*
Dorothy Meyers, Haines	*Raven*
Arlene Henry, Yakutat	*Eagle*

Cooks' Helpers:
| JoAnne O'Brien, Whitehorse | *Crow* |
| Paula Banks, Whitehorse | *Wolf* |

Honorary Dressers:
John Adamson, Shawshe	*Wolf*
Albert Fred, Whitehorse	*Crow*
Charlie Jimmy, Haines	*Raven*
Ed Warren, Klukwan	*Eagle*

Apprentices:
| Harold Johnson, Champagne | *Crow* |
| Calvin Lindstrom, Carcross | *Wolf* |

Figure 3. Pamphlet distributed at the July 2001 funeral service.

Figure 4. Carved box made to hold the cremated remains at the burial site. The glacier discovery site is visible in the centre background. Sarah Gaunt (CAFN) photograph.

Figure 5. Placing the cremation box at the site on the day of the funeral, July 2001. Sarah Gaunt (CAFN) photograph.

Figure 6. Service leaders George Ramos of Yakutat *(front)* and Champagne and Aishihik's Paddy Jim *(right)*, at the Blessing of the Artifacts ceremony, Klukshu, July 2001; the late Joe Johnson of Burwash Landing is on the left. Sarah Gaunt (CAFN) photograph.

completion of the blessing of the artifacts, dancing and storytelling began as preparations for the funeral feast continued. Rose thorns—the rose being a plant used to provide protection in the Yukon—and Devil's Club (from Yakutat)—a plant used for the same purpose—were placed around the feast hall. When the group came back from the mountain, the feast was served; the feeding of the fire ceremony also took place at this time. As per custom, payment was made to all the Cooks for their hard work. The Dressers were not paid for their work, however, as ceremony organizers decided it would be more appropriate for this payment to be made at the final (Headstone) potlatch. The end of the day brought a sense of completion for this part of the cultural ceremonies.

In September 2001 the Forty Day Party for the Kwädąy Dän Ts'ìnchį individual was held at Klukwan (fig. 8), hosted by the Chilkoot Indian Tribe and Chilkat Indian Village. Following tradition, the ceremonial regalia of the Klukwan people was a prominent part of the Forty Day Party. The Kwädąy Dän Ts'ìnchį hat was also brought to the ceremony, its presence symbolizing the spirit of the man who wore it. The Forty Day Party also included a ceremony to recognize and provide support for Champagne and Aishihik First Nations in fulfilling its cultural responsibility as steward of the discovery.

While the group was returning the remains to the discovery site, other events were taking place back at Klukshu. All the Kwädąy Dän Ts'ìnchį man's belongings as well as the other artifacts recovered from the site had been brought to Klukshu for a Blessing of the Artifacts ceremony (fig. 7). Led by George Ramos of Yakutat, assisted by CAFN Elder Paddy Jim, this ceremony took place outdoors, by a fire that had been specifically cleaned and blessed for this purpose. Bags of ashes that had been made by Wolf and Crow women were distributed to all attending. While not part of a typical burial service, this ceremony helped to ease uncertain feelings guests may have had about being in the presence of a deceased person's belongings. With the

2008 Symposium in Victoria

EVENTS AND ACTIVITIES related to the Kwädąy Dän Ts'ìnchį discovery were relatively quiet from 2001 to 2007. In the spring of 2008 the results of the community DNA study were released at the Kwädąy Dän Ts'ìnchį Symposium in Victoria. The two-day symposium brought together Champagne and Aishihik First Nations researchers and community members, as well as international project scientists. Results of the numerous cultural and scientific studies were presented, shedding insight on the last days, lifetime and cultural milieu of the young man who lost his life on the glacier several hundred years ago (Hughes et al. 2009). The symposium

Figure 7. Group witnessing the Blessing of the Artifacts ceremony, Klukshu, July 2001. Sarah Gaunt (CAFN) photograph.

Figure 8. Klukwan representatives *(left to right)* Joe Hotch, Marsha Hotch and Lani Hotch drape Champagne and Aishihik First Nations Chief Bob Charlie *(far right)* with Chilkat blanket at the Forty Day Party held at Klukwan, September 2001. Sarah Gaunt (CAFN) photograph.

Figure 9. Attendees enjoying the post-feast entertainment at the Forty Day Party held at Klukwan, September 2001. Sarah Gaunt (CAFN) photograph.

included 23 verbal presentations and eight poster displays. Co-hosted by the University of Victoria and the Royal BC Museum and held in conjunction with the Northwest Anthropological Conference, the symposium was cross-cultural in nature, including much more than presentations of academic papers. British Columbia's then Lieutenant-Governor, the Honourable Steven Point, officially opened the symposium, and representatives of the host Songhees and Esquimalt First Nations extended welcomes to their traditional territory. A cultural ceremony was also held at the beginning of the symposium to help ensure that the

event went well (chapter 37). A number of the identified living maternal relatives, as well as their families and other community members from the Yukon, were able to attend the symposium. Some were able to examine the Kwädąy Dän Ts'ìnchį robe being curated at the Royal BC Museum (see chapter 37, figs. 1, 2 and 3).

In November of that year, Champagne and Aishihik's Lawrence Joe and Sheila Greer presented an overview of the project and its results at the Transcending Borders conference in Whitehorse. Hosted by Yukon College, Transcending Borders was a public symposium on the Alaska and Yukon land claims.

2009 Winter Event in Haines Junction, Presentation at Clan Conference in Juneau

IN MARCH 2009 a mini-version of the 2008 Kwädąy Dän Ts'ìnchį Symposium in Victoria was presented in Haines Junction, Yukon. At this weekend event, hosted by Champagne and Aishihik First Nations, the various posters from the Victoria symposium were on display, and verbal presentations were given by selected scientists and by Champagne and Aishihik personnel involved in the project. Most of the artifacts being conserved in Whitehorse (i.e., the wood tools and the woven hat), as well as select pieces being conserved in Victoria, were transported to Haines Junction for this event. There was a community feast and cultural entertainment, and a meeting of the identified living maternal relatives was also held in conjunction with the mini-symposium.

Later that month, at Sharing Our Knowledge: A Conference of Tlingit Tribes and Clans[5] in Juneau, Alaska, Marsha Hotch presented a slide show giving an overview of the discovery and results obtained. A resident of Klukwan and a Tlingit-language teacher at the village school, Hotch presented the overview on behalf of Champagne and Aishihik First Nations. She had also participated in spruce-root weaving workshops at Klukwan, which included efforts to reproduce the Kwädąy Dän Ts'ìnchį man's hat. Most of the posters prepared for the Victoria symposium were also displayed at this conference.

What's Ahead

A MEMORIAL POTLATCH for the Kwädąy Dän Ts'ìnchį individual has not been held. One was scheduled for May 2010, at Klukshu, Yukon but was cancelled due to the passing of a Champagne and Aishihik citizen. While the event was not rescheduled, some interest remains within the CAFN community in raising a marker or sign in Tatshenshini-Alsek Park to commemorate the Long Ago Person Found; this might be installed at an appropriate location along the Haines Road.

The belongings and artifacts recovered from the site are presently being curated in a Yukon government facility in Whitehorse. While safely stored, the items are not on public display, and permission is needed from Champagne and Aishihik First Nations to view the belongings and artifacts. No decision on the future disposition of these items has been made.

This paper has summarized how Champagne and Aishihik First Nations fulfilled its role as cultural steward of the Kwädąy Dän Ts'ìnchį discovery, and the involvement of neighbouring First Nations and Tribes from southeastern Alaska, northwestern BC and southern Yukon in such activities. The challenges of bringing together the cultural practices of different groups were addressed to provide cultural ceremonies that not only ensured the young man's spirit was honoured in the proper way but also offered opportunities for younger community members to learn about traditional mortuary practices. The ceremonies held brought our different nations together and highlighted our common heritage and shared cultural practices.

ACKNOWLEDGEMENTS

We acknowledge, with gratitude and appreciation, the contribution of the following individuals and organizations: site discoverers Bill Hanlon, Mike Roch and Warren Ward for reporting the find; researchers Owen Beattie, Al Mackie, Greg Hare, the individuals who had primary responsibility for recovery of the remains of the Kwädąy Dän Ts'ìnchį individual from the glacier, and the Yukon Heritage Resources Unit for temporary storage of the remains, before they were transported to Victoria for the autopsy studies; Delmar Washington of Capital Helicopters, and Alkan Air, Whitehorse, for transporting the remains; Champagne and Aishihik First Nations citizen Harold Johnson, for accompanying the remains as they were transported from Whitehorse to Victoria; the Royal BC Museum, for secure storage of the remains during their period of study, facilitated by Grant Hughes, Jim Cosgrove and Kelly Sendall; the Royal Oak Burial Park and Crematorium in Victoria for cremating the remains of the Kwädąy Dän Ts'ìnchį individual at no charge; and Champagne and Aishihik First Nations staff Sarah Gaunt for accompanying the cremated remains north from Victoria.

The planning committee for the 2001 funeral service included Joe Hotch (Klukwan), George Ramos (Yakutat) and Paddy Jim, James Allen and Gerald Brown (Champagne and Aishihik), with input from Joe Johnson (Kluane First Nation), Bert Adams (Yakutat) and Rosita Worl (Sealaska Heritage) and support from Diane Strand and Sarah Gaunt (Champagne and Aishihik). Champagne and Aishihik citizens Gerald Brown (Crow) and James Allen (Wolf) were clan hosts and masters of ceremony for the funeral. The planning committee for the 2001 Forty Day Party included Joe Hotch, Ruth Kasko and Marsha Hotch (all of Klukwan), Mary Paddock (Haines) and George Ramos (Yakutat), with support from Diane Strand and Sarah Gaunt. Assistance in preparing, transporting and displaying the belongings and other artifacts for the July 2001 ceremonies at Klukshu and the September 2001 ceremony at Klukwan was provided by Kjerstin Mackie (Royal BC Museum), Al Mackie (BC Archaeology Branch) and Valery Monahan (Yukon government).

Sarah Gaunt coordinated communication and consultation efforts in the initial years following the discovery, including the various meetings with neighbouring First Nations, and the 2001 regional meeting.

Co-chairs for the 2008 symposium at the Northwest Anthropological Conference were Grant Hughes (Royal BC Museum), Al Mackie and Sheila Greer (Champagne and Aishihik First Nations). The government of Canada's International Polar Year program was the principal financial supporter for the 2008 symposium. Champagne and Aishihik First Nations provided support for a number of the identified maternal relatives and community members to attend the symposium. Sheila Greer organized the 2009 event in Haines Junction. The government of Yukon's Community Development Fund was the principal financial support of the March 2009 event; supplementary support came from Indian and Northern Affairs Canada as well as Parks Canada's Healing Broken Connections project (Kluane National Park).

—

9

TATSHENSHINI-ALSEK HISTORY

Champagne and Aishihik First Nations, Sheila Greer and Sarah Gaunt

The Indigenous people of this region have long told stories of their ancestors and their connection to this land, reflecting the millennia of connection between people and place. It is appropriate that these stories, which recount how the world began, introduce this chapter because of their importance to the region's Indigenous peoples, and because they would have been fundamental to the identity of the Kwädąy Dän Ts'ìnchį individual.

"That Crow, he's like God. This is how he made the world. Long time ago, animals were all people. This is before they had light …" ("How Crow Made the World", as told by Kitty Smith, in Cruikshank 1991, 14)

"You know Ts'ürki [Crow]—he go through all over the place. First time maybe before, it's like that too, used to be—dark, they say, it used to be dark, and no daylight, nothing, long time ago. So that Ts'ürki, that Crow, he want to get the daylight and the moon and things like that …" ("Crow Creates the World", as told by Marge Jackson and learned from her stepfather, Big Jim Fred; in Jackson 2006)

"From there, they split. Ts'ürki goes to Coast Indian side. Beaver goes to Yukon side. They're going to straighten up this world. No more danger, they say. 'So, I'm going to be Yukon' [Beaver said]. 'Well, I'll come back' [Crow said]. 'I'm going to saltwater side.'" ("How First This Yukon Came to Be", as told by Mrs. Annie Ned, in Cruikshank et al. 1990)

TODAY OUTSIDERS LARGELY VIEW the Tatshenshini-Alsek area as empty wilderness. But it once had a vibrant Aboriginal population with an engaging history. In pre-contact times, Dän (Southern Tutchone) are understood to have lived along the Tatshenshini and Alsek rivers, as far downstream as Dry Bay on the coast of what is now Alaska. Then in the 18th and 19th centuries the population of the middle and lower sections of the basin was heavily affected by epidemic diseases and flooding. During these same centuries, Tlingit from the upper Lynn Canal area of southeast Alaska established their own settlements in the basin. They married into resident Dän groups as well.

When the first non-Aboriginal people entered the Tatshenshini basin in the 1890s, they found a bilingual and bicultural population, speaking both the Dän and Tlingit languages, living in the upper part of the basin, but noted abandoned villages further downstream. During historic times, depopulation of the basin continued as a result of shifts in regional economic and settlement patterns. Today there are permanent residents only in the Yukon and BC headwaters portion of the Tatshenshini River basin.

This chapter presents an overview of the history of far northwest British Columbia, where the Kwädąy Dän Ts'ìnchį discovery was made. The term "ethnohistory" has also been used to refer to the account presented in this chapter, which focuses on the region's Aboriginal past.

Our chronological narrative includes information on the cultural changes that have occurred in this area over the past couple centuries, thereby providing information on the social milieu of the times during which the Kwädąy Dän Ts'ìnchį individual lived, as well as that within which the present project operated. While a detailed ethnographic description of the culture and lifeways of the region's Indigenous peoples is beyond our scope, we highlight content that is relevant to understanding the find and the various studies undertaken. Readers wishing further ethnographic detail on the region's Indigenous cultures may consult the sources cited in this chapter.

Table 1 (page 182) presents a historical chronology of key events that took place in the study area, as well as outside occurrences that would have affected area residents. The approximate temporal boundaries of the pre-contact, early historic (or proto-historic) and historic temporal eras are also indicated in this same table. The historic era, which begins with the arrival of the

Europeans, happened earlier on the coast than it did in the Tatshenshini-Alsek local study area.

Study Area

IN THIS CHAPTER the local study area is defined as the BC portion of the Tatshenshini-Alsek basin and the adjacent portion of the Chilkat River basin that lies within the province. A greater study area is also recognized in order to include those peoples living further afield whose history is nonetheless connected to the local study area. The boundaries of the greater study area extend from the upper Lynn Canal west to the Yakutat area and north to southwestern Yukon (fig. 1, page 580).

The peoples whose history is connected to the local study area include the Champagne and Aishihik First Nations (CAFN), with headquarters in Haines Junction, Yukon; the Chilkat Indian Village (CIV), based in Klukwan and Haines, Alaska; the Chilkoot Indian Tribe (CIT), based in Haines, Alaska; and the Yakutat Tribe, based in Yakutat, Alaska. The Huna Tlingit—who are based in Hoonah, Alaska, but whose traditional homelands include the area now within Glacier Bay National Park—also need to be included in this list.

For convenience, three sections or reaches of the Tatshenshini River are recognized in this chapter. Working downstream these are: the upper basin, which refers to the BC headwaters and the Yukon portion, including the upper and lower canyons of the Tatshenshini River; the middle reaches, from around Detour Creek to below the Tkope River; and the downstream portion, from below the Tkope to the confluence of the Tatshenshini and Alsek rivers (fig. 2).

Ethnohistorical and Ethnographic Sources

THERE IS ABUNDANT ethnohistorical and anthropological (ethnographic) literature on the history and cultures of Indigenous peoples in the study area, including sources that provide relatively

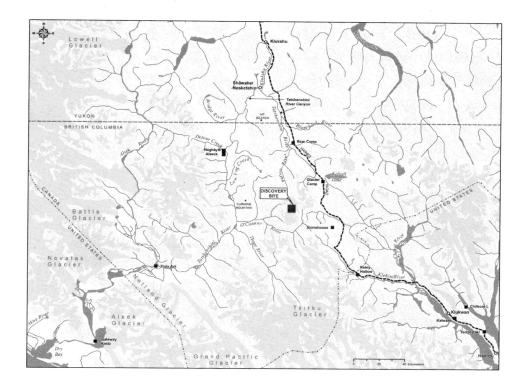

Figure 2. Map showing places in the local study area mentioned in text. CAFN map.

detailed information from the close of the 19th century. There are two key sources. One is a version of a map produced in 1869 by the Chilkat Gaanaxteidi clan Chief Kohklux—also spelled Kloh-Klux or Kloh kutx and also known as Chartrich, Shotrich or Shotridge (Davidson 1901; Johnson 1984; YHMA 1994; see chapter 10 on villages, this volume). The second key source is a series of late 19th-century reports by explorer E.J. Glave, the first non-Aboriginal person known to have travelled down the Tatshenshini River (Glave 1890–91).

The importance of these early documents notwithstanding, it is mid- to late-20th-century research with Aboriginal people based in the upper and Yukon portion of the Tatshenshini basin that established the basic framework for understanding the study area's Aboriginal history. Before the building of the Alaska Highway in southwestern Yukon and the connecting road between Haines Junction and Haines, Alaska, in 1942–43, little was known about the region's Indigenous peoples (see Clarke [1940s]). These highways opened up the study area, allowing

anthropologists Douglas Leechman and Catharine McClellan to begin their work with local Aboriginal people, and anthropologists Julie Cruikshank and Beth O'Leary followed these pioneering researchers.[1] Yukon community members who provided information to these various investigators or to later interview work continued by Champagne and Aishihik First Nations government (CAFN, Gaunt and Greer 1993, 1995) include Johnny Fraser, Lily Hume, Harry Joe, Jessie Joe, Bobby Kane, Bessie Kane, Parten Kane, Lily Kane, Jimmy Kane, Nelly Pringle, Hazel Hume, Maggie Brown, Maggie Jim, Paddy Jim, Kitty Smith, Annie Ned, David Hume, John Adamson, Jimmy G. Smith, John Brown, Ron Chambers, Chuck Hume, Doreen Grady and Marge Jackson. More recently, the latter elder, who has lived in the upper Tatshenshini River basin for much of her life, has told her own story (M. Jackson 2006).

Anthropological studies with Aboriginal people now based outside the study area but whose history is linked to it also include important information about the history of the Tatshenshini-Alsek area. The most important of these sources is the work by Frederica de Laguna in the Yakutat area in the 1950s and later. De Laguna's work, along with that of Goldschmidt and Haas, clarified the cultural makeup of the lower part of the Alsek basin in early historic times (de Laguna 1953, 1972; de Laguna et al. 1964; Goldschmidt and Haas 1998). De Laguna interviewed Dry Bay elders Maggie Dick, Emma Ellis, Frank Italio, Esther Johnson and Jenny White (1972, vii). Nora and Richard Dauenhauer recorded more stories related to the history of the lower Alsek country from Dry Bay Elders Frank Dick and Mrs. Emma Marks (Dauenhauer and Dauenhauer 1987, 1990, 1994). Additional stories about the history of the Dry Bay area can be found on the "Project Jukebox" website at the University of Alaska Fairbanks.[2]

Published sources for information on the Chilkat and Chilkoot Tlingit—or Jilkáat and Lkoot as these names are correctly written—include Dauenhauer and Dauenhauer (1987, 1990, 1994), Emmons (1916, 1991, n.d.), Goldschmidt and Haas (1998), Arthur Krause (1882), Aurel Krause (1956), Krause and Krause (1981, 1993), Larson and Larson (1977), McClellan (1975), Oberg (1973), Olson (1936, 1968), Sackett (1979) and Thornton (2004). Information on Chilkat Tlingit history and culture can also be found on the website of the Sheldon Museum in Haines, Alaska.[3] We know the names of some but not all of the community members these researchers spoke with. Goldschmidt and Haas cite James King, James H. Lee, Charlie James, Mrs. Patsy Davis, John H. Willard, Victor Hotch, Mrs. Sparks, Gus Klaney, Susie Nasook, Mary Williams and Archie Watson (Goldschmidt and Haas 1998). Olson worked with informants Mrs. Paddy Ganat, Mrs. John Benson, Ed Warren, Dan Katseek and others (Olson 1968). Sackett indicates that his report includes information from community historians Paul Phillips, Mildred Sparks, Austin Hammond and Horace Marks (Sackett 1979). Thornton reports that Haines and Klukwan area residents Dave Berry, Charles Brouillette, Ray Dennis, Lee Heinmiller, Evelyn Hotch, Joe Hotch, Edith Jacquot, Hank Jacquot, Charles Jimmie, Cynthia Jones, Ruth Kasko, Diana Kelm, David Light, Kathleen Menke, Albert Paddy, Debra Schnabel, George Stevens,

Margaret Stevens, Minnie Stevens, Gene Strong, Kimberley Strong, Ed Warren, Marilyn Wilson and Paul Wilson contributed to the ethnographic overview of the Chilkoot Trail that he prepared (see Thornton 2004 appendix C).

Other useful sources for information on Tlingit history and culture are Cruikshank (2005), de Laguna (1975, 1989, 1990), Fair and Worl (2000), Hope and Thornton (2000), Kan (1987), Newton and Moss (2005), W. Olson (1997), Swanton (1908, 1909), Thornton (2007) and Worl (1990).

McClellan is the principal source for information on the history and culture of the Tagish. Their language belongs to the Dän (Athapaskan) family of languages (McClellan 1975, 1981c, 2007b).

Key sources for information on the 18th- and 19th-century coastal fur trade are Gibson (1992), Hinckley (1996) and Malloy (1998). Legros (1984) and various works by McClellan (1964, 1975) consider the trade from the perspective of the residents of the interior.

Information related to the local study area in federal (Canada) and provincial (BC) government records was also consulted in the preparation of this chapter. The records searched include those of the North West Mounted Police and federal Indian Agent, as well as provincial trapline and game commissioner records housed in the National Archives of Canada (federal records centres in Ottawa and Vancouver), and the British Columbia Archives in Victoria (see Isaac, n.d.).

Place Names

ABORIGINAL-LANGUAGE PLACE NAMES, also referred to as toponyms, are an important historical data source. Unlike English language place names, which often commemorate individuals, Aboriginal place names have been recognized as mnemonic (recall) devices that encode Aboriginal history (Cruikshank 1990). The Yukon Native Language Centre assembled a considerable body of place names for the study area in the 1980s, based largely on information provided by CAFN Elder Jimmy Kane. Additional toponyms have been documented by

CAFN from Elders John Adamson, John Brown, Marge Jackson and Paddy Jim (CAFN, Gaunt and Greer 1995). While this body of data requires considerably more linguistic, oral history and field study, a few examples are worth highlighting.

Some Aboriginal place names are descriptive, such as Îch', the Tlingit name of the dark-coloured mountain southeast of the mouth of the O'Connor River (Adamson 1993). Nu'ay Chù is the Dän name for Bridge Creek, a tributary that enters the Tatshenshini River around the western crossing of the BC–Yukon border. The river's English name is a direct translation of the Dän toponym, which indicates that some type of bridge—perhaps a log-jam—allowed people to cross the river (Jackson 1993). Similarly, the Tlingit name for the mouth of the Parton River, Jâji Sîdak'u— meaning "little snowshoe crossing"—describes the conditions encountered where the trail crosses this watercourse (Adamson 1993). The Dän name for the Parton River, Tän Chù, means "glacier-fed" or "ice water" and is another example of a descriptive toponym, referring to the glacier at the headwaters of this river. Other place names describe a resource available at the location. For example; Gyu Chù, the Dän name for Low Fog Creek, translates as "king salmon creek" (Jackson 1993). Tínix Âní, which refers to an island located around the confluence of the Tatshenshini and Alsek rivers, means "kinnickinnik land" in Tlingit (J. Kane 1978).

Many Aboriginal toponyms have stories connected with them, such as Ts'ürki'i Kù, a Dän toponym meaning "Crow's house". This landscape feature, discussed later in this chapter, is also known as "Stonehouse" (Adamson 1993). A part of Carmine Mountain, which is located on the west side of the Tatshenshini River south of Low Fog Creek, has the Dän name Ts'ach'an, which refers to the period of seclusion that all young women traditionally experienced (CAFN, Gaunt and Greer 1995). Tli̦ Thia, the Dän name for the northern end of Mount Beaton, which is the mountain block located to the south of Shäwshe village, translates as "dog rock". According to Elder Paddy Jim, the name refers to the traditional practice of using dogs in sheep hunting (Jim 1995).

The pool of place names available for the local study area shows its bilingual and bicultural character. Many features and places have both Tlingit and Dän topoyms (Adamson 1993, 1997; Jackson 1991, 1993; Kane 1978, 1979). Shäwshe, the Dän village located in the upper Tatshenshini basin by Village Creek, was also known by the Tlingit name Nesketahin. By the end of the 19th century this settlement took on a third name, as it began to be referred to as Dalton Post after the trading and police post, established nearby. Klukshu, the northernmost fishing village in the basin, is a Tlingit name meaning "end of the silver salmon run". The Dän name for this settlement, Łu gha, translates as "place where fish are" (McClellan 1975a; Jackson 1991).

Some place names have also been confused, including those of the principal rivers for which the park is named. For example, Âłsêxh, which had various spellings—Alsek, Âłsexyik, Alséi—is the true Tlingit name for the river now officially known as the Tatshenshini on topographic maps. Tlingit scholar Andy Hope translated this name, which he spells Aalseix', as "back towards the mountains" (Hope 2000, 26). Another Tlingit speaker told CAFN staff member Lawrence Joe that the river's name meant "resting place" (L. Joe pers. comm. to Greer, January 2011). There are various spellings of the name Tatshenshini, which is also Tlingit in origin, including Chachànsîni and Tachan shini. The meaning of this toponym is uncertain. The late John Adamson suggested that the "Chachàn" root may translate as "headwaters" (Adamson 1993; Greer 2005). The Tlingit version of the Tatshenshini name refers to the watercourse that is now officially known as the Blanchard River; the latter is a tributary located in the uppermost BC portion of the Tatshenshini basin.

Aboriginal and anthropological sources of even quite recent vintage still refer to what is now known as the Tatshenshini by its Tlingit name, Âłsêxh, if not by its Dän name, Shäwshe Chù. A few sources even refer to the river as the "Dalton Post River". The earliest written document, by Glave (1890–91), as well as sources from around the turn of the 20th century and later (e.g., Davidson 1903; White-Fraser 1901), concur with the Aboriginal toponymic data, showing how the names of the rivers in this area are mixed up, with the Tatshenshini name being applied to the Àłsêxh. A surveyor working in the area in the 1890s is believed to have confused the names. In addition to the Àłsêxh/ Tatshenshini name confusion, the Alsek name has been mistakenly applied to the tributary then known to the region's Indigenous peoples as the Kaskawulsh (Kane 1978; Tero 1973). The confusion of place names may be connected to the fact that the Tlingit toponym, Âlsêxh or Alsek, applies to the river now known as the Tatshenshini as well as to the portion of the Alsek River that is below the confluence of the Tatshenshini.

> That Dalton Post River right down you, we call that Alsek, Âlsêxh … All the way down to salt water Alsek River. (CAFN Elder Jimmy Kane, quoted in Waddington 1975)

In the Aboriginal naming system, the river now known as the Tatshenshini is considered to run all the way down to the ocean. The Kaskawulsh—the portion of the Alsek River above the Tatshenshini confluence—appears to be considered tributary to the latter in the Aboriginal naming system. In Western geography, the hierarchy for naming rivers is based on flow, with the Tatshenshini considered a tributary to the main Alsek River. The Aboriginal naming system appears to be based on other criteria. Perhaps it is based on the presence of salmon. The river now known as the Tatshenshini is the more important salmon river.

Several categories of English-language place names are evident in the Tatshenshini basin (CPCGN n.d.). Some watercourse names were applied by geological survey and exploration parties, such as Low Fog Creek and Debris Creek. Surveyors working on the international boundary named the region's larger glaciers, such as the Netland, Battle and Reynolds glaciers. There is also a growing body of unofficial English-language names being applied to landscape features by the rafting community, such as the "Walker" Glacier.

Chinook jargon was used along the northwest coast from California to the Alaska panhandle in the 19th century (Thompson and Kinkade 1990, 41), and several place names in the Tatshenshini basin, such as Tkope River and Tomahnous Creek, are believed to be of Chinook origin, according to the files of the Canadian Permanent Committee on Geographic Names.

Indigenous Peoples in the Greater Region

OUR DISCUSSION of the Indigenous peoples of the greater region whose history is in some way connected to the local study area requires us to look in four different directions. To the southwest are the Tlingit of the Dry Bay–Yakutat area; to the south, the Hoonah (or Huna) Tlingit of Glacier Bay; to the southeast, the Chilkat and Chilkoot Tlingit of the upper Lynn Canal (Haines-Klukwan) area; and to the north, the Dän and Tlingit of the Tatshenshini basin and adjacent southern Yukon (fig. 1). The discussion could also be broadened to include Tlingit and Dän communities further afield and to incorporate all communities that participated in the community DNA study (see chapter 33, this volume). We know from clan histories that various Tlingit clans have spent time in the local study area in the past, including, for example, the Dakl'aweidí and Deisheetaan ancestors of both the Carcross-Tagish and Teslin Tlingit people (Patsy Henderson in McClellan 2007b, McClellan 1975a: 446; Hope 2000; Angela Sidney, in Cruikshank 1990: 39).

The Yakutat Tribe of today is an amalgamation of the Dry Bay and the Yakutat people, with the latter a mixture of Eyak, Atna Athapaskan and Tlingit, and the former composed of people related to the Southern Tutchone and Tlingit from farther to the southeast. The Hoonah Tlingit are represented by the Hoonah Indian Association. The Chilkat and Chilkoot included the historic villages of Klukwan, Kalwaltu, Yandestaki and Chilkoot Lake (Sackett 1979; Thornton 2004). Today, the political bodies representing these northern Tlingit are the Chilkat Indian Village, the Chilkoot Indian Association and the Skagway Traditional Council. The Dän and Tlingit of the middle and upper Tatshenshini basin are represented by Champagne and Aishihik

First Nations (CAFN). Champagne and Aishihik is an amalgamation of two mid-20th-century Indian "bands" (i.e., the Champagne Band and Aishihik Band).[4] CAFN citizens are primarily of Dän background, though some families are of Tlingit ancestry.

Common Traditions

SIMILAR, BUT DIFFERENT—this phrase best describes the cultures of the Dän, Tagish and Tlingit of the interior of northern BC and southern Yukon, and the Tlingit of southeast Alaska, especially the northern Tlingit of the upper Lynn Canal, Glacier Bay and Dry Bay–Yakutat areas.

These peoples, Tlingit, Tagish and Dän, share many cultural traits and traditions. All reckon descent and inheritance of clan affiliation from one's mother (see chapter 33). Family and clan relationships play a critical role in one's life—defining who one is and what is expected. The languages of these peoples are also recognized as being related. While the peoples are distinct groups with different languages, linguists recognize Tlingit as related to the various languages of the Athapaskan family, including Dákwanjè (Southern Tutchone) and Tagish (Krauss 1980). This suggests a common history for the ancestors of the speakers of these different languages, long ago.

Traditional stories of all of these peoples also have much in common. They begin with Crow (for the Dän and Tagish) or Raven (for the Tlingit) bringing light to the world (M. Jackson 2006; McClellan 2007a, 2007b, 2007c; W. Olson 1997; Sidney in Cruikshank et al. 1990; Smith in Cruikshank et al. 1990). In order to do this, Crow/Raven first had to trick a chief who controlled the sun, moon and stars. As noted in the first two quotations at the beginning of this chapter, Crow is forced to steal these celestial objects in order to bring them into being in the world. Crow/Raven experiences many adventures, outwitting people—occasionally being outwitted himself (Cruikshank 1979, iv).

Back in the beginning of time, Crow/Raven (Ts'urki in Dákwanjè, Yêɬ in Tlingit) not only created the world,

but also people, establishing the kinship rules that humans must live by, as well as the basic duality in society. This was explained by Tagish/Tlingit Elder Mrs. Angela Sidney:

> After he walks around, flies around alone. He's tired— he's lonely—he needs people. He took poplar tree bark. You know how it's thick? He carved it and then he breathed into it. "Live," he said. And he made a person. He made Crow and Wolf too. At first they can't talk to each other—Crow man and woman are shy with each other—look away. Wolf, same way too. "This is no good," he said. So he changed that. He made Crow man sit with Wolf woman. And he made Wolf man sit with Crow woman. So Crow must marry Wolf and Wolf must marry Crow. That's how the world began. (Quoted in Cruikshank 1991, 54)

In addition to Crow/Raven, old-time stories also make reference to Beaverman, or Äsùya, also known as Smartman, who rid the land of dangerous animals thereby making it safe for humans (J. Jonathan in Workman 2000 and Ned in Cruikshank et al. 1990). Beaverman, like Crow, was part of history right from the beginning, as shown in the quote from Mrs. Annie Ned presented at the opening of this chapter.

Both cultures, Dän and Tlingit, recognize the ties that exist between human culture and the natural world. Such ties include the moral obligations to take care of the land and the local resources and to respect and maintain balance between humans and animals (Betts 1994, 15; McClellan 1975a). For these peoples, as anthropologist Cruikshank has noted, nature "is included in human affairs" (Cruikshank 2005, 6).

Interior people have stories about Animal Mother (also referred to as Game Mother, or Moose Mother), the being who gave birth to all the animals (Brown in McClellan 2007a, 154–58; Cruikshank et al. 1990, 44ff.; Sidney in Cruikshank 1991, 29, 48; Smith in Sidney et al. 1979). Some Dän stories include details on glacial surges and ice-dammed lakes (Cruikshank et al. 1990; Cruikshank 1980). Elements of these and other traditional Dän stories suggest that Dän culture has deep roots in the interior country. The Tatshenshini basin's main 19th-century Dän settlement, Shäwshe, is also understood to be an old settlement, occupied for many generations (J. Kane 1978; McClellan 1975a, 26).

Tlingit history, as recounted by Tlingit historians, focuses on the history of individual clans, and how they came to be. As previously mentioned and as discussed further below, various clans have moved through or spent time in the Tatshenshini-Alsek basin at some point in their history (Hope 2000). Notwithstanding these connections to the Tatshenshini-Alsek country, Tlingit clan historians have also noted that clan histories place the ancestral homeland of the Tlingit further afield (de Laguna 1990). This homeland is thought to have been to the south in southeastern Alaska and north-central British Columbia—that is, the area of Prince of Wales Island, Kupreanhof Island and the Stikine and Nass river valleys, according to accounts recorded from Klukwan sources (Larson and Larson 1977; W. Olson 1997; Shotridge 1920; see Sheldon Museum website). Clan histories talk about how the Tlingit expanded northward from these areas. As Chilkat native and Kaagwaantann clan member Louis Shotridge wrote, "Chilkat is not the original home of the Tlingit Indians; they migrated to this region from the south" (1920, 24).

Tlingit scholars Nora Marks Dauenhauer and Richard Dauenhauer have noted that the interior country may have once been the home of at least some of the ancestral Tlingit even further back in time. Stories recount how some of the ancestors migrated downriver from the interior to the coast, where they subsequently dispersed to the current Tlingit homeland (Dauenhauer and Dauenhauer 1990, 3). As noted in chapter 29, these long-ago journeys from the interior to the coast are understood to have involved travel under glaciers.

Traditional Tlingit stories recount how their people began migrating north in the more recent past, eventually establishing villages in the upper Lynn Canal area, as in Haines and Klukwan (Hope 2000, 28; Larson and Larson 1977). Tlingit oral history even records that hooligan (eulachon), a key subsistence resource, was transplanted from the Nass River country to the upper

Lynn Canal area (J. Hotch in Betts 1994, 17). The Russian historian Grinev (2005, 20) points out that the northern expansion of the Tlingit is illustrated in place names. Hoonah, the name of a southeast Alaska community, means "northern" in Tlingit, indicating that this settlement was at one time the northernmost Tlingit settlement. Later, the Tlingit settled even further north, in the Yakutat area.

Anthropologist de Laguna believed that the northward expansion of the Tlingit was related to trade and began about 300 years ago (de Laguna 1990). Apparently when the Tlingit arrived in the Haines area, a settlement already existed at Klukwan. It is for this reason Klukwan has been referred to as the mother or eternal village (Oberg 1973, 56; Shotridge 1920, 24).

As we discuss below, in the 18th century the Tlingit continued to expand northwestward into the lower Alsek River and Dry Bay area, and then in the 19th century into the Tatshenshini basin. Tlingit expansion along the northwest coast and into the interior was eventually interrupted by the arrival of non-Indigenous people at the end of the 19th century. This interruption allowed the interior to remain, on the whole, the territory of Athapaskan-speaking peoples (Dän, Tagish), with the exception of the Tlingit communities established at Atlin, Teslin and Carcross.

Despite different histories and varying cultural traditions, the ties between the Dän, Tagish and Tlingit of the greater study area of southeast Alaska, southern Yukon and northwest BC continue to be strong even today.

The Adventures of Crow/Raven in the Local Study Area

TRADITIONAL STORIES recount Crow/Raven's experiences in the local study area, including the Chilkat Pass area, which is a traditional and current travel route between the upper Lynn Canal area of the coast and the upper Tatshenshini River country. The Kwäday Dän Ts'ìnchị individual travelled through this pass on his last journey, en route to the Fault Creek Glacier where he lost his life. Two versions of the story of Crow's adventures in the Chilkat Pass, both told by Champagne and Aishihik storytellers, are presented here. Additional versions by Lilly Hume and two by Johnny Fraser, former chief of CAFN, can be found in McClellan (2007a). According to Mrs. Susie Pringle,

> Crow discovered the Chilkat Pass. He was the first one to go over it. Somewhere near the [Canadian] Customs—on the other side—Crow fell. He fell so hard his cane pushed a hole right through the mountain. You can still see through this hole. You can't see this place from the road. (As told to McClellan July 27, 1948, at Klukshu; McClellan 2007a, 55)

Jessie Allen described the same event:

> Every summer we hear stories from Big Jim, Johnny Fraser's uncle, my father's uncle.... At Klukwan Crow fell down. His stick went right through the mountain. He got a walking stick with iron, And he made a hole through [the mountain]. (As told to McClellan August 29, 1954, at Klukshu; McClellan 2007a, 105; bracketed notes are McClellan's)

As noted in chapter 32 on travel routes, the hole in the hill made by Crow's walking stick is located in the Rainy Hollow area. It likely would have been a guide post for the Kwädąy Dän Ts'ìnchį individual on his journey inland, as would another landmark fixed by Crow known as Stonehouse—Ts'urki'i Kù, as it is called in Dákwanjè. Referenced in several of the accounts of Crow's adventures in the Chilkat Pass area, Stonehouse is a rock feature known to both CAFN and Klukwan people (Adamson 1993; Emmons 1991; Johnny Fraser in McClellan 2007a). Mrs. Susie Pringle spoke of how

> Crow built a house of stone. It was like a tent. There were big flat stones for the roof.
>
> Crow got married many times. When he got tired of his wives, he just left them. One time when Crow and his wife were coming across the Chilkat Pass, she got tired. She said her eyes hurt from the

smoke. So Crow built a house. Next morning when she woke up she felt good. She felt like a sixteen year old! (As told to McClellan July 27, 1948, at Klukshu; McClellan 2007a, 55)

Jessie Allen discussed it, too:

> He got a house too, a stone house, Ts'ürk'i [had]. [It was] around Mile 60 or 61 [on the Haines Road]. You can't see it from the road. It's way high up, on the right side going down. It's on the other side of Clear Creek. The old foot trail used by the Chilkat people goes to it. The old trail is on the other side of the lake here [at Klukshu]. There is a big stone house on the west side. A big man, he did that. He camps there. [It's an] old one, where the Chilkat used to go. That's [an] old trail, Dalton Trail. Behind [in time] way back of '98! It goes right through to Hutshi. (As told to McClellan August 29, 1954, at Klukshu; McClellan 2007a, 105; bracketed notes are McClellan's)

Stonehouse, located between Rainy Hollow and Clear Creek on the high plateau west of the Haines Road, is in the general area of the route followed by the Kwädąy Dän Ts'ìnchį individual on his last journey (see chapter 32).[5] In the 19th century and perhaps earlier as well, the Chilkats used Stonehouse as a storehouse, a place to cache goods being traded into the interior as well as the furs being brought back out to the coast (de Laguna 1989; J. Fraser in McClellan 2007a; Seton-Karr 1891).

Crow is also known to have "fixed" (i.e., put into their present forms) other places in the local study area. Another Stonehouse-type feature understood to have been made by Crow is said to be located on the north side of the Tatshenshini River below Shäwshe (J. Fraser in McClellan 2007a, 25-26; McClellan 1975a, 33; Swanton 1909). Further downstream and close to the mouth of the Alsek is another feature. In this place Crow is said to have broken open a big box and spilled its contents down the side of a mountain (S. Pringle in McClellan 2007a, 53). At yet another spot on the lower Alsek, a place where the river is 16 kilometres (10 miles) wide, there are big and small stones that represent the place where a woman and her dog crossed the river. Champagne and Aishihik Elder Susie Pringle reported that as a child, likely in the 1880s, she had seen the latter place where these rocks stood up "like Indians" when she travelled downriver to Dry Bay and Yakutat (ibid., 54).

Our chronological overview of the basin's history began with quotations that recount the activities of Crow (Raven, for the Tlingit), reflecting the Indigenous perspective of the region's history. These sources provide insights into the distant human past that the material or archaeological record cannot.

Archaeological Insights

BEING RELATIVELY INACCESSIBLE, the local study area has seen few systematic archaeological field projects and little site survey. The first archaeological research effort here was the 1970s attempt to reach the village known as Núghàyík (see chapter 10). While dense vegetation kept the party from reaching the site (Gates and Roback 1972), in that same decade Champagne and Aishihik citizen Ron Chambers was able to locate the site based on information provided by Elder Jimmy Kane (see chapter 30).

In 1993 CAFN undertook an archaeological survey along the middle and lower reaches of the river (French 1993). The survey team identified a few pre-contact and historic archaeological sites and revisited the Núghàyík site. The party also documented the presence of petroglyphs near the mouth of the Tatshenshini River. A few pre-contact era sites have also been recorded by projects elsewhere in Tatshenshini country. All are located along the Haines Road and in the Yukon portion of the upper basin (Government of BC, n.d.). Though more accessible, this part of the study area has not been well studied archaeologically. In contrast, the Yukon portion of the Tatshenshini basin, north of the provincial-territorial boundary, is better understood archaeologically (CAFN, Gaunt and Greer 1993).

Excepting the Kwäd̠ą̄y Dän Ts'ìnchį discovery site, no archaeological site in the local study area has been investigated in any detail. The few pre-contact era sites that have been recorded remain undated, with their cultural affiliation remaining uncertain. Only the petroglyphs noted near the mouth of the Tatshenshini River have been ascribed to a Tlingit culture (French 1993, 50). We do not know precisely when people began living in or using the Tatshenshini basin. It has been suggested that the region's complex glacial history may have influenced the timing of early human occupation, as glaciers blocking some of the rivers would have prohibited salmon, a critical subsistence resource, from migrating up-river (O'Leary 1992a 49; 1992b; Cruikshank 2005).

The 18th Century

YAKUTAT SOURCES interviewed in the 1940s by US government representatives reported that Dry Bay was once owned by the "Stick Indians" (Goldschmidt and Haas 1998). The Tlingit of southeastern Alaska used this term, as well as another one, gunana (also spelled as *guna.xu*, *gunaax̱u*, *gu nah ho*, *ghunanà*, *gonakho*), the Tlingit word for "stranger" to refer to the Dän and other Athapaskan-speaking people.

> A long time ago, Dry Bay was owned by the Stick Indians. The people there spoke the language of the interior. Now the only people who speak this language are at Cordova, except for one man here at Yakutat named Billy Jackson. These people here learned to speak Tlingit just as now we are changing over to English. (Yakutat Elder Jack Ellis, quoted in Goldschmidt and Haas 1998, 84)

On the basis of stories and place names shared by Aboriginal elders in the Yakutat area in the 1950s, anthropologist de Laguna concluded that prior to the appearance of European trading ships on the Pacific coast in the mid 18th century, the entire Tatshenshini-Alsek basin was likely home to people who spoke an Athapaskan language (1972, 18). The Athapaskan language spoken by the residents of the basin was most likely a variant of Dákwanjè (Southern Tutchone) as we refer to it in this paper. Information from other sources (Swanton 1909, 67), as well as stories from Champagne and Aishihik elders agree that the entire Tatshenshini-Alsek basin was Dän territory in times past, a position supported by the map produced by Chilkat trading Chief Kohklux. Discussed below, this mid-19th-century map similarly indicates only "Stick Indian" villages along the Tatshenshini River (see chapter 10).

The Tatshenshini basin's Dän population in the 18th century is estimated at a few hundred (McClellan 1964, 7). Based on the stories told by Champagne and Aishihik elders, anthropologist McClellan understood that these people spent the majority of the summer at several different fishing camps along the Tatshenshini and its tributaries, travelling downstream to meet the first salmon runs of the season and moving back upstream to harvest the later salmon runs in the Yukon portion of the basin. The rest of the year found them in smaller family groups elsewhere in the Tatshenshini drainage or the adjacent country of the Yukon interior (McClellan 1975a, 1981b). They spread out for hunting large game (such as Caribou, sheep, Moose or goats) to get small game and birds and to fish the region's lakes. They also travelled further afield to trade with neighbouring peoples.

The people who lived in the lower Alsek basin and Dry Bay area at this time are understood to have had regular exchange and marriages with the residents living farther up the Tatshenshini River. Interior goods such as copper, white marble (used for dolls' heads), beautifully tanned skin garments decorated with porcupine quills, ground squirrel robes and other furs, rare feathers, and soapberries in birch-bark boxes were brought down the Alsek (de Laguna 1972, 350). As previously clarified, where de Laguna mentions the Alsek River, she is referring to the watercourse now known as the Tatshenshini.

But before this trade developed, at some point in the far distant past, the people living in the lower Alsek and Dry Bay country did not know about those living upriver. We are aware of this because of a traditional story told by CAFN Elder Mrs. Annie Ned, recounting how the two peoples first met (Ned in Cruikshank 1991, 80). As the story goes, the downriver people noticed chips of wood floating by their camp on the river. Curious about the source of the wood chips, they travelled up the Tatshenshini to the place that would later be known as Núghàyík. Here they found the people who had chopped down the trees and caused the wood chips in the river. Thus began the connection and trade between the upriver and downriver people. As Haines resident and *Dakl'aweidí* clan member Judson Brown (Shaakakóoni) put it, "We traded with the Athabaskan [sic] people from time immemorial." (quoted in Dauenhauer and Dauenhauer 1994, 134).[6]

The lower Alsek and Dry Bay people were not the only ones trading with the people living up the Tatshenshini River in the 18th century. As noted in the foregoing quote from Judson Brown, traditional accounts of the Chilkat people indicate that residents of their area also carried on lively trade with the interior people in ancient times—even before the 19th-century fur trade. Moose hides, highly decorated moccasins, birch-wood bows wound with porcupine gut and thongs and sinews of various kinds for sewing, binding and the making of snowshoes were said to be traded to the coast (Oberg 1973). According to one researcher, the stories shared by Klukwan-area elders indicate that prepared Caribou hides (used to make shirts and trousers) were highly valued items owing to their fine texture and durability, with a ready market in all Tlingit villages (Oberg 1973, 108). Copper was another important material traded from the interior to the coast. In exchange for these goods, the Dän received such things as cedar-bark baskets, fish oil, iron and shell ornaments.

Forged metal was also making its way to the Dän in the interior, even before European trading ships arrived on the adjacent coast. CAFN Elder Jimmy Kane reported that the first metal his ancestors received in trade from their coastal neighbours were pieces recovered from the driftwood of shipwrecks (Kane 1971). Before the trade in metal, obsidian was an item of exchange between the coast and the interior. Obsidian found at three sites in the Aishihik Lake area of CAFN traditional territory in the southern Yukon has been sourced to Sumez Island, which is situated near Prince of Wales Island in southeastern Alaska (Bailey 1998). These archaeological finds are evidence of this ancient trade between the two regions.

Tlingit Expansion

THE MIGRATION of the Tlingit northward to the Klukwan area several hundred years ago has already been mentioned. As they explored new places and developed new trade relationships, the Tlingit continued to expand from the mother village. The traditional Tlingit story known as "The Ghost of Courageous Adventurer" recounts the travels of a party of Shangukeidi clan members who were the first to venture any significant distance northwestward. The party crossed parts of what is now called the St. Elias Range—including a desert of ice—until it reached the Copper River (Shotridge 1920; see chapter 29 on glacier travel, this volume). It is not known exactly when this trip would have occurred, though it was likely before the 19th century.

In the mid-19th century, Tlingit traders from the Hoonah and Chilkat areas began expanding into the Dry Bay and lower Alsek River country (de Laguna 1972, 81). Stories suggest that the Luknax.adi clan was the first to come in contact with the interior tribes of the Alsek, from whom they procured Moose and Caribou hide, which were in demand for clothing (Emmons in Hope and Thornton 2000, 142). The relations between the newcomers and the host Dän population are thought to have been good, though the history of one of the Tlingit clan that settled in the Dry Bay area, the X'at'ka.aayí, makes reference to conflicts with the interior tribes of the upper Alsek country (Emmons in Hope and Thornton 2000, 144). A Luknax.adi clan man named Qakew'wte is credited with establishing the trade in copper.

> He travelled north along the coast to Dry Bay and then up the Alsek River until he met the interior people. He taught them how the Tlingit took

animals and fish, what roots, bark, and plants they ate, and how they prepared food for the winter. Here he first saw copper, in the form of knives, spears and arrow blades, and the copper shield or "copper." Upon leaving, he was given copper which he carried back to his people. From this time, the Hoonah made trading trips to the Alsek country and procured native copper which they traded to the southern coastal tribes. (Emmons 1991, 177)

There are various versions of this story (Thornton 2004; Cruikshank 2005, 37ff.), including some saying that this Tlingit explorer spent two years living with the Gunana up the Alsek, meaning the Tatshenshini River. Possibly he stayed at Shäwshe, where, so the story goes, he taught the Dän how to trap salmon and harvest and prepare other foods. In exchange, he received knowledge and goods from his hosts, including information about sources of native copper. He cemented his trading partnership with the local chief by marrying the chief's daughter (or sister, depending on the version), as well as giving gifts of skins, furs and copper.

Travel along the Pacific coast between the Lynn Canal area, Glacier Bay and Dry Bay areas was dangerous for the 18th- and 19th-century Tlingit. To avoid these open-water journeys, interior travel routes were explored. The Glacier Bay Tlingit used one that took them over the Melburn and Grand Pacific, "through" glaciers to the lower Tatshenshini-Alsek River country (Cruikshank 2005). The route initially taken by the Chilkat Tlingit to the lower Alsek River and Dry Bay country similarly involved travelling over a glacier (de Laguna 1972, 350). Sometime later, the Chilkat people discovered a new route that went from Klukwan up the Klehini, over to the O'Connor River and then downriver (de Laguna 1972, 82, 90; McClellan 1975a, 509). This Klehini-O'Connor route was still in use in the 1890s (Seton-Karr 1891), and it appears that on his last journey, the Kwädąy Dän Ts'ìnchį individual was following at least the initial portion of this route (see chapter 32).

European Trading Ships

IN THE MIDDLE OF THE 18TH CENTURY, the first European (Russian) trading ships visited the northwest Pacific coast (table 1). The arrival of these ships marks the beginning of the European fur trade in this area, with ships—French, Spanish, English and American (belonging to the "Boston Men", as they were known)—subsequently journeying here on a regular basis as the trade developed (Gibson 1992; Hinckley 1996; Malloy 1998). Initially the trade in furs was focused on the Sea Otter, with the northern Tlingit area (upper Lynn Canal and Dry Bay and Yakutat areas) recognized as being the most productive for this species (Davidson 1867–68). When overhunting dramatically reduced the otters' numbers, an increased demand for inland furs made the trade with inland peoples more important. It is believed that inland furs took on greater importance around the turn of the 19th century (McClellan 1964).

Unfortunately, the foreign traders brought more than just new goods to the region. Fatal diseases entered the area. The first smallpox epidemic is thought to have hit southeastern Alaska in 1775, with others following in 1836 and 1862 (Cruikshank 2005, 36, citing Boyd 1999; Sackett 1979, 24). The second (1836) epidemic is believed to have been the most devastating. One scholar estimated that 25–50 per cent of the population of southeast Alaska died at that time. Measles (1800) and typhoid (1819, 1848, 1855) also spread through the area (Cruikshank 2005).

As a young man, CAFN's John Adamson heard stories about the epidemic(s) that resulted in empty villages along the Tatshenshini River, including one account that described how five men came to a village where they found everyone dead, except for one woman.

> They found one that was alive ... Well ... they shouldn't even of come near her, but the way the story goes is that these five guys took her down to the river and bathed her, just had to pull her clothes off, they were just stuck to her.... She was pretty ill, no resistance. Probably she was just giving up living then ... But these guys waited around until she got her health back.... They dressed her with what clothes they had, some they could spare, and took it slow, and brought her to camp. (Adamson 1993)

Mr. Adamson didn't know whether the lady who survived or the men who found her were Tlingit or Dän, nor did he know the village location or when this tragedy occurred. The description of the survivor's clothing being stuck to her body suggests this may have been one of the 18th-century smallpox epidemics rather than any of the other diseases that made their way to the area.

The 19th Century

THE EXPANSION of the Tlingit into the Dry Bay area and lower Alsek basin through marriage and by taking up residency in the area resulted in the mixed Tlingit-Athapaskan group known as the Dry Bay or Guna.xu tribe (de Laguna 1990, 203; Emmons 1991, 57; cf. Dauenhauer and Dauenhauer 1990, 547). There are no accurate population figures for this early-to mid-19th-century mixed Tlingit and Dän population. Their numbers were likely similar to those of the Dän living farther upriver—perhaps a few hundred (de Laguna 1972, 81–90).

The history of the clans of this mixed population is closely tied to the lower Alsek River country. Dry Bay elders told de Laguna in the 1950s that four days' hard journey up the Alsek (but only one day's run down) near a camping spot called Glacier Point, a Cankuqedi boy was accidentally left behind and rescued by the Thunderbirds. This event entitled the clan to use the Thunderbird as a crest and to sing Thunderbird songs (de Laguna 1972, 224). The Dry Bay Tlukwaxadi (spelled Luknax.adi in Hope and Thornton 2000) and their Cankuqedi spouses are also reported to have ascended the Tatshenshini-Alsek to fish, hunt and trade in the interior. The trip upstream in the fall is understood to have been made on foot, when the river was frozen. Cottonwood dugout canoes, which had either been made by or procured from upstream residents, were used for the return downstream journey.

My father's people, Tlukwaxadi, used to go way
up to the head of Alsek, Alsexyik. They would
catch king salmon, slice it and cover it over with
cottonwood branches. They used duq (cottonwood)
leaves (kayani) and put it on top of the dry fish. They
would just leave it there, and when they came down
from the head of Alsek, it was just dried good. Up
at Tinx kayani [Tínix Àní] [Kinnikinik Leaves] at
the head of Alsek, they used to get soapberries and
other kinds of berries and put them in a box. They
used to go up there for all kinds of meat—black
bear meat, and then they come down. That's where
my father's people stay, way up there on an island,
getting soapberries and king salmon. (de Laguna
1972, 87)

In the previous quote, "Alsek" refers to the river
now known as the Tatshenshini. The reference to
Tínix Àní being located at the "head of Alsek" reflects
the coastal orientation of the story teller. Tínix Àní
is actually located on or by an island located near the
confluence of the Alsek and the Tatshenshini River,
rather than at the head of the river. Regardless, the
lower Alsek River people would go to this place every
fall to catch salmon because the weather conditions for
drying fish were better here than they were on the coast
(de Laguna 1972, 87; Kane 1978).

Vivid accounts of the dangers of travelling up
and down the Tatshenshini-Alsek were widely
circulated, told by Tlingit as far away as Sitka, Alaska
(Swanton 1909, 67). Stories tell that for a period a
glacier blocked the Alsek River, requiring the people
heading upstream to portage through a V-shaped
notch. The landmark, now known as Gateway Knob,
is situated below the confluence of the Alsek and
Tatshenshini rivers and has both a Tlingit and a
Dän name (de Laguna 1972, 87).

This is the way it's told about our land. This is the
way my paternal aunt would tell it to me. We would
pole our way up the Alsek River to the interior. Yes,
a cliff of ice ran across this river, the Alsek River.
We did not go under; we went over the mountains.

We would carry our boats over. (Mrs. Emma Marks,
quoted in Dauenhauer and Dauenhauer 1990, 195)

When the river was blocked by glacial ice, travellers
were forced to take their boats under the glacier through
an ice cave. There were also places along the river where
pebbles frequently fell, and if one of these hit a canoe, it
was an omen that someone was going to die. To prepare
for the trip's dangers and the possibility of death, the
people would put on their best clothes. If they made it
through successfully, it is said that they would yell with
joy and sing songs (de Laguna 1972, 87).

While the lower Alsek and Dry Bay area was home
to a mixed Dän/Tlingit population by the early 19th
century, the upstream parts of the basin remained Dän
territory at this time. The exact number of Dän villages
along the river is uncertain. The map produced by Chief
Kohklux in 1869, discussed in further detail in chapter
10 on villages, indicates four settlements: Shäwshe,
located at the river's most northerly point, within what
is now the Yukon; a settlement on the west side of the
river in the area of the Núghàyík site; and two more
villages on the east side downstream from where the
Núghàyík site is located, but north of what is thought
to be the O'Connor River (see chapter 10). Information
from Champagne and Aishihik elders as well as Dry
Bay sources has suggested the possibility of further
villages. However, as the location of such settlements
is unknown, it is not certain if these represent villages
beyond those mapped by Chief Kohklux, or different
names for these same four villages or additional villages
that were in existence at varying points in time.

Relationships between the upriver Dän and the
lower-river and Dry Bay people are understood to have
been positive in the 19th century. It is recorded that the
latter thought very highly of the Dän, admiring their fine
looks, open grassy country, rich furs and beautiful skin
clothing and beadwork (de Laguna 1972, 214). De Laguna
was told that all Dry Bay people of the 19th century could
speak their interior neighbours' language and performed
their songs and dances (de Laguna 1972, 82).

Some lower-river people did take up residency
in the middle reaches of the Tatshenshini basin for a

period. Two different stories have come down to explain why this occurred. Explorer Glave, based on information that was most likely supplied by upriver people, reports the settlement named Tin Char Tlar as a place that the Yakutat people fled to following a late-18th-century coastal battle with the Russians (Glave December 27, 1890, 396). Dry Bay elders told de Laguna that their people escaped to a place named Cix'a'anúwu (meaning Eddy Fort), for protection after a battle with the Yakutat Teqwedi (de Laguna 1972, 89). Whether these are the same or different sites remains uncertain. Neither has been located. Based on information in Glave's account, the old fort site is thought to be on the middle reach of the Tatshenshini River, somewhere between Detour Creek and the Tkope River.

The Dän of the upper basin also made the long journey downstream to visit their relatives and trading partners at Dry Bay. Not only were goods such as copper and furs traded, but also stories, songs and dances (McClellan 1975a, 509). Champagne and Aishihik elders told McClellan that their grandfathers made the two-month trip to Dry Bay in the winter. Women and children also sometimes made the long journey. When she was a child, Wut'ima'—known in later years around Haines Junction as Susie Pringle—walked from Shäwshe to Dry Bay and Yakutat, as did the young man who would later be known in the same interior community as Frank Smith (McClellan 2007a, 57).

Summer trips by dugout canoe were still being made until around the end of the 19th century (Cruikshank et

Figure 3. Dugout canoe crossing Tatshenshini River in front of Shäwshe (Nesketahin/Dalton Post) settlement in the 1890s. Original caption reads "Stick Indian in canoe". H.D. Banks photograph collection (1890s), #3636, Yukon Archives.

al. 1990, 203). Mrs. Kitty Smith—Aboriginal names Kàdùhikh, K'ałgwach and K'odetéena—who was raised at Shäwshe, was born on one of these trips at a place named Gaax'w áa yéi, a fish camp on the coast near Dry Bay (Cruikshank et al. 1990, 201, 366). Mrs. Smith did the trip downriver a second time when she was a child in the 1890s:

> Three boats went down, went to see Yaakwdáat [Yakutat]. They fixed poplar tree boats— big ones; they cut out the inside; then they make hot water by putting hot rocks inside. When this boat tries to open a little bit this way, they [pry it open and] put in cross-pieces. My grandpa made that boat— Grandpa Scottie made that kind of boat.... And that's the kind of boat they used to go down ...
>
> From Dalton Post, three boats went down. Big Jim and his wife [in one boat]; in another one, Grandpa's wife's sister was in the boat with my grandma. Lots of people went! There were lots in each boat, altogether more than ten people.... That trip took about five days.... They just go easy. They camp at Noogaayík [Núghàyík in this volume]. Lots of people living there that time. Tínx Kayaaní [here spelled as Tínix Àní], that's down that way, too. We went down in summertime; Paddy Duncan shot a goat there. (Cruikshank et al. 1990, 203–04)[7]

As previously mentioned, the Dän also travelled downstream each year to the middle reaches of the river, to meet the fish as they migrated up the Tatshenshini River. This practice continued until sometime in the early decades of the 20th century. Hunting and trapping continued in the mid to lower reaches of the river, to at least the confluence of the Tatshenshini and Alsek rivers, until the mid-20th century (Kane 1978).

The Fur Trade and Movement of Tlingit into the Tatshenshini Basin

THE NORTHERN TLINGIT of the upper Lynn Canal area (i.e., the Jilkáat and Lkoot) expanded north into the Tatshenshini basin just as they had into the Dry Bay area. Anthropologist McClellan thought this expansion began around the turn of the 19th century and was similarly connected to the fur trade following the near extinction of the Sea Otter (McClellan 1964; 1981, 388; Krause 1956, 127). The Dän residents had always taken some furbearers, but they now incorporated more trapping into their yearly round in order to procure surplus furs that they could exchange. One source reports that marten and lynx were the most important furs in the trade, likely because they were lightweight and easy to pack to the coast (P. Henderson in McClellan 1975a, 511).

The Jilkáat and Lkoot Tlingit, as well as the Tlingit of the Taku River basin, took on the role of middlemen in this 19th-century fur trade, packing trade goods to the interior, exchanging them for furs that they carried back to the coast, then selling the furs to European traders and explorers. The volume of the trade was significant. In 1867, the year the United States purchased Alaska, more than 2,300 marten sable were exchanged in the Chilkat area (Davidson 1867–68).

While at least one fully Tlingit settlement is known to have been established in the Tatshenshini basin in the 19th century, much of the Tlingit expansion into the basin occurred through marriage. Following Tlingit practice, Chilkat traders married Dän women or arranged the marriage of their sisters or daughters to interior men. The paternal grandmother of Mrs. Kitty Smith (born *ca.* 1890), for example, was one of four Raven clan Tlingit sisters from the coast who married into the interior during this period (Cruikshank et al. 1990, 159, 175). The mother of Big Jim Fred, who was the Crow clan leader at Shäwshe in the early 20th century, was a coastal woman who married into the interior (Jackson 2006). So, too, was the mother of the previously mentioned Frank Smith (McClellan 2007a, 57). The genealogical records of all south Yukon First

Nations have many such examples of coastal-interior marriages (Cooley n.d.; Cruikshank et al. 1990).

The presence of Tlingit clans in Dän culture, including the Raven Gaanaxteidí, and Wolf Kankukedí and Wolf Dakl'aweidí (including K'etl'inmbet, a Dakl'aweidí variant), is attributed to such marriages (McClellan 1975a, 441ff.). We note though that the presence of the moiety system, the basic Crow and Wolf duality of Dän society, is older and predates this close interchange between the two peoples. Still, the intensified relations between the Tlingit and Dän in the 19th century brought many changes. Most community members of the time are thought to have understood both languages, and many had both Dän and Tlingit personal names (McClellan 1975a; CAFN personal names files). It was similar for the Tagish people (Sidney 1983; McClellan 1975a).

Following Tlingit practice, trade with the interior peoples was conducted between trading partners. These were established along kinship lines, hence the importance of cross-cultural marriages to secure the partnerships. According to Yandestuki village resident Joe Wright (also known as Skookum Joe), who was interviewed by anthropologist Ronald Olson in the 1930s, the Lukhaax.ádi clan controlled trade over the Chilkoot Trail, while rights to the Chilkat Trail were shared by the Dakl'aweidí and the Gaanaxteidi of Klukwan (R. Olson 1936).

Shäwshe, the upper basin Dän village, was an important 19th-century regional gathering and trading centre (Davidson 1901; Emmons 1991, 57; de Laguna 1972, 90). Twice a year, in February and again in early summer, large Chilkat Tlingit trading parties arrived at Shäwshe loaded with bundles of European goods that were exchanged for furs and hides. In turn, the Shäwshe people traded these European goods with other Athapaskan-speaking groups, including other Dän, located farther inland (McClellan 1964, 1975a, 509). CAFN elders told McClellan that

> the Chilkat came to Neskatahin [i.e., Shäwshe] in February and again in early summer when the water was low. Good-sized groups of about

a hundred people arrived but usually there were only men in the party. They carried heavy loads of over 100 pounds.... When the Tlingit arrived at Neskatahin [Shäwshe], they stayed two or three days. The head traders and their packers lived in the houses of their trading partners ... the Tlingit brought both native coastal products and white men's goods, such as calico, good quality blankets, kettles, knives, muzzle loaders, shot, powder, axes, mirrors, vermilion, leaf tobacco, flour and coffee ... The Chilkat's own products were Chilkat blankets, baskets, and a fungus which made a red paint useful for protection against sunburn ... sometimes the Chilkat also brought dentalia.... In return for the above articles, the Southern Tutchone bartered the yellow lichen [used as dye], goat skins or hair, sinew, caribou and moose skins that were either ready to tan or already tanned, and furs—especially beaver, lynx and fox. (McClellan 1975a, 505).

Emmons (1991, 55) described the packers: "The average pack of the man weighed one hundred pounds, although some carried over two hundred pounds, in addition to snowshoes and food. All this they carried over three thousand feet of steep mountain, and then several hundred miles beyond, taking advantage of all available water."

Champagne and Aishihik Elder Mrs. Annie Ned heard first-hand from her parents about the visits of the coastal traders.

> Coast Indians got guns, knives, axes. They came on snowshoes. They packed sugar, tea, tobacco, cloth to sew. Rich people would have eight packers each! They brought shells, they brought anything to trade. They traded for clothes.... At first these Yukon people didn't want it.... But people here got crazy for it [trade goods]. They traded for knives, they traded for anything, they say—shells, guns, needles. When you buy that gun, you've got to pile up furs how long is that gun, same as that gun, how tall! Then you got that gun. (Quoted in Cruikshank 1991, 80)

Figure 4. Chilkat Tlingit man ready for trading trip into the interior, taken around the turn of the 20th century. George Emmons photograph. American Museum of Natural History.

Figure 5. Carved tree marker located along one of the old trading trails. This particular trail marker no longer exists, but it was located near the Blanchard River crossing on the Haines Road. Canadian Museum of History, Photograph NO. 98065.

The region's major Aboriginal trails became known to some as "grease trails" because of the importance of eulachon (oolichan or hooligan) oil as a trade commodity (de Laguna 1972, 350). These trading routes are marked on the mid-century maps drawn by Chilkat Chief Kohklux (Davidson 1901; YHMA 1994; see chapter 32). The well-known Chilkat Trail to the interior was actually at least two trails, with the eastern branch of the Chilkat Trail going from Klukwan via the upper Chilkat River over to Kusawa Lake, while the western

Chilkat Trail went from Klukwan up the Klehini River and over into the Tatshenshini basin (see chapter 32).

The Chilkat traders carved elaborate markers at some locations along these trading routes. Several of these carvings, including unfinished examples, have been found along traditional travel routes in CAFN territory. Other locations where tree carvings were once situated are also known (CAFN Heritage program site files, n.d.).

Ron Chambers heard about these carvings from CAFN Elder Jimmy Kane. He learned that they were markers connecting different places along the interior travel routes, allowing the traveller to be directed from one marker to the next (Chambers in Blackstock 2001). Klukwan Elder Ed Warren provided additional information on the tree carvings. He mentioned that the tree carvings were commemorative, recording events that happened at the particular location on the trail, such as the drowning of a trader at a river crossing (CAFN 2001).

While much remains to be learned about the 19th-century trade, its time frame has largely been established. A detailed analysis of the southeastern Alaska trade records from that century shows that the trade in inland furs peaked during the years 1839–67 (Legros 1984; McClellan 1964). The flow of European trade goods to the Tlingit increased in 1839, after the Hudson's Bay Company acquired trading rights in southeastern Alaska from the Russian-American Company. The trade tapered off somewhat when America purchased Alaska a few decades later in 1867 (de Laguna 1991, 209; Legros 1984; McClellan 1964). During the fur trade years there was such a demand for western trade items that the Chilkats had to look further afield for goods that could be exchanged. One Klukwan Elder recounted that the Chilkat took annual spring trading trips by canoe to places as far south as Vancouver to obtain trade goods from the British (M. Sparks in Sackett 1979, 27).

At the height of the trade, the Chilkat Tlingit established one or more villages on the Tatshenshini. The best known of these settlements is Núghàyík. The name of this village translates as "house of the Nua qwa" (or Nu gha; Ritter 2009). Stories told to McClellan by CAFN Elders, closely paralleled by those Glave recorded in 1890, indicate that the Nua qwa traded with the Dän, exchanging seal oil for furs and skins. According to McClellan, some of these people then settled permanently in the interior to avoid the journey back to the coast (1975, 28). Some individuals within the Champagne and Aishihik community, however, understand the Nua qwa to have been a distinct group of mixed Tlingit and Dän, rather than just Tlingit who moved inland. Regardless, there was said to be a

direct trail from Klukwan to Núghàyík, and it has been suggested that the Long Ago Person Found may have been following the route from Klukwan to Núghàyík (see chapter 32).

Another possible Tlingit settlement may have been located somewhere downstream from Núghàyík, on the west side of the river opposite the mouth of the O'Connor River. Champagne and Aishihik elders refer to this place as Kálawà, the name of the Tlingit man from Klukwan who lived there (Adamson 1993; recorded as "Gashurwa" in O'Leary 1992). Big Jim Fred, Crow clan leader at Shäwshe early 20th-century, is understood to have obtained his first metal traps for catching furbearers from this individual (M. Jackson in Greer 2005). The description of the residence of Kálawà as being on the west side of the river suggests that it was not in the same place as the "Sticks Village" that once stood near the mouth of the O'Connor River, as marked on the Kohklux map. The latter is indicated as being on the east side of the Tatshenshini River (see chapter 10). Âloaʔ (or Alar) has been suggested as the name of another possible Tlingit settlement in the basin, but it is likely that this name referred to Tlingit speakers rather than a specific group or village (McClellan 1975a, 28, 32, 581).

Mineral resources were an important trade item in the 19th century, as they likely were in earlier times as well. One of the resources that reportedly brought the Nua qwa people into the Tatshenshini basin was flint. This was obtained from a mountain named Klecea that is supposed to be located near Núghàyík (Glave December 20, 1890, 376). The people of the lower river had also come upstream to obtain marble from a mountain near a river named Djinuwu hini, which means "mountain goats stream" (de Laguna 1972: 89). Now referred to as Tats Creek, according to Jimmy Kane, Djinuwu hini (also spelled Jánwu Hîni) enters the Tatshenshini from the west downriver from Núghàyík (CAFN, Gaunt and Greer 1995).

The Tatshenshini River people were also well known as traders of native copper, and the trade in copper figures prominently in regional oral history (McClellan 1975a, 509; Emmons 1991, 176–77). Raw copper was obtained from neighbours in the White River area, farther inland, and exchanged with the Chilkat and Dry Bay people (see chapter 24).

Finished copper products were traded back into the interior. According to CAFN Elder Johnny Fraser, the house door of the Gaanaxteidi (Raven) clan leader at the Núghàyík settlement featured a copper plate that had been obtained in trade from the Dry Bay Tlingit (McClellan 2007a, 13). The copper trade was of such importance that this clan leader was referred to as Tinneh sarti, a Tlingit name meaning "copper master". When the Núghàyík settlement was abandoned, this copper door plate was moved to Shäwshe, where it was mounted on the door of the Drum Sound House, the residence of Crow (Raven) clan leader Big Jim Fred (CAFN Gaunt and Greer 1993). A replica of this door plate is understood to have been moved to Champagne village sometime in the 20th century (CAFN oral history files).

The 19th-century trade in furs, copper and European goods brought considerable wealth to the Tlingit of the upper Lynn Canal area (Legros 1984).[8] During the height of the trade, the Chilkat Tlingit took measures to protect their trading interests, going so far as to destroy Fort Selkirk, a Hudson's Bay Company trading post located in Northern Tutchone country almost 400 kilometres north of Klukwan. Fort Selkirk was established in 1848 and destroyed in a raid lead by Chief Kohklux in 1852 (Legros 2007; YHMA 1994).

In order to protect their trading interests, the Chilkat Tlingit prohibited non-Aboriginal people from travelling to the interior to trade, and they would not allow Dän or Tlingit living in the interior visiting the coast to trade directly with Europeans (Beardslee 1880; Krause 1956; McClellan 1964, 1975a). Such restrictions are not thought to have existed in the early days of the trade (see chapter 29 on glacier travels, this volume), but once established, it appears they remained in place until at least the mid-1880s.

The trade was highly valued by the residents of the study area; even from the vantage point of the 20th century, it is viewed as a focus of local history (Cruikshank et al. 1990; Cruikshank 2005; CTFN and Greer 1995). Stories told about this period remind us that trade was about more than the exchange of

goods. Songs and stories were also traded, and the family and social relationships that developed have been long-lasting. The trade has also been recognized as an important factor in shaping the ethnic character of the greater region, being seen by outsiders (such as anthropologists) as the mechanism by which Tlingit people, culture and language expanded into the study area and the adjacent Yukon (Cruikshank 1990; de Laguna 1972; McClellan 1964, 1975a).

One could characterize the study area basin in the mid-19th century (the height of the interior fur trade) as bicultural in nature. It was home to both Dän and Tlingit communities, and because marriage between the two groups was common, the region also included multicultural Dän and Tlingit villages. According to McClellan, it is hard to tell just how much of the population at this time may have been Athapaskan (i.e., Dän), and how much may have been Tlingit, or to what degree the two groups had mixed (1975, 24).

Population Losses

THE BASIN'S POPULATION continued to grow until it peaked in the middle of the 19th century. While numbers are impossible to specify, the population density is thought to have been high compared to other areas in the Yukon interior (see Cruikshank 1974, v–12; 2005, 46). In 1890, the explorer Glave estimated that there were 200 people living on the Tatshenshini downstream from Shäwshe (Glave December 20, 1890, 376), but his observations were made after the basin experienced major population losses.

Mid-19th-century disasters first affected the mixed Dän/Tlingit population of the lower part of the basin. Almost all people living at Dry Bay and at the settlement known as Tínix Àní, which is near the confluence of the Alsek and Tatshenshini rivers, drowned in a catastrophic flood. This tragedy occurred after the Alsek River broke through an ice-dam that had been created by an advance of the Lowell Glacier, in what is now Kluane National Park. The glacier ice-dam—known by the Dákwanjè toponym Nàłúdi, meaning "fish stop"—had been created by a shaman who had been offended by the lack of respect shown to him by a youth (Washington quoted in CAFN 2010; Cruikshank 2005, 43–45; Smith quoted in Cruikshank et al. 1990, 205–208). After the ice-dammed lake had been in existence for a period, the shaman ordered the water to break through the ice, which resulted in the downstream flood. The event, dated by scientists to 1852, is remembered in the oral history of both the Dry Bay and Champagne and Aishihik people (CAFN 2010; Cruikshank et al. 1990, 367; de Laguna 1972, 276).

Sometime mid century, perhaps after the flood wiped out the settlement at Tínix Àní, the previously mentioned epidemic diseases swept through the area. According to oral history recorded from CAFN elders, many people died from disease in the middle of the 19th century, including most of the Núghàyík people (McClellan 1975a, 28; 2007a, 13). The survivors from this settlement joined their relatives at Shäwshe or returned to the coast. A few were also reported to have moved permanently to Guse'x on the Akwe River near Dry Bay (spelled as Akoi in Grinev 2005). The latter was a Tlingit town at that time but was formerly known as an Athapaskan settlement (McClellan 1975a, 24; de Laguna 1972, 83). Anthropologist de Laguna suggests that smallpox epidemics may well account for the virtual disappearance of the Dry Bay and lower Tatshenshini-Alsek population (1972, 277).

While the basin's Dän population was affected by the epidemics and perhaps the Alsek River flood, the effect was greater on the Tlingit communities. Up to this point in the 19th century, Tlingit people and culture had been expanding through both migration and marriage into the basin and the adjacent southwestern Yukon. The various epidemics greatly slowed—if not halted—this process, and by the time the first non-Aboriginal people descended the Tatshenshini in 1890, only the villages of the upper basin were inhabited. The lower portion of the basin never recovered from its population losses.

Chapter 10 of this volume discusses the possible 19th-century Tlingit or Dän settlements in the Tatshenshini basin. The basin's residents would have had other settlements and sites away from the river as well. While hunting camps, a Moose or Caribou hunting fence or corral, berry picking camps, and other important resource locales have been mentioned, the locations of only a few of these sites have been identified (de Laguna 1972; McClellan 1975a, 32, 34, 108, 127; CAFN Heritage files).

The Later Contact Period and First Outsiders (1882–1906)

GERMAN EXPLORER ARTHUR KRAUSE is believed to be the first non-Aboriginal outsider to enter the Tatshenshini River basin part of the local study area. Departing from Klukwan and accompanied by Chilkat guides, Krause explored the Chilkat Pass in 1882 (Watson 1948, 11). Krause was most likely following part of the same route that the Kwäday Dän Ts'ìnchį person took. Additionally, though the published English-language reports (Krause 1956; Krause and Krause 1981) do not mention that he travelled as far north as the Blanchard River, his detailed unpublished map of this area was later incorporated into a map produced by geologist George Dawson (Dawson 1887, map 3; see Watson 1948, 11).

The explorer Heyward Seton-Karr was the next outsider to leave a record of his travels in the study area. In May 1890, this adventurer followed the well-used Indigenous travel route that went from Klukwan up the Klehini (Seton-Karr 1891a, b). Like Krause, Seton-Karr was likely following the same route as the Kwäday Dän Ts'ìnchį person, though the latter crossed over and entered the Tatshenshini basin, while Seton-Karr did not. Seton-Karr nonetheless provided key information on this travel route to the Tatshenshini basin via the Klehini and the O'Connor rivers. He noted it was easier travelled in the winter, when snowshoes and sleds could be used. He also mentioned meeting a trading party near the pass between these two drainages. This group, which included a lady from Yakutat, told Seton-Karr that it took seven days to travel from their meeting point to Dry Bay (Seton-Karr 1891, 80–82). Seton-Karr also learned that various non-

Aboriginal people, including prospectors, were in the lower reaches of the Tatshenshini basin around this time, and from one of them he learned that the Chilkat Tlingit had storehouses in the basin where they stored goods that they traded with the Dän (Seton-Karr 1891, 82).

Glave and Dalton's 1890 Trip Down the Tatshenshini

THE UPPER AND MIDDLE REACHES of the Tatshenshini basin were first explored by non-Aboriginals in 1890 by Americans Edward Glave and Jack Dalton, who were part of an exploration party sponsored by *Frank Leslie's Illustrated Newspaper* (Glave 1890–91). A Chilkat man guided their expedition over the first leg of their journey from Klukwan to Kusawa Lake, following the eastern variant of the Chilkat Trail (see chapter 32). At Kusawa, the expedition split into different parties; most of the group travelled north, while Glave and Dalton headed west. At Frederick Lake, a small water body west of Kusawa Lake, they met a Dän family. These people accompanied the explorers to the Tatshenshini basin, visiting several Dän settlements along the way (Glave November 22, 1890, 286) (see chapter 10).

When the Glave and Dalton party reached the main Tatshenshini River settlement of Shäwshe, everyone was away fishing downriver. Accompanied by the Dän family, the explorers continued downstream on foot, as was the practice. Local people normally walked the first stretch of the river downstream from Shäwshe because it was too dangerous to travel by boat through the lower Tatshenshini River canyon (Jackson 1993; Adamson 1993).

Downstream from the canyon the party met the local people at various summer fishing camps. At one of these fish camps they obtained the boat that took them to Dry Bay. It was a cottonwood dugout canoe, roughly six to seven metres in length and one metre wide. Glave reports that two guides were engaged for the trip downstream: Shank, who was reportedly from the Nua qwa people, and an unnamed individual, who is

understood to have been an Aboriginal doctor, that is, a shaman. CAFN Elder Jimmy Kane, who was a young boy and present at the time of Glave and Dalton's trip, reported that his father's brother was the second man that accompanied the explorers down the Tatshenshini River to Yakutat in 1890. Dalton spoke Tlingit, a factor that must have facilitated negotiations for the guiding services; beads were requested and received in payment for the trip (J. Kane in Waddington 1975).

Slightly farther downstream, Glave and Dalton's party stopped at the abandoned village of Núghàyík. Downstream from this settlement they met the chiefs or headmen of the local Crow and Wolf clans, who were in the area with their families for summer fishing. Glave sketched several camp members, including Ick Ars, described by Glave as a Chief; Kooseney, wife of Ick Ars; War Saine, reported as Second Chief; Kaishar; Gunar Arcku; Geunaaya; and two children, Yute Kutu and Jetejoo (Glave December 13, 1890, 352). Most of the individuals sketched by Glave were later identified by the CAFN community. For example, the child Jetejoo would later become known by the English name Jimmy Kane (Waddington 1975; transcribed as Ch'ädäwu in CAFN, Gaunt and Greer 1993, 61). While Kooseney, the mother of Big Jim Fred, was a resident of the study area at the time of Glave's visit, she was originally from the coast and had married into the interior (Jackson 2006, where her name is spelled Kwäsnę). The head of the Wolf clan, whom Glave named as Ick Ars may have been the father of the previously quoted Mrs. Pringle (McClellan 2007a).

Leaving behind the families that were fishing along the middle reaches of the Tatshenshini, Glave and Dalton continued their trip downriver, accompanied by their guides. They reported seeing abandoned structures at the settlement named Tin Char Tlar, with even the trails around this empty village described as being "all grown in" (Glave December 27, 1890, 396). It is noteworthy that Glave reports no other settlements between this place and the confluence of the Alsek and Tatshenshini rivers. He refers to the main stem of the river as the "Kaska Wurlch", noting that there were no Aboriginal people living on its banks. But he adds that "during the winter months, bands of Indian hunters and trappers repair there in search of the beaver, otter, bear and moose" (Glave December 27, 1890, 396). Presumably, his guides provided the latter information.

On the lower reaches of the Alsek, Glave described the river as flowing amidst giant ice fields (Glave January 10, 1891, 438). There is no mention of places where the river was blocked by ice; by this date it appears that the blockages which had caused downstream travellers to run their boats under the ice, as recounted in the stories of the Dry Bay people (de Laguna 1972, 87; Marks in Dauenhauer and Dauenhauer 1990, 195), no longer existed. Eventually, Glave and Dalton's party made its way to Dry Bay and then Yakutat. There is no further account of the travels of the two guides, though it is likely they made their way back upstream on foot, as was the common practice.

The following year, Glave and Dalton returned again to Tatshenshini country on horseback. This time they came directly via the western Chilkat route—that is, from Klukwan via the Klehini River to the upper Tatshenshini basin and Shäwshe (Glave 1892a, 1892b). Two years later, Glave established a trading post in Dän country.

Changes in Clothing and Technology

The arrival of the first outsiders into both the greater and local study areas brought many changes to the lives of the Dän and Tlingit. It is worth reviewing how clothing as well as other items of material culture shifted with the arrival of the historic period. Consideration of these changes provides background information for the chapters that report on the belongings and artifacts recovered from the Kwädąy Dän Ts'ìnchį site and surrounding area.

For both the northern Tlingit and the Dän, skin and fur clothing was the norm in pre-contact times. The Tlingit also made and wore items of clothing woven from the hair of mountain goats, and from plant fibre. Dry Bay area sources told de Laguna that

Men's everyday summer dress consisted of fur capes or shirts of deer and seal skin and sometimes included loincloths.... In summer they usually went barefoot. The head was covered with plaited waterproof hats. (de Laguna 1972, 436)

The clothing worn by the northern Tlingit and the Dän obviously reflects the local climatic conditions they lived in coastal residents had to deal more with rainy conditions, and thus used hats made of spruce root to keep the rain off, or waterproof shirts made from animal gut (de Laguna 1972). Residents of the interior dealt with conditions of extreme cold for many months of the year, and were used to wearing layers of hide and fur clothing.

The only known drawings of local study area people in skin clothing are those in the reports of the explorer Glave (1890–91). These illustrations, based on summertime observations, show women in long skin dresses and men in skin shirts and leggings. From other sources we know that the latter was a one-piece outfit that combined pants and footwear (Van Stone 1982). We note, though, that all of the Tatshenshini River people shown in the lantern photographs taken by the Glave expedition are wearing fabric rather than skin clothing; the former would have been obtained in trade from the coast.

Metal working was not unknown to the Tlingit and Dän before the arrival of Europeans. The latter group had discovered and independently developed its own ways of working native copper (see chapter 24). Living on the coast, the Tlingit started working with iron before the arrival of the Russians; they extracted pieces from driftwood that made its way to their shores (de Laguna 1964, 88–89).

Dalton Post and the Dalton Trail

MORE DIRECT ACCESS to the new items of material culture came in the mid-1890s, when Jack Dalton established the Tatshenshini basin's first trading post. Dalton Post, as it came to be known, was located at the mouth of the Klukshu River, a short walk from Shäwshe, the basin's principal Indigenous village.[9]

Dalton upgraded the western branch of the Chilkat Trail, specifically the part of the route that went directly from Klukwan to Shäwshe, and from this settlement on to other points in the interior. He operated this improved trail as a toll road, with his Dalton Post store serving as a supply point for prospectors moving into the interior in the 1890s (Gates 2010; J. Kane in Waddington 1975; Kane 1979). During the Klondike gold rush of 1897–98, the trail became known as Dalton's Trail (later as The Dalton Trail). It functioned as an important freight route during this period (Gates 2010; Thompson 1905; Waddington 1975; Yukon Archives 1985).

Much of the route of the modern Haines Highway between Klukwan and Shäwshe closely follows that of the Chilkat or Dalton Trail. The northern section, however, Blanchard River crossing to Shäwshe varies significantly from the old route, in that it is situated on the eastern rather than western side of the Tatshenshini River. Regardless, Dalton's route was based on the old foot trail between these two settlements, which elsewhere in this book is referred to as the western variant of the Chilkat Trail, and which Champagne and Aishihik people refer to as Alur Dän Tan—literally "coast peoples trail"—meaning the trail they travel on (see chapter 32).

The Klondike Gold Rush Years and Beyond

The growing influx of non-Indigenous people into the Tatshenshini basin in the late 1890s triggered the arrival of the North West Mounted Police in 1897. The force was sent in to keep the peace, and a detachment was established at the international US–Canada boundary. In 1898 a second post was created by Dalton's store. Winter patrol cabins were also built along the Dalton Trail at Bear Camp, Glacier Camp and Rainy Hollow (Jarvis 1899; Wood 1899; Yukon Archives 1985). By 1906, however, freight and toll traffic over Dalton's trail had decreased dramatically and all of these posts and patrol cabins were closed. The decrease in traffic occurred with the end of the rush to Dawson and the opening in 1900 of the nearby White Pass and Yukon

Railway, which provided access from the coast to the interior via a Skagway-Carcross-Whitehorse routing.

The Shäwshe people are thought to have made the last of their trips down the Tatshenshini River to Dry Bay around the turn of the 20th century (Cruikshank et al. 1990, 159). According to de Laguna (1972, 214), Dry Bay people likewise reported that there was marriage and trading up and down the river until around 1900–10. Another source reports a slightly later date for this visit. Around 1916, Dry Bay resident Frank Dick made the trip upriver to Shäwshe; Mr. Dick went from here on to Whitehorse, and then took the White Pass and Yukon Railway to Skagway, from which point he made the final leg of his journey home by boat (Dauenhauer and Dauenhauer 1987, 450).

In addition to prospectors, government surveyors and geologists also descended on the study area around the turn of the 20th century. They came to determine the area's mineral potential and to formally establish the Canada–US border as well as the Yukon–British Columbia boundary line (ABT 1904; Davidson 1903; de Laguna 1989; Green 1982; McArthur 1898; Tyrrell 1957; White-Fraser 1901; Cruikshank 2005, 222–29). As Cruikshank has noted, no effort seems to have been made to ask the local Indigenous people where they thought these lines should be situated or, indeed, if they should be recognized. As discussed below, these artificial markers would have profound effects on the lives of the local residents.

The presence and activity of the newcomers in the Tatshenshini basin had slowed considerably in a few short years though, causing Dalton to close his store in 1902. This, along with the previously mentioned newly imposed boundaries and ensuing government hunting and trapping regulations that now began to be enforced, severely impacted the families based at Shäwshe. A number moved their headquarters to Champagne because of the resulting hardships (see McClellan 1975a, 25).

It is thought that by this time there were few, if any, people living in the lower and middle portions of the Tatshenshini basin, perhaps only the old Tlingit man named Kálawà, who stayed around the mouth of the

O'Connor River. The upper and Yukon portions of the basin had more inhabitants, but the population of these areas was also decreasing. Families were shifting their residence and land use areas either to the north, into what was now known as the Yukon Territory, or south into the Haines area, as we recount next.

The Early 20th Century

DURING THE EARLY DECADES of the 20th century, the local study area was home to mostly Aboriginal people. The few non-Indigenous people living here were either men involved in mineral exploration or those who had married into the First Nations community (Hume 1979; Watson 1948; McClellan 1975a). The trading post had closed and visits from the North West Mounted Police were rare. Missionaries made only occasional summer visits to settlements such as Klukshu (Jackson 2006). Local residents were nonetheless affected by world events, such as the 1918 flu epidemic, which claimed many lives locally (Greer 2004).

The people continued with their subsistence activities of hunting, fishing and trapping in the surrounding area, but with their houses at Shäwshe functioning as home bases. By this time the last of the Núghàyík people had joined with the Shäwshe community (Clarke [1940s]). The Crow leader of this group from Núghàyík was the man known as Big Jim Fred, who is the stepfather of CAFN Elder Marge Jackson and her siblings (Clarke [1940s]; Jackson 2006). It is understood that Johnny Fraser assumed this leadership role after the passing of Big Jim.

While the Crow and Wolf clan leaders and their families had their headquarters at Shäwshe, families headed by non-Aboriginal men who had married local women lived in another location. Their residences were at the Dalton Post site rather than at the old Shäwshe village. They used the buildings left behind by Dalton and/or the North West Mounted Police as their residences. Some also built their own homes here (CAFN, Gaunt and Greer 1993; CAFN and Greer 1998; O'Leary 1984). Later, particularly once the old village

site ceased to be used as home base by any families, the settlement at Dalton's former trading post location would take on the Shäwshe name.

Many of the families based at Shäwshe in the early 20th century continued with the 19th-century practice of going downriver to meet the first salmon runs of the season. This was in the spring, when they were "hungry for fresh fish" (Jackson 1993; Kane 1978). It is not certain how far downstream they travelled during this early-20th-century period. Stories recorded from elders who were children in the 1920s and 1930s indicate that they are familiar with the country at least as far downstream as the O'Connor River. They still gaffed salmon at Kudwat Creek and stayed at Datl'at, a camp near Núghàyík, during this period (Jackson 2006; O'Leary 1992, 65, 67). Smaller hunting, prospecting and trapping parties went at least to the confluence with the Alsek River (Kane 1978; Brown 1996).

The families who had shifted their headquarters to Champagne still came back to Shäwshe each summer during the first half of the 20th century (CAFN, Gaunt and Greer 1993, 1995; McClellan 1975a). Their visits were timed as they had been in the past, to coincide with the various salmon runs in Klukshu and Village creeks. The connections between the Shäwshe people and the Champagne people were strong throughout this period. In the middle of the century, Johnny Fraser—who described himself as the last of the Noogaayík (i.e., Núghàyík) people—would be recognized as the chief of Champagne/Aishihik Indian Band, as it was then known (McClellan 1975a, 2007a).

Some of the basin's families also moved south to Haines during these early decades of the 20th century. Those that moved in this direction appear to have been drawn there for different reasons from those who shifted their land use area north into the Yukon. For example, one family living at Klukshu in the 1910s moved to Haines so that their daughters could go to school (H. Rappier in Greer 2008). Two families living at the Shäwshe/Dalton Post site—the Claytons and the Browns—headed by former North West Mounted Police men married to First Nations women, were drawn to

Haines by work and schooling opportunities (CAFN and Greer 1998). Sadly, children of another Shäwshe family were placed in a Haines orphanage after the passing of their mother (Adamson 1993).

In the early 1920s, tragedy struck the Shäwshe community when three residents of the old village site died of poisoning from eating tainted fish heads (see chapter 10). The deceased included the wife, daughter and sister-in-law of Crow clan headman Big Jim Fred. As a result of this unfortunate event, people stopped staying at the old village site. Instead, they shifted their place of residence either to the new village site of Dalton Post-Shäwshe or to nearby Klukshu village (CAFN, Gaunt and Greer 1995; Jackson 2006; F. Joe in CAFN 2010). The settlement, once the basin's main village and the central gathering place for trade and visiting, was now no longer home to any families. Individuals still came here seasonally to catch salmon when they were running, but no one lived at the old settlement.

During this period, people continued to use the trail improved by Dalton to travel back and forth between the Klukwan-Haines area and Shäwshe, as well as to other points in the Yukon interior (Adamson 1993). A wide array of goods from the "outside" was brought into the area, and other points further north, on horseback via the Chilkat Trail/Dalton Trail route. Lumber from Haines—used in the construction of the Wolf and Crow clan houses at Shäwshe, as well as numerous grave fences—was brought in. Even carved headstones shipped north from Victoria came in by this route (CAFN Heritage site files).

Early 20th-Century Mining Activities

WITH AN INTIMATE KNOWLEDGE of the basin's resources, including mineral resources, many local men became actively involved in prospecting in the early 20th century (Kane 1978; Hume 1979). This practice was consistent with past traditional land use activities. The early 20th century saw Shäwshe men investigating the mineral potential of creek basins throughout both the Yukon and BC portions of the Tatshenshini-Alsek

drainage (Hume 1979; Gaunt 1994; Mandy 1992). Copper had first been discovered at Rainy Hollow in 1898, and more finds were reported in 1908–09 (Watson 1948, 12). In 1914 gold was found on the Klehini. While these discoveries attracted as many as 1,000 prospectors to the Dalton Trail area, by 1916 the claims were inactive. These would be reactivated later in the century (ibid.).

Local Aboriginal men transported goods for many of these prospecting efforts, continuing to play an important role in the regional transportation system—just as they had in the late 19th century. Using the old trading trails, including the one upgraded by Jack Dalton, the local men moved prospectors' goods into the Tatshenshini country from the Klukwan and Haines area (Adamson 1993; Kane 1990; Gaunt 1990). Packtrains operated along the route now followed by the modern highway. The routes from the Klehini River to the Parton River and then north toward Shäwshe, or west to upper Low Fog Creek and the middle reaches of the Tatshenshini River, were also used during this period.

The year 1927 saw an influx of people into the Tatshenshini basin. That year local resident Paddy Duncan discovered gold about 19 kilometres (twelve miles) south of Shäwshe in the British Columbia portion of Squaw Creek. With practically all the staking of the Squaw Creek rush done by local residents (Mandy 1933; Mandy 1992), the next few years saw many Champagne and Aishihik families shifting the focus of their land use activities back to the basin. The men worked the mining claims in the creeks and operated the pack trains that brought in supplies. A store also operated for a period on Squaw Creek (Adamson 1993; Gaunt 1990, 1991; Jackson 1979; Kane 1990; Mandy 1933; Mandy 1992). While the men carried on with these activities, the women and children stayed at Shäwshe/Dalton Post to put up fish and carry on the other regular activities of their land-based, subsistence-oriented way of life (CAFN, Gaunt and Greer 1993).

Government Involvement in the Area

IN THE 1930S, the British Columbia government instituted a system of registered traplines for this area of the province. Registered traplines were in conflict with the Aboriginal system of land tenure, which saw the clans as having stewardship responsibility for specific resources and resource areas. The Indigenous system also recognized different family land use areas. The provincial system penalized trappers for following the traditional practice of not trapping an area for a period in order to allow the animal population to rebound. Administered out of an office on the Stikine River—more than 400 kilometres distant as the crow flies (and over twice that distance by road)—the registered trapline system brought about additional problems.

The earliest government map showing registered traplines in the BC portion of the Tatshenshini basin and adjacent Chilkat River basin indicates two group trapping areas. The country west of about 137° longitude is indicated as belonging to the "Alsek Indian Band", and east of this meridian, to the "Dalton Post Indian Band" (Stikine Agency, November 30, 1933). In 1937, the BC Game Commission cancelled the trapline belonging to the "Alsek Indian Band" because "these Indians are residents of Yukon Territory" (E. Martin, BC Game Department letter to Indian Agent H. Reed, Telegraph Creek, November 2, 1937). As previously noted, by this date Shäwshe and Klukshu, both located within the Yukon, were the only Tatshenshini basin villages being used as home bases by the Indigenous community. The Indian Commissioner for the province, D.M. Mackay, tried to have this decision overturned, writing:

> … it is well known that many of the Indians of the Stikine Agency are nomads and much of their time is spent in Yukon Territory, but this is hardly a sufficient reason for the action taken by Dr. Martin and I would ask that this line be reinstated in the name of the Alsek Indian Band. (Mackay letter to BC Game Commission, December 14, 1937)

Mackay's efforts were in vain, and the following year, the trapping grounds belonging to the Dalton Post people were similarly cancelled for the same reason (Indian Agent H. Reed to D.M. Mackay, January 18, 1938). This meant that it was now illegal for the Shäwshe people to trap in the BC part of their traditional territory. Over the next few decades, these traplines were allotted to non-Aboriginal people. Despite these regulations, CAFN people such as Jimmy Kane and David Hume continued to catch furbearers in this part of their homeland, often accompanied in their travels by younger CAFN community members (Adamson 1993; Banks 1996; Brown 1996; Gaunt 1996; Greer 1995). Hunting also continued in this part of the traditional territory, with Moose, Caribou and sheep being caught and processed for storage and eventual distribution within the community.

Within the decade, further changes were forced upon the First Nations community with the building of the Alaska Highway and the connecting Haines Highway (between Haines Junction, Yukon, and Haines, Alaska). The creation of these new travel routes resulted in the establishment of Kluane Game Sanctuary, which eventually became Kluane National Park (McCandless 1985, 80). People were not allowed to hunt in these protected areas—a major blow for the Shäwshe residents, as it deprived them of a key hunting and trapping area situated within a few kilometres of the home village.

By the 1950s the impact of government-imposed subsistence and trapping regulations within BC and Yukon made it more difficult for Shäwshe people to live off the land full-time as their ancestors had done. In addition, government-provided school, medical and housing services, stores and other private businesses along the Alaska Highway, as well as opportunities for wage labour (Cruikshank 1982), motivated the last of the families living at Shäwshe to move to Haines Junction or Klukshu. The latter settlement now became the principal Indigenous fishing village of the upper Tatshenshini River basin, a role that it continues to play today (CAFN, Gaunt and Greer 1993; CAFN 2010). This residence shift, however, further reduced

Figure 6. Glave sketch of late 19th-century Tatshenshini River village resident. The woman has been identified as Kwäsnę, mother of Big Jim Fred (Jackson 2006). Kwäsnę was born on the coast and married into Nu gha hit settlement; after the latter was abandoned, she lived at Shäwshe/Nesketahin. *Frank Leslie's Illustrated Newspaper*, December 13, 1890.

Indigenous use of the middle and lower portions of the Tatshenshini basin.

Throughout the early decades of the 20th century, Klukwan residents continued to hunt and trap in the local study area. Goldschmidt and Haas mapped the boundary of early 20th-century Chilkat land use as going beyond the Canada–US border, up the Kelsall and Chilkat rivers to the area of Kelsall Lake on the Haines Road, and up the Klehini River. They report that this area was primarily used for hunting and trapping (Goldschmidt and Haas 1998).

Through the last half of the 20th century, Champagne and Aishihik citizens continued to hunt, fish, trap and prospect for minerals in the BC portion of their traditional territory, never giving up their rights to this part of their country. In the 1990s, CAFN, the Province of British Columbia and the Government of Canada attempted to negotiate a treaty that would address CAFN's Aboriginal claim to these lands and rights following therefrom. As discussed elsewhere in this volume (see chapter 7), when these negotiations failed, efforts shifted to reaching a co-management agreement for the newly created Tatshenshini-Alsek Park. Today, many families have camps that they use on a regular basis, and harvesting activities continue in this part of the traditional territory, as they have for generations (Greer 1995; O'Leary 1992).

Conclusions

TODAY MANY VIEW the northwest corner of British Columbia as untouched, uninhabited wilderness, but this is not the case. Since time immemorial and well into the 20th century, the Tatshenshini-Alsek basin's rich salmon resources have supported families who had an intimate knowledge of not only the region's fish and wildlife, but also a wide range of resources that included minerals like flint, copper and marble. All parts of the study area—highland and lower elevations, glacial and non-glacial—were used in times past. An extensive network of foot trails and travel routes criss-crossed the region, including a known travel route that went from the upper Klehini River area northwestward to the Tatshenshini River. The Kwäday Dän Ts'ìnchį person may have been following at least a portion of this route when he lost his life.

Radiocarbon dates indicate that the Kwäday Dän Ts'ìnchį individual lived *ca.* 1720–1850 AD (Richards et al., chapter 6). It is uncertain, though, just when, during this 130-year interval, he actually lived—that is, whether in the decades before the arrival of the European exploring ships in 1741; during the early days of trade with the newcomers (i.e., 1740s–1800); or in the period when the trade in inland furs was fully operational (i.e., 1800–67). Other evidence, including his belongings (see chapter 6) and the presence of trade metal in the man's knife, suggest that he likely lived after 1740, when the foreign trading ships sailed in nearby coastal waters.

The historic era brought many changes to the lives of the Indigenous peoples of the greater study area. Assuming he lived during early historic times, there are reasons that the Long Ago Person Found was travelling in the area where he died. He may have been

Figure 7. Tatshenshini basin Dän/Tlingit, taken at Shäwshe/Nesketahin, 1890s. CAFN Heritage Collection.

a messenger heading to one of the settlements, such as Núghàyík, which the Chilkat Tlingit established in the basin. Or perhaps he was heading to a Dän settlement for other reasons. It's also possible that he was travelling to one of the Tatshenshini River settlements that was devastated during the 18th and 19th centuries by disease or flooding. The latter situation might explain why his remains were never retrieved, as occurred in other circumstances where individuals disappeared while crossing glacial landscapes (see chapter 32).

Connections to the Shäwshe (Nesketahin) People

Change is clearly a theme in the story of the Long Ago Person Found, but so, too, are continuity and connections. Such connections were demonstrated in part by the results of the community DNA study (see chapter 33), which showed that the Kwädąy Dän Ts'ìnchį individual is not a person of uncertain identity, but rather one connected to individuals of Aboriginal ancestry living in both Alaska and Yukon, if not communities further afield. Even closer connections can be recognized in the pool of identified maternal relatives. Six of the seventeen individuals identified as being related to the Long Ago Person Found through the maternal line are descended from people that lived at Shäwshe around the turn of the 20th century (fig. 7).

The six Shäwshe descendants who are related to the Long Ago Person Found can all be placed on the same extended family tree chart. This same chart includes Big Jim Fred and Paddy Duncan, who were the villages, Crow and Wolf clan leaders, respectively, during the early 20th century. As we do not know precisely when the Kwädąy Dän Ts'ìnchį individual lived and died, it is impossible to establish the number of generations that separate him from people like Paddy Duncan, or any of the people residing at the village at that time. There may have been as few as two or as many as seven generations between him and people of the historic Shäwshe (Nesketahin) population.

It is possible that the individuals shown in the figure 7 photograph knew the story of the young man who was lost while travelling in the mountains between the Klukwan area and the middle reaches of the Tatshenshini River. The connections between the tragedy of the past—a young man's death on the glacier in Tatshenshini-Alsek Park several hundred years ago—and the present descendants, who trace their history to this area, are very real. So, too, are the traditional stories that tell of the shared history between the Tlingit and the Dän of the Tatshenshini-Alsek country. The Kwädąy Dän Ts'ìnchį discovery brings this history and all these stories to life.

Postscript

IN THE DAYS before the Haines Road, people travelled from Shäwshe down to the Haines and Klukwan area by foot. CAFN Elder Mrs. Annie Ned recalls that when travel parties got above treeline, they would sing a south wind song, to bring this warming breeze (in Cruikshank et al. 1990, 331). It is more than likely that the young man who lost his life on a glacier would have known a similar type of travel song. The Dákwanjè version goes as follows:

Nis'i dhal tl'e
(wind blows warm)
Nis'i dhal tl'e
(wind blows warm)
Nighra ke kwätu däjela
(your sons summit they went)
Nighra ke kwätu däjela
(your sons summit they went)
Nis'i dhal tl'e
(wind blows warm)
Nikaghwa kudejela
(for you they are travelling)
Nis'i dhal tl'e
(wind blows warm)

ACKNOWLEDGEMENTS

The present chapter is an updated version of a report on the ethnohistory of the Tatshenshini-Alsek area prepared in the mid-1990s by authors Sheila Greer and the late Sarah Gaunt (CAFN, Gaunt and Greer 1995). The 1995 report built upon an earlier effort by these same two researchers (CAFN, Gaunt and Greer 1993), which drew heavily on work by pioneering researchers whose data are reported in this chapter. We thank as well the numerous First Nations and tribal elders who provided information to researchers Emmons, de Laguna, McClellan, Cruikshank and O'Leary, and those who shared information with Gaunt and Greer. We cite in particular Elders Marge Jackson and the late John Adamson. We also wish to thank past and present Chiefs of Champagne and Aishihik First Nations Paul Birckel, Bob Charlie, Diane Strand, James Allen and Steve Smith for their support for the research presented here. Lawrence Joe, former director of Heritage, Lands and Resources for CAFN, has provided guidance through this and related studies.

Margaret Workman and Jeff Leer assisted with the Dän and Tlingit place name data presented in the 1995 report; they are not responsible, however, for any shortcomings in how these data have been used in this chapter. We also thank John Ritter of the Yukon Native Language Centre for the translation of the Núghàyík place name. We wish to acknowledge as well the numerous colleagues and friends who have shared resources and provided insight on the history of the study area. These include Paula Banks (formerly with CAFN Heritage, Lands and Resources); Mike Crawshay (a Haines Junction resident who trapped in the Tatshenshini basin); Lance Goodwin (a Haines Road resident who currently traps in the Tatshenshini basin); Marsha Hotch (Klukwan Indian Village); Wayne Howells (Glacier Bay National Park); Mike Jim (CAFN Tatshenshini-Alsek Park warden); Judy Ramos (Yakutat Tlingit Tribe); Al von Finster (Whitehorse resident, formerly with Department of Fisheries and Oceans Canada), and Linaya Workman (formerly with CAFN Heritage, Lands and Resources).

CAFN has provided support for this research over the two decades it has been ongoing. Additional funding in support of the 1996 research came from the British Columbia Heritage Trust, and for research post-dating that effort, from the International Polar Year research program (Government of Canada). Other agencies whose funding contributions have supported various stages of the work reported in this chapter include Communications Canada (Access to Archaeology Program); Environment Canada (National Historic Sites; Canadian Heritage Rivers Secretariat); Energy, Mines and Resources Canada (Research Agreements Program; Place Names); Yukon Parks and Outdoor Recreation; Yukon Heritage Branch; Northern Research Institute, Yukon College; and Yukon Foundation. Support in kind has also been provided by Kluane and Glacier Bay national parks.

Table 1. Historical chronology, Tatshenshini basin.

Period	Date	Event or Situation
Prehistoric or Pre-contact	Ancient times	Dän living in basin. Some trading with coastal peoples.
Protohistoric	1741	First Europeans visit the northern northwest coast, trade in Sea Otter pelts starts. Tlingit begin expanding from Hoonah and Chilkat country into the Dry Bay/Alsek area.
	1775	First smallpox epidemic hits southeastern Alaska
	1794	Chilkat Inlet first charted by Europeans.
Historic	*ca.* 1800	Trade in inland furs begins. Chilkats begin expanding into the Tatshenshini basin. Measles epidemic. Conflicts between Chilkats and visiting traders on ships.
	1819	Typhoid epidemic.
	1836	Smallpox epidemic. It is estimated that 25–50% of the population of southeast Alaska may have died.
	1839	Hudson's Bay Company obtains trading rights in southeastern Alaska and more trade goods enter area. All trade is from boats, as Chilkats prohibit non-Aboriginal men from landing.
	1840–64	Height of the Aboriginal fur trade. 1848 typhoid epidemic. 1852 Chilkat Tlingit destroy Fort Selkirk. 1852 Flood destroys village at confluence of Alsek and Tatshenshini rivers. 1855 typhoid epidemic. 1862 smallpox epidemic.
	1867	US purchases Alaska from Russia
	1869	Chilkat trading chief Kohklux draws map showing interior trading routes and Dän villages in the interior.
	1879	Chilkats remove their prohibition of non-Aboriginal people in their territory.

Period	Date	Event or Situation
Historic	1882	Explorer Arthur Krause is first non-Aboriginal person to enter local study area.
	1890	Explorers Glave and Dalton descend the Tatshenshini River, meeting Dän and Tlingit families. They note abandoned settlements.
	1894–96	Jack Dalton establishes first non-Aboriginal trading post in the basin and upgrades the trail from Klukwan to the interior.
	1897–1900	Klondike stampeders move through area. Dalton Trail operates as toll road. NWMP presence in area. Mineral exploration.
	1898–1903	Survey and determination of the international border and provincial-territorial boundary.
	1902–06	Dalton's store closes. Police pull out of area. Some families move to Champagne.
	1910s	Some Aboriginal families move to Haines.
	1918–20	1918 flu epidemic.
	1923	Poisonings occur at old Shäwshe village.
	1927–35	Gold discovered on Squaw Creek; some Dän and Tlingit families move back to Shäwshe.
	1942–43	Haines Road built by US army. Kluane National Park and Game Reserve established. Last of the Aboriginal families based at Shäwshe move to Klukshu or Haines Junction. Hunting and trapping continues.
	1950s–80s	Prospecting, hunting and trapping throughout area.
	Late 1980s–90s	Large mine development proposed. Recreational river rafting begins.
	1993–95	BC government establishes park in Tatshenshini-Alsek basin.
	1999	Kwädąy Dän Ts'ìnchį remains discovered.

10

WHERE MIGHT HE HAVE BEEN HEADED? THE TATSHENSHINI RIVER VILLAGES

Champagne and Aishihik First Nations, Sheila Greer and Sarah Gaunt

Various lines of evidence have given us a reasonably good idea of the general area where the Long Ago Person Found started his last journey (Dickson et al. 2004; Mudie et al. 2005). Based on this, we know his probable direction of travel, but our understanding of his trip's purpose and intended destination is less certain. The abundance of artifacts found in the discovery site area, more than would be transported by a single individual (see chapter 23), suggests some of these pieces may have been left by other travellers, and therefore that the Kwäday Dän Ts'ìnchį individual was on a regularly used travel route. We also know that Tlingit from the Klukwan and Haines area were taking up residence in the Tatshenshini River basin during the Long Ago Person Found's time, by either marrying into Dän communities or establishing their own settlements (see chapter 9). It is therefore reasonable to suggest that the Kwäday Dän Ts'ìnchį individual's intended destination might have been one of the settlements on the Tatshenshini River or its tributaries (fig. 1).

TYPICAL HABITATION ON THE ALSECK FISHING-GROUNDS.

Figure 1. Nineteenth-century sketch of a Tatshenshini River village (Glave 1890–91).

THIS CHAPTER SUMMARIZES archival and community oral history information on the 19th-century Tatshenshini River villages. While we do not know whether any of the settlements discussed might have been visited by the Long Ago Person Found, the presence of these villages, as well as the density of occupation that existed along this river in times past, must be acknowledged. Today the Tatshenshini-Alsek Park area is considered by many to be wilderness, devoid of human presence. Yet the area has a rich human history that is represented in part by these old settlements. They provide a necessary context for understanding and interpreting the Kwädąy Dän Ts'ìnchį discovery.

We know these villages existed for at least part of the 19th century, but it is uncertain if they were also occupied in earlier times. This is an important consideration, since it is understood that the Kwädąy Dän Ts'ìnchį individual lived sometime between *ca.* 1720 and 1850 AD (Richards et al. 2007; see chapter 6). Insight into the time depth of the occupation of these villages can be gained through archaeological field research and oral history accounts, as recorded from First Nations and tribal elders. Stories shared by elders suggest that at least some of the sites discussed in this chapter do predate the 19th century. This chapter, therefore, relates what was learned about these settlements from archaeological investigations and from the available archival and oral history information. We review and compare these two lines of information and identify potential locations to which the Long Ago Person Found may have been travelling. The various geographic locales referred to in this chapter are indicated in figure 2.

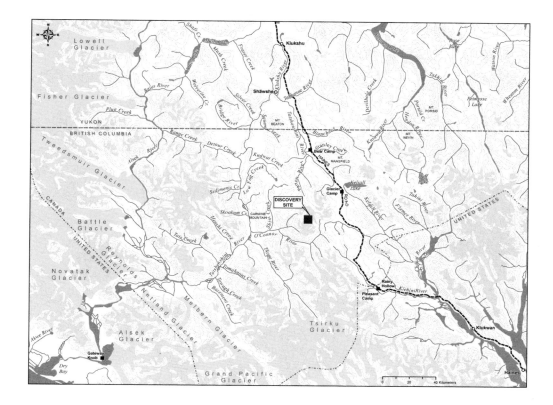

Figure 2. Map of regional setting and location of discovery site, Tatshenshini basin.

Our Sources and Their Limitations

THE SOURCES FOR INFORMATION about the Tatshenshini River villages include the various versions of the Kohklux map (Davidson 1901; Cruikshank 1991; YHMA 1995), and the reports of writer/adventurer Edward J. Glave (Glave 1890–91). Most of the information in stories Glave published in *Frank Leslie's Illustrated Newspaper* also appeared in other magazines the following year (Glave 1892, 1892b). Glave's unpublished journal, which describes his 1891 trip, is housed at the University of Alaska Fairbanks Archives (Davey n.d.). This manuscript, understood to contain information omitted from the published accounts, was unfortunately not available during the preparation of this volume. It has subsequently been published (see Glave 2013).

Oral history and traditional accounts from First Nations and tribal sources are critical in learning about the Tatshenshini River villages. These include published compilations from the local area (McClellan 1975, 2007; Cruikshank et al. 1990), unpublished transcripts in the files of the Champagne and Aishihik First Nations (CAFN) Heritage program and from other Yukon sources, and published sources related to the southeast Alaska Tlingit such as Krause (1956), Emmons (1991), Olson (1967) and de Laguna (1972). As noted elsewhere (see chapter 9), the region's Tlingit and Dákwanjè place names (toponyms) are important sources for information on the old villages.

Little archaeological fieldwork has been done in the Tatshenshini-Alsek area. The most important effort was the 1993 survey for the old villages undertaken by Champagne and Aishihik First Nations. Directed by Dr. Diana French (presently at the University of British Columbia Okanagan), this project spent two weeks in search of signs of the old villages in the BC portion of the basin. Starting out below the main Tatshenshini River canyon at Silver Creek, the crew worked their way downstream in an inflatable rubber raft. They encountered very challenging working conditions; river navigation proved difficult, leaving the crew limited time to spend on shore searching for the sites. The nature of the terrain, the dense vegetation and natural erosion processes presented additional challenges to finding sites (French 1993). This pioneering effort demonstrated that the basin has significant archaeological potential, but also that it would not easily give up the secrets of its past.

We present the information on the Tatshenshini River villages in chronological order. The sources considered strongest and most reliable are the 19th-century first-person accounts from Chief Kohklux and his wives (1859), and from Glave (1890–91), and the information on settlements from CAFN sources, particularly that which can be tied to a specific location.

As travel is an underlying theme not only of this chapter, but also of overall story of the Kwädąy Dän Ts'ìnchį individual, we offer a few comments on travelling in Tatshenshini-Alsek country (see also chapters 9 and 29).

The Kohklux Map(s), 1869 and 1901

THE OLDEST SOURCE showing villages along the Tatshenshini is the Kohklux map, produced in 1869 by Kohklux, a Chilkat Tlingit chief who lived at Klukwan. The map was produced at the request of visiting US scientist George Davidson, who wanted Kohklux to show the routes used by his people to reach settlements visited for trading purposes—including Fort Selkirk, far to the north on the Yukon River, which was the farthest point inland to which they travelled.

There are actually three versions of the Kohklux map. The first version of the map was produced by the chief working on his own but with Davidson's assistance (Johnson 1984). This version, first reproduced in Cruikshank (1991, 114) and later in YHMA (1995) does not indicate any village locations and is somewhat difficult to decipher owing to its small but unspecified size; the landmarks shown are less easily matched with landscape features as we know them today.

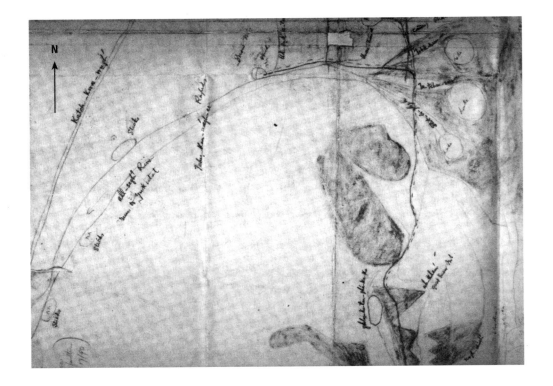

Figure 3. Section of the large 1869 version of the Kohklux map. On the left is the All-segh River (i.e., the Tatshenshini River) which "runs to E of Yuk atat" (i.e., Yakutat). Also indicated is a village near the mouth of the Unahini (i.e., the Klukshu River), which is Shäwshe. Below the latter are the "rapids" and, downstream of the rapids, three additional "Sticks Villages". Adapted from YHMA 1995.

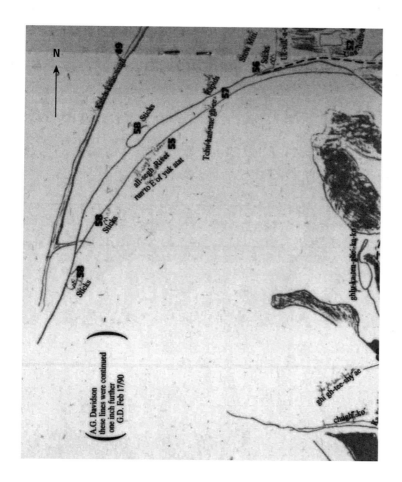

Figure 4. Section of YHMA annotated version of the large 1869 version of the Kohklux map. #49 is Katch-kwae-wugh (i.e., Kaskawulch River), now known as the Alsek River; #52 marks the Una-heen-ae (i.e., Unahini River), now known as the Klukshu River; #55 marks the All-segh River, now known as the Tatshenshini; #56 indicates the "Sticks Village" of Shäwshe; #57 marks the rapids; #58 indicates the three "Sticks Villages" located downstream from Shäwshe. Adapted from YHMA 1995.

The second version of the Kohklux map, drawn that same year by the chief but with assistance from his two wives, is larger in size (109 by 68 cm) and more informative. It has been reproduced and published on an 84 per cent scale (YHMA 1995). In the lower left area of this larger version, the river that we now call the Tatshenshini is clearly indicated. Figure 3 provides a close-up of a section of the Tatshenshini River from the large version of the Kohklux map, and figure 4 shows an annotated version of this same area.

On the large version of the original Kohklux map, the settlements along the Tatshenshini River are identified by the label "Sticks"—a designation referring to Dän settlements (see chapter 9). The settlements are also marked by upside-down Vs, possibly indicating structures, and in one case by a small rectangular roofed

structure. The latter is situated in the approximate location of the Núghàyík site.

The most northerly Dän village indicated on the river is located where the Tatshenshini turns from flowing northwestward to flowing southwestward, around the mouth of a watercourse labeled as the Una-heen-ae. The latter, also written as Unahini, is the Tlingit-language toponym for the Klukshu River (McClellan 1975; McArthur 1898). According to CAFN Elders Moose Jackson and Marge Jackson, the Unahini place name is a corruption of Gunana hini, Gunana being the Tlingit-language word for the Dän, meaning "stranger". The name therefore translates to "Indian" or "Stick Indian" Creek (Greer 1995, 1997). The village indicated here is known in Dákwanjè as Shäwshe; its Tlingit name is Nesketahin (also spelled Neskataheen).

Figure 5. Davidson's 1901 version of the Kohklux map. From Davidson 1901.

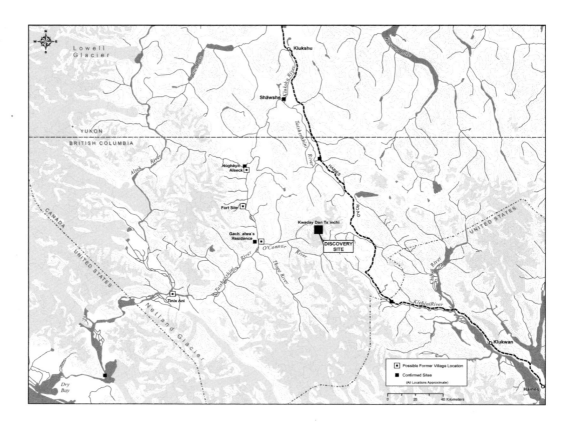

At the end of the 19th century this settlement also began to be referred to by the English name Dalton Post after the trading post established here.

The major canyon and rapids of the Tatshenshini River are shown on the large Kohklux map downstream from Shäwshe, with three "Sticks Villages" indicated below the rapids (see figs. 3 and 4). One settlement is situated on the west side of the river, around where it begins to flow largely southward. This settlement might be the village known as Núghàyík, which is located on the west side of the Tatshenshini River just south of the BC–Yukon boundary near the mouth of Detour Creek. Alternatively, it could be the village known as Alseck (or Alsek), which was reported by the explorer Glave as well as by Klukwan and Champagne and Aishihik oral history sources. Both of these settlements are discussed further below.

Downstream from this village, Kohklux and his wives marked two additional Dän settlements, both on the east side of the river. No additional information is provided to further pinpoint these settlements, other than their location above the confluence of the Tatshenshini and Alsek rivers.

In 1901 Davidson published a variant of the map, which is here considered the third version (fig. 5). Davidson's annotated version includes additional information not indicated on the two versions produced in 1869. For example, beside the "Sticks Village" marked on the west side of the river (below the rapids) is a place labelled "Log Hut". The name "Tin char har" appears as well in this area of the map. The log hut identifier also appears in Glave's report (1890–91), described below, as does a similar place name: Tin Char Tlar. We therefore might presume that Davidson's additions to the map were informed by Glave's published account of 1890–91 or by sources familiar with the same information. The possible location of the fort site—Tin char har or Tin Char Tlar—is considered further below.

The Davidson (1901) annotated version of the Kohklux map also provides more information on the villages situated further downstream, on the east side of the river. The northernmost of these two villages is on the north side of a major tributary that enters the Tatshenshini from the east in a more or less direct line. This tributary—indicated as a travel route overland to the Klehini River—must be the O'Connor River, the Tatshenshini tributary whose headwaters abut those of the Klehini (see fig. 2).

Champagne and Aishihik sources concur that there used to be a settlement around the mouth of the "Boundary River", which is a local name for the O'Connor (McClellan 1975, 581; Adamson 1993; Jim 1996; Moose Jackson in Greer 1995; Watson 1948, 15). McClellan reports, citing unspecific sources, that the name of the settlement by the O'Connor was Si'diq, and that in earlier times it had been occupied in the spring for the King salmon harvest.

The 1993 archaeological survey found a stratified pre-contact-era archaeological site with stone tool remains close to the mouth of the O'Connor; remains representative of a 19th-century village site were not detected at this locale, however (French 1993). Nonetheless, a large area of flat ground on the north side of the river mouth has been recognized as a possible setting for a village, and well-weathered axe-cut tree stumps have been observed here (M. Jim, communication to Greer, August 2009). The location has been wetland due to beaver activity for some time, making it difficult to undertake any kind of serious archaeological assessment.

On the Davidson version (1901) of the Kohklux map, the more southern village on the east side of the river is plotted in the area where the Tatshenshini turns from flowing south to flowing west. This places the settlement somewhere between the mouth of the O'Connor and the confluence of the Alsek and Tatshenshini rivers, perhaps in the area of Ninety-Niner and Melt creeks (French 1993). The source(s) for Davidson's information regarding the two villages on the river's east side remain(s) unknown.

The Davidson version of the Kohklux map includes another detail that is not indicated on the 1869 map versions. This is a trail going overland from the Klukwan area to the Tatshenshini River through the general area where the Kwädąy Dän Ts'ìnchį discovery site is located. We note that while the large Kohklux map has a line marked in this same general area, it is far from certain that this line marks a trail or travel route. Again, Davidson's information source for the trail between Klehini and Tatshenshini is uncertain. Perhaps he learned of the existence of a trail in this area from either the explorer Heyward Seton-Karr (1891) or the original reports of the Krause brothers (see Krause and Krause 1993).

Edward J. Glave, 1890

EDWARD J. GLAVE AND JACK DALTON are thought to be the first non-Indigenous people to have descended the Tatshenshini River. Glave's published account of their 1890 journey includes a record of his own observations as well as information he received from the local Dän and Tlingit whom they met or who guided the party through uncharted territory (Glave 1890–91).[1] The account includes reference to settlements and village remains they passed while travelling on foot from Klukshu to the Tatshenshini River and then, by boat, down the Tatshenshini.

The first Tatshenshini basin settlement Glave visited in 1890 was Klukshu, situated on the Klukshu River at the outlet of Klukshu Lake. Klukshu is now the basin's principal fishing village, though now occupied primarily in the summer and used as a Champagne and Aishihik gathering place. In the first half of the 20th century, Klukshu was a year-round residential base for some CAFN families and though a detailed history of this settlement has not been prepared, a considerable amount of oral history exists about it (CAFN 2010; Greer 1995, 1996, 1997; Jackson 2006; McClellan 1975; O'Leary 1992a).

As Glave continued his trip on foot down the Klukshu River to the Tatshenshini River, he passed an abandoned settlement. Information provided by CAFN sources suggests several possibilities for this site. These include the locale today referred to as Old Klukshu; a site called Khwalk'u, also known as Klukshu Crossing; and a place called Dadnlen, also known as Klukshu Flats.

The site known as Old Klukshu, called Na she uh (approximate spelling) in Dákwanjè according to Elder Marge Jackson, is thought to be located about 1 kilometre from the current Klukshu settlement, west of the confluence of the Klukshu River and Wolf Creek. This location was identified by Mrs. Jackson based on structures seen in the area in the late 1920s. The latter include dadzils, the rectangular-shaped log ground caches in which salmon caught in the fall are frozen whole and stored for later use. Beaver activity in this area has resulted in flooding, which has made it difficult to search for the remains of Old Klukshu and thereby pinpoint its location (Greer 1997, 1998b, 1998c).

Khwalk'u, located further downstream, is reported as a site long used for gaffing fish (Jackson 1993; O'Leary 1992). Information recorded from mid-20th-century CAFN sources led anthropologist McClellan to suggest Khwalk'u was the abandoned village seen by Glave on the river south of Klukshu (McClellan 1975, 27, 581, where it is spelled Kqal kù). While detailed archaeological investigations have not been completed at Khwalk'u, stone tools have been found here in disturbed surface context and the locale is a documented archaeological site, as recorded by the Canadian Museum of History's archaeological site records database.

The third option for the site noted by Glave downstream from Klukshu, Dadnlen (a Dän k'e toponym), is located about 10 kilometres south of Klukshu Village, on the west side of the river downstream from Klukshu Crossing. Dadnlen is similarly known as an "old site" (Jackson 1993) but has not been archaeologically tested to establish if there is evidence for remains of subsurface materials. Nonetheless, a carved wooden box has been

documented here. Community sources have reported that it once held cremated human remains (Greer 2008).

After the abandoned settlement on the Klukshu River, Glave came to the basin's principal settlement, Shäwshe (Dän k'e), which is downstream of Klukshu River and at the mouth of a smaller tributary named Village Creek. Glave referred to the village by its Tlingit name, Nesketahin, which he spelled Neska Ta Heen. Some CAFN sources also use the Dän k'e toponym for Village Creek, Thet'à Chuà, to refer to this village (J. Kane 1978, 1979; P. Kane 1979; Jackson 1979, 1993), and as noted, the name Dalton Post has also been used for the settlement (see chapter 9). There is considerable oral and documentary history on the old village as well as the Dalton Post part of this site or site complex (McClellan 1975; O'Leary 1979, 1984, 1992a, 1992b; CAFN, Gaunt and Greer 1993; CAFN and Greer 1998).

Shäwshe is understood to have been perhaps the largest and most permanent of the Tatshenshini River's many Aboriginal fishing villages. At the end of the 19th century, it was a home base for local residents of the basin, both Dän and Tlingit, as well as a summer fishing centre for Dän from as far north as Hutchi and Aishihik. Perhaps one of the settlement's most distinctive traits is that it has access to two salmon spawning streams, Village Creek and the nearby Klukshu River. Traditionally, traps for salmon were installed in both streams, with one stream managed by the Crow Clan and the other by the Wolf Clan. Salmon were also gaffed in the main river (CAFN, Gaunt and Greer 1993; CAFN and Greer 1998).

At the height of the trade between coastal and interior peoples in the mid-19th century, and quite possibly in the preceding century as well, Shäwshe functioned as a regional trading centre, a place where Tlingit from the coast exchanged goods with Dän from across the southern Yukon and northern BC (CAFN, Gaunt and Greer 1993, 6; CAFN and Greer 1998). As previously reported, this settlement was well known outside the region, being mentioned by Dry Bay, Yakutat and Klukwan area sources as places they or their ancestors visited (de Laguna 1976; Emmons 1991). The Dry Bay people, for example, reported that they obtained native copper at Shäwshe (de Laguna 1972, 90). At the end of the 19th century, Jack Dalton established his trading post by this settlement because of its importance as a regional gathering and trading place (see chapter 9). It is likely that Shäwshe would have also been an important trading centre during the time that the Kwäday Dän Ts'ìnchį individual lived.

Some archaeological investigations have been undertaken at Shäwshe and the surrounding area, including the Dalton Post part of the site (CAFN and Greer 1998; Gates 1974, 1976; Gates and Roback 1972; Gotthardt, 1992; O'Leary 1979, 1980, 1992b; Rutherford 1992). Archaeological materials have been recognized over a wide area, and it is suggested that Shäwshe is best described as an archaeological locality, or site complex, rather than a single site. At the old village site, part of the settlement, the graveyard and various structures remain, including the remains of two clan houses. Various family residence cabins, as well as structures attributed to the North West Mounted Police occupation of the site, are still present at the new (Dalton Post) part of the Shäwshe site (CAFN and Greer 1998). Photographs of the site around the turn of the 20th century also exist.

Leaving Shäwshe, Glave continued his journey downstream by boat. Less than a day's travel on, and after going through the canyon and rapids, he came to the next fishing settlement:

> We reached the first of the Gunena encampments at noon.... The fishing-camps are pitched at intervals of three and four hundred yards along the western bank of the river, extending for about a mile and a half. The settlement is known as Alseck and is a well-selected position. (Glave 1890–91, 332)

CAFN Elders Parten Kane (P. Kane 1979), Stella Jim (1996) and Jessie Joe (Lawrence Joe, pers. comm. to Greer, 2008) all referred to a Tatshenshini River fishing village by this name. Chilkat Tlingit sources also mentioned the settlement named Alsek, adding that it was a place they went to trade in the 19th century (Emmons 1991, 57; Olson 1967).

The location of the Alseck settlement remains uncertain. Based on one CAFN source (P. Kane 1979) and Glave's report, we understand Alseck to be somewhere between the mouth of the Bridge River and the mouth of Detour Creek. This is the same general area where the next settlement visited by Glave, Núghàyík, is located. The status of and relationship between the two sites remains uncertain. Additionally, we do not know if Alseck was a seasonal village, occupied primarily when the salmon were running, or a more permanent settlement. The fact that the Chilkat traders came here suggests it may have been the latter.

Below Alseck, Glave mentions passing several small fishing camps, around which there were some older structures:

> There are a few dilapidated log buildings in the vicinity of the present camp which were once permanent dwellings and belonged to an old coast tribe called Nua Quas.... At present there are a few members, offspring of the once big tribe who are now living with the Gunena (Glave 1890–91, 396).

The Nua Quas were Tlingit from the Haines and Klukwan area who had moved into the Tatshenshini basin. Various spellings of the name of this settlement have been used, including Núghàyík, Nuqwa'ik', Nuqwaik, Nùghàyik, Noogaayík and Nu gha hit (CAFN, Gaunt and Greer 1995; Cruikshank 2005; French 1993; Jackson 2006; McClellan 2007; 1975, 32). The place name is Tlingit in origin, and translates as "house of the Nu gha (or Nua Qua) people" (John Ritter, pers. comm. to Greer 2009). The Nu gha, or Nua Quas, as McClellan spelled their name, are thought to have moved to the interior during the height of the fur trade (McClellan 1975; 2007, 13). By the time of Glave's descent of the river in 1890 they had abandoned the settlement, however. CAFN Elder Johnny Fraser, born ca. 1883, reported that most of the Nu gha people either returned to Klukwan or joined the Dän following a devastating flu epidemic in the latter part of the 19th century (McClellan 2007, 13). Fraser was himself descended from those who had joined the Dän, and he identified himself as the "last of the Noogaayík" (ibid.).

Information provided by Champagne and Aishihik sources has allowed us to pinpoint the location of the Núghàyík site on the west side of the Tatshenshini, north of Detour Creek (Jackson 1993; Kane 1978). It remains the best known of the 19th-century settlements situated in the BC portion of the basin, and most contemporary CAFN people refer to this settlement by the Núghàyík name. Nonetheless, the locale also has a Dän k'e name, which McClellan translated as "drying fish camp" (1975, 32). Moreover, it is understood that a Dän settlement existed in this same area before the Nu gha Tlingit settled here. The location of this earlier settlement is unknown, whether in the same physical spot as the house or houses of the Nu gha people, or somewhere else in the same general area, that is, around the location of the settlement known as Alseck, on the river near the mouth of Detour Creek.

Our understanding that a Dän settlement once existed in the area of the Núghàyík site comes from two sources, the first being the large version of the Kohklux map (YHMA 1995), which shows a "Sticks Village" in this area. The second is a traditional story told by CAFN elders. The story recounts how Dän stopped staying at this place because a copper clawed owl was terrorizing and eating their people. An old lady and some children, left behind when the village was abandoned, are said to have tricked and trapped the owl and burnt the place up, saving the owl's claws as proof (Jackson 1993, 2009; McClellan 1975, 32, 74). The mountain to the north of the settlement (which was where the owl lived) is said to represent the ashes of this monster owl; its Dákwanjè toponym, Madzin' kezunwàt, translates as "owl burn up", commemorating the deed.

Dän families, or mixed Dän and Tlingit families, certainly camped in the general area of Núghàyík when the Nua gha people were based there, according to one source (Jackson 1993). They also used the site after the Tlingit had abandoned it (Jackson 1993; Shadow and Jackson 1997). Yukon Elder Kitty Smith reported that "lots of people" were living at Núghàyík when she camped there with her family as a child (Cruikshank et al. 1990, 204). This would have been in the 1890s, by which point in time the Nu gha people were no longer

staying here. According to CAFN Elders Marge Jackson and Paddy Jim, one place at which the Dän stayed in the Núghàyík area was known as Datl'at (Greer 1996; Greer 2007). This fishing camp is located about one kilometre southwest of the Núghàyík site, and about the same distance back in from the main Tatshenshini River channel, beside a back stream channel now referred to as Red Creek. We suspect that Datl'at may not have been the only auxiliary fishing campsite in the Núghàyík area that was used in times past. Quite likely a number of different locales around the mouth of Detour Creek were places where fish were caught, or where fish processing or other activities took place.

The information provided to McClellan by CAFN sources suggests that Núghàyík was a permanent settlement, albeit one that eventually failed, and that the settlement's history could be measured in decades rather than centuries (McClellan 1975, 26; 2007, 13). Detailed investigation of both surface and subsurface archaeological remains in the area of the Núghàyík settlement is needed in order to explore the relationship between the Nu gha occupation and earlier use of this locale by the Dän.

Based on information provided by Elder Jimmy Kane, CAFN citizen Ron Chambers located the actual Núghàyík site, with its building remains, in the early 1970s. The site was also visited briefly in 1973 by researchers working for Kluane National Park, while another party had attempted to reach it by travelling overland the previous summer (Gates and Roback 1972). Over the years, the CAFN Heritage program has received reports of sightings of several types of traditional structures, or possible structural remains, at various locales in the Núghàyík site area. The reports come from Champagne and Aishihik elders and from other Haines Junction area residents who have trapped furs in this area (Greer 2007). The 1993 archaeological crew examined the main Núghàyík area; they confirmed the presence of features that might represent the remains of the buildings from the Nu gha occupation, and found evidence for other uses of the site area post-dating this occupation. They were not able to reach the Datl'at location, though, due to high water conditions in the marshland between the main river channel and areas where the latter is understood to be situated (French 1993).

Downstream from Núghàyík, Glave reports another abandoned but once important settlement, "apparently of great age", which his guides told him was named Tin Char Tlar (1890–91, 396). He described this settlement as Russian in style, "very strongly built" (ibid). Glave's guides reported that this settlement was established by the Yakutat people after a late-18th-century battle on the coast with the Russians. Dry Bay elders also told de Laguna about a place in the interior that their people fled to after a conflict. It was reported as Cix'a'a nú wu, a Tlingit name meaning Eddy Fort (de Laguna 1972, 89). The latter researcher suggested that the two stories may be referring to the same historical event. We do not know for certain where on the river this fort site was located, though as previously mentioned, Davidson (1901) mapped it as being on the west side of the river, in between Núghàyík and the mouth of the O'Connor (fig. 2).

Where the original 1869 version of the Kohklux map had shown one Dän settlement on the west side of the river below the canyon, Glave's 1890–91 account reported one settlement (Alseck) and two abandoned sites—that is, the settlement of the Nua Qua people, today referred to as Núghàyík, and the log fort named Tin Char Tlar. Glave's account does not mention any additional villages downstream of the fort site, such as the two plotted on the east side of the river on the 1869 Kohklux map. We presume Glave's guides didn't point out these settlements because they were either situated on tributaries away from the main river or they were not in use at this time. The sketch maps prepared by Glave do not show any settlements downriver from Shäwshe.

Klukwan Area Sources, *ca.* 1880s–1930s

LESS THAN TWO DECADES after Davidson's mapping work with Chief Kohklux, the ethnographer Aurel Krause recorded information from Klukwan sources regarding interior settlements visited by their people. His report

was released in German in 1885 and only later translated into English (Krause 1956). While the translated report suggests that Krause met a Tlingit trading party (or parties) at the settlement named Alseck, (ibid., 136) other documentation from the Krause brothers dating from 1881–82 (Krause and Krause 1981, 1993) does not suggest that either Aurel or his brother Arthur actually visited any of the Tatshenshini River villages.

The early ethnographer George Emmons, who also worked with Klukwan sources around the turn of the 20th century, similarly recorded information on interior villages that were visited for trade purposes (Emmons 1991, 57). Emmons' Chilkat Tlingit informant mentioned three settlements in the Tatshenshini River basin: Klukshu, Shäwshe (which they reported as Nesketahin) and Alseck.

In the 1930s Klukwan sources told anthropologist Ronald Olson that their people had traded to the settlement named Alseck in the previous century. These sources noted that there was an overland trail which led from Klukwan up the Klehini River to a place called Tcanwukå'h, and from there on to Alseck (Olson 1967, 26). We do not know the location of Tcanwukå'h or the meaning of this name. As noted below, it is similar to one documented by de Laguna from Dry Bay area sources (1972, 90).

Yakutat/Dry Bay Area Sources, *ca.* 1950

DRY BAY PEOPLE interviewed by anthropologist Frederica de Laguna in the period 1949–54 mentioned a number of settlements up the Tatshenshini River that were visited by their ancestors. One of these places was Shäwshe, which de Laguna referred to as Weskatahin (i.e., Nesketahin, the settlement's Tinglit name); her sources and described it as a "great trading centre" (1972, 90). Another settlement mentioned to de Laguna was Cix'a'a nú wu, Eddy Fort, the place to which the Dry Bay people fled for protection after a battle with the Yakutat Teqwedi clan. Because of the similarity of the stories, de Laguna (1972, 89) wondered if Eddy Fort might be the fort settlement that Glave recorded as Tin Char Tlar.

Tcanukwa, likely the same place as that recorded by Olson as Tcanwukǎ'h, is another Tatshenshini River village reported by Dry Bay sources. The meaning of this place name is unknown. Dry Bay sources also reported a place named T'aketci, which means "king salmon bone" or "dried king salmon". There is no locational information on either of these settlements, but it is possible that T'aketci might have been located near Low Fog Creek, a Tatshenshini tributary located on the east side of the river, upstream from the O'Connor. The Dákwanjè toponym for this tributary, gyu chù, translates as "king salmon creek" (Adamson 1993; Jackson 1993). Making such a linkage is far from certain, although it would account for one of the "Sticks Villages" located on the east side of the Tatshenshini River on the Kohklux map, as described earlier.

Other than Shäwshe, the only Tatshenshini River settlement mentioned by Dry Bay sources, for which we know the approximate location, is the place called Tínix Àní. As discussed elsewhere (see chapter 9), Tínix Àní is situated on or near an island at the confluence of the Alsek and Tatshenshini rivers. While the 1993 archaeological survey recorded petroglyphs in this area, no village remains were detected. This is not surprising, considering that oral history records show the settlement was wiped out by a flood.

Additional Information from Champagne and Aishihik Sources

MUCH OF THE INFORMATION from Champagne and Aishihik sources about the old settlements in the Tatshenshini basin has already been mentioned. These settlements include Klukshu, Old Klukshu, Khwalk'u and Dadnlen, all on the Klukshu River; Shäwshe on the main river by the mouth of the latter; Alseck and Núghàyík, both on the main river in the area of Detour Creek; Si'diq, the village around the mouth of the O'Connor; and Tínix Àní, by the confluence of the Tatshenshini and the Alsek. CAFN sources reported two other settlements in the basin that are not mentioned in any other sources. The first of these is Âloaᐟ (or

Alar; McClellan 1975, 28, 32, 581). It is uncertain if this name refers to a specific group of Tlingit or a Tlingit settlement; if the latter, we have no information on where it might be located.

The other settlement mentioned by CAFN sources goes by the name of the Tlingit man that lived at this place. Variously spelled Gach'alwa, Kàlawà, Kuchlua and Gashurwa (Adamson 1993; Jackson 1993, 2006; O'Leary 1992a), this place is believed to have been a single residence rather than a village site, dating perhaps to the beginning of the 20th century. Elders John Adamson (ibid.) and John Brown (1996; see also Banks 1996) had also heard stories about Gach'alwa or seen the remains of his house.[2] The available information suggests that this man and his wife may have been the last people to live on a full-time basis in the middle reach of the basin. Gach'alwa's house was on the west side of the Tatshenshini River, opposite the mouth of the O'Connor, in the vicinity of Skookum Creek; the Tinglit name for the latter is Shànáxh Tlèn (Adamson 1993).

Data Review and Discussion

THIS SECTION REVIEWS the distribution and size of the reported settlements; and the limitations faced in learning more about these sites.

The locations of Shäwshe and of the other fishing sites on Klukshu Creek to the north of this settlement have been established. This is not the case for the villages located in the BC portion of the basin. Of the various sites mentioned, the location of only one—Núghàyík —was confirmed by the 1993 archaeological survey.

All village locations, both confirmed and suggested, are situated near good fishing sites, particularly near salmon spawning creeks (Joe 1990). Three species of Pacific salmon spawn in the basin: Sockeye, Coho and Chinook (also known as King) salmon (see chapter 7). The availability of fish appears to have been an important criterion in determining where people settled in the basin, though we do not believe this would have been the only factor.

Historically, people are known to have caught some salmon in the main Tatshenshini River with gaffs (Adamson 1993). Gaffs are long poles with a sharp hook on the far end (see chapter 9 and chapter 23; O'Leary 1977, 1992a). Gaffing involves considerable skill, as one has to locate the salmon, often by feel, then align the gaff hook parallel to the body of the fish and quickly pull back toward oneself in order to catch the fish on the hook. The technique works well in the fast-flowing, turbid waters of rivers such as the Tatshenshini and the O'Connor, where the salmon can't be seen (Adamson 1993). But it is thought that a greater quantity of salmon was caught not in the main river but in smaller watercourses, where these fish spawn. In these clear creeks the salmon can be seen, allowing them to be either gaffed or caught with spears (Joe 1990). Self-operating fish trap devices could also be set up in the gentler tributary waters but not in the streams that are directly fed by glaciers; the extreme flows of the latter would destroy traps (see O'Leary 1977, 1992a; Joe 1990).

The suggested correlation between salmon spawning creeks and villages suggests an approach to locating the old sites. Locales in close proximity to known or suspected former salmon spawning creeks are identified as targets for detailed archaeological investigation and subsurface testing. This approach was at least partly followed by the 1993 archaeological survey, though at the time, there was an incomplete understanding of which tributaries were spawning streams. Ongoing inventory work by the CAFN and federal Department of Fisheries and Oceans biologists, as well as observations by Tatshenshini-Alsek Park rangers and others, is improving our knowledge of where salmon spawn in the basin. For example, tagging studies during the 2001, 2002 and 2003 seasons demonstrated the strategic location of the Shäwshe settlement, which is located between Klukshu River and Village Creek. Between 22 and 32 per cent of the Sockeye Salmon spawning in the entire Alsek basin in those years were making their way to Klukshu River, and another 5 to 14 per cent were spawning in Village Creek (Smith et al. 2009, table 8). Spawning has also been observed in the Bridge River; the Detour/Debris, Red, Bear Bite, Kudwat, Shini and

Sediments creeks; the Tkope River; various tributaries around the confluence of the Tatshenshini and Alsek rivers; and in the Alsek River below where the two rivers join (L. Workman and M. Jim, pers. comm. to Greer, 2009; Chambers in Greer 1995).

While we know that salmon was a key resource for the basin's residents, the 1993 site survey effort had limited success in relocating the settlements (excepting Núghàyík) that once existed along the BC portion of the river. The environment of the Tatshenshini basin is very dynamic; as glaciers advance and retreat, new water channels are established, mudslides occur, salmon colonize and subsequently abandon watercourses. These changes would impact the pattern of human settlement in the basin. A location that was ideal for occupation at one time may have been later abandoned because conditions were less favourable.

The basin's dynamic environment also impacts our ability to learn more about the region's human history. The same natural forces that presented challenges to those living here are continually removing signs of past human presence, as noted by the 1993 effort to locate the old village sites (French 1993). Some of the places where villages are understood to have been situated are now wetlands flooded by beaver activity. Because there is very limited trapping of the species along most parts of the river, beaver now flourish in the basin, and shallow creeks that once provided prime salmon spawning habitat now have deeper waters owing to the presence of beaver dams. Active erosion in other places is suspected to have removed evidence of past human presence. These conditions make it very difficult to recognize the old village sites. Our knowledge of the Tatshenshini River settlements has many gaps, leaving many opportunities for future research.

We have also lost local knowledge of former village locations due to human agency, as a result of shifting land use patterns. Sometime in the middle of the 20th century, the Champagne and Aishihik people stopped spending extended periods of time in the Tatshenshini basin downstream from Shäwshe (see chapter 9). The last spring trips downstream to "meet the fish" or for winter trapping may have occurred in the 1930s or early 1940s.

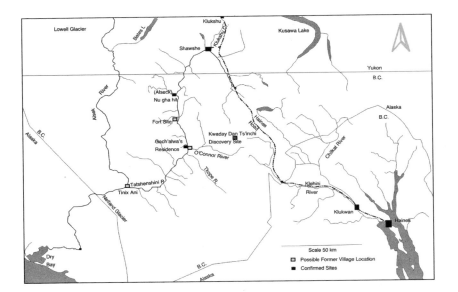

Figure 6. Map showing discovery site and possible village destinations of the Long Ago Person Found, with all locations approximate. Dark boxes indicate the more likely destinations.

When people stop using an area, they are no longer observing the signs that indicate former human presence or passing on such knowledge to younger generations. Fortunately the last individuals to travel to the O'Connor and the confluence, born in *ca.* 1890–1920s, shared their stories, allowing at least some information on the Tatshenshini villages to be documented.

Shäwshe and Núghàyík (Alseck) are the best known of the Tatshenshini River villages; both are understood to have a long occupation history. CAFN Elder Jimmy Kane, who was born in the early 1880s, said of Shäwshe: "That's the oldest village, long time ... just the same place, we stay there" (Kane 1978). The Núghàyík settlement, or at least the area of this site, is similarly understood to have been used since ancient times. These major settlements also drew visitors, travellers and/ or traders from outside the basin. The oral history data suggest that Shäwshe and Núghàyík were occupied more or less on a year-round basis. They likely would have been among the most permanent of any of the villages in the basin, and we suspect that during their zenith, the two settlement clusters were home to the bulk of the basin's population.

The size and permanency of the other reported sites are poorly understood. Perhaps the fort settlement was also permanently occupied during its period of use, whenever that might have been. While there is little information on Si'diq, the settlement that was once located in the area of the O'Connor, what is known suggests that it might have been a seasonal, rather than permanent, community. We do not know if people lived year-round at Tínix Àní, the settlement located at the confluence of the Alsek and Tatshenshini. Some of the reported settlements may not have been occupied at the time the Kwädąy Dän Ts'ìnchį individual lived.

Possible Destinations

HAVING REVIEWED THE POSSIBLE LOCATIONS of the basin's 19th-century settlements, and the limitations of our data, we now take a more detailed look at which settlements might have been destinations for the last journey of the Long Ago Person Found (fig. 6). We also review the possible routes from the Kwädąy Dän Ts'ìnchį discovery site to these sites or site clusters.

The strongest and most reliable information for possible destinations comes from Glave (1890–91) and from the Kohklux maps of 1869 and 1901. Glave reported three settlements on the river below Shäwshe:

Alseck, Núghàyík and the fort site (Tin Char Tlar), all on the western side. The maps show only one site west of the river—perhaps one of the above settlements, or a site complex comprising all three—but they also show two additional settlements on the eastern side, one near the mouth of the O'Connor and one at the confluence of the Alsek and Tatshenshini. We suggest that this last site may have been Tínix Àní.

Based on the reliability of these sources, and the criteria of proximity and permanence, three settlements or settlement clusters are considered the best candidates for likely destinations. These are the settlement located near the mouth of the O'Connor River; the sites known as Alseck and Núghàyík, located on the west side of the river near the Yukon–BC boundary; and Shäwshe. The fort site is a fourth possible destination. While the limited available information on this site indicates that Dry Bay people fled there and built a substantial residence for safety, nothing is said about how long they stayed at this place, and there are no stories that might suggest they had a longer presence in the area. One is left with the impression that this was a relatively short-term occupation. For this reason, we suggest that the fort site is less likely as a candidate for possible destination. We note, too, that while it is possible that the Long Ago Person Found was heading to one of the settlements located downstream from the O'Connor, he would likely have stopped first at the settlement there, and we therefore consider only the latter in our discussion of possible destinations. For a similar reason, we dismiss the settlements located on Kluk shu River upstream from Shäwshe.

Shäwshe, the first candidate destination, is located in the Yukon, near the northernmost point of the main Tatshenshini River. It is understood to be an old site, and also the most permanent of any in the basin. It is about 49 kilometres from the site of the Kwädąy Dän Ts'ìnchį discovery to Shäwshe, with the most likely route requiring circumnavigation of the Mt. Beaton mountain block which lies directly to the south of Shäwshe. Historically, people travelling to Shäwshe from Klukwan did not go past the discovery site, but well to the east, on one of the variants of the Chilkat Trail route, which the present-day Haines Road approximates, and which the commercial trail operated by trader Jack Dalton in the 1890s followed (fig. 3; see also YHMA 1995; chapter 32 this volume). Based on the routing factor alone, Shäwshe is considered less likely as the possible intended destination of the Long Ago Person Found than either of the other two possibilities.

The former village site, possibly the King Salmon camp named Si'diq near the mouth of the O'Connor River, is the second possible destination. Consideration of this site as a destination is based on its proximity (18 kilometres) to the Kwädąy Dän Ts'ìnchį discovery site (figs. 1 and 6). Oral history sources on this settlement are extremely limited. While Champagne and Aishihik people were known to be travelling and trapping in this area in the early 20th century, few details have been passed down about this place. We therefore do not know its precise location or anything about its size or permanency.

The area of the confluence of the O'Connor and the Tatshenshini village is easily accessed from the discovery site. The route goes west, down the middle tributary of Fault Creek, then to the main branch of this tributary, which enters the O'Connor about 10 kilometres upstream from the river mouth. From there, one would proceed west along the north side of the O'Connor. The only major water crossing on this route is Shini Creek (unofficial name), which enters the O'Connor about three kilometres below the mouth of Fault Creek. A possible old foot trail was also observed in this area during a mid-1990s helicopter flight. While now clearly in use as a game trail, the possible old travel route was observed downstream from the mouth of the Shini Creek on the high-terrace bench overlooking the lowermost reach of the O'Connor River (Greer 1995). We note, too, that a couple of years after the 1999 discovery, the CAFN Heritage program received a verbal report of a sighting of possible village remains in the area of the mouth of the O'Connor River (Gaunt 2001). To date, this lead has not been pursued due to the logistical complexity and the high costs of conducting archaeological surveys in this area.

The two settlements known as Núghàyík and Alseck are the third destination candidates. They are located in the same general area—on the west side of the river, in the area of Detour Creek or slightly upstream, near the BC–Yukon boundary. Núghàyík is approximately 33 kilometres from the Kwädąy Dän Ts'ìnchį discovery site. Since the exact location of Alseck is not known, its distance from the discovery site can't be specified, though it would be similar. These two sites are together considered possible destinations because they may have an interrelated history or, broadly speaking, they may be the same place. That is to say, we think it possible that the Nu gha Tlingit built their houses at a location that was previously a former Dän village site. After, and possibly even during, the time when the Nu gha stayed at this place, the Dän returned seasonally to catch salmon.

Two possible routings are suggested for travel from the Kwädąy Dän Ts'ìnchį discovery site to the Núghàyík-Alseck site area. One would be via Fault Creek to Shini Creek, then north up the latter valley to its headwaters (figs. 2 and 6). The alternative, suggested to be the easier route by those familiar with the local terrain, would be via the north Fault Creek tributary, then over the height of land into the uppermost Parton River basin (Lance Goodwin, pers. comm. to Greer, 2011). From either of these routes it is an easy hike west over to the headwaters of Low Fog Creek, north of Carmine Mountain, and from that point, west down to the Tatshenshini River in the area opposite the mouth of Detour Creek, where the river can be easily crossed (ibid.). Remnants of possible old foot trails have also been seen from the air in the headwaters country, between Shini, Low Fog and Kudwat creeks (Greer 1995).

Conclusions

- -

IN THIS CHAPTER we have reviewed information on the settlements thought to have existed in the Tatshenshini River basin in the 19th century. The extent of our knowledge of the past villages along the river varies considerably, with those villages located in the Yukon portion of the basin being best understood. This is

not unexpected, since this region has had continuous use and occupation into the 21st century. The number and location of the settlements located in the British Columbia portion of the basin are uncertain. The most reliable sources for information on these villages are first-hand accounts from the 19th century. The reports of the Chilkat Tlingit chief Kohklux, who mapped Dän villages on the river in 1869, show two settlements on the east side of the river and one on the west. Based on information from CAFN sources, we suggest the settlement on the west side of the river is the possibly related sites Núghàyík and Alseck. The settlements on the east side of the river would be the one located around the mouth of the O'Connor River and the one around the confluence of the Alsek and Tatshenshini rivers. E.J. Glave, who descended the river in 1890, reported one active settlement (Alseck) and two abandoned ones (Núghàyík and the fort site) in this part of the basin.

Promising as these first-hand accounts are, many of the settlements located in the BC portion of the basin that are mentioned in archival and oral history sources have not yet been located. The 1993 archaeological survey that searched for signs of the old villages encountered difficult conditions, showing that the basin is reluctant to give up the secrets of its past. Consequently, we know very little about many of the reported settlements, particularly whether they were permanent villages or only seasonally used—that is, occupied only or primarily for harvesting salmon. We also do not know which sites may have been in use at the time the Long Ago Person Found lived.

While three 19th-century sites or site clusters are suggested as possible destinations for the Kwädąy Dän Ts'ìnchį individual, the first of these, Shäwshe, located 49 kilometres to the north in the area, is considered a less likely candidate based on travel direction and distance. The Núghàyík–Alseck site area 33 kilometres to the northwest, just south of the BC–Yukon boundary, and Si'diq, the site that is reported to have existed near the mouth of the O'Connor River, roughly 18 kilometres from the Kwädąy Dän Ts'ìnchį site, are suggested as the most likely destinations. Our combined sources of first-hand 19th-century accounts, contemporary research and oral history from peoples in the area have allowed us to make a significant start in understanding the past human presence in the study region, but many opportunities for future research remain.

ACKNOWLEDGEMENTS

Our understanding of the Tatshenshini-Alsek villages has drawn heavily on the investigations of researchers who have preceded or worked with us, including George T. Emmons, Frederica de Laguna, Catharine McClellan, Julie Cruikshank, Beth O'Leary, John Ritter, Daniel Tlen, Jeff Leer, Margaret Workman and Bess Cooley. Their contributions are gratefully acknowledged, along with those of the First Nations elders who have shared their knowledge with these individuals or with the CAFN personnel who have worked on the Kwädąy Dän Ts'ìnchį project or on earlier efforts to document the history of the Tatshenshini-Alsek area. Direct funding in support of the research reported in this paper came from the British Columbia Heritage Trust, the government of Canada's International Polar Year program and CAFN. Thanks to CAFN staff member Paula Banks and contractor Gordon Allison for editorial assistance and insight on matters related to Tatshenshini-Alsek history, and to Haines Road resident and local outfitter Lance Goodwin for insights on old travel routes in Tatshenshini-Alsek Park. Darcy Mathews provided assistance with maps.

3

LESSONS FROM A SHORT LIFE

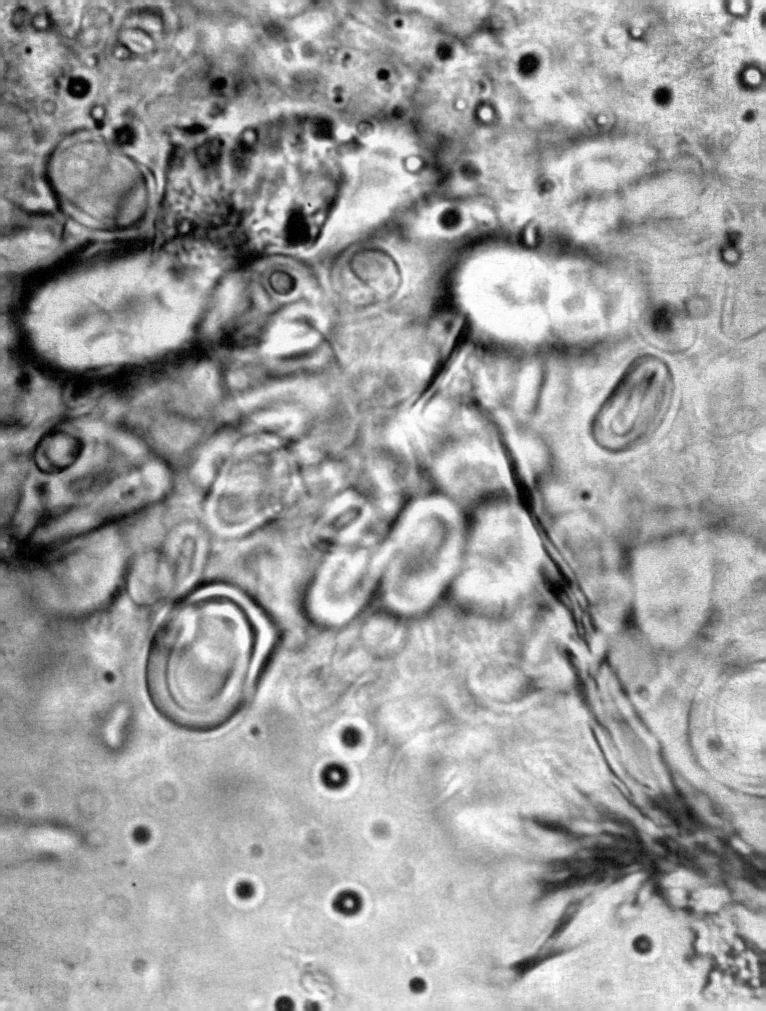

INTRODUCTION

Richard J. Hebda

The remarkable preservation of the Long Ago Person Found's remains has enabled a range of analyses based on traditional techniques and new technology. Following exceptionally careful transport from the find site, the remains underwent a conventional autopsy. Studies of microorganisms, parasites and insects were undertaken, and the human DNA, as well as that of associated microorganisms, provided important knowledge. New technology allowed for two studies investigating the mineral content of hair and preservation of DNA, and a study of carbon and nitrogen provided vital insight into the Kwädąy Dän Ts'ìnchį man's daily diet.

STRICT CONTROL AND CONSERVATION PROTOCOLS were used at all stages of the project to remove, transport and conserve the human remains. Those working on the project both adopted and developed standards and methods based on advice from experts on the study of glacier remains. This approach showed respect for the person and ensured the best possible material for scientific study.

The autopsy revealed the state of the young man's physical health and provided an understanding of the history of the remains following death. He was about 18 years old when he died, in good health, and his body was mostly very well preserved. The torso, however, was likely broken into pieces once frozen in the glacier.

Fourteen different microscopic organisms were detected through a transmission electron microscope (TEM) search of skin, muscle and bone of the left arm. These organisms were likely acquired from outside the body after death, possibly during a freeze-thaw event. Cysts of microbes in the heart, lymph nodes and lungs strongly suggest a pre-death infection, supporting the conclusion that the Kwädąy Dän Ts'ìnchį individual had tuberculosis, despite no obvious physical signs.

Studies of insects and parasites reveal much of a person's life and conditions of the body after death. In an examination of the skin, body tissues, gut contents and hair, no parasites and insects typically associated with humans were discovered. Nor was any evidence found of mites, lice or trichinosis. However, unambiguous remains of the fish tapeworm common in salmon from the region were discovered, showing that the young man ate fish.

DNA evidence for disease was also sought —specifically, the bacterium *Helicobacter pylori*, a cause of gastric ulcers, and *Mycobacterium*, responsible for tuberculosis. The variation in the DNA sequences of these organisms provided important insight into the movements of ancient human populations. *Helicobacter* was detected with sequences that are unique, and thus it shed light on the young man's relationship to other people in the region. *Mycobacterium* DNA was present, but there were no visible symptoms of the disease; interestingly, its sequence was identical to European lineages.

Scanning electron microscopy and energy dispersive x-rays revealed that the human hair had been degraded by chemical and physical processes but not by microorganisms. The surface of the hair was rich in iron, silica and aluminum. High iron and zinc concentrations likely came from red ochre, but silica, aluminum and some of the iron may have been precipitated by groundwater onto the hair surface.

Recent advances in the study of human DNA make it possible to relate people long dead to modern human populations, provided that the DNA can be extracted. In the case of frozen remains, soft tissue may yield DNA, as was the case with the Kwädąy Dän Ts'ìnchi man. In a supplementary study, the preservation of soft and hard cellular microstructure was examined and important new information was gained regarding the relationship between preservation and DNA yield, especially in soft tissues. The investigation also revealed which parts of the remains had experienced the best preservation.

The results of the DNA sequences indicate that within the broad western North American context, the Kwädąy Dän Ts'ìnchi man's remains appear to be more closely related to Athapaskan DNA types than Tlingit types. The genetic connections with Canadian Inuit and Chukchi people of northeastern Asia (far-eastern Russia) are even stronger. The genetic connections within the smaller regional community are explored later in this book.

"We are what we eat" is an expression with a strong scientific basis, and it has special meaning in the case of the Long Ago Person Found man. The ratios of the atomic weights of elements, notably carbon and nitrogen, helped reveal that while the Kwädąy Dän Ts'ìnchi man ate mostly marine foods for much of his life, his intake of land foods went up noticeably in the last year of his life. As you will read further on, the contents of his stomach revealed more precise evidence for his diet just before the time of death. In each chapter in this section, modern science brings us an intimate understanding of a person who lived and died long ago.

11

GENERAL OBSERVATIONS ON THE CONSERVATION AND MONITORING OF THE KWÄDĄY DÄN TS'ÌNCHĮ HUMAN REMAINS DURING THE ANALYSIS PERIOD

Jim Cosgrove, Owen Beattie, Kelly Sendall and Nicholas Panter

When the Royal British Columbia Museum was notified that the body of a young man found frozen on a glacier, as well as associated artifacts, would be coming to our Class II biological containment facility (*Biosafety* 1999), there were a number of issues to address. Immediate steps were taken to ensure that a) proper and prudent protection, security and preservation measures were in place when it all arrived and b) the man's body would be handled in a respectful way, as defined by the agreement between the BC Archaeology Branch and the Champagne and Aishihik First Nations.

THE CONSERVATION TEAM consulted with a number of experts on cold environments and the medical aspects of deep freezing. The recommendations of the team responsible for the preservation of the 5,300-year-old body found in the Tyrolean Alps in 1991 were followed (Gaber et al. 1992), but minor modifications were made to fit the specific context—that is, the conditions of the discovery site and the research time frame. Further recommendations and suggestions were provided by Dr. Locksley McGann and Dr. James Russell, University of Alberta, and Prof. Konrad Spindler, University of Innsbruck.

One of the achievements of the University of Innsbruck researchers was to develop a system to replicate, as closely as possible, the temperature and humidity where the body was found. For the Kwädą̈ Dän Ts'ìnchį remains, the recommended conditions were -17°C (the temperature of the glacial ice below the man's body at the discovery site) and a minimum relative humidity of 80 per cent (Hardy 1998). In consultation with Brian Apland and Al Mackie of the BC Archaeology Branch and Dr. Owen Beattie, the lead scientist of the research team, three goals were set for the conservation of the remains.

- Protection of the remains from contamination that might interfere with the results of research projects being considered. This meant controlling the environment where the body would be exposed for testing.

- Minimization of any possible health risk to both conservation staff and researchers, such as spore-forming bacteria and viruses that could be present with the body, by using biological safety measures for handling human remains.

- Provision of a stable, consistent environment—one mimicking the high humidity and temperature of the glacier from which the man's body was collected—optimal for the preservation of the remains. Temperature would be controlled by altering the walk-in freezer and by keeping the

remains in the original chest freezer (used for transport) within the walk-in freezer. Relative humidity would be kept as high as possible by packing the body with ice within the chest freezer.

Collection and Transfer to the Museum

THE DETAILS OF THE DISCOVERY SITE and the recovery of the Long Ago Person Found's body are described in other chapters in this book. On August 23, 1999, the human remains were transferred directly from the ice into three separate bundles: the body, which had been divided by glacial ice into two parts, plus the separate exposed tibia. These were each wrapped in sterile, non-absorbent hospital fabric sheeting to protect them from contamination. Two bundles were placed into a new, clean, portable plastic camping cooler and the other into a Coroplast box, all carried to a waiting helicopter for the two-hour trip to Whitehorse (fig. 1).

Upon arrival, the bundles were transferred to a new, clean one-metre^3 chest freezer in the conservation section of the Yukon Heritage Branch, awaiting the outcome of negotiations regarding disposition of the human remains and artifacts. Once the decision had been made to transfer the remains to the Royal BC Museum in Victoria, BC, the body was carefully packed—secured against shifting and thawing with ice packs—in preparation for loading into a twin-engine airplane for the six-hour direct trip on September 2, 1999 (fig. 2).

At all times the man's body was accompanied by a Champagne and Aishihik First Nations representative chosen by the elders. Due to the airplane's small size, reconfiguration and removal of non-essential furniture and ducting from the plane interior was required to accommodate the freezer (for more on the airplane transfer, including issues of fitting the freezer into the plane, see chapter 4).

The conservation strategy during the recovery and transport of the remains to the museum was to keep the remains frozen and minimize contamination by material that could affect the research process. While temperature ·

Figure 1. The wrapping and transfer of the remains in preparation for transport to Whitehorse, Yukon. Al Mackie photograph.

Figure 2. Loading the chest freezer into the airplane at the Whitehorse airport in preparation for transport to Victoria, BC.

Figure 3. The Class II containment facility located in the Royal BC Museum's Fannin Building. The walk-in freezer is on the right; the access doors to the examination room are on the left.

was not monitored along the way, the remains arrived in Victoria still frozen. Samples of the glacial ice from below the body, which had travelled along with it, were clearly unthawed, suggesting that the body also remained frozen in transit. The temperature of the glacial ice containing the body at the time it was recovered from the discovery site was -17°C. The coldest temperature encountered en route was in the chest freezer in Whitehorse, where it could have reached -20°C. Accordingly the temperature of the body's exterior may have fluctuated between -20°C and close to freezing during transport.

Preparing the Walk-in Freezer

WHILE THE BODY WAS IN TRANSIT from the recovery site to Victoria, the 30-metre³ walk-in freezer (fig. 3) housed in the Royal BC Museum's Fannin Building was prepared, creating a stable environment to store the remains for study.

The freezer is normally used for storing natural history specimens before they are readied for the museum's research collections. Out of respect for the cultural importance of the young man, and in order to protect the remains from contamination, it was necessary to relocate most of the contents of the walk-in freezer, leaving only a small number of specimens on fixed shelving on one side of the freezer. These were covered with clean polyethylene sheeting in order to completely isolate them from the rest of the freezer. Then, since there was insufficient time to disinfect the rest of the freezer by steam cleaning, the interior surfaces were thoroughly cleaned with isopropyl alcohol.

To ensure a continuous electrical supply to the walk-in freezer, a backup system was built with connection to an on-site emergency generator. A padlock was installed on the freezer door, with the keys held at all times at the museum security desk. Access to the room and the freezer was controlled by security through a strictly limited list of those approved to sign out keys.

A system for tracking temperature with a high/low alarm at plus or minus two degrees Celsius was already in place, and this was monitored continuously from two locations staffed at all times. Opening the door of the walk-in freezer immediately registered as a temperature drop at the two locations, and if the door was left open for even a few minutes, a high-temperature alarm would sound.

A new temperature control system with greater sensitivity—able to hold the walk-in freezer temperature to within plus and minus one degree Celsius—was installed to ensure the accuracy of the thermostat and to minimize the fluctuations in temperature due to the freezer's frost-free system. A four-channel ACR Systems data logger was installed on the exterior of the walk-in freezer. One channel was to connect to three ACR low-temperature probes or sensors, and the other three channels were to connect to ACR combination external temperature and relative humidity probes. The leads were routed under the freezer door into the interior for placement when the man's body arrived.

The Royal BC Museum's walk-in freezer was set at -17°C and was successfully stabilized at this temperature for a two-day period prior to the body's arrival. Before the remains (in the chest freezer) arrived from Whitehorse, a decision was made to place the chest freezer directly into the museum's walk-in freezer. This was proposed for three main reasons.

- It would further reduce the possibility of contamination from contemporary influences by providing another physical barrier.

- It provided a convenient and adaptable enclosure for the use of crushed ice (made from sterile, distilled water) that would be packed loose around the wrapped bundles in the cooler, providing further protection from temperature and relative humidity fluctuations of the walk-in freezer.

- In an emergency, the remains could be rapidly transported out of the museum and the chest freezer plugged in at an alternative location. To speed up the chest freezer's removal in such an emergency, a wheeled platform was built for installation under the chest freezer.

Preparing the Museum Containment Facility

THE ROYAL BC MUSEUM'S bird and mammal preparation lab and associated freezers were determined to be the most suitable area to use as a containment facility and lab.

However, the facility and lab had to be prepared for autopsy and research. The ventilation system in the museum's Class II containment facility was designed to bring outside air in and then to have that air exhausted, creating a negative pressure, so that air would not escape into the building from the lab. To reduce the amount of outside airborne contamination from modern pollens and pollutants, it was necessary to reduce the amount of outside air entering the facility and to direct the airflow away from the autopsy area. All non-essential materials were removed from the lab, and the facility was thoroughly cleaned with Betadine disinfectant and isopropyl alcohol.

Sterile hospital techniques for all access and interaction with the remains were implemented in order to protect the body from contamination by outside materials and also to protect the researchers and other project participants from any potential communicable diseases from the body. Hospital-type sterile and disposable gowns, gloves, masks, protective footwear and caps were to be worn by all participants (fig. 4) anytime the body was out of the freezer for inspection or sampling. All implements and containers to be used were either new and sterile or to be autoclaved before use. Any waste wrappings and used hospital wear would be autoclaved and stored for disposal.

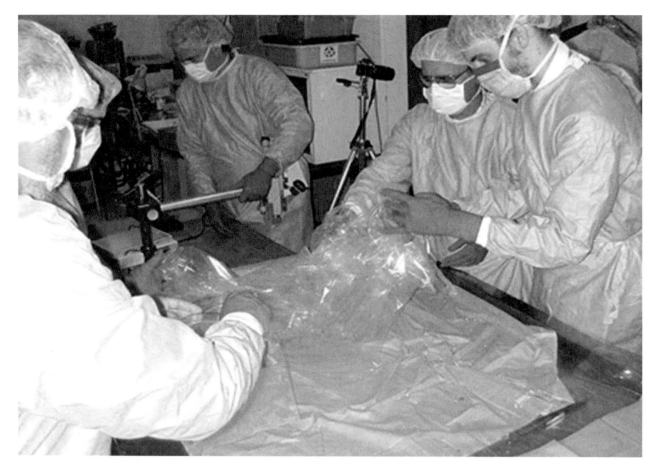

Figure 4. Examination of the wrapped remains in the examination room.

Conservation Treatment of the Body

THE HUMAN REMAINS arrived at the museum on September 2, 1999, and were allowed to stabilize at -17°C in the walk-in freezer overnight. The next day, Beattie and Mackie briefed selected museum staff on the procedure for Beattie's initial examination and documentation of the body's condition. On a large, sterilized, stainless steel table within the containment facility, the frozen remains—in their three separate bundles—were unwrapped, examined, documented for condition and weighed. They were then rewrapped in two layers of sterile plastic sheets. Ethanol-sterilized low-temperature/ combined-temperature and relative-

humidity sensors were placed between these layers, but not in direct contact with the body. Sterile crushed ice made from distilled water was loosely packed around each of the wrapped bundles, in contact with them, and sterile, synthetic, non-absorbent surgical fabric sheeting was wrapped around the bundles to hold the ice in place. Finally, another sterile plastic sheet was used to cover each bundle. The three bundles were then returned to the chest freezer, which also contained plastic buckets filled with sterile crushed ice.

After the body was returned to the chest freezer within the walk-in freezer, the sensor probes were connected to the data logger. Sensor data from within the temperature- and humidity-stabilized chest freezer were examined to verify the desired conditions. As expected, the interior of the chest freezer showed less temperature and humidity fluctuation than the walk-in freezer.

The man's body was brought out of the freezers and examined in the lab directly adjoining the freezer room five times over the seven-month period before research samples were taken. During these examinations, complete videotape and still photographic records were made, allowing for monitoring of the body's physical condition over time. The weight of the three body components was recorded after each examination to detect any changes that could be associated with moisture loss or accumulation. Also after each examination, new wrapping material was prepared in order to reduce the risk of contamination, and the body was rewrapped as described above. Prior to returning the body to the chest freezer, the freezer's inside surfaces were disinfected with ethanol, which was allowed to dissipate. The time to complete this procedure ranged between 20 and 35 minutes, and the remains were carefully monitored to ensure no thawing occurred. Researcher access was carefully scheduled to coincide with this monthly procedure (fig. 5).

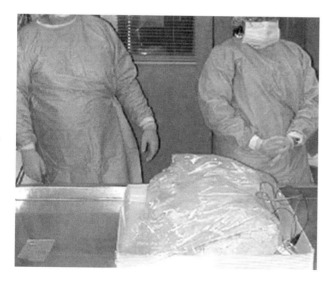

Figure 5. The rewrapped remains in a Coroplast tray after the monthly examination.

Observations and Comments

USING THE PROCEDURES DESCRIBED ABOVE, the total weight change of the remains over the first six-month period was kept to less than 0.5 per cent (table 1). When the body was later autopsied and samples removed for research purposes, the remains were allowed to thaw for those procedures, and fluid was lost as well. It therefore became too difficult to continue to relate the weight of the remains to weight loss due to dehydration in storage. After the sampling was complete, the Kwädąy Dän Ts'ìnchį man's body was cremated at a Victoria crematorium and returned to the discovery site for reburial (see chapters 4 and 8).

The greatest percentage of weight loss was noted in the torso, which had the highest proportion of tissue to bone. Most of the weight loss occurred during the first two months, followed by general stability. Notably, no visible changes in the body condition or tissue colours were observed, except for some very slight bands of darkening; these measured less than 0.5 centimetres wide along the edges of some exposed muscle tissue and were due to desiccation.

The conclusion of the conservation team was that this treatment of wrapping and freezing was a relatively simple and very effective procedure, especially for working within a specific time frame. There were several major components of this successful treatment aimed at minimizing dehydration of tissues by keeping relative humidity high during freezer storage.

- Creation of a stable environment through modifications to the walk-in freezer and addition of an inner chamber (the chest freezer) with sufficient insulation to act as a buffer during the defrost cycle.

- Storage of the remains at a higher temperature than the normal -20°C of most freezers.

- Creation of a local environment for the remains with higher humidity by adding sterilized crushed ice to the chest freezer.

- Placing crushed ice in direct contact with the body by wrapping it around the separate parts.

Situations involving longer-term preservation and stability would definitely bring additional challenges.

A more detailed review of the preservation history and conditions of the Kwädąy Dän Ts'ìnchį man's remains, including detailed sensor data and an evaluation of its long-term effectiveness, is planned for a future publication.

————

Table 1. Weights of the man's remains during storage prior to sampling (September 10, 1999, to March 30, 2000).

	Date	Weight (g)	Change (g)	% Change (last)	% Change (total)
Torso	10-Sep-99	9,898			
	01-Oct-99	9,868	-30	-0.30	-.030
	29-Oct-99	9,851	-17	-0.17	-0.47
	14-Jan-00	9,857	6	0.06	-0.41
	30-Mar-00	9,855	-2	-0.02	-0.43
Pelvis	10-Sep-99	9,987			
	01-Oct-99	9,976	-11	-0.11	-0.11
	29-Oct-99	9,977	1	0.01	-0.10
	14-Jan-00	9,982	5	0.05	-0.05
	30-Mar-00	9,982	0	0.00	-0.05
Tibia	10-Sep-99	408.1			
	01-Oct-99	408.7	0.6	0.15	0.15
	29-Oct-99	408.6	-0.1	-0.02	0.12
	14-Jan-00	408.3	-0.3	-0.07	0.05
	30-Mar-00	408.5	0.2	0.05	0.10

12

A REVIEW, DISCUSSION AND INTERPRETATION OF THE HUMAN REMAINS FROM THE KWÄDĄY DÄN TS'ÌNCHĮ DISCOVERY

Dan Straathof and Owen Beattie

The frozen remains of a young man and a number of directly associated artifacts were discovered in August 1999, partially embedded in glacial ice in northwestern British Columbia, Canada (Beattie et al. 2000). The discovery has been given the name Kwäday Dän Ts'ìnchį (Long Ago Person Found) by the local First Nations community, and the man is thought to have lived between 1720 and 1850 AD (see chapter 6). The objectives of this chapter are a) to provide a detailed description of the physical condition of the recovered and examined human remains and b) to review the findings of the medical autopsy and osteological examination. The remains were sufficiently well preserved to allow analysis of soft tissue elements that are infrequently available for study in remains of this age. The focus here will be on anatomical observations, providing a context for the additional specialized analyses dependent on tissue and other samples collected from the body (e.g., detailed analysis of microbial flora and gastrointestinal contents) that are presented and discussed elsewhere in this publication.

The discovery of the human remains consisted of four events:

1. the original discovery by the sheep hunters in August 1999;

2. the primary recovery efforts in August 1999, with the discoveries flown from Whitehorse, Yukon, to the Royal British Columbia Museum in Victoria, BC, for interim preservation and detailed examination;

3. and 4.

    secondary discoveries related to site visits in August 2003 (the discovery of most of the cranial bones and maxillary dentition, the right scapula and right hand bones) and August 2004 (the discovery of the left parietal bone, two teeth and small amounts of soft tissue). Materials discovered in 2003 and 2004 were photographed and examined in the field but not removed from the site.

Examination of the soft and skeletal tissues identified the individual as a young man, approximately 18 years of age at the time of his death, appearing to be in good health, of average body build and with a living stature of approximately 169 centimetres.

Interpretation of the discovery area on the glacier and the manner in which the man was positioned in the ice suggests that just prior to his death, he lay down—or fell—in a partially prone position, with his right arm outstretched, his right forearm flexed and his head possibly resting on his right hand (Beattie et al. 2000). Here he was frozen and perhaps covered by snow, then ice. Due to the dynamics of the glacier, his body

was later transversely bisected at a point approximately 15 centimetres below the umbilicus (i.e., separated roughly horizontally across the lower abdomen). At some point following, the upper body slipped into a crack that had formed in the ice, ending up in a vertical orientation.

A Review of the Discoveries

THE REMAINS RECOVERED IN 1999 consisted of three major components (see Beattie et al. 2000): the partially fleshed torso with attached left upper extremity, weighing approximately 10.1 kilograms; the largely fleshed pelvis and proximal lower extremities to the knees, weighing approximately 10 kilograms; and the left tibia with a small amount of soft tissue at the proximal and distal ends, including a short segment of fibula attached at the proximal end, weighing approximately 0.4 kg. Found on the surface in association with the other remains, and with a combined weight of approximately 0.04 kilograms, were three right metatarsals (I, IV and V); five right carpals (lunate, pisiform, trapezium, triquetral and capitate); one proximal phalanx; three intermediate phalanges; and one distal phalanx.

With most of the skeleton in relatively good condition and a great deal of the soft tissues preserved, researchers had a nearly unprecedented opportunity to examine and interpret the individual's state of health, determine his general physical characteristics and collect tissue and auxiliary samples (e.g., gastrointestinal contents)—all with the potential to provide an even deeper understanding of what this person's life was like. Critical to achieving these goals was the appropriate, effective and successful protocol for conservation of the body during the research access period at the Royal BC Museum, which was modelled after the procedures established at the University of Innsbruck for application to the 5,300-year-old body recovered in 1991 from the Tyrolean Alps (Gaber et al. 1992). Details of the Victoria protocol are described in chapter 11 (Cosgrove et al.), this volume.

Five day-long medical examinations were performed at the museum, in April, June, September, November and December of 2000. These examinations involved the collection of specimens for DNA and other analyses, the medical autopsy and the osteological observations. Medical imaging of the remains was conducted separately, in June 2000, at Victoria General Hospital.

Human remains discovered on the glacier in 2003 included small portions of hair, small fragments of adipocere tissue, the skeletonized right scapula, several hand bones, seven loose teeth and skeletonized portions of the cranium. The findings were described, examined, measured and photographed on site. And in 2004, more new-discovered remains included small portions of soft tissues, two teeth and the left parietal bone. These findings were also briefly examined, described and photographed on site.

Post-Mortem Examination of Soft Tissues

1. Gross Examination

The following description relates only to the discoveries of August 1999, as the minute soft-tissue fragments identified in 2003 and 2004 could not be examined in detail. Some general observations about the skeletal elements are included in this section, but a more detailed description, including of those bones discovered in 2003 and 2004, is presented below, under Osteological Observations—specifically, estimation of age, sex and stature, analysis of dentition and evidence of healed injuries.

The upper-body segment (torso) included the vertebral column from the second cervical to fifth lumbar vertebrae and was flattened in the anteroposterior plane. Much of the skin of the neck, chest and abdomen was intact, extending inferiorly to approximately 15 centimetres below the umbilicus. Posteriorly, there were patches of skin, with interspersed fields of exposed subcutaneous tissue, which displayed adipocere change. The skin of the chest and abdomen was relatively pliable, whereas that of the arm appeared somewhat desiccated and displayed a few longitudinally oriented, rigid folds or creases. None of the tissues could be described as completely mummified.

Examining the skin of both the torso and pelvis with conventional and alternative light sources (i.e., ultraviolet light) revealed no cutaneous markings (e.g., tattoos), nor was there evidence for any animal scavenging activity.

Plain radiographs and computed tomographic (CT) scans were performed after DNA samples were collected. Other than a transverse fracture of the proximal end of the left fibular shaft, of possible perimortem or post-mortem origin, no fractures or other pathologic findings were identified. The internal structures had a compressed, layered appearance with intervening air-like density.

The internal organs were exposed for examination by a vertical incision through the anterior midline of the neck and chest, extending onto the abdomen to the right of the midline. Adipocere formation was noted, involving the subcutaneous tissues of the chest and abdomen to a variable degree, with the usual accompanying odour; however, fascial planes were preserved, and recognizable fibrillar skeletal muscle was seen in many areas, including the pectoralis musculature and rectus sheath. The tissue planes in fact appeared to be accentuated, in keeping with the x-ray. The tissues were only loosely attached to the ribs and sternum, and they were easily removed by blunt dissection. The tissues of the left side showed patchy discolouration, being relatively darker than those of the right—features likely due to post-mortem change.

The bones were, unexpectedly, very soft and pliable. This was most noticeable in the ribs, which were quite flexible and easily retracted laterally from the anterior midline to expose the thoracic cavity. The costochondral junctions were distinct and easily detached. Similarly, the clavicles were appropriately positioned, but the ligamentous attachments had deteriorated so that the clavicles were easy to remove. There was no evidence of antemortem injury in the axial or appendicular skeletal elements examined.

The heart, lungs, liver and gastrointestinal tract were present and flattened in the anteroposterior plane. The lungs had a uniform spongy consistency, without masses or calcifications. Gastrointestinal contents were abundant. The intestinal mass was sharply transected horizontally across the lower abdomen, corresponding to where the body was bisected. The torn ends of individual loops were slightly ragged, but the overall pattern was distinctly linear. The small intestinal loops were otherwise normally positioned and did not protrude from the defect. The spleen and retroperitoneal organs were not recognizable.

The odontoid process of the second cervical vertebra was intact, as was the rest of the vertebral column. No degenerative changes or other abnormalities were identified. The intervertebral discs were fleshy and pliable, and the spinal cord was visible in the spinal canal both rostrally and caudally (i.e., at both upper and lower ends of the vertebral column). Further examination of the caudal portion of the spinal cord (approximately T10-lumbar) showed flattening, with an intact dura and recognizable *cauda equina*.

The medullary cavity contents, visible upon sampling the proximal left humerus, had a homogeneous grey-white appearance with a pasty texture.

The lower body segment (pelvis and thighs) contained the urinary bladder, identifiable as a hollow viscus anterior to the rectum. The prostate gland was palpable, and male external genitalia were present (flattened scrotal tissue). Most of the large intestine was attached to the upper body segment.

2. Microscopic Examination

The tissues were, on the whole, poorly preserved due to autolysis and putrefaction. But as table 1 (page 222) illustrates, many organs were sampled successfully for histological examination. One sample from each tissue was fixed in formalin, and another in Ruffer's solution; there was no appreciable difference in tissue preservation or staining quality between the two solutions. Sections from each block were stained with hematoxylin and eosin, and with Masson's trichrome. In sections containing muscle tissue, no striations were visible. Moderate amounts of anthracotic pigment were detected in the lung tissue. Elastin stain performed on selected blocks (lung, aorta) showed no detectable elastic fibres. Oil Red O stain for fat (fixed, unembedded lung and adipose tissue) was negative. Despite many successful samplings, the histologic examination was severely limited by decomposition; most tissues had numerous birefringent crystals typical of adipocere. In general, the trichrome stain was useful in highlighting tissue architecture, which was otherwise obscured for most tissues.

Osteological Observations

1. Inventory

Most of the skeleton was present (including part of the hyoid) *except* the following:

- mandible and mandibular dentition (except at least one tooth)
- right and left nasal bones
- nasal conchae
- auditory ossicles
- first cervical vertebra
- right humerus, radius and ulna
- right scaphoid, trapezoid and hamate carpals
- right metacarpals I, IV and V
- right manal sesamoids
- three manal phalanges

- left foot
- part of left fibula (distal two-thirds)
- right tibia and fibula
- all seven right tarsals
- right second and third metatarsals
- right pedal sesamoids
- all right pedal phalanges

2. Determination of Sex

Though sex was confirmed through soft-tissue criteria (male genitalia), it was possible to evaluate a number of anatomical features that demonstrate various degrees of sexual dimorphism. Observed skeletal features generally associated with males included narrow greater sciatic notches, a narrow subpubic angle, a triangular pubic bone shape, absence of a ventral arc, the robust and foreshortened shape of the pubic tubercle and crest, the S1 body/alae proportions (body larger than each ala), moderate brow ridge development, rounded superior orbital margins and large mastoid processes. The nuchal features were more ambiguous, with a lack of cresting and an indistinct inion.

3. Age Assessment

The excellent preservation of bone detail allowed us to evaluate a number of skeletal features valuable in interpreting age at time of death, leading to an age estimate of approximately 18 years. These features include:

- pubic symphyses: *ca.* 18.5 years, range 15–23 years (Suchey-Brooks casts and criteria; Phase 1 of Brooks and Suchey 1990);
- fourth ribs: *ca.* 19 years (features of both Phases 1 and 2; Iscan et al. 1985).

Other elements showing morphological consistency with the age estimate derived from the symphyses and fourth ribs are the spheno-occipital synchondrosis (open), clavicles (unfused medial epiphyses), anterior iliac crests (fully fused), clavicular notches, sacrum, sternal body, acromion, epiphyseal rings, proximal humerus, proximal femur and cranial sutures (unfused vault sutures).

4. Stature Estimation

Stature was approximated using the research of Trotter (1970, formulae for "white" and "mongoloid" males) and applied to the left femur, left tibia and left humerus. All bones produced estimates between 169 and 170 cm ± 3.80 cm.

5. Evidence for Healed Injuries

The right parietal bone was marked by two depression features that appeared to be healed depressed fractures. The larger measured 1.7 centimetres in diameter and was located 3.2 centimetres from the sagittal suture and 6.5 centimetres from the right coronal suture. The smaller feature measured 0.8 centimetre in diameter and was located 4.2 centimetres from the sagittal suture and 5 centimetres from the right coronal suture. The distance between the centre points of both features was 2.3 centimetres. The lack of surface damage to the features, and the nature of the appearance of the outer table, indicated that they were not the result of a post-mortem process—for example, pressure in the ice, or forceful or movement-related contact with rocks during the thawing process. Also, their close proximity suggests that they may relate to a single event, though this is not absolutely clear.

6. Dental Features

Even though some teeth have considerable post-mortem damage to the enamel, including loss of enamel tissue, all discovered teeth show signs of occlusal attrition or abrasion. The third and second molars have enamel wear, while signs of dentin exposure are visible on the first molars. Premolar occlusal wear appears to have reduced the crown heights by one-half, and the anterior six teeth show progressively greater signs of wear to the midline. The central incisors show nearly complete loss of the crown. The attrition is significant in an individual so young and is reminiscent of dental wear patterns seen in prehistoric peoples found in other parts of coastal British Columbia.

Though the anterior dentition has suffered pulp exposure, there does not seem to be evidence in the alveolar bone for periapical abscessing; however, post-mortem damage to the outer surface of the alveolus for the incisors may have masked any indications of infection.

7. Other Observations

There was a distinctive flattening of the posterior part of the skull involving the occipital squama and posterior portions of the parietal bones, and incorporating the osteometric point lambda. These observations are strongly suggestive of lambdoidal cranial deformation. This interpretation is problematic in that we would need to know that the bone softening seen to varying degrees in all parts of the skeleton does not account for the deformation. The vault flattening could also have been caused by unequal pressure applied to different parts of the cranial vault when it was encased in ice, and also during the release at the time of thaw. But the symmetry of the lambdoidal flattening, and its location in the appropriate part of the skull, seems too coincidental to be explained as a consequence of a random natural process. This interpretation is supported by the field examination, which revealed that the cranial bones were soft but did not demonstrate any flexibility. The left parietal bone discovered in 2004 was in considerably poorer shape than the cranial bones found and examined in 2003—probably as a result of being exposed shortly after the field visit in 2003—and was showing the effects of exposure, movements and contact with rocks over a longer period than the other recovered remains. These effects range from damaged and shredded outer table bone and pericranium to fractures, cracking and warping from post-mortem movement and desiccation.

Comments, Interpretations and Conclusions

THE KWÄDĄY DÄN TS'ÌNCHĮ DISCOVERY is one of the most well-preserved frozen historical bodies ever recovered from a North American glacier. Though rare, the few frozen bodies that have been found in archaeological contexts worldwide are relatively well known—the most relevant precedent to the Kwädąy Dän Ts'ìnchį discovery being the 5,300-year-old body recovered from a Tyrolean glacier in 1991 (for past examples, see Sudtiroler Archaeologiemuseum 1999; Rollo et al. 2000; Fleckinger 2003; Samadelli 2006). The Kwädąy Dän Ts'ìnchį find has provided a remarkable opportunity to examine not only the skeletal elements but major organs and other soft-tissues from an individual who lived more than 150 years ago. As detailed elsewhere in this publication, these soft tissue observations and analyses have yielded a large quantity of high-quality information about this man's life, including diet, travel patterns and family connections. Based on analysis of the body and recovery site, it appears that the Long Ago Person Found died on the glacier while pursuing regular daily activities and that the body was not disturbed by others prior to its discovery in 1999.

The post-mortem examination of these ancient remains was limited by several factors. For one, the remains were incomplete—that is the head was skeletonized, and portions of the extremities were not recovered. This means that possible injuries or other findings affecting these missing components cannot be excluded. And though it was remarkable for most of the tissues to be preserved and recognizable, the condition of the tissues was generally poor. The kidneys, spleen and pancreas were not recognizable, and pathologic findings in the identifiable organs may have been obscured by the post-mortem changes. Histologic examination was severely compromised by these post-mortem changes.

Several post-mortem effects deserve further comment. The presence of adipocere (demonstrated both grossly and microscopically) indicates prolonged exposure to a damp environment in the post-mortem period. This phenomenon represents a transformation of lipid substances to stable, indigestible molecules by bacteria. Once formed, adipocere may persist without further deterioration for many years. Also, the skeletal remains were very soft and somewhat flexible; this was most evident in the ribs. The precise cause of such a condition is uncertain but is possibly related to calcium salts leaching from the mineralized bone matrix. The exposed ribs and those still covered by soft tissue appeared equally affected. Histologic examination of the bone tissue did not reveal any demonstrable pathologic abnormality that would explain these findings.

As described above, the recovered remains had been sharply separated through the lower abdomen. The disruption of the internal organs along this separation line, particularly the intestinal loops, indicates that this separation likely occurred after the remains had been frozen solid. Shearing forces applied by shifting layers of ice seem a reasonable explanation. A similar phenomenon may be responsible for detachment of the head.

The body was markedly flattened in the anteroposterior dimension, and the internal organs were similarly flattened. Despite this deformation, there were no fractures or other traumatic indicators. Gradual compressive force on an increasingly pliable skeleton is the most likely explanation for this finding.

With due consideration for the limitations of the post mortem examination, we found no evidence of natural disease. Aside from a perimortem/post-mortem fracture of the left fibula, and possible well-healed depression fractures of the right parietal, there was no evidence of trauma. There was also no pathologic evidence of infectious disease or parasitic infestation. Examination of the lungs did reveal accumulation of black particulate material consistent with inhaled carbonaceous material. Such accumulation is most likely due to chronic exposure to smoke or soot. This material was not quantitated.

No cause of death was identified. The circumstances of death, combined with the absence of demonstrable natural disease or significant perimortem injury, are most consistent with an accidental death due to exposure or other environmental factors. Additional possibilities, including fatal head trauma or immobilization due to injury of the extremities, could not be excluded. The body's remarkable preservation over many decades (i.e., persistence of soft tissues, including many internal organs) can be attributed to the freezing conditions on the glacier and the fact that the remains seem to have been spared from any significant animal activity. Several of the chapters in this volume could not have been written, their information never discovered, if the post-mortem conditions had not allowed for this degree of preservation.

In 2001, after completion of all examinations, the remains recovered in 1999 were repatriated to the Champagne and Aishihik First Nations. These remains were cremated in Victoria and then returned to and interred at the discovery site (see chapters 4 and 8). The human remains discovered subsequently in 2003 and 2004 were interred with the others at the site.

ACKNOWLEDGEMENTS

We would like to acknowledge the contributions of the following agencies, organizations and individuals who provided invaluable support for and input into this part of the investigation of the Kwäday Dän Ts'inchi discovery: Champagne and Aishihik First Nations; the Kwäday Dän Ts'inchi Management Group; Royal BC Museum staff; Wax-It Histology Services; Royal Columbian Hospital Histology Department; Victoria General Hospital; the Faculty of Arts, University of Alberta; Drs. R.J. Hebda, M. Noble, G. Anderson and J. Dickson.

Table 1. Samples submitted for histologic analysis.

Samples submitted
Thyroid cartilage
Left fourth costal cartilage
Right fourth rib
Trachea
Lung (left upper, left lower and right lower lobes)
Right and left ventricular myocardium
Left anterior descending coronary artery
Mediastinal tissue
Left hemidiaphragm
Region of spleen (left upper abdomen)
Liver
Gallbladder
Abdominal aorta
Region of pancreas
Small intestine
Proximal descending colon
Skin (lower left abdomen, anterior left thigh)
Spinal cord (caudal)
Prostate gland

13

ANALYSIS OF MICROORGANISMS IN BONE AND MUSCLE TISSUES IN THE KWÄDĄY DÄN TS'ÌNCHĮ FIND

Maria Victoria Monsalve, Elaine Humphrey, David C. Walker, Mike Nimmo, Jacksy Zhao, Claudia Cheung and Paul Hazelton

The Kwädąy Dän Ts'ìnchį man was approximately 18 years old when he died between 300 and 150 years ago in a remote part of northwestern British Columbia. There was no evidence of a violent or traumatic cause of death (Straathof and Beattie 2008). It is possible that, at such elevation, he was caught in an unexpected storm and froze to death. The severing of his trunk was attributed to the slumping of the glacier he was buried in. All of this makes it interesting to learn as much as possible about the general state of his health—a topic especially worth pursuing in this case, given the quality of preservation of some of his soft tissues (Monsalve et al. 2008a). We were further motivated to look specifically for pathogens because of the observed presence of DNA from *Mycobacterium tuberculosis* in his myocardium, mediastinal lymph node and lung (Monsalve et al.

2008b; Swanston et al. 2008). As in other ancient frozen corpses, it is reasonable to look for evidence of microorganisms and parasites (Leles et al. 2008). Therefore, during our microscopic examination of all the Kwäday Dän Ts'ìnchį individual's tissues, we made images of any objects resembling microorganisms—the subject of this chapter. These observations are discussed in terms of their possible contribution to the state of the Kwäday Dän Ts'ìnchį man's health at the time of his death.

Looking for Microorganisms

MICROORGANISMS have been identified in ancient remains buried in permafrost (Cano et al. 2000; Rollo et al. 2000). Analysis of the Tyrolean Ice Man's tissues by transmission electron microscopy (TEM) demonstrated that bacteria were present in colon and stomach tissues (Cano et al. 2000). While the presence of bacteria in the Tyrolean Ice Man's gut is not remarkable, it is perhaps remarkable that they were structurally recognizable. In addition, the DNA of bacteria was found in the Tyrolean Ice Man's skin and muscle (Rollo et al. 2000). As there were two distinct populations of bacterial DNA, skin versus muscle, it was suggested that those from the skin were of environmental origin while those from the muscle were of the corpse (Rollo et al. 2000). This difference and the presence of adipocere were consistent with an interval of thawing, both of which may have contributed to the subsequent difficulty in isolating DNA. In addition to bacterial DNA in the gastrointestinal tract researchers identified eggs of parasites (whipworm) in the body's sacro-gluteal region (Aspöck et al. 1996). Such information not only helps us gain an overall picture of the individual's health but it may suggest possible contributing factors in his death (Cano et al. 2000; Rollo et al. 2000).

There are various possible explanations for the presence of bacteria in and on post-mortem tissue (Morris et al. 2006). During life, formerly commensal bacteria can become pathogenic through the invasion of organs or body fluid following injury. Alternatively, bacteria and other microorganisms may also enter the tissues of a corpse post-mortem, prior to mummification. In the case of the Kwäday Dän Ts'ìnchį person, where we assume that preservation by freezing began quickly, the time available for bacteria to enter the body would be limited. Since care was taken to avoid contamination when removing the body from the glacier, it seems likely that the microorganisms we report here were not introduced at the time of recovery. The presence of adipocere in the back and arm suggests the Kwäday Dän Ts'ìnchį man's remains were exposed through an episode of thawing (Straathof and Beattie 2008). The microorganisms found in these tissues were most likely acquired post-mortem from the environment during a thaw. It is possible, though, that internal organs of the thorax, contained by the diaphragm, may have avoided the introduction of microorganisms from thawing events. Therefore, the microorganisms or parasites we observed in the internal organs of the thorax could well have been present in life and had the potential to affect the young man's health. All of these factors must be considered for any microorganisms located on or within the Kwäday Dän Ts'ìnchį man's remains in order to assess possible roles in the state of his health.

In 1999 the body was recovered from the glacier and transferred to a controlled, sterile lab at the Royal BC Museum in Victoria, BC. In April 2000 biopsies from the arm, lung and heart tissue were taken from the frozen corpse. Samples were kept at -80°C and shipped in a container with dry ice to the Kuvin Centre for the Study of Infections and Tropical Diseases at the Hebrew University Hadassah Medical School. Attempts to culture bacteria from these tissues were made using standard techniques and culture media by Dr. Mark Spigelman's research teams at the International Health, Department of Infections, Windeyer Institute of Medical Sciences, UCL, UK, and the Kuvin Centre. This team's attempts to culture microorganisms using standard techniques-media for fungal spores were unsuccessful. The negative results suggested that viable bacteria were not present in the frozen remains (Monsalve et al. 2008b). At the same time, given advances in the ability to identify bacteria through amplification of DNA by polymerase chain reaction (PCR), Dr. Treena Swanston, currently Assistant Professor of Anthropology and Biological Sciences at MacEwan University who was at the time a PhD student in the Archaeology and Anthropology department at the University of Saskatchewan succeeded in amplifying DNA of *Mycobacterium tuberculosis* from the myocardium and a mediastinal lymph node of the frozen tissues (Monsalve et al. 2008b; Swanston et al. 2008). Through PCR product reamplification from lung tissue, Swanston and her team detected *Mycobacterium tuberculosis* DNA (Monsalve et al. 2008b; Swanston et al. 2008). In addition, Dr. Spigelman and his collaborators at the Koret School of Veterinary Medicine, Hebrew University of Jerusalem, were able to confirm by PCR the presence of *Mycobacterium tuberculosis* in lung tissue (Monsalve et al. 2008b).

The microscopy techniques we used to analyse microorganisms present in tissues of the Kwäday Dän Ts'ìnchį man have been previously described (Monsalve et al. 2008b). Here we extend our previous study of the heart for microorganisms by using light microscopy, histochemistry and TEM of the lung. Since the Kwäday Dän Ts'ìnchį man died around the time of European contact (see chapter 6), and since we have identified *Mycobacterium tuberculosis* DNA in internal organs

(Monsalve et al. 2008b), we include a summary of the global distribution of *Mycobacterium tuberculosis* in both Old World and New World ancient remains detected by DNA amplification (tables 1, page 232 and 2, page 236). In this study we report morphological evidence of microorganisms, acquired both pre- and post-mortem, that have not been previously reported in the Kwäday Dän Ts'ìnchį man's tissues.

Material and Methods

SOME ELECTRON MICROSCOPIC IMAGES of microorganisms observed in the arm muscle and humerus bone were sent to the College of Medicine, Medical Microbiology, University of Manitoba, for identification. A detailed description of conventional, microwave and freeze substitution techniques has been published elsewhere (Monsalve et al. 2008a).

Distinctive characteristics, such as subcellular content and organization, must be taken into consideration when attempting to identify the biological agent of an infectious disease in any organism (Mims et al. 2004). Bacteria as prokaryotes contain no organelles other than ribosomes and mesosomes for protein synthesis. Prokaryotes do not have a nucleus. Eukaryotic organisms, on the other hand, contain membrane-bound organelles (nuclei, mitochondria, endoplasmic reticulum, Golgi apparatus and lysosomes). To classify the microorganisms we observed in the arm muscle, humerus, lung and heart tissues from the Kwäday Dän Ts'ìnchį body, we applied the standard criteria of identification used to differentiate prokaryotes from eukaryotes (Mims et al. 2004).

To differentiate between Gram-positive and Gram-negative organisms, we applied the criteria of Rogers and Perkins (1968) and Seltmann and Holst (2002). Gram-positive organisms are characterized by the following:

1. A cytoplasmic membrane enclosed by a thick peptidoglycan layer.
2. A thick layer of carbohydrates composed mainly of polysaccharides. Together the peptidoglycan and

polysaccharide layers form the cell wall. A lamellar effect is apparent in favourable planes of section.

3. A thick layer of glycosaminoglycans, frequently hyaluronic acid, forms the capsule.

Gram-negative organisms feature other characteristics:

1. A cytoplasmic membrane, surrounded by a periplasmic space, and a second phospholipid bilayer, comprising an outer membrane.
2. Lipopolysaccharides are inserted into the outer membrane and extend into the environment.
3. A thin layer of peptidoglycan lies between the cytoplasmic membrane and the outer membrane.

Samples of the Kwädąy Dän Ts'ìnchį man's lung and heart tissues were processed and mounted on slides for light microscopy. The tissue slides were stained with H & E and also Ziehl-Neelsen (ZN) to detect *Mycobacterium tuberculosis* and stained with Grocott-Gomori methenamine-silver to detect for evidence of fungi. Positive controls were used for both of the additional stains.

Results and Discussion

OUR KWÄDĄY DÄN TS'ÌNCHĮ HUMAN REMAINS investigative team was assembled in order to provide a multidisciplinary approach to explain the presence of microorganisms observed in various tissues (Monsalve et al. 2008b). Our team reported 1) successful PCR amplification of *Mycobacterium tuberculosis* complex DNA from mediastinal lymph node, myocardium and lung tissues by team members at the University of Saskatchewan; 2) confirmation of PCR amplification of *Mycobacterium tuberculosis* complex DNA by team members at the Hebrew University; and 3) identification of a prokaryote in heart, bicep muscle and humerus tissues by team members at UBC using electron microscopy. Spigelman and his team were unable to culture microorganisms from Kwädąy Dän Ts'ìnchį tissue samples. An autopsy of the remains provided

no clinical evidence of tuberculosis in pulmonary and extrapulmonary tissues (Straathof and Beattie 2008). In light of this, we suggest that the presence of *Mycobacterium tuberculosis* complex DNA and the presence of cellular ultrastructures of microorganisms are indicative of the presence of subclinical evidence of tuberculosis. These findings led us to search previous studies of ancient remains for reports of clinical or subclinical evidence of tuberculosis and the simultaneous presence of *Mycobacterium tuberculosis* complex DNA (tables 1 and 2). We found that the presence of *Mycobacterium tuberculosis* DNA correlated with the presence of tuberculosis in 8 of the 37 studies of Old World ancient remains. Eight samples of bones—obtained from York (England), Egypt and Hungary—with clinical findings suggesting the presence of tuberculosis failed to amplify for *Mycobacterium tuberculosis* (table 1, page 232; reference notes 2, 6, 7, 11). Four New World studies have provided evidence of tuberculosis associated with *Mycobacterium tuberculosis* complex DNA. Our team's efforts suggest that the Kwädąy Dän Ts'ìnchį man is another example of ancient remains in which no clinical evidence of tuberculosis was observed; however, there was clearly evidence of *Mycobacterium tuberculosis* complex DNA (Monsalve et al. 2008b; Swanston et al. 2008).

Zink et al. (2001) studied 37 skeletal tissue samples from cadavers in the necropolis of Thebes-West, Upper Egypt (2120–500 BC) and the necropolis of Abydos (3000 BC). Their study reported the presence of *Mycobacterium tuberculosis* complex DNA in 5 out of 13 non-specific cases and 2 of 14 cases without pathological bone changes. In a study by Spigelman and Lemma (1993), the presence of tuberculosis DNA was found in one bone with suspected Yaws disease in a pre-European contact individual from Borneo.

Tables 1 and 2 (page 236) present demographic characteristics, dates and clinical findings of ancient remains with *Mycobacterium tuberculosis* found in both Old and New Worlds, respectively. In our study of the Kwädąy Dän Ts'ìnchį man's remains, subclinical presentation was consistent with presence of *Mycobacterium tuberculosis* complex DNA.

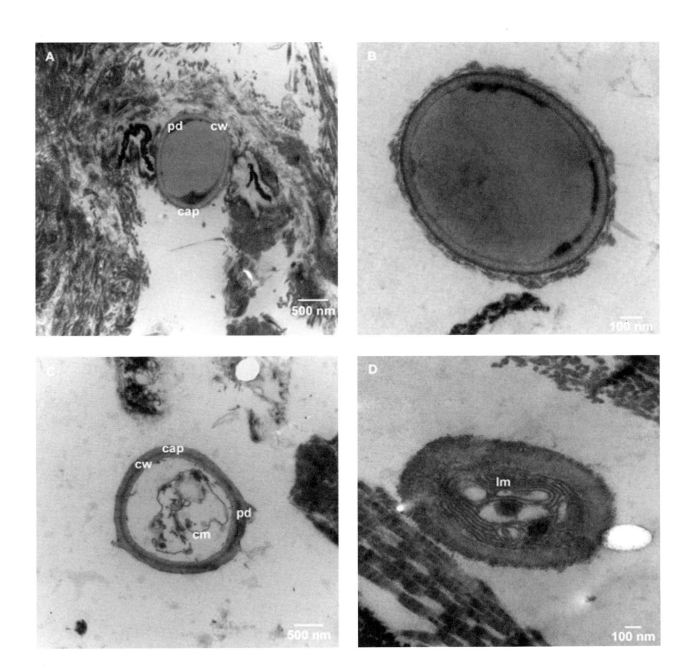

Figure 1. Electron micrographs of microorganisms in arm muscle. *A*, Single Gram-positive cell in start of division: *cap*, capsule; *cw*, cell wall; *pd*, plane of division. *B*, Bacterium or fungal spore. *C*, Gram-positive coccus. Note the layered effect seen in the peptidoglycan layer of the cell wall. The cytoplasmic membrane has collapsed during processing. *cap*, capsule; *cm*, cytoplasmic membrane; *cw*, cell wall; *pd*, plane of division. *D*, Gram-positive spore-forming bacterium: *lm*, lamellar mesosome.

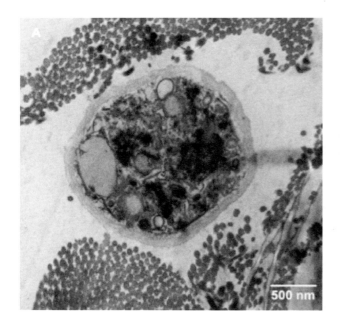

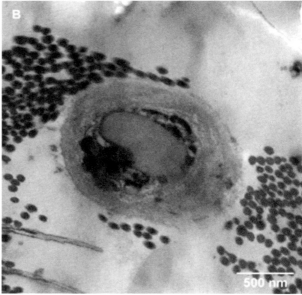

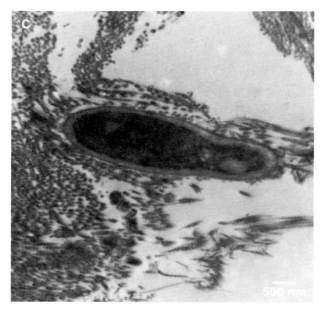

Figure 2. Electron micrographs of microorganisms in arm muscle.

Light microscopy studies using the ZN stain for *Mycobacterium tuberculosis* and the Grocott stain for fungi both failed to provide positive results in the Kwädąy Dän Ts'ìnchį man's heart and lung tissues. On the other hand, TEM demonstrated the presence of microorganisms in heart and lung tissue samples.

A variety of microorganisms was observed in the skin, muscle and humerus of the upper arm. In the arm muscle we observed some microorganisms that were clearly prokaryotes (fig. 1), but other microorganisms contained intracellular material, suggesting that they were eukaryotic (fig. 2).

Microorganisms observed in samples from the humerus appeared to be a mixture of Gram-negative and Gram-positive organisms (fig. 3). The designation of Gram-positive or Gram-negative here is based solely on the ultrastructural appearance of the cell walls of these prokaryotes. The presence of these microorganisms in essentially superficial tissues suggests that they are of environmental origin. Environmentally introduced microorganisms could come from either contamination at the time of discovery or a thawing/refreezing episode that occurred sometime between death and discovery (Monsalve et al. 2008a). We suspect the latter to be the case here, since great care was taken to avoid contamination at the time of recovery and since our team was unable to culture microorganisms from the tissues. The presence of adipocere in muscle tissues

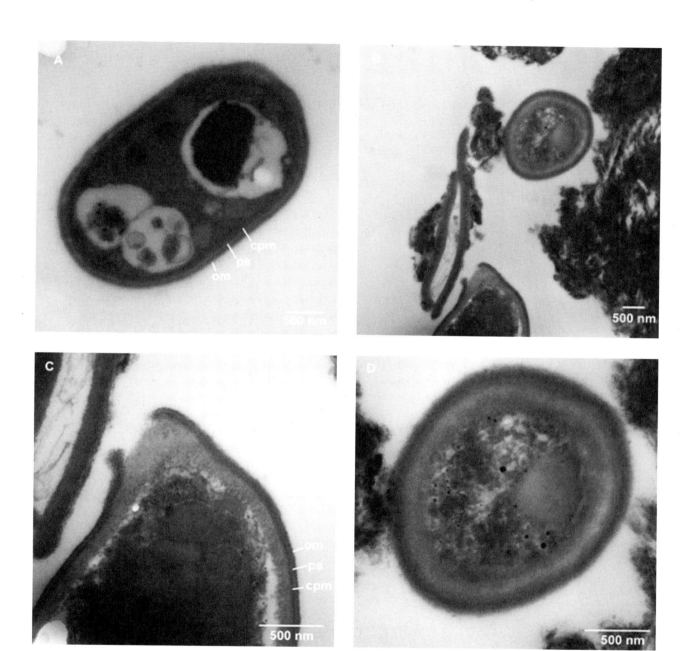

Figure 3. Electron micrographs of microorganisms in humerus. *A,* Gram-negative bacillus with double membrane: *om,* outer membrane; *ps,* periplasmic space; *cpm,* cytoplasmic membrane. *B,* Gram-positive bacterium, *top,* and Gram-negative bacterium, *bottom. C,* Higher magnification of Gram-negative bacterium in *B* with double membrane: *om,* outer membrane; *ps,* periplasmic space; *cpm,* cytoplasmic membrane. *D,* Higher magnification of Gram-positive bacterium in *B.*

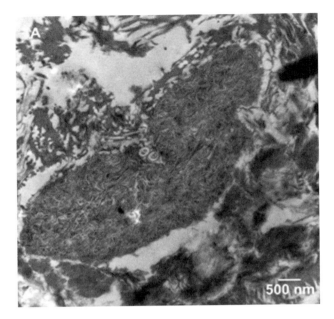

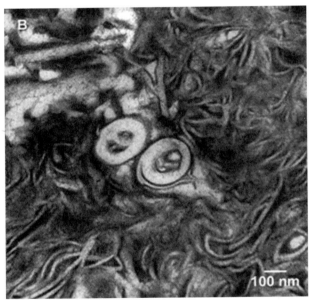

Figure 4. Electron micrographs of microorganisms in lung tissue. *A*, low power of microorganisms. *B*, Higher magnification of microorganisms in *A*. We believe this conglomerated material containing two microorganisms represents a mass of surfactant that originated in alveoli. Note connective tissue at the right an alveolar tissue at the upper left.

from the arm and shoulder also clearly supports the proposition that a thaw/freeze episode did occur (Liu et al. 2010; Christensen et al. 2010). The severing of the torso by glacier slumping (Beattie et al. 2000) also adds weight to the argument for an environmental origin for these microorganisms. All of these findings reinforce the position that microorganisms probably had no role in the young man's health.

The autopsy of the Kwädą̈y Dän Ts'ìnchį individual did not reveal any obvious cause of death. The pathological examination of his lungs revealed only evidence of the accumulation of carbon, presumably from smoke (Straathof and Beattie 2008). Despite this, the observation that *Mycobacterium tuberculosis* complex DNA was amplified from myocardium, mediastinal lymph node and lung tissue suggests the presence of subclinical pathogenesis. These findings are further supported by our observation of apparently encysted microorganisms in the heart (Monsalve et al. 2008a) and in the lung (shown in fig. 4).

Conclusion

THROUGH MICROSCOPY, we identified microorganisms in the Kwädą̈y Dän Ts'ìnchį man's remains that were probably acquired both pre- and post-mortem. We found 14 different microorganisms (eukaryotes and prokaryotes) in an extensive TEM search of skin, muscle and humerus tissues of the left arm. The finding of microorganisms along with small amounts of adipocere suggests these microorganisms were acquired post-mortem, as a result of at least one thawing/refreezing episode. The chemical identification of fatty acids further confirms the presence of adipocere in the arm and back tissues (Liu et al. 2010; Christensen et al. 2010). The proposition that the microorganisms (prokaryotes) observed in the lung and heart tissues were acquired pre-mortem and not during the recovery process is supported by the fact that attempts to isolate microorganisms in the internal organs were not successful. Molecular, cellular biological and bacteriological tools helped to confirm the presence of

Mycobacterium tuberculosis in the Kwädąy Dän Ts'ìnchį individual's remains. Our multidisciplinary approach has made it possible to consolidate the available knowledge concerning the Kwädąy Dän Ts'ìnchį man's state of health at the time of his death.

ACKNOWLEDGEMENTS

We thank the Champagne and Aishihik First Nations members for their active interest and support, and the Kwädąy Dän Ts'ìnchį committee—Al Mackie in particular—for its help in the execution of this project. We thank the Royal BC Museum for facilitating access to the samples. We are grateful to Drs. Maria Juana Aristizabal, Julian Davies, Brian Leander and Corine Ong at the University of British Columbia; Dr. Vladimir Yurkov at the University of Manitoba; Dr. Mark Spigelman at the Centre for Infectious Diseases and International Health Department of Infection, Windeyer Institute of Medical Sciences, UCL, UK, and the Kuvin Centre for the Study of Infections and Tropical Diseases, the Hebrew University of Jerusalem; and Dr. Oliver Dutour at the University of Marseille for their reviews of our findings. As well, we thank Jennifer Alsfeld at UBC for her editorial contributions. We acknowledge the financial support received by Claudia Cheung through the UBC Faculty of Medicine Summer Student Fellowship Program (2006) and the Summer Student Research Program (2007), and by Jacksy Zhao through the Summer Student Fellowship Program (2008).

Table 1. Presence of *Mycobacterium tuberculosis* in Old World ancient remains.

Archaeologic/anthropological data	Sample origin/clinical findings	DNA confirmation*
Predynastic necropolis of Adaïma, district of Esna, Upper Egypt; 3400–3300 BC; child, 12–14 years old[1]	Bone samples from rib and vertebral fragment of skeleton exhibiting classic kyphotic signs of Pott's disease and containing lesions suggestive of tuberculous involvement	(+) for *Mycobacterium* spp.
Thebes-West, Upper Egypt; *ca.* 1550–1080 BC; (?) adult[2]	Fusion L4-S1, lumbosacral tuberculosis	(+)
Thebes-West, Upper Egypt; *ca.* 1550–1080 BC; male, < 35 years old[3]	Samples from both lungs with extensive pleural adhesions to chest wall in right thoracic cavity suggestive of pulmonary tuberculosis, and a macroscopically unremarkable left lung	(+) in right lung, (–) in left lung
Thebes-West, Upper Egypt; *ca.* 1550–1080 BC; male, 20–30 years old[2]	Bone samples from lumbar (L4/L5) spine showing signs of advanced-stage vertebral tuberculosis	(+)
Thebes, Egypt; *ca.* 700–600 BC; female, 40–50 years old[4]	Samples from lung tissue and femur with possible tuberculosis	(+) in lung and femur
Negev desert, Karkur; *ca.* 600 AD; male, 35–45 years old[5]	Sample from calcified tissue, possibly lung pleura, in chest cavity suspected of tuberculosis	(+)
Turkey; Byzantine; not described by author[6]	Bone sample from lumbosacral spine suspected of tuberculosis	(+)
York, England; Medieval; not described by author[6]	Bone sample from lumbar vertebra suspected of tuberculosis	(–)
England; Medieval; not described by author[6]	Bone sample from talus suspected of leprosy	(+)
Egypt; 4th–12th dynasty; not described by author[6]	Bone sample from lumbar vertebra suspected of tuberculosis	(–)

Archaeologic/anthropological data	Sample origin/clinical findings	DNA confirmation*
Sükösd cemetery, Hungary; 7th–8th century AD; female, young adult[7]	Bone samples from collapsed and fused thoracic (T3-T6) vertebral bodies consistent with advanced-stage tuberculous spondylitis	(+)
Bélmegyer cemetery, Hungary; 7th–8th century AD; female, advanced-aged[7]	Bone samples from collapsed and fused thoracic and lumbar vertebral bodies consistent with advanced-stage tuberculous spondylitis	(+)
Bélmegyer cemetery, Hungary; 7th–8th century AD; female, advanced-aged[7]	Fusion of L5-S1 and destruction of the acetabulum and femoral head suggestive of tuberculosis	(−)
Pitvaros cemetery, Hungary; 7th–8th century AD; male, middle-aged[7]	Samples from vertebra of bamboo-spine exhibiting bilateral lumbosacral fusion, and those from calcified pleural fragments consistent with ankylosing spondylitis and possible pulmonary tuberculosis	(+)
Pitvaros cemetery, Hungary; 7th–8th century AD; male, young adult[7]	Lumbar vertebrae and acetabulum suspected of tuberculosis	(+)
Pitvaros cemetery, Hungary; 7th–8th century AD; male, middle-aged[7]	Ankylosis of the left ankle suspected of tuberculosis	(−)
Wharram Percy, England; 900–1400 AD; male (?), 50+ years old[8]	Bone samples from thoracic and lumbar spine and ribs affected by possible tuberculous lesions	(+) in all samples
Wharram Percy, England; 1060–1170 AD; female, 25–35 years old[8]	Bone samples from thoracic and lumbar spine, ribs, ulna and ilium affected by possible tuberculous lesions	(+) in all samples
Sant Cristòfol de la Castanya, Montseny, Spain; 12th–13th century AD; adolescent, 14–16 years old[9]	Bone sample taken from anterior edge of superior metaphysis in one of the knees affected by tuberculosis gonarthropathy	(+)
Wharram Percy, England; 1270–1410 AD; female, 35–45 years old[8]	Bone samples from thoracic vertebrae and ribs affected by possible tuberculous lesions	(+)

Archaeologic/anthropological data	Sample origin/clinical findings	DNA confirmation*
Black Death cemetery, London, England; 1350–1538 AD; male, 45 years old[10]	Bone sample from fused wrist suspected of tuberculosis	(+)
Black Death cemetery, London, England; 1350–1538 AD; male, 15–25 years old[10]	Bone samples from two lumbar vertebrae (L1 and L2) with large lytic lesions in the bodies of both	(+) in both specimens
Yangiu, Korean Peninsula; 15th century AD; child[11]	Collapsed thoracic cavity and nodules of the liver suggest tuberculosis	(+)
Kraziai, Lithuania; 15th–16th century AD; male, 18–20 years old[12]	Bone samples from thoracic (T12) vertebra and right femur with lesions suggestive of tuberculosis	(+) in both specimens
Alytus, Lithuania; 15th–17th century AD; male, 45–50 years old[12]	Bone samples from thoracic (T12) vertebra suggestive of tuberculosis and left upper second molar	(+) in both specimens
Alytus, Lithuania; 15th–17th century AD; female, 50–55 years old[12]	Bone samples from lumbar (L1) vertebra suggestive of tuberculosis and left lower second molar	(+) in both specimens
Alytus, Lithuania; 15th–17th century AD; male, 50–55 years old[12]	Bone samples from lumbar (L4) vertebra suggestive of tuberculosis, left upper third molar, and soft tissue	(+) in all specimens
Bácsalmás cemetery, Hungary; 17th century AD; female, juvenile[7]	Bone samples from thoracic and lumbar vertebrae exhibiting periostitis and hypervascularization consistent with possible early-stage tuberculosis	(+)
Bácsalmás cemetery, Hungary; 17th century AD; male, 16–18 years old[7]	Bone samples from thoracic and lumbar vertebrae exhibiting periostitis and hypervascularization consistent with possible early-stage tuberculosis	(+)
Bácsalmás cemetery, Hungary; 17th century AD; male, middle-aged[7]	Samples from three right ribs (fused) and from pleural plaques consistent with probable pulmonary tuberculosis	(+) in bone and pleural samples

Archaeologic/anthropological data	Sample origin/clinical findings	DNA confirmation*
Bácsalmás cemetery, Hungary; 17th century AD; male, young adult[7]	Periostitis and erosive lesions in nine right ribs suggestive of tuberculosis	(–)
Bácsalmás cemetery, Hungary; 17th century AD; male, young adult[7]	Osteolytic change of lumbar vertebrae suggestive of tuberculosis	(–)
Bácsalmás cemetery, Hungary; 17th century AD; male, young adult[7]	Osteolytic changes of right rib suggestive of tuberculosis	(–)
Scotland; 17th–18th century AD; not described by author[6]	Bone sample from lumbosacral spine suspected of tuberculosis	(+)
Dominican Church, Vác, Hungary; 1731–1838 AD; male, 56 years old[13] (++)	Tissue sample from right lung with evidence of tuberculosis	(+)
Historical pathological collection from the files of the Institute of Anthropology, University of Gothenburg; end of 19th century to mid 20th century AD[14]	Bone samples from two femora and one skull affected by tuberculosis	(+) in all three specimens

*(+) is positive for *Mycobacterium tuberculosis* DNA; (–) is negative for *Mycobacterium tuberculosis* DNA.

(++) is positive for acid-fast intracellular bacteria; (--) is negative for acid-fast intracellular bacteria.

[1]Crubézy et al. 1998; [2]Zink et al. 2001; [3]Nerlich et al. 1997; [4]Minnikin et al. 2007; [5]Donoghue et al. 1998; [6]Spigelman and Lemma 1993; [7]Haas et al. 2000; [8]Mays et al. 2001; Mays et al. 2002; [9]Baxarias et al. 1998; [10]Taylor et al. 1996; Taylor et al. 1999; [11]Donoghue et al. 2007; [12]Faerman et al. 1997; [13]Fletcher et al. 2003; [14]Baron et al. 1996.

Table 2. Presence of *Mycobacterium tuberculosis* in New World ancient remains.

Archaeologic/anthropological data	Sample origin/clinical findings	DNA confirmation[*]
Chiribaya Alta, Osmore Valley, Peru; 910 AD ± 44 years; female, 40–45 years old[1]	Tissue samples from calcified right hilar lymph node and calcified lesions in upper lobe of right lung consistent with pulmonary tuberculosis	(+) in both lymph node and lung tissue
Arica, Chile; *ca.* 1000 AD; female, 12 years old[2] (⁺⁺)	Bone samples from vertebrae with multiple lytic lesions and exhibiting kyphosis consistent with Pott's disease	(+)
Schild Cemetery, Illinois, USA; 1020 AD ± 110 years; female, 21–22 years old[3]	Bone sample from thoracic (T11) vertebra with extensive central vertebral body necrosis consistent with osseous tuberculosis	(+)
Uxbridge Ossuary, near Toronto, Canada; 1440 AD ± 30 years; unknown gender, adult[3]	Bone sample from two fused, partially collapsed lumbar vertebrae with lesions highly suggestive of tuberculosis	(+)

[*](+) is positive for *Mycobacterium tuberculosis* DNA; (–) is negative for *Mycobacterium tuberculosis* DNA.

(⁺⁺) is positive for acid-fast intracellular bacteria; (⁻⁻) is negative for acid-fast intracellular bacteria.

[1]Salo et al. 1994; [2]Arriaza et al. 1995; [3]Braun et al. 1998.

14

THE PARASITOLOGY AND ENTOMOLOGY OF THE KWÄDĄY DÄN TS'ÌNCHĮ FIND

Bruce J. Leighton, Gail S. Anderson, John M. Webster, Niki Hobischak and Michael Petrik

The presence of parasites and insects in human remains provides clues to diet, state of health before death, living conditions of the host and even the cause, time or season of death. We examined skin, muscle tissue, gut contents and hair rinsate from the frozen body (recovered from a BC glacier in 1999) for insects and common parasites that would furnish information about the life, times and death of the Kwädąy Dän Ts'ìnchį man.

SKIN FROM THE LEFT AXILLA was examined for the itch mite, *Sarcoptes scabiei*. The microscopic mites live in burrows in the skin—usually the skin of the hands, elbows, axillae, groin and buttocks—and the burrows may contain adults, eggs and fecal pellets of the mites. These mites are transmitted by human contact and spread rapidly between people in crowded conditions.

Tissues from the diaphragm and the intercostal muscles were examined for cysts of the nematode *Trichinella* spp., which causes trichinosis. Humans are infected by *Trichinella* when they eat raw or inadequately cooked pork, bear, walrus or horse meat containing infective larvae of the parasite. The worms mature in the intestinal mucosa and deposit motile larvae that can penetrate any tissue, preferring the active, striated muscles of the diaphragm, larynx, tongue and ribs (intercostals). Once in a muscle cell, a larva will form a small cyst (400 x 260 μm). Five larvae per gram of body muscle can cause death in humans (Garcia 2007).

Gut contents were examined for helminth eggs, including those of trematodes, cestodes and nematodes. These eggs are often thick-shelled, armoured against digestion and desiccation. Helminth eggs have previously been isolated from ancient, frozen human remains—for example, whipworm eggs from a frozen Incan body in Chile (Beaver et al. 1984) and in Ötzi, the neolithic glacier mummy found in the Alps (Aspöck et al. 1996).

Rinsate from the hair mass of the Kwädąy Dän Ts'ìnchį man was examined for insects, including lice (*Pediculus humanus*), plant material or any artifacts. Lice have been found on natural mummies from Greenland and the Aleutian Islands (Hansen et al. 1991).

Methods

SAMPLES WERE ACQUIRED during the autopsy of
the thawed body in a lab at the Royal BC Museum in
Victoria, BC, and examined at Simon Fraser University
in Burnaby. In order to search for the itch mite, *Sarcoptes
scabiei*, skin from the left axilla of the frozen body was
scraped and macerated with a scalpel, then suspended
in mineral oil for microscopic examination. Oil was
used instead of saline for two reasons: it enhances
visualization of the mites because of the greater
refractive difference between the mite and the oil,
and it does not dissolve fecal pellets.

In order to study tissue from the diaphragm and the
right side intercostal muscle for *Trichinella* spp., thin pieces
of tissue were pressed between glass plates and examined
in transmitted light for the capsules or unencapsulated
larvae of *Trichinella*. Both tissue types were also artificially
digested in a mixture of pepsin and hydrochloric acid, and
the sediment was searched under the microscope for the
digestion-resistant, first-stage larvae.

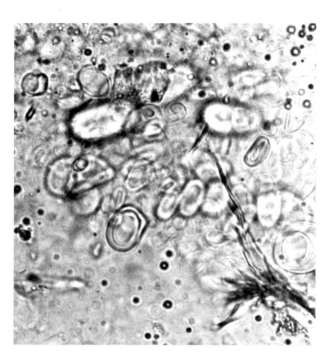

Figure 2. Colourless and incompletely formed eggs of *Diphyllobothrium* sp.
in a matrix of tissue that may be the decaying body of the tapeworm in the
Kwädąy Dän Ts'inchį man's small intestine.

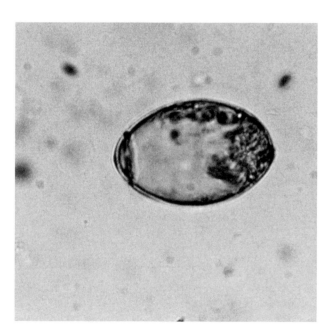

Figure 1. Egg of the tapeworm *Diphyllobothrium* sp., from the Kwädąy Dän
Ts'inchį man's small intestine (length 56–66 µm, width 38–45 µm).

A series of samples taken from parts of the gut
ranging from the stomach to the rectum was examined
for helminth eggs using two methods. Each sample was
suspended in saline and filtered through cheesecloth,
with additional saline washes of the cheesecloth. Part of
each filtrate was transferred to a petri dish and searched
for eggs under a dissecting microscope. The other part
of the filtrate was centrifuged in zinc sulfate solution to
float the eggs out of the sample. After centrifugation,
the fluid at the top of the centrifuge tube was collected
with a pipet, transferred to a petri dish and searched for
eggs with the dissecting microscope.

The wash-water rinsate from the hair mass was
examined under the dissecting microscope.

Results and Discussion

NO EVIDENCE OF PARASITES was found in the skin from the axilla or in the muscle tissue of the diaphragm or intercostal muscle.

Eggs of the fish tapeworm *Diphyllobothrium* sp. were found in large numbers in the contents of the small intestine and in lesser numbers in the descending colon and the rectum (fig. 1) (Dickson et al. 2004). The cestode eggs varied in colour from yellow-brown, tanned eggs to almost colourless, and there were many asymmetrical, colourless objects that may be eggs in an earlier stage of formation. Some of these objects and colourless eggs were found in clumps of a colourless matrix, which may be the decaying body of an adult worm (fig. 2). Hilliard (1972, 586) describes developing eggs from the uterus of diphyllobothriid cestodes as "asymmetrical, thin shelled, cohesive and plastic". This description fits the objects found in the samples from the small intestine. Mudie et al. (see chapter 31) reports finding the rectangular segments (proglottids) of the tapeworm in samples from the colon.

The measurement of 50 tanned eggs from the small intestine showed the following dimensions: mean length 61.6 (SD = 1.99) μm; mean width 41.7 (SD = 1.71) μm; a range of 56–66 μm length and 38–45 μm width, with a length by width ratio of 1.48 (SD = 0.68). These means are smaller than the mean dimensions for *Diphyllobothrium latum* eggs given by Rausch and Hilliard (1970) or Anderson and Halvorsen (1978), and the length by width ratio is higher than that reported for *D. latum* by Anderson and Halvorsen (1978).

Identification of *Diphyllobothrium* species from egg dimensions or form is uncertain. Meyer (1966) questioned the validity of egg dimensions and form for species identification because of the intraspecies variability of these factors. Anderson and Halvorsen (1978) showed a large, interspecies overlap in the dimensions of the eggs, and that egg size varied with the definitive host species, with the intensity of infection and with the age of the mature worm over the first 10 days of egg production. Reinhard and Urban (2003) found that eggs of *D. pacificum* from Chinchorro mummies in Chile (4,000–5,000 years ago) were smaller than those from contemporary samples, suggesting that the measured eggs were immature eggs released during post-mortem decomposition of the worms.

Some sources have reported that *D. latum* can be identified by a characteristic knob on the abopercular end of the egg's shell. While this characteristic was seen on some of the eggs from the Kwäday Dän Ts'ìnchi body, its presence was inconsistent. Using scanning electron microscopy to examine the external surface of the eggs of three species of *Diphyllobothrium*, including *D. latum*, Anderson and Halvorsen (1978) found no differences of taxonomic value at the species level.

While *D. latum* is the most common species infecting humans in Alaska, Rausch and Hilliard (1970) considered it possible it was introduced there by Europeans in the 18th century, although evidence from coastal middens indicates that the genus *Diphyllobothrium* pre-dates first European contact in North America (Bathurst 2005).

The presence of cestode eggs indicates that raw or undercooked fish was a part of the Kwäday Dän Ts'ìnchi man's diet. It can take 7–14 days from the time that infected fish is eaten until the tapeworm matures and begins to release eggs in a human host. However, these worms can survive and continue producing eggs in humans for 25 years (Schmidt and Roberts 1989), so the infection could have been acquired at any time during the Kwäday Dän Ts'ìnchi man's approximately 20 years of life.

Auger soil samples from coastal shell middens in British Columbia have shown evidence of several intestinal parasites, including *Diphyllobothrium* spp., as well as *Ascaris lumbricoides* and *Nanophyetus salmincola* (Bathurst 2005). The evidence from these middens suggests that such parasitism dated back at least 5,500 years.

These tapeworms have many fish hosts. McDonald and Margolis (1995) list the Canadian records of the fish hosts for *D. dendriticum*, *D. ditremum*, *D. latum* and *Diphyllobothrium* sp. Of the 32 species of freshwater and anadromous fishes of the north coast rivers of BC listed by Troffe et al. (chapter 20), 14 are listed by McDonald and Margolis as hosts of one or more *Diphyllobothrium* species. These include Sockeye Salmon (*Oncorhynchus nerka*) and other Pacific salmon. Pieces of fish found with the Kwäday Dän Ts'inchi man's body were identified as *O. nerka* (chapter 20).

Examination of the hair rinsate yielded an insect wing, a plant leaf and fish scales. No evidence of lice was seen. The insect wing was cross-referenced against a known insect wing catalogue (Borror 1989) and was identified as belonging to the insect family Asilidae, suborder Brachycera. Identification was possible by examination of the rounded arch venation pattern of the wing, typical of the Asilidae. The family Asilidae is a large group composed mainly of robber and grass flies, with nearly 1,000 North American species (Borror 1989).

Fish scales were also found in the hair rinsate. These were cross-referenced against a field manual of known North American fish scales and showed many characteristics of Salmonid scales. Although the species was not determined from the scales in the hair rinsate, subsequent DNA studies confirmed the fish found with the body to be *O. nerka* (chapter 20). The plant leaf from the hair rinsate is from a Dwarf Willow commonly found in high alpine areas (Mathewes, pers. comm.). The rounded shape and veinous structure of the plant are characteristic of the Dwarf Willow species *Salix stolonifera*, presently found only in high alpine areas extending far north into Alaska, well beyond the British Columbia boundary.

The fish scale provides evidence of the man's diet and suggests he might have had fishing skills. The salmon meat in his possession indicates that he used fish as a food staple when travelling great distances.

There was no evidence of blow fly (Diptera: Calliphoridae) or other carrion-feeding insect colonization seen, either in the form of insect remains or insect-related tissue damage. Carrion insects are attracted to remains shortly after death, when conditions are suitable for colonization. Such insect colonization and development on a body is commonly used to estimate the time elapsed since death (Byrd and Castner 2001). The lack of insect colonization of the Kwäday Dän Ts'inchi man's remains could mean that conditions were not suitable or that the body was inaccessible. For insects to colonize remains, a number of criteria must be met, including a suitable temperature, season, geographical area, habitat and level of exposure of the remains (Anderson 2005). Season and temperature are the most limiting factors for insect colonization. Palynological evidence suggests that the Kwäday Dän Ts'inchi man's death occurred when the plant *Salicornia* was producing pollen, which would suggest the end of July and all of August (Mudie et al. 2005). In temperate climates today, this would suggest very suitable conditions for insect colonization. Climatic conditions at the discovery site at the time of death are unknown, but even today there is the possibility of early snowfall in those months, or even snow on the ground persisting from previous years (Mackie, pers. comm.). The high altitude alone (1,600 metres, Beattie et al. 2000) would not limit insect colonization, as several studies have collected large numbers of carrion insects at much higher altitudes—albeit in very different and much warmer geographical areas (Martinez et al. 2007; de Jong and Chadwick 1999; Smith and Heese 1995; Baumgartner and Greenberg 1985). Had snow been present, cold temperatures alone could have prevented insect colonization. If conditions were warm enough for colonization, then it is possible the remains were inaccessible to insects or other scavengers, suggesting that the remains were covered rapidly—although insects will colonize shallowly buried remains when other conditions are suitable (VanLaerhoven and Anderson 1999).

Overall, these findings help us verify that the Kwäday Dän Ts'inchi man was indeed a native inhabitant of the northwest coast of North America and that he consumed local fish for food, which he carried with him on his long journey in what is now British Columbia.

ACKNOWLEDGEMENTS

The authors thank Owen Beattie for his invitation to join the study and Al Mackie for his work on the organization of the studies and symposium. We thank Steve Halford, Zamir Punja, Rolf Mathewes and Mark Skinner of Simon Fraser University, and James Dickson of the University of Glasgow for their assistance.

15

MICROBIAL DNA ANALYSIS OF THE KWÄDĄY DÄN TS'ÌNCHĮ INDIVIDUAL'S TISSUES

The Identification of Helicobacter pylori *and* Mycobacterium tuberculosis *DNA*

Treena M. Swanston, Monique Haakensen, Harry Deneer and Ernest Walker

In 1999 human remains were discovered eroding out of a glacier in Tatshenshini-Alsek Park in northern British Columbia. The find was on the traditional land of the Champagne and Aishihik First Nations (CAFN) and given the name Kwädąy Dän Ts'ìnchį, meaning "Long Ago Person Found". Radiocarbon dating of bone samples and artifacts associated with the Kwädąy Dän Ts'ìnchį individual resulted in an estimated time frame of 1720–1850 AD (Richards et al. 2007; see also chapter 6, this volume). An agreement was made between the CAFN and the BC government allowing the remains and artifacts to be studied. Our research team proposed to extract and amplify the microbial DNA associated with various tissue samples in order to identify any pathogens (disease-causing microorganisms) or normal microbial flora that may have been present.

AS DNA STAYS INTACT LONGER in cold environments (Lindahl 1993), there was a high potential for the amplification of microbial DNA due to the individual's encasement within the glacier. Enzymatic amplification of DNA using polymerase chain reaction (PCR) has been successfully used by researchers of ancient DNA for several decades, and the field of ancient DNA continues to evolve (Kaestle 2002; Pääbo et al. 2004). It is important to note that strict precautions and protocols are necessary for ancient DNA work due to the constant presence of modern DNA (Roberts and Ingham 2008).

DNA is amplified by combining a number of components, including template DNA (extracted from the sample tissue); a thermostable DNA polymerase; deoxynucleotide triphosphates (basic DNA building blocks); $MgCl_2$; buffers; and primers (short strands of known DNA that allow for the amplification of specific regions of template DNA). In our analyses, we began by using universal bacterial primers that are complementary to conserved regions in all bacterial DNA. This approach was unsuccessful, however, because the universal primers amplified all bacteria that were present in the sample. It was difficult to determine if the amplified DNA was originally associated with the tissue samples from the Kwädąy Dän Ts'ìnchį individual or if it was a modern contaminant. As a way of controlling for modern DNA contamination, primers were chosen that would only amplify specific bacteria—namely, *Helicobacter pylori* and *Mycobacterium tuberculosis*. In the following section, we expand on the reasons those two particular bacteria were chosen. Additional information on this microbial DNA analysis can be found in Swanston (2010).

Helicobacter pylori

Helicobacter pylori are Gram-negative, helical bacteria that infect approximately half of the world's population (Carroll et al. 2004). The bacterium *H. pylori* was first isolated from stomach tissue in 1984 (Marshall and Warren 1984) and has since been recognized as the primary cause of gastric ulcers and gastritis, although only 15 per cent of infected individuals present clinical symptoms (Atherton 2006). This infection is common in developing countries but is on the decline in industrialized countries (Ghose et al. 2002). Transmission of *H. pylori* most often occurs within families, thereby connecting a particular *H. pylori* strain to a specific geographical area (Kersulyte et al. 2000).

Scientists have sequenced the complete genome of *H. pylori*, and polymorphisms have been found between the DNA of many strains. The nucleotide sequences of two of the virulence-associated genes, *vacA* and *cagA*, are commonly used for typing *H. pylori* strains. The *vacA* gene contains two variable regions: the signal (s) region, which encodes the signal peptide for post-translational information, and the middle (m) region, which is related to the toxicity of the bacterium. These variable regions are present in several forms, also referred to as alleles, which allow for further classification of *H. pylori* strains. Studies have shown that variations in the *vacA* m and s regions correspond with different populations because limited horizontal transmission has resulted in the co-evolution of this pathogen with its human host, and this information has been used to trace ancient population movement (Kersulyte et al. 2000; Falush et al. 2003; Carroll et al. 2004). The *vacA* s region consists of either the s1a, s1b, s1c or s2 allele, whereas the m region is composed of the m1a, m1b or m2 allele. It has been shown that type m1 is found more often with *H. pylori* infections that cause disease (Aviles-Jimenez et al. 2004).

Scientists have debated if *H. pylori* bacteria were present in the New World prior to the arrival of Europeans. Phylogenetic analyses based on the nucleotide sequences of the *vacA* s and m regions of modern strains indicate that this bacterium was in the New World for millennia (Yamaoka et al. 2002). Antigens associated with *H. pylori* have been identified in South American mummies (Allison et al. 1999), but scientists have suggested that it is possible for antigens from other bacteria to cross-react with *Helicobacter*-specific antibodies used in the tests (Kersulyte et al. 2000).

Mycobacterium tuberculosis

The paleoepidemiological study of tuberculosis is an area that is currently receiving widespread attention (Spigelman and Lemma 1993; Donoghue et al. 2004). This disease is still a worldwide issue, despite advances in chemotherapeutic treatments (Sacchettini et al. 2008). Tuberculosis (TB) is the result of an infection by bacteria from the *Mycobacterium tuberculosis* complex (MTBC), which is a group of rod-shaped bacilli that includes the following species: *M. tuberculosis, M. africanum, M. bovis, M. canettii, M. microti* and *M. caprae.* Since one in three individuals globally is estimated to be infected with MTBC bacteria (Bishai 2000), we used PCR to test the Kwädąy Dän Ts'ìnchį man's tissue samples with primers for MTBC and *M. tuberculosis* DNA even though a pathological and histological examination was negative for any signs of illness.

Mycobacteria are different from other bacteria because their cell walls contain a high concentration of lipids. Scientists identify the bacteria using a Ziehl-Neelsen stain that demonstrates the acid-fast characteristic of their cell walls. But cell deterioration in archaeological samples means this stain is often unsuccessful when testing them. Other identification methods include the analysis of the mycobacterial DNA. Although post-mortem DNA degrades, it has been suggested that the amplification of MTBC DNA fragments from a deceased individual who had an infection is more probable. The high concentration of lipids in the MTBC cell wall may protect the bacterium from enzymatic attack, and MTBC DNA has a high percentage of guanine and cytosine that may aid in DNA stabilization (Donoghue et al. 2004).

Materials and Methods

TISSUE SAMPLES WERE OBTAINED from the September 2000 autopsy of the ancient individual at the Royal BC Museum in Victoria. The pathologist used sterile surgical tools during the sampling to reduce the risk of contamination, and all members of the autopsy team were dressed in appropriate protective clothing. The samples were frozen and placed on ice in an insulated container for the journey to the University of Saskatchewan, where they were stored in a -70°C freezer upon arrival.

Tissue extractions were set up in a separate laboratory with a biological safety cabinet that was surface-cleaned with 10 per cent (v/v) bleach. A sterile scalpel was used to mince the stomach tissue into small fragments, and the DNA was extracted using the tissue protocol with the QIAamp DNA Mini Kit (QIAGEN Inc., Mississauga, Ontario). DNA was also extracted from additional tissues, including samples from the lung, a mediastinal lymph node, myocardium, descending colon, small intestine, liver, cecum, rib and skin using a modified protocol for a silica-based method (QIAquick PCR Purification Kit, QIAGEN) that was developed for ancient DNA work (Yang et al. 1998).

Three sets of previously published PCR primers were used for the amplification of the *Helicobacter pylori vacA* variable regions (table 1). For the *Mycobacterium tuberculosis* study, primers were chosen to amplify regions within the IS*6110* insertion sequence, and *Rv3479, gyrB* and *katG* genes (table 1). PCR products were sequenced directly at the National Research Council Plant Biotechnology Institute (Saskatoon, Saskatchewan).

The DNA sequences were compared with reference sequences in the National Institute of Health (NIH) GenBank database, and phylogenetic analysis was used for the determination of evolutionary relationships. Sequences were aligned with the ClustalX software program (Thomson et al. 1997), and DNA alignments were subsequently visualized and manually edited using the GeneDoc software program (Nicholas et al. 1997). All phylogenetic trees were produced and visualized using MEGA4 (Molecular Evolutionary Genetics Analysis software version 4.0, Tamura et al. 2007). Tree topology was evaluated using Minimum Evolution, Maximum Parsimony, Neighbor-Joining, and Unweighted Pair Group Method of Arithmetic Means algorithms. A bootstrap test (Felsenstein 1985) of 1,000 replicates was performed.

Table 1. PCR primers for the amplification of regions within *Hp vacA, MTB* IS6100, *Rv3479, gyrB* and *katG.*

Region	Primers		Product size	Annealing temp. / Number of cycles	Source
vacA s	VA1F	ATGGAAATACAACAAACACAC	s1 176 bp	50°C	Atherton et al. 1995
	VA1XR	CCTGARACCGTTCCTACAGC	s2 203 bp	40 cycles	van Doorn et al. 1998
vacA m	MF1	GTGGATGCYCATACRGCTWA[a]	m1 107 bp	50°C	van Doorn et al. 1998
	MR1	RTGAGCTTGTTGATATTGAC[a]	m2 182 bp	40 cycles	van Doorn et al. 1998
	y98vacAmF	CCTTGGAATTATTTTGACGC	m1 479 bp	58°C	Yamaoka et al. 1998
	y98vacAmR	ATCCATGCGGTTATTGTTGT	m2 488 bp	45 cycles	Yamaoka et al. 1998
IS6110	P1	CTCGTCCAGCGCCGCTTCGG	123 bp	68°C	Eisenach et al. 1990
	P2	CCTGCGAGCGTAGGCGTCGG	40 cycles		
Rv3479	MTB1	ATGTGTAGCAGACCAGCGAT	156 bp	62°C	This study
	MTB2	GGCAAGTTGCGTCAAGGT	40 cycles		
gyrB	TSgyrBF	CACATCAACCGCACCAAGAAC	203 bp	64°C	This study
	TSgyrBR	TTGTTCACCACCGACGTCAG	45 cycles		
	TSgyrB2F	ACACCATCAACACCCACGAG	209 bp	64°C	This study
	TSgyrB2R	CAACTTGGTCTTGGTCTGGC	45 cycles		
	TSgyrB3F	AGGTCAGCGAACCGCAGTTC	200 bp	64°C	This study
	TSgyrB3R	CACCAACTCTCGTGCCTTAC	45 cycles		
katG	katGF	TCAGCCACGACCTCGTCGG	163 bp	68°C	Zink and Nerlich 2004
	katGR	AGGCGGATGCGACCACCGTT	45 cycles		

[a] R is A or G, W is A or T, and Y is C or T

Results

Helicobacter pylori

We identified *Helicobacter pylori* DNA in the stomach tissue of the Kwädąy Dän Ts'ìnchį individual and typed the bacterial strain as *vacA* s2-m2a/m1d (fig. 1). Interestingly, the s region of the Kwädąy Dän Ts'ìnchį man's *H. pylori vacA* gene shares a remarkably high per cent identity (98 per cent at the nucleic acid level) to two previously identified Alaskan strains (Yamaoka et al. 2002; see fig. 2). Unfortunately, GenBank does not contain sequences of any British Columbia or Yukon *H. pylori* strains for comparison with the Kwädąy Dän Ts'ìnchį man's strain.

Mycobacterium tuberculosis

Mycobacterium tuberculosis complex (MTBC) DNA was initially identified in the Kwädąy Dän Ts'ìnchį man's lung tissue through the amplification of an IS6110 insertion sequence DNA fragment. Sequencing of the amplicon confirmed that the PCR product was identical to reference MTBC DNA sequences. Because one of the biggest concerns when working with ancient DNA is the possibility of contamination with modern DNA, it was necessary for these results to be confirmed in a second laboratory. A sample of the lung tissue was re-tested in the laboratory of Dr. Mark Spigelman (Hebrew University of Jerusalem, Israel), and their PCR and sequencing results were also positive for MTBC DNA.

Further characterization of the bacterial strain included the successful amplification of 156 bp of the Rv3479 gene that is complete in *M. tuberculosis* but contains a deleted region in *M. bovis*, which is one of the MTBC members known to cause human tuberculosis. Additional single gene analyses of the katG and gyrB genes were also included to determine that the infection was due to *M. tuberculosis* and not one of the other MTBC members. Interestingly, the 163 bp amplified with katG primers was found to be 100 per cent identical to DNA sequences from *M. tuberculosis* strains associated with European populations.

Discussion

Helicobacter pylori

The Kwädąy Dän Ts'ìnchį man's *H. pylori* strain possesses a *vacA* gene type s2 with a hybrid m2a/m1d region. Although hybrid m regions have been previously described (Pan et al. 1998), the hybrid m region possessed by the *H. pylori* strain isolated from the Kwädąy Dän Ts'ìnchį individual does not match any known DNA sequences in GenBank. Moreover, the combination of an s2 and m1 type has rarely been seen in other phylogenetic studies of *H. pylori* (Morales-Espinosa et al. 1999; Letley et al. 1999). The m1d-type *vacA* has previously been shown to be an indicator of an original indigenous strain (Yamaoka et al. 2002), but it has never before been found in conjunction with

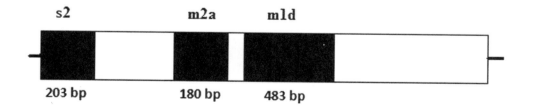

Figure 1. Schematic diagram of the Kwädąy Dän Ts'ìnchį man's *H. pylori vacA* gene. Shaded regions indicate areas that were amplified by PCR in this study.

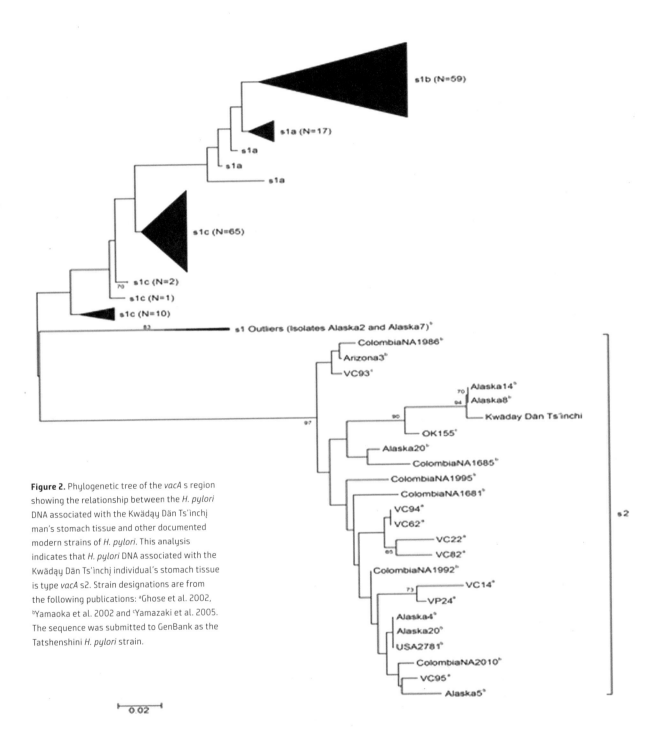

Figure 2. Phylogenetic tree of the *vacA* s region showing the relationship between the *H. pylori* DNA associated with the Kwädạy Dän Ts'inchị man's stomach tissue and other documented modern strains of *H. pylori*. This analysis indicates that *H. pylori* DNA associated with the Kwädạy Dän Ts'inchị individual's stomach tissue is type *vacA* s2. Strain designations are from the following publications: [a]Ghose et al. 2002, [b]Yamaoka et al. 2002 and [c]Yamazaki et al. 2005. The sequence was submitted to GenBank as the Tatshenshini *H. pylori* strain.

an s2-type s region. These findings suggest that the *H. pylori* strain present in the Kwädąy Dän Ts'ìnchį man's stomach tissue is an ancient strain that has evolved (or become nonexistent) over the years that have passed since his death.

Mycobacterium tuberculosis

It was surprising to discover that *M. tuberculosis* DNA was associated with the lung tissue. Since there was no evidence that the Kwädąy Dän Ts'ìnchi individual was ill, we propose that he may have had a latent tuberculosis infection. Individuals with a latent infection have no symptoms and are not infectious to others, but there is a small chance that they may develop active tuberculosis at some point in their lifetime.

TB is an ancient bacterial disease that still affects modern populations. It is the result of an infection by *M. tuberculosis* complex bacteria that are inhaled and take up residence in the lungs. This infection may not cause symptoms, but activation of the disease can occur years later if there is a change in the individual's immune system (Roberts and Buikstra 2003). According to the World Health Organization (2008), a third of the world's population is infected by the TB bacilli, one in 10 people who have been infected will develop the active disease, and the current death rate is a staggering 1.7 million people per year. In the past, the study of ancient tuberculosis mainly involved the recognition of the disease process on dry bone. Only five to seven per cent of diseased individuals in the pre-antibiotic era had any skeletal involvement, and most often the evidence is apparent only on the vertebral bodies (Aufderheide and Rodriguez-Martin 1998).

It was once controversial to say that TB was present in the New World prior to the arrival of Europeans. Researchers thought a large, sedentary population was required to promote the spread of the disease through overcrowding and poor ventilation (Morse 1961). Disease origins, especially of tuberculosis, have been the focus of many studies. The first documented case of TB in the New World was reported in 1973 (Allison et al. 1973). Evidence of the disease was found in a mummified child, dated 700 AD, from the Nazca culture of southern Peru. White nodules or tubercles were identified on lung and kidney tissue, and Ziehl-Neelsen staining indicated the presence of acid-fast bacilli. Skeletal involvement included part of the lumbar spine.

In the past few decades, paleopathologists have documented more skeletal and soft tissue evidence of tuberculosis in the Americas (Buikstra and Williams 1991). Critics continued to raise concerns that the evidence was not conclusive and that other disease processes, such as osteomyelitis and septic arthritis, have the same effect on tissue. It was noted that even the staining technique used by Allison for acid-fast bacilli was not specific enough to rule out nonpathogenic soil mycobacteria. The first molecular analysis of New World TB was published in 1994, when Salo and colleagues reported on their discovery of *M. tuberculosis* DNA in lung tissue from a female Peruvian mummy dated 1000 AD (Salo et al. 1994). It is now generally accepted that TB did exist in pre-contact Aboriginal populations.

In North America the diagnoses have been based mainly on examination of skeletal material, but molecular technology has been used on samples from two North American sites. Scientists have confirmed the presence of MTBC DNA using PCR in samples from the Uxbridge Ossuary near Toronto (1440 AD) and the Schild site in Illinois (1020 AD) (Braun et al. 1998). Interestingly, an extinct Bison (17,000 BP) was recovered in Wyoming and identified as having an MTBC infection (Rothschild et al. 2001).

There is no conclusive evidence of pre-contact TB on BC's northwest coast or in Alaska, but two cases have been reported with soft tissue involvement due to a possible TB infection. In the 1970s, human remains (400 AD) recovered from St. Lawrence Island, Alaska, were autopsied, and a calcified lymph node was found (Zimmerman and Smith 1975). And near Barrow, Alaska, a female (1510 ± 70 AD) was found with calcified lymph nodes and calcified granulomas in the lungs (Zimmerman 1998). Both cases were linked with histoplasmosis (a fungal disease) because, at the time, it was thought that tuberculosis was not present in the area until after the arrival of the Russians in 1741 (Fortuine 2005).

There are probable TB cases from the historic period in northwest coast BC and Alaska. Ortner (2003) discusses the likely diagnosis of tuberculosis of the hip associated with an Aboriginal skeleton recovered near the Yukon River, Alaska. The female's estimated age was 20, based on epiphyseal union and morphology of the pubic symphysis. The left innominate and femoral head were deformed, and the acetabulum had evidence of healing and remodelling. While the vertebrae are the most common area to be affected in skeletal TB, the hip is the second most likely area.

A second historic-period TB case based on skeletal evidence was identified on Barkley Sound, Vancouver Island (Schulting and McMillan 1995). Archaeologists located and mapped a burial cave with skeletal evidence of nine individuals. Although the remains were not intact, the excavators' goal was to limit disturbance, so they only briefly assessed the skeletal remains. Nine fused thoracic vertebrae were identified with evidence of kyphosis due to the collapse of vertebral bodies T3 to T6. There was no evidence of reactive bone, and the neural arch and spinous processes were not affected—leading to the conclusion that the individuals had suffered from spinal tuberculosis.

Historical records indicate that the first documented case of tuberculosis in a northwest coast Aboriginal person occurred in 1793 at Nootka Sound, where the Spanish had established a post that was occupied between 1790 and 1795. By the 1840s tuberculosis had reached the Tlingit communities near Sitka, an area occupied by the Russians since 1799 (Menzies 1793 cited in Boyd 1999).

Conclusions

THROUGH THE MICROBIAL DNA ANALYSIS of the Kwädąy Dän Ts'ìnchį man's stomach tissue, we have determined that *Helicobacter pylori* bacteria were present in the northwestern coast of North America prior to 1850 AD. Sequence variation in the *H. pylori vacA* gene

is common in the different strains, and we have found the Kwädąy Dän Ts'ìnchį individual's *H. pylori* strain to be a unique hybrid *vacA* type s2-m2a/m1d. This rare s2/m1d combination is an indication of strain evolution, which is further supported by the m2a/m1d hybrid region. We hope that, over time, other ancient *H. pylori* DNA sequences become available in GenBank, as it would be optimal to perform analyses using other ancient DNA samples in order to determine the relationship between the Kwädąy Dän Ts'ìnchį man's *H. pylori* strain and ancient strains from different geographic areas.

The analysis of the Kwädąy Dän Ts'ìnchį man's lung tissue resulted in the discovery of *Mycobacterium tuberculosis* DNA. It is not clear if this individual lived during the late pre-contact or the early historical period, based on the radiocarbon dates (see chapter 6). But the presence of *M. tuberculosis* DNA in his lung tissue lends itself to two theories. It is either the first conclusive evidence of pre-contact TB on the northwest coast, or it might indicate that the Kwädąy Dän Ts'ìnchį individual lived in a time after European contact. Interestingly, the amplified katG DNA sequence contained a single nucleotide polymorphism that suggests the strain is related to European MTBC strains.

ACKNOWLEDGEMENTS

We would first like to thank the Champagne and Aishihik First Nations for their permission to learn more about their ancestor. Also, thanks go to the Kwädąy Dän Ts'ìnchį Management Group, especially Al Mackie, whose advice was always greatly appreciated. In addition, we would like to thank Dr. Vicky Monsalve and Dr. Mark Spigelman, both of whom were instrumental in the confirmation of the TB results. Finally, we thank the University of Saskatchewan for its financial support.

16

ORIGINS OF THE KWÄDĄY DÄN TS'ÌNCHĮ MAN INFERRED THROUGH MITOCHONDRIAL DNA ANALYSIS

Maria Victoria Monsalve

The genes of each organism are composed of deoxyribonucleic acid (DNA) in combinations of four different bases: adenine (A), thymine (T), cytosine (C) and guanine (G). In the nuclei of human cells, some three billion of these bases are configured to form the human genome. The bases are arranged in such a way that each individual has a unique DNA sequence. When individuals are more closely related, we expect more similarity between their sequences.

OUTSIDE THE NUCLEUS OF THE CELL, within the mitochondria, is other DNA known as mitochondrial DNA (mtDNA). The mitochondrion is an organelle localized in the cytoplasm of the cell, and its function is to generate energy (ATP) for the cell. The human mtDNA genome is a closed circular molecule of 16,569 base pairs. It contains a control region of one kilobase base pair composed of hypervariable region 1 (HVI) and hypervariable region 2 (HVII).

Transmission of Mitochondrial DNA (mtDNA)

THE MTDNA GENOME is inherited matrilineally—since sperm carry their mtDNA in their tails and necks, it is left behind during fertilization. Embryos therefore only inherit the mitochondria in the mother's eggs, and it can be traced from individuals back through generations to common ancestors.

Advantages of Using mtDNA in Archaeological and Forensic Samples

EACH CELL CONTAINS many mitochondria and, therefore, many identical copies of the mtDNA. There is only one copy of the nuclear DNA in each cell. This is particularly important since DNA degrades when exposed to environmental factors such as heat, solar radiation and humidity. In older remains, relatively more copies and larger fragments are likely to have survived for mtDNA than for nuclear DNA.

We previously investigated whether DNA could be extracted from the ancient remains of the Kwä̀dą̀y Dän Ts'ìnchį man and be used to identify mtDNA origins (Monsalve et al. 2002). To this end, nitrogen and protein content in the collagen of the hard and soft tissue was quantified to establish protein preservation (Monsalve et al. 2002, 2003). The remains were characterized as belonging to haplogroup A, one of the Indigenous North American mtDNA lineages (Monsalve et al. 2002). In this paper we report the genetic relationship between the mtDNA lineage of the Kwä̀dą̀y Dän Ts'ìnchį person and groups determined by sequencing analysis to belong to haplogroup A. These groups are central and northeast Asian, and Indigenous North American, Central American and Greenland populations.

Material and Methods

METHODS OF EXTRACTION, amplification and sequencing of mtDNA fragments from different Kwä̀dą̀y Dän Ts'ìnchį tissue types in independent laboratories are described elsewhere (Monsalve et al. 2002). Styrofoam boxes containing dry ice were used for sample storage and transportation between different locations, and stringent precautions were taken to avoid contamination by external sources. In an effort to minimize external DNA contamination, samples for DNA analysis were selected from different areas of the remains before any other tissues were removed. Arm muscle and humerus tissues—obtained at the Royal BC Museum in Victoria, BC, where the body was being stored in a secure, sterile containment facility—were divided into pieces for analysis at several other institutions. The mtDNA analysis was done at the University of British Columbia (UBC), Canada; quantification of proteinaceous material at the University of Oxford, UK; and the replication of mtDNA analysis at the University of New Mexico (UNM), US.

Macromolecules Analysis

THE EXTENT of macromolecule preservation was determined by measuring the carbon (%C) and nitrogen (%N) content in two samples of the same bone and a sample of the muscle (Monsalve et al. 2002, 2003) according to the method described by Ambrose (1990).

DNA Analysis

THE SMALL SAMPLE of humerus bone was divided in two, and each piece was cleaned with a Dremel tool fitted with a sanding disc in order to remove surface contamination. Sections of the cleaned bones were sealed in a sterile grinding vial, then placed inside a Spex 6700 Freezer Mill containing liquid nitrogen and pulverized. This powdered humerus was kept at -20°C until DNA extraction. An arm muscle sample was ground in a mortar and pestle with liquid nitrogen.

Three separate DNA extractions were made. One extraction from the humerus was made in April 2000, following the protocol described by Hagelberg and Clegg (1991). A piece of Moose antler found at the same site as the Kwä̀dą̀y Dän Ts'ìnchį man was used as a control for the DNA extraction and amplification of mtDNA. QIAquick columns were used for DNA purification. A second DNA extraction from the humerus was made in May 2000, according to the method described by Yang et al. (1998). The third extraction was from the arm muscle in August 2000, following the protocol by Guhl et al. (1999).

Polymerase Chain Reaction (PCR)

PCR IS AN IN-VITRO PROCESS that enables the replication of a specifically targeted segment of DNA. A standard reaction mixture contains the DNA sample, thermostable Taq DNA polymerase, deoxynucleotide triphosphates and two oligonucleotide primers (Innis and Gelfand 1990). The primers complement opposite ends of the targeted segment. The mixture is heated to denature the template DNA and cooled to allow the primers to anneal to the targeted DNA segment. The primers are then extended by the Taq DNA polymerase. The resulting extension products are complementary to and capable of binding the two oligonucleotide primers. Through repeated cycles of heating, cooling and extending primers, an exponential accumulation of the targeted DNA segment is achieved. Primers, annealing temperatures and amplified products for amplification of the mtDNA control region appear in table 1.

DNA was amplified by PCR in a different laboratory at UBC to prevent transfer of amplified DNA products into unamplified samples. All the instruments were cleaned with bleach or irradiated with ultraviolet (UV). Filter tips were used for PCR. Similarly, extraction of DNA from the humerus was performed in a different laboratory from the amplification at University of New Mexico (UNM).

Restriction Enzyme Analysis

A TECHNIQUE called restriction enzyme analysis allows the detection of variants (called polymorphisms) of the DNA sequences through the presence or absence of specific sites. If a site is present, the DNA fragment submitted to the restriction enzyme digestion will be cut into two pieces. Using a combination of restriction enzyme analysis and the presence of other genetic markers, Indigenous Americans have been clustered into six mtDNA haplogroups (Torroni et al. 1992; Malhi et al. 2007). Haplogroup A is characterized by the presence of a HaeIII site at position 663; haplogroup B by a nine-base pair deletion (an Asian genetic marker);

haplogroup C by the absence of a HincII site at position 13259; haplogroup D by the absence of the AluI site at position 5176; haplogroup X by the absence of the DdeI site at position 1715; and haplogroup M by the presence of the AluI site at position 10397.

At UBC and UNM, DNA was screened for the major haplogroups found in Indigenous Americans using restriction enzymes (Monsalve et al. 2002). Overlapping fragments of the hypervariable region 1 (HVI) of the mtDNA amplified using Taq DNA polymerase were sequenced at UBC and UNM. The amplification consisted of 40 cycles. Details of the procedures are described in Monsalve et al. (2002).

Results

THE QUANTIFICATION of proteinaceous material by determining nitrogen and carbon levels indicated a high likelihood of successful DNA extraction (Monsalve et al. 2002, 2003). Moreover, extraction and analysis of DNA was also successful at UNM.

To validate DNA sequences, we adhered to standard recommendations on the following aspects: 1) physically isolated work area and control amplifications; 2) appropriate molecular behaviour; 3) reproducibility; 4) cloning; 5) independent replication; 6) biochemical preservation; and 7) associate remains (Monsalve and Stone 2005). DNA was extracted in a laminar flow hood. Blanks were used as controls in two DNA extractions of bone and one extraction of soft tissue. DNA sequence data obtained from the three extracts were the same, and repeated amplification of these samples gave consistent results. Direct sequencing obtained at UNM showed the same nucleotide substitutions found by direct sequencing of PCR products and cloning at UBC.

Comparison of a 362-base pair fragment of the control region HVI with the reference sequence (Anderson et al. 1981) showed substitutions at six sites. These six substitutions were found in 1 of 41 Haida analysed (Ward et al. 1991) and 5 of 280 Amerindians from Central and South America (Monsalve et al. 2002). The presence of the 16111T, 16223T, 16290T, 16319A and

16362C polymorphisms and +663 HaeIII indicated the A2 root type (Forster et al. 1996). The high frequency of type A2 in Inuit and Na-Dene peoples is a feature that distinguishes them from Amerindians (Forster et al. 1996). Recently, the presence of haplogroup A2 has been found in pre-Clovis human remains in south central Oregon (Gilbert et al. 2008). A fourth sequence within this group, defined as A2-16189C, has been found in North, Central and South America and in the Kwädąy Dän Ts'ìnchį man's remains (Monsalve and Stone 2005). In a GenBank database survey, we found that 12 Indigenous North Americans presented additional substitutions with the A2-16189C sequence (table 2). The substitutions are different among the Athapaskan, Bella Coola, Emberá and Wounaan groups, indicating that these lineages are characteristic of certain groups. These "private polymorphisms" have been suggested as a result of tribalization in early Indigenous North American prehistory (Lorenz and Smith 1996).

Results of mismatch analysis from sequences of some Asian and North American Aboriginal peoples belonging to haplogroup A (defined as having 16223T, 16290T, 16319A and 16362C substitutions), and sequences that did not include information about nucleotide at position 16362, were described in Monsalve and Stone (2005). Table 3 presents information on the genetic distance retrieved from the GenBank database on Canadian Inuit, central Yupik, Navajo, and Tlingit from North America and Inupiaq, Siberian Yupik from northeastern Asia and west Greenland Inuit not previously reported. We also included the genetic distance (previously published) of the Athabaskan, Bella Coola (Nuxalk), Haida, Nuu-Chah-Nulth and Oneata from North America; Emberá, Kuna, Ngöbé and Wounaan from Central America; and Chukchi from northeastern Asia (Monsalve and Stone 2005). The genetic affinities between the Kwädąy Dän Ts'ìnchį man and the groups in both the previous study (Monsalve and Stone 2005) and this study were investigated by mismatch and intermatch analysis using Arlequin software (Schneider et al. 2000).

The Kwädąy Dän Ts'ìnchį man was found near the present-day Champagne and Aishihik First Nations, who speak Tlingit and Athapaskan languages of the Na-Dene linguistic group. Measurement of genetic distances, or corrected intermatch distances, showed that the Kwädąy Dän Ts'ìnchį man is more closely related to the Athapaskan (distance 0.60) than the Tlingit (distance 0.77) people. The Kwädąy Dän Ts'ìnchį man also shares close affinities with the Canadian Inuit and Chukchi (0.50; see table 3). The Kuna are the most distantly related (distance 0.86). More precise genetic connections to the Kwädąy Dän Ts'ìnchį man were established through a recent analysis of mitochondrial DNA in 240 people living on the coast of BC, southeastern Alaska and Yukon. Of these, 17 were found to share the mtDNA lineage of the Kwädąy Dän Ts'ìnchį man (Strand et al. 2008).

UBC's pathology lab was selected to have first access to the Kwädąy Dän Ts'ìnchį man's tissues in order to place his mitochondrial DNA lineage. It is important to note that a key aspect of the research was our proceeding with full disclosure to the Champagne and Aishihik First Nations and the Kwädąy Dän Ts'ìnchį Management Group. The work was characterized by mutual respect and support throughout—between researchers, First Nations representatives and the project committee—and it serves as a model for future projects involving Indigenous peoples.

ACKNOWLEDGEMENTS

We are grateful to the Champagne and Aishihik First Nations for making this project possible. We thank the Kwädąy Dän Ts'ìnchį Management Group and, in particular, Al Mackie from the BC Archaeology Branch and Jim Cosgrove from the Royal British Columbia Museum for facilitating the execution of this project. We also thank Dan Straathof from the Royal Columbian Hospital and Owen Beattie from the University of Alberta for their constant support. We acknowledge the editorial contribution of Jennifer Alsfeld from the University of British Columbia.

Table 1. Oligonucleotide primers for PCR amplifications of the mtDNA control region of the Kwädąy Dän Ts'ìnchį man.

Primers*	Product Size (bp)	Reference
L15996 – H16218[1]	222	Ward et al. 1991, Handt et al. 1996
L16192 – H16401[1]	209	Higuchi et al. 1988
L16131 – H16303[1]	172	Handt et al. 1996
L16055 – H16142[2]	87	Handt et al. 1996, Stone and Stoneking 1998
L16132 – H16218[2]	86	Handt et al. 1996
L16287 – H16356[2]	69	Handt et al. 1996
L16347 – H16410[2]	63	Handt et al. 1996

[1]UBC PCR amplifications; [2]UNM PCR amplifications

* L and H refer to the light and heavy strand, respectively of the human mtDNA genome, while the number identifies the base at the 3' primer end according to the numbering of the reference sequence (Anderson et al. 1981).

Table 2. A2 – 16189C sequence distribution in the Kwäday Dän Ts'inchi man remains and other Native American groups.

| | Cambridge Reference Sequence Position | | | | | | | | | | | | | | |
	16111C	16126T	16129G	16168C	16186C	16189T	16192C	16218C	16223C	16290C	16319G	16355T	16362T	Frequency	Source
*Group A															
KDT	T					C			T	T	A		C		1
Brazilian	T					C			T	T	A		C	3 of 247	2
Haida	T					C			T	T	A		C	1 of 42	3
Maya	T					C			T	T	A		C	1 of 3	4
Quiche	T					C			T	T	A		C	1 of 30	5
**Group B															
Athapaskan	T					C	T		T	T	A		C	1 of 21	6
Bella Coola	T					C			T	T	A	C	C	6 of 40	3
Emberá	T	C				C			T	T	A		C	2 of 44	7
Wounaan	T					C		T	T	T	A		C	1 of 31	7
Bella Coola	T		A	T	T	C			T	T	A		C	2 of 40	3

*Identical sequence with the A2 - 16189C sequence; **additional substitutions with the A2 - 16189C sequence.

Base positions are compared with the reference sequence (Anderson et al. 1981). Nucleotide positions shown in bold are the substitutions shared with the Kwäday Dän Ts'inchi man remains.

Sources: 1. Monsalve et al. (2002); 2. Alves-Silva et al. (2000); 3. Ward et al. (1993); 4. Torroni et al. (1993); 5. Boles et al. (1995); 6. Shields et al. (1993); 7. Kolman and Bermingham (1997).

Table 3. Genetic distances between mtDNA control region sequences of the Kwädąy Dän Ts' ìnchi man and North and Central American, Greenland and northeast Asian populations.

	n =	Genetic Distance	Geographic Location	Reference
Athapaskan	15	0.60	North America	2
Bella Coola	25	0.60	North America	2
Canadian Inuit	44	0.50	North America	1
Central Yupik	23	0.68	North America	1
Chukchi	37	0.50	Northeast Asia	2
Emberá	8	0.57	Central America	2
Haida	35	0.70	North America	2
Inupiaq	15	0.56	Northeast Asia	1
Kuna	45	0.86	Central America	2
Navajo	9	0.79	North America	1
Ngöbé	31	0.60	Central America	2
Nuu-Chah-Nulth	14	0.57	North America	2
Oneata	11	0.55	North America	2
Siberian Yupik	22	0.84	Northeast Asia	1
Tlingit	8	0.77	North America	1
West Greenlander Inuit	11	0.70	Greenland	1
Wounaan	7	0.64	Central America	2

The genetic distance is given as the intermatch distances (distance between two populations minus the mean of the mismatch distances within each population) (Nei and Li 1979). 1 This study; 2 Monsalve and Stone (2005).

17

MINERALIZATION OF THE KWÄDĄY DÄN TS'ÌNCHĮ MAN'S HAIR SAMPLES

A Confounding Factor in Interpretation of Metal Content

Ivan M. Kempson and Ronald R. Martin

This study presents an analysis of hair, water and ice samples associated with the remains of the Kwädąy Dän Ts'ìnchį man (Long Ago Person Found). The hair samples include human hair and Arctic Ground Squirrel hair from the robe found with the remains. Structurally, hair is remarkably stable. Due to the incorporation of endogenous xenobiotics from the blood stream, the analysis of hair can reflect an individual's diet or even disease status due to metabolic dysfunction. Hair's slow growth rate (about 1 centimetre per month) means a temporal record can be identified in the length of the hair shaft from root to tip. Hair samples are generally easy to obtain, and chemical analysis needs little sample preparation. Plants, mammals and fish all have different isotopic distributions, and details of dietary intake may be inferred from the isotopic distribution of carbon (C), oxygen (O), nitrogen (N) and sulphur (S) in an individual hair. Indeed, the most useful dietary information from hair is probably obtained from stable isotope analysis (White 1993; O'Connell and Hedges 1999; Macko el al. 1999).

In principle, the metal content of human hair could provide further information regarding environmental exposure and other aspects of dietary intake; in practice, though, this information is obscured by the impact of external contamination. The work presented here is limited to the more problematic trace elemental distribution and potential diagenesis of the Kwädąy Dän Ts'ìnchį man's hair. This is a critical consideration if any attempt to imply dietary habits based on trace elemental concentrations is to be made.

THREE ANALYTICAL TECHNIQUES WERE EMPLOYED. 1) Scanning electron microscopy with energy dispersive x-ray analysis (SEM/EDX) was used to determine gross surface morphology and to obtain a rough estimate of the principal element composition. This technique provides high-resolution images of the hair with the microscopy, as well as concentrations of the minor and bulk elemental composition. 2) Inductively coupled plasma mass spectrometry (ICP/MS) was used to determine the trace element composition of the hair and the water in which the hair was submerged when discovered. This is a destructive technique that consumes the sample in the analysis, but it is capable of detecting virtually all elements, down to trace concentrations. Only a small amount of material was available in this study, though, so the results must be treated with caution. 3) Time-of-flight secondary ion mass spectrometry (ToF-SIMS) was also used, which is a surface-sensitive analytical technique that provides compositional information of the first three or four atomic layers on top of the sample. This technique identifies elements at concentrations as low as trace levels and, more importantly, provides surface distribution information on each element. Some of the results that will be briefly reviewed in this paper have been reported elsewhere (Kempson et al. 2010).

Materials and Methods

HUMAN HAIR associated with the Kwädąy Dän Ts'ìnchį man, the Arctic Ground Squirrel pelts comprising his robe and samples of modern human head hair (from undergraduate students at the University of Western Ontario) were analysed. All hair samples were washed in Triton-X, a non-phosphate detergent, for two minutes in high-density polyethylene bottles—except those from a small part of the pelt, which were analysed as received by ToF-SIMS only. The samples were then rinsed with water and air-dried. All water was of ultra pure Milli-Q grade.

Scanning electron microscopy was carried out on a Hitachi S-4500 Field Emission instrument with an accelerating voltage of 1.0 keV for imaging and 15 keV for element identification. Element identification and quantification were carried out using software supplied with the instrument. Samples were coated with a thin layer of gold (to minimize surface charging) and mounted on carbon stubs for analysis.

Trace element analysis was carried out at the Trace Elements Laboratory, a biochemical facility at London Health Sciences Centre, using a Finnigan MAT Element ultra-trace ICP/MS. Washed hair was dissolved in freshly distilled nitric acid and diluted with 18 mega-ohm water prior to analysis. Two water samples originating from the Kwädąy Dän Ts'ìnchị discovery site were also analysed: meltwater collected at the recovery site, and ice that had been found in contact with the Kwädąy Dän Ts'ìnchị man's hair samples.

ToF-SIMS data were acquired using a Physical Electronics Inc. (PHI) TRIFT II instrument with a 25 kV liquid metal gallium (Ga) ion gun (20 ns, 600 pA DC pulses at 11 kHz). Both negative and positive secondary ions were collected in the mass range from 1–500 amu (atomic mass unit).

Results and Discussion

WHILE THIS WORK was primarily concerned with trace element analysis, we do note that interpretation of the trace metal content of hair is both difficult and problematic (Potsch and Moeller 1996; Radosevich 1993). As well, the small sample mass (12 human hairs) and low trace element concentrations present additional problems for interpretation. Accordingly, we have adopted a conservative approach in interpreting the results, and in general, only data showing consistent variability are reported.

Scanning Electron Microscopy and Energy Dispersive X-ray Analysis (SEM/EDX)

FIGURE 1 presents a comparison of the Kwädąy Dän Ts'ìnchị man's hair and modern hair samples. Two features of note are the poorly defined cuticle scales on the Kwädąy Dän Ts'ìnchị man's hair and little or no evidence of biodegradation (Rowe 1997). The deterioration in the hair therefore appears to be due to oxidation or physical processes rather than fungal activity or the action of microbes that consume hair. The latter processes would present evidence of hyphae or distinct pocks and holes radially penetrating the hair.

The EDX results are shown in table 1. At the accelerating voltage used, analysis effectively probes only the top two to three microns of material, with detection limits little better than 0.1 atomic per cent. The decrease in sulphur concentration of the Kwädąy Dän Ts'ìnchị man's hair is consistent with loss of the sulphur-rich cuticle scales from the outermost region.

The results are consistent with surface oxidation, resulting in a decrease in sulphur and loss of exchangeable ions such as sodium. The Kwädąy Dän Ts'ìnchị man's hair and the robe hair surfaces also appear to be enriched in iron, silicon and aluminum, suggestive of alumino-silicate material.

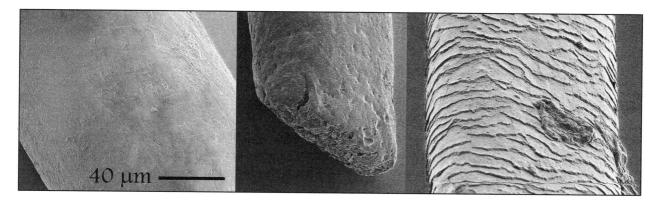

40 μm ⎯⎯⎯

Figure 1. SEM imaging of the surface of the Kwädąy Dän Ts'ìnchị man's hair, *left* and *centre*, showing marked reduction in surface scaling due to cuticle erosion—evidence for environmental degradation. A modern hair is shown on the right for comparison, with a well-structured and intact cuticle surface.

Table 1. SEM/EDX results listed in order of increasing atomic number, atomic per cent. Bold numbers in the KDT column emphasize the hair concentrations that differ markedly from modern hair, as discussed in the text.

	KDT (N=3)	Modern (N=10)	Robe hair
C	59.66	68.39	66.44
N	17.95	13.73	13.79
O	21.66	16.78	18.62
Na	00.00	00.17	00.05
Mg	00.03	00.00	00.00
Al	**00.24**	00.07	00.15
Si	**00.14**	00.01	00.17
P	00.00	00.00	00.00
S	**00.27**	00.85	00.75
K	00.00	00.00	00.00
Ca	00.00	0.00	00.00
Fe	**00.06**	00.00	00.04

Bulk Compositional Analysis by Inductively Coupled Plasma Mass Spectrometry (ICP/MS)

THE ICP/MS RESULTS for a limited suite of elements are shown in table 2. The selection is limited because useful interpretation was not possible for some exotic species, and a software failure resulted in the loss of the results for sodium.

The results for iron are consistent with those obtained for manganese, showing significant enrichment in both the robe and the Kwädąy Dän Ts'ìnchį man's hair, with very low concentrations in the meltwater, ice water and modern hair samples.

Iron and manganese have similar chemistry and would be expected to behave alike. Their anomalous concentrations in the material from the ancient remains may be attributed to the ochre used to decorate the robe (Mackie 2005).

The enrichment in aluminum and silicon, again in the robe hair and human remains, is striking and consistent with mineral enrichment of the hair surfaces. Under the circumstances in which the Kwädąy Dän Ts'ìnchį individual was preserved, bacterial lithification was not unlikely (Wierzchos et al. 2003); however, direct precipitation from groundwater would provide a viable mechanism.

Table 2. Selected elemental concentrations as detected by ICP/MS, listed in order of increasing atomic number. Bold numbers in the KDT column have been used to emphasize hair concentrations that differ markedly from modern hair—in particular, being consistent with alumino-silicate deposition (Al and Si), ochre (Fe) and possible dietary impact (Zn).

Element	KDT hair (N=3) (in ppm)	Robe hair (N=5) (in ppm)	Modern hair (N=10) (in ppm)	Meltwater (in ppb)	Ice (in ppb)
Al	**366**	484	7.2	43	0.8
Si	**172**	195	11.7	2169	97
S	35.1	17.2	51.1	0.4	0.2
K	**447**	994	28.7	3261	193
Ca	247	316	232	2964	161
Mn	**15.6**	16.7	0.071	1.82	0.089
Fe	**3567**	1974	7	305	6.9
Zn	**36.5**	72.3	179	0.56	4.7
Se	1.06	1.47	0.69	1.88	0.13

The high concentration of potassium is surprising since, as a readily exchangeable ion, it might be expected to be lost to the groundwater from the hair matrix (as the scanning electron microscope results suggest for sodium). The potassium may be incorporated in the alumino-silicate minerals, which appear to be present as a result of the overall surface mineralization of the samples.

The pronounced lower zinc concentration relative to modern hair deserves passing mention, as it may result from decreased dietary zinc. Alternatively, the modern population may consume significantly greater amounts of zinc as compared to the diet of a historical individual. Zinc concentrations in hair do appear to be relatively stable, suffering less from exogenous processes (Kempson et al. 2007). The squirrel hair also contained lower amounts of zinc compared to the modern human population, but it was still approximately double that of the Kwädąy Dän Ts'ìnchį man's hair. No firm conclusions can be drawn from this observation.

Surface Analysis by Time-of-flight Secondary Ion Mass Spectrometry (ToF-SIMS)

THE TOF-SIMS RESULTS are reported elsewhere (Kempson et al. 2010) and show regions of surface deposition of alumino-silicate material—again, consistent with both the SEM/EDX and ICP/MS data. Shown in figure 2, however, is a sample of ToF-SIMS images from the robe hair. These samples were analysed "as-collected", with no washing or pre-treatment. The features in these images highlight the extent of mineral

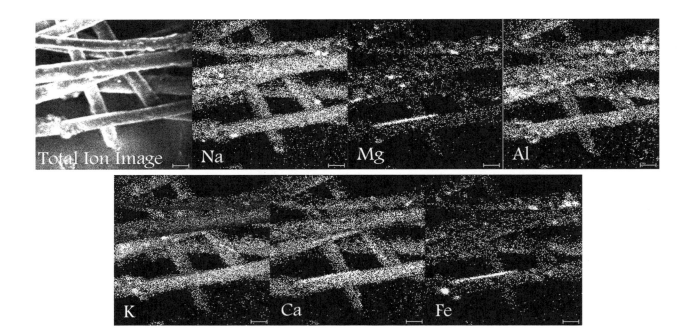

Figure 2. Ion images from the Time-of-flight
Secondary Ion Mass Spectrometry (ToF-SIMS),
demonstrating the mineral components
accumulated on the outside of the robe fur. Bar =
10 microns.

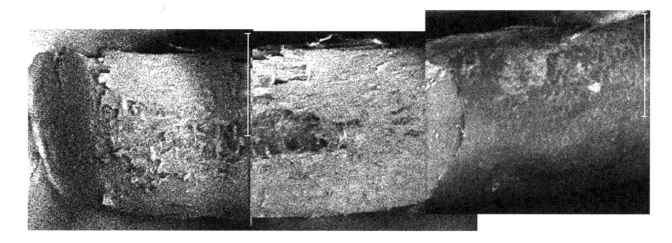

Figure 3. Time-of-flight Secondary Ion Mass Spectrometry (ToF-SIMS) total
ion image of an oblique section of a hair from the Kwädąy Dän Ts'inchį man.
The outer surface had regions of mineral encrustation, as seen in the lower
right portion of this mosaic. Bar = 100 microns.

accumulation on the outside of the robe fur. A number of the elemental components demonstrate a high degree of co-localization, confirming the alumino-silicate mineralogy. The proteinaceous composition of the fur is comparable to that of the human hair and is expected to behave analogously in a biochemical sense.

Figure 3 exhibits an oblique section cut through the Kwädąy Dän Ts'ìnchį man's hair and imaged with ToF-SIMS, revealing the internal structure. The structure's integrity is remarkably good, but the structure of the external cuticle layer is relatively subtle, as seen by SEM. Mineral deposition is exhibited on the outside of the hair, in the lower right portion of the image. The mineral components were most prevalent on the hair's external surface, with no distinct mineral accumulations inside the hair.

Overall, the results must be interpreted with caution. The ICP/MS data are suspect, because the available sample material was very small, and in the case of the water samples, handling may cause contamination (though the low concentrations of dissolved ions make this unlikely). Given these constraints and the concordance between the human hair and robe hair data, the following inferences are possible. Both materials are enriched in iron, which is consistent with close contact with the red ochre found on the robe (Mackie 2005). Both the robe hair and the human hair are also markedly enriched in aluminum and silicon. There is some evidence for reduced zinc, which could possibly reflect endogenous deficiency. For this conclusion, though, we must assume that the small sample size analysed here sufficiently and accurately represented all of the Kwädąy Dän Ts'ìnchį man's hair.

Conclusions

THE SEM AND EDX DATA show the surface of the hair to have undergone considerable environmental degradation, as evidenced by a marked reduction in the scale ranges on the hair strands. This would appear to be mainly the result of oxidation or physical abrasion, as evidenced by reduction in surface sulphur.

The most realistic hypothesis is that both the human and robe hair had been degraded by exposure to the environment—specifically, an environment dominated by freeze-thaw cycles in the glacier. The meltwater associated with the hair and the water obtained from melting adjacent ice show different ICP/MS signatures, probably because the meltwater has been in contact with local rock and airborne dust. Freshly melted ice has a much lower concentration of dissolved material. We hypothesize some loss of mobile ions from the hair and precipitation of alumino-silicates onto the hair surface. Due to the minimal biological degradation of the hair, these observations could be consistent with the very first stages of fossilization of the material. ToF-SIMS data reported elsewhere are consistent with this interpretation. Clearly, this surface mineralization represents a confounding factor, limiting interpretation of the trace element composition in terms of diet or disease.

18

THE USE OF CELLULAR STRUCTURE IN ANCIENT FROZEN HUMAN REMAINS TO PREDICT DNA RETRIEVAL

Maria Victoria Monsalve, Elaine Humphrey, Wayne Vogl, Mike Nimmo, Jacksy Zhao, Claudia Cheung and David C. Walker

It has been estimated that the Kwädąy Dän Ts'ìnchį individual was about 18 years old at the time of his death, which occurred between 300 and 150 years ago. We used mitochondrial DNA (mtDNA) analysis to establish his genetic relationship with Aboriginal peoples from the Americas as well as east Siberian, Greenlandic and northeast Asian populations (chapter 16). This was successfully accomplished through molecular genetic analysis of mtDNA from tissues dissected shortly after the body's discovery. The DNA analyses showed that the Kwädąy Dän Ts'ìnchį man shared origins most closely with the Haida people of British Columbia, Canada, and other groups from Central and South America (Monsalve et al. 2002). After testing, the young man's remains were returned to the Champagne and Aishihik people for burial. The current presence of Champagne and Aishihik First Nations in the discovery region made it critical to try to identify the Champagne and Aishihik First Nations people to whom the ancient man was

most closely related. Subsequently, Strand et al. (2008) compared
our mtDNA sequences with those of contemporary people from
northern British Columbia, Yukon and Alaska (see also chapter 33).
They demonstrated that the Kwädąy Dän Ts'ìnchį man was most
closely related to people of northern BC, Yukon and Alaska
(Strand et al. 2008).

TISSUES THAT CAN PROVIDE DNA SAMPLES suitable
for establishing the identity of ancient remains are not
always easily identified, and not all tissues are available
for sampling in incomplete remains. This is an especially
interesting issue for remains that were frozen and may
or may not have remained so until discovery. Therefore,
along with our tissue sampling for DNA analysis, we
assessed samples of the Kwädąy Dän Ts'ìnchį man's
hard and soft tissues with light and electron microscopy
to determine their state of preservation 150–300
years post-mortem. We found that the humerus and
biceps tissues from which we had extracted DNA were
preserved well enough to recognize their characteristic
architecture, and we also learned that characteristic
tissue cell architecture was recognizable in heart, lung
and vertebra (Monsalve et al. 2008).

While sampling of mummies for DNA analysis is
often done on hard tissues like bone and teeth, frozen
bodies may also provide suitable soft tissue sources
not otherwise available. Despite evidence of ice crystal
damage in soft tissue samples from the Kwädąy Dän
Ts'ìnchį man's body, we observed regions of reasonably
well-preserved ultrastructure in both hard and soft
tissues (Monsalve et al. 2008).

Recognizable tissue and cell ultrastructure have
also been reported in the 5,300-year-old Tyrolean Ice
Man (Hess et al. 1998). DNA retrieval was successful in
that corpse (Handt et al. 1994), and DNA extraction has
also been successful for four frozen mummies found in
the Andes (Wilson et al. 2007). Observations of tissue

preservation and DNA extraction from frozen mummies
around the world are summarized in table 1 (page 275).

The purpose of our present study was to extend our
survey of both hard and soft tissues from the frozen
remains of the Kwädąy Dän Ts'ìnchį man to determine
the full extent of reasonable tissue preservation. A
better understanding of the frozen tissue's state of
preservation may increase options for sampling DNA.

Material and Methods: Histological Analysis

AN EXTENSION of the initial research agreement
was obtained from the Royal British Columbia
Museum—where the body was being stored—to include
a study of cellular architecture and microorganisms in
these remains. Methods of structural and ultrastructural
analysis of cellular components from different tissue
types from this body are described elsewhere (Monsalve
et al. 2008). In the present study, both light and electron
microscopy were used. The hard tissue was decalcified
and embedded in paraffin wax prior to sectioning and
staining for light microscopy with haematoxylin and
eosin (HE stain) and Masson's trichrome (Monsalve et
al. 2008). Observations of hard and soft tissue using light
microscopy showed sufficient architecture preservation to
proceed with electron microscopy.

Electron microscopy was used to determine the
state of preservation of several parts of the remains.
Soft and hard tissues were obtained as follows:

1. April 2000: muscle samples from the biceps brachii, the gluteal region and the thigh, and a bone sample from the humerus, were excised at the first sampling of the body at the Royal BC Museum, where the remains were kept at -19°C.

2. May 2000: a bone sample from a lumbar vertebra was provided from a second sampling, also obtained at the Royal BC Museum.

3. February 2006: samples of lung tissue were received from the University of Saskatchewan, derived from material collected at the Royal BC Museum in September 2000.

4. February 2006: heart tissue was provided by the Royal BC Museum.

Biceps brachii, gluteal and thigh muscles, as well as bone samples from the humerus and lumbar vertebra, were processed at the UBC Bioimaging Facility, using standard conventional and microwave tissue processing techniques. Lung and heart tissues were processed in a Leica automatic freeze substitution unit through acetone at -85°C and osmium at -20°C. They were embedded in an Epon-Spurr, 50:50 resin mix, and 70- to 90-nm thick sections were viewed on a Hitachi H7600 transmission electron microscope (TEM) with an AMT digital camera, following staining with lead and uranyl acetate. A detailed description of the electron microscopy methodology appears in Monsalve et al. (2008).

Results and Discussion

FOR THE PURPOSE OF DISCUSSION BELOW, we provide results of our previous analysis (Monsalve et al. 2008) and our most recent observations in order to show that in frozen ancient remains, a variety of soft tissue may provide for the successful extraction of DNA.

Bone

Since the most common element of ancient human remains is skeletal tissue, most attempts to extract DNA use bone. Well-preserved architecture of cellular components in skeletal tissue has also been documented in other ancient remains. The findings of Colson et al. (1997) and Haynes et al. (2002) indicate that if bone is well preserved, DNA extraction is more likely to be successful. We have successfully extracted DNA from the Kwädąy Dän Ts'ìnchį man's humerus (Monsalve et al. 2002). Hedges et al. (1995) propose criteria for assessing the degree of bone preservation in ancient remains using a scale from 1 to 5, category 5 showing the best state of preservation (see fig. 1). Light and electron microscopic assessment showed that the Kwädąy Dän Ts'ìnchį man's humerus fell into category 2 and that his vertebra fell into category 3. From this observation, we would predict that the lumbar vertebra could also serve as a site for successful DNA extraction from the Kwädąy Dän Ts'ìnchį body.

We sampled the spinous process of a lumbar vertebra for compact bone. Analysis revealed the presence of preserved Haversian canals, concentric rings of lamellae and lacunae (figs. 1A and 1B).

As one might expect, we successfully isolated DNA sampling from the Kwädąy Dän Ts'ìnchį man's skeletal tissues that showed a reasonable degree of preservation.

Muscle (Skeletal and Cardiac)

In addition to well-preserved skeletal tissue, an advantage of ancient frozen remains is the apparent preservation of some soft tissues. For example, in 2000 we successfully isolated DNA from the Kwädąy Dän Ts'ìnchį man's biceps brachii. We have since sampled a variety of soft tissues from the frozen body, and our analysis is presented below.

Previous analysis of biceps brachii muscle tissue by light microscopy revealed the presence of skeletal

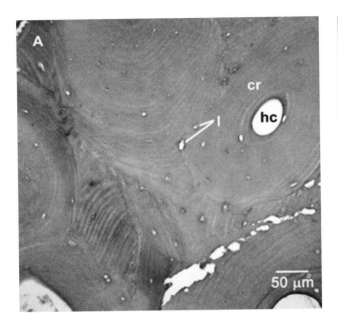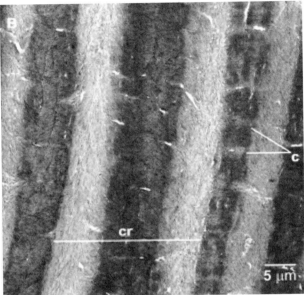

Figure 1. Lumbar vertebral bone. *A*, Light micrograph of transverse section through a lumbar vertebral bone, Haematoxylin- and Eosin-stained: *hc*, Haversian canals; *cr*, concentric rings of lamellae; *l*, lacunae. *B*, Electron micrograph of concentric rings (*cr*) and canaliculi (*c*) of vertebral bone in longitudinal section. Microwave processed. From Monsalve et al. (2008). Courtesy of John Wiley and Sons, Inc. *A*, colour version on page 588.

muscle striations. Re-examination of this muscle sample in 2006 confirmed the presence of striations (fig. 2*A*). We also observed remarkably well-preserved collagen in this soft tissue. Electron microscopy showed this to be type I collagen that retained its characteristic periodicity of banding and fibril diameters (fig. 2*B*).

In the Tyrolean Ice Man, poor muscle tissue preservation was observed, and electron microscopic images were therefore not provided (Hess et al. 1998). While skeletal muscle striations were seen by light microscopy, ultrastructural examination only showed evidence of the internal architecture of myocytes from the Kwädąy Dän Ts'ìnchị man's biceps brachii muscle, but not Z lines or banding (fig. 2*C* and *D*).

The cellular architecture of gluteal and thigh skeletal muscles was not as clear as that of the biceps brachii samples. In the gluteal tissue a spherical organelle was recognizable (fig. 3*A*) and it was possible to identify cell borders (fig. 3*B*), but it was not possible to determine whether these organelles were mitochondria, lipid droplets or peroxisomes. In the thigh muscle we

observed no evidence of cell boundaries or identifiable inclusions (fig. 3*C*).

When compared, skeletal muscle from the upper part of the body (biceps brachii) appeared better preserved than muscle tissue from the lower parts of the body (gluteal and thighs). Arm muscle had well-preserved type I collagen, while gluteal and thigh muscles did not.

Better muscle preservation was observed in the tissue samples from the heart. In more recent heart samples we have observed fibres showing a repeating pattern of banding, similar to that seen in sarcomeres of cardiac myocytes (fig. 4). At only 0.1 μm in length, the repeating units of this banding pattern were shorter than a tenth of the length of normal, relaxed striated muscle. This shortening of the banding pattern's periodicity might be explained by shrinkage or contraction of the tissue through freezing or death. We also observed well-preserved type I fibrillar collagen in this heart sample. The periodicity of the type I collagen banding is distinctly smaller than the banding pattern seen in the cardiomyocytes (fig. 4; cf. fig. 2*B*). In the cardiomyocytes,

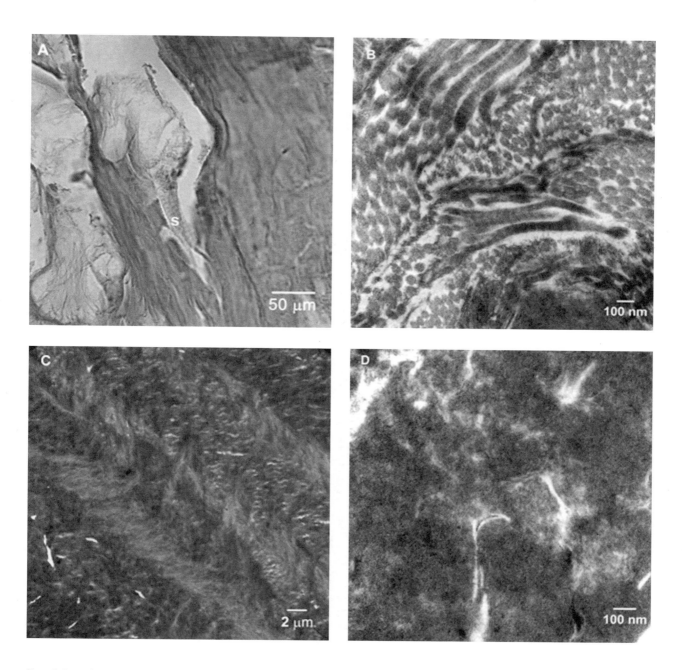

Figure 2. Biceps branchii muscle. *A,* Light micrograph in longitudinal section showing good preservation with cross striations (*s*). *B,* Electron micrograph of tissue in cross section showing good tissue preservation with type I collagen. Microwave processed. From Monsalve et al. (2008). Courtesy of John Wiley and Sons, Inc. *C* and *D,* Electron micrographs showing myocyte architecture at the ultrastructural level. Microwave processed. *A,* colour version on page 588.

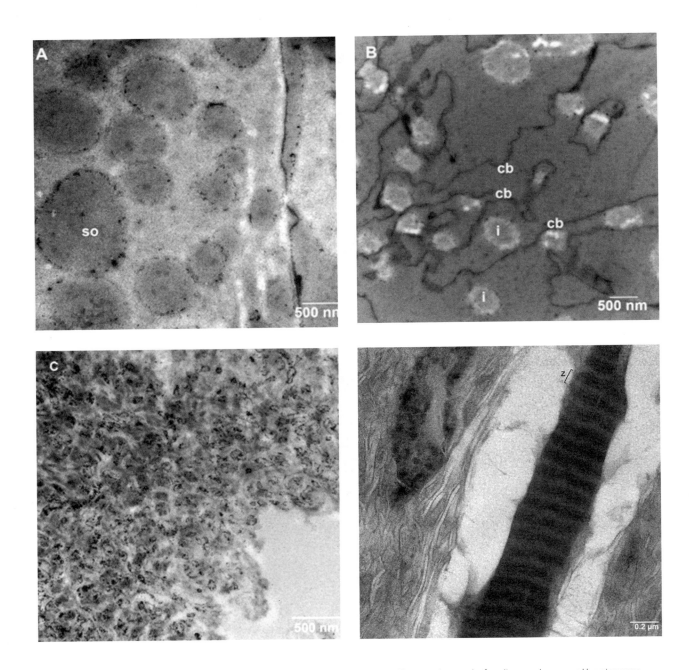

Figure 3 (A, B, C). Electron micrographs of gluteal and thigh muscles prepared by microwave processing showing poor preservation. *A*, Gluteal myocyte with spherical organelles (*so*), osmophilic particles at the periphery of organelles. *B*, Gluteal myocytes with cell borders (*cb*) and unidentifiable inclusions (*i*). *C*, Thigh muscle with ice crystal damage.

Figure 4. Electron micrograph of cardiac muscle prepared by microwave processing. Repeated banding pattern suggestive of Z band.

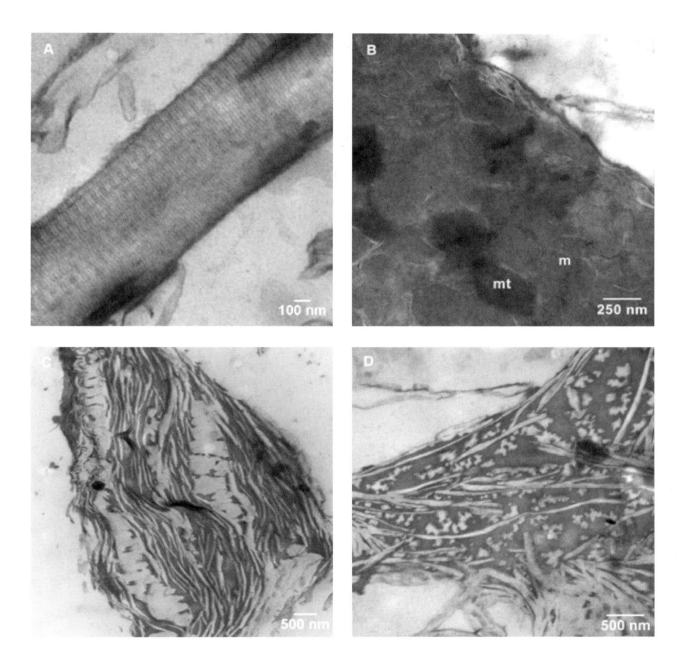

Figure 5. Electron micrographs of heart tissue prepared by freeze substitution. *A*, Fibrillar collagen from heart. Courtesy of John Wiley and Sons, Inc. *B*, Presumed mitochondria (*mt*) and myofibrils (*m*). Courtesy of John Wiley and Sons, Inc. *C* and *D*, Cardiomyocytes showing damage from ice formation and adipocere.

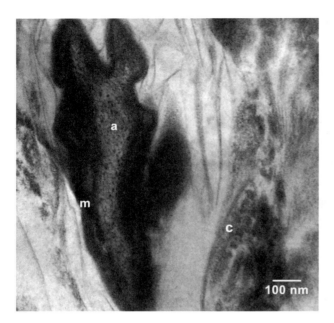
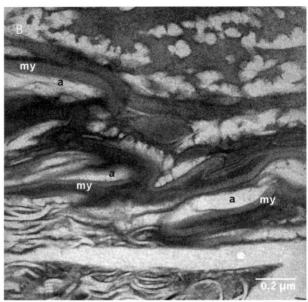

Figure 6. Electron micrograph of a nerve. *A,* Myelinated axon of a peripheral nerve. The single axon (*a*) with possible cytoskeletal elements is enveloped in membranes of the myelin sheath (*m*). Fibrillar collagen (*c*) still surrounds the myelinated axon. Microwave processed. From Monsalve et al. (2008). Courtesy of John Wiley and Sons, Inc. *B,* Three possible myolinated axons in the heart. Axon (*a*), myelinated axon (*my*). Microwave processed.

mitochondria and myofibrils could be distinguished in cross-section by electron microscopy (fig. 5B).

Two types of crystalline damage to cardiomyocytes were observed. One was of a linear nature, and a second was globular (fig. 5C and D). These patterns could result from either ice crystal formation or adipocere, or both (Hess et al. 1998).

Bereuter et al. (1996) describes adipocere as a waxy lipid mixture formed by post-mortem biochemical conversion of fat. It has been found in the Tyrolean Ice Man and other bodies released by glaciers (Bereuter et al. 1996). Adipocere formation was found in the Kwädąy Dän Ts'ìnchį man's spine and in soft tissue in his back (chapter 12; see also Beattie et al. 2000). Microscopic examination indicated some areas of cell and tissue deterioration (Monsalve et al. 2008). Further infrared analysis determined the presence of collagen, protein and lipid spectra in the ancient tissue sample (Christensen et al. 2010). In collaboration with Liu et al. (2010), we have confirmed the presence of adipocere in the Kwädąy Dän Ts'ìnchį man's tissues.

We previously reported the presence of a single myelinated axon of a peripheral nerve in the arm muscle (fig. 6A). Recently, we have observed another example of peripheral nerve: in the heart muscle sample (fig. 6B). Nervous tissues have also been described in 550-year-old mummies found in west Greenland (Kobayasi et al. 1989) and in the 5,300-year-old Tyrolean Ice Man (Hess et al. 1998). These findings suggest that reasonable preservation of nervous tissues in frozen remains expands the possibility of successful DNA extraction for studies in forensic medicine and anthropology.

Lung (Respiratory Tract)

In our previous analysis of lung samples, we observed remarkably well-preserved cell ultrastructure (Monsalve et al. 2008), and we easily identified examples of type I collagen in these samples (fig. 7A). Evidence of shrinkage was also observed. A periodicity of 64 nm, which is usually observed in freshly fixed lung tissues, was not found in the lung sample; in the Kwädąy Dän Ts'ìnchį

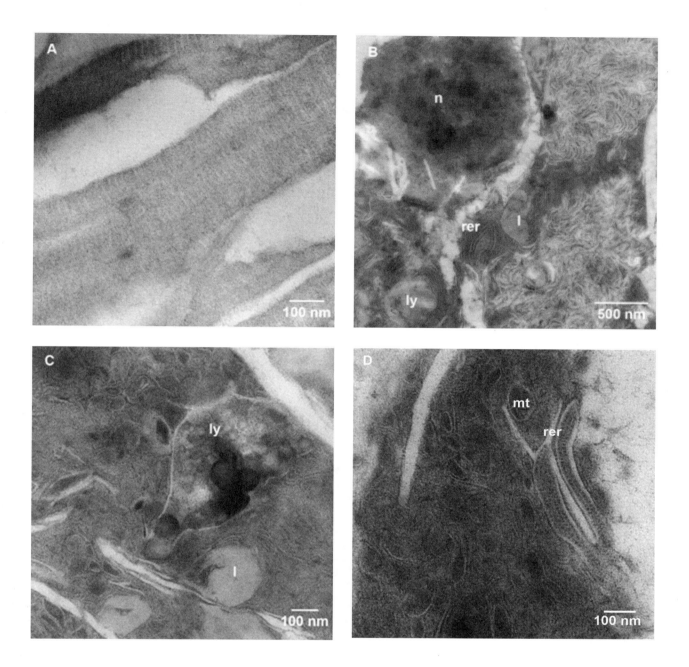

Figure 7. Electron micrographs of lung tissue processed by freeze substitution. From Monsalve et al. (2008). Courtesy of John Wiley and Sons, Inc. *A,* Fibrillar collagen from lung. *B,* Possible type II pneumocyte with condensed heterchromatin in the nucleus (*n*). Rough endoplasmic reticulum (*rer*), lipid drops (*l*) and lysosome (*ly*) are also seen. *C,* Lysosome (*ly*) and lipid drops (*l*). *D,* Mitochondria (*mt*) and rough endoplasmic reticulum (*rer*).

man's tissue, it was measured at approximately 20 nm (Monsalve et al. 2008).

A possible pneumocyte with condensed heterochromatin in the nucleus, rough endoplasmic reticulum, a lipid droplet and a lysosome are shown in figure 7B (Monsalve et al. 2008). Mitochondria were observed in the lung sample with electron microscopy (fig. 7C and D), and we observed the remarkable preservation of what appears to be rough endoplasmatic reticulum with ribosomes still attached (fig. 7D).

Conclusions

THE COMBINED RESULTS of our electron microscopy analysis suggest that the internal organs of this ancient frozen body proved to be sites of recognizably preserved tissues and cells. In contrast, parts of the body not embedded deep in the ice—therefore more susceptible to periods of thawing and freezing—did not show recognizable preservation of tissues and cells. Our observations thus lead us to suggest that the internal organs of frozen ancient remains that stay fully embedded in ice may offer alternative sites for successful DNA extraction.

ACKNOWLEDGEMENTS

We are grateful to citizens of the Champagne and Aishihik First Nations for their support in the research. Thanks also go to the Kwäday Dän Ts'ìnchį committee, Al Mackie in particular, for facilitating the execution of this project. We thank the Royal BC Museum for providing the samples, and Jennifer Alsfed and Syndy De Silva from the University of British Columbia for the editorial contributions. We acknowledge the financial support received by Claudia Cheung through the UBC Faculty of Medicine Summer Student Fellowship Program (2006) and the Summer Student Research Program (2007), and by Jacksy Zhao through the Summer Student Fellowship Program (2008).

Table 1. Histological and/or DNA analysis of frozen human tissues.

Demographic characteristics and age (radiocarbon-dated)	Location found	Year found	Histological/pathological findings
Eskimo female body, 53 years old ± 5 years, *ca.* 400 AD[1]	St. Lawrence Island, Alaska	1972	Tissues desiccated, coronary atherosclerosis, fibrous adhesions to chest wall and diaphragm in lower lobes of lungs, heavy deposits of anthracotic pigment in lungs, calcified carinal lymph node
Eight mummified bodies, 1475 AD ± 50 years[2]	Qilakitsoq, Northwestern Greenland	1972	Heavy anthracosis of lungs (possibly due to frequent exposure to open fires), collagen better preserved than other components of skin, nodule on a finger showed outlines of tissue components such as keratinocytes, melanocytes, vessels, nerves and histiocytes as well as collagen and elastic fibres, *Clostridium perfringens* found in tissues in pelvic cavity (suggests possible gas gangrene)[†]
Female body, 45 years old, *ca.* 1475 AD ± 50 years[3*]	Qilakitsoq, Northwestern Greenland	1972	Not discussed by author
Two female bodies, 1520 AD ± 70 years[4]	Utquiaqvik, Alaska	1980	i) Northern body: lungs and hilar lymph nodes black (possibly due to prolonged smoke inhalation from heating lamps), right-side dilatation of heart, osteoporotic bones ii) Southern body: lungs and hilar lymph nodes highly anthracotic, atherosclerosis of aorta and coronary arteries, focal calcification of mitral valve (possibly as a result of bacterial endocarditis), pleural adhesions (suggesting previous pneumonia)
Tyrolean Ice Man, 4546 BP ± 17 years[5*]	Ötztal Alps	1991	Soft tissues displayed signs of shrinkage, collagen and elastin well preserved with 64/67 nm periodicity, osteocytes not visible except for their outlines, local structural breakdown of myelin sheaths and rearrangement of myelin constituents, epithelial and reticular connective tissue transformed into amorphous to crystalline material (adipocere)

Demographic characteristics and age (radiocarbon-dated)	Location found	Year found	Histological/pathological findings
Kwädąy Dän Ts'ìnchį man, calibrated 1720–1850 AD[6][*]	Tatshenshini-Alsek Park, northern British Columbia	1999	Presented in this study

[1]Zimmerman and Smith 1975

[2]Ammitzbøll et al. 1989; Hart Hansen 1989; Kobayasi et al. 1989; Hart Hansen and Gulløv 1989; Hart Hansen and Nordqvist 1996

[3]Thuesen and Engberg 1990

[4]Zimmerman and Aufderheide 1984; Zimmerman 1985

[5]Hess et al. 1998

[6]Monsalve et al. 2002; Monsalve et al. 2008; Richards et al. chapter 6

[*]Retrieved DNA for mtDNA analyses

[†]Not enough information provided to claim whether findings are of one body or of several bodies

19

STABLE ISOTOPE ANALYSIS TO RECONSTRUCT THE DIET OF THE KWÄDĄY DÄN TS'ÌNCHĮ MAN

Michael P. Richards, Sheila Greer, Lorna T. Corr, Owen Beattie, Alexander P. Mackie, Richard P. Evershed, Al von Finster and John Southon

I **n this chapter** we report on the paleodietary information from stable isotope analysis of the Kwädąy Dän Ts'ìnchį man's body tissues (originally reported in Richards et al. 2007) and explore how this dietary information helps us understand the ancient individual's life history—notably, his diet, location and geographical movements in the last year of his life.

STABLE ISOTOPE MEASUREMENTS of body tissues, especially bone collagen, are a powerful tool for reconstructing past diets. The isotope ratios can be used as signatures or tracers that allow us to determine the foods a person was consuming in the past, particularly in terms of marine versus terrestrial foods (see general reviews by Lee-Thorp 2008 and Sealy 2001). Specifically, bone collagen carbon isotope ($\delta^{13}C$) values can be used to indicate dietary protein, carbohydrate and lipid sources, while nitrogen isotope ($\delta^{15}N$) values indicate dietary protein sources over a long time period, on the order of 10–30 years (Ambrose and Norr 1993). Muscle tissue bulk $\delta^{13}C$ and $\delta^{15}N$ values also reflect dietary protein sources, but over the much shorter time frame of a few months.

Sample Preparation and Measurement Methods

Bone and Muscle Tissue

The methods of sample preparation are described in chapter 6 on radiocarbon dating, this volume. The carbon and nitrogen contents and stable isotope values of the solid proteinaceous matter from two bone aliquots and the single muscle sample were measured using an elemental analyser and continuous flow mass spectrometer at the Oxford Radiocarbon Accelerator Unit (ORAU), University of Oxford (see table 1). The C:N ratios of the extracted collagen-like material from

Table 1. Carbon and nitrogen contents and stable isotope measurements of bone collagen, muscle tissue and segments of two hair strands from the Kwädąy Dän Ts'ìnchį individual. The % extract column indicates the amount of solid proteinaceous extract compared to the starting sample bone mass. The bone values are slightly less than the observed average modern bone value of *ca.* 20 per cent, while the muscle tissue is lower, likely due to high water content in the original muscle sample. The %C and %N columns are the amounts of carbon and nitrogen in the proteinaceous extract. The $\delta^{13}C$ values are measured relative to the vPDB standard, and the $\delta^{15}N$ values were measured relative to the AIR standard. Muscle and bone fragments were prepared and measured at the Research Laboratory for Archaeology and the History of Art, Oxford, UK (OX). The clothing and additional bone samples were extracted and measured at the Rafter Radiocarbon Laboratory, New Zealand (RFT). The hair was prepared at the Department of Archaeological Science, University of Bradford, UK, and the isotope measurements were undertaken at Iso-Analytical Ltd., Cheshire, UK (ISO). Measurement errors are approximately ± 0.3 ‰ for the muscle and bone and ± 0.2 ‰ for the hair.

Sample	Material	$\delta^{13}C$	$\delta^{15}N$	C:N	%C	%N	% Extract	Source
	Muscle	-14.8	19.9	3.5	42	14	3.7	OX
	Bone-1	-13.8	18.0	3.2	47	17	18.7	OX
	Bone-2*	-13.7	17.9	3.1	42	16	19.2	OX
	Bone	-14.0	17.6	3.2	52.2	19.5		RFT
KDSHA1	Hair	-14.5	17.5	3.6	45.6	14.8		ISO
KDSHA2	Hair	-15.0	17.5	3.6	40.3	13.0		ISO
KDSHA3	Hair	-15.4	17.8	3.7	38.2	12.1		ISO
KDSHA4	Hair	-16.3	16.8	3.8	31.0	9.6		ISO
KDSHA5	Hair	-16.1	16.6	3.8	39.3	12.0		ISO
KDSHA6	Hair	-16.1	16.6	3.7	52.4	16.5		ISO
KDSHB1	Hair	-15.2	17.4	3.9	44.7	13.5		ISO
KDSHB2	Hair	-14.7	17.6	3.6	42.3	13.8		ISO
KDSHB3	Hair	-15.0	17.0	3.9	47.5	14.1		ISO
KDSHB4	Hair	-14.5	17.6	3.8	55.9	17.3		ISO
KDT-31-30a	Fur	-21.2	5.0	3.3	43.5	15.3		RFT
KDT-31-30b	Skin	-20.4	5.6	3.3	38.2	13.6		RFT
KDT-Robe-24-8-7	Skin	-21.9	2.2	3.2	18.1	6.7		RFT

*With a chloroform-methanol pre-treatment

the bone samples is within the range observed for well-preserved "collagen" (DeNiro 1985). There is no statistical difference between the stable isotope values and C and N contents for the two aliquots of bone.

Hair

The hair was rinsed in 2:1 chloroform-methanol for 30 minutes, rinsed twice in distilled water, then dried at 60°C overnight. Two hair samples were used for isotopic analysis. Sample KDSHA (KDS2-8) was 13 centimetres long in total and was cut into 2-centimetre sections, resulting in six sections for isotopic analysis. The second hair, KDSHB (KDS15-6), was 9 centimetres long and was also cut into 2-centimetre sections, resulting in four segments (the last segment was 3 centimetres long). Isotope measurements on the hair were undertaken at Iso-Analytical Ltd., Cheshire, UK.

Single Amino Acids

Approximately two micrograms of bone collagen and skin was hydrolyzed under vacuum in Young's tubes (6 M HCl, 500:1 v/w; 100°C, 24 h). On cooling, the samples were transferred to screw-capped vials with double distilled water (3 × 0.5 mL) and methanol (3 × 0.5 mL). The free amino acid solutions were evaporated under a gentle stream of N_2, transferred to screw-capped test tubes and dried under N_2. γ-Amino-n-butyric acid (40 μl of 0.2 mg mL^{-1} solution in 0.1 M HCl) was added to each tube as an internal standard. Amino acids were then derivatized to their trifluoroacetyl isopropyl (TFA-IP) esters, as described in Corr et al. (2005).

Cholesterol

Cholesterol extractions were performed using chloroform-methanol (2:1, v/v, 3 × 1 h, ultrasonication). An aliquot of each total lipid extract was transferred into screw-capped culture tubes, dried under N_2 and saponified with 2 mL 0.5 M methanolic NaOH (70°C 1 h). On cooling, the neutral fraction was extracted into hexane (3 × 2 mL). Cholesterol was then converted into its trimethylsilyl (TMS) ether, as described in Stott et al. (1997). Amino acid and cholesterol δ^{13}C values were measured in triplicate following derivatization to TFA-IP esters and TMS ethers, respectively, using an HP 6890 gas chromatograph coupled to a Finnigan MAT DELTA^{plus} XL isotope ratio monitoring mass spectrometer via a Finnigan MAT GC III combustion interface, as described in Howland et al. (2003).

Isotope Results

THE KWÄDĄY DÄN TS'ÌNCHĮ PERSON's bone and muscle tissue δ^{13}C values (table 1; fig. 1) indicate that the majority of his dietary protein (approaching 100 per cent) was from marine sources. The δ^{15}N value indicates the consumption of higher trophic level marine protein, such as piscivorous fish and marine mammals (Peterson et al. 1985; Peterson and Fry 1987; Richards and Hedges 1999). Previous human isotopic studies of bone collagen in BC coastal contexts also found carbon isotope values suggestive of a heavily marine-based diet (Chisholm et al. 1982, 1983). As might be expected, marine food consumption was found to be much less at inland sites (Lovell et al. 1986). The evidence of a mainly marine-based diet for the Kwädąy Dän Ts'ìnchį individual is significant because it indicates that, despite being far from the coast at the time of his death, the young man lived either on or near the coast for most of his life—or at least in an area where salmon constituted the dietary mainstay, such as one of the villages reported to have existed in the 19th century on the middle to lower Tatshenshini-Alsek basin (McClellan 1975; see also chapter 10 on villages, this volume).

Hair carbon and nitrogen isotopic values also reflect dietary protein, but over a much shorter timescale than bone, with one centimetre of hair reflecting approximately one month of growth (O'Connell and Hedges 1999). Hair isotopic values take some time to equilibrate with a change of diet (i.e., weeks), although the exact length of time is not known (O'Connell and Hedges 1999). The isotopic data from two strands of the

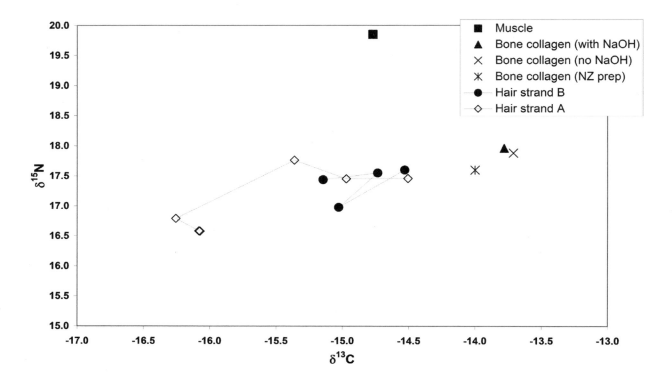

Figure 1. δ¹³C and δ¹⁵N values of bone collagen, muscle tissue and two hair strands from the Kwädąy Dän Ts'ìnchį man. Each hair segment was two centimetres long and reflects approximately two months of diet. The bone indicates the source of dietary protein over a much longer term (e.g., 10 years or longer), while the muscle tissue likely indicates dietary protein sources over the last year or so of life.

Kwädąy Dän Ts'ìnchį man's hair are presented in table 1 and figures 1 and 2. The hair has somewhat different δ¹³C values than the bone collagen. This may be simply due to a difference between bulk hair and collagen δ¹³C values as a result of the different amino acid compositions of the two tissues. This difference was observed by O'Connell et al. (2001), where the average offset in their study was 1.4 ‰. Regardless, the hair isotopic values show that although there was still a strong marine component in the diet, there is an increasingly terrestrial isotope signal along the lengths of the two hair samples. Unfortunately, we could not confidently determine which end of the hair was from the scalp. Therefore, we cannot definitively say that the observed dietary change was from a mixed terrestrial/marine to mainly marine diet (i.e., if section 1 of both strands were from the scalp) or from a marine diet to one with

increasing amounts of terrestrial foods—that is, if segment 4 (hair strand B) and segment 6 (hair strand A) were nearest to the scalp. Based on the isotopic data from compound-specific studies discussed below, we believe it is the latter. Either way, the hair isotopic data show that during the last year of the Kwädąy Dän Ts'ìnchį man's life, his diet changed, with terrestrial foods introduced into a predominately marine diet.

We were also able to provide a more refined picture of his diet, and especially a dietary change within the last months of life, when compound-specific δ¹³C values of amino acids and cholesterol of the Kwädąy Dän Ts'ìnchį man's bone (average last 5–10 years of life) and skin (last months of life) were compared (fig. 3). Bone collagen single amino acid δ¹³C values confirmed a diet rich in marine food sources (as indicated by the bulk collagen values), shown particularly in the extremely

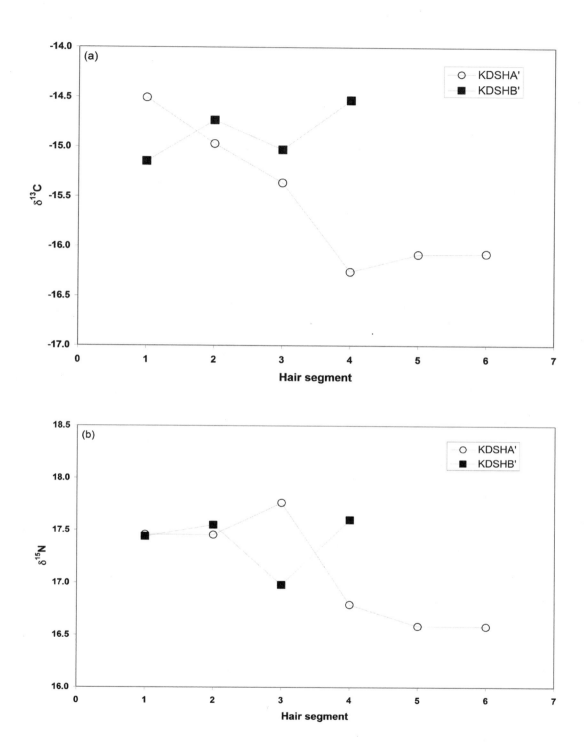

Figure 2. $\delta^{13}C$ (a) and $\delta^{15}N$ (b) values of two hair strands found in association with the Kwädąy Dän Ts'ìnchį man: KDSHA (KDS2-8) and KDSHB (KDS15-6). Hair segments are two centimetres long, except for segment 4 of KDSHB, which is three centimetres long.

high glycine $\delta^{13}C$ (-4.6‰) and $\Delta^{13}C_{\text{Glycine-Phenylalanine}}$ (15.6‰) values; the latter has been shown to be an indicator of high marine protein consumption (Corr et al. 2005). Skin amino acid $\delta^{13}C$ values were less positive, with lower glycine $\delta^{13}C$ (-9.1‰) and $\Delta^{13}C_{\text{Glycine-Phenylalanine}}$ (12.7‰) values, which suggests a greater abundance of C_3 terrestrial dietary sources in the final months of life.

Bone cholesterol $\delta^{13}C$ values can provide useful information on the isotopic composition of the whole diet over a relatively short period prior to death, as its turnover rate is faster than collagen's (Jim 2000). Studies of human and animal bone cholesterol (from archaeological sites and modern studies) have shown that humans with mainly C_3 terrestrial diets had $\delta^{13}C$ values of greater than or equal to -23.5‰, while humans with some marine dietary sources had higher $\delta^{13}C$ values, and those consuming mainly marine foods had cholesterol $\delta^{13}C$ values of between -19‰ and -20‰ (Stott et al. 1999). The Kwädąy Dän Ts'ìnchį man had a bone cholesterol $\delta^{13}C$ value of -21.6‰, which is characteristic of a mixed C_3 terrestrial and marine diet and seems to indicate that he was consuming more C_3 dietary sources in the last several months of life. The exact turnover time of human bone cholesterol has not been determined. But the 99 per cent turnover time of rat bone cholesterol is 287 days (Jim 2000), and so turnover is likely to be at least a year in humans. Interestingly, we observed a more enriched skin cholesterol $\delta^{13}C$ value of -19.7‰ (fig. 3), which may indicate that in the last period of his life, the quantity of marine foods in the Kwädąy Dän Ts'ìnchį man's diet had again increased.

Other Dietary Evidence

OTHER SOURCES OF DIETARY EVIDENCE, such as stomach content analysis, can tell us about the last days—even hours—of diet. Analysis of the Kwädąy Dän Ts'ìnchį man's stomach contents (Dickson et al. 2004) indicated the presence of coastal foods—specifically, a marine crustacean (unidentified to species), pollen from an intertidal salt marsh plant (*Salicornia* species) and fish bones (not identified to species level; see chapter 31).

While the pieces of fish and the fish scale found with the man's remains were originally thought to be chum salmon (Dickson et al. 2004), more detailed studies re-identified the fish he was carrying as sockeye (see chapter 20).

Travel and Residence Patterns

THE DATA DERIVED from the stomach contents of the Kwädąy Dän Ts'ìnchį man informs us about his diet during the final days of his life. As noted elsewhere (Dickson et al. 2004; chapter 5), fish bones, a marine crustacean and a coastal plant (likely *Salicornia*, or beach asparagus) were found in his gut. Salmon, identified as sockeye (*Oncorhynchus nerka*) through DNA analysis, was also found in association with his remains (see chapter 20). Sockeye is found both in the Haines, Alaska, area and the Tatshenshini River basin. Because salmon was often dried for later use, however, the point of origin for this food source is less informative than that of the other two food items found in his gut. Shellfish are typically not preserved for later use but are eaten fresh (Emmons 1990), and beach asparagus is consumed while in flower (Mudie et al. 2005). The presence of these two "beach foods" (Newton and Moss 2005; Emmons 1990) strongly suggests that the Kwädąy Dän Ts'ìnchį individual had been on the coast or had been in contact with someone who had recently been there. The discovery of pollen and the remains of plants from coastal biotic zones on his clothing, as well as in his gut, supports the first interpretation (Dickson et al. 2004; chapter 28).

Interpretation of the man's lifetime dietary signature is a bit more complex, because full-time coastal residency is not the only pattern that could result in a marine signature. This pattern could also occur if the protein consumed consisted solely of piscivorous fish (salmon); a person living in the interior, whether year-round or on a temporary basis, could potentially have a marine dietary signature if salmon was the individual's sole source of protein.

There is little evidence, however, to support the idea that the man's diet was based entirely on salmon. As

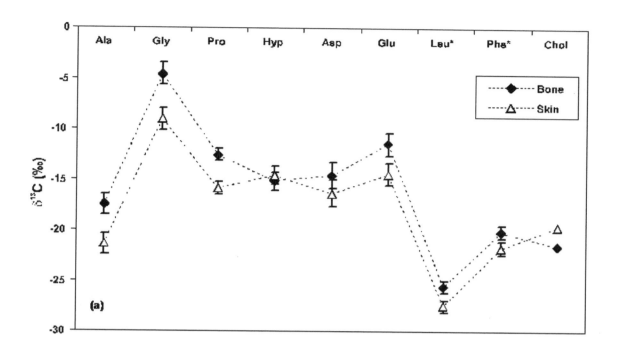

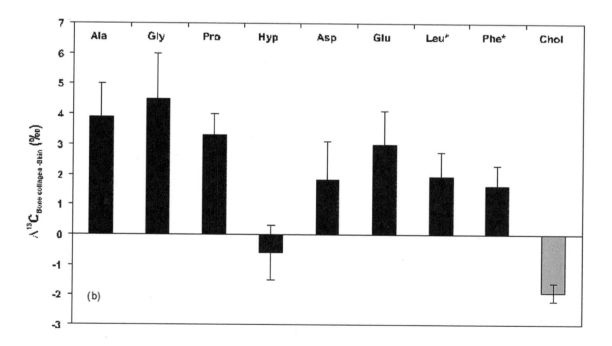

Figure 3. (a) Bone collagen and skin amino acid and cholesterol $\delta^{13}C$ values, and (b) difference between bone collagen and skin amino acid and cholesterol $\delta^{13}C$ values ($\Delta^{13}C_{\text{Bone collagen-Skin}}$).

previously noted, the isotopic pattern detected at inland pre-contact sites located further to the south in BC suggests a mixed diet (Lovell et al. 1986). Oral history accounts suggest that a mixed diet was also the norm in this far northwestern part of the province. The Dän and Tlingit residing in the upper Tatshenshini in the early 20th century reported that in addition to salmon, they ate both big game, such as Caribou, Moose and sheep, and small game (McClellan 1975). Anthropologist Beth O'Leary, seeking to discover why a mixed diet would be followed when a seemingly abundant resource like salmon was available, found that the salmon runs reaching the upper Tatshenshini, the part of the basin with the most abundant runs (cf. Smith et al. 2009), differed in size significantly from year to year, and that the timing of these runs also differed appreciably. O'Leary concludes that the mixed diet made sense when these timing and abundance factors were considered in light of the human labour requirements for catching and processing fish (O'Leary 1992).

While we do not know if the Tatshenshini basin residents had a mixed diet during the lifetime of the Kwädąy Dän Ts'ìnchį individual, it is difficult to argue that these people would have eaten only salmon when other high-protein foods such as Caribou, Moose, sheep and goat were locally available. The simplest and most plausible explanation for the lifetime dietary signature of the Kwädąy Dän Ts'ìnchį individual is therefore that he lived on the coast and that his diet was based on salmon, other seafood and possibly marine mammals.

The nearest coastal area where he might have lived is the Haines and Klukwan area to the south. Haines, about 90 kilometres south of the discovery site, is situated at tidewater, and Klukwan, about 60 kilometres south, is accessible to tidewater by watercraft. The sediment and plant evidence found on the man's robe suggest that he was travelling from the south when he arrived at the place where he lost his life (see chapter 31), which leads us to suspect that he started his journey at a village in the Haines and Klukwan area.

While the botanical evidence suggests where his last journey began, it does not establish whether this location was his home. It is possible that he lived in a different coastal location, perhaps in the area of Dry Bay or Glacier Bay, and that his final voyage was one of many in a period of travelling that involved multiple journeys, eventually arriving in the Haines and Klukwan area. The Dry Bay area is located about 140 kilometres to the southwest and is accessible via the Tatshenshini and Alsek rivers. The Glacier Bay region is somewhere between 60 and 100 kilometres from the lower Tatshenshini basin. Travel routes connecting both of these locations to the Tatshenshini basin were known to be in use in the 19th century and may have been in use during his lifetime (see chapter 9).

The isotopic data also indicate that there were changes in the diet of the Kwädąy Dän Ts'ìnchį individual during the last months of his life, with sufficient food from land mammals being consumed to result in the terrestrial dietary signature recorded in his hair, skin and bone. The explanation for this dietary shift is more difficult to pinpoint. It could have resulted from a cultural practice (for example, adhering to a special diet when participating in a period of ceremony). The variation could also have occurred because of a shift in residency and hence in the nature of the food available for consumption.

Given the cultural exchange and trade taking place during his lifetime (see chapter 9), the most readily available explanation for the increase in terrestrial foods in his diet in the months before his death was that he left his home location with its seafood-based diet and began travelling. He may have journeyed inland where he personally hunted land animals for food, or spent time with people whose diet included a significant component of protein derived from big or small game such as Caribou, Moose, sheep, marmots or ground squirrels.

Conclusions

OUR ISOTOPIC ANALYSIS of bone and tissue indicate that, although he died far from the ocean, the Kwädąy Dän Ts'ìnchį person, like most pre-contact people from the Northwest Coast, had a predominantly marine diet for most of his life. The dietary analysis cannot specify anything further about the exact geographical area(s) where this individual may have resided during his 18 years of life, beyond indicating that he probably spent most of his time on or near the coast, in a place where seafood and marine mammals were consumed as the primary protein source. The isotopic data also showed that his food consumption patterns changed in the year before his death. Although he still had a predominantly marine-based diet during this interval, sufficient food from land mammals was consumed to result in the terrestrial dietary signature that was recorded in his hair, skin and bone.

Based on our understanding of the cultural changes and exchanges that were taking place during the times in which he lived, the lifetime dietary evidence could be interpreted to suggest that the Kwädąy Dän Ts'ìnchį person was a traveller: a visitor to the inland area where he lost his life. The shorter-term isotopic dietary signature could similarly be interpreted to suggest that in the last year of his life, the Kwädąy Dän Ts'ìnchį individual may have taken one or several trips away from the coast, possibly making extended stays in the interior, where he consumed food from the land rather than from the ocean.

ACKNOWLEDGEMENTS

We would like to thank the Champagne and Aishihik First Nations for allowing the isotopic studies reported here. The Kwädąy Dän Ts'inchį project is jointly managed by representatives of the province of British Columbia and Champagne and Aishihik First Nations, and we thank project management group members Grant Hughes and Jim Cosgrove (Royal British Columbia Museum), and Lawrence Joe and Diane Strand (Champagne and Aishihik First Nations) for their ongoing support of the project's scientific studies. We also thank the Heritage program of the Yukon government and the Royal British Columbia Museum for their support of the Kwädąy Dän Ts'inchį project, noting the ongoing contributions of artifact conservators Valery Monahan (Yukon) and Kjerstin Mackie (Royal BC Museum), as well as the assistance—particularly in the project's early days—of archaeologists Ruth Gotthardt and Greg Hare (Yukon). We also thank Brian Apland (Royal BC Museum) for his opinions and recommendations during the preparation of this paper. Wayne Howell (Glacier Bay National Park, Gustavus, Alaska) provided information on relevant oral history texts.

4

THE BELONGINGS AND THE ARTIFACTS

INTRODUCTION

Alexander P. Mackie and Richard J. Hebda

The artifacts found with people frozen in glaciers provide direct insight into the daily lives of these individuals. In comparison, burial objects are placed by others after a person's death and may not be connected to the deceased's activities in life. Several belongings were found with and near the body of the Kwädąy Dän Ts'ìnchį individual, and other items lay scattered around the discovery site. These belongings may have been his or were perhaps left behind by other travellers. All of the items are exceptional because they consist of perishable organic material that rarely survives in typical archaeological sites. Like the human remains, the belongings open a window into this person's daily life as he made his way on foot through an icy mountain landscape.

THE LESSONS OF THE BELONGINGS were investigated in many ways. The objects underwent conventional and cutting-edge physical, chemical and biological analyses, and they were also compared to similar items in museum collections. In an effort to bring to light the artifacts' role in the man's life, consultations were held with community elders, who provided insight into the objects' function and cultural connections.

The study of the discovered belongings began with their conservation, the methods and challenges of which are described in several of the following chapters. This careful handling and sampling permitted the recovery of microscopic materials such as pollen, seeds, DNA and insects—all key to understanding the objects' history and use.

Several items of clothing were found with the Kwädąy Dän Ts'ìnchį man, including a basketry hat. One chapter describes its conservation, and another provides an analysis of the hat's form, revealing remarkable details of weaving, repair and techniques. This study points to a Tlingit style of manufacture.

Fur objects directly associated with the human remains include a robe or blanket and a bag. One chapter describes in detail the conservation and deciphering of the robe's structure, showing that it was manufactured from 95 Arctic Ground Squirrel (gopher) skins meticulously sewn together. In contrast, the bag consisted of beaver skin. DNA analyses revealed that many other animals were involved in the manufacture of these fur and skin artifacts—in particular, in the case of the sinew or thongs used to hold them together and repair them. Among the most remarkable discoveries was the use of Blue and Humpback whale sinew to repair the robe.

The fur items, especially the robe, trapped microscopic plant fragments and pollen grains. The last chapter in this part of the book traces the robe's exposure to many different environments and thus sheds light on the young man's experiences before his death.

Numerous stick-like wooden objects were also found with and near the human remains. Most featured carving or paint, clearly linking them to human use. Consultation with First Nations elders suggests that some of these objects may have been used to gaff salmon, while others may have been involved in setting traps for ground squirrels. A detailed study of the fish remains found with the frozen human body reveals that the Kwädąy Dän Ts'ìnchį man carried Sockeye Salmon on his journey.

Two metal objects were apparently among the Kwädąy Dän Ts'ìnchį man's possessions as well. A tiny bead attached to sinew fragments consists of nearly pure native copper, which raises stories of the well-established and wide-ranging copper trade in the region. A second composite object involves a piece of iron inserted into wood and secured with bone or antler. Comparisons to ethnographic collections and accounts reveal that he was carrying a knife with a wooden handle.

From these studies, we learn many lessons about geography, as the tools reflect not just activity but specific areas where those activities might have occurred. The discoveries tie together many widespread places, from inland mountains with copper to distant seas, as well as the people that manufactured and used the objects.

Equally important are the lessons in cultural history derived from these objects. We learn about the Tlingit style of spruce-root weaving; the important role that native copper played in regional Aboriginal history, as well as the introduction of European trade metals; the significance of salmon as a local subsistence resource; and how clothing can carry a record of one's travels. Together, the chapters in this section show how the discovery of one person in one place connects a huge region and many people.

20

IDENTIFICATION OF SOCKEYE SALMON FROM THE KWÄDĄY DÄN TS'ÌNCHĮ SITE

Peter M. Troffe, Camilla F. Speller, Al von Finster and Dongya Y. Yang

The study of the Sockeye Salmon (*Oncorhynchus nerka*) remains found with the Kwädąy Dän Ts'ìnchį man began in January 2000, immediately after the frozen remains and associated research materials were transported to the Royal BC Museum by the Kwädąy Dän Ts'ìnchį Management Group. Among the frozen samples were fish remains collected throughout the discovery site. These included scattered, loose scales and numerous pieces of skeletal tissue complete with skin displaying intact rows of translucent scales (fig. 1).

THE NORTH COASTAL RIVERS of British Columbia support 32 species of freshwater fish, with 18 of these species being anadromous, like salmon, which means they make seasonal movements between fresh water and the ocean on foraging or reproduction migrations (see appendix 1). Drainages to the upper Lynn Canal include the Chilkat and Chilkoot rivers, their tributaries and a number of smaller streams. These waters tend to be short, with steep gradients, high flows and significant bedloads. This results in a relatively high degree of channel instability and associated groundwater-fed side and back channels. Such conditions provide significant spawning and rearing habitat for Sockeye, Coho and Chinook salmon, and spawning habitat for Pink and Chum salmon. These salmon populations seldom migrate far from the ocean to spawn. The current use of habitats in the upper Lynn Canal reflects over a century of industrial harvest and associated salmon management, including enhancement of habitats and stocks. Land use activities in the area that may have affected salmon distribution include forest harvesting, civil and military infrastructure, and mining. Independent of human activities, the landscape of the area and the network of streams draining it

Figure 1. One of several pieces of intact Sockeye Salmon flesh and skin (sample 105), as found in situ at the Kwäday Dän Ts'ìnchį discovery site. Al Mackie photograph. Colour version on page 588.

are intrinsically dynamic. Stream reaches may be inaccessible or unsuitable for salmon under one set of conditions but may be highly productive a century or a decade later. Colonization of suitable habitats tends to be rapid.

The Tatshenshini River watershed is entirely in Canada and is currently used by Chinook, Sockeye, Coho and Steelhead salmon. Most tributaries to the lower river receive drainage from glaciers and are of limited value to spawning salmon (Waugh et al. 2004), although associated groundwater-fed areas are highly utilized. Tributaries to the mid and upper river drain the dry interior of the

Yukon and are now little affected by glacial melt. All four species utilize habitats accessible to the virtual headwaters of the river for spawning, rearing and overwintering. There are no records of Chum or Pink salmon in the river. There has been very little enhancement of stocks or habitats, and the drainage has been essentially unaffected by mineral or forestry development. Infrastructure construction has been limited.

The physical characteristics of the mid and upper Tatshenshini River and its tributaries reflect a more stable landscape than the coast, and salmon habitats tend to endure for extended periods of time. The

Figure 2. A traditional type of salmon fishing weir and trap located in the Klukshu River, a tributary of the upper Tatshenshini River. This image was taken on August 28, 2003, when Sockeye were present in the river near the trap and the adjacent smokehouses (beyond view) were full of food fish. Al Mackie photograph.

most productive of these habitats are located in small tributaries, particularly those with lakes in their upper watersheds. The Klukshu River and Village Creek are examples. Both enter the Tatshenshini River at the northernmost extent of the river's course, and both are generally clear and have flows buffered by the upstream lakes. Sockeye Salmon return to the Klukshu River in two distinct runs, with a small "early run" in June–August and a larger "late run" in August–September (Fillatre et al. 2003). Weirs and fish traps can be easily installed and safely operated, with salmon drying quickly as compared to coastal areas (fig. 2). It is worth noting that during the Kwädąy Dän Ts'ìnchį man's lifetime, these streams probably supported larger stocks than at present.

Analysis of Salmon Remains

WE USED TWO PRIMARY METHODS to analyse the fish remains associated with the Kwädąy Dän Ts'ìnchį man: a morphometric scale analysis and an ancient DNA analysis. At the beginning of the project in 2000, the first attempt at DNA analysis was undertaken using multiple fish tissue and scale samples. The results, however, were inconclusive, and we focused our efforts solely on the morphometric analysis until 2008, when ancient DNA analysis techniques were reapplied. The morphometric analysis of the fish scales and their internal structures provided information about the number of individual fish present, an estimate of age and seasonality at time of fish death, and species identification through comparison of the ancient fish scales to dichotomous keys in the contemporary fisheries literature and fish museum specimens. The ancient DNA molecular technique involved digesting tissue or scale samples and comparing the highly diagnostic, species-specific DNA sequences in a portion of the mitochondrial genome to those available in genetic databases (such as GenBank) and DNA sequence repositories focused on fisheries resources.

Scale Analysis

FISH SCALES show sufficient pattern variation to allow for species identification in archaeological remains. Out of necessity, a number of dichotomous keys have been produced for scales of freshwater, anadromous and marine fish species for archaeologists, paleontologists and fisheries biologists (e.g., Casteel 1972, 1973; Bilton et al. 1964; Koo 1962; McAllister 1962). The majority of these keys are used in cases where the number of species present at a site is only identifiable through the use of scale signatures. Unlike vertebral or skeletal elements—for which there are only one or few unique elements per animal (e.g., atlas, vertebrae)—there are numerous scales per animal, requiring the scale researcher to use frequency analysis (rather than a singular approach) to make species-specific identifications. Scale features used for species identification are generally surface or internal features, such as circuli (individual growth rings) or annulus (collection of annual growth rings) patterns. In contrast, positive species identification can be difficult when using differences in scale shape or size since the dimensions of scales from different regions can vary among individual fish (Casteel 1976).

Scale Samples

SAMPLES from the Kwädą̄y Dän Ts'ìnchį find thought to contain fish remains had been sorted into five sample bags at the discovery site. On January 4, 2000, these bags were sequentially removed from the storage freezer at the Royal BC Museum and allowed to thaw for inspection under a dissection microscope of between three and six times resolving power. In two of the five sample bags (8, 34-22) scales were removed from

the surface of tissues comprised of intact epidermis, dermis and body wall muscle tissue. The scales from these samples were arranged in immaculate rows and appeared to be in excellent condition, despite being encrusted in fine brown-black sedimentation; they were supple when compared to scales removed from fresh contemporary specimens. Scales found freely scattered among the other sample bags (7, 32-1, 34-4, 34-31) were covered in sedimentation and appeared less flexible than those associated with tissue.

In total, 116 cycloid fish scales were removed from the remains stored at the Royal BC Museum (table 1). Once removed from the remains with sterile forceps, the scales were mounted between two glass microscope slides that were taped together, labelled and placed in a freezer maintained at -17°C until cleaning. In a method similar to that used by Whaley (1991), individual scales were cleaned by suspending no more than five scales at a time in a shallow petri dish filled with distilled water which was floated in an ultrasonic cleaning device partially filled with water. The scales were subjected to ultrasonic treatment for five to ten minutes and were intermittently inspected under a dissecting microscope. Once the majority of the sedimentation had been removed, the remaining dirt was cleaned mechanically under a dissecting scope. The clean scales were then mounted between two new microscope slides and the edges were sealed with a film of wax for final storage.

Age and Seasonality

THERE ARE SEVERAL EXAMPLES of archaeological and fisheries studies estimating the season of death or age from fish scale circuli patterns (Follett 1967; Struever and Thurmer 1972; Brinkhuizen 1997). Fish are cold-blooded, with their growth rate dependent on the temperature of the surrounding water. Evidence of seasonal variation in fish growth is represented in several calcified tissues, including vertebrae, fins, otoliths and scales, and these features are commonly employed in aging techniques (Hogman 1968; Pearson 1966; Watson 1964; Deelder and Willemse 1973).

Figure 3. A scale from a contemporary Sockeye Salmon illustrating four seasonal annular bands (circuli) of ocean growth (Ocean 1–4) and a period of freshwater growth at the centre (focus) of the scale.

Early Life History

THE INTERNAL PATTERN of the scales found at the Kwädạy Dän Ts'ìnchị site suggests they belong to the family Salmonidae (trout, whitefish, char and salmon), and the presence of marine growth patterns identified the scales as belonging to a salmon of genus *Oncorhynchus*, the Pacific salmon. Generally, Pink Salmon (*O. gorbuscha*), Chum Salmon (*O. keta*) and some populations of Chinook Salmon (*O. tshawytscha*) migrate out of freshwater environments and into estuaries quickly after emergence from their natal redds. These species will typically rear in the ocean for one to five years before returning to fresh water on natal spawning migrations. Other salmon species, such as Coho (*O. kisutch*), Sockeye (*O. nerka*) and most Chinook populations, often (but not always) spend significant portions of their early life history in fresh water before maturing in the ocean. Growth rates for salmonids are much slower in freshwater environments than in marine environments, and as a result, the interval between the scale focus and the first-year annulus is much smaller (rarely exceeding 1.10 mm) for species that spend one or more years in fresh water, while those that enter the ocean quickly after emergence have a large (rarely less than 1.10 mm) first annulus diameter (Bilton et al. 1964). The scales associated with the Kwädạy Dän Ts'ìnchị find did not exhibit a

Generally, fish scales grow rapidly in the summer and the circuli spacing is wider, with greater distance between them than in periods of slower growth in late summer and fall. The scales of north temperate freshwater fish generally show annuli as a function of seasonal variations in water temperature, represented by a series of rows of thin, closely packed circuli separated by thick, widely spaced circuli (see, for example, fig. 3). In the coolest seasons, periods of slow or zero growth are represented by broken "crossing over" circuli in the annular region (fig. 3).

strong freshwater annulus pattern, and the relatively large diameter of the first annulus (average 1.78 ± 0.15 mm) suggests the scales might have originated from a salmon population that spent very little time rearing in a freshwater environment before out-migrating to the ocean.

Adult Life History

AGE ESTIMATES of the scales found with the Kwädąy Dän Ts'ìnchį remains suggest that the fish had completed four seasonal growth periods in a marine environment after a short freshwater life history phase (fig. 4). The number of circuli beyond the last annulus at the periphery of the scales ranged from three to six, suggesting the fish was into its fifth year and was most likely on spawning migration at the time of death (Bilton and Ludwig 1966).

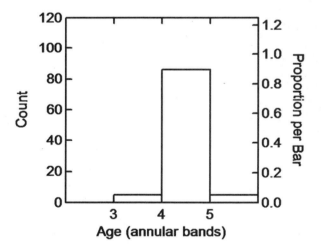

Figure 4. Frequency distribution of the annular age band counts estimated from cleaned and mounted Sockeye Salmon scales found associated with the Kwädąy Dän Ts'ìnchį remains.

Number of Individual Fish Present

THE PROCESS FOR ASSESSING the minimum number of individual fish present at an archaeological site usually uses the frequency count of various skeletal elements in an average individual. In cases like the Kwädąy Dän Ts'ìnchį find, where there are no ossified remains, scales can be used. After preliminary identification, frequency plots of various characters can be made to determine if the scales fall into one or more groupings based on similar characteristics. Variation in the grouping of character counts or measurements can be expected if the scales come from more than one individual. If scales have come from a single individual or the same population, one would expect to find variation among continuous characters (e.g., relative size of the scales) but not among discrete characters (e.g., number of first-year circuli). A multivariate analysis (Principal Component Analysis, or PCA) was conducted on discrete scale characters and included the number of complete central circuli, the number of circuli after the last annulus and the number of circuli present in the first annulus. Continuous scale characters used in the PCA included scale diameter, first annulus diameter and distance from focus to scale edge.

Due to the unique nature of the remains, there was some variability in preservation, and only a portion of the scales sampled exhibited good-quality characters. There appeared to be some differences among the continuous scale shape characters examined, suggesting two different shapes of fish scales present in the samples. A scatter plot of the component scores from the PCA revealed two groupings, with one dominating the positive ordinal plane of Factor (PC) 1 and the other group occupying the negative ordinal plane of Factor (PC) 1 (fig. 5). The total explained variance in the PCA was 89.7 per cent, with 68.6 per cent and 21.1 per cent variance accounted for by the first and second principal components respectively. Post-hoc labelling of the PCA scores suggests that the groups fall out based on their sample numbers: 32-1 scales are in the positive plane of Factor (PC) 1, while samples 34-22-1 occupy the negative plane of Factor (PC) 1 (fig. 5). The difference in the PCA scatter plot suggests there are two different scale shapes from samples found with the Kwädąy Dän Ts'ìnchį remains, although the scales all exhibit the same discrete character patterns. The overall morphology suggests that the narrow, elongate scales from 34-22-1 are from the ventral surface of the fish, while the presence of lateral line canal pores and the

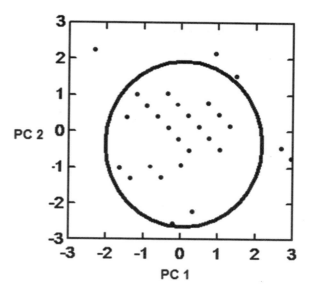

Figure 6. Principal component plot of discrete scale counts based on the number of complete central circuli, the number of circuli after last annulus and the number of circuli present in the first annulus does not demonstrate any distinct grouping patterns, suggesting that the salmon scales among the Kwädąy Dän Ts'ìnchį remains originated from an individual salmon.

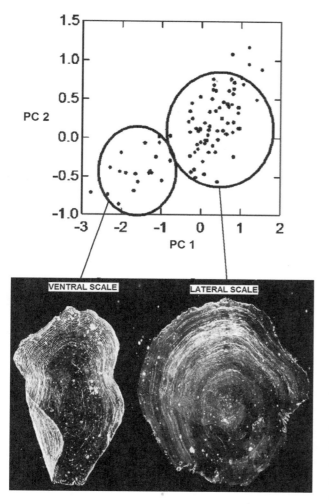

Figure 5. Principal component plot of continuous scale measurements demonstrating the presence of two different shapes of salmon scales among the Kwädąy Dän Ts'inchį remains. The ovate scale *(left)* is typical of scales on the ventral (belly) surface, and the circular scale *(right)* is typical of scales from the lateral side of salmon.

overall circular appearance of samples 32-1 imply they originated from the lateral surfaces, near the lateral line (fig. 5).

The results from the discrete character PCA analysis suggest that the scales found with the Kwädąy Dän Ts'ìnchį man's remains are closely related and most likely all from a single fish. The total explained variance in the discrete character PCA was 75.7 per cent, with 41.8 per cent and 33.8 per cent variance accounted for by the first and second principal components respectively, and post-hoc labelling of sample series 32-1 and 34-22-1 suggests there are no differences among the scales. A scatter plot of the component scores of the PCA reveals one major cluster arranged around the origin, suggesting there is little variation among the discrete scale characters (fig. 6).

Species Identification from Scales

GIVEN THE PRESENCE of at least four years of marine growth and the life history profile of the salmon scales, we eliminated several salmon species with different freshwater or marine life history phases (e.g., Coho, Chinook and Pink). Further tests in 2000, using the salmonid keys of Bilton et al. (1964) and Koo (1962), provisionally—and incorrectly—identified the fish remains as belonging to Chum Salmon. This was based on the fact that for Pacific salmon, most of the circuli at the focus of the scale terminate abruptly as they pass into the posterior portion of the scale, usually leaving some complete circuli at the focus. For Sockeye, Coho and Chinook salmon, the number of complete circuli ranges from four to fourteen (Bilton et al. 1964; Koo 1962), while Pink Salmon have two to nine (Koo 1962) and Chum the fewest, with zero to three complete focus circuli (Bilton et al 1964; Koo 1962). Given that the scales found at the Kwädąy Dän Ts'ìnchį site exhibited a range of two to three unbroken focus circuli, we determined that the remains most closely resembled Chum Salmon scales. We have since learned that due to the cool temperature profile of the high-latitude north coastal rivers, many Sockeye Salmon populations in the discovery region spawn in the mainstem of headwater tributaries, not in rearing lakes, and emergent juveniles quickly out-migrate to the ocean without a significant freshwater rearing phase—similar to the life history strategy of Chum Salmon throughout their species range. While the morphological analysis of the scales failed to correctly identify the salmon remains to the species level (2008 DNA evidence strongly suggests Sockeye Salmon), it yielded important information about the number of fish present at the site and described the early life history and age at maturity of the Sockeye Salmon remains associated with the Kwädąy Dän Ts'ìnchį find.

Molecular DNA Analysis, Introduction

GIVEN THAT THE SCALE ANALYSIS was based largely on comparison to contemporary specimens from published fisheries literature and the fact that this type of analysis can be subjective, we thought it prudent to attempt modern molecular genetic techniques to provide another line of evidence for identifying the Pacific salmon species associated with the Kwädąy Dän Ts'ìnchį discovery. At the beginning of the project in 2000 there was not sufficient expertise in ancient DNA analysis readily available, and so the first attempts at species identification using DNA were inconclusive.

2008 Molecular Analysis

THE 2008 MOLECULAR ANALYSIS was completed once the project teamed up with members of the dedicated ancient DNA laboratory in the Department of Archaeology at Simon Fraser University (SFU) in Burnaby, BC. The lab used strict precautions for contamination controls—including a positive pressure laboratory and Tyvek coverall suits, masks and gloves—with techniques designed to efficiently extract and amplify degraded DNA from ancient archaeological samples (Yang et al. 1998, 2004). Ancient DNA analysis was conducted on five separate pieces of salmon from the samples archived at the Royal BC Museum. Samples were incubated with proteinase K overnight at 50°C, in 3 mi of lysis solution (0.5 per cent SDS, 0.5 mg/ml proteinase K, 0.05 M EDTA), and DNA was extracted using a modified silica-spin extraction protocol (Yang et al. 1998). Two different fragments of salmon mitochondrial DNA were targeted: one 249 base pair (bp) fragment was from the salmon D-loop, and the other 169 bp fragment was from the cytochrome b gene. Detailed information on the PCR condition and the primer sequences can be found in an earlier publication (Yang et al. 2004).

Four of the samples (1-1, 34.22, 32.5, 105) yielded positive DNA amplifications, and one sample failed to amplify (8-6) (table 1, fig. 7). The latter sample was a piece of salmon skin, while the other samples came from

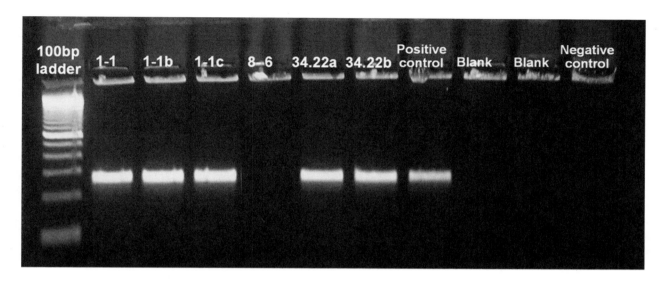

Figure 7. This electrophoresis gel image shows some of the successfully amplified D-loop DNA samples (the bright band shows the presence of a 249 bp DNA fragment; some of the samples were amplified multiple times to replicate results). Two of the samples (1-1 and 34.22) yielded DNA, and one sample failed to amplify (8-6). Colour version on page 589.

Figure 8. Part of the amplified cytochrome b sequence aligned with reference sequences retrieved from GenBank: Rainbow Trout (NC_001717), Chum (AJ314561), Coho (AJ314563), Sockeye (AJ314568), Pink (AJ314562), Chinook (AJ314566) and Atlantic salmon (NC_001960). Dots indicate identical bases with Rainbow Trout. The ancient sequence recovered from the Kwädạy Dän Ts'ìnchị discovery samples is represented by a DNA consensus and matches identically with the Sockeye Salmon reference. Colour version on page 589.

salmon flesh, which may explain the difference in DNA yield (fig. 7). For all four samples that yielded DNA, the D-loop and cytochrome b sequences from amplifications match DNA sequences of Sockeye Salmon (*Oncorhynchus nerka*) from GenBank and other publications (Yang et al. 2004). Figure 8 shows the alignment of the amplified cytochrome b sequence with reference sequences of all five Pacific salmon species; the species identity of Sockeye can be readily determined from the sequence comparison (fig. 8).

The D-loop and cytochrome b sequences from all extracted samples matched each other exactly, indicating a single individual. Because the mitochondrial genetic diversity of Pacific salmon is low compared to that of other fish (Nielsen et al. 1994; Billington and Hebert 1991), the presence of an identical sequence in all of the Kwädạy Dän Ts'ìnchị find samples does not irrefutably support the scale analysis results that indicated all samples came from the same fish, but the sequence analysis does not disprove the conclusion. Additional analyses to identify whether the ancient salmon DNA is related to any contemporary Sockeye Salmon populations were unsuccessful due to a lack of sufficient nuclear DNA templates.

Table 1. Results of ancient DNA analysis performed on salmon remains from the Kwädąy Dän Ts'ìnchị discovery.

Sample number	Sample type	D-loop sequence	Cytochrome b sequence	Replicated extraction
1-1	Flesh	Sockeye	Sockeye	Yes
8-6	Skin	No DNA	No DNA	No
34.22	Flesh	Sockeye	Sockeye	Yes
32.5	Flesh	Sockeye	Sockeye	No
105	Flesh	Sockeye	Sockeye	No

Blank extractions and negative controls were included in every step of DNA extraction and amplification. None of these controls produced unexpected DNA amplifications (see fig. 7), indicating that systematic lab contamination was not an issue for this study. In the past five years, the ancient DNA laboratory at SFU has processed more than 300 ancient salmon remains from all five Pacific salmon species recovered from various archaeological sites. No systematic contamination has ever been detected, also demonstrating the effectiveness of the lab's contamination controls.

Conclusions

THE EXACT ORIGIN of Sockeye Salmon found with the Kwädąy Dän Ts'ìnchị man is difficult to pinpoint, as the freshwater habitat in the drainages of Lynn Canal and the upper Tatshenshini River are intrinsically variable. But the identification of edible Sockeye Salmon remains with the Kwädąy Dän Ts'ìnchị man's body highlights the historically intimate relationship coastal peoples of western North America have long had with Pacific salmon.

At the beginning of the project in 2000, initial identification of the fish remains associated with the Kwädąy Dän Ts'ìnchị man was conducted through a comparison of the ancient scales to scales removed from museum specimens and to dichotomous keys from the literature. The patterns near the centre of the ancient scales suggest that as a juvenile, the anadromous Pacific salmon had spent limited time rearing in a freshwater environment and later went on to spend at least four years maturing in an ocean environment. The presence of growth after a fourth winter of ocean rearing suggests the salmon might have died close to spawning time during late summer or early fall. Further analysis was conducted by measuring discrete and continuous characters on all 116 scales found throughout the Kwädąy Dän Ts'ìnchị remains. The scales varied in size and shape, but the arrangement of internal characters suggests they all originated from a single salmon, and evidence suggests the two tissue pieces were from two body regions: lateral (side) and ventral (belly).

Two molecular DNA trials were applied to all the pieces of salmon tissue associated with the Kwädąy Dän Ts'ìnchị remains for the purpose of species identification. The first survey was conducted in 2000, concurrently with the scale analysis, and was inconclusive. However, the 2008 trials, conducted in a laboratory dedicated to ancient DNA analysis, successfully recovered mitochondrial DNA from four of the five archived samples, with sequences from two markers matching Sockeye Salmon (*Oncorhynchus nerka*).

ACKNOWLEDGEMENTS

We are indebted to the following people and groups for their enthusiastic support on this unique project: Champagne and Aishihik First Nations Elders' Council; the Kwädąy Dän Ts'ìnchį Management Group; Sheila Greer; Al Mackie; and, at the Royal BC Museum, Jim Cosgrove, Nick Panter, Val Thorpe, Grant Keddie, Kelly Sendall, Grant Hughes, Andrew Niemann and David Gillan. Dr. E.B. Taylor and Dr. J.D. McPhail offered support from the University of British Columbia. The ancient DNA analysis performed at SFU was conducted with support from the BC Archaeology Branch, the Royal BC Museum, a Canada Graduate Scholarship (Speller) and a Social Science and Humanities Research Council of Canada grant (Yang).

21

TECHNOLOGICAL AND STYLISTIC ANALYSIS OF THE KWÄDĄY DÄN TS'ÌNCHĮ BASKETRY HAT

Kathryn Bernick

This chapter discusses a basketry hat that had been preserved in a glacier in northwestern British Columbia. It was recovered in August 1999, along with other artifacts found with human remains that elders of the Champagne and Aishihik First Nations named Kwädąy Dän Ts'ìnchį, or Long Ago Person Found (Beattie et al. 2000). While other authors in this volume describe the circumstances of the discovery, the range of finds and the results of analytical studies, this chapter is limited to a technological and stylistic analysis of the basketry hat.

THE CHAMPAGNE AND AISHIHIK FIRST NATIONS (CAFN) retained me in 2001 to analyse the hat and prepare a descriptive report; this chapter is a revised version of that original report (Bernick 2001a). One topic of interest was determining whether the finds represented coastal (Tlingit) or interior (Southern Tutchone) communities. The hat seemed to be a good candidate for clarifying the issue of cultural origin, and this potential guided the research design. Uncertainty about the hat's final disposition provided impetus for detailed descriptive documentation.

Additionally, the details reported here may be of interest to basket weavers and to scholars analysing basketry from other regions. The hat also has the potential to contribute to scientific knowledge in other ways—for example, the identity of the plant species from which it is made, the environmental residues embedded in the fabric and the composition of applied paint. Cultural inferences can be drawn based on associations with other finds and on the hat's function as an item of apparel. However, these topics lie outside the scope of my study. To avoid possible confusion, I relate the history of Gregory Young's wood species determination for the hat materials, as his identification changed after it was cited in my previous report (Bernick 2001a). Valery Monahan (chapter 22) presents an account of the hat's conservation treatment and relates details about the paint she observed on the hat. Sheila Greer (chapter 36) relates the hat's role as a catalyst for the revival of weaving in local communities.

Research Framework

BASKETRY HAS GREAT POTENTIAL to inform us about the past. An additive manufacturing process, complex structure and stylistic sensitivity make basketry artifacts particularly useful for investigating questions of cultural attribution. Studies of archaeological assemblages from the Northwest Coast show that basket types and styles reflect particular cultural groups. (Northwest Coast refers to an anthropological "culture area" extending from southeastern Alaska to northern California along the Pacific coast of North America. The Tlingit are an ethnolinguistic group on the northern Northwest Coast; see Suttles 1990.) In addition to giving clues about the ethnicity or cultural identity of the basket-makers, characteristics of basketry artifacts may provide chronological markers (Bernick 1987, 1998a, 1998b, 2003; Croes 1977, 1987). These observations pertain to all culture areas with basketry, as well as to ethnographic collections and archaeological assemblages.

Basketry hats have the same research capacity as baskets, plus the potential to reflect aspects of the individuals for whom they were made. But hats are seldom recovered from archaeological contexts and have not been studied as much as baskets. The small sample sizes, which constrain interpretation, can be increased by including appropriate aspects of basketry containers. Technological similarities allow basketry hats to be analysed and described as though they are a type of basket (in one sense, a basketry hat is an upside-down basket).

This detailed description of the hat at the Kwäd̠ay Dän Ts'ìnchį site will add significantly to the body of information now available about basketry hats from the Northwest Coast and surrounding areas. Studies of post-contact headgear from the region tend to address hats with decorative embellishments, emphasizing the artwork over details of construction (Devine 1980, 1981, 1982; Hodge 1929; Laforet 2000; Watkins 1939). Although painted designs and appendages signifying social status clearly have important interpretive potential, comparative studies can be enhanced by incorporating archaeological specimens. While archaeological artifacts tend to be utilitarian and relatively plain, they are suitable for considering technological aspects of manufacture and structural ornamentation. Archaeological specimens also provide a temporal dimension from eras for which museum collections and written or pictorial accounts do not exist.

The find site's location suggests that the Kwäd̠ay Dän Ts'ìnchį hat was made either in the Tlingit area of the northern Northwest Coast or in the adjacent interior. I am unaware of any information about traditional basketry hats from that part of the interior or of any record that the Southern Tutchone made basketry hats. On the coast, people made basketry hats in different styles corresponding to ethnolinguistic groupings (Croes 1977; Laforet 1990). If the Kwäd̠ay Dän Ts'ìnchį hat was made on the coast, it should resemble Northwest Coast basketry hats in general. If it was made by a Tlingit basket-maker, it should display diagnostic characteristics of Tlingit basketry and of Tlingit basketry hats in particular—taking into account that some of the characteristics may have changed through time.

Analysis of the Kwäd̠ay Dän Ts'ìnchį hat offers a case study for exploring cultural attribution of Northwest Coast archaeological basketry hats. The research framework involved identifying attributes of the hat that indicate a general Northwest Coast style and those that suggest a specifically Tlingit style, determining how the hat differs from those made by

Figure 1. The basketry hat partially reshaped. Scale in centimetres. Valery Monahan photograph. Colour version on page 590.

other Northwest Coast ethnolinguistic groups, and distinguishing chronological markers indicated by the hat's technological and stylistic characteristics.

Methods

I ANALYSED the Kwädąy Dän Ts'ìnchį basketry hat over a three-day period in May 2001, in the conservation section of the Yukon Heritage Branch in Whitehorse, Yukon. Valery Monahan, the Heritage Branch conservator, arranged workspace and facilities, assisted with handling the object and provided copies of conservation treatment records. The details and descriptions in this chapter document my observations, which were enhanced by discussion with Monahan, with individuals from the Champagne and Aishihik First Nations who stopped by to talk about the hat, and with Whitehorse resident and traditional weaver Anne Smith, who was planning to make a replica.

The hat was dry, stiff and extremely fragile when I examined it. Flattened when found, it had been partially reshaped in preparation for conservation treatment (fig. 1). The hat rested on a custom-made flip mount that permitted examination alternately from top and bottom perspectives without handling it. The compactness of the weaving, as well as the misshapen condition of the hat and its lack of pliability, constrained observation of some details.

I selected attributes to record in keeping with conventional archaeological practice (Adovasio 1977) and my previous experience conducting stylistic analyses of Northwest Coast basketry (Bernick 1998a, 1998b, 2000). In addition to notes of visual observations, measurements were taken using dial calipers and a flexible tape. I did not remove samples from the hat.

Terminology used here conforms to that in scholarly archaeological studies of basketry from the Northwest Coast (Bernick 1983, 1998a, 1998b; Croes 1977). "Warp" refers to a passive element (vertical in this case) and "weft" to an active element (horizontal). These terms may refer to single elements or to the entire set. Figure 2 defines weave types. Other technical terms are defined in the text upon first usage.

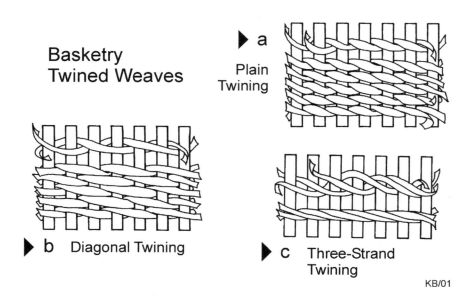

Basketry Twined Weaves

▶ **a** Plain Twining

▶ **b** Diagonal Twining

▶ **c** Three-Strand Twining

KB/01

Figure 2. Weave types: *A*, two-strand twining, often called plain twining; *B*, diagonal twining, sometimes called twilled twining; *C*, three-strand twining. Kathryn Bernick illustration.

For comparisons, I consulted published material and drew on my familiarity with archaeological specimens. The main sources for information on Tlingit basketry technology in the post-contact era are Emmons (1903, reprinted 1993) and Paul (1981). Pre-contact baskets that have been recovered from the Tlingit area of Alaska include 5,000-year-old specimens from Baranof Island (Bernick 1999) and a 6,000-year-old basket from Prince of Wales Island (Fifield and Putnam 1995).

Several archaeological sites on the Northwest Coast have yielded basketry hats for which Croes (1977) gives basic descriptive information. To my knowledge, only one other ancient basketry hat has been recovered from the northern Northwest Coast or the adjacent interior: an isolated find from archaeological site HdTw y, referred to here as the Katete-Stikine hat. It was found in May 2000, on a sandbar at the confluence of the Katete and Stikine rivers in northwestern British Columbia, very near the border with Alaska (Bernick 2001b). That location is south of the Kwädąy Dän Ts'ìnchį find site but is similarly at the conjunction of coastal and interior areas. The Katete-Stikine hat yielded a radiocarbon date estimate of 180 ± 30 uncal BP (Beta-146265), indicating a late pre-contact or early

post-contact age (Bernick 2001b). The Katete-Stikine hat is in the Royal BC Museum's archaeology collections, where I examined it in detail in 2000 and 2001.

Descriptive Results

THE KWÄDĄY DÄN TS'ÌNCHĮ BASKETRY HAT is skilfully woven from thin wood fibres using several variants of twining. A basketry headband is firmly attached to the inside, and remnants of a hide chin strap are attached to the headband.

The hat is complete but damaged. Abrasion along the bent edge at the top, horizontal slits on the sides (where warps have broken) and rips in the headband suggest that it had been worn over a long period or in adverse conditions. Mends testify to maintenance for continued use.

Materials

The weaving elements used to make the Kwäday Dän Ts'ìnchị hat are thin, narrow, longitudinally split sections of withes or roots, generically referred to as wood fibres. No traces of adhering bark were visible, suggesting the bark had been removed prior to splitting. The warps comprise "inner splints" with rectangular cross-sections and two flat surfaces, whereas most of the weft elements—including the overlay strands—appear to be very thin "outer splints" with semicircular cross-sections (one flat surface and one curved surface). The curved outer surfaces are smoother than the flat inner surfaces. When freshly peeled and split, the curved surfaces were probably somewhat shiny. These characteristics of weaving element composition correspond to descriptions of Tlingit spruce-root basketry (Emmons 1993; Paul 1981) and also to 5,000-year-old baskets from Baranof Island in the Tlingit area of Alaska (now in the collections of the Alaska State Museum in Juneau; personal observation, March 2009).

Due to the compactness and nearly intact condition of the weaving, it was impossible to confirm whether the Kwäday Dän Ts'ìnchị hat's construction follows the Tlingit practice of backing the warps with additional strands, as described by Emmons (1993, 243) and Paul (1981, 30). The few exposed warp ends that I observed and the apparent introduction of new warps as single elements suggest that the warps may be single elements (rather than doubled or tripled), at least in some areas of the construction.

Table 1 summarizes element sizes. Notably, the warps are somewhat wider near the start (centre top of the hat) and become narrower as the weaving progresses. Also, the wefts are slightly narrower than the warps, which is common for many types of archaeological basketry throughout the Northwest Coast culture area. The headband elements are essentially twice as wide as those used for the hat itself, but they otherwise appear similar. Although some elements may have shrunk during the drying process, there was no visible evidence of distortion or loosening of the weave that would indicate a significant change in dimensions.

Table 1. Element width and weaving gauge.

Hat part	Element width, in mm		Weaving gauge, per 10 centimetres	
	Warp	Weft	Warp[1]	Weft[2]
Top, two-strand twining	2	1.5	45	55
Side, three-strand twining	2	1.0–1.5	60	55
Side, two-strand twining	1.5	1.0–1.5	50	60–65
Headband, diagonal twining	3	3	16	28

[1] Number of stitches across the warp (horizontal)
[2] Number of weft rows (vertical)

Determining plant part and species requires microscopic analysis of thin sections with reference to comparative specimens. In 2000 Gregory Young, a scientist at the Canadian Conservation Institute in Ottawa, examined samples from the hat brim and the headband. He determined that they are the same species, possibly *Juniperus* sp. (Juniper) or *Chamaecyparis* sp. (Yellow Cedar, also known as Alaska Cedar) (Young 2000). These identifications are unexpected for a Tlingit-style basketry object such as the Kwädąy Dän Ts'ìnchį hat, as traditional Tlingit basketry is woven from split roots of *Picea sitchensis* (Sitka Spruce) (Emmons 1993; Paul 1981). Moreover, Sitka Spruce appears to have been a common basketry material for the past several thousand years in the Tlingit area of southeastern Alaska, though other species may also have been used (Bernick 1999; Fifield and Putnam 1995). At the request of the Champagne and Aishihik First Nations, Young re-examined samples from the hat in October 2001 and changed his identification to "*Picea* sp., likely *Picea sitchensis* (Sitka Spruce)" (Young 2001).

Form and Size

Prior to being flattened under glacial ice, the Kwädąy Dän Ts'ìnchį hat was shaped like a truncated cone, with a round, nearly flat top and straight, flared sides. Its original form, which I reconstructed from its partially reshaped state using geometrical calculations, would have stood 10.5 centimetres high and measured 93 centimetres in circumference at the brim edge. For detailed dimensions and a schematic sectional view of the hat's form, see figure 3.

Figure 3. Schematic view of the original shape of the Kwädąy Dän Ts'ìnchį basketry hat and the sequence of weave types. Kathryn Bernick illustration.

The hat's sides do not exhibit distinct crown and brim sections, though the bottom portion functions as a brim. Neither the top nor any part of the hat's sides would have rested on the wearer's head. Instead, the hat has an inner band that fits the head, analogous to a crown band on European-style hats (fig. 4). One edge (long side) of the headband is firmly attached to the inside surface of the hat about 10 centimetres above the brim edge. The band tapers in circumference from about 55 centimetres at its opening to 44 centimetres along its top edge (where it is attached). The headband circumference corresponds to a very small size by present-day Canadian hat standards.

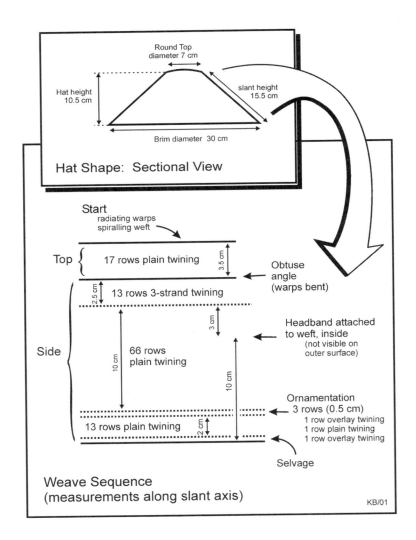

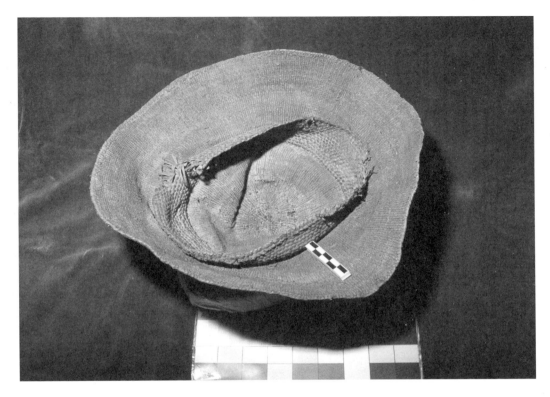

Figure 4. Inside view of the basketry hat showing attached headband. Loose ends of splices and inserted warps give the inside surface a slightly rough appearance. Scale in centimetres. Valery Monahan photograph. Colour version on page 590.

The Kwädąy Dän Ts'ìnchį hat mirrors traditional Tlingit-style work hats in shape and composition—and, with some variations, hats made by other Aboriginal peoples on the Northwest Coast (Emmons 1993, 256; Laforet 1990). Museum collections and archaeological finds document conical basketry hats with inner headbands for at least the past 2,500 years on the Northwest Coast (Croes 1977, 234–43). Special-occasion basketry hats had the same basic shape and composition as the everyday hats but with decorative or symbolic appendages and often painted designs.

Construction Techniques

The Kwädąy Dän Ts'ìnchį hat is woven entirely by twining, with stitches leaning up to the left (\). The top of the hat and most of the sides are woven in two-strand plain twining (fig. 2A). The upper part of the side features a band of three-strand twining (fig. 2C). There are also two rows of overlay twining near the brim

(fig. 5). Figure 3 illustrates the weave sequence, including the number of weft rows and their relative placement. The headband is discussed under a separate heading.

The hat's top, seven centimetres in diameter, was circular and nearly flat. It comprises 17 weft rows (all two-strand twining), with numerous inserted warps that expand the circle while maintaining a consistent tightness in the weave. The warps were then bent at an obtuse angle to form the sides of the hat, and the first 13 rows are woven in three-strand twining. Thus, the uppermost part of the hat side consists of a 2.5-centimetre-wide band woven in a technique that provides greater strength than the two-strand twining used to make the other parts of the hat. The rest of the flared sides consist of 82 weft rows (12.5 centimetres) woven in two-strand plain twining. Two of those plain-twined weft rows, positioned 2 centimetres from the brim edge, have a twisted overlay that strengthens the construction. The overlay technique is further discussed in the ornamentation section. Plain twining with an up-to-the-left stitch slant as the primary weave in combination with three-strand twining in the upper area of the sides characterizes Tlingit hats as described by Emmons (1993), Laforet (1990) and Paul (1981). But their examples have more complex weave combinations (and more ornamentation) than the Kwädąy Dän Ts'ìnchį specimen.

Weaving gauge, calculated as the average number of stitches per 10 centimetres, is given in table 1. This measurement represents relative fineness or coarseness of the weave. As expected for consistently compact weaving, areas with narrower elements have a finer weave (more stitches per 10 centimetres). To my knowledge, there are no published statistical data on weaving gauge for Tlingit hats. The Tlingit-style Katete-Stikine archaeological hat has a somewhat coarser weave than the Kwädąy Dän Ts'ìnchį hat.

The weave start on the Kwädąy Dän Ts'ìnchį hat is a clockwise spiral with radiating warps, which is typical for northern Northwest Coast baskets and hats (Emmons 1993; Paul 1981). During construction, warps were added to expand the form. Most appear to be single elements inserted into the weave. Some straight-cut ends of new warps show on the inside surface of the weaving, but most ends are covered by weft stitches.

Construction proceeded with a spiralling weft. The hat's inside surface shows numerous cut ends of weft elements, indicating that splices were introduced from the inside. Weft splices are ubiquitous in basketry and have little diagnostic potential. However, at the point where weaving techniques change, the direction of the spiral can be ascertained by noting the relative position of the weft rounds. At the jog point on the Kwädąy Dän Ts'ìnchį hat (looking at it with the top up), the stitches on the right are lower than those on the left. Assuming that the weaver worked from left to right, which was usual, this indicates that the hat was held in an upside-down position when it was being woven—that is, with the top facing the ground—and that the weaving proceeded in an upwards direction. This orientation is typical in Tlingit basket weaving and is opposite to the Haida practice (Laforet 1990, 291).

The brim edge is finished in a strong, elaborate manner that involves folding the warps to the inside and securing them with a braid-like twining sequence. The precise method of this selvage (edge finish) is difficult to confirm due to its tightness and complexity. It appears to be the Tlingit-style edge called "Border 12" by Emmons (1993, 248) and is consistent with Laforet's depiction of the usual Tlingit-style hat brim finish (1990, fig. 59b). The Katete-Stikine archaeological specimen has a similar selvage.

Headband

The inner headband is an integral component of the Kwädąy Dän Ts'ìnchį hat (fig. 4), and is made from the same (or very similar) materials as the rest of the construction. The weft is oriented parallel to the sides of the band; it is not woven on a bias.

The 4.5-centimetre-wide band has 13 weft rows plus the selvage. All but the first two rows, which are plain twining, are woven in the twilled twining technique that is also known as diagonal twining (fig. 2B). The twining stitches lean up to the left (\). The headband appears to be as skilfully woven as the rest of the hat, though its wider elements and the weave type make it coarser (table 1). The band is broken in two places, apparently where a chin strap had been attached. In both instances, the rips extend 1.5 centimetres inward from the finished edge.

The headband was affixed during construction, while the hat sides were being woven. A particularly stiff weft row discernible on the outside of the hat coincides with the location of attachment, suggesting that the headband is tightly secured to the weft stitches. It was impossible to ascertain the precise method of attachment, since that would have required examining both sides of the headband along its attached edge—an examination precluded by the stiffness and friability of the material. The open edge of the headband weave was finished by folding over the warp ends to the outside (the side facing the hat wall) and securing them to the respective adjacent warps with a row of plain twining (\). This is a relatively simple but strong selvage technique that Emmons refers to as "Border 7" (1993, 247) and Paul as "Border 4" (1981, 28). Both note its ubiquity on articles intended for use, which presumably includes hat headbands.

Comparative information about headbands on Northwest Coast basketry hats indicates that they were always present, but details of construction, attachment and size are lacking. The Katete-Stikine hat headband resembles the Kwädąy Dän Ts'ìnchį specimen. It, too, is woven in diagonal twining, though considerably coarser. The method of attachment appears to differ. Croes (1977, 432) reports that archaeological spruce-root hats from the southern Northwest Coast have cedar-bark headbands. These differ from the northern specimens in weave type, but band width and placement on the hat (i.e., distance from the brim edge) are similar. He does not give circumference dimensions.

Mends

Two slits in the side of the Kwädąy Dän Ts'ìnchį hat had been repaired in antiquity. Each consists of a 2.5-centimetre-long horizontal line of broken warps held together by sewing stitches. One, located 4.5 centimetres from the brim edge, was clearly mended from the inside, though the stitches emerge onto the outside surface. The mending thread consists of two-strand twisted cordage, Z-laid, 0.5 millimetres in diameter, and appears to be sinew. The cord is anchored on the inside with a knot slightly beyond one end of the slit (see fig. 4 in chapter 22). Vertical stitches close the slit and are fully visible on the inside surface. On the outside the stitches appear to be diagonal but are obscured by a ravelled weft. The second mend, located 7 centimetres from the brim edge, appears to be identical in technique, but the repair job is partially obscured by the headband. A third slit on the hat, 3.5 centimetres from the brim edge, has not been repaired.

Both the stitching and the cord differ from those used to construct and repair the garment found with the Kwädąy Dän Ts'ìnchį remains (see chapter 26). Compared to the garment, the mends on the hat have more widely spaced and less even stitches and are sewn with thinner cord of a different colour (personal observation, May 31, 2001).

Ornamentation

The Kwädąy Dän Ts'ìnchį hat displays two kinds of structural ornamentation that involve variations in weaving technique. In addition, the hat was likely painted—but for weatherproofing rather than decoration.

As previously mentioned under construction techniques, the uppermost portion of the hat's side features a 2.5-centimetre-wide band of three-strand twining. This band contrasts structurally and visually with the rest of the hat side, which is woven in two-strand twining. The combination of these particular weaves, and their relative placement, is common on Tlingit hats. In addition to providing strength at a stress-bearing location, the band of three-strand twining has a decorative quality.

A second kind of structural ornamentation occurs near the bottom of the hat's side. This consists of two rows of overlay twining made prominent by being raised and having a stitch slant opposite to the weaving on the rest of the hat (figs. 5 and 6). Although the two rows of overlay twining comprise a continuous spiral, they have a row of plain two-strand twining between them. It is not clear how this was accomplished; perhaps two weft rows were woven at the same time.

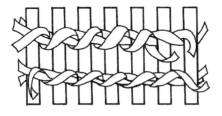

Figure 5. Overlay twining consisting of two-strand twining with an added element twisted around the weft stitches on the outside surface only. This is called false embroidery when executed with coloured overlays.
Kathryn Bernick illustration.

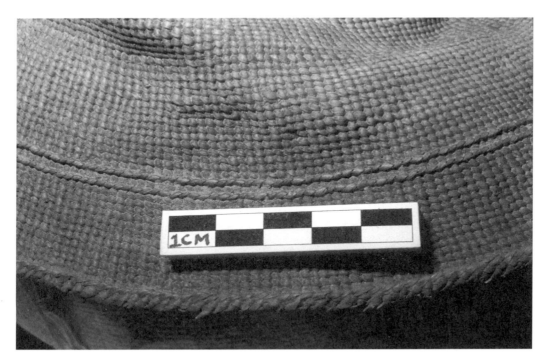

Figure 6. Lower portion of the hat's side, showing two rows of overlay twining (above the scale). Scale in centimetres. Valery Monahan photograph.

The method of construction referred to here as overlay twining is exactly like Tlingit-style false embroidery executed with coloured grass overlay elements. Emmons (1993, 242) describes the variant that occurs on the Kwädąy Dän Ts'ìnchį hat, with an overlay strand identical to a regular weft strand, as a "twist" used to strengthen or reinforce the construction. The Katete-Stikine archaeological basketry hat has two rows of overlay twining in a similar location, as well as two rows higher on the hat's side and one on the top.

There are traces of a reddish substance on the Kwädąy Dän Ts'ìnchį hat. The faded pigment, barely noticeable without magnification, is most apparent on low-relief portions of the weaving elements and as a broad smear on the inside of the headband on one side. Whether the pigment represents a residual painted design cannot be ascertained by visual inspection, but there is no indication that it does. After my examination of the hat, Valery Monahan examined the pigment in detail. She observed it under a thin film of silt that covered much of the hat's outer surface and submitted a sample for identification. Monahan (chapter 22, this volume) reports that the pigment is red ochre, and she concludes that the hat had been covered with paint. Emmons (1993, 256) notes that conical Tlingit hats were painted for protection from the elements and that hats with painted designs, worn by wealthy people on festive occasions, had a different shape. The Katete-Stikine hat also has traces of a red substance.

Appendages

A hide strap was attached to the headband when the Kwädąy Dän Ts'ìnchį hat was recovered. With the exception of one short piece anchored to the headband with a knot, the strap was detached and broken when I saw it. The detached pieces had been treated with polyethylene glycol and dried in a freezer (see chapter 22). They were extremely fragile, dry, stiff and somewhat distorted. The material appears to comprise at least two types. No species determinations have been attempted and the method of tanning, if any, remains unconfirmed. Because they had been attached to the headband, these hide pieces can be identified with confidence as parts of a chin strap that would have secured the hat while it was being worn.

From a technological perspective, one interesting aspect of the strap concerns the way in which the pieces were combined. Two pieces, 15.8 centimetres and 9.4 centimetres long respectively, are combined end to end with a "slit-join" and knots. The slit-join, which has an associated stopper knot, resembles features on other items associated with the Kwädąy Dän Ts'ìnchį remains (K. Mackie, pers. comm., May 31, 2001; see chapter 26, this volume). All of the knots—one anchored in the headband and three on the detached pieces—appear to be variants of overhand knots, though this is difficult to confirm as they are tightly snugged with broken ends. Two pieces, 2.8 centimetres and 1.1 centimetres long, are apparently tag ends broken off the knots. The maximum extant width of the strap is 5 millimetres, but the amount of distortion is unknown.

Discussion

THE KWÄDĄY DÄN TS'ÌNCHĮ BASKETRY HAT is typically Tlingit in style and construction technique. This conclusion rests on comparison with published studies of North American basketry technology. The hat displays numerous attributes identifying it as Tlingit: longitudinally split spruce root materials; conical shape with a nearly flat, rounded top and straight sides; the particular set of weaving techniques, including up-to-the-left stitch slant, spiral start with radiating warps, construction executed in an upward direction, three-strand twining near the top of the hat's sides, type of brim selvage and headband woven in diagonal twining. Although some of these attributes also characterize basketry hats made by other Northwest Coast peoples, the combination indicates a Tlingit style.

Radiocarbon assays confirm a late pre-contact or early post-contact age. There are two date estimates on plant fibre from the hat: 500 ± 30 uncal BP (Beta-133765) and 197 ± 23 uncal BP (OxA-16690) (Beattie et al. 2000; Richards et al. chapter 6, this volume). In their discussion of the radiocarbon dating, Richards

et al. suggest that the younger date for the hat is more accurate. Their conclusions indicate that the hat was made in the 18th century or early 19th century AD.

While stylistic characteristics of the basketry hat have potential for comparative dating, that would require appropriate samples with associated dates and detailed descriptions. Ethnographic collections in museums may provide pertinent comparative information for the post-contact era, but as stated in the methods section, only published descriptions of specimens in ethnographic collections were reviewed for this study. Archaeological basketry hats are scarce in assemblages from the Northwest Coast, and none are reported with full descriptive information. The only archaeological Tlingit-style basketry hat recovered to date is the Katete-Stikine specimen. Although I analysed it thoroughly, the reports (Bernick 2000, 2001b) do not include all of the technological details that might be useful for comparisons.

The Kwädąy Dän Ts'ìnchį hat bears considerable resemblance to the Katete-Stikine specimen, especially in weaving techniques and selvage construction. The two hats are also roughly similar in age, both dating to late pre-contact or post-contact times. The Kwädąy Dän Ts'ìnchį hat is shorter and more broadly flared, and has less ornamentation. The Katete-Stikine hat lacks contextual information, thereby reducing the insight it might contribute toward attribution of the Kwädąy Dän Ts'ìnchį specimen. Some of the shared attributes, such as the ornamental "twist" overlay, might be chronological markers, but that remains speculative until more specimens with a wider range of dates become available.

Some construction details of the Kwädąy Dän Ts'ìnchį specimen that might be significant for comparisons were not observable due to the integrity of the hat and its stiffness. Had the hat's woven fabric been (somewhat) pliable when I examined it, additional information would have been ascertainable. For example, it would probably have been possible to determine the exact method by which the weaver attached the headband—a characteristic with potential for embedded cultural variability and an essential detail for fully accurate replication. Determining the composition of completely covered weaving elements was not possible. In this regard, ragged, broken specimens that reveal the inner structure have advantages for the analyst intent on documenting intricacies of construction. Yet the outstanding preservation environment produced a complete article that is unmistakeably recognizable as a hat.

Conclusion

TECHNOLOGICALLY AND STYLISTICALLY, the Kwädąy Dän Ts'ìnchį hat conforms to Tlingit basketry—specifically, to Tlingit basketry hats. It shares some basic attributes with hats made by other ethnolinguistic groups on the Northwest Coast and, in that respect, represents a cultural tradition associated with the Northwest Coast culture area. At the same time, the hat displays other attributes clearly distinguishing it from hats made by non-Tlingit Northwest Coast groups. These conclusions rest on comparison with published information, most of which concerns objects made in the 19th and 20th centuries. How far back in time these cultural distinctions prevail cannot be confirmed from available information. The scant archaeological record suggests that on the Northwest Coast, basketry hat types have the same spatial and chronological distribution as basketry containers. With the caveat that sample sizes are small, Tlingit-area spruce-root basketry appears to be a craft tradition that is several thousand years old.

Analysis of the Kwädąy Dän Ts'ìnchį hat points to possibilities for further research. Potentially fruitful research directions include a comprehensive study of Northwest Coast basketry hats and a search for Pacific Rim associations. A thorough review of historical documents and photographs might help answer the question of whether the Southern Tutchone and their interior neighbours wore basketry hats, and if so, whether these were coastal imports. More and larger archaeological assemblages would be highly useful. Detailed descriptions of existing specimens should precede regional synthesis and comparative research; this description of the Kwädąy Dän Ts'ìnchį hat is a first step. Wherever it leads, the present study confirms that detailed technological and stylistic analysis can inform cultural attribution.

ACKNOWLEDGEMENTS

I sincerely thank Champagne and Aishihik First Nations (CAFN) and the late Sarah Gaunt for the opportunity to analyse the Kwädąy Dän Ts'ìnchį hat and to visit Whitehorse. I also greatly appreciate the congeniality and logistical assistance of the Yukon Heritage Branch staff, particularly Valery Monahan. Many people generously shared ideas and information, among them Jerry Cybulski, Sarah Gaunt, Sheila Greer, Al Mackie, Kjerstin Mackie, Valery Monahan, Anne Smith and the proprietor of Roberta's Hats in Victoria, BC. My thanks to Sheila Greer for editorial advice on revising this article from my previous report. Selected results of my analysis were presented in a poster produced by the CAFN for the Northwest Anthropological Conference in April 2008.

22

CONSERVATION OF THE KWÄDĄY DÄN TS'ÌNCHĮ WOVEN HAT AND WOODEN ARTIFACTS

Valery Monahan

The 1999 Kwädąy Dän Ts'ìnchį discovery included a woven plant-fibre hat with an animal-hide strap, a composite tool with metal blade and wooden handle, and a variety of cut and carved wooden artifacts. These artifacts were brought to the Yukon government museums conservator for treatment and storage. Additional wooden pieces were found in 2001, 2003, 2004 and 2005, during annual monitoring at the Kwädąy Dän Ts'ìnchį discovery site (see chapter 23). These artifacts were added to the original collection, along with a roll of modified birch bark found in 2005 (Greer 2006). The initial goal of treating the artifacts in Whitehorse was to make them stable enough for culturally relevant study and scientific analysis. Initially, the Champagne and Aishihik First Nations identified a one-year study period for the Kwädąy Dän Ts'ìnchį human remains and any related finds. It was also their wish that the artifacts remain in the Yukon. These factors limited conservation treatment options. When it was decided that the Kwädąy Dän Ts'ìnchį artifacts would be kept beyond that initial study year, long-term storage and community access became important considerations for their care. Conservation of the Whitehorse artifacts has consisted of assessment and documentation, treatment to allow them to be dried safely, post-treatment storage and handling, and the removal of samples in support of a variety of approved scientific studies. The recovery of archaeological artifacts from glaciers and other ice contexts is relatively rare worldwide. Hopefully, experience gained from the assessment and treatment of the Kwädąy Dän Ts'ìnchį collection will help researchers working with future frozen finds.

Condition of the Wooden Artifacts

THE CHOICE OF CONSERVATION TREATMENT for archaeological wood depends upon its condition at the time of recovery. This, in turn, is determined by environmental factors at the site. Archaeological sites in the boreal forest zone of Canada do not support organic preservation, so it is presumed that the wood found at the Kwädąy Dän Ts'ìnchį site survived because it became encased in a glacier. The Kwädąy Dän Ts'ìnchį site consists of an ice ridge at the margin of an active glacier, at an altitude of approximately 1,600 metres. Seasonal snow is absent from the site for only a few weeks in warm summers, but the glacier experienced enough melting in the recent past for running water to create a channel. In 1999 wooden artifacts were found exposed on gravel slopes some distance from the glacier as well as on ice surfaces and, in one case, protruding from the ice (Beattie et al. 2000). This suggests that the artifacts melted out of the glacial ice at different rates, possibly in different years. While frozen in the glacier, artifacts were likely subjected to physical forces from ice movement. As they emerged from the ice, the artifacts may have been exposed to additional physical forces as a result of ice break-up and transport by running water. Once out of the ice, the artifacts would have been exposed to the action of both running and standing

oxygen-rich water, to ultraviolet light and to the attack of cold-tolerant microorganisms.

Just after recovery in 1999, the wooden artifacts were described as "damp" and "weathered" (Komejan et al. 1999–2008). Now dry, the artifacts range from brown, dense pieces of wood with considerable structural integrity to lightweight, brittle ones with soft, grey surfaces (fig. 1). While cracks and dimensional distortion can be seen on many of the pieces, wood deterioration is not universal. A few wooden artifacts have reddish areas that resemble deposits of pigment. One artifact (IkVf-1:125) has what appears to be a brown surface coating.

In some artifacts broken into several fragments, condition varies from fragment to fragment. In others, condition varies along the length of one piece. Some have sound, brownish cores with grey, soft, reduced or otherwise distorted exteriors. Running water removes soluble lignin from wood, while sunlight causes it to first darken, then turn grey (Cronyn 1990). The results of these processes, commonly referred to as weathering, are familiar to most people from driftwood—which many of the Kwädąy Dän Ts'ìnchį wooden pieces resemble—and the uncoated wooden surfaces of historic structures.

Wood begins to weather within weeks or months of exposure to the elements (Feist 1990). The shrunken

Figure 1. Range in wood condition. Wood on the left is lightweight, grey and fibrous. Left to right, condition improves. The fragment at the far right exhibits physical damage in the form of cracks, but the wood is dense, with crisp surface details. All scales are in centimetres. Government of Yukon. Colour version on page 591.

exteriors of the most deteriorated wooden artifacts in the Kwädąy Dän Ts'ìnchį collection suggest that they became at least partially waterlogged after melting and had dried partially, or completely, before recovery. Their weathered surfaces suggest exposure to the alpine environment during or immediately after this process. During microscopic examination for wood identification, the sample from one of the more deteriorated wooden artifacts (IkVf-1:123) was observed to contain structures consistent with attack by a brown rot fungus (Florian 2005). In contrast, the more robust collection pieces seem to have had little water exposure. The water within glacier ice may not impact the artifacts until melting is actually underway. Even then, there may not be sufficient exposure for organic artifacts to become fully waterlogged. Permafrost and short

summer seasons at Arctic sites can preserve buried wood and other organic materials without waterlogging. Wood exposed at the surface of these sites may survive for longer than would be expected in other climates (Scott and Grant 2007).

While all the wooden artifacts have breaks, cracks or other evidence of damage caused by physical force, evidence of wood deterioration varies at this site. The artifact with the best-preserved wood (IkVf-1:106, see fig. 2) was the one recovered directly from the ice (Beattie et al. 2000), confirming the role that continuous freezing has in organic preservation. The kinds of wood deterioration present in collections pieces—weathering, partial waterlogging, fungal attack and distortion from uncontrolled drying—are all consistent with surface exposure in an alpine

Figure 2. Wooden artifact IkVf-1:106, after drying. The left fragment was recovered from ice. It is stained, but surface details like wood grain are visible. The right fragment has weathered and lost surface details. Dark scratches show that weathering extends only a few millimetres at the surface. Scale in centimetres. Government of Yukon. Colour version on page 591.

environment. Over time it has become clear that the wooden artifacts in the Kwädąy Dän Ts'inchį collection resemble cold-climate-weathered wood both in appearance and condition.

Treatment of Wooden Artifacts

AT THE TIME OF RECOVERY, it was assumed that all the wooden artifacts were at least partially waterlogged and would require some kind of bulking agent if they were to be dried without structural damage. While there are a variety of recognized treatment options for waterlogged wood (Cronyn 1990; Hamilton 1998), cost and health considerations have meant most Canadian conservation laboratories treat waterlogged archaeological wood by immersion in polyethylene glycol (PEG) and water, followed by freezing and sublimation of excess water. Depending on the size and condition of the wood and

the availability of a purpose-built freeze-dryer, the process of polyethylene glycol impregnation and freeze-drying can take between several months and several years to complete, with artifacts generally inaccessible to researchers during that time. This treatment option was rejected for the Kwädąy Dän Ts'inchį wooden artifacts because it was in conflict with the year-long study framework initially established for the collection (Beattie et al. 2000).

All but one section of one of the wooden artifacts were placed into available chest-style freezers until an alternative treatment could be designed and approved. A section of the largest wooden artifact (IkVf-1:106) did not fit into any of the available freezers, and two months after it was recovered, it was observed to be splitting as it dried. It was brushed with a polyethylene glycol and water solution, in an attempt to stabilize the wood against further physical change, and then wrapped in layers of polyethylene foam to slow the drying process

Figure 3. The underside of the woven hat at the discovery site. It has breaks in the headband and it is crushed, but it holds its shape with limited support. Photograph courtesy of Champagne and Aishihik First Nations Heritage program.

(Komejan 2000). One of the visible cracks grew six centimetres, but the section dried in one piece, without noticeable additional distortion.

The remaining wooden artifacts were removed from the freezer periodically for inspection. They were observed to be losing unbound water to sublimation without shrinking or becoming more distorted, so they were allowed to dry completely in the freezers without further treatment. By December 2000 all were dry enough to be removed from the freezers. This technique of slow-drying archaeological wood (and other organic materials) without chemical impregnation has been used successfully where organics have been preserved in frozen sites without waterlogging (Scott and Grant 2007; Monahan 2004). The man's knife was kept frozen until

treatment and associated analysis could be undertaken systematically (see chapter 25).

Condition of the Woven Hat

IT SEEMS LIKELY that the hat melted out of permanent ice between the August 1999 discovery and the arrival of the recovery team three days later. The hunters who first discovered the remains did not see the hat (Beattie et al. 2000), but it was disturbed by the air currents generated by the recovery team's helicopter (Greer 2006). At recovery the hat appeared complete. It had considerable structural strength but was flattened, suggesting that it had been crushed (see fig. 3).

The main portion of the hat had folds and distortions, and the headband was broken in two places. The tie strap was broken on one side, and the hat was coated with fine silt. Its overall good condition, with preserved hide tie strap and delicate sinew thread repairs, suggests it spent little time exposed to sun and water. Semi-tanned animal hides and sinew are rarely preserved archaeologically, except in the context of dry burial or under conditions where they remain continuously frozen (Cronyn 1990). Once exposed on the surface, these materials deteriorate so rapidly that their discovery has been used as proxy evidence of glacier retreat resulting from climate change (Grosjean et al. 2007).

Conservation of the Woven Hat

IN DECEMBER 1999 two small fragments of the woven plant fibre were removed from the wet hat. One was sent for radiocarbon dating (Beattie et al. 2000) and the other for plant identification (Young 2000). Like the wooden artifacts, the hat was assumed to be at least partly waterlogged, requiring chemical impregnation before it could be safely dried. But unlike the wooden artifacts, the hat was flexible (while wet, at least) and so had the potential to be reshaped—an opportunity which could have been lost if the hat had dried without intervention.

While treatment decisions were being made, the hat was kept immersed in changes of water. This is a common short-term storage method for untreated, waterlogged wood (Cronyn 1990), but it is not risk-free. Wood will degrade slowly during wet storage, but this is seen as a reasonable compromise when weighed against the greater risk of structural damage that occurs when waterlogged wood is allowed to dry without impregnation. In this case, the water had an adverse effect on the hat's animal tissue components, which began to degrade. Three small slits in the woven plant fibre were recorded in 1999. Two of the slits retain fine sinew thread, apparently used to repair them (see fig. 4). Similar thread was thought to have been lost from the third slit during wet storage (Komejan, pers. comm. 2008). As well, the tie strap began to break into pieces, becoming detached from the hat.

Since wet storage was proving damaging to the hat and conventional PEG/freezing was incompatible with the one-year study period, it was decided that slow-drying should be attempted, as this treatment option would allow more or less continuous access for research. In preparation, the deposits of silt on the hat's exterior were reduced by gentle brushing. The hat's original shape was clear prior to treatment, but the general rigidity, the broken headband and many distortions suggested that the kind of manipulation required to restore the hat's shape would result in irreparable structural damage to both the headband

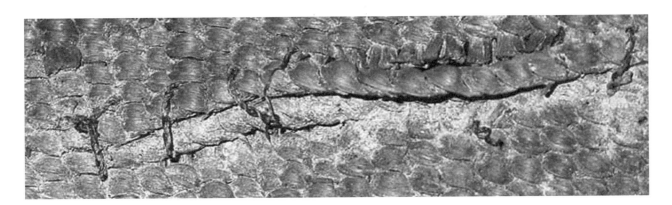

Figure 4. Twined sinew has been used to repair a slit in the woven hat. Stitches are 0.4–0.7 centimetres apart. Government of Yukon. Colour version on page 592.

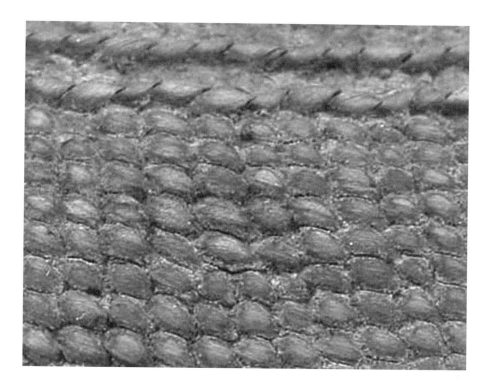

Figure 5. Gentle cleaning of the hat's exterior reveals bright ochre in recesses of the weaving. Government of Yukon. Colour version on page 592.

and the hat (Komejan 2000). Therefore, the wet hat was cautiously and partially reshaped, placed on a Tyvek-covered Ethafoam temporary support and then into a chest-style freezer. After drying, the hat retained some flexibility at the brim, but the headband and adjacent exterior areas remained quite rigid.

With the silt reduced and excess water gone, the hat was lighter in colour, many details were revealed, and the fine quality of the weaving could be appreciated. Even the glossy surfaces of the spruce root (Young 2001) were preserved. Under magnification, small deposits of red material were seen distributed all across the hat's exterior, and samples were removed for analysis. To date, only preliminary work has been done, identifying the red material as iron oxide, otherwise known as red ochre (Mager 2001; Kate Helwig, pers. comm. 2007). Future studies could include characterizing the ochre, looking for associated binding media and identifying any media found, if possible. Removal of silt from small areas across the hat's exterior revealed remnants of vivid orange-red pigment in recesses of the weaving

(fig. 5). This pigment suggests that the hat was originally brightly coloured. There is also a smear of red pigment on the bottom edge of the headband. The remaining two sinew thread repairs to the woven spruce root also became clearly visible. A number of dark, straight hairs were seen in the interior of the hat—some caught in the recesses of the weaving and others woven into the hat, either by accident or design.

The detached sections of the tie strap were kept frozen until treatment could begin. After the tie strap was thawed, a collagen fibre sample was removed and sent to the Canadian Conservation Institute for thermal analysis. The Kwädąy Dän Ts'ìnchį tie strap sample returned a result within the normal shrinkage temperature range of mammalian hide (Young 2005). This result was unexpected. The strap had already broken into pieces, which seemed evidence of deterioration. This apparent contradiction may stem from the structure of hide: a dense network of interwoven collagen fibres. As hide deteriorates, weak collagen fibres may dissolve or drop out of the network.

Figure 6. Hide tie strap. This section is unevenly cut with a clear pattern of hair follicles. Government of Yukon. Colour version on page 592.

Stronger fibres may then be removed for analysis, returning a result that does not reflect the overall condition of the hide (Young 2005).

With the tie strap breaking into pieces, there was concern that it would become too fragile to handle and too fragmented to study. Risk that chemical treatment would prevent future analysis had to be weighed against the loss of the entire tie strap. Since most of the strap was no longer attached to the rest of the hat, chemical treatment could be limited to the strap with no impact to the rest of the hat. The largest and most fragile pieces of the tie strap were immersed in a solution of polyethylene glycol and water for several days, blotted dry and then allowed to dry in a freezer. One end of the strap remained attached to the spruce root of the headband and was left to dry there along with the hat. After treatment, the tie strap was in eight pale brown fragments that were quite brittle.

Details of the hide and its processing were revealed. The central section of the strap, between two knots, retains hair follicles, indicating that it had minimal processing (fig. 6). The edges of this section are cut quite unevenly, while the remainder of the strap has parallel and relatively smooth edges. This difference suggests the central section may have been added to the tie as a rough repair during the hat's use. Also revealed were small streaks of what appears to be red pigment on several of the tie strap fragments. With successful DNA analysis on the hide components of the Kwädąy

Dän Ts'ìnchį robe completed (see chapter 27), it would be interesting to attempt identification of the animal source(s) of the tie strap hide.

Storage of the Whitehorse Artifacts

ONCE DRY, the wooden artifacts were removed from the freezer. Due to their length and variable sizes, commercially made archival boxes were not an option. Initially, simple boxes were made using lightweight Coroplast sheet. After the decision was made to keep the artifacts indefinitely, more thought was put into their long-term storage needs. With additional finds being added to the collection and the identification of adjoining fragments increasing the size of some artifacts, a reorganization of storage was required. Currently, the artifacts are housed in Coroplast storage/travel boxes, each with a custom-made Ethafoam liner, itself lined with Tyvek, cut out for the artifacts.

ACKNOWLEDGEMENTS

I would like to thank the Champagne and Aishihik First Nations and particularly the staff of their Heritage office—Sheila Greer, Frances Oles and the late Sarah Gaunt—for the opportunity to care for and examine these precious artifacts. Tara Grant, Kate Helwig and Gregory Young of the Canadian Conservation Institute have provided much helpful advice and expertise over the years.

23

WOODEN ARTIFACTS FROM THE KWÄDĄY DÄN TS'ÌNCHĮ SITE AND SURROUNDING AREA

An Analytical Catalogue

Champagne and Aishihik First Nations and Sheila Greer

This chapter presents an analytical catalogue of the artifacts related to the Kwädąy Dän Ts'ìnchį discovery that are presently being curated in Whitehorse, Yukon. This includes all artifacts collected except for the robe and other hide/skin fragments (chapter 27) and the copper bead (chapter 24), which are being curated at the Royal British Columbia Museum in Victoria. The hat made of spruce root, while part of the Whitehorse collection, is reported on in chapters 21 and 22 of this volume.

WE DESCRIBE the individual wooden artifacts along with the results of various specialized and comparative studies that have been completed on these pieces. In an attempt to learn more about these unique pieces, including what they may have been used for, the wood artifacts are also compared to objects known from relevant archaeological and ethnographical contexts. In total, 23 wooden artifacts have been identified (table 1). This count is based on

33 fragments recovered, some of which have been refitted together. Other fragments, although not refitted, are assumed to be part of the same artifact since they are of similar morphology and were found in close proximity.

The wood artifacts were recovered within an area of approximately one square kilometre around the Kwädąy Dän Ts'ìnchį discovery site, and were found either during the initial discovery and recovery efforts of 1999, or during subsequent annual monitoring visits. Except for two pieces dug out of the snow, all artifacts were surface finds. A single piece of cut birch bark, recovered from nearby site IkVf-2, which is located more than a kilometre from the discovery site, is also reported in this chapter. Readers are reminded that the discovery site is located in alpine tundra above the treeline, and that all pieces of wood found must have been brought here by some person. This means that even unmodified sticks are recognized as artifacts.

The location data on all wooden artifacts are also reviewed in this chapter, in order to illustrate which wooden pieces were found in association with the Kwäday Dän Ts'ìnchi individual. Contrary to the initial report (Beattie et al. 2000), only two wooden artifacts recovered from the site area can be definitively associated with the Long Ago Person Found and are therefore considered his belongings.

Methods

Catalogue System

The Kwäday Dän Ts'ìnchi discovery site has been entered into the British Columbia registry of archaeological sites. This database uses a national system, referred to as the Borden system (after the archaeologist who developed it), to establish a unique identifier or label for each site based on the site's latitude and longitude coordinates (Borden 1952). The identifier for the Kwäday Dän Ts'ìnchi site is registered as IkVf-1.

A piece of cut and rolled birch bark was also recovered in a valley bottom more than a kilometre southeast of where the Kwäday Dän Ts'ìnchi individual was discovered. This find has been registered under a separate site number, IkVf-2.

The wooden artifacts reported here have each been assigned their own catalogue number, which is a sub-number of the site reference number. For example, catalogue number IkVf-1:102 is for the man's

knife.[1] Table 1 presents an abbreviated version of the detailed catalogue for the artifacts that are being conserved in Whitehorse.[2]

Conservation

Unless otherwise noted, all artifacts reported on in this chapter have been under the care of the Yukon government staff conservator since they were recovered from the discovery site area.

As reported by Valery Monahan (see chapter 22), conservation treatments have generally been non-interventionist. Only one artifact fragment—the #115 part of refitted artifact IkVf-1:104, interpreted as a gaff pole/walking stick—was surface-treated with a preservative. All the other wood pieces were slowly freeze-dried until their moisture content was sufficiently reduced so that they could be safely stored, with supports as necessary, at room temperature.

Artifact Provenience

The provenience of the wood artifacts is reported in table 1. All pieces are surface finds with the exception of fragment #115 of artifact IkVf-1:104, which was dug out of the snow. It should be noted that the glacier's shrinkage over the years since the discovery means that the ground surface has, in many places, changed from snow or ice to rock or colluvium.

Some items were found in either uncertain or only approximately known locations. This is particularly true for artifacts (especially the less shaped ones) collected during the August 1999 site visits, when time was extremely limited and fieldwork was focused largely on recovery of the human remains. More accurate locational data, including GPS coordinates, are available for most finds from 2002 and later. Figure 1 shows the approximate find location of all pieces for which we have locational data.

Most of the wood artifacts were found east of and uphill from where the human remains were discovered melting out of the glacier, around the landscape feature referred to as the Cairn Knob (fig. 1). The site

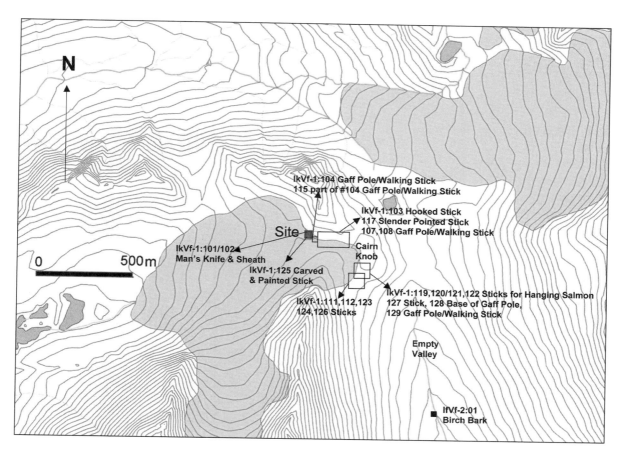

Figure 1. Map showing location of artifact finds at site and in surrounding area.

discoverers—Bill Hanlon, Mike Roch and Warren Ward—reported that they approached the find area from the Cairn Knob to the east (Mackie 2000; Champagne and Aishihik First Nations 2000b). They picked up various pieces of wood along this route before encountering the human remains, at which point they put down the bigger wood pieces they were carrying—with the exception of the IkVf-1:103 hooked stick artifact, which they subsequently carried out to their vehicle and took to Whitehorse. Consequently, some of the wooden artifacts recovered were not in their original location when collected on the site visits of August 17 and 23.

After reviewing all available provenience information (Mackie 2000; Champagne and Aishihik First Nations 2000b), we concluded that only two artifacts can be clearly associated with the Long Ago Person Found: IkVf-1:101/102, the man's knife in its fur sheath, and IkVf-1:125, the carved and painted stick. See figure 1 for their find locations.

One other artifact was found in relatively close proximity to the human remains and therefore might be associated with the human remains. This is IkVf-1:104, the gaff pole/walking stick, which consists of four fragments refitted together. Ward reported that the notched fragment of this artifact (#104) was found off the ice in the rocks below the body (Mackie 2000). Another fragment, the long #115 piece, was dug out of the snow at the base of the glacier ice in which the body was found. The provenience of the other two fragments of this artifact, #106 and #116, is uncertain, though it's possible the #116 fragment was found in the same general area as the #115 piece (Champagne and Aishihik First Nations 2000b).

Analytical Studies

Given the unique characteristics of the site collection, the broadest possible approach was used to research the artifacts. The methodological approaches and analytical tools have varied over the years since the discovery, depending on the individual directing the work. It should also be mentioned that the financial resources available for the artifact analysis were limited, which impacted the studies completed. None of the artifacts collected after 1999, for example, have been radiocarbon dated.

Initial specialized study of the artifacts—for example, radiocarbon dating and wood identifications—was supported by the Yukon government's Heritage Branch and facilitated by Dr. Ruth Gotthardt, staff archaeologist. CAFN staff heritage planner Sarah Gaunt was actively involved in organizing research on the artifacts and doing community consultation work during the first few years following the discovery. Author Greer assisted Gaunt with some of this work and assumed responsibility for coordinating the specialized studies, the community consultation work and the completion of the comparative studies after Gaunt's passing in 2003.

Specialized studies completed on the wooden artifacts were limited to radiocarbon dating and wood species identification. Initial work on the latter was completed by Gregory Young of the Canadian Conservation Institute (Young 2001), with later, more detailed identifications conducted by Mary-Lou Florian (Florian 2005, cf. Greer 2005a).

In addition to basic descriptive work, compiled in Greer (2005b), the comparative studies reported here involved two approaches. One included interview work with elders knowledgeable about components of Aboriginal technology traditionally used in this biogeographic context, and the other was a comparative study of relevant museum collections and text materials.

Due to financial limitations, both approaches used easily accessible sources such as web-based or online museum catalogues, as well as consultation of elders residing in the local south Yukon community. During a meeting of the Kwädąy Dän Ts'ìnchį Management Group in Victoria, Greer also consulted the collections of the Royal BC Museum, and through correspondence with the curator of collections at the Alaska State Museum in Juneau, Greer was able to clarify details on holdings of possible interest in that institution.

Published source materials with significant content on relevant traditional Aboriginal material culture were consulted. These sources were those available either in Greer's personal library or in the library system of the University of Alberta, Edmonton, or through inter-library loan. Examples of sources consulted for the subarctic culture area include Clark 1974; Johnson and Raup 1964; McClellan [1975] 2001; McClellan and Denniston 1981; McClellan et al. 1987; Nelson 1983a; O'Brien 1997; Osgood 1936, 1937, 1940, 1971; VanStone and Simeone 1986; Thompson 2008; and VanStone 1974. Examples of the materials consulted for the Tlingit culture area included Davis 1990, 1996; de Laguna 1960, 1972, 1990; de Laguna et al. 1964; Emmons 1991; Jonaitis 1988; Krause 1956; Krause and Krause 1981, 1993; Langdon 1986; McClellan 1981; Oberg 1973; Olson 1936; Sackett 1979; Shotridge 1920; Stewart 1977; Thornton 2004; and Wood 1882.

Relevant museum collections were also investigated (cf. Lohse and Sundt 1990). Of particular interest were museum collections holding artifacts obtained from the Tlingit culture area by George T. Emmons around the turn of the 20th century. Emmons, a US naval lieutenant posted in southeastern Alaska, collected and sold artifacts to many museums; a listing of these institutions can be found in Emmons 1991 (see also Low 1977). The online collections of the American Museum of Natural History (Anthropology Collections) in New York, the Burke Museum of Natural History and Culture in Seattle and the Alaska State Museum in Juneau were found to be the most useful for the work reported here.[3]

Historic photographs taken in Aboriginal communities in the late 19th and early 20th centuries were also consulted to see if similar artifacts could be found in period images (Scherer 1981, Lohse and Sundt 1990). Many images of possible interest are available in the online collections of the Alaska State Library.[4] The state collection includes historic images from numerous sources, including the noted photographic teams of Case and Draper, and Winter and Pond (cf. Wyatt 1989). Unfortunately, this research approach was not particularly fruitful. Period images taken in camp settings often show various long poles or sticks around a camp, but the function of the items cannot typically be identified.[5]

Interview work was also undertaken with Champagne and Aishihik First Nations elders to learn more about the wooden artifacts recovered from the site area. The methodological approach was inspired by a project that documented Yupik elders' insights into museum collections (Fienup-Riordan 2005) and by work done with Gwich'in Elder Rev. David Salmon (O'Brien 1997). While the work with Yupik elders brought people to the museum where the artifacts were housed, this project brought the artifacts to people in their home community, Haines Junction. Similarly, whereas the research with Rev. Salmon involved interviewing one particular knowledge-holder about the various types of traditional tools he had personally made, the elders interviewed for this project had not made the tools being discussed. Nonetheless, it was hoped that they had seen these tool types around their family homes or camps when they were children, or heard about such implements from their ancestors, and could therefore provide input on what the artifacts might have been used for or part of. The intent was that the artifacts would be a catalyst

for discussions that would highlight the substantive knowledge about such items and the beliefs and values held around them—their cultural importance. The interviews were conducted by Greer and were informal in nature, documented with notes (Greer 2005c; Greer 2005a; Greer 2006b; Greer 2006d; Greer 2008).

Source materials providing Tlingit and Dákwanjè vocabulary information were also consulted. For Tlingit, these included Leer et al. (2001), as well as the online Alaska Native Languages Dictionary.[6] For Dákwanjè, sources included Tlen (1993) and Southern Tutchone Tribal Council (1999).

The Wood Artifact Collection

TABLE 1 lists all wooden artifacts recovered from the Kwädạy Dän Ts'ìnchị site. Descriptions are provided for those items recognized as artifacts simply because they are "manuports," having been brought to the site area from elsewhere. The table also provides descriptions and results of specialized analyses for artifacts lacking defining characteristics that would allow them to be recognized as any type of formalized tool or artifact.

One of the objectives in the present analysis was to determine the activities represented by the artifacts in the collection. Nine pieces in the Whitehorse collection (shown in fig. 2) are clearly recognizable as artifacts and are, therefore, the focus of the chapter. Detailed description and discussion of these more shaped and recognizable artifacts follow below the table. Even these

Kwädạy Dän Ts'ìnchị Belongings

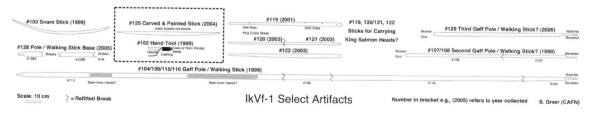

Figure 2. Schematic drawing of selected artifacts collected at IkVf-1, to scale. The objects are illustrated with the distal end of the tool on the right and the proximal or base end on the left. The pieces inside the box were found in direct association with the Long Ago Person Found. Sheila Greer illustration. For larger version, see appendix 3.

Table 1. Artifacts conserved in Whitehorse, Yukon, listed in catalogue number order. Items with names in boldface are further discussed in the section titled Detailed Artifact Descriptions. Items with underlined names count toward the total artifact count.

Catalogue #	Name/Description/Provenience
IkVf-1:101	Sheath from man's knife Provenience: found by Hanlon, Roch and Ward, near human remains. Found: August 14, 1999
IkVf-1:102	**Man's knife** (previously referred to as the hand tool) Provenience: found by Hanlon, Roch and Ward, near human remains. Found: August 14, 1999 See chapter 25 for a detailed study of the knife.
IkVf-1:103	**Hooked stick** Provenience: found by Hanlon, Roch and Ward between Cairn Knob and human remains—i.e., uphill from the body. Found: August 14, 1999
IkVf-1:104	(First) **Gaff pole/walking stick** (part of) Provenience: found by Hanlon, Roch and Ward between Cairn Knob and human remains—i.e., uphill from the body. Found: August 14, 1999
IkVf-1:105	Pointed stick with bashed end: bag label "Bag 5"; a short (61-centimetre) fragment of a sapling, with one carved and pointed end, the other broken; the surface still has some bark on it. Wood identified as a hardwood, possibly *Betula* species (Birch) (Young 2001). Interpreted as a "pointed stick with bashed end". Provenience: found by Ron Chambers; exact location uncertain, but not near the body. Found: August 23, 1999
IkVf-1:106	Refitted part of first gaff pole/walking stick; see #104 Provenience: uncertain, but found uphill from the body. Found: August 14, 1999
IkVf-1:107	Part of second gaff pole/walking stick; see #108 Provenience: found by Hanlon, Roch and Ward off the ice, uphill from the body. Found: August 14, 1999

Catalogue #	Name/Description/Provenience
IkVf-1:108	(Second) **Gaff pole/walking stick** (part of) Provenience: found by Hanlon, Roch and Ward off the ice, uphill from the body. Found: August 14, 1999
IkVf-1:109	Fragments of a small sapling: the combined length of the four fragments is 30 centimetres. The wood has been identified as *Betula* species (Birch) (Florian 2005). One of the fragments was radiocarbon dated to 120 ± 40 years BP (Beta-140637). It was originally suggested that these fragments were possibly sections of an arrow (Beattie et al. 2000). More detailed examination indicates a lack of defining characteristics to support such an interpretation, and the fragments are now thought to be little more than pieces of a small broken stick. Provenience: no information on where these were found. Found: August 17, 1999
IkVf-1:110	Stick with carved end: a slender sapling (77 centimetres long) carved on one end, while the other end is broken. Wood identified as a *Picea* species (Spruce) (Young 2001). This was perhaps used as an informal/impromptu walking stick. Provenience: no information on where this was found. Bag label "Wood A". Found: August 23, 1999
IkVf-1:111	Stick with cut end: a sturdy sapling (36 centimetres long) that shows a right-angled cut break at one end, with the other end broken and some carving on the face. Wood identified as a *Picea* species (Spruce) (Young 2001). Provenience: found by Erik Blake, overlooking Empty Valley. Bag label "Wood 2". Found: August 23, 1999
IkVf-1:112	Stick, unaltered?: a stick *ca.* 14 centimetres long, broken at both ends and too weathered to tell if modified. The wood was identified as *Picea* species, probably Sitka Spruce (Young 2001). Provenience: found by Blake, overlooking Empty Valley. Bag label "Erik Wood". Found: August 23, 1999
IkVf-1:113	Hat (made of woven spruce root; see chapter 21, this volume) Provenience: hat blown around by helicopter landing; assumed to be associated with human remains. Not seen by hunters on August 14. Found: August 17, 1999
IkVf-1:114	Carved bark pieces: refers to four small fragments of bark that have been carved. The bark is identified as *Picea* sp., probably Sitka Spruce (Florian 2005). Provenience: no information on where these were found. Found: August 23, 1999

Catalogue #	Name/Description/Provenience
IkVf-1:115	Refitted part of more or less complete gaff pole/walking stick; see #104 Provenience: found protruding from snow/ice at base of solid ice about 10 metres up-slope from where the body was found. "Bag 15". Found: August 17, 1999
IkVf-1:116	Refitted part of more or less complete gaff pole/walking stick; see #104 Provenience: uncertain where this portion of the artifact was found. A photo taken on August 17 shows a stick on the ice in the area close to but uphill from where the body was found; it may be this stick. "Bag 16". Found: August 17, 1999
IkVf-1:117	<u>Slender pointed stick with carved end:</u> #117/118 (two refitted pieces, totalling 169 centimetres in length) is a slender sapling carved to a bevelled end, while the other end is broken. The wood identification on the #118 portion was inconclusive, possibly Alder or False Huckleberry (Florian 2005). The #117 portion of the artifact was radiocarbon dated to 360 ± 40 BP (Beta-140634). The bevelled end shows a depression, which was previously interpreted as representing the dimple of a dart (Beattie et al. 2000), but this dimple was subsequently recognized as a natural pith depression. As a result, the functional identification for the #117/118 piece is revised to a "possible walking stick". Provenience: uncertain where this stick was picked up, possibly near where the helicopter always landed, which is between the Cairn Knob and where the human remains were found. Found: August 17, 1999
IkVf-1:118	Part of slender pointed stick with carved end; see #117 Provenience: uncertain where this stick was picked up, possibly near the spot where the helicopter always landed, which is between Cairn Knob and where the body was found. Found: August 17, 1999
IkVf-1:119	**Stick for hanging salmon** Provenience: found by Chambers, on southwestern side of Cairn Knob. Found: July 21, 2001
IkVf-1:120	(Second) **Stick for hanging salmon** Provenience: found by Chambers on southwestern side of Cairn Knob. Found: August 28, 2003

Catalogue #	Name/Description/Provenience
IkVf-1:121	Part of second stick for hanging salmon; see #120 Provenience: found by Chambers on southwestern side of Cairn Knob. Found: August 28, 2003
IkVf-1:122	(Third) **Stick for hanging salmon** Provenience: found by Chambers on southwestern side of Cairn Knob. Found: August 28, 2003
IkVf-1:123	<u>Pointed stick fragment:</u> a short (73-centimetre) fragment of a carved stick, with one carved, pointed end, the other broken. The wood was identified as *Picea* sp., probably Sitka Spruce (Florian 2005). Perhaps this is a fragment of an informal/impromptu walking stick. Provenience: found by Sheila Greer, overlooking Empty Valley, southwest of Cairn Knob. Found: August 28, 2003
IkVf-1:124	<u>Fragments of a small stick:</u> refers to many fragments of a short (25-centimetre) stick; the pieces are too badly weathered to tell if the surface is modified. Wood identified as *Picea* sp., probably Sitka Spruce (Florian 2005). Provenience: found by Greer, overlooking Empty Valley, southwest of Cairn Knob. Found: August 28, 2003
IkVf-1:125	**Carved and painted stick** Provenience: found by Greer, on ground surface in area covered by ice in previous seasons and less than three metres from where the last skull fragment was found in 2004. Found: August 20, 2004
IkVf-1:126	<u>Stick unmodified?:</u> refers to a stick *ca.* 25 centimetres in length, with both ends broken and crumbly; no modification seen on the intact parts of the stick. Provenience: found by Greer, overlooking Empty Valley, southwest of Cairn Knob. Found: August 20, 2004
IkVf-1:127	<u>Stick with carved end:</u> the end fragment (26 centimetres long) of a slender carved stick; the carved end shows shallow bevelling similar to the end of the #118 piece. Provenience: found by Greg Eikland, among rocks, draw on southwest side of Cairn Knob. Found: August 12, 2005

Catalogue #	Name/Description/Provenience
IkVf-1:128	Fragment, **base portion of a gaff pole/walking stick**. Note that this may be part of artifact #129. Provenience: found by Eikland, among rocks, draw on southwest side of Cairn Knob. Found: August 12, 2005
IkVf-1:129	Distal portion of (third) **gaff pole/walking stick**. Note that #128 may be part of this same artifact. Provenience: found by Eikland, among rocks, southwest side of Cairn Knob; recovered September 22. Found: August 13, 2005
IkVf-1:130	<u>Fragments of a carved stick with a pointed end</u>: refers to some badly weathered fragments of a carved stick that tapers to a carved, pointed end. The largest fragment is 27 centimetres long. The second largest piece is carved to a flat surface on one side, with the rounded exterior surface of the stick also showing carving. Provenience: no information on where it was found. Found: August 23, 1999
IkVf-1:131	<u>Stick with carved end</u> (other end broken): a badly weathered stick (22 centimetres long) showing a cut on one end, with the other end broken. One side shows the exterior surface; the other is the interior of a sapling. Provenience: found by Greer, overlooking Empty Valley, southwest of Cairn Knob. Found: August 9, 2002
IkVf-2:001	Cut and rolled birch bark. Length of piece is *ca.* 58–78 millimetres. It is estimated that unrolled, the piece would be *ca.* 70–100 millimetres wide.

items, though, proved challenging to understand, in part because they appear to be everyday objects rather than ceremonial, and it is ceremonial objects that are most commonly featured in museum collections and existing photography records.

Detailed Artifact Descriptions

IN THE FOLLOWING ARTIFACT DESCRIPTIONS, the proximal end is considered the end closest to the person holding or using the tool, or the end closest to the ground surface. The distal end is that furthest from the individual holding or using it, furthest from the ground—that is, the working or "business" end of the tool form. The functional interpretation of the artifact is indicated after the catalogue number of the piece—for example, IkVf-1:102 (man's knife). A more detailed discussion of the likely use or purpose of each artifact is presented later in the chapter, under the heading "Shaped Artifacts Discussion and Analysis".

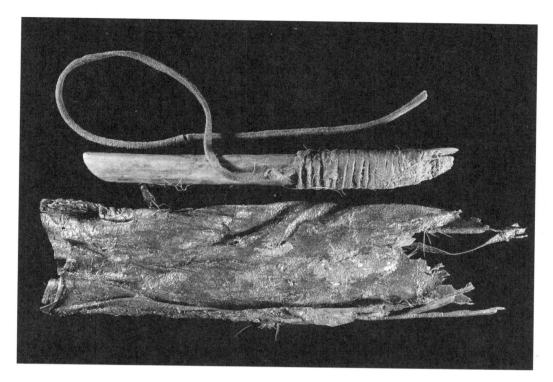

Figure 3. Photograph of IkVf-1:102, man's knife *(upper)* with its fur sheath *(lower)*. The man's knife has been radiocarbon dated to 150 ± 50 c14 years and was found in association with human remains. Ruth Gotthardt (Yukon government) photograph. Colour version on page 593.

IkVf-1:102 (man's knife—fig. 3)

Dimensions: *ca.* 14 centimetres long by 15 millimetres by 12 millimetres.

Description: This small composite tool, analysed in more detail in chapter 25 (Helwig et al.), consists of a wooden handle into which a short metal blade has been inserted. The blade is held in place by hide lashing wound around the wooden handle. A piece of bone or antler is inserted underneath this lashing. The function of the latter appears to have been to secure and stabilize the blade; perhaps it also prevented the metal from abrading the lashing. The highly corroded metal has been confirmed as non-meteoritic, meaning it is a trade metal rather than from an indigenous source. Note that the man's knife was found inside its own hide/fur sheath, not described here.

Previous label or reference number: Referred to as the "Hand Tool" in Beattie et al. (2000).

Year collected: 1999

Associated with human remains: Yes; found inside its sheath, on top of the human remains.

Curation/study history: In 2000 an initial examination of this artifact was completed by Dr. M. Wayman, Department of Metallurgy, University of Alberta. The artifact was then returned to the Yukon, where it remained until late 2006, when it was transferred to the Canadian Conservation Institute in Ottawa for specialized studies, as reported by Helwig et al. (chapter 25). Since 2009 the artifact has been curated at Yukon Heritage facilities in Whitehorse.

Radiocarbon date: 150 ± 50 c14 years [CAMS#71937]

Wood species: The handle is made of *Tsuga* sp., Hemlock, possibly Western Hemlock or Mountain Hemlock (Florian 2005).

Other specialized studies: See Helwig et al. (chapter 25) for specialized studies completed on this artifact at the Canadian Conservation Institute.[7]

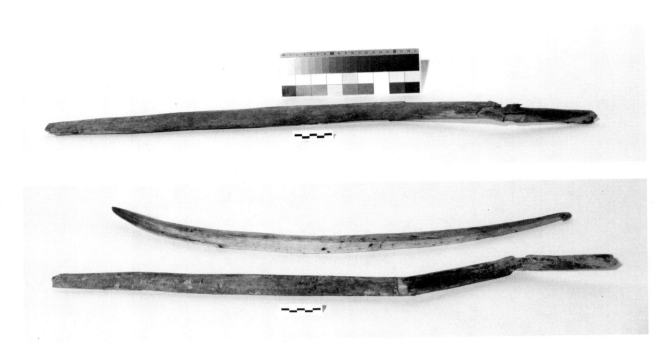

Figure 4. *(top)* Photograph of IkVf-1:125, carved and painted stick. It has not been radiocarbon dated. Function is unknown; wood not identified. Found in association with human remains. *(bottom)* Photograph showing both IkVf-1:103 (fig. 5) and IkVf-1:125. Robin Armour (Yukon government) photograph. Colour versions on page 593.

IkVf-1:125
(carved and pointed stick—fig. 4)

Dimensions: *ca.* 75 centimetres long, width ranges from 20 to 27 millimetres, thickness from 12 to 22 millimetres.

Description: This medium-length, mostly straight stick is almost complete, with small portions of both ends broken off. The artifact is carved over all faces, and shaped into a plano-convex cross-section. The artifact's surfaces are covered with a red-brown paint. Specialized analysis has not been completed to identify the chemical composition of the pigment and binder.

Previous label or reference number: None.

Year collected: 2004

Associated with human remains: Yes; found less than two metres from the last skull bone found.

Curation/study history: Piece has been curated at Yukon Heritage facilities since it was collected.

Radiocarbon date: This artifact has not been radiocarbon dated.

Wood species: This piece has not been identified by an expert in wood identifications. To the non-specialist, the wood looks and feels similar to the wood artifacts that have been identified as Hemlock.

Other specialized studies: None.

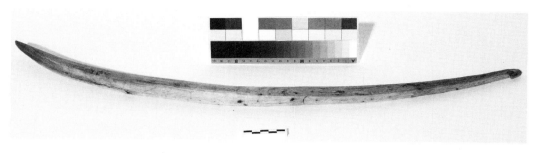

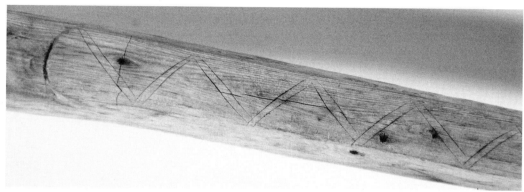

Figure 5. Photograph of IkVf-1:103, hooked stick, radiocarbon dated to 230 ± 40 BP. Hemlock wood; suggested function is for setting marmot snares. Not associated with human remains. Ruth Gotthardt (Yukon government) photograph.

Figure 6. Photograph, close-up, showing carved decorative detail on inside face of IkVf-1:103, hooked stick. Robin Armour (Yukon government) photograph.

IkVf-1:103 (hooked stick—figs. 5 and 6)

Dimensions: 72 centimetres long, mid-point diameter *ca.* 26 millimetres.

Description: This curved, slender artifact features a distinctive hook, similar to a crochet hook, at its distal end. The opposite proximal end is flattened and comes to a rounded point. The latter displays extensive use polish, similar to what might result from the object being carried inside a hide bag. A shallow, horizontal, intentionally placed cut is situated on the implement's inside surface near the proximal (i.e., hooked) end. It is suggested that the cut may have been used to secure some type of thong or string. There are other cut marks on this same inner surface that are scattered in their orientation and distribution. These appear to be haphazard damage, not intentional cuts or use damage from a specific repeated activity. The distal portion of the somewhat flattened inside face of the artifact displays an incised zig-zag decoration (fig. 6). Red pigment, though very faint and difficult to see, is also present on the piece's outside surface. Specialized analysis work has not been completed to confirm if this red pigment is ochre and whether or not a binding agent was employed. The colour is visually similar to ochre identified on wooden hunting artifacts from the Yukon ice patch sites (cf. Helwiget al. 2004; Hare et al. 2004).

Previous label or reference number: In Beattie et al. (2000) it is noted as being similar in morphology to throwing board or atlatl devices, and also to a Southern Tutchone snare-setting stick.

Year collected: 1999

Associated with human remains: No; this piece was found several hundred metres east of and uphill from the human remains.

Curation/study history: The piece has been curated at Yukon Heritage facilities since it was collected.

Radiocarbon date: 230 ± 40 BP [Beta-140633]

Wood species: *Tsuga* sp., Hemlock, possibly Western Hemlock or Mountain Hemlock (Florian 2005).

Other specialized studies: None.

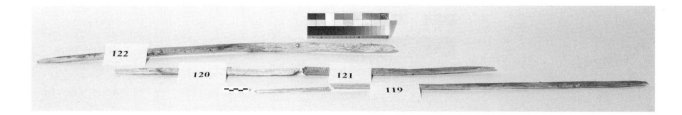

(Above) Figure 7. Photograph showing the three sticks for hanging/carrying salmon, IkVf-1:119, 190/121 and 122. These have not been radiocarbon dated. #119 is Western Red Cedar; #120/121 probably *Picea* sp., Sitka spruce; #122 wood not identified. Not associated with human remains. Robin Armour (Yukon government) photograph.

(Facing page) Figure 8. The four sections of refitted artifact (IkVf-1:104, 105, 115, 116) that make a complete gaff pole/walking stick. The #104 fragment is radiocarbon dated to 290 ± 40 BP and the #106 fragment to 140 ± 40 BP. The #106 fragment is identified as *Picea* sp., probably Sitka spruce. Association with human remains uncertain but unlikely. Robin Armour (Yukon government) photograph.

(Facing page) Figure 9. Close-up photograph of the distal end of the complete gaff pole/walking stick (IkVf-1:104). Robin Armour (Yukon government) photograph.

IkVf-1:119, 120/121 and 122

(sticks for hanging salmon—fig. 7)

Dimensions: The three artifacts range in length from about 80–85 centimetres, with maximum width *ca.* 22–24 millimetres and maximum thickness *ca.* 11–15 millimetres.

Description: These three wooden objects (one in two pieces, another cracked and almost in two pieces) were found in close association. The wood of the three is extremely weathered. All are made from staves and are similar in shape, size and manufacture. They feature carved surfaces and are spatulate (flattened) at one end, going to a more tapered shape, oval in cross-section, at the opposite end. One of the three shows compression ring depressions, presumably from having been lashed to something at one time.

Previous label or reference number: The #120 and #121 pieces refit together into one complete stick.

Year collected: The #119 piece was found in 2001. The #122 pieces as well as the #120 and #121 fragments that refit together were all found in 2003 within a few metres of each other, near where the #119 piece was picked up the previous year.

Associated with human remains: No; these pieces were found uphill from and east of the human remains.

Curation/study history: These pieces have been curated at Yukon Heritage facilities since they were collected.

Radiocarbon date: None of these pieces have been radiocarbon dated.

Wood species: The #119 piece was identified as Western Red Cedar, *Thuja plicata* (Florian 2005). The #120 and #121 pieces were identified as *Picea* sp., probably Sitka spruce (Florian 2005). The wood species of the #122 piece has not been identified.

Other specialized studies: Tiny holes were noted on two of the three pieces and examined microscopically by conservator Valery Monahan. The holes contained mineral sediment and appeared to be natural in origin. It is suggested that they were naturally occurring knotholes.

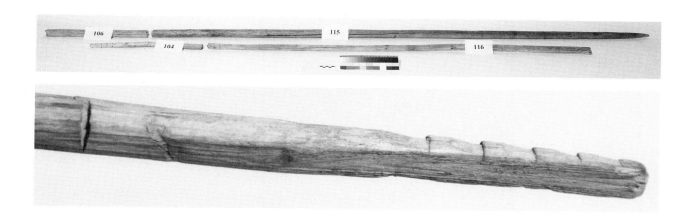

IkVf-1:104

(gaff pole/walking stick—figs. 8 and 9)

Dimensions: Total length of refitted artifact is *ca.* 3.8 metres (13 feet). Diameter ranges from maximum 36 millimetres close to the base to a delicate 20 millimetres near the beginning of the notches toward the far end. The lengths of the various fragments are as follows: #104 fragment, 39 centimetres; #116 fragment, 141 centimetres; #106 fragment, 35 centimetres; #115 fragment 195 centimetres.

Description: This artifact is composed of four broken and refitted fragments (IkVf-1:104, 106, 115, 116) of a robust carved sapling. The pieces were not originally recognized as parts of the same object because of their differences in colour. It is likely that the fragments melted out of the ice at separate times and thus experienced a different weathering history that caused their varying hues.[8] The artifact is carved over its entire surface and is circular in cross-section, except for the notched distal end (fig. 9). The proximal end is carved to a sturdy pointed base, while one face of the distal end has been carved into a series of five stepped notches, which are spaced over an eight-centimetre distance. The reverse face of the distal end is bevelled and flat, as if to "receive" or hold another part of this composite artifact. As discussed further and illustrated below, the notches would have allowed lashing to secure whatever item was intended to rest on the flat surface. Red pigment, thought to be ochre, shows around some of the notches on the distal end. Surface staining is present at *ca.* 53–70 centimetres from the base and at *ca.* 125–183 centimetres from the base.

Previous label or reference number: The #104 piece was referred to as the multi-notched stick, while the #106 and #115 pieces were referred to as a large walking stick, similar to sticks used in bear hunting, in Beattie et al. (2000).

Year collected: 1999

Associated with human remains: Uncertain, but unlikely. While one fragment was dug out of the ice in the general area where the human remains were found, the other three portions were found some distance away, east of and uphill from the remains.

Curation/study history: These pieces have been curated at Yukon Heritage facilities since they were collected in 1999. The #115 fragment was treated with preservative; the others are untreated.

Radiocarbon date: The #104 fragment was dated to 290 ± 40 BP [Beta-140636]; the #106 fragment was dated to 140 ± 40 BP [Beta-140635]; the #115 and #116 portions have not been radiocarbon dated.

Wood species: The #106 fragment was identified as *Picea* sp., probably Sitka Spruce (Florian 2005). The other fragments of this refitted artifact have not been analysed by a wood specialist.

Other specialized studies: None.

(***Top***) **Figure 10.** Photograph of the two sections of refitted artifact (IkVf-1:107, 108) that comprise the distal end of the second gaff pole/walking stick. The #107 fragment is radiocarbon dated 500 ± 40 BP and identified as *Picea* sp., probably Sitka Spruce. Not associated with human remains.

Robin Armour (Yukon government) photograph.

(***Bottom***) **Figure 11.** Close-up of the distal end of the second gaff pole/walking stick (IkVf-1:107).

Robin Armour (Yukon government) photograph.

IkVf-1:107
(gaff pole/walking stick—figs. 10 and 11)

Dimensions: Length of existing fragments is 122 centimetres; diameter (except for bevelled end) ranges from 24–28 millimetres.

Description: This incomplete artifact is made of two refitted fragments (#107 and #108), with an intact distal end. The base end is broken off and missing. Made from a robust sapling that was carved over its entire surface, the resulting cross-section is circular except for the carved distal end. One side of the latter has been shaped into a flat, right-angled bevel, providing a place to "receive" or hold another part of this composite artifact (see fig. 11). Unlike IkVf-1:104 (above) or IkVf-1:129 (below), the reverse face of the distal end is not notched in any way to provide surfaces that would help secure lashing; this is discussed further and illustrated below.

Previous label or reference number: In Beattie et al. (2000), the #107/108 piece was referred to as the thick stick.

Year collected: 1999

Associated with human remains: No; these pieces were found east of and uphill from the remains.

Curation/study history: The piece has been curated at Yukon Heritage facilities since it was collected.

Radiocarbon date: The #107 fragment has been dated to 500 ± 40 BP [Beta-140638].

Wood species: The #108 fragment was identified as *Picea* sp., probably Sitka Spruce.

Other specialized studies: None.

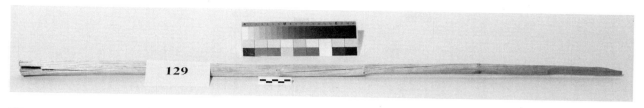

(Top) Figure 12. Photograph of IkVf-1:129, the distal end of the third gaff pole/walking stick. Not radiocarbon dated; wood not identified and not associated with human remains. Robin Armour (Yukon government) photograph.

(Bottom) Figure 13. Close-up photograph of the distal end of the third gaff pole/walking stick (IkVf-1:129). Robin Armour (Yukon government) photograph.

IkVf-1:129
(gaff pole/walking stick—figs. 12 and 13)

Dimensions: Length (incomplete) is 99 centimetres. The diameter of the pole, away from the notches, ranges from 18 to 25 millimetres.

Description: This incomplete artifact is made from a robust sapling that was carved over its entire surface. The resulting cross-section is circular, except at the uniquely shaped and finely carved distal end. One side of the distal end shows a series of three notches spread over a distance of five centimetres (fig. 13). Gouging marks that created the three notches are distinctly parallel, suggesting they may have been made with a metal chisel. The cuts that created the notches are also considerably finer than those on the IkVf-1:104 piece (compare figs. 9 and 13). The stick face opposite to these notches has been carved to a flat surface, providing a place to "receive" or hold another part of this composite artifact; this hafting arrangement is discussed further below.

Previous label or reference number: None.

Year collected: 2005; found near IkVf-1:128, discussed below.

Associated with human remains: No; this piece was found east of and uphill from the remains.

Curation/study history: This piece has been curated at Yukon Heritage facilities since it was collected.

Radiocarbon date: this artifact has not been radiocarbon dated.

Wood species: This piece has not been examined by an expert in wood identifications. It is visually similar to pieces that have been identified as probably Sitka Spruce.

Other specialized studies: None.

Figure 14. Photograph of IkVf-1:128, the base portion of a gaff pole/walking stick. Not radiocarbon dated; wood not identified and not associated with human remains. Robin Armour (Yukon government) photograph.

IkVf-1:128
(base of gaff pole/walking stick—fig. 14)

Dimensions: Length of largest fragment is 28 centimetres; diameter is *ca.* 26 millimetres.

Description: The "a" and "b" pieces, plus other small fragments, are parts of a badly weathered but carved artifact.

Previous label or reference number: None

Year collected: 2005

Associated with human remains: No; these fragments were found east of and uphill from the remains.

Curation/study history: This artifact has been curated at Yukon Heritage facilities since it was collected.

Radiocarbon date: This piece has not been radiocarbon dated.

Wood species: This artifact has not been examined by an expert in wood identifications. It appears similar to the pieces that have been identified as probably Sitka Spruce.

Other specialized studies: None.

Figure 15. Photograph of cut birch bark found at nearby site IkVf-2. Not radiocarbon dated and not found in association with human remains. Sheila Greer (CAFN) photograph.

Artifact from Site IkVf-2

During a visit to the site and surrounding area in 2005, Bill Hanlon found a piece of birch bark southeast of the discovery site in the so-called Empty Valley (fig. 1)—a treeless (vegetationless) valley bottom setting. Owing to the distance from the original find area, the find was recognized as a separate site and registered as IkVf-2. This site location is further described in chapter 4 of this volume (Mackie and Greer).

No other artifacts were found in the same area. The IkVf-2 site collection therefore consists of the single piece of cut birch bark. The cut bark is *ca.* 58–78 millimetres long and rolled on both ends (fig. 15). It is estimated that unrolled, the piece would be *ca.* 70–100 millimetres wide. No specialized studies have been completed on this piece of bark, and it has not been radiocarbon dated.

Overview of the Wood Artifact Collection

THE COLLECTION of wood artifacts from the Kwädąy Dän Ts'ìnchį site is not typical of finds from pre-contact or early-contact archaeological sites in either the coastal or subarctic interior region (cf. Davis 1990, Clark 1981). For both regions, sites dating to pre-contact times tend to be dominated by stone tools or the by-products of stone tool manufacture. For sites dating to the post-contact era, expected finds include metal, ceramic or other trade items. Wooden artifacts are not typically preserved at sites dating to either era, though have been found (e.g., Davis 1996 for a coastal context). In the adjacent southern Yukon, examples of pre-contact-era wooden hunting artifacts have recently been recovered from the ice patch archaeological sites (see Hare et al. 2004, 2012; Carcross-Tagish First Nation et al. 2002, 2005).

Half of the wood artifact collection is composed of incomplete pieces—that is, stick fragments, sticks that are minimally modified or sticks that are too badly weathered to establish if they had once been shaped into distinct artifacts. Twelve wood artifacts were thus classified (table 1). These pieces include a group of longer (>30 centimetres) sticks such as IkVf-1:105, pointed stick with bashed end; IkVf-1:110, stick with carved end; IkVf-1:111, stick with cut end; IkVf-1:117/118, slender pointed stick with carved end; and IkVf-1:123, pointed stick fragment. They are clearly artifacts, but being minimally modified, little information can be extracted from them.

Some fragments of carved bark pieces are also part of the collection (IkVf-1:114), as well as four fragments of a small sapling (IkVf-1:109). The latter were initially thought to possibly be parts of an arrow (Beattie et al. 2000), but morphological details are lacking to substantiate this interpretation.

The collection also includes six wood artifacts that represent shorter (<30 centimetres) pieces of wood which display minimal or no modification (see IkVf-1:112, 124, 126, 127, 130, 131 in table 1). These fragments are either too small to establish if they might once have been part of something, or they display no modification and are only interpreted as artifacts because someone carried them to the treeless landscape of the discovery site.

We suggest that some of the longer sticks were possibly brought to the IkVf-1 site area as informal (i.e., minimally prepared) walking sticks. Some of the other sticks may have been brought there for firewood.

Shaped Artifacts: Discussion and Analysis

IkVf-1:102—Man's Knife

In shape, design, size and construction materials, this artifact matches specimens in museum ethnographic collections from the southeastern Alaska Tlingit culture area which are identified and labelled as knives or, more specifically, "men's knives". Over 20 examples of this particular type of artifact are in the holdings of the

Burke Museum in Seattle; three originated in Klukwan.[9] Other examples of this tool type, similarly originating from Klukwan or other Alaskan Tlingit communities, can be viewed in the online museum collections of the American Museum of Natural History.[10]

All of these specimens labelled "knife" or "man's knife" were obtained from the Tlingit culture area by collector George Thornton Emmons around the turn of the 20th century. We do not know if Emmons personally coined the "man's knife" label for this shape or type of knife based on his observations or if his informants referred to these knives in that manner. Regardless, the label suggests that the intention was to distinguish this type of knife—which features a linear, straight blade—from that used by women. A Tlingit "woman's knife" features a broad curved blade, similar to the Inuit ulu or the Athapaskan tabular scraper (thechel; see chapter 35, fig. 12). Examples of Tlingit "women's knives" made of shell, bone and metal can also be seen in the Burke and American Museum of Natural History online artifact collections. An online Tlingit dictionary identifies the word for the curved knife as *wéiksh*, and for the straight knife used in carving as *t'áa shuxáshaa*.[11]

The metal on this piece has been identified as non-meteoritic, which means it is trade metal and not of local native manufacture. Most likely it is either Asian or European in origin. Given the artifact's age, as indicated by the wooden handle being dated to 1750–1850 AD, the trade metal could have come from a variety of different sources. It could have been obtained in trade from any of the numerous Spanish, Russian, British and American trading ships that were operating along the northwest Pacific coast throughout the late 18th century (Cole and Darling 1990; Emmons 1991, 183ff.). The first land-based trading post to open in Tlingit territory, established by the Russians, opened in 1799. Alternatively, the trade metal could have come from shipwrecks of Asian origin that washed up on the Pacific coast via the Japanese current even before the maritime fur trade developed (Emmons 1991, 183). Wike (1951, 34, as cited in Cole and Darling 1990, 120) reports that the Tlingit were doing their own limited forging of iron by 1786.

Identification: We identify IkVf-1:102 as a man's knife.

IkVf-1:125—Carved and Painted Stick

At present no functional or typological identification is offered for this piece. We do not know what it might have been used for or been part of. While vaguely similar in width and thickness to a simple bow, its configuration, including its considerably shorter length, is different from any historic or pre-contact period bows known from the southern Yukon, northern BC and southeastern Alaska.

The pigment on this artifact, while reddish-brown in colour, is dissimilar to ochre Indigenous pigment from this region of northwestern North America. Its hue is more brownish than the ochre colour present on IkVf-1:103, the hooked stick, and on the notches of the #115 portion of IkVf-1:104, the gaff pole/walking stick. Possibly the pigment is non-local in origin; technical studies would be required to confirm this.

Identification: We identify IkVf-1:125 as a carved and painted stick, function unknown.

IkVf-1:103—Hooked Stick

In shape, this large crochet-hook-type wooden artifact is similar to the implement traditionally used by the Dän to set the eagle feather snares used to catch gophers (also known as Ground Squirrels, *Spermophilus parryii*), as discussed in chapter 35 on the gopher robe. The inland Tlingit term for the item used to set gopher snares is *tsâts kât'i* (Leer et al. 2001). The Dákwanjè term for what is referred to as a gopher stick in English is *thèmèl gät* (Greer 2008) or more simply *gat* (Southern Tutchone Tribal Council 1999, 42). Figure 16 depicts a contemporary Southern Tutchone gopher stick. Illustrations can also be found in Harp (2005, 57) and Johnson and Raup (1964, 194–95, plate 53-e); the latter also provides a description of how the stick was used to set the snares.

The IkVf-1:103 artifact is noted as being more curved and robust than gopher sticks examined by Greer and other staff of the Champagne and Aishihik First Nations Heritage program. Generally, gopher sticks are not known to be decorated, as is the IkVf-1:103

artifact, and the wood it is made from, Hemlock, is also much denser and heavier than the spruce used to make most gopher sticks.

Although clearly different, the distinctive hooked end of the IkVf-1:103 implement suggests it may have been a tool that fulfilled a purpose similar to the gopher stick. It is proposed that this artifact may have been used to set the snares used to catch marmots (*Marmota caligata*), locally referred to as groundhogs. This species is common in the alpine of Tatshenshini-Alsek Park and is reported to have been similarly snared in times past (Greer 2005c). Very little documentation exists, though, about the practice of marmot snaring. One Champagne and Aishihik citizen said he had been told that marmot snares were set in a manner similar to those used for catching gophers, adding that often "the marmots had to be dug out when snared, because they would back down and fill the hole with dirt as they struggled" (Greer 2005c). This same source also indicated that feather quill snares used to catch gophers were not employed for marmots. The ethnological and ethnographic literature from the western subarctic (e.g., McClellan and Denniston 1981, O'Brien 1997) and the coastal Tlingit culture area (e.g., Emmons 1991) unfortunately provides no information on the technique or equipment used. There are no known museum examples of Southern Tutchone or Tlingit devices that may have been used for setting groundhog snares.[12]

The similarity of this wooden artifact to an atlatl, or throwing-board or -stick type of implement, has been previously noted (Beattie et al. 2000, 38). For a number of reasons, we consider this functional identification problematic. New archaeological data from the adjacent southern Yukon suggest that the throwing-stick and dart system of hunting stopped in the northwest interior around 1,250 years ago (Hare et al. 2004). Wooden artifacts from the nearby Yukon ice patch archaeological sites indicate that around this time, this kind of hunting tool was replaced by the bow and arrow (ibid.). More than a millennium separates the IkVf-1:103 hooked artifact from the last recorded use of a throwing board at the southern Yukon ice patch sites, making it unlikely that the IkVf-1:103 artifact was used as an

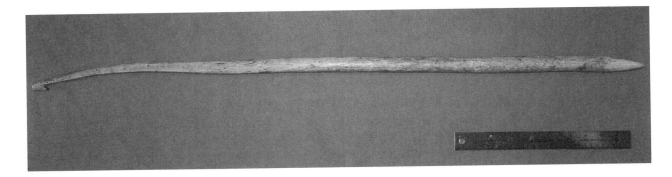

Figure 16. Dän (Southern Tutchone) tth'émèl (gopher stick); compare with artifact IfVf-1:103, the Hooked Stick (fig. 5). Champagne and Aishihik First Nations Heritage program collection. Sheila Greer (CAFN) photograph.

atlatl or throwing-board device. We realize that the throwing board or throwing stick continued to be used in the coastal Tlingit culture into historic times (Brown 2000, 178). In the latter context, though, the tool was used solely for sea mammal hunting. Historic Tlingit throwing boards, of which many examples exist in museums, also differ significantly in morphology from the IkVf-1:103 artifact.

Identification: We consider it unlikely that the IkVf-1:103 artifact was used as a throwing-stick (atlatl) type of device. Rather, we suggest that it may have been a tool used to set marmot snares.

IkVf-1:119, 120/121 and 122—*Sticks for Carrying/ Hanging Salmon*

When shown these three wooden artifacts, a Champagne and Aishihik First Nations elder immediately identified them as the kind of sticks that were traditionally used to carry King Salmon heads (Greer 2005c). Other elders later confirmed that these same sticks, known by the Dákwanjè term *£u kal*, were also used to hang salmon to dry after it received the "second cut" (Greer 2008). A detailed description of the Dän method of salmon processing, including the purpose of the different cuts, can be found in O'Leary (1992).

Identification: We identify these three artifacts as sticks for carrying/hanging salmon.

IkVf-1:104, 107, 129—*Three Gaff Poles/ Walking Sticks or Staffs*

Four pieces, representing a minimum of three objects, are identified as gaff poles/walking sticks (or staffs)—a combination and multi-purpose tool form. We suggest that this multi-purpose tool may be unique to the local geographic area. More detailed discussion of the tool as a whole follows the individual artifact identifications.

IkVf-1:104—*(First) Gaff Pole/Walking Stick*

This largely complete artifact is tentatively identified as part of an artifact here referred to as a gaff pole/walking stick, albeit one lacking the working or "business" part of the composite gaffing tool. The artifact is nearly 13 feet long, with a pointed but sturdy base. It is suggested to have been used not only for catching salmon but also as a walking stick or staff in the manner of an alpenstock—the long, wooden pole with an iron-spike tip that has been used by shepherds for travel on icefields and glaciers in the Alps since the Middle Ages.[13] The earlier "Walking or Bear Stick" descriptor (Beattie et al. 2000) has thus been dropped.

While further research is needed to confirm the functional interpretation of the IkVf-1:104 artifact, the tentative identification of this piece suggests that three other wooden artifact pieces (two distal and one proximal, or base, portions) are possible parts of the same or a similar type of artifact.

IkVf-1:107—(Second) Gaff Pole/ Walking Stick (Distal End Fragment)

The earlier "thick stick" descriptor (Beattie et al. 2000) has been dropped in favour of the tentative gaff pole/ walking stick identification, as discussed above, noting that the artifact is similarly lacking the working or "business" part of the composite gaffing tool.

IkVf-1:129—(Third) Gaff Pole/ Walking Stick (Distal End Fragment)

Although incomplete, this artifact is tentatively identified as part of the combination tool here referred to as a gaff pole/walking stick, owing to its similarity in size and general design characteristics, but noting that it is similarly lacking the working or "business" part of the composite gaffing tool.

IkVf-1:128—Base/Proximal End Fragment of a Gaff Pole/Walking Stick

While highly fragmentary, this piece is definitely similar in shape, size and construction to the base of the more-or-less complete IkVf-1:104 artifact. This piece is therefore suggested as possibly being the base portion of what we call a gaff pole/walking stick combination tool. Note, too, that this piece was found near IkVf-1:129 and so might be part of the same artifact.

Gaff Poles/Walking Sticks

In 2000 Klukwan Elder Joe Hotch noted that long sticks played

> a big part in our culture.... They used a stick to walk across creeks, rivers so they know where it's deep. When it goes way down, then that's where they're not going to step because there's a hole there. That's why they use the long stick. (Champagne and Aishihik First Nations 2000a, 1)

Other sources confirm the importance of walking sticks in the northern Tlingit area. Lieutenant C.E.S. Wood, who visited Glacier Bay in 1877, mentions being outfitted with one by the local (i.e., Hoonah) Tlingit in preparation for a hunting and climbing trip to the interior (Wood 1882, 334).

Champagne and Aishihik elders similarly reported that walking sticks were an important part of travel for the Southern Tutchone in times past, when travel was largely by foot. Elder Marge Jackson mentioned that women always travelled with a walking stick, noting that one's walking stick was also used for bear protection (Greer 2005c). Long walking sticks were identified as being especially important when travelling in the avalanche-prone mountainous environment of the Tatshenshini-Alsek country. Elder Paddy Jim mentioned that a sturdy walking stick can be used as a rudder of sorts, helping one to keep upright under (presumably light) avalanche conditions (Greer 2006b). Elder John Adamson, who had travelled extensively in the Tatshenhini-Alsek country as a young man, mentioned that the long sticks were used to test for dangerous snow and ice. He noted that creek overflow ice would often form under high-water conditions. Later, when the water level dropped, one ran the risk of breaking through this thin, snow-covered ice layer, only to be trapped on the ground surface of the creek bed, a couple of metres below the surface (Adamson 1993). Traditional Southern Tutchone stories report that even the creator, Crow, used a walking stick (McClellan 2007).

While it is suggested that some of the minimally modified or shaped sticks (such as IkVf-1:105) may have been simple walking sticks, the IkVf-1:104 artifact appears to be more than just a walking stick. The distal end of this piece features a series of notches that allowed something to be attached with lashing, which was held more securely because of the notches. It is possible that a gaff or spear hook (now missing) might have been secured to the distal end. We are thus proposing that this artifact represents a multi-purpose, multi-component tool form.

(Left)
Figure 17. Klukshu
resident gaffing for
salmon in Klukshu
Creek in the 1970s.
CAFN photograph.

(Below)
Figure 18. Well-used
path alongside
Village Creek, Yukon,
where salmon have
been gaffed for
generations. Sheila
Greer (CAFN) photograph.

It is worth noting that similar-sized gaff poles were
historically used by Champagne and Aishihik people to
catch salmon in the Tatshenshini River tributaries (figs.
17 and 18), and gaffs continue to be employed in salmon
fishing today.

Ethnographic information suggests that
Tatshenshini basin salmon were traditionally caught by
one of two means: either by traps, not discussed here, or
by gaffing (O'Leary 1992). The Yakutat Tlingit, who live
in the area near the mouth of the Tatshenshini-Alsek
River, are also reported to have traditionally gaffed
salmon (de Laguna 1972, 386). Krause (1956, plate II)
provides an illustration of a Tlingit gaff pole, noting that
it was used for catching both salmon and eulachon.

In streams featuring a heavy sediment load, such as
those of the Tatshenshini-Alsek basin, fish can't be seen,
making a spear or leister (which requires the fisherman
to see his prey) less effective. In such settings, the gaff is
preferred. Secured on the end of a long pole (fig. 19), its
large barbless hook does not toggle like harpoons or fish
spears (see Stewart 1977).

Gaffing is a highly skilled activity. It involves feeling
the fish with the distal end of the long pole and then, at
precisely the right moment, quickly pulling the gaff back

346

toward oneself to catch the fish with the hook. The latter usually lodges in the back section of the fish, in the area of the dorsal fin or the tail (O'Leary 1992, 64).

The late Champagne and Aishihik Elder John Adamson, who spoke both Tlingit and Dákwanjè languages, referred to this style of fishing as "poking" salmon. He added that the Tlingit language place name for the Takhane River, which is a tributary in the upper basin of the Tatshenshini River, was derived from the Tlingit root word *tak* meaning "to poke" (Greer 2005d). The Southern Tutchone term for the gaff pole is *k'ek'a käw*, and the term for the gaff hook is *k'ek'a*. The term for a walking stick is *tü* (Greer 2008). In the interior Tlingit dialect, the term for gaff pole is *kùxìdà* or *khuxìdà*, while a walking stick or staff is known as *wùtsàghâ* or *yùtsàghâ* (Leer et al. 2001).

While de Laguna mentions that salmon were gaffed from drafting boats by the Yakutat people, in the Tatshenshini basin gaffing seems to have been done from the shore or by standing in the stream (figs. 17 and 18). O'Leary reported that contemporary Dän gaff poles varied in length from 8 to 12 feet (1992, 64). Champagne and Aishihik Elder Marge Jackson, who had more than 80 years of experience in gaffing for Tatshenshini basin salmon, indicated that the length of a gaff pole depends on the size of stream where it will be used (Greer 2006b).

Another detail of the IkVf-1:104 artifact supports the gaff pole interpretation. The artifact shows stains in two places along its length. The location of these stains (shown in fig. 2) matches where one puts one's hands when using a Tatshenshini-style gaff pole (fig. 17). It is therefore suggested that the stains on the IkVf-1:104

Figure 19. Contemporary Southern Tutchone gaff pole mounted on vehicle for transport to salmon-bearing stream. Sheila Greer (CAFN) photograph.

artifact are grease or sweat stains from the tool being similarly held and used.

The Distal or Working End of Gaff Poles

Contemporary Tatshenshini River gaff poles, as well as the late 19th-century example illustrated by Krause, are outfitted with metal gaff hooks (see fig. 19). Usually these hooks are made from a bent nail or spike that has been heated and hammered into the desired shape (O'Leary 1992, 64). Southern Tutchone informants reported that before metal became available, gaff hooks were made of sharpened bone, such as the nose bone of a Moose (ibid.).

An example of a gaff hook made of antler rather than bone was collected by Emmons in the Haines area over a hundred years ago and is in the collections of the American Museum of Natural History.[14] This J-shaped specimen, 17.5 centimetres long and spreading over 9.5 centimetres at its maximum width, is similar in size and configuration to the modern metal hooks. It is possible that something like this antler gaff hook specimen was once attached to the IkVf-1:104, 107 and 129 gaff poles.

The three proposed gaff poles collected from the Kwädąy Dän Ts'ìnchį site area each feature a different design for securing the missing gaff hook that would have been mounted on the distal end. The simplest design, a flat and bevelled surface, is evident on the IkVf-1:107 artifact (fig. 11). As mentioned above, the flat surface would have held the missing gaff hook, which would have been secured with lashing.

The IkVf-1:104 and IkVf-1:129 specimens display a slightly more complex design. Both artifacts have the same flat surface where the gaff hook would have been placed, but these two artifacts also have a series of notches on the face opposite (figs. 9 and 13). The lashing would have been wound around these notches, thereby holding the hook more securely in place.

As previously mentioned, the carving at the distal end of the IkVf-1:129 specimen is more delicately executed than that of IkVf-1:104. We suggest that the notches on the former may have been made with a metal knife, rather than one of stone. The notches on the IkVf-1:104 artifact are definitely crude by comparison,

but whether this item was shaped with a stone knife is uncertain. Unfortunately, the weathered wood of both artifacts makes it impossible to firmly establish what type of technology was used to construct the pieces.[15]

Wood Types

Analysis by wood specialists demonstrated that a range of wood types are represented in the wood artifact collection (Florian 2005, Young 2001). These include *Picea* sp. (unknown spruce); *Picea* sp., probably Sitka spruce; *Betula* sp., (birch); *Tsuga* sp., (hemlock, possibly Western Hemlock or Mountain Hemlock); *Thuja plicata* (Western Red Cedar); and possible Alder (*Alnus* sp.) or False Huckleberry (*Menziesia ferruginea*).

Excepting the unknown spruce wood, all of the woods identified to the species level are native to the coastal rainforest biogeographic zone (chapter 2). The settings closest to the discovery site featuring this zone are the lower Tatshenshini River area, *ca.* 20 kilometres to the southwest, or the Rainy Hollow to Klukwan region, *ca.* 25 kilometres to the southeast of the discovery site.

Dates for the Wood Artifacts

Six of the wooden artifacts, including four of the ten more recognizable artifact pieces, have been radiocarbon dated. Since dates were obtained on two fragments that were later refitted together (the #104 fragment and the #106 fragment of IkVf-1:104 gaff pole/walking stick), seven dates in total are available.

The earliest date, 500 ± 40 BP (or *ca.* 1409–41 AD), is on IkVf-1:107, second gaff pole/walking stick, and dates to the pre-contact period (see chapter 9). The next youngest date, 360 ± 40 BP (or *ca.* 1550–1630 AD), is on IkVf-1:117, slender pointed stick with carved end, and is likewise pre-contact in age.

The two dates obtained on the IkVf-1:104 gaff pole/walking stick artifact are 290 ± 40 BP (or *ca.* 1620–1700 AD) and 140 ± 40 BP (or *ca.* 1770–1850 AD). These are variable and do not overlap; not surprisingly, therefore, they could fall anywhere within the pre-contact or protohistoric periods. Artifact IkVf-1:103, the hooked stick, was dated at 230 ± 40 BP, which displayed a wide range of intersect variability (i.e., 1530–45 AD, 1635–80 AD, 1740–1805 AD, 1930–50 AD). Its date could likewise fall anywhere within the pre-contact or protohistoric periods.

The two other dates obtained fit somewhere within the protohistoric or early historic period for the local and greater region. These include the date on IkVf-1:102, man's knife, at 150 ± 50 BP (or ca. 1750–1850 AD) and the date on IkVf-1:109, fragments of a small sapling, at 140 ± 40 BP (or ca. 1790–1870 AD).

Summary and Conclusions

ANALYSIS OF THE WOOD ARTIFACTS from the Kwädąy Dän Ts'ìnchį site has been challenging. Close to half of the pieces are stick fragments, minimally modified or too badly weathered to establish if they had once been purposely shaped. A lack of information on wooden artifacts traditionally made and used by the region's Aboriginal peoples, including a paucity of comparative museum examples, limits our ability to identify the purpose of objects that are clearly artifacts. We have no idea what the item referred to as the carved and painted stick (IkVf-1:125) was used for. It has been suggested that another tool, the hooked stick (IkVf-1:103), may have been used to set snare traps for catching marmots/groundhogs. This tentative identification is based on the item's general similarity to the hooked sticks that were traditionally used by the Dän and inland Tlingit to set snares for gophers/ground squirrels.

Three long, carved sticks have been tentatively identified as either part of or a whole artifact that is here referred to as a gaff pole/walking stick. This multi-purpose and multi-component tool may be unique to this biogeographic zone, or at least the Tatshenshini-Alsek country—an area of glacier-fed, fast-flowing streams where salmon spawn. The presence of the gaff pole/walking stick artifacts at the site, as well as the items identified as sticks for carrying/hanging salmon (IkVf-1:119, 120/121, 123), remains to be explained, though.

Given that the closest location where salmon might have been gaffed is ca. 17 kilometres away, finds of such artifacts in the high mountain location of the Kwädąy Dän Ts'ìnchį site are somewhat unexpected. Nonetheless, the importance of salmon in the traditional subsistence economy of the basin's Aboriginal inhabitants (see chapters 9 and 10) means finding such artifacts related to salmon fishing is not totally surprising.

Based on comparable museum objects, artifact IkVf-1:102, previously referred to by the more generic label "hand tool," has now been identified as a man's knife. Many examples of this tool type exist in museum ethnographic collections, which suggests that the man's knife was a common item in the tool kit of the late 18th-century southeast Alaskan Tlingit.

While there is uncertainty about the function and use of some of the wooden artifacts from the Kwädąy Dän Ts'ìnchį site and surrounding area, a few things have been established. One is the source area for the wood used in making these tools. It appears that most of the wood types are native to the coastal rainforest biotic zone. The settings closest to the discovery site that feature this biogeographic zone are the lower Tatshenshini River area to the southwest and the Rainy Hollow to Klukwan region to the southeast of the discovery site.

We also note the considerable variation in age of the wooden artifacts that have been dated. The youngest item in the collection is the IkV-1:106 portion of the complete gaff pole/walking stick, which was dated to 140 ± 40 years BP. At 150 ± 50 BP, the date on the man's knife (IkVf-1:102) almost matches this. Both of these age estimates correspond closely with the revised date range on the Kwädąy Dän Ts'ìnchį individual (chapter 6).

The oldest dated wooden artifact is the IkVf-1:107 portion of the distal end of the second gaff pole, which returned a significantly older date: 500 ± 40 BP. The wood artifacts from the site thus appear to span the pre-contact to early historic period for the region (see chapter 9 on ethnohistory, this volume).

Only two wooden artifacts—IkVf-1:102, the man's knife, and IkVf-1:129, the carved and pointed stick—were found in definite association with the Kwädąy Dän

Ts'ìnchį individual and are consequently considered his belongings. The artifact IkVf-1:104, gaff pole/walking stick, was noted as being found in the same general area as the human remains, and therefore it may be associated with them. We hesitate to link this artifact directly with the Long Ago Person Found, however, because of the quantity of other artifacts found at the site. The existence of these other finds opens up the possibility that this particular gaff pole/walking stick was left at the site on another occasion.

The provenience or location of the remaining wooden artifacts strongly suggests that these other pieces are not directly associated with the Kwädąy Dän Ts'ìnchį individual. Their quantity also implies that they ended up at the site by other means. That is, they are considered to have resulted from past uses of or visits to this area by individuals other than the Long Ago Person Found.

Many questions remain about the wooden artifacts, particularly those not considered to be associated with the Kwädąy Dän Ts'ìnchį individual. The functional identification of the pieces, as proposed in this chapter, has not simplified the explanations for these finds. We do not know why these artifacts were brought to the site area. Various scenarios may be proposed, but most pose additional questions. For example, were these wooden items abandoned because they were no longer of use? If so, why were they abandoned at this location? Were these items intentionally cached with the idea of returning for them at a future time? If so, why were they cached here? Alternatively, were some of these wood artifacts left behind over time as markers, possibly to indicate a trail or travel route? Might at least some of them have been left to mark the approximate location where a community member was buried by an avalanche and never seen again?

We conclude by noting that the number of wooden artifacts found in the area demonstrates that the Kwädąy Dän Ts'ìnchį discovery site does not represent a single event site. Rather, the site area was likely visited or used by others in addition to the individual whose remains were recovered here. The fact that these other artifacts (i.e., those not belongings of the Long Ago Person Found) appear related to activities that could not have been conducted in the immediate area also implies that the site is in some way connected to other locations, situated in different ecological settings. We believe this suggests that the Kwädąy Dän Ts'ìnchį site is on some kind of trail or overland travel route connecting these various places. The find of a piece of cut birch bark at the nearby but separate IkVf-2 site further supports such an interpretation.

ACKNOWLEDGEMENTS

Staff of the Yukon Heritage Branch has played a significant role in the project and the study of the artifacts reported here. Their contribution was particularly important in the initial period after the discovery, when staff archaeologists Greg Hare and Ruth Gotthardt assisted with the field assessment and recovery of the remains and artifacts, and with artifact photography and initial analysis. Their program also prepared and paid for radiocarbon samples that are reported here. More recently, Yukon government staff photographer Robin Armour assisted by taking lab photographs of the wood artifacts.

Costs for conservation studies and stabilization work on the wooden artifacts have also been covered by Yukon Heritage Branch, with the exception of the specialized analysis work completed on the IkVf-1:103 specimen, the man's knife, which was covered by the Canadian Conservation Institute. In addition to providing conservation services for the artifacts, Yukon government conservator Valery Monahan must be acknowledged as the individual who recognized that wooden pieces IkVf-1:104, 105, 115 and 116 all fit together into one complete, long artifact. Valery also collected the samples that were subsequently used for wood species identification.

The wood species identifications reported here were done by Mary-Lou Florian in 2005. Thanks also to Gregory Young (Canadian Conservation Institute) for his early insights on the same subject. Dr. M. Wayman (University of Alberta) provided an initial assessment of the metal in the man's knife artifact.

We also wish to thank Dr. Steve Hendrickson, Alaska State Museum, for input on relevant museum comparative collections, as well as Brian Seymour at the Royal BC Museum for assistance in reviewing collections of interest. Wayne Howell, Glacier Bay National Park, alerted us to relevant ethnohistoric information on the Hoonah Tlingit.

Elders Moose Jackson, Marge Jackson, John Adamson, Paddy Jim, Frances Joe and Hayden Woodruff and Chief Diane Strand (Champagne and Aishihik First Nations) provided insights into the technology used by the Dän in harvesting and processing gophers (ground squirrels) and salmon, including information on gopher sticks, gaff poles, walking sticks and sticks used for hanging and carrying salmon. Elder Joe Hotch (Chilkat Indian Village, Klukwan) provided insights on walking sticks in Tlingit culture.

Greer's research work on the wooden artifacts, as well as Florian's wood identification work, was financially supported by CAFN. Supplementary contributions in support of the former were provided by Northern Research Institute (Yukon College) and Canada's International Polar Year program. Greer was also a research associate (status only) of the Canadian Circumpolar Institute, University of Alberta, and gratefully acknowledges the library support provided by this institution.

Staff and citizens of Champagne and Aishihik First Nations have contributed to our understanding of the site and the artifacts recovered from it. These individuals include Frances Oles, Ron Chambers, Micheal Jim, Greg Eikland and the late Sarah Gaunt. The contributions made by the site's discoverers—Bill Hanlon, Mike Roch and Warren Ward—should not be forgotten, and it should also be mentioned that Bill Hanlon found the site IkVf-2 birch bark in 2005.

24

THE KWÄDĄY DÄN TS'ÌNCHĮ COPPER BEAD

H. Kory Cooper, Kevin Telmer, Richard J. Hebda and Alexander P. Mackie

A **native copper artifact,** nearly circular in outline, was found at the Kwädąy Dän Ts'ìnchį site in 1999 (Beattie et al. 2000). Informally called a bead, it is 7.5–8.0 millimetres in diameter, 1.5 millimetres thick, weighs 0.51 grams and has a 2-millimetre diameter hole located slightly off-centre of the object (fig. 1*A* and 1*B*). Two sinew threads (2 centimetres and 1.5 centimetres long) with two-ply S-twists are attached to the bead, which appears to be made from a single piece of coiled copper (fig. 1*A*). Native copper is that which is found naturally in a metallic state, often 99.9 per cent pure, and numerous sources have been noted throughout south-central Alaska and southwestern Yukon (see fig. 2 and table 1; Cooper et al. 2008). Indigenous people in the region were using native copper by at least 1000 AD and continued to use it after industrially produced metal trade goods became available in the 18th century.

Copper Sources and Trade

MANY REGIONAL SOURCES of native copper fall within the traditional territory of the Tutchone and Ahtna Athapaskans (Cooper 2007). Kletsan Creek, a tributary of the White River, was reportedly the main source of native copper used by people of southwestern Yukon at the time of contact with non-Aboriginal people (Hayes 1892; Brooks 1900; Moffit and Knopf 1910; Schwatka 1996). However, native copper would have been available at many other locations in southwestern Yukon, Alaska, and northwestern British Columbia (see table 1 on page 363 and fig. 2). As noted by Bostock (1957, 126), "native copper is found with the gold on Bullion, Sheep, Kimberley,

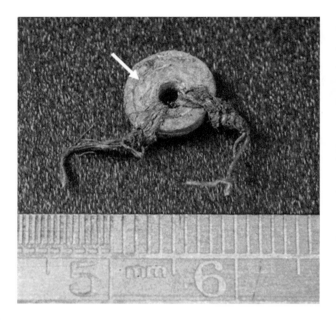

Figure 1. Copper bead and thong fragments found with the Kwädạy Dän Ts'inchị man. *A*, Plan view showing thong attachment, spiral structure and millimetre scale. *B*, Edge view of bead showing flattened edge, same scale as 1*A*. Colour versions on page 594.

Burwash and, in fact, on nearly all the creeks in this portion of the St. Elias range on which any mining has been done".

Kletsan's name comes from one of the northern Athapaskan terms for copper: *tsetsaan'*. This creek was once known as Klet-san-dek, meaning "copper creek" (Orth 1967). While travelling through Tutchone territory, Glave (1892) recorded the name *Eark Heene* (written "*Erk-heen-ee*" on Glave's map, p. 869) for the White River. The same name was used by the Tlingit to refer to the Copper River in Ahtna territory (Emmons 1991). The Tlingit word for copper was *e-ak* or *ik* (Emmons 1991; given as *eq* in Boas 2002), and according to Emmons, the Tlingit referred to the modern-day Copper River as *eek heene* or *e-ak heene*, meaning "water of copper".

Native copper from this region was reportedly traded great distances on the northwest coast, and its exchange figured prominently in the relationship between people of the interior and people of the coast (de Laguna 1972; de Laguna et al. 1964; McClellan 1975). The Haida of the Queen Charlotte Islands (Haida Gwaii) obtained copper from this region via the Chilkat Tlingit

(Acheson 2003; Dawson in Brooks 1900; Emmons 1991), and Meares (in Keithahn 1964) notes that the Nootka had copper acquired from the Tlingit.

In the latter part of the 19th century the movement of White River native copper to the south and east was controlled by a man known as the Copper Chief, who lived near Snag along the White River (McClellan 1981, 1987). For the contact period, McClellan (1964, 1975) lists dentalium, Chilkat dance blankets, medicinal roots and fish oil, as well as trade goods such as pipes, beads, knives, mirrors, leaf tobacco and tea, as some of the items being traded into the interior. In exchange, the Athapaskans were trading native copper, tanned Caribou and Moose skins, furs and tailored clothing to the Tlingit.

Emmons (1991) described some of the routes by which native copper travelled from the interior to the coast: down the Nizina, Chitina and Copper rivers to the mouth of the Copper River; from the Chitina River over the coastal range and glaciers to the coast west of Icy Bay; from the White River to the headwaters of the Chilkat River and down to the shores of Lynn Canal;

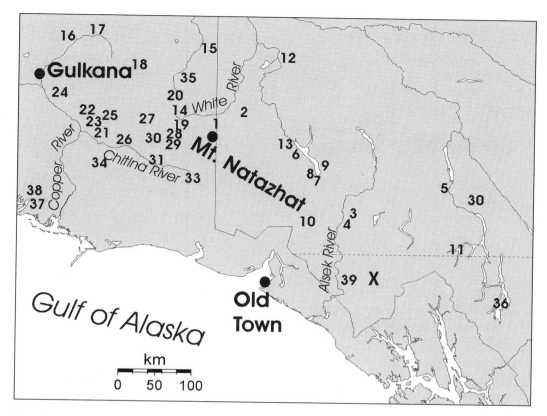

Figure 2. Selected place names mentioned in the text and potential sources of native copper, referenced by number to table 1. The Kwädąy Dän Ts'ìnchį discovery site is marked by an X.

and, particularly relevant to the Kwädąy Dän Ts'ìnchį find, from the White River to the headwaters of the Alsek River and down to Dry Bay.

The Chilkat and Chilkoot Tlingit obtained native copper from the Tutchone as late as the mid 19th century. Though the Tlingit were most interested in furs, other items were obtained in trade from the Tutchone, including native copper. It would appear that they were trading for native copper even though they had access to trade metal from non-Aboriginal fur traders. Indeed, the Tlingit seem to have been trading industrial copper metal—in the form of kettles—to the Tutchone, while simultaneously receiving native copper (Legros 1984).

MICROPROBE: LASER ABLATION INDUCTIVELY COUPLED PLASMA-MASS SPECTROMETRY ANALYSIS (LA-ICP-MS)

Methods

IN ORDER TO VERIFY whether the bead from the Kwädąy Dän Ts'ìnchį site was made of native or smelted (i.e., trade) copper and to obtain trace element data to determine the material source, it was analysed at the Inductively Coupled Plasma-Mass Spectrometry (ICP-MS) Facility at the University of Victoria. The preparation and analysis of the bead followed the methods outlined in Sanborn and Telmer (2003). In simple terms, a laser vaporizes microscopic amounts of the object at which it

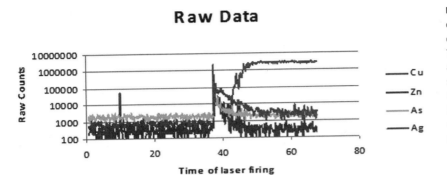

Figure 3. Major element composition of the Kwädąy Dän Ts'inchį bead. Vertical axis is a logarithmic scale of hits received by the detector. The horizontal axis is in seconds. Colour version on page 594.

is pointed. A continuous stream of argon gas carries the vaporized material into a mass spectrometer, where it is separated into its constituent elements by atomic mass. The detectors in the instrument count the number of atoms of each element present in the vapour stream, and this information is used to report the composition of the object (fig. 3). The qualitative composition of the inside of the bead was determined from later times in the analysis (after 60 seconds), when the laser had bored through the outer material and was ablating (vaporizing) only core material.

Results

THE ORIGINAL BEAD MATERIAL (the core) contained mostly copper (Cu) with traces of silver (Ag) and arsenic (As) (fig. 3). Lead (Pb), antimony (Sb), mercury (Hg) and gold (Au) occurred as very minor elements (fig. 4). The bead's outer layer contains higher levels of some trace elements: zinc (Zn), arsenic, silver, antimony, gold, mercury and lead (see figs. 3 and 4). This is likely due to oxidation and formation of a copper oxide or oxidized copper mineral film—probably malachite ($CuCO_3(OH)_2$)—which adsorbed these elements from the environment. This is not uncommon. Oxides, hydroxides and carbonate hydroxides (like malachite) have a high adsorption capacity. Silicon, calcium, aluminum and magnesium can be present in native copper in the form of non-metallic inclusions, most frequently as silicates or oxides (Patterson 1971; Rapp et al. 2000; Wayman 1989).

The data conclusively show that the bead is composed of native copper, which is noted for its purity, often greater than 99.9 per cent copper. The LA-ICP-MS analysis produced data for zinc, arsenic, silver, antimony, gold, mercury, lead and bismuth, though silver and arsenic were the main elements found once the outer corrosion products were penetrated. Silver and arsenic are the most common and significant impurities associated with native copper. Both can be present in solution with copper up to their respective solubility limits as distinct inclusions, and their concentrations can vary widely within a single native copper deposit (Broderick 1929; Wayman 1989). Silver and arsenic are both found in smelted 19th-century copper, but while the amount of silver in both native copper and historic copper can be around 1000 ppm, arsenic, antimony, nickel and selenium are found in amounts two orders of magnitude greater than in native copper (Wayman et al. 1985).

Provenance

THE PURITY OF NATIVE COPPER, which allows it to be distinguished from smelted copper, makes determining provenance difficult, as not all sources of native copper have a unique trace element signature.

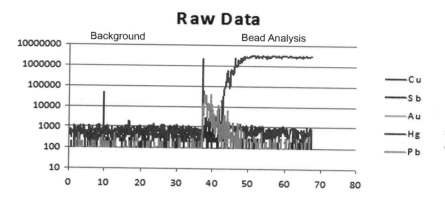

Figure 4. Minor element composition of the Kwädąy Dän Ts'ìnchị bead. Vertical axis is a logarithmic scale of hits received by the detector. The horizontal axis is in seconds. Colour version on page 594.

However, analyses of North American native copper have demonstrated that it is possible to distinguish between sources based on trace elements (Rapp et al. 2000). In a study by Cooper et al. (2008), trace element data were obtained for native copper from three sources: Chititu, Dan and Kletsan creeks (fig. 2, source #28, 29 and 1, respectively) in south-central Alaska. Seventeen samples from Kletsan Creek, seventeen from Dan Creek and seven from Chititu Creek were subjected to Instrumental Neutron Activation Analysis (INAA), and ten samples from Kletsan Creek, six from Dan Creek and four from Chititu Creek were analysed using ICP-MS (solution mode). Dan and Chititu creeks originate in the same ridge system, approximately 10–15 kilometres apart, and flow into the Nizina River, a tributary of the Chitina River. Kletsan Creek heads in Alaska and crosses over into the Yukon before crossing back into Alaska, where it enters the White River.

The source material from Dan and Chititu creeks was identical with regard to its trace element signatures. It was possible, though, to distinguish between the Dan and Chititu creek material and the Kletsan Creek material based on the presence or absence of arsenic, mercury and selenium. The trace element signature for Dan and Chititu creeks included arsenic and mercury but no selenium, while the trace element signature for Kletsan Creek included selenium but neither arsenic nor mercury. The results for the bead more closely match material from Dan and Chititu creeks than that from

Kletsan Creek, but it would be premature to claim that this artifact has been sourced to a specific locale since there are many potential sources of native copper in south-central Alaska, southwestern Yukon and northwestern British Columbia that have not been analysed (see fig. 2; Cooper et al. 2008). It is, however, significant that the LA-ICP-MS analysis produced results for both arsenic and mercury—effectively ruling out Kletsan Creek as the source of native copper used for the Kwädąy Dän Ts'ìnchị bead—because according to ethnohistoric accounts, this was the best-known source of native copper in southwestern Yukon in the 19th century (Hayes 1892, Schwatka 1996).

NATIVE COPPER IN NORTHWEST NORTH AMERICAN ABORIGINAL SOCIETY

Archaeological Evidence

THE MAJORITY of archaeological finds of native copper in northwestern North America fall within the vast, geologically cupriferous zone in southwestern Yukon and south-central Alaska. Archaeological evidence indicates that native copper was used by Athapaskans and other groups in this region both at the time of contact and before, beginning around 1000 AD. This raw material was used primarily for tools such as knives, projectile points and awls, though objects of personal adornment such as rings and bracelets are also known. According to published and unpublished literature, at least 25 archaeological sites in southwestern Yukon contained one or more native copper artifacts, and more have been found by private landowners and miners (Cooper 2006, 2007).

Despite the high value and popularity of copper on the northwest coast, especially during the early part of the maritime fur trade, few pre-contact archaeological sites with native copper have been reported in the southeastern panhandle of Alaska (Cooper 2007). This supports Emmons' statement that it "was never very abundant and increased in value as it was traded southward" (1991, 178). The Old Town site, a late prehistoric-protohistoric Tlingit village in Yakutat Bay consisting of seven house pits, many smaller pits and a midden, produced 48 native copper artifacts (fig. 2). In addition to copper, 19 pieces of iron were recovered. Two radiocarbon dates associated with the site are 136 ± 62 BP and 328 ± 78 BP (uncalibrated 1552–1698 AD; de Laguna et al. 1964). Other than this site, however, there are few examples of native copper in southeastern Alaska. A single copper artifact was recovered from each of the Tlingit fort sites known as SIT-135 and SIT-244, both with midden deposits (de Laguna 1960). The specimen from SIT-135 is a copper band wrapped around a piece of wood. It has what was thought to be a stamped design resembling an eye and was found at the bottom of the uppermost midden layer. The site was considered by de Laguna (1960) to be prehistoric, and a radiocarbon date of 668 ± 86 BP was obtained from the location (Moss et al. 1989). The specimen from SIT-244 is a cone (or "tinkler"). The site has radiocarbon dates of 278 ± 86 BP, 468 ± 67 BP and 905 ± 103 BP (Moss et al. 1989) and is believed to have been occupied both prehistorically and historically (de Laguna 1960). The relationship of the radiocarbon dates to the copper artifacts is unknown.

Both native copper and iron were being used prior to the beginning of sustained contact with Europeans and non-Aboriginal North Americans in the middle of the 18th century. A number of sites in the region contain both native copper and iron, while lacking other obvious indications of trade with non-Aboriginal people, such as beads or ceramics (de Laguna 1956; de Laguna et al. 1964; Shinkwin 1979; Workman 1976). This is not surprising given the availability of iron from Asia and Siberia—whether from shipwrecks or trade across the Bering Strait—to Indigenous peoples in the western Arctic, subarctic and northwest coast as early as the first few centuries AD (as noted by scholars such as Acheson 2003; de Laguna 1956, 1972; Keddie 1990; McCartney 1988; Wayman et al. 1992).

Origin of Copper Use

COASTAL TLINGIT acknowledge learning of native copper from interior Athapaskans—i.e., Ahtna or Tutchone (de Laguna 1972; de Laguna et al. 1964; McClellan 1975). The Yakutat Tlingit credited the Ahtna with the discovery of native copper and the knowledge of how to work it. The northern Tlingit referred to the Ahtna as *ʔiqka* or *ʔiqkaha*, meaning "copper diggers" (de Laguna 1972; de Laguna and McClellan 1981, 662). A story told by the Tutchone and Hoonah Tlingit credits the introduction of copper to a Tlingit man who travelled up the coast to Dry Bay and then up the Alsek River until he met some people (presumably Tutchone Athapaskans). There was an exchange of knowledge, whereby the Tlingit man introduced Tlingit ways of hunting, fishing, gathering and preparing food, and in return, he learned about native copper. It was among these people that the Tlingit man first saw native copper, as it was used for knives, arrowheads, spears and the shield-like Copper. This traveller married into the Tutchone but eventually returned to the coast with his new relatives, who took some copper with them, thus starting the coast-interior copper trade (Emmons 1991; McClellan 1975, 27).

McClellan recorded a story about the origin of copper from Chief Albert Isaac, a Southern Tutchone from Aishihik. He told a story he had heard as a boy from an individual known as the Copper Chief. Long ago a man had two picks, one of iron and the other of copper. He used these picks to climb a wall of ice on a mountain now known as Mt. Natazhat, just south of the Kletsan Creek headwaters. When he reached the top, he could not climb back down because the steps he made using the picks had become smoothed over again. He threw the iron pick toward the coast and the copper pick further inland, saying: "Be copper on this side!" And he sat on top of the mountain until he turned to stone—where he can still be seen, according to some. This story provides an explanation for why people on the coast had iron and people living in the interior had copper (McClellan 1987, 56). The fact that both materials are associated with the Kwädąy Dän Ts'ìnchį site attests to that individual's access to the resources of both regions.

Copper-working

THE PRIMARY METHOD of working native copper in the region was through a combination of cold-hammering and annealing (Franklin et al. 1981). Because of its high purity, native copper is malleable and deforms easily. But native copper can only be worked for so long before it becomes brittle and cracks. This process is known as work-hardening, which can be reversed by pausing to heat the copper for a relatively short period of time at temperatures as low as 200–300°C. This heating of the copper, known as annealing, reorganizes the crystal structure and restores malleability (Wayman 1989). According to information collected by McClellan (1975), native copper was heated and pounded with rocks in the southern Yukon, the nuggets made flat using a hammer stone and stone anvil. These flat sheets were then folded over, as is evident in the Kwädąy Dän Ts'ìnchį artifact, in order to build up bulk (fig. 5). The result of this stage was often a rectangular bar (Franklin 1982; Franklin et

Figure 5. Hypothetical reconstruction of native copper bead manufacture. *A* shows stage 1, the nugget; *B*, stage 2, forming the sheet; *C*, stage 3, finishing the copper sheet; *D*, stage 4, folding the copper sheet; *E*, stage 5, finished folded bar of copper; *F*, stage 6, rolling the bar into a tube/spiral form; *G*, stage 7, forming a bead.

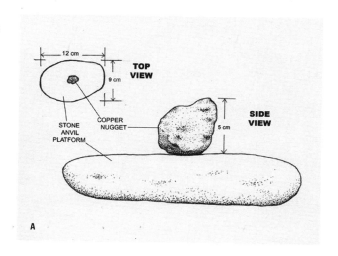

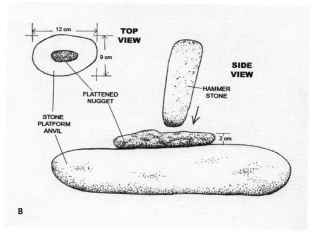

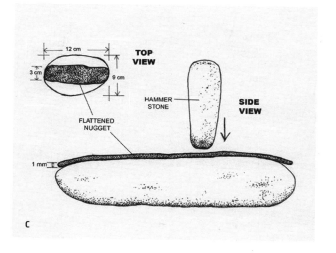

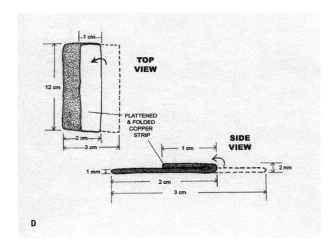

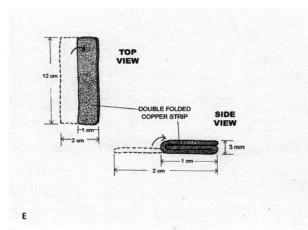

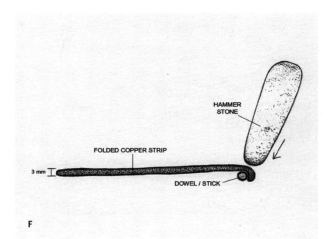

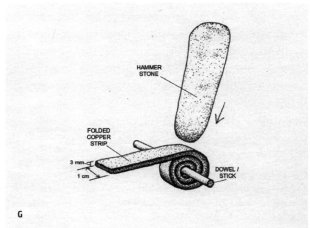

Figure 5. Hypothetical reconstruction of native copper bead manufacture, continued.

al. 1981; Workman 1976). Frequent annealing would have allowed a copper bar to be bent into the spiral shape observed in the Kwäd̠ay Dän Ts'ìnchi̧ artifact without the copper fracturing. Additionally, the coil appears flattened on the surface, probably a result of the entire object being hammered as a final step (fig. 1A and 1B). Thus, the bead conforms to oral history accounts of Indigenous metalworking techniques.

Discussion

THE DATE OBTAINED for the Kwäd̠ay Dän Ts'ìnchi̧ individual, 1720–1850 AD (Richards et al., chapter 6), fits well with what is known about the use of native copper in the region both before and after sustained contact with Europeans. Several non-Aboriginal travellers in the region commented on the use of native copper by Indigenous groups into the late 19th and early 20th centuries (Abercrombie 1900; Allen 1887; Glave 1892; Legros 1984; Powell 1909; Schwatka 1996). While at Fort Selkirk in 1891 in Northern Tutchone territory, Schwatka observed the use of copper by First Nations for items such as bullets, arrow-heads and a "long copper spatula" used to get the marrow out of Moose and Caribou bones. They also had unworked pieces of native copper in their possession, as well as "powdered azurite, or the blue carbonate of copper", used to tattoo the faces of women (1996, 129). Though he did not visit Kletsan Creek, Glave (1892) travelled through Tutchone territory the same year as Schwatka, in 1891, and noted the use of copper, in addition to bone and iron, for arrow points.

Though most native copper artifacts found in the region are tools such as awls, knives and projectile points, oral history often emphasizes copper as a form of wealth, associated with prestige and rank (Cooper 2006, 2007). Native copper also had associations with luck and health. According to information collected by McClellan (1975), copper objects were worn to guarantee health, and during a young girl's confinement while being initiated into womanhood, she might keep copper pieces in her mouth to ensure strong teeth in old age. A tubular piece of native copper was recovered from the Old Town site in Yakutat Bay, Alaska, and de Laguna (1972, 664) was given the following information by a local resident on its potential use as an amulet:

> Some old people got their dope in there—the roots they believe are lucky. Sometimes they use it for hunting sea otter. Sometimes before they eat, they use it. We don't use it much [now], but we hear about it. This time we believe in God, but I hear about it. It [the copper tube] is just to keep it [the root]. They tie it up and make a wish for luck…. Because copper cost so much, they use a little of that. There's lots of things they save out of this to make money on.

The Kwädąy Dän Ts'ìnchį bead has two pieces of two-ply twine tied to it that might have tied around the neck so it could be worn as an amulet. It may also have been attached to another artifact such as the robe or hat. Objects of personal adornment are rare in the region, the Old Town site being a noticeable exception. However, spiral forms in copper—most often associated with knife handles (see Rogers 1965; Wayman et al. 1992)—have been documented in the area. Many of these objects are made of trade metal obtained during the protohistoric and historic periods, but there are also examples from

prehistoric contexts, such as the spiral native copper artifact (fig. 6) from the Gulkana site (fig. 2).

Emmons' (1991) description of the various routes by which native copper reached the coast was based on information he was given in the 1880s and 1890s, and the accounts of European fur traders attest to the movement of native copper from the Tutchone to the Tlingit in the 19th century. The presence of native copper in an archaeological context—that is, at Old Town in Yakutat Bay (de Laguna 1964) and at least two sites further south in the Alaskan panhandle (de Laguna 1960)—indicates native copper was moving to the coast during the late prehistoric period (Cooper 2007). Though native copper was not seen in the immediate vicinity of the Kwädąy Dän Ts'ìnchį discovery, copper mineralization was noted at the site, and native copper is associated with the Windy Craggy mineral deposit (fig. 2, #39; Peter 1989). Anyone familiar with native copper would have recognized the potential for its

Figure 6. Coiled native copper artifact from a late prehistoric context at the Gulkana site. Courtesy of the State of Alaska Office of History and Archaeology.

occurrence in the area surrounding the discovery site. Though we may never know for sure, the presence of copper may have been what brought this man to the area.

Conclusion

THOUGH WE CANNOT CURRENTLY ASSIGN the bead to a specific source, the ICP-MS data have confirmed it to be native copper and not a foreign trade object. Additionally, the data suggest that the copper used to manufacture the bead was obtained from a source other than Kletsan Creek.

Successfully determining the provenance of native copper artifacts requires the collection of quantitative trace element data on numerous samples of native copper from multiple sources. The main difficulty in conducting native copper provenance research in northwestern North America is obtaining material from sources to analyse in order to develop trace element signatures.

Numerous native copper sources have been identified for Alaska, Yukon and British Columbia through both oral history and geological reports, but there are no pre-existing collections of native copper available for analysis. Due to the remoteness of many of the known sources and a lack of detailed location information for even these sources, collecting native copper in the field is a costly and time-consuming endeavour.

The use of native copper did not end with the appearance of iron or industrial smelted copper trade goods; native copper was still being traded to coastal people as late as the mid 1860s. Native copper was used as raw material in the manufacture of a variety of objects, such as tools and weapons, but it was also used for items of personal adornment that signified wealth and prestige or, possibly, were associated with Indigenous concepts of luck and health. The bead at the Kwädąy Dän Ts'ìnchį site is consistent with a long tradition of native copper use in the region, where it served as an important material link between interior and coastal people.

Table 1. Potential sources of native copper in southwestern Yukon Territory, south-central Alaska and northwestern BC.

Map ID#	Potential source	Reference
1	Kletsan Creek	Hayes 1892, Brooks 1900, Moffit and Knopf 1910
2	Generc River	Cairnes 1915
3	Beloud Creek	Kindle 1953
4	Mush Lake	Kindle 1953
5	Grafter Mine	Kindle 1964
6	Burwash Creek	McConnell 1905
7	Bullion Creek	McConnell 1905
8	Sheep Creek, YT	McConnell 1905
9	Fourth of July Creek	pers. comm. LeBarge
10	Kimberley Creek	McConnell 1905
11	Windy Arm	McConnell 1906
12	Nisling River	Dawson 1899
13	Tetamagouche	Muller 1954
14	Middle Fork, White River	Brooks 1911, Moffit and Knopf 1910
15	Nutzotin Mountains	Moffit 1943
16	Chistochina River	Moffit 1954
17	Slana River	Moffit 1954
18	Nabesna River	Rohn 1900
19	Sheep Creek, AK	Capps 1916
20	Chisana Glacier	Mendenhall and Schrader 1903, Moffit and Maddren 1908
21	Strelna Creek	Schrader and Spencer 1901
22	Kluvesna River/Fall Creek	Mendenhall and Schrader 1903
23	Kotsina Drainage	Schrader and Spencer 1901, Mendenhall and Schrader 1903
24	Tsedi Kulaende	Kari 2005

Map ID#	Potential source	Reference
25	Nugget Creek	Moffit and Maddren 1909, Mendenhall and Schrader 1903
26	Bear Creek	Mendenhall and Schrader 1903
27	Nikolai Mine	Moffit and Maddren 1909, BIA 1995
28	Chititu Creek	Moffit and Maddren 1909, Mendenhall and Schrader 1903
29	Dan Creek	Moffit and Maddren 1909
30	Chieftain Hill, Little Chief/Arctic Chief, S and SE of Whitehorse	Craig and La Porte 1972, Wheeler 1961
31	Young Creek	Rosenkranz p.c. 2004
32	Glacier Creek	Moffit and Maddren 1909
33	Chitina River headwaters	Mendenhall and Schrader 1903
34	Hanagita Valley/ Chitina River	Mendenhall and Schrader 1903
35	Bonanza, Bryan and Chathenda creeks	Cobb 1973
36	Goat Island, Atlin Lake	Dawson 1899
37	near Cordova	Grant and Higgins 1910, de Laguna 1956
38	NE of Orca	Grant and Higgins 1910, Moffit 1908, de Laguna 1956
39	Windy Craggy	Peter 1989

25

EXAMINATION AND ANALYSIS OF THE KWÄDĄY DÄN TS'ÌNCHĮ KNIFE

Kate Helwig, Tara Grant, Jane Sirois, Michael Wayman, Gregory Young, Jennifer Poulin and Valery Monahan

The man's knife (previously referred to as the hand tool), shown in figure 1, is one of the wooden items from the Kwädąy Dän Ts'ìnchį discovery in Tatshenshini-Alsek Park (see chapter 23). It was directly associated with the body of the young man melting out of the glacier and was probably one of his belongings. The knife was found inside its sheath, on top of the robe that was on top of the body (Beattie et al. 2000). Radiocarbon dating of the wooden shaft gave a date of 150 ± 50 BP, which calibrates to 1725–1780 AD (33.6%) (chapter 6). This agrees with the most probable date for the man, estimated to be between 1720 and 1850 AD. These dates fall within the pre-contact or early European contact period for the region (Richards et al. 2007). Our examination and analysis of the knife had two main goals: to provide information about the materials and construction of the object and to provide information about its current state of preservation.

Figure 1. The Kwädąy Dän Ts'ìnchį man's knife.

Description of the Man's Knife

THE MAN'S KNIFE is a composite object, consisting of a metal blade hafted to a wooden shaft. The tool is relatively small, approximately 16 centimetres long (just over 6 inches) and about 2 centimetres wide. A thinly cut hide thong is wrapped around the top 5 centimetres of the wooden shaft near the blade, to tighten the wood around the blade and provide a better grip for cutting. The thong is knotted twice, to extend its length. The second knot attaches the wrapped portion of the thong to a 28-centimetre length of loose thong. On the wooden handle beneath the hide lashing there are two areas where lifting wood is visible. These could correspond to partial breaks in the wooden handle or to small, secondary pieces of wood present under the lashing.

It is difficult to see how the blade is attached to the wood, as the hafting is obscured by both the hide lashing and corrosion products that have migrated from the blade onto the surrounding materials. An examination of the wooden ends above the lashing suggests the blade was inserted into an open slot extending across the entire width of the wooden handle. The top ends of both sides of the slot have rounded edges and a carved indentation to hold the lashing in place. The wood on one side of the slot is wider, thicker and several millimetres lower than the wood on the other side. Figure 2*A* shows the wider, lower side of the slot, and figure 2*B* shows the thinner, higher side. One explanation for the different dimensions of the two sides of the slot is that the higher, thinner side could be a secondary piece of wood that was placed on top of the blade. This would have created a slot by sandwiching the blade between the handle and the secondary piece of wood.

A brown, organic piece with a tapered end (marked with an arrow in fig. 2*B*) is wedged between the wood and the hide lashing. Based on visual examination prior to the analysis, the piece was thought to be antler, bone or quill. It probably functioned as a backing for the blade and provided support by closing one side of the open slot. There are blue deposits in several areas on the surface of the organic piece, and spots of similar blue material appear on the hide lashing. Several of these blue deposits are visible in figures 2*A* and 2*B*.

The metal blade is heavily corroded and has a red-brown surface, suggesting it is made of iron. The corroded metal is very fragile and has split into at least three separating layers. The blade has a moderate magnetic pull, indicating that intact metal or magnetic

366

Figure 2. *A*, Detail of blade showing the lower side of the wooden slot. *B*, Detail of blade showing the higher side of the wooden slot. The upper edges of the wood are outlined with dotted lines to illustrate their shapes. The organic backing piece is marked with an arrow.

A

B

corrosion products are present. In its current state, the blade extends about a centimetre beyond the end of the wooden handle. Because the blade is so corroded, it is not possible to determine its original length or shape. It may have had a blunt end, as it appears now, or it may have originally come to a point (and the tip has broken off).

Cultural Context

IRON TOOLS AND WEAPONS were used by First Peoples in northern British Columbia prior to European contact in the late 18th century (Emmons 1991;

Keddie 2006). Although there is no evidence that they smelted metals, people living in this region had several potential sources of iron prior to European contact, including the trade of wrought iron around the northern Pacific Rim (Keddie 1990), trade with First Peoples in eastern Canada (Keddie 2006), trade northward from Mexico (Emmons 1991) or the use of "drift iron" obtained from shipwrecks (Emmons 1991). After the late 18th century, wrought iron obtained through trade with Europeans became common on the northwest coast (Emmons 1991; Keddie 2006).

Given both this context and the knife's likely date of 1725–1780, one would assume the blade is probably wrought iron, which could have been obtained through various trade sources. However, the use of naturally occurring metallic iron from a meteorite cannot be ruled out without detailed elemental analysis (see below). Large quantities of meteoritic iron existed in Greenland, and smaller sources have been found in Yukon (Wayman 1989). There was also trade in meteoritic iron; artifacts containing iron from meteorites have been found across the eastern Arctic and into Hudson's Bay (Wayman 1989).

The Kwäday Dän Ts'ìnchị knife is similar in appearance to items from several locations in Alaska that were collected by Emmons in the early 20th century. Examples in the Burke Museum collection in Seattle, Washington, that are labelled as "knives" or "men's knives" show similar wooden shafts, lashing made of hide or other materials and iron blades with variable sizes and shapes (see chapter 23). It is very likely, then, that the Kwäday Dän Ts'ìnchị knife is an iron-blade knife.

With this in mind, some of the observations made by early European explorers on the northwest coast that have been summarized by Keddie (2006) are relevant to the man's knife. For example, during his 1778 expedition, Cook observed that at Nootka Sound "the chisel and the knife are the only forms, as far as we saw, that iron assumes amongst them. The knives are of various sizes". He goes on to say that most of the blades had "the breadth and thickness of an iron hoop". At Snug Corner Cove he noted that "they had a great many iron knifes; some of which are straight; others a little curved; and some very small ones, fixed in long handles, with the blade bent upward, like some of our shoemakers instruments" (Keddie 2006, 22). Similarly, at Nootka Sound in 1786 Elliot described "small knifes crooked, made of iron hoops or some other thin pieces of iron and good for nothing" (ibid., 25). Early European explorers also mentioned how such iron knives were carried and used. For instance, near Yakutat Bay in 1787 Dixon saw "short daggers in leather cases. A leather thong was tied to the dagger, wrapped around the wrist and looped onto the middle finger so the knife would not be dropped" (ibid., 28), and in Observatory Inlet

in 1793 Vancouver observed that "every man had an iron dagger in a leather neck sheath" (ibid., 32). Such observations suggest that the Kwäday Dän Ts'ìnchị man would have stored the knife in its sheath and may have worn it around his neck or wrist.

Scientific Examination: Methods

SEVERAL METHODS were used to reveal as much information as possible about the knife's materials and construction (details of the analytical methods are given in appendix 3). The examination's goals were to determine a) the structure of the metal blade inside the wooden handle, b) the type of iron used for the blade and the range of corrosion products, c) the composition of the organic backing piece wedged between the lashing and the wood, d) the species of wood used for the handle, e) the level of preservation of the hide thong and f) the identification of any coating that may have been applied to the wood.

The first step was to investigate both the metal and organic components using X-radiography. Because materials absorb X-rays differently depending on their thickness and composition, X-radiography can show the internal structure of an object (Wainwright 1990). X-radiographs were taken of the knife in two orientations: first, with the blade flat (that is, with the broad side of the blade parallel to the image plane; see figure 3A), and second, turned by 90 degrees so that the blade was edge-on (the broad side of the blade perpendicular to the image plane; see figure 4A). The initial X-radiography was undertaken in 2006. A second series of X-rays was taken in 2008, after two years of frozen storage, to determine whether any changes had occurred during the storage period.

Microscopic samples were taken from the corroded metal blade, the wooden handle, the hide lashing, the organic backing piece and the blue material visible on the lashing and backing piece. All samples were taken from unobtrusive areas under magnification using micro-sampling tools. Several different analytical techniques were then used. All the samples were

Figure 3. *A,* X-radiographic detail of the blade in a flat orientation (broad side of the blade parallel to the image plane). *B,* Possible contours of the blade superimposed on the X-radiograph.

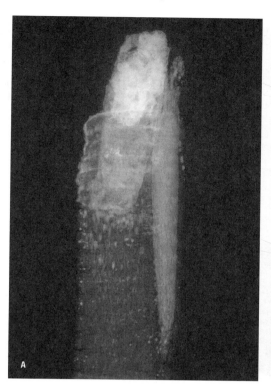

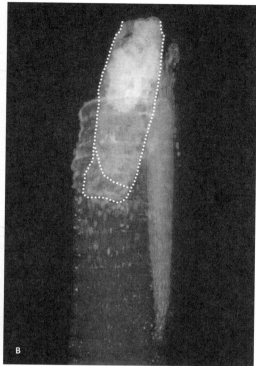

examined by various types of optical microscopy (incident light, polarizing light, transmitted light). Certain materials were also analysed using scanning electron microscopy/energy dispersive spectroscopy, Fourier transform infrared spectroscopy, X-ray diffraction, gas chromatography/mass spectrometry and thermal microscopy.

Scientific Examination: Results and Discussion

Structure of the Metal Blade

The X-radiographs provided significant information about the metal blade.

The X-radiograph of the blade in a flat orientation (figures 3*A* and 3*B*) showed that the radiographic density of the metal blade's corroded area is variable due to differences in thickness and degree of corrosion.

The exposed part of the blade is more opaque in the radiograph than the part under the wrapping. This opaque area has an irregular oval shape and is approximately 1.5 centimetres long. It extends into the lashing by a few millimetres. The opaque zone in the radiograph corresponds to a thick lump of corrosion visible on the exposed part of the blade (see fig. 2*A*). This indicates that the radiographic opacity is partially due to a buildup of corrosion products. It is possible that there is an uncorroded metal core in the opaque zone, but this is not certain. A second X-radiograph, taken after two years of frozen storage, showed very little difference in the corroded blade, indicating that corrosion has not progressed rapidly since its recovery.

The X-radiographs also provided information about how the metal blade is hafted to the wooden handle. Figure 3*A* shows that the portion of the corroded blade inside the handle is approximately 1.5 centimetres long and 1 centimetre wide at its widest point. It appears to have a roughly rectangular outline, but there is also a slightly more opaque area with a curved outline. Due to the extent of corrosion, the original shape of the blade inside the handle is not clear. Figure 3*B* shows two possible contours superimposed on the radiograph. The relatively low radiographic density of the blade under the lashing indicates that it is primarily composed of corrosion products. There is likely no intact metal inside the handle, except for—potentially—the few millimetres of the opaque zone that extend under the lashing. It is interesting to note that the blade is not parallel to the shaft but angled toward the organic piece.

This may be part of the original construction or perhaps the result of pressure applied to one side of the tool during use, which caused the blade to shift in its slot.

An X-radiograph taken with the metal blade on edge (that is, with the broad side of the blade perpendicular to the image plane; see figure 4*A*) clearly shows the exposed part of the blade's separation into layers. The cross-section of the metal inside the handle is also apparent. It was estimated from the X-radiograph that the blade was about two millimetres thick where it fit into the slot. The asymmetrical hafting that was observed visually is evident in the X-radiograph; the wood on the right side of the slot is thinner and comes up higher on the blade than the wood on the left side. The thinness of wood on the right side, producing a significantly off-centre hafting, supports the idea that the slot was formed by placing the blade on one side of the handle (which may have been carved flat to receive it) and lashing a small, secondary piece of wood on top of the blade. In fact, one of the lifting pieces of wood visible under the lashing is at the approximate location where the X-radiograph showed the metal blade terminating. Figure 4*B* shows a hypothetical contour for the slot based on the profile of the exposed portions of wood visible in the X-radiograph. The location of the piece of lifting wood visible under the lashing on the right side is shown with an arrow.

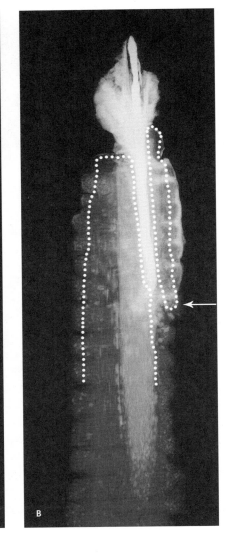

Figure 4. *A*, X-radiographic detail of blade on edge (broad side of blade perpendicular to image plane). *B*, Possible contour of the wooden slot superimposed on the X-radiograph. The arrow shows the location of lifting wood.

Additional evidence supporting the idea that the blade was wedged between the handle and a secondary piece of wood is that some early 20th-century Tlingit knives in the Burke Museum collection show this type of hafting. A number of these artifacts have a significantly off-centre blade, and in a few cases, the end of a secondary piece of wood is visible where it extends past the lashing (Burke Museum 2008).

Composition of the Metal Blade

The chemical elements present in a small fragment of the blade's corroded metal were determined using scanning electron microscopy coupled with energy dispersive spectrometry. The major metallic element identified in corrosion from the blade was iron. Oxygen, carbon and traces of other elements—such as aluminum, silicon, sulfur and calcium—were also found. These elements could be related to dirt or other accretions on the surface of the corroded blade. No nickel was identified in the corrosion products from the blade, indicating that if nickel is present, it is below a concentration of one per cent (the detection limit of the instrument used is in the range 0.1–1% by weight). The absence of nickel in the corrosion products indicates that the iron is not meteoritic. Meteoritic iron contains between five and sixty per cent nickel by weight, with the average being about eight per cent (Wayman 1989). Nickel is also preserved in the corrosion products of meteoritic iron; for example, goethite corrosion on the surface of meteorites from the Antarctic was found to contain between one and eight per cent nickel (Buchwald and Clarke 1989).

Since elemental analysis showed that the iron is non-meteoritic, the blade is probably composed of wrought iron. Several corrosion flakes were mounted as cross-sections for examination using light microscopy and scanning electron microscopy. Relic microstructure preserved within the corrosion layers could provide evidence for the blade being made of wrought iron. Wrought iron is composed of a matrix of ferrite and

sometimes pearlite, with slag inclusions—often largely composed of iron silicate—within the metal matrix. These slag inclusions are one of the distinguishing features of wrought iron (Wayman 1989) and are sometimes preserved in corrosion layers.

A backscattered electron micrograph of a corrosion flake is shown in figure 5. Backscattered electron images show areas with compositional differences; the brightness of the image is related to the atomic numbers of the chemical elements present. The low contrast

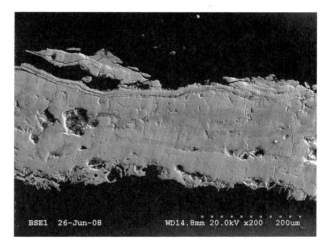

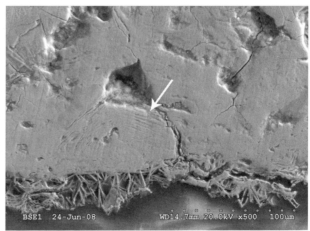

Figure 5 *(top)*. Backscattered electron image of a corrosion fragment from the blade.

Figure 6 *(bottom)*. Detail of backscattered image shown in figure 5.

in the figure 5 image indicates that the sample has a fairly homogeneous chemical composition. Mapping of chemical elements over the cross-section did not show areas with a high concentration of silicon, which would be visible if slag inclusions were present.

There was some evidence of a lamellar constituent in one area of the corroded metal. A higher magnification backscattered electron micrograph of this area is given in figure 6. In the region marked with an arrow, lamellae are visible, running parallel to each other but at several different orientations. Based on their appearance, it is possible these lamellae are relics of pearlite colonies (that is, areas with higher carbon content). But since some of the corrosion products also have a lamellar structure, the identification of pearlite is not certain.

Analysis of corrosion from the metal blade and surrounding areas using X-ray diffraction and Fourier transform infrared spectroscopy showed a range of iron corrosion products: goethite and lepidocrocite (iron oxide hydroxides); siderite (iron carbonate); magnetite (a magnetic form of iron oxide); and vivianite (an iron phosphate). No chlorine was identified in the corrosion, which indicates that active corrosion caused by chloride ion contamination is not currently taking place (Selwyn 2004).

The vivianite was present as blue accretions on both the hide lashing and the organic backing piece. Magnetite was identified in a black area of corrosion under the vivianite on the hide. Siderite, goethite, lepidocrocite and possibly magnetite were found on fragments of corrosion from the blade itself. Goethite and lepidocrocite are formed by oxidation, while vivianite and siderite form in low-oxygen conditions (Selwyn 2004). The presence of both types of corrosion is consistent with the knife having been in a low-oxygen environment under the ice and snow, then subsequently melting out. The presence of vivianite indicates that there was a source of phosphate ions, such as decaying bones or other organic matter, near the iron (Selwyn 2004).

It was originally thought that there might have been intact metal protected from corrosion inside the wooden slot. The fact that the metal has been completely corroded within the slot, and that there is a large buildup of corrosion on the blade's surface (for example, the large lump of corrosion visible in figure 2A), may be due to a process known as crevice corrosion. Crevice corrosion occurs in shielded areas and is based on a concentration difference in dissolved oxygen and metal ions created between the protected and exposed areas of the metal. This produces a so-called concentration cell and accelerates corrosion in the protected area (Callister 2003).

Organic Backing Piece

As well as providing information about the metal blade, the X-radiograph of the knife revealed the structure of the organic backing piece. Figure 3A illustrates that the organic piece is held against the wooden shaft by hide lashing and extends beneath the lashing for about four centimetres. It is tapered to a point at both ends and has a total length of approximately five and a half centimetres. It is relatively opaque in the radiograph for most of its length, with a density similar to portions of the corroded blade. However, both of the tapered ends are more transparent in X-rays. This appearance indicates that the backing piece does not have the same composition along its length. The difference in composition could be due to deterioration at the thinner ends of the piece or, perhaps, to another material filling part of its interior.

Several different areas of the organic backing piece were analysed by Fourier transform infrared spectroscopy. Most samples contained only collagen, which is the major protein found in the connective tissue of mammals. But one sample, from a protected area under the hide lashing, was found to contain a small proportion of hydroxyl apatite (calcium phosphate hydroxide) in addition to collagen. The presence of collagen and hydroxyl apatite confirms that the backing piece is made of bone or antler, as both bone and antler are composed of hydroxyl apatite in a framework of collagen fibrils (O'Connor 1987). Undeteriorated bone contains approximately 70–80 per cent hydroxyl

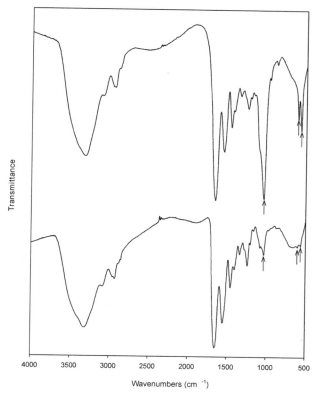

Figure 7. Fourier transform infrared spectra of antler *(top)* and a sample from the backing piece of the man's knife *(bottom)*. The spectrum from the backing piece shows primarily collagen and only a trace of hydroxyl apatite. The major bands for hydroxyl apatite are marked with arrows on the spectra.

apatite, while antler contains about 65 per cent (Chadefaux et al. 2008).

A comparison of the two Fourier transform infrared spectra shown in figure 7 illustrates the very low concentration of hydroxyl apatite in the sample of the backing piece as compared to that of undeteriorated antler. The upper trace is the spectrum of undeteriorated antler, and the lower trace is the spectrum from the backing piece under the lashing. The major bands for hydroxyl apatite (1,037 centimetres⁻¹, and a doublet at 600 and 565 centimetres⁻¹) are marked on the spectra with arrows. The low concentration of hydroxyl apatite indicates that the outer surface of the backing piece is highly demineralized. For the hydroxyl apatite to

have dissolved out of the backing piece, the knife's local environment must have been somewhat acidic (O'Connor 1987).

During sampling of the organic backing piece under the lashing, it was observed that at approximately one millimetre below the surface, the brown collagenous material gave way to a blue, powdery substance. Analysis showed this blue powder to be vivianite—the same iron phosphate compound present in patches on the hide wrapping and on the exterior of the backing piece. This indicates that hydroxyl apatite has dissolved from the backing piece and been replaced by vivianite. The presence of vivianite is not unusual in archaeological bone (O'Connor 1987; Scott and Eggert 2007), and its formation in this case was certainly accentuated by the abundance of iron ions available from the dissolution and subsequent corrosion of the iron blade.

Wooden Handle

The wood of the knife's handle has features consistent with an identification of the softwood hemlock (*Tsuga* sp.). Possible species include Western Hemlock (*T. heterophylla*) and Mountain Hemlock (*T. mertensiana*), which grow in neighbouring regions at lower elevations than the discovery site (see chapter 2, fig. 10). Western Hemlock grows primarily to the west of the coastal mountains but can also occur east of the mountains in major river valleys (Egan 1997). Mountain Hemlock grows at higher elevations; on the northern coast of British Columbia it is found from 400 to 1,000 metres above sea level (Egan 1999).

A scraping from the surface of the wood was analysed using Fourier transform infrared spectroscopy and gas chromatography/mass spectrometry, which showed that both protein and a triglyceride compound (oil or fat) are present. These materials could have been an intentional application to the wood, residues related to the use of the tool or residues from other adjacent materials at the site. No paint or pigment residues were observed on the wood's surface during microscopic examination.

Hide Wrapping

Determination of the animal species used for the hide lashing has not been carried out to date. It would, however, be an interesting avenue for future research, since DNA analysis has been successful for components of the Kwädạy Dän Ts'ìnchị robe (chapter 27, this volume). Measurement of the thermal stability of collagen fibres from the hide wrapping indicated good chemical preservation of individual fibres. The onset shrinkage temperature of the lashing was 45°C. Undeteriorated collagen fibres have an onset temperature of approximately 55°C. With increasing deterioration, the onset drops and lowers to room temperature under conditions of severe chemical breakdown. The 45°C onset for the lashing represents a high level of preservation for a collagenous archaeological material. This result was expected given the site's icy conditions. Unpublished work by the authors showed a similar high level of preservation for sealskin artifacts excavated from the surface layers at Thule occupation sites in the high Arctic. These skins had been exposed to repeated freeze/thaw cycling, but damage was limited to the physical breakdown of the fibre matrix and the shortening of fibre length by ice crystal formation. Similarly, the lashing on the knife shows an opening up of the collagen fibre matrix through the action of water and ice, but little damage appears at the molecular and fibrillar levels.

Conclusions

THE EXAMINATION AND SCIENTIFIC ANALYSIS of the Kwädạy Dän Ts'ìnchị knife determined that a blade made of non-meteoritic iron, likely wrought iron obtained through trade, is held in an asymmetrical, open-sided slot in the wooden handle. The blade is stabilized within the slot by hide lashing and a tapered, organic backing piece made of partially demineralized bone or antler that has been wedged between the wood and the lashing. The off-centre hafting suggests that the slot may have been formed by placing the blade on one side of the handle and lashing a small, secondary piece of wood on top of it. The blade is highly corroded and separating into layers, and X-radiography indicated that there is likely no intact metal inside the wooden handle—perhaps no intact metal remaining at all. The corrosion products found on the knife (goethite, lepidocrocite, magnetite, siderite and vivianite) are consistent with the object having been in a low-oxygen environment under ice and snow and subsequently melting out. The wooden handle is made of hemlock and shows residues of protein and an oil or fat material on its surface.

The Kwädạy Dän Ts'ìnchị knife is significant because it was found directly associated with the young man's body and was very likely one of his possessions. He would have stored the knife in its sheath and may have worn it around his neck or wrist. The information discovered about the materials used to make this object and the technology of its construction adds to the growing body of knowledge about the life and culture of the man who lost his life on a glacier so many years ago.

ACKNOWLEDGEMENTS

Thanks are due to Sheila Greer and Frances Oles of the Champagne and Aishihik First Nations for giving us the opportunity to examine the knife. Several current and former colleagues at the Canadian Conservation Institute contributed to the work: Jeremy Powell and Carl Bigras carried out the X-radiography and photographic documentation; Lyndsie Selwyn provided helpful information about iron corrosion; Judy Logan shared her expertise in interpreting X-radiographs of archaeological materials; and Carole Dignard provided bone and antler reference materials.

26

ANALYSIS, DOCUMENTATION AND CONSERVATION OF THE KWÄDĄY DÄN TS'ÌNCHĮ ROBE, BEAVER-SKIN BAG AND UNDESIGNATED FRAGMENTS

Kjerstin Mackie

The garment found with the remains of the young man discovered melting out of a glacier in Tatshenshini-Alsek Park was a blanket-shaped, rectangular robe that was made by stitching together about 95 ground squirrel pelts (fig. 1). On the robe's fur side there were long fringes originating in the vertical seams, as well as a long, narrow neckband made from a different kind of fur. Two leather thongs were used to tie the robe closed at the chest. On the robe's skin side was a pattern of wide lines of dark red. Evidence of mending is visible—patching, re-stitched seams, replacement furs and mended tears.

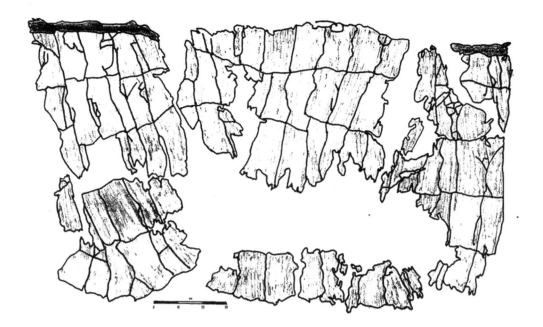

(*Left*) **Figure 1.** Sketch made with tracings of largest robe fragments, arranged in the positions they would have occupied during use.

(*Below*) **Figure 2.** Stitching on contemporary ground squirrel skins showing how two skins were stitched together in the Kwädąy Dän Ts'ìnchi robe. Colour version on page 595.

TO CONSTRUCT THE ORIGINAL ROBE, two ground squirrel pelts were placed fur sides together. The fore and hind limb areas of the pelt were folded in toward the fur, creating a straight edge. Sinew was used to make stitches through all four layers of skin, sewing from the skin side (fig. 2).

Seams were made in a straight line, and approximately 19 pelts were stitched side to side in a row. The excess fore and hind limb areas of the squirrel pelts, still protruding on the fur side of the robe, were then cut into thin fringes. Likely, fringes were present in all vertical seams. Five rows of 19 pelts were stitched together to form a large rectangle. A 5-centimetre strip of unidentified fur was stitched to one long edge of the rectangular robe, from the corners toward the centre. It may originally have extended across the whole length of this edge, but only fragments at the corners remain. When the robe was draped over the shoulders, this fur would have formed either a partial or a complete neckband. Two small patches of unidentified hide, each approximately 5 centimetres square, were stitched to the top neckband of the robe at about 70 centimetres from

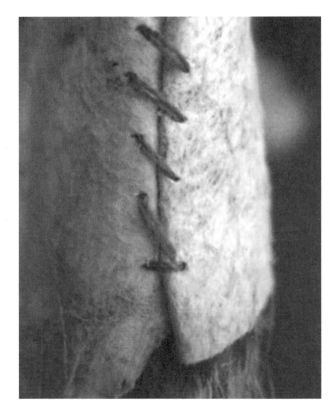

each corner. Each of these patches had a Moose-leather thong looped through it, which could be tied across the chest of the person wearing the robe.

Fur samples were examined by scientists at the Royal British Columbia Museum and the forensic laboratory of Alberta Environment and Parks' fish and wildlife section. Their study results indicate that most of the robe was created using ground squirrel, *Spermophilus* species (Mackie 2005). This was confirmed through DNA analysis (see chapter 27).

The original threads used to stitch the pelts together were made of two-ply, Z-twist sinew with an average diameter of 0.8 millimetres. Most stitches were taken very evenly, about 3 millimetres apart. Several samples taken from these sinew stitches have been identified as Moose (see chapter 27).

Some areas of the robe—for example, in the lower corners and at the centre back—show evidence of it having been repaired during its lifetime with thicker sinew threads, roughly placed stitches and what appear to be patched repairs. All the patches that have been sampled were identified as ground squirrel fur. The repair sinews that were tested appear to originate from a variety of different animals, including mountain goat and at least two species of whale. Records indicate that whale is a known source of sinew used in "sewing and tying" among some First Nations (Boas 1909, 375).

The robe was found frozen into the glacier with the fur side exposed and the skin side folded in toward the centre. Fur robes depicted in historic photographs (for example, plates 19, 45 and 66 in Wyatt 1989) and robes still used in the community (chapter 34) are worn with the fur side facing out and the skin side toward the wearer. The orientation of this robe as found at the site indicates that it was worn or carried with the fur side out.

The remains of a red pigment were present on the skin side of the robe. McClellan (2001, 256) records a Southern Tutchone woman's statement that in the old days, instead of smoke-tanning skins, "they were liberally rubbed with red ochre". On the Kwädąy Dän Ts'ìnchį robe it appears that a pattern of dark red ochre was applied in wide stripes along all horizontal seams, possibly also along the top and bottom edges. The centre

Figure 3. Kwädąy Dän Ts'ìnchį robe fragment from centre back area showing original stitching, patches and red ochre pigment visible from the skin side. Colour version on page 595.

back vertical seam is also coated with a wide stripe of red ochre (fig. 3). These stripes are between four and five centimetres in width. The mineral was identified by Dr. Mary Majer of the University of British Columbia (Mackie 2005), and DNA tests done on the ochre indicate that the carrier mixed with the mineral was salmon (chapter 27).

Some pairs of thin, vertical stripes look as though they could have been made by two fingers dipped into the ochre paste and drawn down from the neck edge to the hem. These stripes are not associated with seams.

Study of the ancient robe has shown numerous similarities with other robes, some preserved in museums (e.g., RBCM 18822, RBCM 2837) and some still in use by the Champagne and Aishihik First Nations and other communities. Comparison between existing garments and the Kwädąy Dän Ts'ìnchį robe has helped to determine the robe's original size and construction.

Arrival at the Museum

THE FUR GARMENT associated with the Kwädąy Dän Ts'ìnchį project arrived at the Royal BC Museum less than three weeks after its initial discovery in 1999. The museum's conservation section was asked to participate in the project in order to forestall deterioration of the materials, provide structural support to the conserved garment, document the garment's construction methods and facilitate analysis of the robe's materials and materials found in association, such as fish remains, pollen, other plant remains and sediment.

Treatment Decisions

TARA GRANT, archaeological conservator at the Canadian Conservation Institute, helped to formulate a plan to clean, freeze-dry and re-house the fragments. After a number of trials using immersion baths, running rinses and a dental scaler, it was decided to proceed with a two-part treatment of rinsing and freeze-drying, followed by analysis and packaging for long-term storage.

First, a running rinse of distilled water was given to each fragment in order to separate surface sediments and potential sample material from the material of the garment. Then the fragments were freeze-dried to enable long-term storage, analysis and preservation. Samples were collected both during and after treatment. The primary goal during treatment was to preserve information that could be learned from studying the robe. No attempt was made to reconstruct or repair the original robe, and treatment decisions reflect this focus on preserving information.

Unfolding and Rinsing

DISTILLED WATER was used to prevent introduction of foreign minerals, pollen and additives. The robe's materials were separated into four discrete, smaller sections for ease of handling and to maintain control over location and context of each fragment. Individual fragments were then separated carefully, using Mylar polyester film to lift sections. As each fragment was eased away from the mass, its location was plotted on a "map" of fragments. Small fragments were placed onto tilted glass plates for rinsing, and larger ones were put into shallow trays of the appropriate size, also tilted at an angle to collect the rinse water without re-depositing soils onto the fragment. Macro-detritus—such as fish remains, plant remains and compacted glacial sediment—was collected by hand when possible and mapped. These materials were placed into sample bags for future analysis. Rinse water was collected and labelled as relating to specific fragments.

Significant wrinkles were eased from the fragments during the rinsing process but were not removed. Sheets of Mylar were used to turn fragments over while wet, reducing the strain on vulnerable wet protein fibres in the skins and sinews. Fragments with adhering soils were brushed gently under water to minimize abrasion. To avoid contaminating the robe with exotic animal species, soft synthetic fibre paintbrushes were used instead of natural fibre brushes. When rinsed, each fragment was lifted clear of the water using a Mylar sheet and then placed skin side down onto a thin piece of spun-bonded fabric tissue. This prevented the surface tension from creating an artificial flatness on the skin side during drying, while leaving the fur to dry with a more natural pile. Some of the innermost fragments were not coated with glacial sediments, and in these cases, the pieces were unfolded without any rinsing.

Freeze-Drying

THE FREEZE-DRYING occurred in the Royal BC Museum's walk-in freezer. No bulking agents were used to pre-treat the fragments, in part to preserve the integrity of any original fats or waxes used in the manufacture of the robe but also because after defrosting, the fragments appeared to be simply wet, not waterlogged. Whether rinsed or not, each fragment was placed onto tissue and then onto a piece of Mylar labelled with its specific number. A tray of

Coroplast (corrugated polypropylene) was cut to size to provide some support for each fragment during handling, and finally, the whole package was wrapped in Tyvek (spun-bonded olefin). This outermost layer provided additional buffering during the freeze-drying process and slowed the rate by creating a micro-layer for the water molecules to pass through. The Tyvek layer also prevented the fragments from being mechanically disturbed during the fan cycle of the freezer and appeared to promote uniform drying over the total surface area of the fragment. Very small fragments were placed on polyester fabric tissue within small boxes constructed of Tyvek paper, which were stacked in the freezer to dry.

The weight and dimensions of each piece were recorded at intervals, removing the packaged fragments and recording the decrease in weight. Approximately six to ten weeks were required for the weight loss to taper off and for the fragments to appear visually dry. Measurements and tracings indicate that the fragments experienced between 2.9 and 5.9 per cent shrinkage during the process, within expectations for freeze-drying (Grew and Neergaard 1988, 139).

Documentation

THE CONSERVATION PROCESS was thoroughly documented using notes, sketches, photography and, in some cases, video and digital recording. Careful and accurate documentation of the robe fragments was vital in trying to understand the appearance and construction of the original robe. Tracings made during the separation of the fragments were essential to orienting the fragments in their original positions in the robe and to documenting shrinkage rates. After treatment, four more tracings were taken of each piece. Each clean, dry fragment was traced to show the following attributes: fringe location; direction of fur; stitch lines and exact length and location of stitches; wrinkles and folds; torn or cut edges; and presence of red ochre pigment.

Using the tracings, it was possible to build up an accurate picture of the shape and size of the entire robe without having to subject the actual fragments to excessive handling. Tracings showing the fur's direction helped in initial orientation and alignment of the fragments. The position of each fragment relative to the robe's outside edge could be determined using the tracings that showed cut, or outside, edges. Fringe location tracings helped build a picture of the number and position of fringes. A clear idea of the original pattern of ochre was understood only after assembling the tracings showing pigment.

Sample-taking

SAMPLES OF SKIN, fur, sinew and wash water were taken from the robe in various locations and noted on the tracings. The samples have been examined by researchers who used them to build a more complete picture of the environmental context of the robe, its use and origins.

Plant materials such as leaves, needles and seeds were found attached to and entangled in the robe, and wash water samples provided pollen. Examination of these materials has revealed evidence of plants characteristic of both coastal and interior biogeoclimatic zones. Botanical remains were studied by Richard Hebda, paleobotanist at the Royal BC Museum; Petra Mudie, Natural Resources Canada; and James Dickson of Glasgow University (see chapter 28).

Fur samples have been examined by Dave Nagorsen of the Royal BC Museum; Mary-Lou Florian, conservator emerita at the Royal BC Museum; and Thomas Packer of the forensic laboratory at the fish and wildlife section of Alberta's Ministry of Environment. These analyses determined the species used for the majority of the robe, and DNA samples analyzed by Dongya Yang and Camilla Speller (see chapter 27) added greatly to the knowledge of other animal species in the robe through testing skin, fur, and original and repair threads and thongs. It was not feasible to sample every fragment or every thread, and not all samples taken yielded conclusive results.

Storage

THE MEDIUM-SIZED FRAGMENTS are housed in a map cabinet, with the larger pieces on temporary shelving in the Royal BC Museum's textile conservation laboratory and the very small pieces in padded Ziploc bags. Each bag is labelled so that individual fragments can be retrieved easily. The laboratory's ambient temperature is stable at 19°C, and the relative humidity is 45 per cent.

After conservation, the robe fragments, beaver-skin bag and undesignated fragments remain in individual pieces. Each piece is clean and dry, resting on, but not attached to, individual storage mounts. The dry robe fragments are less vulnerable to tearing than when they were wet. All the robe fragments are presently stored on individual custom-sized trays of Coroplast, each in a polyethylene bag. They are easily viewed and handled within their storage containers. All the robe fragments, even the larger pieces comprising several animal pelts, can be removed from their storage containers for close examination. The smaller pieces can be turned over easily, while larger pieces incorporating sinew stitching threads are a little more complicated. Sinew is stiff and brittle when dry and presents areas of weakness along all lines of stitching. Two people are required to safely turn the larger pieces over.

Beaver-Skin Bag

A SINGLE PIECE OF BEAVER SKIN was found frozen into the ice next to the left side of the young man's torso. It is the only object found in direct contact with the body—all the other objects were found partly or completely melted out and exposed to the air.

This fragment appears to have consisted of a single piece of fur, at least 28 by 35 centimetres, folded in half longitudinally and stitched up one side, forming a long, narrow rectangle. The bottom may have been knotted or stitched. There were folds near the top, and a thong tie found in association with the fur was probably looped around this area.

The bag appears to be constructed with the fur on the inside of the bag and the skin on the outside. The fur's nap runs from bottom to top, so that placing something inside the bag would require pushing against the fur. Inside the bag were two small balls of material, apparently plants of different species, including two kinds of moss, deciduous plant leaves and possibly lichen.

In general, the skin of the bag was almost completely disintegrated (fig. 4). Only fur remained in situ to show the form and dimensions of the bag (fig. 5). It was extremely difficult to separate the layers enough to identify the shape and size of the bag without causing further damage and distortion, so all measurements are approximate.

Rinsing was not considered, due to the extreme fragility of the beaver-skin bag. Sediments visible on the bag's surfaces had to be removed with forceps to prevent damage to the underlying fur. Numerous samples were removed from the bag in 2001 and 2006, including meltwater; fish remains; beaver hairs; beaver skin tissue; thong; plant materials from inside the bag; material for radiocarbon and isotope testing; and fur, skin, thong and sinew for DNA testing.

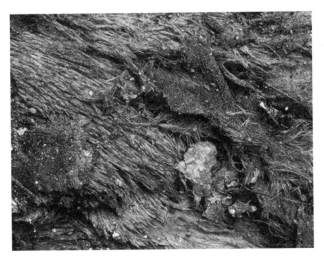

Figure 4. Beaver-skin bag: skin side showing loss of skin; hair follicles are visible in situ. Colour version on page 595.

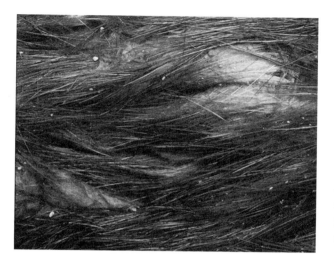

Figure 5. Beaver-skin bag: fur side.

The bag was freeze-dried without rinsing, following the same procedures developed for the robe fragments (described above), and the treated bag is stored on a Coroplast support in a polyethylene bag.

Undesignated Fragments

A NUMBER OF SMALL FRAGMENTS were collected from the glacier surface in 1999, at the same time the robe was removed from the glacier, and over several years of revisiting the site. Some of these fragments may have originally been part of the robe or the bag. Of the more than 1,000 individual fragments, only three were sampled for DNA testing. These materials were all rinsed and freeze-dried, just like the robe fragments. They include several fragments similar to, and possibly part of, the ground squirrel robe; small lengths of seam with sinew stitching; some loose fragments of sinew; numerous clumps of beaver fur (in most cases, missing the skin); large numbers of detached fringes; fragments of leather thong in various thicknesses and lengths; and some very short lengths of what appears to be spun fibres, in one case threaded through a fragment of leather. Most of these fragments are very small, and many are quite deteriorated, due to their exposure and

decomposition on the glacier's surface. They are all dissociated from other pieces, making it very difficult to determine whether they originally belonged to the robe or were once part of other items. It will be difficult to determine how these fragments might be connected to one another or to identified objects (e.g., the robe or beaver-skin bag), but they form a reservoir of information that may, at some point, be studied.

The numerous small fragments are currently stored in polyethylene bags with Tyvek paper supports and Ethafoam (polyethylene foam) padding, while the larger fragments are stored in polyethylene bags with a Coroplast support tray.

Conclusions

THE KWÄDĄY DÄN TS'ÌNCHĮ ROBE would have been soft and flexible during its use. When the masses of fragments arrived in the lab, they were frozen solid. Upon thawing, the fragments were flexible but wet and vulnerable to tearing. After freeze-drying, the fragments were stiff and slightly smaller than their original dimensions. The sinew stitching threads and thongs were brittle and inflexible. The intensity of the red ochre markings was reduced after conservation, though recorded in images and on tracings.

Nevertheless, the unfolded fragments can now be understood as a garment and not as an unidentifiable mass of fur. The freeze-drying process minimized shrinkage and resulted in robe fragments that can be stored without complex equipment or continual monitoring. Handling is now possible by a single person, except for three very large pieces, which require two people to lift and turn. Cleaning has allowed for viewing and research of the original garment's pattern, construction methods, materials and repairs without the necessity for specialized handling. Any original oil, fat or wax residues have been unadulterated and retained.

Conservation has forestalled deterioration of the robe, beaver-skin bag and undesignated fragments, but they remain in pieces, with no attempt to reconstruct them into a complete state. The fragments, stored in containers that support and protect them, are available to people wanting to study and understand the original creator's materials and methods. Maintaining stable storage conditions should preserve the fragments indefinitely. This material is scheduled to be returned to Champagne and Aishihik territory in the autumn of 2017.

Documentation, sampling and study of the robe and the bag have revealed new insights into their history—and into the life of the young man who carried them. The people who made the Kwädąy Dän Ts'ìnchį robe and bag have given us new knowledge about an ancient way of life.

ACKNOWLEDGEMENTS

I would like to acknowledge the Champagne and Aishihik First Nations for their continuous support and interest in the project, the Royal BC Museum for facilitating the work, and the individual people who worked with me. I am so grateful for the privilege of collaborating on this project, I have learned to see much more than I was taught to look for in school.

27

ANCIENT DNA ANALYSIS OF THE KWÄDĄY DÄN TS'ÌNCHĮ ROBE AND SEWN BAG

Camilla F. Speller, Kjerstin Mackie, Alexander P. Mackie and Dongya Y. Yang

S ince the 1999 discovery of the Kwädąy Dän Ts'ínchį man (Long Ago Person Found) in Tatshenshini-Alsek Park, a wealth of studies have been undertaken to learn as much as possible about this individual and his physical and cultural environment. The incredible preservation of both the body and the associated belongings have allowed for detailed archaeometric and molecular research to investigate this young man's short- and long-term diet, place of residence, population affinity and even clan identity, as well as the approximate age of the remains (Dickson 2004; Martin 2001; Monsalve et al. 2002; Richards 2007; chapter 37). Our study follows a similar vein, by applying molecular analysis to several items associated with the Kwädąy Dän Ts'ínchį individual, including a robe and bag made from animal pelts.

BASED ON HAIR AND PELT MORPHOLOGY, the robe and bag had been tentatively identified as being constructed from ground squirrel and beaver pelts, respectively. Much work has already gone into understanding the robe's construction, materials and original appearance (Mackie 2004, 2005, 2008). The main aims of this DNA-based study were to determine precisely which animal species were used in the construction of the robe and bag and to explore how these items were made and repaired.

The discovery site, located in Tatshenshini-Alsek Park, is within the traditional territory of the Champagne and Aishihik First Nations (CAFN). The Tatshenshini basin has been home to both Athapaskan speakers (Southern Tutchone) and, closer to the ocean, the Tlingit. There has historically been a close relationship between the Tutchone and Tlingit groups, and this continues today. Although the ancient body was discovered 80 kilometres inland, botanical studies of his intestinal tract, as well as isotopic analysis of bone, muscle and hair, have indicated a mixed reliance on both marine and terrestrial resources at different times in his life (Dickson 2004; Richards 2007; chapter 6). Thus, an additional aim of this study was to determine whether the animal species used to make the robe and bag included both coastal and inland species—a further means to explore the young man's connections to these communities.

The Robe and Bag

WHEN THE REMAINS of the Kwädą̈y Dän Ts'ínchi individual were discovered and excavated in August 1999, numerous hide and fur fragments were found in close association with the body. The main mass of the robe was collected on the initial site assessment visit, with minimal forensic controls, and was kept frozen in a folded mass in Whitehorse. The other fur and hide items, including many fragments of the robe, were collected by personnel wearing Tyvek suits and latex gloves, and these pieces were immediately wrapped in sterile bags and stored in sub-zero temperatures until they were transferred to the Royal British Columbia

Museum in Victoria, BC (Beattie et al. 2000). The robe's hide and fur fragments were then conserved at the museum by Kjerstin Mackie and other staff members. (For more on treatment and storage conditions prior to sampling, see chapter 26; detailed conservation methods are recorded in Mackie 2004, 2005, 2008).

Many of the large hide fragments were first rinsed with distilled water in order to clean them and to collect any glacial sediments, pollen samples and plant and fish remains that might aid in reconstructing the young man's life history and physical environment before death (chapter 28). In order to avoid contaminating the robe with exotic species, only soft synthetic brushes were used to brush off water and soils. The wet robe samples were then placed on Mylar sheets and wrapped in Tyvek before being freeze-dried in a secure walk-in freezer for approximately 10 weeks. Some smaller robe fragments were not cleaned and dried but stored in sub-zero conditions. The bag fragments were unfolded but not rinsed, due to the disintegrated condition of the skin, and then freeze-dried in a similar manner.

Dave Nagorsen, the Royal BC Museum mammal curator, and Mary-Lou Florian, conservation scientist emerita at the museum, visually examined the robe and bag fragments, as well as hair samples. They visually identified the pelts making up the majority of the robe as ground squirrel (*Spermophilus* sp., now *Urocitellus* sp.) (Florian, n.d.), and the bag pelts were visually identified as American beaver (*Castor canadensis*).

The robe's original appearance would likely have been similar to gopher robes worn and valued by both coastal and interior First Nations groups in historic times (Emmons 1991). Made of around 95 individual ground squirrels (Mackie 2004), the robe would have had a rectangular shape, approximately 204 centimetres by 110 centimetres. It had a five-centimetre-wide strip of fur at the top, tentatively identified as cervid or bovid fur, and a leather thong was threaded through a patch of thicker leather along this top edge. The thong was likely one of two used to tie the robe across the chest. Several areas on the robe have apparently undergone repair; some appear to be patched with thinner, darker squirrel pelts and coarsely sewn with thicker sinew (Mackie 2004). While tentative

Figure 1. Diagram indicating the location of some of the DNA samples taken from the robe.

species identifications were made for the pelts and the border fur, morphological identification of the animal sinew used for both the original and repair stitching, and for the tying thong, could not be done. This was also the case for the beaver-skin bag's thong and stitching.

Ancient DNA analysis has often been used to accurately identify species when traditional macroscopic or microscopic techniques are inadequate (Nicholls et al. 2003; Yang et al. 2004, 2005). The majority of ancient DNA studies focus on bones or teeth, since these hard tissues are most often recovered within the archaeological record and since DNA generally preserves well within the hard tissue matrix of bone or enamel (Burger et al. 1999; Haynes et al. 2002; Lassen et al. 1994). The unusual preservation of the fur garment and bag within the glacier, however, has provided a relatively rare opportunity to apply DNA analysis to ancient clothing. Previous studies have been successful at extracting DNA from archaeological animal-skin products, such as bookbindings and parchments (Burger et al. 2001; Teasdale et al. 2015), as well as some archaeological leather goods (Vuissoz et al. 2007).

Additionally, recent work has shown that animal hair can also be an excellent source of mitochondrial DNA, especially when preserved in sub-zero temperatures (Gilbert et al. 2007, 2004). In this study, DNA analysis provided a unique test to confirm the species identifications for the robe and bag pelts and allowed for identification of the species used for the border fur, sinew stitching and leather thongs.

Ancient DNA Analysis

IN MAY 2006, 17 samples were taken from various areas of the robe and bag using clean scissors or tweezers (the sample types and their Royal BC Museum-assigned numbers can be found in table 1). Sample types included fur, skin, leather thong and sinew from different parts of the robe and bag; the locations of some of the robe samples can be seen in figure 1. These samples were frozen in sterile Eppendorf tubes until they were ready to be processed in the ancient DNA laboratory in the Archaeology Department of Simon Fraser University (SFU).

Table 1. Robe and bag samples procured from the Royal BC Museum for ancient DNA analysis (aDNA).

Lab Code	RBCM #	Item	Sample type	Presumed species
K1	28.34b	Robe	Fur and skin	Ground squirrel
K2	28.27c	Robe – fringe	Fur and skin	Ground squirrel
K3	24.8a	Robe – repaired area	Fur and skin	Ground squirrel
K4	28.27d	Robe	Fur	Ground squirrel
K5	R31b	Bag	Fur	Beaver
K6	R31d	Bag	Fur and skin	Beaver
K7	28.34d	Robe – repaired area	Fur	Ground squirrel
K8	24.8c	Robe	Sinew and fur	Unknown
K9	28.34e	Robe – ochre-covered area	Fur	Ground squirrel
K10	28.34a	Robe	Sinew	Unknown
K11	28.34c	Robe	Thong	Unknown
K12	28.27a	Robe – top strip	Hair	Unknown
K13	28.27b	Robe – top strip	Skin	Unknown
K14	24.8b	Robe – repaired area	Sinew	Unknown
K15	24.8d	Robe	Sinew	Unknown
K16	31a	Bag	Sinew	Unknown
K17	31c	Bag	Thong	Unknown

DNA Extractions and PCR Amplifications

DUE TO THE FRAGILE NATURE of the hair and sinew samples, the vigorous decontamination techniques typically used on bone or other hard tissue samples were not undertaken for initial extractions. Hair, sinew and leather samples were placed in a small weighboat with one millilitre of double distilled, ultra-pure water and then irradiated in a UV crosslinker for 30 minutes. A modified silica-spin column method was used for DNA extraction (Yang et al. 1998). The samples were placed in four to five millilitres of lysis solution designed to release DNA from hair (0.5 per cent SDS, 0.5 mg/ml proteinase K, 10 mg/ml DTT, 0.05 M EDTA) and incubated overnight at 50°C in a rotating hybridization oven. Approximately three millilitres of the lysis solution was then concentrated using a Millipore Amicon centrifugal filter Ultra-4 and purified using a

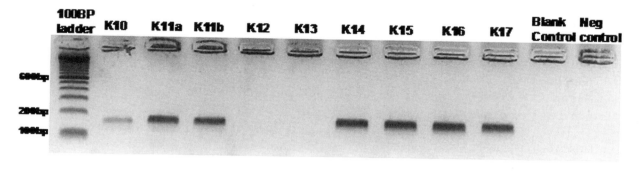

Figure 2. Electrophoresis gel displaying some positively amplified samples using the universal 12S primer set designed to amplify fragments of approximately 120 bp. Note the lack of amplification in K12 and K13; these two samples consistently failed to amplify, despite repeated efforts with a variety of primers. Colour version on page 596.

QIAGEN QIAquick column, from which approximately 100 µl of DNA solution were collected for each extract.

Repeat extractions were performed on five of the samples (K11, K13, K14, K15, K17) using a modified decontamination technique that has been found to be successful for hair extractions (Gilbert et al. 2004). These samples were immersed in a 3 per cent sodium hypochlorite solution (50 per cent commercial bleach solution) for 20 seconds, rinsed thoroughly in double-distilled, ultra-pure water and irradiated in a UV crosslinker for 30 minutes with one millilitre of ultra-pure water. The extraction process was followed using the same protocols as the initial extraction.

The DNA analysis targeted short fragments of mitochondrial DNA. Mitochondrial markers were selected for analysis since mtDNA is present in higher copy numbers than nuclear DNA, making it easier to recover from degraded archaeological remains (O'Rourke et al. 1996). Additionally, some of the samples consisted solely of fur, with no adhering skin or tissue. Although the hair shafts may retain adequate quantities of mitochondrial DNA, nuclear DNA may be severely degraded unless the hair root is present (Higuchi et al. 1988; Nozawa et al. 1999). Samples of ground squirrel fur from the robe and beaver fur and skin from the bag were PCR-amplified using species- or genus-specific primers designed to amplify short cytochrome b gene and/or control region fragments. The samples of sinew, thong, skin and fur from unknown species were PCR-

amplified with a variety of primers targeting pinniped, canid, ursidae, bovid, cetacean, mustelid and ovid/caprid cytochrome b or control region fragments, as well as two universal primers designed to amplify 12S gene fragments from a variety of large mammals.

PCR amplifications were conducted in an Eppendorf Mastercycler Personal in a 30 µl reaction volume containing 50 mM KCl, 10 mM Tris-HCl, 2.5 mM MgCl$_2$, 0.2 mM dNTP, 1.0 mg/ml BSA, 0.3 µM each primer, 3.0 µl DNA sample and 1.5–3.0 U AmpliTaq Gold or AmpliTaq Gold LD (Applied Biosystems). PCR was run for 60 cycles at 94°C for 30 seconds (denaturing), 52–55°C for 30 seconds (annealing) and 72°C extension for 40 seconds, with an initial 12-minute denaturing period at 95°C. Five µl of PCR product were visualized via electrophoresis on a 2 per cent agarose gel using SYBR Green I (Invitrogen) staining. Figure 2 is an electrophoresis gel displaying some of the positively amplified samples. PCR products were purified using QIAGEN's MinElute purification kits and were subjected to direct sequencing.

Species Identifications

DIRECT SEQUENCING was carried out using combinations of forward and reverse primers to maximize sequence information from overlapping primer sets. The obtained electropherograms were visually edited and subjected to GenBank BLAST searches for initial species identification. Multiple alignments of the sample sequences and published mtDNA reference sequences were conducted using ChromasPro and ClustalW through BioEdit software (for more on these technologies, see Thompson et al. 1994; Hall 1999). A confident species identity was assigned to a sample if it matched identically, or was within only a few base pair difference, from published reference sequences and if no other evidence—including reproducibility tests or additional sequencing of the same sample—indicated a different species. When multiple species sequences were recovered from a single sample, a tentative species identity was suggested when alternative lines of evidence pointed to a particular species (see discussion).

Contamination Controls

DUE TO DEGRADATION of DNA over time, DNA from modern sources can very easily overwhelm the low quantities of DNA present in ancient remains and lead to false positive results (Poinar 2003). Comprehensive controls were taken during every step of laboratory analysis to reduce the risk of contamination, including 1) the use of a dedicated ancient DNA lab, where no modern DNA samples have ever been processed and which was equipped with positive pressure and UV sources; 2) the use of protective clothing, including masks, Tyvek suits and disposable gloves; and 3) the inclusion of multiple blank extractions and negative controls for contamination detection.

Species Identities from Ancient DNA Analysis

MITOCHONDRIAL DNA SEQUENCES for a variety of species were obtained from the robe and bag samples, and table 2 (page 393) shows a synopsis of the species identities for each sample. Generally, the DNA seemed to be relatively well preserved in the samples, with a number of samples yielding DNA fragments over 250 bp

in length. Of the 17 samples tested, only two failed to yield reproducible DNA sequences (K12 and K13).

Ground Squirrel Robe

ALL ROBE SAMPLES taken from the squirrel pelts (fur and skin samples K1–4, K7 and K9) returned mitochondrial sequences that matched most closely with an Arctic Ground Squirrel subspecies (*Spermophilus parryii plesius*). Two different mitochondrial loci were tested (cytochrome b and D-loop fragments); both fragments were sequenced for each robe fur or skin sample, with all samples yielding the same or very similar haplotypes.

Sample K8, consisting of both sinew and attached squirrel fur, yielded ground squirrel as well as Moose (*Alces alces*) mitochondrial DNA sequences, revealing that the sinew and the fur pelt came from two different species. The other samples of original sinew from the robe also yielded Moose DNA (K10), and DNA recovered from the tying thong (K11) pointed to the same species identity. These three samples (K8, K10, K11) repeatedly returned *Alces alces* mitochondrial sequences when amplified with the two different 12S primer sets.

As mentioned previously, a portion of the robe seemed to have been repaired with a coarser sinew than that used in the original manufacture. When amplified with multiple 12S and cytochrome b primers, this sample (K14) consistently returned Blue Whale (*Balaenoptera musculus*) mitochondrial sequences. A second extraction of this sample was done, and both 12S and cytochrome b amplification confirmed the original species identity.

One piece of sinew (K15) initially identified as original sinew, demonstrated unusual species identity results. The first extraction of sample K15 also consistently returned 12S and cytochrome b whale sequences, though from Humpback Whale (*Megaptera novaeangliae*) rather than Blue Whale. Like sample K14, this sample was extracted a second time, using a more vigorous decontamination technique, in order to confirm the cetacean species identity. The second extraction also yielded Humpback Whale sequences when amplified with cetacean cytochrome b primers, though Mountain Goat (*Oreamnos americanus*) DNA was also recovered when the sample was amplified with the 12S primers and caprid cytochrome b primer sets.

The species used to create the bordering strip of fur at the top of the robe could not be identified. Two samples from this area, one of hair (K12) and one of skin (K13), were amplified using a variety of primer sets, including pinniped, canid, ursidae, bovid, mustelid, cetacean, ovid/caprid and universal 12S primers, yet they consistently failed to yield any reproducible DNA sequences. This area of the robe was re-sampled (including both fur and skin) and re-analysed. The second set of DNA analyses also failed to produce any successful PCR amplifications.

Beaver-Skin Bag

AS EXPECTED, the two fur samples from the bag that had been visually identified as beaver (K5 and K6) returned cytochrome b sequences most closely related to American Beaver (*Castor canadensis*), confirming the morphological identification of the fur pelts. The sinew sample (K16) and the tying thong (K17) for the beaver-skin bag yielded Moose (*Alces alces*) mitochondrial DNA sequences when amplified with the two 12S primers. Interestingly, when amplified using beaver-specific primers, these Moose sinew and leather thong samples returned *Castor canadensis* sequences, pointing to the unavoidable transfer of beaver DNA from the pelts to the associated sinew stitching and leather thong through direct contact.

Discussion

DUE TO THE EASE by which ancient samples may be contaminated with modern DNA, all ancient DNA results must be carefully scrutinized to ensure they are authentic. This study faces a special challenge in that the modified conditions of the remains means

species identities cannot be readily determined using morphological analysis. Consequently, multiple lines of evidence must be incorporated to test the authenticity of the results.

Several factors support the species identities obtained in this study (each of which will be described in more detail below): 1) a dedicated facility was used for ancient DNA analysis; 2) no PCR amplifications of expected band length were ever observed in blank extracts and negative controls; 3) the hypothesized species identifications of ground squirrel and beaver were confirmed using DNA analysis; 4) multiple species were amplified, corresponding to animals described in ethnographic documentation; 5) species identities were confirmed through repeat extractions and multiple DNA fragments; 6) PCR reactions designed to amplify DNA from common domestic species (e.g., cow) failed to produce amplifications of common domestic species from the ancient samples; and 7) repeat extractions confirmed initial species identities (but see discussion on sample K15).

The Depositional Environment

The excellent preservation of the robe and bag, and their relatively young age (*ca.* 300 years old), suggests that DNA would still be well preserved in the material. As well, several studies have demonstrated that DNA preserves better in cold temperatures (Lindahl 1993) and has been known to persist in permafrost conditions for over 100,000 years (Lydolph et al. 2005, Willerslev et al. 2004). Quite likely, the maintenance of sub-zero temperatures since the items were deposited between 1720 and 1850 AD (chapter 6) would have drastically reduced the rate of DNA degradation. The successful recovery of DNA from the young man's body, yielding a haplotype consistent with First Nations populations from that region (Monsalve et al. 2002), further supports that authentic DNA has been preserved in his belongings. More recent work by Monsalve et al. has also indicated the incredible morphological preservation of the body's soft tissue (see chapter 18).

Samples K12 and K13 did not yield any PCR amplifications. These samples, taken from the bordering strip of fur at the top of the robe, could not be identified, despite the fact that they were amplified using myriad primer sets. Three common reasons for amplification failure are a lack of DNA template, PCR inhibition and non-specific primers. Considering the excellent DNA preservation demonstrated in the other robe and bag samples, it is unlikely that DNA degradation, or a lack of DNA template, was an issue for these samples. However, the tanning processes that render leathers more stable and supple are detrimental to DNA preservation (Vuissoz et al. 2007). It is possible that a tanning process used on this bordering section of fur significantly degraded the DNA. Tanning processes may also increase the likelihood of PCR inhibitors, such as tannins or organic compounds (Pääbo et al. 1989, Wilson 1997). These particular samples were specifically tested for inhibition by spiking positive samples with their DNA extracts; amplifications of the positive control samples were not affected, suggesting that inhibition was likely not a factor in PCR failure. Another possibility is that the primers used in PCR amplification did not target the correct species.

Contamination Controls

Vigorous contamination controls were put into place during every step of this research, beginning with sample collection and conservation and right through the DNA extraction process. The general lack of amplification of human or domestic animal DNA strongly demonstrates that contamination from human handling, or from lab reagents and materials, was very low. Only one sample (K12) produced two separate amplifications—which were found to match dog and human DNA, respectively—when amplified with a universal 12S primer. This sample yielded only these two sequences, despite being included in at least 17 different amplifications with various primers (including canid-specific primers, which failed to amplify). These infrequent amplifications were likely the result of sporadic human and common domestic

animal contamination that may occasionally occur in lab reagents and materials (Leonard et al. 2007). One cannot rule out, though, the possibility that this DNA may have been introduced onto the robe at any time since its manufacture, including during the time of its use. Overall, the extremely low levels of contaminant DNA amplification indicate that systematic lab contamination was not an issue for this study.

Each sample was amplified multiple times, with multiple different primer sets. The majority of the samples consistently yielded sequences with the same species identity, even when targeting different portions of the mtDNA genome. Multiple blank extracts and negative controls were included in the analysis; none of these controls produced DNA amplifications of expected length (some controls produced primer-dimers during PCR amplifications), further demonstrating that systematic lab contamination, or contamination caused by post-PCR carryover and/or human handling, was not an issue for this study.

A few samples yielded sequences from two different species (table 2). Typically in ancient DNA studies, the amplification of several different species from the same sample would point to the presence of contamination. The multiple species identities in this study, however, seem to support the authenticity of the results. For example, sample K8 was made up of both sinew and squirrel fur and was therefore expected to yield both squirrel and Moose DNA sequences. Another example can be seen in samples K16 and K17, sinew and leather taken from the beaver-skin bag. It was not unexpected that they should yield Moose DNA (likely the main constituent of the samples) as well as beaver DNA. Given the close association of the beaver pelt with the stitching and leather thong over the course of several hundred years, it is expected that some DNA transference would take place between samples that are in direct contact with each other. The authenticity of the results is demonstrated by the fact that beaver DNA was amplified in the sinew and leather samples when using beaver-specific primers, while more conservative 12S primers, targeting a wide variety of mammals, preferentially amplified the more abundant authentic Moose DNA.

390

Biogeographic Evidence

The natural ranges and habitats of the species identified in this study also support the species identities determined using ancient DNA. All the identified animals can be found within the general vicinity of Tatshenshini-Alsek Park and the northwest Pacific coast. Arctic Ground Squirrels are distributed across eastern Siberia, Alaska and northern Canada, living in Arctic and alpine tundra. Eight subspecies are recognized: *Spermophilus parryii ablusus, kennicottii, kodiacensis, lyratus, nebulicola, osgoodi, parryii* and *plesius*. These subspecies form fairly distinctive phylogeographic clades, in part due to the glacial history of the Nearctic region and enhanced by the high levels of female philopatry and low levels of gene flow (Eddingsaas et al. 2004). Tatshenshini-Alsek Park lies within the home range of the subspecies *plesius* (the same subspecies identified from the robe samples), which occupies the tundra of southeastern Alaska and the Yukon.

The American Beaver, identified from the bag pelts, is distributed around North America, including the Pacific Northwest coast and interior. Although there are several subspecies of beaver, no genetic studies of modern populations have, to date, characterized their mitochondrial haplotypes, so a subspecies identity for the beaver samples is not yet possible. Moose (*Alces alces*) and Mountain Goat (*Oreamnos americanus*) inhabit the study area, and the two marine species (Blue Whale and Humpback Whale) also have ranges within the north Pacific.

Ethnographic Evidence

A cursory review of the ethnographic evidence finds that the study's species identity results are congruent with some records for the Tlingit and Southern Tutchone-speaking groups (de Laguna 1972; Emmons 1991), and there is documentation for the use of Arctic Ground Squirrel, Moose and whale products. Ground squirrel robes were highly valued by both the Tlingit and Southern Tutchone, and the pelts were often traded either individually or as complete robes. The use of Moose hide and sinew, obtained through trade with interior peoples, has also been recorded in Tlingit groups, and there are indications of the use of whale sinew on the coast—and of its trade to the interior.

Unexpected Results

Although many lines of evidence support the authenticity of the DNA-based species identities, this study also yielded some unexpected results—that is, the recovery of two different species' mitochondrial DNA for sample K15, a piece of robe sinew. This sample was extracted twice. The initial extraction yielded only Humpback Whale sequences, while the second extraction yielded sequences from both Humpback Whale and Mountain Goat DNA. Several possibilities may account for these results: 1) either the Humpback Whale or Mountain Goat (or both) represent contaminant DNA introduced onto the robe during excavation, conservation or DNA extraction; or 2) both of these species identities are correct, and two species are present in this sinew sample.

A few details may support the latter conclusion. First, several lines of evidence (as described above) indicate that there has been no systematic contamination of the samples. Second, sample K15 consistently and repeatedly amplified sequences from these two species, and only these two species. If whale or Mountain Goat DNA had contaminated the robe, one would expect to find it in multiple samples and on multiple areas of the robe. This was not the case, as Mountain Goat and Humpback Whale DNA were only amplified from sample K15. Finally, visual examination of sample K15 indicated that it was made up of two strands of sinew wrapped together. These two strands were extracted separately, one each in the first and second extraction, respectively. It is possible that one of these sinew strands was from Humpback Whale and the other from Mountain Goat, and that the initial DNA extraction targeted a piece of whale sinew while the second targeted a piece of Mountain Goat sinew that had some adhering whale tissue. Although this scenario may seem unlikely, especially considering that the other robe and bag sinew seems to be made of Moose, it should be noted that the K15 sinew sample came from near the bottom of the robe—an area that would have been subject to greater wear and tear and would therefore have more repair stitching than other areas.

In order to clarify these ambiguous species results, an additional piece of sinew was obtained from a portion of the robe that had not been cleaned and freeze-dried but remained frozen and untreated since its recovery from the glacier. This subsequent sample yielded Humpback Whale DNA (using both 12S and cytochrome b primer) and did not yield DNA when amplified with Mountain Goat primers. This result strongly suggests that Humpback Whale sinew was indeed used in the construction or repair of the robe. That Mountain Goat DNA was not obtained from this particular piece of sinew does not preclude the presence of goat sinew used on other parts of the robe.

Summary

THE OVERALL AIM OF THIS STUDY was to identify which species were used in the construction and repair of the Kwädąy Dän Ts'ìnchį robe and bag. The DNA analysis yielded a variety of animal species, some of which are highly supported by biogeographic and ethnographic evidence. The DNA analysis confirmed that the bulk of the robe was made up of Arctic Ground Squirrel pelts (subspecies *plesius*). The thong used for tying the robe was likely made from Moose hide, with at least some of the original stitching made with Moose sinew. Some repair stitching, observably thicker than the original stitching, was likely made from Blue Whale sinew, while other stitching (either original or repair) may have been made using Humpback Whale sinew. Additional pieces of sinew from the robe may have been made from Mountain Goat, though this species identity needs to be repeated and confirmed. The results also confirmed that the bag was indeed composed of American Beaver pelts and that its stitching and

thongs were also made of Moose sinew and hide. The identification of unexpected species, such as whale, will allow researchers to re-analyse how the robe was constructed and eventually enable us to piece together a high-resolution picture of the robe's manufacture and repair technique.

A further aim of the study was to determine if the animal species used in the construction of the robe and bag could indicate any relationship to the coast or interior communities. The robe's original construction—ground squirrel pelts, Moose hide and sinew—incorporates animals typically found in the interior. Species results from the bag (Moose and beaver) support these conclusions. Thus, based on the species used in the original manufacture of the robe and bag, a reasonable conclusion is that these items were made in the interior. The Blue Whale sinew (and perhaps the Humpback Whale as well) used in the robe patches suggests it was repaired on the coast or by someone that had sinew obtained from the coast.

The ancient DNA analysis of these items has helped determine the species used in manufacturing the robe and has reinforced the conclusions drawn by other researchers indicating the young man's close ties to both coastal and interior communities.

ACKNOWLEDGEMENTS

Many thanks to the Champagne and Aishihik First Nations members and representatives, in particular to Frances Oles, and to Sheila Greer for her constructive comments on the manuscript. We are also grateful for the contributions of the Kwädąy Dän Ts'ìnchį Management Group. Our thanks go to Ursula Arndt, Krista McGrath and Sarah Padilla for technical assistance in the SFU ancient DNA laboratory and for discussion and assistance during the design of the project. We also thank the anonymous reviewers of this paper for their constructive comments. This study was supported by the BC Archaeology Branch, the Royal BC Museum, SSHRC Canada Graduate Scholarship (PhD, Speller) and a research grant from the Social Sciences and Humanities Research Council of Canada (Yang).

Table 2. Species identities of the robe and bag samples.

aDNA#	Item and sample type	Species
K1	Fur and skin (robe)	Arctic Ground Squirrel (*Spermophilus plesius plesius*)
K2	Fur and skin (robe fringe)	Arctic Ground Squirrel (*S. p. plesius*)
K3	Fur and skin (robe repair area)	Arctic Ground Squirrel (*S. p. plesius*)
K4	Fur (robe)	Arctic Ground Squirrel (*S. p. plesius*)
K5	Fur (bag)	American Beaver (*Castor canadensis*)
K6	Fur and skin (bag)	American Beaver (*Castor canadensis*)
K7	Fur (robe repair area)	Arctic Ground Squirrel (*S. p. plesius*)
K8	Sinew and fur (robe)	Moose (*Alces alces*) and Arctic Ground Squirrel (*S. p. plesius*)
K9	Fur (robe with ochre)	Arctic Ground Squirrel (*S. p. plesius*)
K10	Sinew (robe)	Moose (*Alces alces*)
K11	Thong (robe)	Moose (*Alces alces*) – tentative
K12	Hair (robe border)	Consistent PCR amplification failure
K13	Skin (robe border)	Consistent PCR amplification failure
K14	Sinew (robe repair area)	Blue Whale (*Balaenoptera musculus*)
K15	Sinew (robe)	Humpback Whale (*Megaptera novaeangliae*)* Mountain Goat (*Oreamnos americanus*)*
K16	Sinew (bag)	Moose (*Alces alces*)
K17	Thong (bag)	Moose (*Alces alces*)

* The species identity for sample K15 could not be confidently identified.

28

FORENSIC BOTANY OF THE KWÄDĄY DÄN TS'ÌNCHĮ GROUND SQUIRREL ROBE

Richard J. Hebda, James H. Dickson and Petra J. Mudie

The Kwädąy Dän Ts'ìnchį site included a number of cultural artifacts, the largest being a ground squirrel–fur item informally called a robe (Mackie 2005) (fig. 1). This object may not be a traditional subject for botanical analysis, but it has a large surface area and the potential for a complex history. It is also an ideal material for retaining plant remains, especially microscopic pollen, spores and diatoms. Many small plant fragments, including conifer needles and moss stems and leaves, can adhere to or become tangled in textile fibres and the fur fibres on animal hides. Even the smooth, inner surface of prepared ground squirrel skins has oils and fats to which plant parts will stick. Items of clothing such as the robe also come into direct contact with soils, mud and water, which may serve as sources of adhering material, each with a distinctive pollen, spore and microalgal signature.

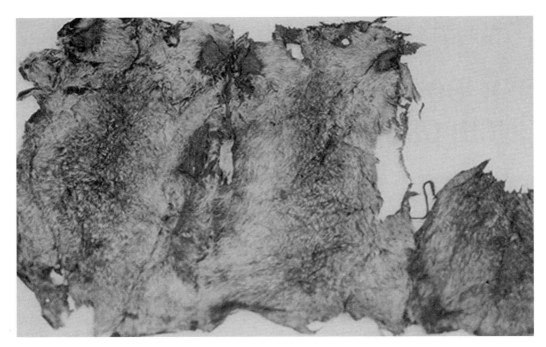

Figure 1. Robe fragment showing general conditions of preservation and Arctic Ground Squirrel fur. Colour version on page 596.

THROUGH ITS VARIOUS EXPOSURES to the natural environment and, in some cases, to plants being used directly by the wearer or bearer, the robe becomes a kind of archive of the environments to which it has been exposed—and, therefore, of the environments to which the object's user has been exposed as well.

Botanical remains are now being more widely used in forensic studies (Coyle 2005; Bryant and Jones 2006; Mathewes 2006; Oeggl 2009). Botanical studies of clothing are largely concerned with identifying what the garments are made of. Typically, wood sections of artifacts are studied in order to learn what constituent tree and shrub species were used, and plant fibres are examined microscopically to identify herbaceous plants (see Bernick chapter 21 as an example). Studies of adhering plant material are far less common, despite their potential to reveal much about the history of the item, its use and its user(s).

In the case of Ötzi, the Tyrolean glacier mummy from 5,200 years ago, pollen from the clothes was studied by Groenman-van Waateringe (1993, 1998). She extracted and counted pollen from "hairs belonging to the clothes" (Groenman-van Waateringe 1993, 121). The pollen spectrum led her to conclude that Ötzi's cultural background had to be found in the cultural groups north of the Alps—a conclusion that very few, if any at all, would now accept.

Dickson et al. (1996) listed about 30 species of mosses recovered from Ötzi's clothes (jacket, leggings and cap). Some could have derived from the immediate area of the archaeological site, but others certainly came from low to moderate altitudes. Most notably, *Neckera complanata* is one of many indications of Ötzi having a domicile to the south and of his last journey having come from the south (Dickson 2000, 2003; Dickson et al. 2005). Heiss and Oeggl (2009) demonstrated that many of the plant remains found with or near Ötzi were from non-local sources.

Here we describe the results of forensic studies of the botanical remains associated with the Kwädąy Dän Ts'ìnchį robe. We report on both microscopic and macroscopic remains, interpreting them with respect to the history of the robe and, by inference, to the history and travels of the man. The plant remains are particularly instructive with respect to his last days, just before he perished on the glacier.

The number of microfossil samples examined is unprecedented for a study of this type. Our work demonstrates how abundantly plant remains may occur in association with prehistoric clothing and shows the potential of these plant remains for use in a range of forensic studies, whether of ancient or contemporary items.

Methods

The Kwädąy Dän Ts'ìnchį Robe and Its Recovery

In any forensic investigation, particularly one involving surface microscopic remains, an account of the context of occurrence and recovery from the study site is central to the interpretation of results. Description of the original environment of occurrence and subsequent handling provides details of a study item's most recent history of exposure to sources of the botanical material found on it. In a way, it allows investigators to peel away the layers or ascribe remains to known environmental exposures, leaving behind the residual material to be explained by the inferred history of the object. In this way, it may be possible to answer questions such as where the item had been and how it was used before it came to rest at the discovery site.

The robe was observed immediately at the time of the site's discovery, being both melted out of and frozen to the ice (Beattie et al. 2000) (fig. 2). It was positioned near the upper end of the man's torso, where the head would have been. The object, with some mud adhering to it, was collected in a mass from the glacial ice and placed in a box. It included several fragments, some of which became detached later when thawed (Mackie 2005). When recovered, the robe mass fragments were tightly bunched and folded, so only the outside surface of the robe mass was exposed to the ambient atmosphere; other parts were not exposed.

Figure 2. Robe mass before conservation. Colour version on page 597.

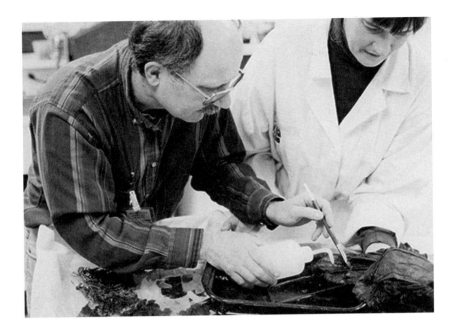

Figure 3. Richard Hebda and Kjerstin Mackie at the Royal British Columbia Museum sampling the smooth inside surface of the robe using distilled water and a fine brush. Colour version on page 597.

Some fragments of the robe were taken to the Yukon Heritage Branch lab in Whitehorse, Yukon, and eased apart. The remainder of the robe was packaged for transportation and storage while still on the glacier and was not exposed to outside ambient conditions until after it was brought to the Royal British Columbia Museum in Victoria, where it was kept frozen except for brief exposure for sampling and conservation. Sampling for forensic botany progressed in tandem with conservation treatment. It entailed removing samples of adhering mud, moving from the outside inward, and retaining samples of distilled water from gentle rinses for conservation purposes as the robe was unfolded over a long series of conservation and sampling episodes (Mackie 2005) (fig. 3).

Macrofossil Sampling and Preparation

Macrofossils were picked from the robe surface or, under a dissecting microscope, from debris and fur samples before preparation for pollen analysis. The sediment from robe rinse samples was similarly examined under a dissecting microscope for plant remains, which were picked out and stored in vials. Using a compound microscope, pollen and spore slides were examined for large plant fragments, such as moss remains, and five subsamples of these pollen preparations were studied with an environmental scanning electron microscope (ESEM) (see chapter 31). In an ESEM, specimens are not subject to a large vacuum and do not need to be coated for electrical conductivity. Preparation of specimens for ESEM imaging does not require the freeze-drying or prolonged dehydration needed for typical scanning electron microscope (SEM) imaging. It is therefore easier and much faster to prepare extremely small samples like those available for the robe study, and there is less risk of losing material in the preparation process. With ESEM imaging, the particles' natural surfaces are also more readily visualized than in SEM studies. Identifications of vascular plant materials were carried out with reference to herbarium specimens at the Royal BC Museum. The remains of bryophytes and lichens were identified by visual inspection, microscopy and comparison with voucher specimens from collections in herbaria, primarily from the University of British Columbia.

Samples were obtained from the conservation treatment, involving gentle distilled-water rinses and removal of adhering debris, at the Royal BC Museum's conservation laboratory. Rinsed and discrete samples were collected directly into large plastic bags, which were then sealed and immediately frozen. A scalpel was used to remove samples of washed fur, which were straightaway placed in small plastic bags and either frozen or taken to the museum's pollen laboratory, where they were examined for macro-remains and then prepared for pollen analysis. Samples from the robe's inside surface were removed by gently scraping it with a scalpel and washing the residue into a small beaker with distilled water. The beaker's contents were then placed in a vial before freezing or processing. To the extent possible, the specific location of all samples was noted with respect to robe fragment and position on the fragment.

During the sampling process, care was taken to limit exposure to ambient pollen sources. The Royal BC Museum's conservation laboratory is isolated from the outside by three doors. The robe's surface was exposed only during the interval of sampling and was either refrozen immediately or packed for conservation purposes. Samples were usually taken from freshly exposed surfaces as the robe mass thawed, and only regularly cleaned or pristine tools (washed in distilled water between samplings) were used. Once placed in sterile containers, the samples were transferred to the pollen preparation lab, which is also isolated from the outside by three doors and by an air filter.

Overall, four different types of samples were obtained from the robe (table 1). It is important to keep in mind that the type, number and size of samples were constrained by the artifact's conservation needs (see chapter 26).

The four sample types included mud and other material adhering to or having detached from the robe surface as it was unfolded and treated for conservation; water and debris samples obtained at the Royal BC Museum conservation lab as the ground squirrel fur was rinsed; small clusters of fur cut by scalpel from within the fur mass of the robe; and scrapings with washing from the smooth, inside hide surface.

Table 1. Robe pollen and spore sample numbers, their sources and sample materials. The first number in the "Sample number" column refers to the robe fragment number and can be related to the reconstructed map of the robe (K. Mackie, pers. comm.).

Sample number	Sample description
Outside surface washes	
R-23	Wash
24-1-1	Wash, ground squirrel pelt
24-5-1	Wash
24-6-1	Wash
24-7-1	Wash, abundant mineral-scattered fibre clumps
24-8-5	Wash
24-12-1	Wash
25-9-1	Wash
R-25-10-2	Wash
25-10-1	Wash
26-4-1	Wash
26-6-2	Wash
26-8-2	Wash
28-5-4	Wash, mineral with some fibre
Inside surface samples	
R-26-6-1	Ochre, scraped inside surface
R-28-10-3	Inside wash
28-28	Inside wash/scrape
28-25	Inside wash/scrape, inside a fold
28-27	Inside wash/scrape
28-30	Inside wash/scrape, near collar?

Sample number	Sample description
R-28-31	Inside wash/ scrape
R-28-32	Inside wash
28-34-6	Skin washing, inside, 15 cm from tie
28-34-7	Skin washing, inside, 15 cm from tie
25-8	Sediment, inside surface
Sediment from outside surface	
25-6	Sediment, robe 25 tray
26-9-3	Sediment lump
28-9-2	Adhering amorphous debris
28-11	Detached sediment
28-12	Sediment matted into fur inside mass
28-13	Mud with fur fibres
28-20	Sediment in fur
28-21	Mud with fur fibres
28-22	Mud with fur fibres
28-29	Mud, a few fur fibres
28-33	Mud with fur fibres
R-26-9-1	Surface fur with sediment
Fur from outside surface	
23-24-6	Fur
51-12-1	Fur, inside of a fold
25-7	Fur from pelt
28-10-1	Mostly fur
28-10-2	Uncontaminated fur, no attached debris
28-14	Masses of fur
28-16	Fur mass
28-19	Uncontaminated fur, inside fold

Sample number	Sample description
28-24-5	Fur, some mineral
28-24-7	Fur
28-34-2	Fur
28-34-3	Fur

A few additional samples of snow and ice were collected on site and analysed either by conventional transmitted light microscopy or ESEM imagery to provide comparative context for the pollen and spore assemblages.

At the pollen preparation lab, pollen and spore samples were prepared in the conventional manner using standard acetolysis (Faegri and Iversen 1989). Glassware was thoroughly cleaned and rinsed in distilled water, and the fume hood used for processing was regularly washed. The working surface consisted of a plastic base of fresh paper towel removed after each day, or more frequently. In all cases, lab- or reagent-grade chemicals were used. Samples were largely prepared in 15-cc conical glass tubes and, if very small, were removed with a glass pipette and placed in 1.5-cc snap-top plastic vials after the last water wash. No mineral acids (hydrofluoric and hydrochloric) were used, so as to limit potential damage to pollen and spores and to keep the number of washes and centrifugation to a minimum. The fur and inside scraping samples were extremely small, and losses of microfossils through decantation and adherance to glassware had to be limited. Residue was mounted in glycerin jelly on glass microscope slides.

Pollen and spores examined by light microscopy were observed, identified and counted in a Nikon Biophot microscope at 400x and, if necessary, 1,000x magnification under immersion oil. Counts of 300 or more were made where possible, but in several small samples all the residue had to be mounted in glycerin jelly slides. Pollen grains and spores were easily moved by applying a heated dissecting probe to the cover slip immediately adjacent to the microscope objective lens;

this melted the jelly and allowed the microscopic object to be manipulated and examined. The five subsamples examined by ESEM imagery were mounted in water on one-centimetre-wide aluminum stubs, coated with gold-palladium and examined at a voltage of 17 KeV at magnifications of 5,000 and 13,000 times. The very high resolution of the ESEM allowed identification and counting of various microscopic algae, including diatoms and chrysophytes, as well as pollen grains, microscopic animal remains and plant fragments.

The relatively large number of samples facilitates a statistical analysis to support interpretation based on the obvious differences in percentages of pollen and spores. Principal components analysis (PCA) helps recognize data patterns by plotting them with respect to derived numerical axes, such that the samples with greatest differences are placed furthest apart and samples having the greatest similarities cluster together on the graph. Where there are several patterns in the data, PCA helps show those patterns as principal components displayed with the trends of the patterns at right angles to each other. The first principal component shows the strongest pattern or structure and, thus, the greatest difference among samples. The second principal component shows the next strongest pattern, and so on. PCA was applied to the percentage values for the pollen data, including pollen and spore types occurring at more than one per cent and in three samples or more.

Results

Pollen and Spores

One of our first observations is the amount and good preservation of pollen and spores in and on the robe, regardless of the type of sample (fig. 4). After acetolysis, surface wash and sediment/mud samples consist predominantly of mineral debris, but pollen occurs in abundance; with scans of only one or two slides, a sum of 300 is readily achieved. Fur samples consist mainly of fur fibres, and pollen and spores occur sparsely. Many transects and several slides were required to obtain a

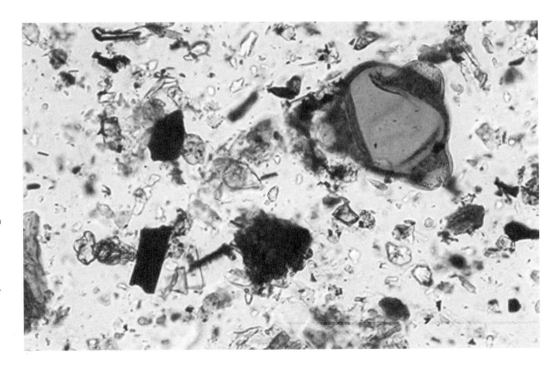

Figure 4. Image of a typical inside surface of robe pollen sample showing a large pollen grain of fireweed (*Epilobium* sp.) at upper right and charcoal (angular black objects) at centre left. Colour version on page 598.

Figure 5. Pollen and spore assemblages from robe samples showing percentages of major types. *Artemisia* and *Salicornia* pollen is included in the "non-trees" value. Analyst: Richard Hebda. Colour version on page 598.

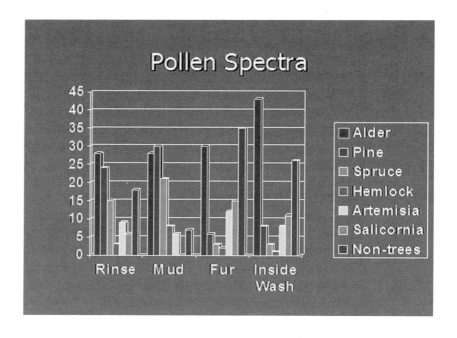

sum of more than 100 pollen and spores, and in the case of a few samples even that total was not reached. The pollen and spores are, however, very well preserved and easily identifiable. Skin-side samples were the smallest, consisting of a limited amount of scraped and washed material on the edge of a scalpel. These yielded mixed residues of mineral material and organic debris such as charcoal, pollen and spores (fig. 4). In most samples a total count of 300 was readily reached.

Few pollen grains were found in the samples examined by ESEM, using subsamples of the prepared pollen residues, in robe samples R23, R28, 28-2, 28-31 and 28-34. The largest number, 29, was from R23 (see table 1, Dickson and Mudie 2008). The composition of the pollen spectra in the ESEM samples was the same found with the light microscope. It is important to note, though, that the very high resolution of pollen morphology provided by the ESEM studies made it possible to distinguish among the genera and some of the species of chenopod pollen (Mudie et al. 2005; chapter 31). A few (1–4) *Salicornia perennis*–type grains were found in samples R28-34 and R23, in contrast to 20–30 grains in samples of surface mud taken from salt marshes at Pleasant Island, Juneau and Hoonah. One other kind of chenopod pollen grain was found in sample R23.

The four different types of samples from the robe yield, broadly, four different pollen and spore assemblages (fig. 5). These assemblages differ with respect to the dominant pollen types and the combination of pollen types. They further differ with respect to predominance of arboreal (tree) compared to non-arboreal (non-tree) pollen and spore types.

The fur-side wash samples, which presumably represent recent surface material, are dominated either by pine or alder. Spruce is usually a secondary type, though it sometimes occurs more abundantly than alder. Fern spores occur in relative abundance as well. Non-arboreal types occur up to 40–46 per cent, with *Artemisia* (sage or wormwood) and beach asparagus being the most abundant.

Samples of sediment adhering to the fur-side surface yielded pollen and spore assemblages somewhat similar to fur wash samples, in that conifer pollen is abundant, but the two types of assemblages are nevertheless distinct. Pine pollen clearly predominates in samples of adhering sediment, but spruce pollen occurs abundantly as well, usually at greater relative values than alder. Alder percentages are much less than in fur wash samples. Western Hemlock pollen is a notable element of the assemblages, reaching values higher than in the other sample types. Non-arboreal pollen occurs in low amounts, consisting mainly of fern spores and *Artemisia*. A most striking feature of the assemblages is the near absence of beach asparagus pollen.

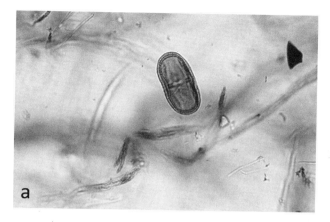 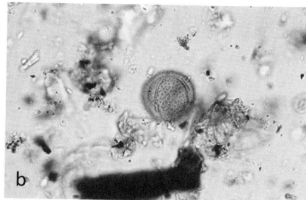

Figure 6. Microscopic pollen of two relatively common non-arboreal (non-tree) pollen types recovered from robe samples. *6A Heracleum maximum* (Cow Parsnip), *6B Artemisia* (sagewort). Colour versions on page 599.

Samples from within the fur of the robe yielded distinctive assemblages easily differentiated from wash and sediment samples. At first glance, the most obvious difference is the low relative values for conifer types pine and spruce and, instead, the abundance of alder pollen. Striking, too, is the abundance and diversity of herbaceous types, including *Artemisia*, grasses, aster family and, in most samples, beach asparagus. Cow Parsnip (*Heracleum maximum*; fig. 6) and burnet (*Sanguisorba*) pollen also occurs in notable quantities.

Scraped and washed samples from the smooth, inside surface are most like the samples from within the fur of the robe. Alder pollen predominates, and beach asparagus pollen grains occur abundantly, reaching values of 10–13 per cent (fig. 5). Pine and spruce pollen occurs at relatively low values. As in the fur samples, there is a diversity of non-arboreal pollen types in addition to beach asparagus. The percentages of *Artemisia* (fig. 6), grasses and aster family are notable. Some single samples include notable percentages of willow (*Salix* species), Sweet Gale (*Myrica gale*), Cow Parsnip and burnet.

Comparison pollen and spore counts from glacial ice and snow are strongly dominated by alder pollen, with low to modest pine and *Artemisia* and little else (table 2).

Principal Components Analysis

The first component axis of the principal components analysis explains 22 per cent of the variance (differences) in the samples, and it clearly separates robe wash and adhering sediment samples from those obtained from the inside of the robe and within the fur (table 3). With the exception of a single sample, the two groups are completely discrete (fig. 7). The main contributing components to this separation are the conifers pine and spruce, representing forest environments, versus herbaceous types such as grasses, asters, *Salicornia* and *Heracleum*, representing a diversity of open habitats.

As interpreted from visual inspection of percentage values (fig. 5), the surface wash and sediment samples, both characterized by abundant conifer pollen, are also nearly discrete from each other. The most distinct samples in this separation (PCA component 2) are those with relatively high alder values from wash samples (table 3).

The non-arboreal cluster of samples (inside fur and inside surface) is highly mixed (see fig. 7), suggesting that it is not possible to distinguish different open habitats, such as seashore salt marsh from alpine meadow. This may be because the two assemblage types accumulated over a long interval, eventually mixing

Table 2. Percentages of major pollen and spore types from snow and glacier ice samples collected near the remains of the Kwädąy Dän Ts'ìnchį man.

Pollen and spore type	Ice sample KDS 57	Snow sample (melted to 600 ml of water) KDS 68:
Pine	6	2
Spruce	<1	–
Western Hemlock	2.	–
Alder	65	94
Grasses	<1	1
Asters	<1	–
Sagewort	10	2
Cow Parsnip	<1	–
Sedges	<1	–
Club-moss	1	<1
Ferns	14	3
Total number of pollen and spores	400	320

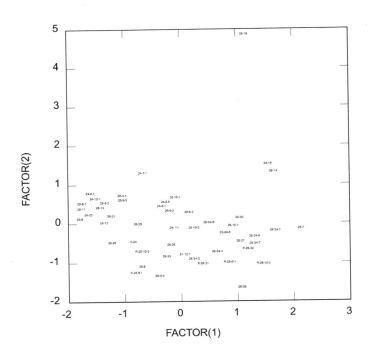

Figure 7. Principal component analysis (PCA) diagram for pollen and spore assemblages from the Kwädąy Dän Ts'ìnchį ground squirrel robe. Sample numbers correspond to those in table 1. Samples to the right originate mainly from inside the fur and the inside surface of the robe. Samples to the left originate from attached sediment and outside surface washes. The term "factor" on x and y axes means the same as the more traditional term "component". Ken Marr analysis and image.

Table 3. Principal component analysis (PCA) component scores for first three components of percentages of pollen and spores in robe assemblages.

Component loadings

Pollen and spore type	1	2	3
Pinus	-0.865	0.204	0.154
Picea	-0.845	0.157	0.179
Abies	-0.315	0.224	0.315
Tsuga heterophylla	-0.736	0.166	0.091
Tsuga mertensiana	-0.582	0.285	0.253
Cupressaceae	0.066	0.042	0.188
Alnus	0.236	-0.578	-0.451
Betula	0.662	0.176	-0.099
Populus	0.345	-0.284	-0.297
Salix	0.620	-0.291	0.005
Myrica	0.361	0.101	-0.071
Ericaceae	0.442	0.588	-0.362
Rosaceae	0.542	0.276	-0.274
Shepherdia	0.388	-0.034	0.405
Caprifoliaceae	0.101	0.748	-0.512
Cornus	0.381	-0.002	-0.001
Ribes	0.127	-0.248	-0.144
Poaceae	0.781	0.034	-0.112
Asteraceae	0.739	-0.048	0.304
Artemisia	0.417	0.082	0.323
Chenopodiinae	0.694	-0.043	0.036
Caryophyllaceae	0.215	0.021	0.383
Thalictrum	0.191	0.687	-0.384
Onagraceae	0.432	0.157	0.454
Valeriana	0.299	0.716	-0.233
Apiaceae	0.303	-0.062	0.273
Heracleum	0.634	0.301	0.382
Liliaceae	0.449	-0.063	0.284
Lysichitum	0.199	0.226	0.367
Lamiaceae	0.192	0.090	0.241
Campanula	0.148	-0.065	0.034
Sanguisorba	0.562	0.104	0.387
Unknown	0.719	-0.102	0.078
Cyperaceae	0.578	0.133	-0.298
Selaginella	-0.094	0.133	0.286
Lycopodium	0.162	0.590	0.296
Cryptogramma	-0.274	0.118	0.112
Equisetum	-0.064	0.192	0.214
Fern spore rugulate	-0.389	0.034	-0.037
Fern indeterminate	-0.568	0.313	-0.065

Percent of total variance explained

1	2	3
22.289	8.730	7.295

pollen from various open sources together. An alpine-subalpine source (assemblage) might also be obscured by a strong recent exposure of both inner and outer surfaces to beach asparagus pollen. Exposure to non-seashore open environments is clearly indicated by relatively abundant pollen from insect-pollinated Sitka Valerian and aster family (excluding *Artemisia*), neither of which occurs in any abundance in the shoreline zone.

Macroscopic Plant Remains

Plant remains recovered in association with the Kwädąy Dän Ts'ìnchį robe include the following types (see table 4).

Cf. *Cladina* (Reindeer Lichen). There is only one small piece of lichen, two millimetres long with four short branches, resembling the tips of a *Cladina*. A single species of *Cladina* grows at location 2.

Distichium sp. or spp. These very short pieces of short-leaved stems are most likely to be *Distichium capillaceum*, usually growing on base-rich soil and rock.

Ditrichum flexicaule. There are three short pieces of leafy stems from this moss that favours base-rich substrates.

Encalypta/Syntrichia/Tortula. The single specimen consists of two very small leaves tight together, in which it is very difficult to see cells. With such poor material it is hard to be sure of generic—far less species—identity, so this and similar material from non-robe samples is assigned to this broad category.

Hylocomium splendens. A single short piece of leafy stem. Around the discovery site there were only two small patches adhering closely to the ground. This moss can grow profusely at much lower altitudes.

Pleurozium schreberi. Sample 10-3 was a "Piece of leather or hide", so is likely to have been part of the robe. There are two short pieces of leafy stem. This species was not found in the vicinity of the discovery site, nor was it noticed elsewhere during the expedition between the discovery site and the Haines Highway. It is often associated with *Hylocomium splendens* and can be abundant at low altitudes.

Polytrichum/Polytrichastrum sp. or spp. There are two short pieces of leafy stem. These two genera, and also *Pogonatum*, grow at the discovery site.

Racomitrium canescens s.l. This is the most common moss recovered from various samples, not just those from the robe. It grows from sea level to high in the mountains.

Juncus sp. A single seed of a rush, which could perhaps be identified to the species level but only after very detailed study.

Osmorhiza berteroi. This half a fruit of Mountain Sweet Cicely was removed from the lower left-hand corner of the robe. It would have become attached to the robe in the coastal zone. The fruits have downward-directed, brittle hairs that are an adaptation for briefly clinging to fur. It was discussed and illustrated by Dickson et al. (2004, 484), who state: "To brush through a stand of *Osmorhiza* plants is to inadvertently gather on your clothing many fruits, most of which are soon knocked or shaken off".

Salix sp. or spp. Small leaves and leaf fragments, some obviously round, with prominent veins and short petioles. Derived from the small, very low-growing species such as those that inhabit alpine environments very near the area of the discovery site now.

Tsuga mertensiana (Mountain Hemlock). This is a single needle (leaf), discussed and illustrated by Dickson et al. (2004). Like the Mountain Sweet Cicely, it would have become attached to the robe in the coastal zone.

Table 4. Macroscopic plant remains from the Kwädąy Dän Ts'ìnchį robe.

Plant remains from the robe							
Plants	Sample numbers						
Lichen							
Cf. *Cladina*	+	24-14-1					
Mosses							
Distichium sp. or spp.	+	28-11		28-29			
Ditrichum flexicaule	+	18-5-1		20-7-1	28-11		
Encalypta/Syntrichia/Tortula	+	24-14-1					
Hylocomium splendens	+	28-21					
Pleurozium schreberi	–	10-3					
Polytrichum/Polytrichastrum	+	24-14-1		51-11			
Racomitrium canescens s.l.	+	12-2-1		20-7-1	28-11		
Vascular plants							
Juncus sp.	+	25-11-1					
Osmorhiza berteroi	–	24-8-3					
Salix sp. or spp.	+	28-51	28-23	28-66	28-7	28-5-2	51-11
Tsuga mertensiana	–	24-14-4					

+ means taxon grows in the near vicinity of the discovery site.

– means taxon absent from the near vicinity of the site.

Diatoms

The high resolution provided by ESEM imagery is a powerful tool for forensic studies using diatoms as environmental indicators because the characteristics of minute diatom species (less than 10 microns) can be clearly discerned. Very small pennate diatoms dominate the ESEM samples from the ponds and snow near the discovery site, as illustrated and discussed by Dickson et al. (chapter 5). The kinds of diatoms and other non-pollen microfossils found in ESEM samples from the robe and in samples of red- and green-coloured snow at the discovery site in 2002 are shown in table 5 and figure 9. For comparison, table 4 also lists the kinds of diatoms found at other locations: in two inland saline lakes near Whitehorse; in Sediment Creek delta on the Tatshenshini River; and in seven salt marshes of southeastern Alaska, between Young Island, Glacier

Figure 8. The finely divided and layered branches of *Hylocomium splendens* (Step Moss), a lowland forest moss found in association with the robe. Jim Dickson photograph. Colour version on page 599.

Bay (YI) and Bostwick Inlet (Bos), near the border of BC. Several of the diatoms on the robe are large (>20 μm), thick-walled cylindric and pennate species that are characteristic of the coastal region diatom assemblages, while others resemble a small centric species found in the snow and ponds near the discovery site. The robe samples also included the broken shell of an ostracod (tiny crustaceans called seed shrimp) and a rhizopod (testate amoeba). These microfaunal remains were not found in the alpine and inland ponds or in river water samples, and they were rare in one inland saline pond. Testate amoebae were only found in the coastal salt marsh samples.

Interpretation and Discussion

RECOVERING AN ABUNDANCE of plant remains associated with the robe has provided a wide range of important insights into not only the robe's recent history but the inferred travel of the Kwädąy Dän

Ts'ìnchį individual and, more broadly, the application of microscopic analyses of clothing to forensic studies as well.

First, as mentioned earlier, we noted excellent preservation and diversity of pollen and spore types. The preservation is not surprising for surface samples; organic material was frozen in the ice and, therefore, well preserved. The preservation quality of samples from within the fur and in the smooth, inner surface is perhaps surprising. These two environments would have been exposed to body temperature and ambient temperatures for many days, perhaps even months, sufficient for pollen to have decomposed. The degree of preservation and the abundance of material indicate tremendous potential for pollen and spore analysis not only of artifacts but of mammals' fur and perhaps bird feathers (see also Bryant and Mildenhall 1998).

Second, the occurrence of unambiguously different assemblages, more distinctive than those of pollen zones representing different ecosystems (Hebda 1995), is also notable. The differences are evident even at the most basic level of the proportion of arboreal (tree) to non-arboreal (non-tree) types. With a cultural object, especially an item of clothing, one might expect differences from different areas on the object, according to what it was in contact with. In this case, the systematic differences according to the type of sample is a surprising result.

We could expect surface washes to be different from assemblages inside the fur. Wash samples presumably represent the most recent exposure, with fur samples perhaps representing earlier exposures. Indeed, this hypothesis underlay our sampling strategy. The distinctiveness of the inside scrapings and washings was not expected but makes sense if the robe was worn during the journey from one environment, where the inner surface was exposed to a particular plant community, through another plant community where the inside was not exposed (because it was being worn) but the outer surface was. Such an interpretation assumes the article was being worn, presumably draped over the shoulders. It might not apply if the man had been carrying it folded or rolled up.

If inside scrapings and samples from within the fur represent environmental encounters prior to the last few days or day of the journey, as we assume, then they strongly indicate that the robe and, presumably, the person carrying/wearing it spent much time in the open along the seashore and possibly in the alpine—both environments with strong non-tree components. The relatively low values for tree species would not be expected on an object (and, by inference, a person) that spent much time in the forest. The high diversity of non-arboreal types, which produce relatively little pollen compared to trees, also suggests exposure to plants of open habitats. The relatively high percentages of the low-growing beach asparagus pollen type strongly suggest that the robe had been in direct contact with these plants either over a long time or, possibly, not long before the trip to the glacier.

The distinctiveness of the sediment/mud samples would not normally be expected. Such debris on the surface of an object recovered from within a glacier would presumably have originated from the glacier itself, unless it had adhered from a previous environmental exposure before arrival in the glacier's alpine environment. Abundant conifers in mud and in surface washings suggest recent exposure to forest environments.

The assemblages of the mud samples do not reflect what might be expected as contamination from glacial mud. Snow and ice spectra show that very high alder pollen values should be expected (table 2; chapter 31). Instead, there is a strong suggestion of coastal forest exposure in the relatively high values of Western Hemlock (cf. Cwynar 1990). Based on visual impression, the spruce pollen grains are relatively large, suggesting the source might be Sitka Spruce rather than White or Black spruce. A comprehensive analysis of spruce pollen in the samples may be worth undertaking (see Bagnell 1975). In any case, the strong conifer component of the sediment samples strongly suggests that the mud was picked up in travels through coastal and adjacent boreal forest after leaving the marine shoreline, en route to the glacier.

Charcoal on the robe's inside surface suggests many hours around fires, such as a house fire, campfire or those used for processing fish and other meat in smokehouses or at outdoor smoking racks (see also chapter 12). The relative absence of charcoal on the outside is interesting, and a quantitative study of charcoal might be informative.

It is important to note that our results are related to the exposure of the robe to various environments and plants, not necessarily the person with whom it was found. We do know that the robe was directly associated with the man and must have been carried there by him, presumably worn by him (Beattie et al. 2000). We also know that he had eaten beach asparagus, so the strong pollen signal of the chenopod pollen and marine salt marsh diatom assemblages on the robe is presumably the result of his direct contact with the salt marsh environment (chapter 31). The sediment and surface wash samples must reflect exposures following those to the beach asparagus site because they were recovered from surfaces on top of the beach asparagus-dominated samples (surface of fur versus inside of fur). It seems reasonable to infer that the robe's pollen assemblages

Table 5. Diatoms and other algal remains from the robe and regional environments.

Sample number	Robe samples					Snow 2002	
	28-31 28-2 28-34	21-1	23-1	R 28-21 R 28-21	R23	Red snow	Green snow
Percent abundance							
Centric diatoms <10µ	20	10	10	0	15	trace	0
Centric diatoms 10–20µ	2	0	5	5	5	trace	0
Centric diatoms >20µ	0	0	0	0	0	0	0
Pennate diatoms <10µ	15	15	5	0	1	0	0
Pennate diatoms >10µ	2	5	5	2	5	trace	0
Other diatoms	3	10	2	1	5	trace	0
Ratio centric:pennate	1.1	0.3	1.25	1.6	1.8	0	0
Chrysophyte – smooth	5	5	1	1	1	1	trace
Chrysophyte – spiny	0	2	0	0	1	5	5
Red snow alga	1	0.5	2	1	1	30	60
Trochiscia aspera alga	0	1	1	1	0	0	5
Artemia brine shrimp	0	0	0	0	0	0	0
Dinoflagellate cyst	0	0	0	0	0	0	0
Foraminifera	0	0	0	0	0	0	0
Ostracode	0	0	1	0	0	0	0
Rhizopod	0	0	1	0	0	0	0
Silicoflagellate	0	0	0	0	0	0	0
Thecamoeban	0	0	0	0	0	0	0

Yukon			Alaskan Coastal Salt Marshes							
Sed Ck	Takl	RP	Hain	Jun	Hoo	Swa	YI	GH	GI	Bos
1	0	4	0	1	1	0	0	0	0	0
0	0	0	12	0	trace	2	2	0	0	0
0	0	0	0	2	8	8	7	7	2	3
0	1	50	2	2	1	0	0	4	0	2
4	0	1	5	5	5	8	4	11	2	9
0	0	0	0	0	0	0	0	0	0	1
0.25	0	0.08	1.7	0.28	1.5	1.25	2.3	0.46	1.01	0.3
0	0	4	0	0	0	0	0	0	0	0
0	0	28	0	0	0	0	0	0	0	0
0	0	0	0	0	0	0	0	0	0	0
0	0	0	0	0	0	0	0	0	0	0
0	0	3	0	0	0	0	0	0	0	0
0	0	0	0	0	0	1	1	1	0	0
0	0	0	1	0	0	1	0	0	0	0
0	0	0	0	0	0	0	0	0	0	0
0	0	0	0	0	0	0	0	0	0	0
0	0	0	0	0	0	0	0	0	0	0
0	0	0	0	0	0	0	0	0	4	0
0	0	0	0	0	0	0	2	0	0	0

provide a record of travel from the marine shoreline through coast and inland forest, presumably in the day or days immediately before the Kwäday Dän Ts'ìnchi man's death on the glacier.

Exposure to the forest is clearly indicated by adhering or associated plant macrofossils such as the Mountain Hemlock needle, the Sweet Cicely seed and several mosses (table 4). The time at which these may have been attached to or associated with the robe is difficult to ascertain from their occurrence alone. In combination with the pollen and spore data, however, it is possible to suggest that the Kwäday Dän Ts'ìnchi man encountered these plants just before arriving at the glacier, as he passed through the forest zone.

Diatoms are well known as a powerful tool in forensic cases involving drowning (Hurlimann et al. 2000), as the species found in the lungs and stomach may reveal if the victim was drowned in the ocean, a lake or a bathtub. We were not expecting to find numerous diatoms on the robe of the Kwäday Dän Ts'ìnchi man, except for the two small, thin-walled species (*Aulacoseira glacialis* and *Pinnularia viridus*) found in the snow samples at the discovery site. Finding three species of large centric diatoms and two species of large pinnate diatoms in the five robe samples

was therefore surprising, and it clearly supported the pollen evidence showing that, in the past, the robe had been laid on the ground in a very different location than the glacier. Fortunately, our fieldwork in Yukon and Alaska had enabled us to collect water samples from a wide range of inland and coastal aquatic environments (chapter 31), and so we had a reference collection of diatom and microfaunal ESEM images from southwest Yukon inland pond and Alaskan salt marsh samples (table 5).

In the Northwest Territories the ratio of centric (Centrales) to pennate (Pennales) diatoms (C:P) has been used as an index of the ecological affinity of diatom populations (Lotter et al. 1999). Assemblages dominated by centric types indicate boreal forest lakes, whereas assemblages dominated by small pennate types indicate lakes and ponds characteristic of Arctic tundra regions. Diatom studies in the Mackenzie Delta region of Yukon

Figure 9.

A–I, ESEM images of diatoms from the robe samples and representative inland and saltmarsh sites. Robe sample KDS28-34. *A–C,* Relatively large centric diatoms: *Cyclotella* sp. *A,* Valve view. *B,* Girdle view. *C, Porosira glacialis. D,* Poorly preserved large pennate diatoms. *E,* Small centric *Stephanodiscus* sp. *F,* Swanson Harbour large centric *Thallasiosira* sp. *G,* Takhini salt pond small pennate diatom. *H, Stephanodiscus* resting spores from discovery site red snow sample. *I,* Pleasant Island large pennate diatom and *Salicornia perennis* pollen grain. Analysis and image: Petra Mudie.

Territory (Campeau et al. 1999) show that centric diatoms are more abundant in marine waters, whereas pennates dominate the shore zone and freshwater ponds. The C:P ratio range is 0.3–1.8 for the robe samples and from 0.3 to more than 2.5 for the coastal salt marshes. In contrast, the snow and inland lake or river samples have C:P ratios of less than 0.3. The ESEM diatom evidence therefore tends to support the pollen and macro-plant evidence for contact with a coastal salt marsh environment.

Other possible evidence for contact with the non-glacial (possibly coastal) environment is the broken ostracod shell and rhizopod. These microfaunal remains were absent or very rare in the inland pond and river water samples. But our general knowledge of these groups in this region is so poor that further study is needed to assess the significance of these occurrences.

Conclusion

OUR STUDY DEMONSTRATES that the Kwäday Dän Ts'ìnchi robe acted like a sampling device of the environments in which the man travelled and stayed. The garment seems to carry traces of many days,

perhaps even months, of exposure to a variety of plant communities, yet the pollen and plant macrofossil assemblages remained distinct. If the robe had been gathering and integrating pollen, spore and diatom assemblages over years of exposure to many environments, we might have expected assemblages to be similar throughout—but they are not. Our study therefore suggests that botanical studies of ancient clothing, especially micro-botanical studies, may be powerful forensic tools in understanding the lives of ancient people. We thank Long Ago Person Found for teaching us this important lesson.

ACKNOWLEDGEMENTS

Thanks go to Kjerstin Mackie and Alexander Mackie for providing data concerning the robe, and to Nicholas Panter for facilitating access and artifact/sample recordkeeping. Thomas Munson helped with preparing pollen samples, and Frank Thomas (Geological Survey of Canada Atlantic) ran the ESEM. Ken Marr of the Royal BC Museum helped perform the PCA. Rolf Mathewes and Vaughn Bryant provided valuable reviews of the manuscript. We also thank the Champagne and Aishihik First Nations for allowing us to be part of this important project and for their continued interest in the study.

5

JOURNEYS

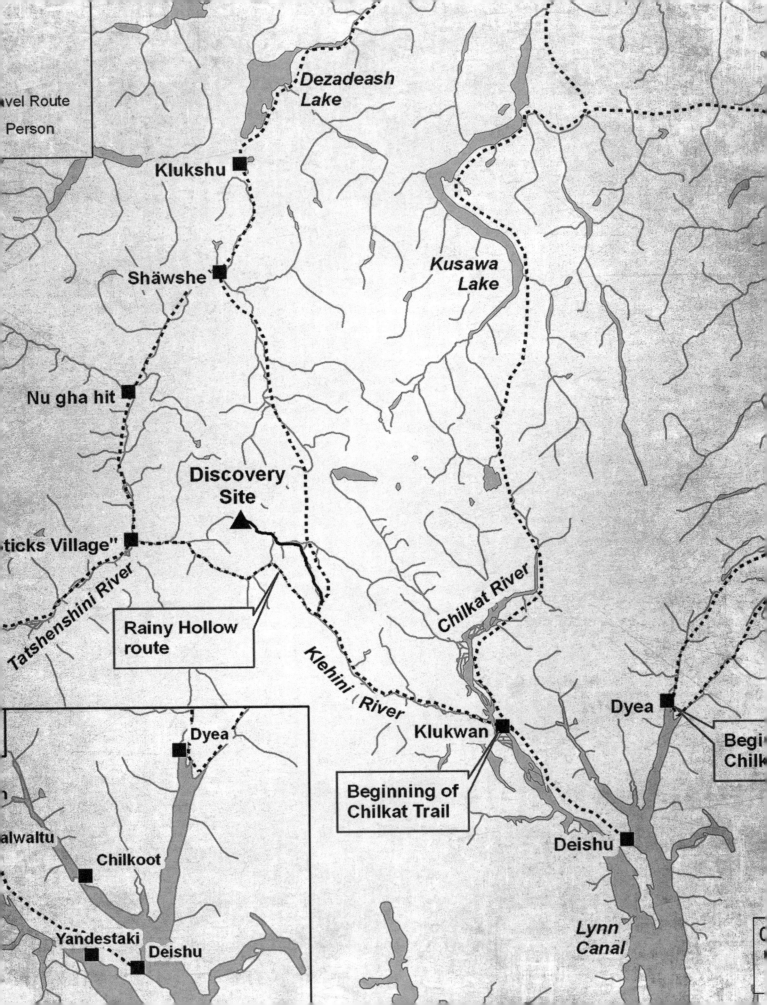

INTRODUCTION

Sheila Greer

Travel is a major theme in the overall story of the Kwädąy Dän Ts'ìnchį man. Contributors to this section describe past and present travels in the region's glacier-dominated landscapes, including on the glaciers themselves. Forensic and other evidence is used to reconstruct the journey that the Kwädąy Dän Ts'ìnchį man himself may have taken before he died.

TRADITIONAL STORIES REVEAL that travel through and over glaciers was a long-standing tradition for the region's Tlingit, Dän (Tutchone) and Tagish peoples. Oral history accounts, including previously unpublished stories from community elders, tell us that glacier journeys involved risk, with travellers sometimes failing to make it to their destinations. These traditional stories bring to life the beliefs and glacier experiences of the region's Indigenous cultures, thereby providing context for understanding the Kwädąy Dän Ts'ìnchį discovery.

The first-hand account from Champagne and Aishihik citizen Ron Chambers, from Haines Junction, provides a modern view of journeys in the icy mountains. A well-known guide and outdoorsman who was directly involved in the recovery of the Kwädąy Dän Ts'ìnchį remains, Chambers explains his culture's belief that glaciers are "absolutely alive" and shares knowledge of the dangers of glacier travel.

Small details can yield incredible information, particularly when the latest scientific equipment is used to unlock their secrets. To learn more about the diet and travels of the Long Ago Person Found, a high-powered electron microscope and an associated electron probe device were used to study minute samples taken from the Kwädąy Dän Ts'ìnchį man's digestive tract. These samples of food eaten during the days prior to death reveal that he had travelled widely.

As well, we learn that he was an omnivore, eating mammals, crab, fruit and seashore plants. Microscopic plant and mineral remains even identify where he likely drank water only hours before he died. His meals, as revealed by the digestive tract samples, establish a strong recent connection to the ocean shoreline many kilometres away from the discovery site, and even suggest the route taken to reach the glacier. The microscopic remains of one plant, beach asparagus

(*Salicornia perennis*; in Tlingit *Suk K'ádzi*), provide especially valuable information about his last journey, including the time of year it took place.

Knowing that the Kwädąy Dän Ts'ìnchį man's final journey involved foot travel from somewhere in the lower Chilkat River country (Haines, Klukwan area), a field party of two retraced the likely route in 2010. In doing so, the duo travelled over a glacier located south and east of the discovery site, as well as through a valley that was likely full of ice at the time of Kwädąy Dän Ts'ìnchį man's journey.

The chapters in this section show that using both traditional knowledge and scientific research to study aspects of the Kwädąy Dän Ts'ìnchį man's journey across an imposing landscape means we get a more complete picture of his life, culture and travels.

29
TRADITIONAL STORIES OF GLACIER TRAVEL

Champagne and Aishihik First Nations, Sheila Greer, Sarah Gaunt, Diane Strand,
Sheila Joe and John Fingland, in collaboration with Ron Chambers
and the late John Adamson, Wilfred Charlie, Moose Jackson and Jimmy G. Smith

There was a lot of fear of glaciers.... People didn't go anywhere near them unless they had to, and trade said you had to come through these places. It's part of what our life was, part of what we had to deal with.

In the spring, before the melt, the weather is quite nice but there are snow bridges on those crevasses.... They look like perfect flat snow ground and all of a sudden, it drops through.[1]

—RON CHAMBERS

LANDSCAPES DOMINATED by mountains and ice are central to the homelands of the Tlingit of southeastern Alaska and the Tagish, Dän (Southern Tutchone[2]) and other peoples of the northwestern interior of British Columbia and southwestern Yukon (fig. 1). Called *tan shi* in Dákwanjè (Southern Tutchone) and *sít'* in Tlingit, glaciers are common in Tatshenshini-Alsek Park, where the Kwädąy Dän Ts'ìnchį man lost his life, as they are throughout the greater region. This area has been designated a UNESCO World Heritage Site (fig. 2; see also chapter 2) and features not only extensive glaciers but the world's largest non-polar icefield and expanses of dramatic wilderness; and some of the tallest peaks on the continent, including the Kluane, Wrangell, St. Elias

and Chugach mountain ranges. In addition to British Columbia's Tatshenshini-Alsek Park, the World Heritage Site comprises three other protected areas: the Yukon's Kluane National Park and Alaska's Wrangell-St. Elias and Glacier Bay national parks.

The Aboriginal peoples whose homelands are now included in this World Heritage Site include the Ahtna, Upper Tanana, Eyak and Tlingit (Wrangell-St. Elias National Park); the Tlingit (Glacier Bay National Park); the Southern Tutchone and Tlingit (Tatshenshini-Alsek Park); and the Southern Tutchone, Tlingit and Upper Tanana (Kluane National Park).[3] Today these various peoples are represented in Canada by three Indigenous governments: Champagne and Aishihik First Nations,

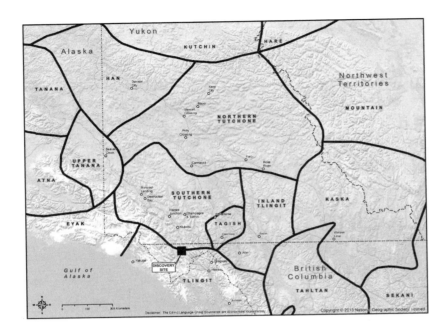

Figure 1. Distribution of Aboriginal cultural/language groups in the region. Boundaries indicated are not political and do not correspond with the traditional territories of recognized tribes and First Nations. Adapted from McClellan (1975).

Figure 2. The UNESCO World Heritage Site comprising Wrangell-St. Elias National Park, Glacier Bay National Park and Tongass National Forest (Alaska); Kluane National Park (Yukon); and Tatshenshini-Alsek Park (BC). White areas indicate glaciers. The Kwädąy Dän Ts'ìnchį discovery site is located in the lower right of the map. Courtesy of the USGS.[10] Colour version on page 600.

Figure 3. Guide Ron Chambers at the Kwädąy Dän Ts'ìnchį site, August 2003. Sheila Greer (CAFN) photograph.

Kluane First Nation and White River First Nation. The Ahtna, Eyak and Sealaska regional native corporations represent the various tribes in the United States.

Reference to glaciers and travels over them are found in the traditional stories of all these Indigenous peoples, as well as in the stories of neighbouring groups, including other Athapaskan-speaking peoples of the interior like the Tahltan (northwestern BC) and the Northern Tutchone (Yukoners located further inland and to the north of the Southern Tutchone). Our focus is on the traditional stories of the Tagish, Dän, and northern Tlingit peoples, whose traditional territories include or are adjacent to Tatshenshini-Alsek Park, where the Kwädąy Dän Ts'ìnchį discovery site is located. Traditional stories are referred to as *kwändür* or *kwädąy kwändür* in Southern Tutchone (Workman 2000) and *tlàgú* in Tlingit (Leer et al. 2001). We present published and previously unpublished traditional stories of glacier travel, highlighting those referring to people who almost or actually lost their lives while crossing these dangerous frozen landscapes. Some background is also presented on the trade between peoples living on the northern northwest coast and those in the northwest interior. Most of the stories recounted here are thought to be related to journeys undertaken for purposes of trade.

The stories in this chapter show that travel mishaps did indeed occur in times past and that the dangers associated with glacier travel were very real for people living in these lands. Perhaps the greatest hazard was the risk of falling down a crevasse—or crevice, as they are sometimes called. Crevasses can reach depths of hundreds of metres and often have vertical or near vertical walls. Fall into a deep one, and it is impossible to get out by yourself. Crevasses are particularly dangerous not just because they are deep but because they are often well hidden, covered by a fragile bridge of snow. The opening quotation from guide Ron Chambers (see fig. 3) describes just this situation. When crossing a snow-covered glacier, one has no idea where crevasses may be, hence the contemporary practice of roping people together to cross a glacier.

Our elders' words remind us that just as these landscapes warranted respect in times past, they must be respected today. Following this spirit, our approach here is to emphasize the elders' words so that knowledge of these past events, as well as the respect these landscapes warrant, will continue to be passed on to younger generations.

After a brief review of our sources, we present stories about glacier travel that have not been previously published. These are followed by a summary of other stories involving glaciers and a brief overview of the theme of glaciers in the history of local Indigenous peoples.[4]

Methods and Sources

STORIES ABOUT GLACIER TRAVEL appear in some of the earliest ethnohistorical and anthropological literature from the Tlingit, Tagish and Dän culture area (Davidson 1901; Krause 1956; Krause and Krause 1981, 1993; Swanton 1909). Later, detailed studies by researchers Frederica de Laguna (1972) and Catharine McClellan (1950, 1964, 1975) highlighted the importance of glacier travel in the history of specific Tlingit clans. Their work also documented the importance of trade and intercultural relations between the coast and the interior, noting that some of this travel involved passage either over glaciers or down rivers going under glaciers.

Subsequent to McClellan's research with southern Yukon First Nations communities, anthropologist Julie Cruikshank collaborated with Yukon First Nations elders Angela Sidney, Kitty Smith and Annie Ned to record their personal life histories (Ned 1984; Sidney 1982; Sidney et al. 1979; Smith 1982). Among the stories shared by these women were accounts of people who became lost while travelling over glaciers between the Pacific coast and the BC/Yukon interior. Cruikshank's 1990 study looked at these stories in the context of individual life histories. A more recent publication by this same researcher contrasts the glacier stories of First Nations elders with those of non-Aboriginal explorers, providing insight into the worldviews of these very different societies (Cruikshank 2005).

Immediately after the Kwädąy Dän Ts'ìnchį discovery was reported, accounts of glacier travel were being brought to our attention (Banks 1999;

Charlie 1999). The Champagne and Aishihik Heritage program began assembling traditional stories about glacier travel to augment those already known, and we began combing our research and interview transcript files for relevant content. We reviewed the aforementioned sources as well as collections of published traditional stories—such as those by Allen and Allen (2006), McClellan (2007), Nyman and Leer (1993) and Workman (2000)—and unpublished transcripts of interviews with community elders.

Our efforts to document relevant glacier stories beyond our own community were limited. Rather, we focused on stories brought to the attention of the Champagne and Aishihik First Nations Heritage program by Yukoners or Alaskans who considered the accounts relevant to understanding the Kwädąy Dän Ts'ìnchį discovery. Stories shared during consultation meetings that Champagne and Aishihik held with its own citizens and with neighbouring First Nations and tribes (see chapter 8) are among those presented here.

It was also beyond the scope of this paper to assemble glacier toponymic data, though it is known that at least some glaciers within the Tlingit, Tagish and Dän area had Indigenous-language place names (cf. Cruikshank et al. 1990; Cruikshank 2005).

Storyteller Wilfred Charlie: Qwantuk's Story

THE FIRST PREVIOUSLY UNPUBLISHED STORY we share is that of a Northern Tutchone man named Qwantuk.[5] Storyteller Wilfred Charlie (1937–2005), traditional First Nations name Kakúwät, was the son of Sarah and George Charlie (fig. 4). An elder of the Northern Tutchone Little Salmon-Carmacks First Nation based in Carmacks, Yukon, and a member of the Wolf clan, Wilfred was a respected hunter, fisherman and trapper, very knowledgeable of the old ways. He was raised in the traditional Northern Tutchone manner, travelling with his family as they made their living off the land (Kakúwät 2005).

Mr. Wilfred Charlie's Northern Tutchone ancestors were heavily involved in trade with the coastal Tlingit during the 18th and 19th centuries, just as their southern neighbours and relatives were (Legros 1984, 2007). Fort Selkirk, located on the Yukon River at the mouth of the Pelly River, is the most well-known Northern Tutchone village and the furthest point inland visited by coastal Tlingit traders (Davidson 1901; Legros 2007; YHMA 1994; see map in chapter 9). The first European trading post to be established in Tutchone territory, Fort Selkirk, was built by Robert Campbell of the Hudson's Bay Company (HBC) in 1848 (Wilson 1970; Wright 1976). Only four short years later, the Chilkat Tlingit from Klukwan destroyed Fort Selkirk because it interfered with their trade monopoly with the people of the interior (Wilson 1970; cf. YHMA 1994).

Qwantuk was a Tutchone leader, and his home area of Little Salmon was one of the places the coastal Tlingit traders visited in Northern Tutchone territory (Charlie 1999). Qwantuk's story is long and sometimes troubling, since it records details about the difficulties that occurred between the Tlingit and Dän during what is assumed to be the early years of trade between them (likely before HBC established Fort Selkirk). It is clear that multiple revenge killings took place for a time, but eventually the parties established or agreed upon trading protocols and operating procedures so that trade could proceed for the mutual benefit of both parties.

Only a portion of Qwantuk's story is presented here—the part referencing a member of a Tutchone (Dän) trading party becoming lost on a glacier en route home to the Yukon from the coast. The man who never made it home was an older brother of Qwantuk. Wilfred Charlie shared Qwantuk's story about a month after the Kwädąy Dän Ts'ìnchį find was reported and suggested that the body might be that of this person lost on his trip home. Present at the time of the telling were Dawn Charlie (Wilfred's wife) and Yukon government archaeologist and family friend Dr. Ruth Gotthardt, who had assisted with the initial assessment of the Kwädąy Dän Ts'ìnchį remains. The full story, from which the following is excerpted, was later shared with Champagne and Aishihik Heritage program staff.

The second oldest brother froze up on a glacier near the coast.

The chief stopped his people from killing all of Qwantuk's party. They killed people before and they didn't come down to trade for a long time.

[After a while] people [i.e., the Tutchone] started to run out of things so they decided to go to the coast. They met a little wolverine who stood up like a man and told them to go back. Qwantuk told his dad: "Let's turn back, something wrong, wolverine bad luck." People say when a wolverine meets you, you got to kill him, bad luck....

His four sons keep going that time. The coast Indians can't kill Qwantuk's father because it is not his war. The war [was] between the other clan but which, I don't know.

Coast Indian cut the packs off the guys and they killed two of them.

Then three boys run back and one got killed for good and two made it back to the Yukon.

One of them climbed up on an icy mountain. He got so scared he climbed up on a high place and sat down. In the morning they hollered at him to come down so they could get going but he never moved. *They went up [to him] but he had frozen.* (Charlie 1999, 1; emphasis added)

The story continues, and we learn that the Northern Tutchone trading party, reduced in number but including the wounded Qwantuk, eventually made its way back to Little Salmon. But the party had been forced to abandon the frozen body of Qwantuk's brother. The remainder of the story recounts how Qwantuk was ready and prepared for a revenge strike when the coast traders next came to Little Salmon. A fort-like structure had been built, including a wall with holes in it to shoot through. Other Tutchone intervened, however, and peace was negotiated with the Tlingit, avoiding further conflict. The parties paid each other for the killings that had occurred, and after this the trade is understood to have proceeded on more peaceful terms.

Storyteller John Adamson: Glacier Rescue Story

JOHN ADAMSON (1917–2008), traditional First Nations names Kha~ Guxh- (Tlingit) and Ká Go (Dákwanjè), was a Champagne and Aishihik First Nations elder and member of the Killerwhale clan (Eagle moiety) (*Kha` Guxh-* 2008; see fig. 5). John Adamson's father was from Sweden, and his mother, Anna Duncan Adamson, was a Tlingit woman raised at Shäwshe, the Champagne and Aishihik traditional settlement also known as Dalton Post and by the Tlingit name Nesketahin (see chapter 10). Adamson's mother died when he was a child, and he then spent a number of years in an orphanage in Haines, Alaska. When he was about seven years old his maternal uncle, Paddy Duncan, the Wolf (Eagle) clan leader at Shäwshe, brought him back home to stay, though his sisters remained in Alaska. After serving in the Canadian army during World War II, Adamson and his Dän wife, Irene Slim Adamson, raised a family in the Whitehorse area. During his later years, Adamson lived at Takhini, a small Champagne and Aishihik community west of Whitehorse.

Adamson was the last Champagne and Aishihik elder to speak both Tlingit and Southern Tutchone in addition to English. He noted that when he was a young man travelling in Tatshenshini-Alsek country with his uncle, he was "forbidden to travel over glaciers" (Adamson 2006, 1). From the earliest time of the Kwädąy Dän Ts'ìnchį discovery, John Adamson understood its significance as well as the obligations it entailed. He supported the Champagne and Aishihik Heritage staff in trying to learn more about the Long Ago Person Found and continued to be a resource person on the project up until the time of his passing.

Adamson shared two stories that he believed were relevant to the Kwädąy Dän Ts'ìnchį discovery. The first is the story of a Tlingit trader who falls down a crevasse on the return journey to Klukwan. The man is taken for dead, and his travelling party returns to Klukwan, where a funeral is held for him. However, when a party returns to the glacier to retrieve the body, the trader is found alive. It is not known if this story refers to the same historical event as that recounted by Annie Ned or Kitty Smith, considered later in this chapter. The second story Adamson shared is that of the escaped slave named Ch'eet (or Ch'eet Tay), the "Rock Climber". We present two accounts of the Ch'eet story, one by Adamson and the second by another Champagne and Aishihik elder, Jimmy G. Smith.

Adamson's first story, of the trader rescued from the glacier crevasse, was recorded in two versions—one in video format (Smith 2001) and the second in audio format (Adamson 2006). There are also notes from a third, unrecorded telling (Greer 2005). Adamson advised us that he learned this story from his uncle, Paddy Duncan, who also told him that the rescued man was Adamson's maternal great-grandfather.[6] Adamson noted that his wife also grew up knowing this story and was able to add some additional details to the account.[7] As the story's central figure was rescued, we know he is not the Kwädąy Dän Ts'ìnchį individual.

The two recorded versions of this story vary in their details, which is expected given that Adamson was presenting to different audiences. The video version was recorded by Whitehorse-based Northern Native Broadcasting in the Dákwanjè language, with Adamson's daughter, Hazel Bunbury, in the role of interviewer (Smith 2001). English-language subtitles were then developed for this video recording, which was broadcast on APTN (Aboriginal Peoples Television Network) that same year. The audio version was recorded solely in English, with a non-family member asking the questions in a private office and no audience in attendance. The differences in the details make both versions worth sharing here. Text from the video version is presented first, followed by text from the audio version.

> Back then, no one really knew much about that man. But I was told the true story about it. It was my great, great, great-grandfather. It was him who jumped across the crevasse and fell into the ice. There was nothing anyone could do, so they left him behind. The river had frozen this way, and then this way. He fell in this way and this is where he landed. There were Tlingit packers from north and he was one of them. Fur was what he was packing,

not food. Furs, beaver fur, lynx fur and fox fur, all tied together. When he finally landed, he could see a little. It must have been light in the daytime. Then he untied his pack and made a bed. He put some on top of himself too, because it was cold in the ice.

The Tlingit men left him there. How can they bring him out? Then they had potlatches for him at GeKa'an, that place they call Klukwan. This is where they had potlatch for him, but he had to stay there without food. How can he get out? While at the potlatch, they were serving food in his honour. They would put out a plate of food for him too. It seemed like it was really for him. That is what he lived on.

"Hurry!" said the people. "Let's hurry and search for him." Now it was eight days and eight nights had passed while he was there. Finally the Tlingit men came back for him there. Then they lowered down one Tlingit man where he fell in. They went to save him. Right straight down there is where they lowered the Tlingit man, lowered him down. He was listening, he heard them lowering down that Tlingit man. They lowered the man down right beside him. He yanked on the rope and they stopped letting it down. All of a sudden his foot caught on something; it seemed to him it was fur. Then he said: "Are you alive?" "Are you human?" He said to him, "Really," sure enough and then he spoke, "Are you human?" "Am I dreaming?" "No, I am human, I come for you. They lowered me down." Then under here, under his arms, he tied a loop around there. Tie it good so that it wouldn't come undone. Then he yanked on the rope. It was going up like this, so they pulled up the man that had fallen into crevasse. "Hold on like this here, hold the rope," he instructed him. Here he was still alive when they pulled him out. Then he closed his eyes against the sun. He had been there for a long time in the ice. It was dark down there. Then he closed his eyes like this.

They untied him, untied the rope from his waist and tied a rock to it. They lowered the rock down there. Meanwhile, he gathered up the things that the man had been packing. Those that were not frozen in his pack he tied it together and yanked on the rope. They pulled it up. Then, he was standing down there. Wasn't he scared? After they tied his pack up above there … they tied the rope around the rock and lowered it back down. He could hear it hitting the sides. Finally, it got to the bottom. He untied the rock and tied the rope around himself. Then he yanked on the rope again. He was finally pulled out then. After that, that man who fell in … they first pulled out was given a piece of tallow in his mouth. Then it took a long time for them down there. Slowly, as he was staggering around. He had no strength, starving.

Then he finally came back to his home. They camped many times because he was staggering. Finally, after he came home, a long time later. It may have been five days later, before he finally asked: "What all was spent on me?" He said: "It was that, it was that, it was that." All of them he paid, he paid them all. Paid all the people that worked over him too. Because of that, they respected him. He was highly respected, my grandpa, who they saved.

Interviewer: What was the name of that man? Do you know?

No, I don't know. But they told me it was my grandpa. (John Adamson in Smith 2001)

In the audio version, which follows, we learn that earlier in his trip this trader had been approached by a Tutchone woman who wanted to buy his metal knife. This knife was exchanged for dry meat, and it was the latter that kept him alive during his time down inside the glacier. We also learn that after the rescue had taken place, the trip home to Klukwan was a slow one because the trader was so weak from his time down in the glacier. It took the group two days to reach Rainy Hollow, which is about 40 kilometres northwest of Klukwan on today's Haines Road.

Now there is another story, it was packers, packing from Selkirk and whatnot, Tlingit people packing. He fell through a crevice. Only thing, he had too much of a load on him there, he probably was just so wide, and he just couldn't make it, you see, he slipped and fell. Well, he's leapt though, over a crevice, was quite a ways down, the crack in the glacier, that just was beyond his hand....

Well, he landed but, gently you see, he was probably trying to hang on, but he didn't land side, he was stuck this ways. So he never got hurt. Now, how can he get out. Well, he was seen by the rest of the packers, but what could they do? They couldn't do anything there, they didn't have a rope. They continued, but they couldn't do anything. They, they marked the place.

Now, what this guy was packing, well furs, they trade there. That was during the time of, I guess they were trading with Russia, Russians, in Sitka. Russians at one time I guess you know that too, they, they owned Alaska. Well, our Uncle [i.e., Uncle Sam, the USA], he bought it from them. And they are kicking themselves yet for ever selling it, Alaska [laughter].

But anyway, well, he had a load of furs. And, recently, my wife, I was telling her that story, and she corrected me on something there. When he was coming up, uh, some women, she seen him, a knife. He had a knife in his belt. She stopped and was talking to him. "I like that knife, but I got no furs. I will give you so much dry meat."

He didn't want to let that knife go either, because in those days knives were [rare].... Wherever they got the steel from, he agreed [to the trade]. She gave him a whole bunch of food. And he give her the knife. That's what my wife's story is, there. I don't know where she heard it.

So, where he, where he landed down there, now he had that food, that dry meat, and some dried berries, stone berries, I think they call that. You can dry them. And they will keep. And he undid his pack of furs, laid it on the ice down there, covered himself up with furs, maybe he had Moose skins too.

Well, he was nice and comfortable, but [laughter] nowhere to go. He can't go up, he can't go this way, he can't go where. He just laid there for about, oh, maybe a week or 10 days. Just laying there, pretty well, he think: "How am I going to get out? I am just going to die?"

Well, his relatives had to give up, they went down to Klukwan, had a funeral for him. Here he was alive, and very much alive, and they, there was no cash money floating around those days. It was trade goods.

A rescue party was formed and returned to the place where the man was last seen. Adamson continues:

Now, uh, now, my uncle Paddy is the one that told me this story. There was a special reason. When they hired a different clan, you see, there are several different clans among the people, you got to work on, you see, your dead people, you got to get different clan [to help with the deceased].... Yeah, like we call ourselves here, Eagle clan, there's Crow clan, there's Wolf clan, and so on. You got to get somebody not in the same clan.

So they hired a guy, explained everything. Well, he knew it, he was in the packing [group] when this happened anyway. He say: "I know, I will volunteer." Now, he's doing it pretty well, and they braided up a bunch of, uh, you see, that babiche that they use for lacing snowshoes? Well, they braided all that, a bunch of that, making a heck of long rope. Now how, how big, did they know they are going to need, well they just had to guess. So, they sent the guy down, and they told him, they said: "Just swing the rope and holler, it will echo, and we can hear, echo." You can make a noise down there, and, well the echo, it goes right up. "And, if you find the corpse, and his pack, tie him, tie a rope on him"....

Now, I guess, the guy down there, he was down there, well, I guess so long, was just about going half nuts, I guess. Nothing to do. There was nothing wrong with him when he fell in there, and it's loneliness I guess and, when somebody come, he could hear something. Scratching or, he thought maybe he was dreaming, he just listen....

Well, this guy [the rescuer], he could feel it, uh, furs, what was furs that was dry doing down the glacier? He could feel a body. Warm. And, he [says], "How are you?" because he was scared. People are scared of corpses. Because he move, eh. They don't bother me none though, I don't know. The guy, I guess he was awake, somebody hollered at him. Now, he heard somebody coming down, but he wasn't thinking right....

And he answered him, "Are you human?" He said: "Yes." He says: "How are you?" "Well," he says, "I am all right but I am getting bored, getting bored." "How did you get here?" "Well, I was let down by a rope." "How?" Now, I guess it was lack of doing anything, he got so darn weak, just laying down. And he tied that rope around him there, got him sat up or, more or less, and tied a knot there, and he says "Hang on to the rope."

Yeah, rope around, and under the arms. Yeah, hang on, so you won't tip over. Now, he give an "Ahhhh" you could hear the echo, and shaking the rope. A whole bunch of guys, start dragging him up. The guy is supposed to be sending someone there, and he couldn't stand it, you see living in the dark for so darn long and then enters the sunlight, he's just [squinting] and just hanging on. [laughter]

Well, it must have been terrible. And the guys untied him, they were hanging on to him because he's staggering just like I stagger, just the same way now, getting weak. And, they give a chunk of tallow. My aunt used to do that to me too when I am tired, when I come in from hunting. She always insist there that I have a chunk of tallow. I don't know why, like.

And, so they tied a rock on him, banging into the gear there, and pulling. In the meantime, well, they sent the guy up. He gather up all his furs, and tied them up with rope, tied the rope onto it, and up went his pack.

Now he must have been a pretty brave man. I sure would have wanted to get out of there first. He was either a brave man or he was a damn fool. One of the two. And, so, up goes the guy's pack.

Well, another rock on the rope, hanging on and on, tied it to himself. Up he went. And they say, you see, it ain't very far from Rainy Hollow to timberline. And there's two guys helping him,

they camped two days, it took them two days ... to Rainy Hollow. Two days. That's how weak he was, he couldn't walk. But he was coming out of it because he was starting to eat. Yeah.

Now, uh, that's the way my uncle was telling me the story. He told us the story, and how, when that guy got back to his home, he called his relatives in. All of them, that looked after his funeral. Spent, whatever they spent. Paid them back. Paid them back, everything. And well, paid for his rescue party too, he paid them. So, he was considered a bigshot.

And then my uncle, when he finished the story, he told me: "Do you know why I am telling you this story?" "Well," I said, "it's a good story." "Well," he said, "that person who fell in that glacier was your great-grandfather," he said. (Adamson 2006, 4–6)

As told by Adamson, this story of a man lost in a glacier crevasse but later rescued is similar in many details to a glacier rescue story told by Annie Ned (Cruikshank et al. 1990) and to the story told by Kitty Smith (1982), referred to later in this chapter. It is not known if the three stories refer to the same historical incident.

The Story of Ch'eet (or Ch'eet Tay), the Rock Climber

THE STORY OF CH'EET, the escaped slave, was first brought to the attention of the Champagne and Aishihik Heritage program by Champagne and Aishihik Elder James Gilbert (Jimmy G.) Smith (fig. 5). Smith (1919–2005), First Nations name Gooch Sháa, was the son of Jake (Jack) and Mary Smith. A Champagne and Aishihik elder and member of the Wolf clan, Smith was raised in the traditional way, trapping and fishing with his family in the Shäwshe, Kusawa and Hutchi areas (James Gilbert Smith 2005).

In 2000 Jimmy G. Smith suggested that the Kwädąy Dän Ts'ìnchį individual might be Ch'eet, the slave who escaped to the interior from the coast. While the story does not directly refer to someone who became lost while travelling over a glacier, the landscape involved is similarly challenging, and the story is known to have taken place in country where there are lots of glaciers.

The story of Ch'eet was also brought forward when a Champagne and Aishihik First Nations delegation travelled to Yakutat to discuss the Kwädąy Dän Ts'ìnchį discovery with Yakutat tribal representatives later that same year. At that gathering, Yakutat Elder George Ramos made reference to the Ch'eet story, noting its similarities to particulars of the Kwädąy Dän Ts'ìnchį discovery (Champagne and Aishihik First Nations 2000b). Champagne and Aishihik First Nations Elder John Adamson, who was present at the Yakutat meeting, recognized the story and was able to provide additional details. His telling follows Smith's account.

The Ch'eet story can be summarized thus: being held as a slave in the Yakutat area after a conflict with another group, Ch'eet is urged by his captors to demonstrate his rock-climbing skills. He takes up the offer, negotiating for his freedom if he can scale a certain rock face. As the slave is about to start, he requests items to be added to his gear, in order to provide "balance" for the climb. The climb is successful, and the slave escapes his captors, heading to the interior, never to be seen again.

Details and emphasis vary in the two accounts, and therefore both are worth sharing. We begin with Smith's telling. Note that the term "dope" refers to the man's medicine power.

> This story was told by my grandfather and my mother from coast. My mother name Mary Smith.... They used to live down there in Dalton Post. That's the way, just before that, the time that Indian war started up in Klukwan. I guess there must be some kind of a trouble against one another or something. Klukwan-tan and the Shangukeidis start fighting. That's when this guy here was the last man left that time with the two of them. The other one he went the other way, they got no way to get out and he hide around in the bush....
>
> I think that's the man that was found. It's gotta be that one because he's got a design and signs and everything on there. He's got a coast Indian, the way

they used to make hats, he's got one of those. Yeah, that's the man.

> So they chase him up, a bunch of boats at the bottom spear right into him. They drive him up a cliff. Cliff run right down to the salt water. Bunch of boats at the bottom for one man to kill. But this man he's got a different idea too. Long time ago people, don't matter where they are—coast and everything—they using this dope [i.e., medicine or power]. They call that, some kind of message come to them. They call them Indian daughter, something like that.
>
> So this man he had something like that in his mind and he did it right on that cliff, had nowhere to go. He works, so he tells those guys, "I'm gonna come down to you fellas," "I'm gonna come down, I'm gonna give up, I'm gonna give myself to you fellas." "But the way I'm gonna make it myself. And give me two and a half dry salmon and a little bit of grease, oil. I'm gonna eat that before I start coming down. And a pair of snowshoes." So they haul that up to him and he's sitting there ready to fall on that cliff there. So this guy tell him, he says, "We don't want to do it for you but we're gonna save you." "You're telling me lies," he says, "I know that. I'm gonna give up on you guys when you give me that pair of snowshoes and stuff like that." He say: "Thank you for that."
>
> So he eats some of that giant salmon and he's ready to come down. Everybody's got a spear for them to kill him. This man, I think I've got his name, I'm quite sure of how the coast people know that man. His name "Cheat-tay", that what his name. This man he's got doped up with everything. So they started up. He eat now, he says, "Gonna come down," he put on his snowshoes.
>
> And this dope work on him, the ptarmigan come on him and he started going with his snowshoes. The snowshoes stuck right to the rock. He make a noise like a ptarmigan: "Uk uk uk uk uk uk uk." He went right straight up on that cliff and he went right over the top with his snowshoes. That's why they call him Chee-tay.

That's what I heard when I was young, I would listen to this story from my grandfather, from my mother, all of them. That's why my mother they got sign from the coast, you see, that's where they come from.

So he went over the hill. They can't do nothing. All they send a bunch of young guys after him. Deep snow up there too. So he starts putting on his snowshoes and put his packsack on and he took off into the Yukon. I believe that's the man that they found. So ... that's the way the story go. So he went up halfway and he sees that they're starting to catch up to him by the sound length, so many people talking here, so he take his snowshoes and he turn around the other way around. So he start walking backwards with it. He set it up to go ahead but his snowshoes were backwards. He got close to the top there where you can get a solid drift now. And he can walk without snowshoes from there easy so he took it off there and run away from there. And this man come on up, he walk up this way. Go head the other way, he's looking around for his track again, and when he turn his snowshoes around to find that's trail going backwards, going back down to the sea. You know everybody then is coming back down this way, went down back to the sea, so everybody follow him. He turn his snowshoes around, pretty tricky too himself I guess....

Ch'eet. That's what his name, that man, supposed to be. Went over that rock. That's why they call him Ch'eet-tay. Ch'eet-tay mean "Ch'eet Rock". Tay is rock. Ch'eet is the man name.

Interviewer: And both of the people that were fighting in the war were coastal people from Alaska?

Yeah. All people from Alaska yeah. They two different relations though.

Interviewer: And this person Ch'eet tay, they chased him inland?

Yeah. (Smith 2000, 1–4)

John Adamson similarly suggested that the Long Ago Person Found may have been Ch'eet, the escaped slave. Adamson noted that the items Ch'eet requested of his captors for balance—items such as the man's knife and possible arrow shafts (see Beattie et al. 2001; chapter. 12)—were similar to those found with the Kwäday Dän Ts'ìnchį individual (Adamson 2006; Greer 2005). Adamson related the story as follows:

Well during the time, you see, they war. In the old days, there was no law in the country. There wasn't no law. People war, and they used to come up and land, and revenge some[thing], or make trouble or some darn thing. Well, Ch'eet was his name. Now, when the war party landed, of course, they had heard about him, how good he was.... His name is mountain ptarmigan, that little ptarmigan, they call mountain ptarmigan.

Well, I guess, they killed other people, for some revenge or over some trouble. But him, they grabbed him, and they explained to him, "We hear how good you are, climbing." "Well," he says, "I do climb, who doesn't?" he says. They went easy on him. They took him to a cliff, and well, "Climb that." He tried. He says, "I can't do it." They said: "Why?" "I haven't got the balance." "What do you need for balance?" "You just pick up a rock or something." "No," he says, he wish for something you see today in a museum, he say, "I have to have a packsack, and in that packsack, I have to have a knife, I have to have lunch," and he described the lunch, that's all in the story.

"And bow and arrows, I got to have that for balance." So, so they give it to him. And they figure he's going to pay anyway, so they gave him that stuff, so that's what you see in a museum now. Because I heard about it. That's in the story. That's what we see.

Well, he went up that wall, and he climbed to a height where they couldn't get at him. Well, really, they didn't use bow and arrows, but spears, they couldn't get near him with spears. "Well, okay, come on back, come on back, you win, you win." "Goodbye." [John does hand motion, hand to nose, waving, to indicate goodbye]....

Yeah, goodbye. Well, there is no danger, and nobody is going to follow him up that cliff. So, this

happened down around Yakutat, and here he come
up the river, he come up that cliff someplace, and
then on level ground, and thereʼs a pass. Well,
where was he going, thatʼs where he was going.
He fell through the, he fell through, he dipped over
a crevice, I guess. Thatʼs my guess.

Interviewer: So you are thinking, wondering,
if that is the man they found [i.e., Kwädąy Dän
Tsʼìnchį]?

Thatʼs the one. Thatʼs got to be! (Adamson
2006, 2–3)

While the Chʼeet term is Tlingit, referring to the
seabird known in English as a murrelet, both John
Adamson and Jimmy G. Smith reported the name as
meaning "ptarmigan". Given that both men had lived
most of their adult life away from the Pacific coast where
murrelets live, it is possible that neither was familiar
with the species. Perhaps the ptarmigan reference was
because of its known association with travel over snow;
it was traditional Dän practice to put ptarmigan on a
childʼs feet or snowshoes, in order that the child would
be fleet-footed when travelling over snow. Regardless,
it is noteworthy that a "Chʼeet Hit", meaning "Murrelet
House", was one of six Dak̲lʼaweidí clan houses that
formerly existed at Klukwan (cf. Hope and Thornton
2000, 150) and that both storytellers had ancestors that
came from this Chilkat Tlingit village.

Migrations, Trade and Glacier Travel

TURNING FROM RECENTLY RECORDED STORIES, we
now consider previously published accounts referring
to glacier travel. It is important to note that the
traditional stories of the Dän and the Tagish, as well
as their fellow inland neighbours who likewise spoke
Athapaskan languages, do not mention their people
having migrated to the region from elsewhere. Rather,
the *kwändür* (traditional stories) of these peoples imply
that they have deep ties with their present territories,
going back in time to when the "Creator" fixed the land
to be as we know it today (de Laguna 1972, McClellan

1975, McClellan et al. 1987). In contrast, migration
stories are a dominant theme in Tlingit history, and
certain Tlingit clans are understood to have migrated
to their present homelands by travelling *under* glaciers
(de Laguna 1972, 1990; Hope and Thornton 2000). As
the migrating groups travelled from the interior to the
coast down rivers such as the Stikine, Taku and Alsek,
they found the way blocked by glacial ice, forcing them
to travel via watercraft under the ice. The journeys
were understandably terrifying, but the travellers
eventually reached the coast and their new homeland.
These migration stories are thought to have taken
place in ancient times, perhaps thousands of years ago,
when glaciers were still receding from their maximum
coverage of the land during the Pleistocene.

Stories of Tlingit migration *over* glaciers appear to
refer to more recent events. The story of a migration
that is thought to have taken place in the 18th century
was one of the first accounts of Tlingit glacier travel
to be published. Recorded in Sitka by researcher John
Swanton, the story was told in Tlingit by clan member
Kʼáadasteen while Don Cameron, another Tlingit,
assisted with translation (Cruikshank 2005, 33; Swanton
1909). The clan migration mentioned in the story was
triggered by a conflict that occurred after the death
of a Raven clan chief. It resulted in the Kwáashkʼi
Kwáan clan, who were Copper River Ahtna-speaking
Athapaskans, migrating to the Icy Bay and Yakutat
Bay areas on the coast via a route that took them over
a glacier in what is now Wrangell-St. Elias National
Park. The travellers eventually married into the coastal
Eyak and Tlingit people. Nonetheless, not long after
the group first reached their new homeland, a party
returned to the interior with goods to exchange, thus
beginning a trade that lasted for many generations
(Cruikshank 2005, 33).

While inter-group conflict was the original motive
for travel in the Kʼáadasteen story summarized above,
the end result was the establishment of a trade network
between interior and coastal residents.

Trade was also the motivation in the story of the
war knife known as "Ghost of Courageous Adventurer"
(Shotridge 1920). This weapon was made to honour or

commemorate the travels of a group of Tlingit men who went, under the leadership of a man named Eagle Head, from Klukwan overland in a northwest direction to seek out new peoples to trade with. The story was recorded at Klukwan by Louis Shotridge, a Chilkat Tlingit anthropologist, from an unknown Chilkat man who was a member of the Shungukaydi clan (Thunderbird House), Eagle moiety.

The "Ghost of Courageous Adventurer" story is the only glacier travel account we are aware of that definitely refers to passing through Tatshenshini-Alsek country. The story recounts how, after about six days of travel heading northwest from Klukwan, the group came to the Alsek River (Shotridge 1920, 16), thus placing at least the first part of the story in the Tatshenshini-Alsek region. But the group's most arduous glacier travels occurred after they crossed this river and continued westward, where they crossed a large glacier: "In the cool of dawn underfoot of man was firm, but as higher rose the sun that firm surface began to melt, hence to the feet came fear.... It was slow, they say, for crevices were many...." (Shotridge 1920, 17, 18). The risks taken when travelling over a glacier, including the possibility of falling down a crevasse, were ever-present.

The party eventually reached Yakutat Bay and continued on, over the Malaspina Glacier, which was even more dangerous. Two young men were lost in crevasses on this portion of the journey. The remaining travellers continued on to the Copper River, from which point they turned back toward home, bringing new trade goods such as ivory and driftwood iron (Shotridge 1920).

Trade and Culture Change

THE QWANTUK STORY, the K'áadasteen story and the "Ghost of Courageous Adventurer" story are all accounts from the early days of trade between the coast and the interior. They relate to a pattern of exchange between the two regions that would develop considerably. This trade, and the culture change associated with it, became the defining theme of the region's history in the 18th and 19th centuries.

Our understanding of this trade and its importance in shaping local culture and history comes largely from oral history accounts as documented by researchers such as McClellan (1975), de Laguna (1972), Emmons (1991) and Cruikshank et al. (1990). Oral history makes it clear that the trade is long-standing, predating the arrival of European trading ships on the northwest Pacific coast in the mid 18th century. Using archival records, one researcher has tracked details on the scope and volume of the exchange (Legros 1984). The trade increased in volume, reaching its height from roughly 1750 to the 1890s. It slowed significantly in the latter decade when the interior Yukon country was opened up to non-Indigenous people.

The trade between the coast and the interior had its roots, as systems of exchange do, in peoples residing in environmentally different regions exchanging goods from their home territories for materials not locally available. Coastal peoples traded things like dentalia shells, obsidian, dried seaweed, fish oil, spruce-root baskets, cedar boxes, fungus for paint, medicinal roots and native tobacco to the interior. Items such as raw copper, sinew, Mountain Goat wool, tanned Moose and Caribou hides, as well as finished hide and fur clothing, made their way in the opposite direction (McClellan 1964, 1975).

With the arrival of Russian and, later, British and American traders on the northwest Pacific coast in the mid 1700s, the Tlingit living there were able to offer the newly available European trade goods to the interior people with whom they traded. Inland furs also became especially sought after by European traders once the Sea Otter, the initial focus of their trade, almost became extinct from overharvesting around the turn of the 19th century (de Laguna 1990).

Like all tribes along North America's western coast, the Tlingit of the Alaskan coast consolidated their position as middlemen in this trade, transporting goods from the European trading ships to the interior, where they were traded to the Dän, Tagish and others. These routes to the interior were strictly controlled, and the

European newcomers were not allowed to access them (de Laguna 1990; Emmons 1991).

The Chilkat Tlingit of the Klukwan and Haines area were the most aggressive about protecting their position as middlemen in the trade. They jealously guarded the travel routes, such as the Chilkat and Chilkoot trails, which they used for trade access to the interior (CTFN and Greer 1995; de Laguna 1990; see fig. 2 in chapter 9). Europeans were not permitted to establish trading posts in this area. All trade in the north Lynn Canal area took place from ships. This blockade continued until the United States purchased Alaska from Russia.

As told in the Qwantuk story, difficulties arose during the early years of trade with interior peoples until such time as trading protocols and trading partnerships between Tlingit and Dän were established. Such partnerships were often formalized by arranged marriages—sisters of coast Tlingit men married to their interior trading partners.

We learn from Emmons (1991, 803) that many of the trading routes to the interior used by the southeastern Alaska Tlingit traversed glaciers. Despite the dangers of crossing glaciers, swift rivers and rugged terrain, trade between the coastal and interior people continued for generations. Eventually, what had started out as a trade in luxury goods became a vital part of the regional economy, a necessity that continued regardless of the travel dangers involved.

More Storytellers

ACCOUNTS OF PEOPLE that went missing when crossing glaciers were among those documented from various Yukon women storytellers by anthropologists Cruikshank and McClellan (Cruikshank et al. 1990; McClellan et al. 1987). Cruikshank worked with storytellers Angela Sidney (1902–1991) (of Tagish and Tlingit descent and a member of the Deeisheetaan Tlingit clan), Kitty Smith (1890–1999) and Annie Ned (1880s–1995) (of Tlingit and Dän ancestry).[8] These three storytellers shared accounts of Chilkoot or Chilkat Tlingit traders who fell in crevasses while traversing

glaciers. Angela Sidney's story talks about her father's rescue of his trading partner while they were travelling between Chilkoot Lake and Robinson, which is situated north of Carcross (Sidney 1982). In this account, the man who fell down the crevasse was rescued right away by his trading partner, Tagish John (Sidney's father), who was able to pull him to the surface with rope.

In the two other accounts recorded by Cruikshank (K. Smith 1982; Ned 1984), we learn of a trader who fell into a glacier crevasse. Unable to make contact with him or rescue him, the others in the party continue home to Klukwan, where a funeral is held for the lost man. In both Ned's and Smith's telling of the story, a group returns to the glacier to retrieve the lost man's body only to find him still alive. These details are similar to the account from John Adamson, related previously, but it is not known if the three stories refer to the same or different historical events. Smith's account does not specify a location or a direction of travel, while Ned and Adamson both refer to an incident that took place as travellers were heading south on the route between Kusawa Lake and Klukwan. Mrs. Smith's version of the story provides a name for the man who fell down the crevasse and was later rescued: Ḵanaẋ (Smith 1982, 95).

McClellan was told a story of glacier travel in 1949 by Teslin Tlingit elder and storyteller Kitty Henry (McClellan et al. 1987, 219). Unlike the accounts related to trade that are mentioned above, this story has an unhappy ending and is also understood to be very old. The story was paraphrased as follows.

… a man who had gone hunting … had slipped down into a crevasse in a glacier. He could not get out, and he did not know what to do. He had no food with him, nor any way to make a fire. Finally he curled up and went to sleep. Because he was wearing snowshoes and there was fresh snow on the glacier, his people were able to track him to the crevasse into which he had fallen. He was still alive when they found him, though he should have been dead. He had lived because the family had put food in the fire for him at home. The man could not be rescued, and eventually he died—but not before he had told the people how to burn the bodies of the dead, how to put things at people's graves, and how to give feasts. He reminded them how important it was to put food into the fire so that the dead would have something to east. (Henry, as paraphrased in McClellan et al. 1987, 219)

Mrs. Henry's story shows that long-ago events involving glacier travel help teach lessons in behaviour, such as the correct way to treat the dead. The story even explains why memorial potlatches are held (see chapter 8). In fact, the practice of "feeding the fire" continues today at funeral services across the southern Yukon, northwestern BC and southeastern Alaska.

Glaciers Must Be Respected

> You travel on a glacier as long as you don't cook nothing there. Glacier you can walk on it, but you got to watch for crack. You got to have stick [i.e., walking stick] all the time.... A lot of people fall on that kind of place I heard. (Moose Jackson in Jackson and Brown 1997, 13)

THE STORIES PRESENTED IN THIS CHAPTER, shared by First Nations and tribal elders, provide insight into the traditional Dän, Tagish and Tlingit way of understanding the world. In this traditional way of life, all plants, animals and even mountains are understood to be animate or alive. Researchers McClellan and de Laguna were the first to help non-Indigenous people understand this worldview, through their writings *My Old People Say* (McClellan 1975) and *Under Mount Saint Elias* (de Laguna 1972). Later, Cruikshank further explained the Indigenous way of understanding the world. Through detailed study of traditional southern Yukon narratives she pointed out that, for these cultures, glaciers are similarly viewed as living entities (Cruikshank 2005).

Traditional Dän, Tagish and Tlingit accounts about individuals who ran into trouble on or near glaciers also communicate the need for appropriate conduct around

them. The stories tell of negative events arising when proper care, respect and precautions were not taken (Cruikshank et al. 1990, Cruikshank 2005), and they contain warnings about the need for extra respect for these special places (Emmons 1991, 803). Perhaps the most commonly cited prescription is the one against cooking grease in close proximity to a glacier (cf. Ned in Cruikshank et al. 1990, 332–36; Smith in Cruikshank et al. 1990, 207; Shadow and Jackson 1997). In addition to reminding us of specific dangers, contemporary elders remind us that the natural world and the old teachings must be respected, and care must be taken when crossing these unique frozen landscapes.

Identity and Responsibility

WHO WAS THE LONG AGO PERSON FOUND? Was he Qwantuk's brother, Ch'eet the escaped slave, an unknown trader who ran into trouble, or perhaps simply someone travelling between villages? Will we ever know his name?

In the years since the Kwädąy Dän Ts'ìnchį discovery, there have been glimmers of hope that we might one day learn the young man's name. One such hint came during an informal interview session with John Adamson (Greer 2005), where he recalled a visit he had when he was a young man with Little Shorty, a Dän man who lived with his wife at Bear Creek, just west of Haines Junction. During this visit conversation touched on mountain passes in the Tatshenshini country. Adamson was regularly travelling in Tatshenshini country with his uncle, Paddy Duncan, at that time and was therefore familiar with it. He recalled that during these discussions, Mrs. Shorty mentioned that "there's a person missing there", but after saying these few words, she offered no further details.

Adamson didn't bring up this exchange with Greer until 2005, a number of years after the Kwädąy Dän Ts'ìnchį discovery. Perhaps he had forgotten it, a comment that had been offered in passing some 60–70 years previous. Or perhaps he was waiting for the right time to mention Mrs. Shorty's words. Three years later, in the spring of 2008 and only months before his passing, Adamson brought up Mrs. Shorty's comment again, this time with Champagne and Aishihik staff Lawrence Joe and Frances Oles when they visited him to advise on the results of the community DNA study (chapter 33; Lawrence Joe, pers. comm. to Greer, April 2008). On that occasion he also mentioned that even the name of the missing individual had been stated by Mrs. Shorty, and while he struggled to recall that name, it failed to come to him.

From today's perspective, one might think that Mrs. Shorty's comments represented unfinished business, something that the community hadn't taken care of. Why hadn't they gone back to retrieve the body? Or had they tried but been unsuccessful? With the passing of elders such as John Adamson, Jimmy G. Smith and Wilfred Charlie, the hope of ever having answers to these questions wanes, as does the possibility of learning the name of the Long Ago Person Found.[9]

Conclusion

THE TLINGIT, TAGISH AND DÄN oral narratives about glacier travel and the people who encountered difficulties crossing these frozen landscapes have been passed on for generations. These powerful stories teach about the perils of such travel and serve as warnings to travellers in these dangerous areas. The stories remind us that glaciers must be respected and that necessary precautions and preparations must be taken when journeying near or over them.

But the stories are more than just teaching devices or simple lessons to be learned. They document past events and processes of societal change—such as clan migrations and the history of trade and interconnections between coastal and interior peoples. For those familiar with even one of the stories presented in this chapter, the discovery of a man frozen on a glacier in Tatshenshini-Alsek Park was not totally unexpected.

Storytellers Wilfred Charlie, John Adamson and Jimmy G. Smith had direct ancestors who were travellers and traders, people who took journeys across glaciers and similarly dangerous landscapes. So while

Figure 4. Wilfred Charlie at camp. Photo courtesy of the Charlie family.

these stories represent aspects of the shared histories of the Tlingit, Tagish and Dän, it is important to recognize that they are also *family stories*, accounts of events and experiences that the storytellers' ancestors actually encountered. Charlie, Adamson and Smith knew these stories because they had been passed down within their respective families. They, in turn, shared their family stories with Champagne and Aishihik First Nations because they knew doing so would help others better understand the Kwädąy Dän Ts'ìnchj discovery. They honour their ancestors by remembering and passing on these accounts.

The Kwädąy Dän Ts'ìnchj discovery clearly verifies the experiences of the Tlingit, Tagish and Dän, as recounted in traditional narratives. It illustrates that our traditional stories are indeed our history, a record

Figure 5. Storytellers Jimmy G. Smith (*left*) and John Adamson at Kusawa Lake, August 1999. Sarah Gaunt (CAFN) photograph.

of our peoples' lives and past experiences. These stories help community members and others understand and potentially interpret many aspects of the Kwädąy Dän Ts'ìnchį discovery. For instance, they give us insight into the social milieu in which the Long Ago Person Found lived, how he might have come to be in the area where he lost his life and the dangers inherent in these landscapes. These stories also illustrate the traditional values of the region's Indigenous peoples, providing insight into worldviews regarding responsibility, duty and obligations to take care of each other, including seeing that the dead are given the respect that our cultures prescribe.

Although few contemporary Dän, Tlingit or Tagish now experience walking atop a glacier as their ancestors did, they nevertheless continue to learn the importance of respecting the natural world through their peoples' past encounters with glacial landscapes. These stories provide the younger generation a sense of who they are as Dän, Tagish and Tlingit.

ACKNOWLEDGEMENTS

Greatest credit must go to the storytellers featured in this chapter: the late John Adamson, Jimmy G. Smith and Wilfred Charlie, who came forward to share stories that they knew were relevant to understanding the Kwädąy Dän Ts'ìnchį discovery. This chapter builds upon the efforts of oral historians such as Frederica de Laguna, Catharine McClellan, Julie Cruikshank and others whose work was possible thanks to the many Aboriginal storytellers who shared their knowledge, including the late Angela Sidney, Kitty Smith, Annie Ned, Kitty Henry and Moose Jackson. We are very grateful that these individuals worked together to document the traditional stories of the region's Aboriginal peoples. We also thank summer student Megan McConnell for searching Champagne and Aishihik First Nations oral history transcript files for relevant content.

30

LIVING IN AND TRAVELLING IN GLACIER AND MOUNTAIN LANDSCAPES

Ron Chambers

Ron Chambers, a Champagne and Aishihik First Nations citizen and Wolf clan member, lives in Haines Junction, Yukon. He is a well-known guide and outdoorsman with considerable experience around glaciers. At the time of the Kwädąy Dän Ts'ìnchį discovery in 1999, when he was Deputy Chief of Champagne and Aishihik First Nations, Ron provided logistical advice and assistance to the recovery team. He assisted on subsequent visits to the discovery site as well and is responsible for finding many of the additional artifacts recovered from the surrounding area. Ron was also part of the group that in 2001 returned the Long Ago Person Found's remains to the mountain where he lost his life.

Wearing his cultural regalia and headdress, Ron began his presentation at the April 2008 symposium in Victoria by singing and drumming for the Long Ago Person Found. The Southern Tutchone song he shared is called Shadäla, the traditional name for the settlement known today as Champagne. The song is about a hunter who asks permission from his family before he goes out. He says: "I have to go out to hunt. I will be gone for a while, but I will be back. Hopefully I will be successful." It is his goodbye song until he comes

back. Ron explained that Shadäla is an appropriate song to
honour the Kwädąy Dän Ts'ìnchį person, the Long Ago Person Found,
who came back and has brought us so much. What follows is the
text of Ron's symposium presentation, as edited by Sheila Greer
and Gordon Allison.

GOOD MORNING. *Gwänischis.* Thank you for allowing us
on your traditional territory. *Gwänischis* to Champagne
and Aishihik First Nations, to Chief Diane Strand and
to the people who have put this conference together.
It's a very powerful thing to be here and to have the
opportunity to come up and sing for you.

We talk about our ancestors and our history. We
need our history to carry on our future. We have many
of our people in the last number of years trying to
bring back some of the stories and relate our history. At
this time in our history it is appropriate for us to open
up, to share our history and our stories. The Kwädąy
Dän Ts'ìnchį discovery has brought our history to the
attention of Canada, North America and many parts of
the world. This makes us feel very good, but it also gives
us a challenge: to take this information back to our
people so we can all benefit by it.

I was involved with the Kwädąy Dän Ts'ìnchį
discovery at the beginning, when the find was
first made. At the time, I was Deputy Chief, which
meant dealing with protocol matters and developing
agreements with the British Columbia government. I
am very proud to say that we had the people in place in
our organization that could handle these matters and
allowed us to get to the place we are today. Our people
have been well represented in this matter, and I believe
the audience here today recognizes that—and that we
have a message to give to people.

I appreciate all the hard work that has been done,
the information that has been gathered and shared at
this symposium. This means a lot to us and has opened
up a lot of interest. There have been some tough times

for us in handling this discovery, as some of our people
have expressed concerns related to spirituality—that we
might not be doing the right thing. But I feel the positive
parts, the inspiration, outweigh the negative things.

I'll give a little background on myself to help
explain my interest in my people's history and culture.
When I was quite young the elders used to get us to
dance almost before we could walk. You might cross
a room and the elders would start to sing and dance,
and without even realizing it, you would be doing it
too. When I got to be a teenager I wasn't so keen to
show this side of myself, but when I got a little older I
went to Haines, Alaska, and ended up being a Chilkat
Dancer. I was one of the original members of the
Chilkat Dancers dance group. The headdress that I am
wearing, which shows my clan and crest, dates from
that time, made in 1962.

Being involved in the Kwädąy Dän Ts'ìnchį
discovery is a continuation of what has been a lifelong
pursuit for me—to be aware of what my culture is. I've
also had the real fortune of being at the right place
at the right time. When I was a teenager growing up
in southwestern Yukon, I worked with archaeologists
Scotty MacNeish and George MacDonald; some of
you may have known these men. George MacDonald,
a student at the time, was more like my understudy
because I had worked with Dr. MacNeish the year before
he did. Scotty MacNeish taught me about those little
stone tools, called microblades, which were used by our
ancestors thousands of years ago. Dr. MacNeish had lots
of nicks and cuts on his wrists from demonstrating to
me, and lots of other people, just how sharp those little

tools were. The Band-Aid company must have made lots of money from him.

These are the types of experiences I am talking about, including meeting the right people at the right time, people who opened my eyes to seeing things. Such experiences have taught me to be looking for old things almost my whole life. Finding things is not luck but keeping your eyes to the ground and appreciating what you see—and, of course, recording what you've found. I am grateful for those people who took the time to talk to me and open up my eyes about my people's history.

I spent a lot of my life working as a park warden for Kluane National Park and working on and around glaciers. I've been involved in search and rescue for people who have dropped in crevasses, and I know what they are like. Now, the Kwäday Dän Ts'ìnchi individual travelling in glacier country—his people had to understand and deal with the dynamics of glaciers. It's not hard to see why our peoples had stories and songs about glaciers, to help teach them about these special places. They are very dangerous places, especially when a blanket of snow covers the glacier so you can't see where a crevasse is. Today's mountaineers still recognize glaciers as one of the biggest hazards they deal with.

In 2003, just a few years after the Kwäday Dän Ts'ìnchi discovery was made and not that far from the discovery site, just such an incident happened. A couple of Royal Canadian Mounted Police (RCMP) officers were snowmobiling on a glacier in Tatshenshini–Alsek Park when one fellow got off his snowmobile, took a couple steps and dropped into a crevasse. He went down far enough that the other person didn't have enough rope to reach him, but he landed on a ledge that stopped him from dropping further down. He spent a whole night down in the crevasse before a full rescue could be mounted to save him. After that, this policeman ended up being referred to as "Not So Long Ago Person Found" in our community.

This incident showed that even today this kind of accident can happen. But during the times the Kwäday Dän Ts'ìnchi individual lived, things were even more dangerous. This is because crevasses are V-shaped, and

if you fall into a crevasse with a pack on your back, the chances are good that you will get jammed so bad that you can't be pulled out. I suspect that this may have been what happened to the Kwäday Dän Ts'ìnchi person, who might have been carrying a heavy load.

He was also found with a gopher-skin robe; these robes were treated with such respect that I think it might have been part of his load rather than wearing it if he was packing. So I am speculating that they may have tried to hook him and pull him up but weren't successful. Perhaps that's why the rest of his goods and clothing were lost.

The people of olden times did ceremonies when they would be travelling across glaciers, especially when they travelled down in May, when there would still be a skiff of snow over the ice. On a cloudy day it can be especially difficult to see when you are travelling across a glacier; there is white all around you, making it difficult to determine just what is in front. You may still know the direction because you can kind of see the mountains, but you can't see what you are walking on.

Despite the risks, the old-time people still had to go there, to cross glaciers. They had a purpose in mind. But there's lots of spiritual feeling about glaciers, because glaciers sing; they hum. I've heard what I thought were helicopters and airplanes and outboard motors when I've been camped by a glacier. Now, 200 years ago if they were camped anywhere near a glacier, they would also be hearing these kinds of sounds, and those things had not even been invented yet. These sounds come from those glaciers. I've heard a booming sound that started a kilometre up the ice and travelled down the ice right past our camp and beyond. When you are out there, by a glacier, you realize that it's alive, absolutely alive. So you can see the spiritual feeling that our people had toward glaciers when you travel in these places. Yet our people still crossed the glaciers because trade was that important to them.

The dynamics of a glacier are fantastic, and I just wanted to add my knowledge of that because glaciers are a big part of the Kwäday Dän Ts'ìnchi picture. Glacier stories are global kinds of things that tie into the Kwäday Dän Ts'ìnchi discovery.

I have just a couple more things I'll comment on about the Tatshenshini-Alsek country, where the Long Ago Person Found lost his life. In this country it's very hard to travel in the summer, because of the brush and alder and everything. Trails are extremely important, and over time we've lost a number of trails as the trees fall down and nobody is there to remove them. These kinds of cultural resources are being affected, with no one taking care of the trails in this country.

I'd also like to mention one of the stories that has come down through our people. One of our former chiefs, Johnny Fraser, was known as the last of the Núghàyík people. He was raised at Shäwshe, but his parents were from further downriver at Núghàyík, and Johnny—like his parents—had travelled up and down the Tatshenshini and Alsek rivers during his lifetime. One day I was talking with Johnny and he mentioned the "roacher". "The roacher?" I said, not knowing what he was talking about, if it's a church group or what. "You know," he said, "they lived in Alaska a long time ago." "Oh, the Russians," I said. "Yeah, Roacher," Johnny said. He then went on to tell me that the Tlingit had told our people about these strangers that were around before the white man came here. The Tlingit in this area are big people, powerful people. They were traders and drove hard bargains, pretty impressive in their own right. Johnny described the Russians as "big, they got white skin like snow, white skin, big ears, red hair." As Johnny told me this story, he was getting pretty excited, describing what he had heard about the Russians. It made me think that stories like the abominable snowman might be like this, coming from somebody describing somebody else.

The explorer Glave, who was one of the first non-Native people to travel through our area, met some of our people living along the Tatshenshini River. One of these people was Jimmy Kane, who was a young boy at the time in 1890. I knew Jimmy as an old man. I could still talk to people such as Jimmy who was there when the first contact was made. This is how recent our contact with the outside world was. The Kwäday Dän Ts'ìnchį discovery has brought back that part of our history that seems just like yesterday.

For many people, Yukon history started with the gold rush of 1898, but for First Nations, the gold rush is really kind of recent. That was within Jimmy Kane's memory, and we are lucky we have had people like him and Elder Annie Ned share so many stories about our history. We've come a long way, but we've got a lot to do yet to put the stories together.

One other thing I'll mention is that people from many different parts of the world have joined our people. My grandmother married a white man, who was a member of the North West Mounted Police. When he died, she remarried another white man, but he, too, died after just a couple years. She was in her 80s by then and said: "I'm not going to marry no more white man; they die too easy."

I close by saying that the reason we are here and what we are doing here made me want to sing. So I sang you a song, and we have some other people here from the Yukon that also want to sing to show our appreciation to everyone for taking part in this event and to those who put it on. *Gwänischis.*

31

ENVIRONMENTAL SCANNING ELECTRON MICROSCOPY

A Modern Tool for Unlocking Ancient Secrets About the Last Journey of the Kwädąy Dän Ts'ìnchį Man

Petra J. Mudie, James H. Dickson, Richard J. Hebda and Francis C. Thomas

In 1999 the frozen body of an ancient Indigenous man called the Kwädąy Dän Ts'ìnchį man, meaning Long Ago Person Found, was found at an altitude of 1,600 metres on the Samuel Glacier in northwestern British Columbian (fig. 1). This find raised important questions about the location of his home, his ancestry and his lifestyle, including his diet. Archaeobotanical studies, ethnobotanical research and forensic palynology[1] studies (Dickson et al. 2004, Mudie et al. 2005, Dickson and Mudie 2008) have been carried out to address the following questions: was his diet mainly of marine or inland origin? Can his diet tell us if he had travelled from the Alaskan coast or from southeastern Yukon on his last journey across the St. Elias Mountains? What was the season of his travel? And does his diet tell us anything about his health?

Figure 1. Map of the study region, showing location of the discovery site and the main places mentioned in this chapter.

BY EXAMINING the plant, animal and microscopic sediment particles in samples from the man's stomach, intestines and clothing, and by comparing the tiny remains with reference samples collected at inland and coastal localities during five years of regional fieldwork (2001–05), we have been able to clarify our earlier results, which indicated that pollen from the chenopod beach asparagus (*Salicornia perennis*) pointed toward his last journey beginning in a salt marsh on the coast of southeastern Alaska. Our work on the life and death of Kwäday Dän Ts'ìnchį is unique in that we look at the life of *one* person, on a scale of months, days and hours. The only other known example of such detailed

archaeobotanical work is the study of the European frozen man, Ötzi, who lived near the border of Austria and Italy about 5,200 years ago (Dickson et al. 2003, 2008; Oeggl et al. 2007; Oeggl 2009; Heiss and Oeggl 2009).

Our work differs from the archaeobotanical studies on Ötzi because we used an Environmental Scanning Electron Microscope (ESEM) and Energy Dispersion Spectrometer (EDS), which allowed us to make routine studies of the smallest pollen grains, algal spores, and organic and mineral particles at very high magnifications (down to 0.1 micron = 10,000th of a millimetre). We were also able to determine the elemental composition of silt particles found in the digestive tract of the Kwädąy Dän Ts'ìnchį individual. At the start of our study in 2002, this ESEM-EDS system was the most powerful microscope system available—one of only three in Canada. Essentially, it uses a high-voltage electron beam to trace out the smallest features of the particle surfaces and to capture these as 3-D images rather than the broader beam of light rays and the mirrors used to examine translucent particles with conventional transmitted light microscopes). The ESEM system has an additional advantage in that it allows observation of the particles in their natural state, wet or dry, without requiring special preparation such as acid digestion, freeze-drying or embedding in a mounting medium. The attached EDS enables further analysis of the elements in an observed particle from the distinctive energy signature of the X-ray particles produced when the electron beam interacts with the scanned object. The element composition is then displayed as a spectrogram showing the relative abundance of the chemicals in the particle, even at very low levels of parts per million.

In this chapter we review the laboratory and field methods used to resolve some of the key questions about the Kwädąy Dän Ts'ìnchį man's life and last journey, and we provide the final results of earlier ESEM studies of his last meals, as initially outlined by Dickson and Mudie (2008). We also present data on pollen, algae and mineral particles in snow samples and local streams as part of our effort to trace the geographic sources of his food and water. All the ESEM studies of pollen, algae and other particles smaller than 125 microns are the work of the first author (Petra Mudie), while the archaeobotanical studies of the macroscopic plant and animal remains larger than one millimetre are the work of Jim Dickson. Field data regarding pollen content of snow in 1999 and the flowering time of salt marsh chenopod pollen in southern British Columbia were provided by Richard Hebda, while data on the distance chenopod pollen travels across the salt marsh were provided by Mudie. Results of ESEM studies of samples from the squirrel-skin robe are reported elsewhere (Dickson and Mudie 2008 and, particularly, Hebda et al. chapter 28).

Methods

Laboratory

The laboratory methods used were conventional light microscope and special ESEM studies of tiny samples (see table 1, about 0.5–1.0 ml volume) saved from the frozen body during the autopsy in Victoria in 2000. ESEM studies of the organic particles were made with pinhead-sized subsamples from the stomach and intestines. The particles studied by ESEM included pollen, microscopic plant and animal remains, and silt-sized mineral grains. The methods for preparing these minute samples are given by Mudie et al. (2005); notably, most of the washed sample residues were not treated with acid but were mounted in water after repeated washing to remove the preservative solution. Methods for preparing the macroscopic plant and animal remains are given in Dickson and Mudie (2008).

The ESEM samples came from the following parts of the digestive tract: the stomach (gaster), representing what was eaten in the last three hours; the duodenum, containing what was eaten about four to five hours previously; the ileum (small bowel), with food remains from six to seven hours previous; the ascending and descending parts of the large colon, containing residue from up to a day ago; and the rectum, with remains from two to three days previous. The snow samples and other water samples were collected at the discovery site in 1999, 2002 and 2004, as described in Dickson et al. (chapter 5).

Field

The discovery site is at an altitude of 1,600 metres, about 600 metres above the treeline, in a glacial alpine setting with steep, scree-covered barren slopes and sedge meadow around the morainal ponds (fig. 1). To collect botanical reference data, we walked about 60 kilometres from there down to the boreal forest near the upper reaches of the Tatshenshini and Chilkat rivers documenting the vegetation along the altitudinal gradient—particularly noting plants of known food or medicinal value—and sampling likely drinking water sources (see chapter 5).

To identify the most common pollen grains found in the Kwäd̲ay Dän Ts'ìnchį man's intestines to species level, it was also necessary to make numerous collections of crucially important chenopod pollen from a variety of habitats and elevations throughout the region. These range from the alkali and salt ponds of the interior Yukon Plateau, near Whitehorse, to the coastal marshes of the Chilkat River estuary and Lynn Canal, and the tidal marshes of the northwest Pacific coast between Yakutat, southeastern Alaska, at 59.56° N, southward to Victoria, at 46.48° N (fig. 2; see also table 3). We collected fresh pollen from chenopod species at most of these sites, and we measured the surface water salinity associated with the plants using a Goldberg refractometer calibrated to a standard seawater salinity of 35‰ (~3.5 grams total dissolved salt). Dickson and Mudie also travelled down the Tatshenshini-Alsek River basin by raft, going from Dalton Post on the Klukshu River to the fish plant at the head of Dry Bay—255 kilometres and an elevation drop of 550 metres—in search of chenopods and to sample the water of rivers draining into the basin from the St. Elias Mountains near the Samuel Glacier. Likewise, Dickson and Mudie spent several days in Yakutat and one day in eastern Dry Bay searching for chenopods, particularly the elusive beach asparagus (*Salicornia perennis*) reported to have grown there in ancient times.

Herbarium material in collections at Whitehorse (Bruce Bennett) and herbaria at Glacier Bay, Skagway, Juneau, Ketchikan, the University of British Columbia and the Royal British Columbia Museum were also examined as references for chenopod pollen morphology and their flowering time. Details of our collections up to 2004 are given in Mudie et al. (2005), where several Yukon First Nation elders and Indigenous Alaskans report their knowledge of traditional uses for chenopods. Here we add further information given to us by the inland and coastal Tlingit people from Skagway, by elders at Yakutat and by Wayne Howell and Carolyn Martin at Glacier Bay National Park, Alaska.

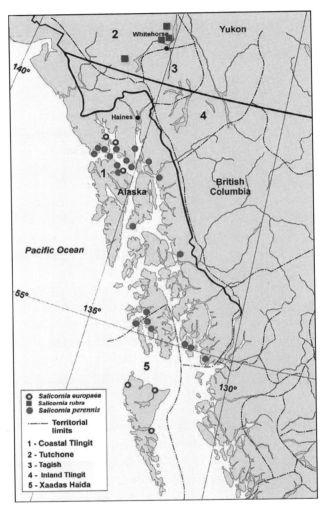

Figure 2. Map showing distribution of *Salicornia* sites in Yukon Territory, southeastern Alaska and adjacent British Columbia. Dotted broken line shows the rough boundaries of Aboriginal territories according to the source map from the Vancouver Museum of Anthropology, 2002. Colour version on page 601.

Results

Laboratory Studies

Table 1 shows the results of the detailed stomach (gaster) pollen analyses made independently by Mudie at the Geological Survey of Canada (sample T30) and by Dr. Susan Ramsay at the University of Glasgow (sample T27). Both analyses reveal the presence of high percentages of chenopod pollen. The ultra-high resolution detail provided by the ESEM for sample T30 further shows that the chenopod pollen is that of *Salicornia perennis* (now called *Sarcocornia perennis*). Other pollen from the gaster sample T30 (fig. 3) includes relatively high percentages of alder, grass, pine and *Artemisia* (wormwood) pollen that could be derived either from local subalpine plants or from long-distance wind transport.

For comparison of what we might normally expect in the wind-transport pollen load from the Samuel Glacier, table 1 shows the kinds of pollen recovered from snow samples taken at the discovery site. No chenopod or Rosaceae pollen grains were found in any of the five snow samples that we collected over three years, between 1999 and 2005. Jocelyn Bourgeois (pers. comm., 2007) also found no chenopod pollen in annual

records of snow and ice core samples from Mt Logan in the St. Elias Mountains northeast of Yakutat.

Table 2 shows the proportions of the main types of organic particles found in stomach (gaster) sample T30-1 examined by ESEM. Oil droplets and a kind of meat we have called "striped chyme" make up about half the volume of this sample. The striped chyme is thought to be seal meat (see Dickson and Mudie 2008). The shellfish is most likely a marine crab (see Dickson et al. 2004 and Dickson and Mudie 2008 for details). Figure 4 also shows the particles identified by light microscope as *Salicornia*, anther fragment (figs. 4.1 and 4.2), and an ESEM image of some plant epidermis that matches a reference sample of blueberry (*Vaccinium* sp.; fig. 4.5). The algae are the red snow species *Chlamydomonas nivalis* and *Troschiscia*, illustrated in chapter 5. Numerous small spheres are thought to be fungal spores (fig. 3.3).

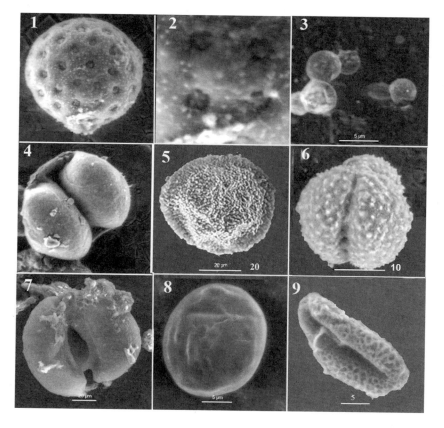

Figure 3. ESEM images of pollen from digestive tract.

1 and 2. Beach asparagus grain (1) and pore details (2), from sample T30-1.

3. Round spores, possibly of fungal origin (T30-1).

4. Pine (*Pinus* sp. *P. banksii* type).

5. Hemlock (*Tsuga heterophylla*) grain.

6. Sagebrush/wormwood (*Artemisia*).

7. Spruce (*Picea* sp.).

8. Small grass grain (*Poaceae*).

9. Willow (*Salix* sp.).

Table 1. Pollen counts for snow samples (KDS-68 and -84 and Snow 2002), the Kwädąy Dän Ts'inchį robe (KDS-R) and intestinal samples (KDS-T). KDS-68 and -84 were 600-ml samples collected in 1999 and analysed by Hebda. Sample KDS T-27 was analysed by Dr. S. Ramsay, and the remainder of the results are Mudie's.

Sample name	KDS-68	Snow 2002 SEM	KDS-84	KDS R28-31	KDS R23	KDS R28-2	Mineral Lake	KDS T27	KDS T30	KDS P14-1
Sample origin	snow	snow	ice + soil	robe	robe	robe	mud	chyme	chyme	feces
Pollen and spores										
Abies	-	-	-	-	-	-	3	-	-	1
Pinus	2	8	9	6	23	-	4	2	2	-
Picea	-	1	2	1	10	-	11	-	5	-
Tsuga heterophylla	-	4	+	1	3	2	3	-	-	-
Alnus	92	5	82	46	31		2	21	10	2
Betula	-	2	+	3	1		-	3	6	1
Salix	-	-	-	2	+	1	-	1	-	1
Poaceae	1	2	1	5	1	-	3	9	14	-
Asteraceae w/o *Artemisia*	-	4	1	3	1	-		-	6	1
Rosaceae	-	-	-	-	-	-	1	-	-	-
Artemisia	2	1	-	5	8		4	1	-	2
Chenopodiinae	-		-	16	3	1		43	25	1
Caryophyllaceae	-		-	-	-			-		
Apiaceae	-		-	2	1		1	3	-	
Cyperaceae	-		-	-	-		-	-	4	
Liliaceae	-	-	-	-			1			1
Ranunculaceae									2	
Sanguisorba						2			2	1
Lycopodium	+		-	-	1			4	-	2
Polypodiaceae	3	1	8	2	16		2	3	6	1
Other	+		1	4	+			7	-	
Unknown pollen	-	2	1	4	1		1	13	14	6
Moss fragments							60			4
Desmids							17			-
Lycopodium spike							4			14
Σ pollen + spore	**327**	**31**	**358**	**366**	**375**	**60**	**36**	**90**	**44**	**21**

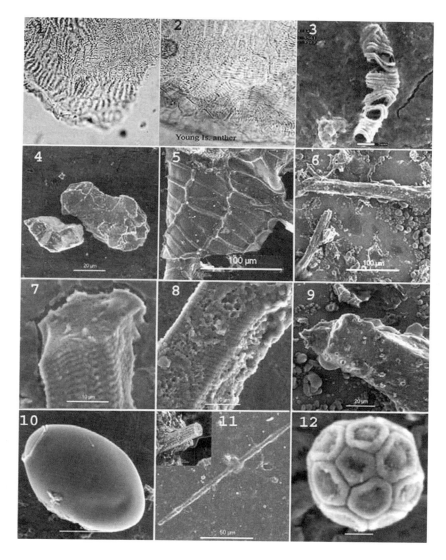

Figure 4. Light microscope (1 and 2) and ESEM images of non-pollen particles from digestive tract and meat/fish reference samples.

1 and 2. *Salicornia* tissue fragments. 1. Stomach sample, T30-1; 2. Anther fragment from Young Island, Glacier Bay.

3. *Salicornia* sclereid, from sample P14-1.

4. T30-1 feldspar silt grains.

5. Blueberry epidermal tissue.

6–9. Meat fibres. 6. Meat fibre and fat in ileum sample T38-11; 7 and 8. Striped meat fibres of sockeye salmon reference sample; 9. Fibre and fat in reference sample of Moose meat.

10. Egg of *Diphyllobothrium latum*–type with lid in place, from sample P14-1.

11. Spines of sea urchin; inset shows a broken spine (from T38-24) where the complex, ribbed, semi-porous structure can be clearly seen.

12. Coccolithophorid with pentagonal calcareous plates, from sample P14-1.

The stomach sample contained the highest proportion of silt-sized mineral particles seen in the ESEM studies. All the silt grains were very angular (fig. 4.4) and characteristic of freshly plucked glacial fine sediment (glacier flour). The elemental composition of the mineral particles (fig. 5E) shows a predominance of silica, aluminum and potassium that is characteristic of alkali feldspar (orthoclase), and a trace amount of copper is also present. Quartz grains (with only silica and oxygen peaks, such as that in fig. 5G) were also common.

The proximal duodenum sample (T38-3-1) comprised mainly chunky meat particles that look like fibres of Bison or Caribou meat when compared to reference samples of partially digested meat (figs. 4.6 and 4.9). Globules of fat were also very common. There was no recognizable plant material except one unknown pollen grain and a stalked spore, possibly that of another fungus. Several tapeworm eggs were found, as well as short, rectangular worm segments like those of *Diphyllobothrium latum*. The few mineral grains in the duodenum were well-rounded or delicate crystalline

Table 2. Dietary components of gaster and intestinal samples from remains of the Kwädąy Dän Ts'ìnchį individual. Approximate numbers of grains, spores or hairs counted and per cent volume of organic matter such as meat tissue, oil and fat observed in ESEM images.

Contents	Gaster T30	Duodenum T38-3-1	Distal ileum T38-11	Small bowel T38-24	Large bowel T-46	Rectum T14-1
Striped meat (1)	25	0	0	0	0	0
Salmon (17)	0	0	0	50	70	55
Oil (2)	25	0	0	20	0	5
Fat (11)	0	25	20	0	0	0
Meat (12)	0	75	55	3	0	0
Hair (3)	1	0	0	5	0	0
Shellfish (4)	10	0	0	1	5	0
Salicornia (5)	10	0	0	0	2	5
Blueberry (6)	5	0	0	0	0	0
Cloudberry	0	0	0	0	2	0
Waterlily (18)	0	0	0	5	0	0
Sedge pollen	0	0	0	0	0	5
Mountain Hemlock pollen	0	0	5	0	0	0
Plant tissue (7,13)	1	3	3	10	23	0
Snow alga (8)	2	0	0	0	0	0
Chrysophyte (14)	0	2	20	5	0	5
Fungi (9)	2	3	3	0	0	2
Charcoal (18)	0	0	0	1	1	0
Bast fibre (15)	0	0	10	10	0	0
Moss (19)	0	0	0	2	0	5
Worm eggs (16)	0	6	1	2	7	10
Worm segments (20)	0	1	0	1	2	8
Minerals (10)	15	3	0	10	1	3
Total particles	96	118	118	124	113	103
Sample size, g	4.5	0.1	0.1	0.4	0.3	0.1

fluorite and calcitic grains (fig. 5D) that suggest a water source in a quiet, carbonate mineral lake, such as those located west of Stonehouse Creek—about 300 metres and 7 kilometres below the discovery site.

The distal ileum sample (T38-11) also largely comprised about 50 per cent meat chunks and 20 per cent fat globules. The pollen included some Mountain Hemlock (*Tsuga mertensiana*), which is a common component of the salt marsh pollen assemblages in the Glacier Bay region, suggesting a location closer to the coastal forest treeline at about 1000 metres above sea level. Animal hairs and some flat plant fibres of unknown

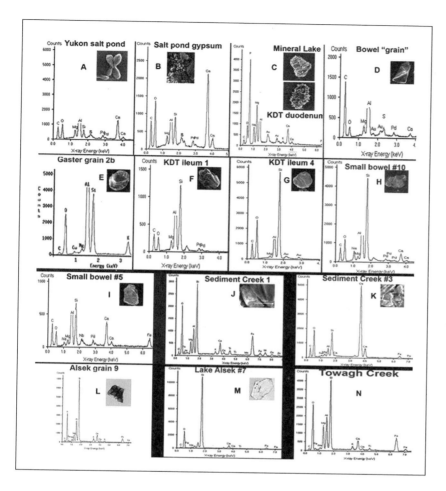

Figure 5. Mineral composition of silt grains in ESEM samples from the digestive tract of the Kwädąy Dän Ts'ìnchį individual (*D–I*) and in field reference samples (*A–C, J–N*), as indicated by energy dispersion spectroscopy (EDS). Each graph shows the relative amounts (vertical axis = counts/area distributions) of X-rays that are emitted by mineral elements in the silt grain being scanned versus their X-ray energy (in KeV, horizontal axis), and the letters indicate the elements present at more than about 0.1 per cent. Note that an aluminum holder was used; therefore, counts of less than 8000 for Al have no importance. Also, because the samples were sputter-coated with gold-palladium, the presence of trace amounts of Au and Pa have no significance. The elements are Al (aluminum), Au (gold), C (carbon), Ca (calcium), Cu (copper), F (fluoride), Fe (iron), K (potassium), Mg (magnesium), Mn (manganese), Na (sodium), O (oxygen), Pa (palladium), S (sulphur), Ti (titanium) and Zn (zinc).

origin were present. Silt particles were sub-rounded grains of quartz (fig. 5*G*) and alkali feldspar (fig. 5*F*).

The ascending (T38-24) and descending (T46) colon samples are both dominated by striated meat fibres resembling that of our smoked salmon reference sample obtained in Klukshu (figs. 4.7 and 4.8). Both of the colon samples contained coiled plant fibres (fig. 4.3) resembling sclereids of *Salicornia*. Wood charcoal was present, perhaps as expected for a meal of smoked fish, although no charcoal was seen in the reference sample of smoked fish. Silt particle grains included both rounded calcium carbonate grains and sub-rounded feldspar (fig. 5*H*), with a coating of a powdery carbonate containing sodium and indicating a marine origin. In contrast, the EDS spectra for the inland salt pond

gypsum deposits (fig. 5*B*) contained only magnesium and traces of sulphur.

Both colon samples also contained chains of the rectangular segments (L:B ratio <0.5) of the *Diphyllobothrium* tapeworm and its oval eggs (as illustrated in Dickson et al. 2004). Two egg shapes and morphotypes were observed by ESEM imagery. Some smaller (less than 40 microns), more rounded eggs were seen in the duodenum and ileum, and these seem to have a smooth, rounded base. However, most of the eggs in the small and large bowel and in the rectum were oval with a short basal projection, presumably representing the mature egg form as described by Leighton et al. (chapter 14). Most of the eggs were open in all the ESEM samples, and smooth, concave, circular lids were common. One egg with the lid still in place was found in the rectum sample P14-1 (fig. 4.10).

The ascending colon (small bowel) contained about 20 per cent oil, while the descending colon sample contained no visible oil but included calcareous spines like those of a sea urchin (fig. 4.11). In the small bowel sample T38-17, light microscope studies (by Dickson) revealed a fragment of *Fontinalis* moss (see chapter 5) and a small piece of spongy (trabecular) leaf tissue from a water lily, *Nuphar* cf. *N. polysepala* (see Dickson and Mudie 2008, fig. 2.4); both of these plants only grow below the treeline and were not seen beyond the coast in our field studies. ESEM studies of the small bowel sample T38-24 also showed the presence of various other vascular plant tissues, including uncarbonized softwood fibres (see Dickson and Mudie 2008, fig. 2.6) similar to those in the bast tissue (inner bark) of spruce (*Picea*) or larch (*Larix*). Silt particles included subrounded alkali feldspar and quartz, and some rounded calcite grains. The large bowel contained pollen of *Salicornia* and a grain like that of Cloudberry (fig. 4.11), but unlike the small bowel, the silt grains were all of carbonate mineralogy.

The rectum sample was also dominated by salmon fibres and oil particles, and it included some *Salicornia* pollen and the distinctive water-storage tracheids found in this halophyte's stem tissues as well as fungal spores and a spiny desmid (*Staurastrum* cf. *S. sexangulare*) not found in any of the glacier lakes or mountain streams (see chapter 5). The leafy stem of a bog moss belonging to *Sphagnum* section Acutifolia was found among the macroscopic plant remains. This section has many species in southern Alaska and northern British Columbia, where the very wet coastal zone favours the profuse growth of *Sphagnum* in general. A calcareous, scaly microfossil resembling a coccolith with pentagonal plates (fig. 4.12) was found, resembling *Braarudosphaera bigelowii* (Gran and Braarud) but with a larger number of more rounded scales. If correctly identified, this microfossil definitely indicates an outer coastal marine environment because coccoliths do not live in water with a salinity of less than 16‰. The only mineral particles found were clusters of round (framboidal) pyrite crystals (FeS) and

angular particles resembling magnesite (fig. 5D), which is a secondary deposit in limestone rocks.

Field Studies

Our field studies focused primarily on obtaining a detailed inventory of the terrestrial and microscopic aquatic plants that grow near the discovery site, and on determining the geographical distribution of chenopods in southwestern Yukon, northwestern British Columbia and southeastern Alaska. The main results of fieldwork up to 2004 are noted by Mudie et al. (2005), and the discovery area flora is listed in a separate chapter (chapter 5). Here we report that we found no chenopod species on the 60-kilometre hike from the alpine tundra at the discovery site down an altitudinal gradient of about 1,000 metres to the valley boreal forest along the Haines Highway—despite the fact that the introduced species *Chenopodium album* (Lamb's Quarters) is a common weed around Whitehorse and grows on the shore of the Chilkat Estuary, and thus might be expected to have been accidentally transported into the lower elevations of the discovery area by early miners. The chenopods we saw closest to the discovery site were *Atriplex gmelinii* and *A. alaskensis*, which grow in the high marshes of the Lynn Canal near the Haines ferry terminal, where the water salinity was 4‰ (~0.4 g/l) at high tide (<2‰ at low tide). *A. alaskensis* also grows at Mud Bay, off the Lynn Canal near the entrance to the Chilkat River, and is listed for the Katzehin River delta north of Berner's Bay (Parker 2000), where the surface salinity in September is 11‰.

At the intertidal marshes of Dyea, near the mouth of the Chilkoot River, the salinity is <1‰ in early August, and we found no chenopods in the tidal marsh—although the Skagway Parks herbarium contains a few small specimens of *Atriplex gmelinii* and two samples of *Chenopodium album*. At the Skagway Traditional Council, Marion Twitchell, a Tlingit elder, told us that about 30–40 years ago she used to pick beach asparagus from the marsh by the pier at Dyea.

During public meetings and an interview with Elaine Abraham at Yakutat, we were informed that jarred beach asparagus was a common item at potlatches and that beach asparagus grew at east Dry Bay. Searches in different areas of Dry Bay were made over several years, between 2001 and 2006, by Wayne Howell and his colleagues at Glacier Bay National Park, and we made an extensive search around Khantaak Island in Yakutat Bay and in eastern Dry Bay during late August 2005. The only chenopods we found, however, were some rare occurrences of *Atriplex gmelinii* in the low marsh on Khantaak Island, where the water salinity was 22–24‰, following two days of very heavy rain (25 centimetres). In eastern Dry Bay, near the entrance to the Alsek River, the channel water salinity was about 1‰ in late August. It is also notable that no chenopods have been reported for either Lituya Bay, about 70 kilometres south of Dry Bay (Post and Streveler 1976), or in Icy Bay, about 60 kilometres northwest of Glacier Bay. Likewise, no chenopods were found at the Copper River delta, 300 kilometres northwest of Yakutat (Thilenius 1995), although there are some chenopods (*Atriplex* spp., *Suaeda depressa* and *Salicornia europaea*) at China Poot Bay, near Homer, another 300 kilometres further west (Crow and Koppen 1977).

Species of the genus *Salicornia* are obligate halophytes, meaning that they cannot survive without some sodium chloride salt in their environment. Therefore, in order to pinpoint possible sources of *Salicornia* pollen found in the digestive tract of the Kwäday Dän Ts'ìnchị man, we focused most of our field studies on areas of saline soils. At the inland sites, where only *Salicornia rubra*—the annual Red Glasswort or Red Samphire—grows, we measured groundwater salinities of up to 80‰ and salt pond surface salinities of 4–80‰ along the Old Dawson Trail and near Cracker Creek, east of Haines Junction (table 1). *Salicornia* was absent, however, in the gypsum salt pond near Carmacks, where the salinity was <1‰. In the salt ponds along the Dawson Trail, algae and other organisms (fig. 6) associated with the glasswort include several morphologically distinct species of chrysophytes, diatoms and brine shrimp (*Artemia*), none of which were found in the intestinal or robe samples.

Along the coast of southeastern Alaska the annual glasswort or samphire *Salicornia europaea* (*S. depressa* in the *Flora of North America*; see Ball 2003a) was usually found as scattered plants on the open mud flats of the upper intertidal zone, or in salt pans within the high marsh, while the perennial beach asparagus (*Salicornia perennis*, previously called *S. virginica*) formed a dense meadow of succulent vegetation covering the middle and high marshes below the extreme high-water line. The salinity at most of the *Salicornia* sites we visited was 20‰, with a range of 15–28‰ (mean 23.4‰, see table 3). We do not have a salinity measurement for the marsh at St. George's Bay, but it is probably similar to that given for Portage Bay, near Haines: 14–28‰ at 3.7 metres of water depth (US EPA 1995). Details of the intertidal marsh zonations and associated chenopod and other salt marsh species have been given elsewhere (Mudie and Dickson 2004).

In August these two Alaskan *Salicornia* species can easily be distinguished by the shape of their inflorescence (fig. 7) and from the longer, exerted stamens in *S. perennis* (now *Sarcocornia perennis* [Miller]; A.J. Scott; see Ball 2003b), in addition to its perennial stoloniferous habit. Therefore, it was possible to reliably collect reference pollen from small, young plants of the perennial before it took the mature form of a loose rope on the beach (*suk k'ádzi*). We had less certainty about the possibility that there are two perennial *Salicornia* (*Sarcocornia*) species at the southern end of the Alaskan distribution. These are *Salicornia pacifica* Standley—now *Sarcocornia pacifica* [Standley] A.J. Scott; see Ball 2003b—and *S. perennis*. The possible *S. pacifica* morphotype seen at George's Bay and Bostwick Inlet near Ketchikan (fig. 7) has the same stoloniferous habit and ecological niche as *S. perennis* but has a notably longer inflorescence, with eight or more rows of flowers (figs. 7A and 7C). Similar morphotypes were observed at Kinsmen Park and elsewhere in Esquimalt, BC. ESEM studies of *S. pacifica* show that its pollen grains appear to be slightly larger than those of *S. perennis* (figs. 6.11 and 6.15) and often wrinkled. However, it is not clear that the pollen of these two perennial *Salicornia* species can always be distinguished reliably, even with the high magnification of the ESEM.

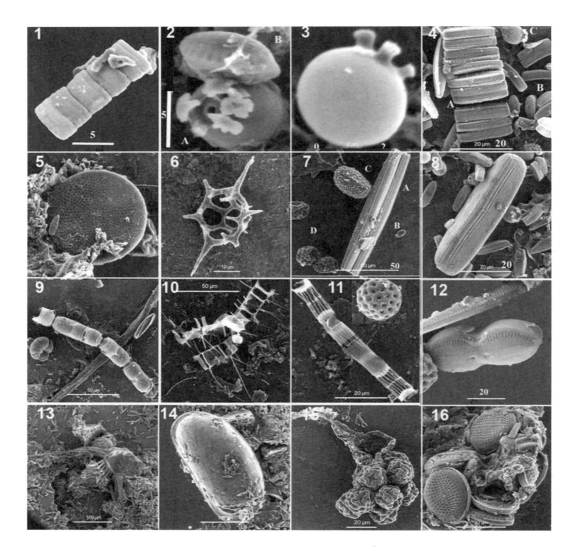

Figure 6. ESEM images of various Yukon salt pond and Alaskan coastal salt marsh algae and microorganisms. Scale bar is in microns.

1–4, 8, 13 and 14. "Ranunculus" salt pond, Old Dawson Trail. 1. Chain of small centric *Stephanodiscus* sp.; 2 and 3. Unknown small, smooth, trispinate stomatocyst (2*A*) and small diatom (2*B*). 2*A*. Apical view of stomatocyst pore with low collar, surrounded by short, thick spines with flared tips. 2*B*. Small pennate diatom. 3. Side view of stomatocyst without flared tips, presumably an immature stage. 4. Diatom-chrysophyte assemblage. 4*A*. Chain of rectangular pennate diatoms; 4*B*. Small pennate diatoms; 4*C*. Trispinate statocyst. 8. Girdle view of rare, large pennate diatom, surrounded by abundant small pennate and centric species. 13. *Artemisia* brine shrimp surrounded by abundant small pennate diatoms. 14. Ostracod and small pennate diatoms.

5, 6 and 15. George Inlet salt marsh, Chichagof Island. 5. Valve of large centric *Coscinodiscus* diatom, abundant small pennates and grain of *Salicornia pacifica*. 6. Valve of silicoflagellate *Dictyota*. 15. Anther from *Salicornia pacifica*, with enclosed pollen grains.

9 and 12. Haines salt marsh. 9. Chain of large centric diatoms. 12. Large pennate diatoms and pollen grain of the chenopod *Atriplex gmelinii*.

7 and 16. Groundhog Bay salt marsh. 7. Very large rectangular (*A*) and small naviculoid (*B*) pennate diatoms, dinoflagellate cyst (*C*) and Salicornia perennis pollen grains. 16. Large pennate diatom valves and frustule fragments.

10 and 11. Young Island salt marsh, Glacier Bay. 10. Large chain of centric *Chaetoceros* sp. 11. Chain of centric diatoms and pollen grain of *Salicornia perennis*.

Figure 7. Inflorescences of *Salicornia* species and varieties.

A and *C. Salicornia pacifica* plants from the southernmost study area, showing the relatively long flowering spikes with more than eight rows of bracts. *A.* George Inlet; *C.* Esquimalt Lagoon.

B and *D. Salicornia perennis*, showing the shorter, more knobby flower spikes with fewer than eight rows of bracts, found on plants from the northern islands and mainland. *B.* Swanson Harbor; *D.* Pleasant Island.

E. Salicornia europaea, Hoonah salt marsh, showing the small annual coastal herb with pointed flowering spike tip and barely visible stamens.

F. Salicornia rubra, showing the distinctive colour, small size and almost sunken anthers of the annual saltwort (Red Samphire) from the Dawson Creek Trail near Whitehorse.

Colour version on page 602.

Our fieldwork also provided the opportunity to observe the flower phenology of *Salicornia* in the region. Red Glasswort was not yet flowering at Takhini when visited by Dickson in early July 2001 or Mudie in late May 2005, and it was near the end of its flowering by August 16, 2004. On the Alaskan coast both *S. europaea* and *S. perennis* were flowering at sites from Hoonah to Taylor Bay between August 9 and 19, 2004, but few plants were still in flower at either Bostwick Inlet on August 30 or Victoria in early September. This suggests a window of pollen production for Alaska and southern Yukon Territory between late July and late August. For comparison, we show chenopod pollen rain and surface sample data collected by Hebda (1977) for the *Salicornia perennis* (possibly including *S. pacifica*) salt marsh and adjacent meadow at Boundary Bay, Fraser Lowland, BC (fig. 8*A*). Ambient pollen rain was collected throughout the growing season at ground level, using shallow jars containing glycerine gel. A rapid rise in percentage of chenopod pollen occurs over the last week of June and first week of July, followed by nearly 100 per cent values during flowering time in July and August. Comparison with data collected in a bog about five kilometres from the *Salicornia* salt marsh shows a different pattern of

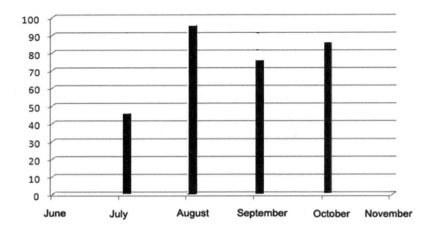

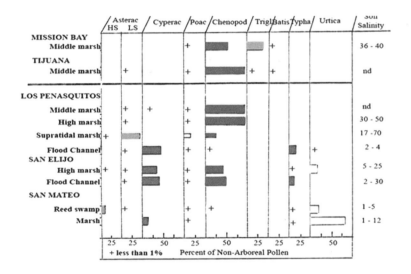

Figure 8. Chenopod flowering time and marsh distribution.

A. Percentage of *Salicornia* pollen in air trap samples of *Salicornia perennis* marsh at Boundary Bay, southern BC (from Hebda 1977).

B. Relative abundances of herb pollen in two intertidal salt marshes (Mission Bay and Tijuana), in the salt marsh and adjacent flood plain freshwater marsh of Los Penasquitos Lagoon, and two more brackish water marshes at San Elijo and San Mateo lagoons, San Diego County, California (Scott et al. 2014). Asterac., Asteraceae; HS, high spine; LS, low spine; Cyperac., Cyperaceae; Poac., Poaceae; Chenopod., Chenopodiaceae; Trigl., *Triglochin*; *Batis*; *Typha*; *Urtica*.

+ less than 1%; percentages are based on total non-arboreal pollen.

Colour version of 8B on page 603.

chenopod pollen output, with a short duration peak in April and two small peaks of up to 30 per cent in early and late July. This indicates that the *Salicornia* pollen does not travel far. Pollen percentages in surface sediments from a *Salicornia virginica* marsh near San Diego (fig. 8B) also show a very rapid decrease in chenopod pollen from the salt marsh to the freshwater marsh in the floodplain about one kilometre inland.

Discussion

USING ESEM IMAGERY to study microscopic particles in samples from the stomach and intestines of the Kwädąy Dän Ts'ìnchį man has been key to unlocking secrets regarding the direction, speed and time of year of his last journey. Both light and electron microscope studies show that pollen of the chenopod family makes up about 25 per cent of the total pollen in samples from his stomach. The ultra-high-resolution imagery provided by the ESEM makes it possible to further distinguish between the pollen of five chenopod genera that grow within a circumference of about 250 kilometres around the discovery site, and to distinguish between inland and coastal species of *Salicornia*. From our collection of fresh chenopod pollen reference samples (Mudie et al. 2005, fig. 6), we recognized the chenopod pollen from the Kwäday Dän Ts'ìnchí individual's stomach and large intestine as that of beach asparagus (*Salicornia perennis*), which grows only in intertidal areas of coastal salt marshes. This locates the start of his last journey near the Alaskan coast. The presence of abundant *Salicornia* pollen on the robe (chapter 28) supports this conclusion.

The presence of *Salicornia* tissue fragments and the specialized, spirally thickened water-storage cells of this succulent halophyte in the stomach and intestines also confirm that he specifically ate this plant, because similar cells are not recovered in either windblown sediment on the Samuel Glacier or in salt marsh surface samples. The occurrence of these *Salicornia* tissues in the food residues from both the first (bowel) and last (stomach) meals eaten on his final journey indicates that he carried this succulent plant as a travel food or a medicine.

Other critically important ESEM evidence for a diet of marine food, in addition to the dried salmon or grease traditionally carried on journeys (Krause 1885, 135; Swanton 1909), is the presence in the rectum and lower colon samples of sea urchin spines. Moss (1993) discusses the role of shellfish in the diet of the Tlingit and notes: "The Tlingit associated shellfish with poverty, laziness and ritual impurity, and those who sought to be 'ideal' persons avoided shellfish". De Laguna (1972, 405) also comments, in the context of a shared meal of sea urchins at Yakutat, that "a shaman and members of his immediate family ... were not allowed to eat any beach food except during one month of the year(February? March? April?) ...". However, Dmytryshyn and Crownhart-Vaughan (1976, 37) quote the 1817–32 reports of Khlebnikov at Sitka, southeastern Alaska, mentioning that sea urchins (*Strongylocentrotus* spp.) were in demand for their "delicious flavor" and "healing properties". Krause (1885, 107) also said of the Tlingit in

the Chilkat region that "the principal dish of the day is always fish, boiled, roasted, dried, but never raw. Next in importance is the meat of land and sea mammals, fowl, crabs, squid, shellfish, sea urchins".

There are several species of sea urchins in Alaska, but only one, the Green Sea Urchin (*S. droebachiensis*), seems to be tolerant of the low salinity of Alaska's sheltered bays and fjords. It lives in rocky shorelines near both Haines (US EPA 1995, Lawrence 2006) and Yakutat (Alaska Department of Fish and Game, n.d.). Holmberg ([1855–63] 1985) also noted that around Sitka, mussels and "starfish" (i.e., sea urchins) were eaten raw, and Emmons (1991, 49) says of the Tlingit in general that "sea urchins were eaten where they were found, by scooping out the insides with the forefinger (or wooden spatula). It was the ovaries that were eaten". De Laguna (1972, 405) also reports that at Yakutat, the orange-yellow ovaries of sea urchins (*niś*) "were eaten raw without seasoning". In contrast, Oberg (1973, 107–08) wrote that the Tlingit on the islands of southeastern Alaska gathered large quantities of marine food, including clams, mussels, sea urchins and seaweed, some of which was dried and traded to mainland Tlingit villagers in exchange for eulachon oil and furs. Thornton (2004) lists *Strongylocentrotus purpuratus* (the Purple Sea Urchin) as a Tlingit food (*Neés'*) that was gathered in spring, summer and winter. The listing of crab specifically for the Chilkat Tlingit may be important because it was part of the Kwäday Dän Ts'ìnchí man's last meal, although it was not a favoured Tlingit food in ancient times (Earlandson and Moss 2001).

At the simplest level of qualitative interpretation, our ESEM study of the dietary diary recording the approximately three last days of the Kwäday Dän Ts'ìnchí man's life reveals that his last meal consisted of a fatty meat resembling seal, some crab, a fruit such as blueberry, and beach asparagus—and it was probably eaten on the glacier or by the morainal ponds at the toe of the glacier (his stomach is the only sample with glacier flour and snow algae). Three to six hours earlier, he ate a relatively small meal of meat resembling that of Bison or Moose, with some fat but apparently without any vegetables or fruit. The silt minerals associated with that meal indicate water from a subalpine pond or small lake with a high calcium carbonate and fluoride content, such as the Mineral Lakes, about five kilomtres southwest of the discovery site. The remains of the previous meal (the ileum sample) are similar in composition. The meals for up to two days before that predominantly consisted of salmon and oil, with some beach asparagus and fragments of other plants that are either confined to or common only at lower altitudes in the coastal zone (see also Dickson and Mudie 2008). The presence of a coccolith in the bowel and sea urchin fragments in the bowel and lower colon, together with the chenopod pollen and *Salicornia* tissue fragments, strongly suggest that the last journey began at or near a tidal *Salicornia* marsh in a rocky coastal area where sea urchins might be gathered easily.

Alternatively, but with more difficulty, it could be argued that all the food carried on this last journey was dried and had been obtained by trading. However, it is uncommon for either beach asparagus or sea urchin to be dried as a travel food. In particular, the perennial *Salicornia perennis* contains woody tissue that would make it unpalatable when dry. On the other hand, we have a report from Carolyn Martin, a Tlingit elder who is a tour guide at Glacier Bay National Park, who told us that when beach asparagus grew in abundance along the Homelands shore near Groundhog Bay (see table 3), it was commonly harvested and some was dried, then later soaked in water before being eaten. Because of the very small particle sizes of the pollen, coccolith and sea urchin spines, it is also possible that these are just debris trapped in another very commonly traded and highly valued sea food like black seaweed (Emmons and de Laguna 1991, Turner 2003, Thornton 2004).

We have also previously discussed the possibility that *Salicornia* was used as a medicine by the Kwädąy Dän Ts'ìnchį person (Mudie et al. 2005). The annual inland species, *S. rubra/borealis*, was used by the Teetl'in Gwitchin for salt flavouring, and several Champagne and Aihishik elders suggested that it was a man's medicine. At Haines Junction in July 2004, Rosalie Washington told Mudie and Dickson that she had spoken with Fred Brown, of Canyon Creek, who knew that Red Glasswort grew in saline areas along the Aishihik Road and said that it might have been used for medicine a long time ago (Mudie et al. 2005). Along the coast, Dr. Menzies, the ship's surgeon of Captain Vancouver's expedition in 1794, reported the use of chenopods as an anti-ascorbic in the Icy Strait region (Olson and Thilenius 1993). In the Ketchikan area beach asparagus is noted as a good low-calorie vegetable and a source of salt, iodine and vitamin A (Garza 2007). In a booklet on Haida foods from land and sea (Cogo and Cogo 1974) a drawing of a shoot of *Salicornia* is called Wild Asparagus, which is recommended as a late spring (June) salad or steamed vegetable.

In India an annual species of glasswort, *Salicornia brachiata* Miq., is commercially grown on a large scale as a source of low-sodium vegetable salts and to produce a nutritious oil rich in lineolic acid (CSMCRI 2015) Extracts of the mature green plant, dried and pulverized, have also been used as an anti-tubercular agent in tests to suppress *Mycobacterium tuberculosis*, the TB bacterium (Choudhary et al. 2015). However, there appears to have been no further use of this drug since 2005. In Egypt a perennial glasswort, *S. fruticosa* (now *Sarcocornia fruticosa*), has been tested as an antioxidant for cancer tumors but was found to be only mildly effective (Radwan et al. 2007). In Brazil there are very high chenopod pollen levels found in worm-infested coprolites, and it has been suggested that some chenopod species were used as a medicinal treatment for worms (Chaves and Reinhard 2006). *Artemisia* (wormwood) is also a well-known vermifuge and was used by the Tlingit at Yakutat in historical times for treatment of worm infestations (Elaine Abraham, pers. comm., August 2005), possibly accounting for some of the pollen in the Kwädąy Dän Ts'ìnchį man's stomach. Also, crushed sea urchin shell is claimed to be a particularly good source of calcium with high oral bioavilability (Kumagai and Kubo 1997), but it is doubtful that spines like those in the intestinal samples could have survived the crushing process if this was the way the sea urchin was consumed.

Another important aspect of the *Salicornia* pollen associated with the Kwädąy Dän Ts'ìnchį man relates to the season during which he travelled. In the 1930s Olson (1936) reports that pre–gold rush trading trips over the Chilkat and Chilkoot trails were usually in mid winter, when "the deep snow of the passes was packed hard by high winds". But Thornton (2004, 94) cites other sources as reporting more substantial trips in spring (April–May), summer (before the heavy salmon fishing) and fall, especially October. The fact that the Kwäday Dän Ts'ìnchí man was very lightly clothed makes it more likely that he travelled during a warm season. Our field studies show that *S. perennis* blooms mainly from July to late August, perhaps starting later (mid to late July) in the Alaskan and Yukon region than in the Fraser Delta. The peak beach asparagus pollen production is from late July to mid August, which is therefore the most likely time of the Kwäday Dän Ts'ìnchí man's last journey. Finally, our knowledge of the geographical range and ecology of *Salicornia perennis* provides some important clues as to the direction of the trail the Kwäday Dän Ts'ìnchí man took from the coastal region inland. In particular, our data allow us to ask: was the start of this trek to

the west (e.g., Dry Bay-Alsek Lake) or east (e.g., Lynn Canal) of the Samuel Glacier?

Figure 2 shows the present geographical range of *Salicornia perennis* from Taylor Bay in Icy Strait southward to just north of Haida Gwaii (Queen Charlotte Islands), near the border between Alaska and British Columbia. Table 3 shows some localities for the perennial *S. virginica/S. pacifica* in southern British Columbia, and there are scattered localities for the annual *S. europaea* around Homer and Anchorage, at 61–62°N. The annual glasswort also grows on Haida Gwaii (fig. 2) and in the Fraser Delta. It is clear, however, that at present there is no *Salicornia* in the tidal marshlands and deltas between Icy Strait and Homer. In fact, there are virtually no chenopods in the marshes throughout this region, which strongly indicates that the climate is too cold and foggy for these plants with the Crassulacean C4 metabolism commonly associated with tropical plants. The low salinities of the estuaries (<16‰) and the very high rainfall of the region probably also result in sediment salinities too low for growth of perennial *Salicornia*. The short-lived annual glasswort may be

Table 3. Pacific Northwest *Salicornia* locations, compiled from citations provided by Carolyn Parker, University of Alaska Fairbanks; Olivia Lee, UBC Herbarium; and our field observations, 2002–05. The asterisk indicates field locations reported in this study.

Taxon	State	Quad	Locality	Latitude N	Longitude W	Salinity	Habitat
S. europaea	AK	Kenai	Kalgin Isl., Cook Inlet	60 28	151 55	nd	Tidal mud flats
S. europaea	AK	Kenai	Redoubt Bay, Big R. Delta	60 39	152 02	nd	Silt-clay mud flat
S. europaea	AK	Anchorage	Potter Pt. State Game Refuge	61 06	149 55	nd	Coastal marsh, wet silt
S. europaea	AK	Anchorage	Coffee Pt., Duck Flats area	61 30	149 22	nd	Intertidal mud flats
S. europaea	AK	Anchorage	Potter Marsh at Seward Hwy	61 04	149 47	nd	Tidal zone

Taxon	State	Quad	Locality	Latitude N	Longitude W	Salinity	Habitat
S. europaea	AK	Anchorage	NW Eagle River Flats	61 19	149 46	nd	Mud flats
S. europaea	AK	Anchorage	Matanuska	61 36	149 06	nd	Flats near Matanuska Village
S. europaea	AK	Seldovia	Homer, East Road, Swift Creek	59 45	151 15	nd	Back beach mud flats
S. europaea	YT	Whitehorse	E of Takhini R., N of Alaska Hwy	60 51	135 41	50	Salt flats
S. europaea	YT	Whitehorse	E of Takhini R., N of Alaska Hwy	60 51	135 41	nd	Sodium sulphate (mirabalite)
S. europaea	YT	Whitehorse	Mi 945 Alaska Hwy, 2 km S Takhini R. bridge	60 52	135 42	nd	Alkali lakes, spruce-pine forest
*S. borealis**	YT	Whitehorse	Old Ranch, SW L. Laberge	61 15	135 15	35	Alkali flat & vernal pool
*S. borealis**	YT	Whitehorse	Dawson Rd, near L. Laberge	61 15	135 15	nd	Alkali flats, rare
*S. borealis**	YT	Whitehorse	Dawson Trail, N of Takhini R.	60 54	135 42	35.4	Salt pond margins, alkali flats
*S. borealis**	YT	Haines Jctn	Cracker Creek, 42 km E of Haines Jctn	60 45	137 30	80	Salt pond in spruce forest
*S. europaea**	AK	Juneau	Lars Isl., Berg Bay, Glacier Bay	59 29	136 08	20	Low & mid marsh
*S. perennis**	AK	Juneau	Lars Isl., Berg Bay, Glacier Bay	59 29	136 08	20	Mid marsh
*S. perennis**	AK	Juneau	St. James Bay, Lynn Canal	58 37	135 10	nd	Common in muddy salt marsh

Taxon	State	Quad	Locality	Latitude N	Longitude W	Salinity	Habitat
S. europaea*	AK	Juneau	Gustavus, dock area	58 25	135 44	10	High intertidal zone
S. perennis*	AK	Juneau	Gustavus, dock area	58 25	135 44	15	Intertidal salt marsh
S. europaea*	AK	Juneau	Bartlett Cove, Glacier Bay	58 27.5	135 52.5	23	Low & middle marsh
S. perennis*	AK	Juneau	Bartlett Cove, Glacier Bay	58 27.5	135 52.5	23	Middle & high marsh
S. perennis*	AK	Juneau	Young Isl., Beardslee Gp, Glacier Bay	58 27.3	135 58	30	Low & middle marsh
S. europaea*	AK	Juneau	Young Isl., Beardslee Gp, Glacier Bay	58 27.3	135 58	30	Low marsh, sandy, bouldery
S. perennis*	AK	Juneau	Dundas Bay, Icy Strait	58 21	136 20	26	High marsh, very rare
S. perennis*	AK	Juneau	Pleasant Isl., Icy Strait	58 21	135 31	31	Middle & high marsh, common
S. europaea*	AK	Juneau	Pleasant Isl., Icy Strait	58 21	135 31	31	Low marsh, common
S. perennis*	AK	Juneau	Fern Harbour, Taylor Bay, Icy Strait	58 18	136 38	26	Middle marsh, occasional
S. perennis*	AK	Juneau	Groundhog Bay, Icy Strait	58 14	135 15	18	Bouldery high-mid marsh
S. perennis*	AK	Juneau	Swanson Harbr, Port Couverden, Lynn Canal	58 12	135 05	20	Low-high marsh, abdt in middle
S. perennis*	AK	Juneau	Cannery Bay, near old pier, Chichagof Isl.	58 07.8	134 27.5	20	Disturbed area, low marsh
S. perennis"	AK	Juneau	Hoonah wharf, Chichagof Isl.	58 06.2	134 26	26	High marsh, scattered
S. rubra*	AK	Juneau	Dalton Creek, Hoonah	58 06	134 25.8	28	High marsh, occasional

Taxon	State	Quad	Locality	Latitude N	Longitude W	Salinity	Habitat
*S. perennis**	AK	Juneau	Dalton Creek, Hoonah	58 06	134 25.8	nd	High-water tidal panne
*S. europaea**	AK	Juneau	Burnt Pt Cove, Port Frederick, Chichagof Isl.	58 02	134 30	26	High marsh, gravelly mud flat
*S. perennis**	AK	Juneau	Burnt Pt Cove, Port Frederick, Chichagof Isl.	58 02	134 30	26	Low marsh, gravelly mud flat
S. virginica	AK	Juneau	Neka Bay, Pt Frederick, N Bay, Chichagof Isl.	58 02.0	135 38	28	Gravelly muddy beach
S. virginica	AK	Juneau	Eight Fathom Bight, Pt Frederick, Chichagof Isl.	58 00	135 44	nd	Mud flat
*S. perennis**	AK	Juneau	Juneau	58 18.0	134 24	nd	Disturbed area
*S. perennis**	AK	Juneau	Eagle Beach, NW of Juneau	58 18.0	134 25	15	Middle salt marsh & channels
S. virginica	AK	Sitka	Carroll Isl., off Admiralty Isl. cove 3 km E of Tyee	57 01.7	134 28.5	nd	Salt marsh
S. virginica	AK	Petersburg	Etolin Isl., Steamer Bay	56 08.0	132 40.0	nd	Mud flats
S. virginica	AK	Craig	Hydaberg, Prince of Wales Isl.	55 12.3	132 49.5	nd	Rocky point on beach
S. virginica	AK	Ketchikan	Annette Isl.	55 15.0	131 31.0	nd	Beach
*S. pacifica**	AK	Ketchikan	Cannery pier, George Inlet, Revillagigado	55 22	131 30	16	Gravelly salt marsh, very rare
*S. pacifica**	AK	Ketchikan	Bostwick Inlet, Gravina Isl.	55 14.5	131 44	23	High and middle salt marsh, abdt
S. virginica	AK	Craig	Soda Bay, Prince of Wales I.	55 15.0	133 02.0	nd	W coast

Taxon	State	Quad	Locality	Latitude N	Longitude W	Salinity	Habitat
S. virginica	AK	Craig	Port Refugio, Suemez Isl.	55 18.0	133 18.0	nd	Mud flat
S. virginica	AK	Craig	Port San Antonio, head of bay	55 21.0	133 37.0	nd	Beach, intertidal mud-gravel
S. virginica	BC	Prince Rupert	Dog Isl., 40 km SSE of Ketchikan	54 59.0	131 19.0	nd	Along beach
S. virginica	AK	Prince Rupert	Kirk Point village beach and slough	54 59.97	130 59.7	nd	Sandy flats of estuary
S. pacifica*	BC	Victoria	Esquimalt, Van. Isl.	46 27	123 25	nd	Sandy shore, EHW
S. pacifica*	BC	Victoria	Pond in Esquimalt Gorge Park, Gorge Inlet, Esquimalt, Van. Isl.	46 27.7	123 25.2	28	Sandy mud
S. pacifica*	BC	Victoria	Esquimalt Gorge Park, Gorge Inlet, Esquimalt, Van. Isl.	46 27.7	123 25.2	26	Sandy beach with fine gravel

able to survive in drier intertidal regions to the north because the plants can grow rapidly during the short period of higher salinities in summer. Hutchinson (1988) also notes that in British Columbia the range of water salinity is important, with *Salicornia* being confined to coastal marshes where the salinity and temperatures are consistently high. In contrast, it is absent from strongly stratified fjords with a low salinity surface layer and outflow of very cold glacier meltwater, and it is not found in the main channels of the Fraser River delta.

Overall, the known distribution of beach asparagus suggests that the start of the Kwäday Dän Ts'ìnchį man's last journey was in the Icy Strait–Lynn Canal region, rather than at Yakutat or Dry Bay. The closest point of *S. perennis* today is at St. James Bay on the Lynn Canal, about 100 kilometres south of Haines. Its densest distribution in the northern part of its range is at the mouth of the Lynn Canal, at Swanson Harbor and around Hoonah.

The EDS studies of silt grains consumed or breathed in by the Long Ago Person Found also give us some important information about his travel direction. The mineralogy of the silt grains in his digestive tract shows that the water he drank at the start of his journey contained a mixture of carbonate and sub-rounded feldspathic grains, followed by increasing amounts of carbonate grains until the last hours, when he consumed some sharply angular feldspathic silt grains as found today in the meltwater of the Samuel Glacier. This changing pattern of mineral chemistry is what might be expected for a journey up the Chilkat River trail, along Stonehouse Creek, where carbonate bedrock and lakes or ponds are common. But our sampling of sediment in Alsek Lake (fig. 5L) and tributaries draining westward into the Tatshenshini basin showed that these streams were laden with mica and very angular grains of alkaline hornblendes, with high magnesium and iron content as well as traces of sodium. In the Tatshenshini-Alsek basin even carbonate grains (for instance, at Sediment Creek; fig. 5K) and silica grains (fig. 5M) had a coating of iron oxide and so remained distinguishable from those we studied from the digestive tract samples.

The sedimentological evidence from the shape and chemistry of the silt in the stomach and intestines therefore points to a journey from low to high altitude in the limestone carbonate-rich valleys of the Chilkat basin, rather than a journey up the lower Tatshenshini-Alsek valley and O'Conner River.

To our knowledge, this is the first study of frozen remains to use a combination of ESEM and EDS to evaluate the probable route followed by an ancient person, and it complements the investigation by Helwig et al. (chapter 25) of the man's knife's mineralogy, which also showed the absence of chlorine (from sodium chloride in seawater) in the metal residues. The petrological study of minerals in Ötzi's intestines (Müller et al. 2003) was made using the more precise, but very expensive, Argon isotope method and was therefore confined to the study of 12 white mica grains, measuring 100–400 microns, that were associated with cereal fragments in his intestine. The EDS system allowed the rapid investigation of a much wider range of mineral particles associated with the Kwädąy Dän Ts'ìnchį man and his environment, and this electronic probe also helped in confirming the visual identification of microscopic calcareous particles such as the sea urchin spine.

With these clues about the location and time of year for his last journey, we can speculate on the speed with

which the Kwädąy Dän Ts'ìnchį man seems to have travelled from the coast to the Samuel Glacier, where he died. The modern distribution of *Salicornia* species in Yukon, southern Alaska and northern British Columbia shows that the closest location to the Chilkat River where fresh beach asparagus can be collected is at St. James Bay, on the north shore of the Lynn Canal, about 120 kilometres west of Klukwan, and the furthest north that beach asparagus grows today is at Taylor Bay, near the entrance to Icy Strait. The salinity and soil conditions at Yakutat, Dry Bay and the Tatshenshini-Alsek River estuary are not suitable for growth of *Salicornia* species. In fact, they are generally unsuitable for any other obligate salt-requiring plants, and *Atriplex* appears to be the only chenopod in the tidal marshes between Icy Strait and Homer. Assuming that the Kwädąy Dän Ts'ìnchį man started from the closest *Salicornia* location, his dietary diary indicates that he must have walked or run a distance of about 100 kilometres (63 miles) over mountainous terrain within two or three days—i.e., about 33 kilometres (~20 miles) per day or 3.3 kilometres (2 miles) per hour for a 10-hour day. For comparison, it took the Tlingit traders four days to travel up the Chilkoot Trail from Dyea to Bennett Lake (Thornton 2004, 94), over a distance of *ca.* 50 kilometres as the crow flies, giving an estimated speed of 12.5 kilometres per day. The journey from Yakutat to the Chilkat Pass took Tlingit traders about seven days (Mudie et al. 2005) over a linear distance of about 150 kilometres, giving a travel speed of approximately 20 kilometres per day. The estimated 33 kilometres per day for the Long Ago Person Found therefore seems high, but it is not unreasonable for a very young man who was trained as a runner (Mudie et al. 2005) rather than a heavily laden member of a trading party.

Conclusions: What, Where and Why

THE VERY FINE DETAILS provided by ESEM imagery make it possible to identify the pollen of the five chenopod genera that grow in the region from Whitehorse, southern Yukon, westward to the Alaskan coast, from Yakutat to Juneau. The ESEM details also allow us to distinguish between pollen produced by the inland species and two or three coastal species of *Salicornia*. This background information allows us to conclude that the chenopod pollen in the Kwäday Dän Ts'ìnchi man's stomach and intestines is the same as the pollen produced by beach asparagus (*Salicornia perennis*). The Tlingit name for beach asparagus is *Suk k'ádzi* in the Hoonah area, and this plant has a long-standing tradition as a food plant and potlatch gift item throughout the northwest Pacific region.

The EDS system, used in conjuction with the ESEM, provided chemical profiles of the mineral grains in the Kwäday Dän Ts'ìnchi man's stomach and intestines. Analysis of these EDS profiles shows that the minerals are either carbonates, like those in the lakes of the Stonehouse Creek and Chilkat River valleys, or feldspars lacking iron and sodium, like those found at the discovery site. In contrast, the EDS profiles for mineral grains in silt collected from the Tatshenshini-Alsek basin are heavily iron-coated or are iron-rich hornblendes. Mineral grains of this type were not found in intestinal samples we examined.

Light microscope and ESEM studies of six samples from the stomach and intestines show that the young man was omnivorous but that he mainly ate salmon and meat with oil or grease during the last three days of his life. He probably also ate shellfish, including sea urchin, when he was close to the coast, and later, just before attempting to cross the glacier, he ate some marine crab. His meals also included some fruit (blueberry or Cloudberry) and vegetable (beach asparagus).

The pollen of *Salicornia perennis* puts the food source for his last journey on the coast, south of Icy Point, with the closest modern day locality being at St. James Bay on the west side of the Lynn Canal, about 120 kilometres from Klukwan. The EDS profiles support the hypothesis that the young man travelled up the Chilkat River and Stonehouse Creek route, and the presence of *Salicornia* pollen puts the time of year at July or August, which is the flowering season for beach asparagus in southern Alaska. Our salt marsh pollen studies, and the absence of chenopod pollen in surface samples of the

St. Elias Mountains, show that this kind of pollen is not carried very far by wind in northern regions, and it is very difficult to account for so much chenopod pollen in the Kwäday Dän Ts'ìnchi man's alimentary canal if his last journey was at another time of year.

Salicornia is a low-calorie food rich in vitamins A and C, niacin and other micronutrients, and it was known by the coastal Tlingit as a remedy for scurvy. Elsewhere in the world, it is known to have medicinal values as an antioxidant and anti-tuberculosis treatment, and it may have some value as a vermifuge. The occurrence of *Salicornia* plant tissues in the food residues from both the first and last meals eaten by the Kwäday Dän Ts'ìnchi man on his final journey indicates that he carried this succulent, salt-rich plant as a travel food and/or a medicine. As his death was apparently non-violent (see chapter 12 for details), we can surmise that the young man's varied diet may have contributed to his generally good state of health as he undertook his final journey across the glacier.

ACKNOWLEDGEMENTS

Many thanks go to Peter Ball (University of Toronto) for help with the tricky and shifting taxonomy of *Salicornia*, and to Reinhard Pienitz (Université Laval) and Dan Selbie (Queen's University) for help with the diatoms. For invaluable field study aid we are indebted to Wayne Howell, Greg Streveler and Judy Brakel in the Glacier Bay–Gustavus region, who fearlessly rowed us across Icy Strait; to Bruce Bennett and David Murray of the Yukon Department of Resources, Whitehorse; to Judy Hall of Alaska Nature Tours, Haines, Alaska, who sailed alone down to St. James Bay to discover the beach asparagus; to Meg Hahr of Skagway Parks; to Cathy Conner and Carol Thelenius, University of Alaska Southeast at Juneau; to Cathryn Pohl of the US Department of Natural Resources, Juneau; to Geno Cisneros of the USDA Forests Service at Hoonah; to Robert Johnstone, US Department of Fish and Game at Yakutat; and to James Llanos, Department of Forestry, Ketchikan. For help with interviewing elders, our grateful thanks go to Sheila Greer for her consistent help with various matters over five years; to Rosita Worl, director of Sealaska at Juneau; to Marsha Hotch, director at Klukwan (*Gunasheesh*); and to Elaine Abraham and Judy Ramos, tribal planners of the Yakutat Tlingit Tribe.

32

HIS TRAVEL ROUTE

Retracing the Footsteps of the Kwädąy Dän Ts'ìnchį Man

Darcy Mathews, Sheila Greer, Richard J. Hebda and Alexander P. Mackie

The remains of the Kwädąy Dän Ts'ìnchį man were found on a glaciated and windswept pass between the Samuel Glacier and the headwaters of Fault Creek, a tributary of the O'Connor River (fig. 1). While seemingly remote, hostile and inaccessible in today's terms, the mountainous homelands of the Dän (Tutchone), Tagish and Tlingit are full of glaciated landscapes, and the oral history of these peoples recounts the use of glaciers as travel routes (Cruikshank 2005; see also chapter 29). One hypothesis is that the Long Ago Person Found was walking a travel route between the lower Chilkat River and an inland destination when he died, with his body then becoming incorporated into perennial snow and glacial ice.

BY CONNECTING THE KNOWN POINTS of his route as determined through scientific analyses, and considering information from ethnographic and archival sources, we construct the Kwädąy Dän Ts'ìnchį man's most probable route, from his starting point somewhere in the Chilkat River estuary or the lower Chilkat River valley to the site of his death in the high alpine. These data, coupled with terrain analysis from topographic maps and aerial photos taken by the recovery team, allows us to illustrate, in a top-down way, the corridor of travel on his last journey. But this was not how the route was experienced by the Kwädąy Dän Ts'ìnchį individual or by the many other ancient peoples who travelled through this area. Experience gained by actually walking through these places offers invaluable insight into aspects of the landscape that are only discernible on the ground. While we may never know why the Long Ago Person Found chose his route, researchers today can experience the terrain he walked—both on the large scale of glaciers, mountain passes and weather patterns and at the smaller, step-by-step scale of vegetation cover and animal habitat. This close-up perspective reveals a

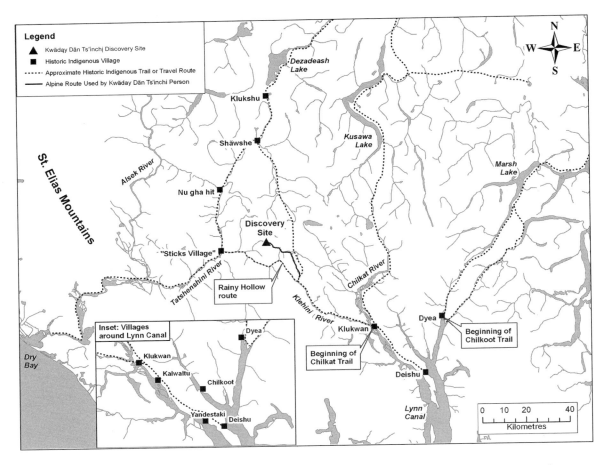

Figure 1. The location of the Kwädąy Dän Ts'ìnchį discovery site in relation to the historic Indigenous villages and the approximate trails and travel routes discussed in the text.

nuanced landscape of micro-topography, full of berries and other plants, game and minerals. The traveller is immersed in an unfolding story, gathering details and impressions while moving within these places.

To better understand the Kwädąy Dän Ts'ìnchį man's travel route, the alpine portion of his last day was retraced on foot. Photographs from that effort illustrate the route he would have seen. As the discovery site and surrounding area is officially closed to visitors, our reconnaissance effort ended one kilometre before the actual site where the body was found, but we nonetheless traversed a major lobe of the Samuel Glacier, the more difficult section of his journey.

On Travelling in Far Northwestern British Columbia

BOTH TLINGIT AND DÄN oral history sources make it clear that prior to the 1940s, when the Haines Highway was built through northwestern British Columbia, most journeys in this area were made overland and on foot. Dugouts and rafts were used to cross the Tatshenshini River and its tributaries; the former were also used to travel down the Tatshenshini to the Pacific coast. Horses first came into this country at the end of the 19th century; a few decades later, motorized vehicles followed. But throughout history, foot journeys have been the predominant mode of travel here, and walking sticks common equipment for the journey (see chapter 23).

People living in the interior, or those who visited Tatshenshini-Alsek country from the coast, were used to travelling hundreds of kilometres by foot. Wherever you came from or were heading to, significant distances were involved. As the crow flies, the distance from the discovery site to Klukwan is 65 kilometres; to Haines, 96 kilometres; to Dry Bay, 115 kilometres; to Yakutat, 152 kilometres; to Klukshu, 64 kilometres and to Haines Junction, 123 kilometres.

Coastal dwellers respected the interior residents as travellers—one of de Laguna's Dry Bay sources reported that the Dän taught their people "how to travel" (1975, 351). Travel through the region in general and the Tatshenshini-Alsek Park area in particular has never been considered easy: Champagne and Aishihik Elder Jimmy Kane described it as "young man's country" (Kane 1978; see chapter 7). Besides the rugged terrain, at the lower elevations the vegetation is thick, particularly when compared to that of the dry open country of the Yukon interior to the north (see chapter 2). The deep accumulations of snow mean that snowshoes are absolutely necessary for winter travel. Despite the severe temperatures, travel here can be somewhat easier in winter as frozen waterways become open highways and there is less need to use maintained trails. Summer travel, particularly below the tree line, can be difficult unless one has access to a network of maintained trails. While sections of old foot trails have been documented in Tatshenshini-Alsek Park (Greer 1995), these probably represent only a fraction of the trails that existed here during the 19th century and earlier, when the basin was more heavily populated (see chapter 9).

Aboriginal Travel Routes as Known Through Ethnographic and Archival Sources

THERE IS CONSIDERABLE INFORMATION within Indigenous communities today, as well as in ethnographic and archival sources, regarding different travel routes from the coast to inland destinations. This knowledge adds insight as to the Kwäday Dän Ts'ìnchį man's chosen route and situates our understanding of his journey within a larger social, geographical and historical context. Historic travel routes leading inland from the area of modern-day Haines, including Klukwan and the lower Chilkat River valley, have commonly been referred to as the Chilkat Trail—a routing of significance in regional Aboriginal history equivalent to the nearby Chilkoot Trail of gold rush fame (see Neufeld and Norris 1996). There are various routings of the Chilkat Trail; in Dákwanjè (Southern Tutchone) these routes are known as Alur Dän Tän. While the literal meaning of the phrase is "Coast Indians Trail", a more accurate translation would be "trail the coast Indians travel on".

The Chilkat and Chilkoot trail routes—so named by non-Aboriginal outsiders who visited this area at the north end of the Lynn Canal in the late 19th century—were associated with the Chilkoot and Chilkat Tlingit regional groups that used and controlled outsider access to these travel routes to the interior (Scidmore [1898] 1973). The history of these two regional Tlingit groups, or *kwans* as they are known in the Tlingit language, is closely related (Thornton 2004). In the late 19th century the Chilkat and Chilkoot *kwans* included the main villages of Klukwan, Kalwaltu, Yandestaki and Chilkoot; smaller villages such as Dyea, Deishu and others (fig. 1); and numerous camps and activity sites (Sackett 1979; Thornton 2004).

The development of these trails as trade routes to the interior is recorded in numerous Tlingit, Tagish and Dän traditional stories, such as the well-known account of Khaakeix'wtí, also known as "the man who killed his sleep" (see Thornton 2004, 90; see also chapter 29, this volume). Although all Tlingit clans seem to have participated in the trading that took place via these routes, the Chilkat Trail was controlled by the Gaanaxteidi Ravens and the Dak̲l'aweidí Eagles of Klukwan by the late 19th century, with the L'ux naxdi Ravens of Chilkoot and Yandestaki having authority over the Chilkoot Trail (Sheldon Museum, n.d.; Larson and Larson 1977). By *ca.* 1890, however, Indigenous control of these trade and travel routes was waning, as the Chilkoot Trail was opened to outsiders (see CTFN and Greer 1995) and the entrepreneur Jack Dalton developed the Chilkat Trail into a commercial route. By the time of the Klondike gold rush of 1898, the latter route was referred to as the Dalton Trail (Gates 2010, Yukon Archives 1985).

Both the Chilkat and Chilkoot trails included numerous routing variants, the general locations of which are shown in figure 1. There appear to have been two general variants of the Chilkat Trail, with the eastern (or Kusawa) route heading from the area of Klukwan village in a northeastward direction up the Chilkat River valley. From the uppermost reaches of the latter drainage basin, this route went north toward Kusawa Lake.[1] Here the trail split, with one route heading down the lake and northeastward via the Takhini River to Lake Lebarge, and then further down the Yukon River (Adamson 1993, Jim 1993, Krause 1956). The other variant of the eastern Chilkat Trail went from Kusawa Lake east via the Primrose River to Robinson and the Marsh Lake area in Carcross-Tagish First Nation traditional territory (CTFN and Greer 1995).

Whereas the eastern variant of the Chilkat Trail headed inland via the middle to upper reaches of the Chilkat River, the western routing went inland via the Klehini River, a tributary of the Chilkat, with headwaters rising in the mountains to the northwest (fig. 1). There are various sources of information on the western routing of the Chilkat Trail. We know, for example, that trade goods and furs were cached on the route at the landscape feature known as Stonehouse (Johnny Fraser in McClellan 2007, 24, 25). Understood to have been created by the being known as Crow (Raven), Stonehouse is located in the basin of a creek by the same name (Adamson 1993). Today this waterway is officially known as Clear Creek.[2] In 1905 the early ethnographer George Emmons, accompanied by guides from Klukwan, travelled to Stonehouse via this western routing of the Chilkat Trail (Emmons [1905?]).[3]

Available sources indicate that the western or Klehini River variant of the Chilkat Trail leading to the Tatshenshini River also had several different routings, depending on the intended destination. The best documented variant of the western Chilkat Trail proceeded in a generally northward direction toward the portion of the Tatshenshini River that now lies within Yukon Territory, heading to the Dän village of Shäwshe (fig. 1)—also known as Nesketahin and later referred to as Dalton Post (see chapter 10). From the Klehini/Rainy Hollow area, this route went via Stonehouse/Clear Creek, and then the Nadahini, Parton and Blanchard rivers, all tributaries of the uppermost Tatshenshini River. The modern-day Haines Road follows a good portion of this route.

Another variant of the western Chilkat Trail may have gone from the Klehini River over a height of land to the headwaters of the O'Connor River via the Klehini tributary located in Rainy Hollow (fig. 1). A possible reference to this route (i.e., westward from the Klehini/Rainy Hollow area through the mountains to the Tatshenshini River) appears in a story told to anthropologist Catharine McClellan by Johnny Fraser, a mid 20th-century chief of the Champagne and Aishihik people (McClellan 2007, 24). In recounting the adventures of Crow in the Chilkat Pass region, Fraser explained that Crow dropped his cane (or walking stick) when he fell in the Klehini River area, and as it slid away, it created a hole through the mountain. Fraser referred to this mountain as "Sunshine Hill", and CAFN Elder Mary Deguerre has identified it as being to the southwest of Rainy Hollow (Deguerre 2006). The story about Crow's walking stick making a hole in the mountain continues to be told in the Champagne and Aishihik community today (Lawrence Joe, pers. comm. to Greer, March 13, 2009).

The Rainy Hollow route over to the O'Connor River area of the Tatshenshini basin may also have been the route used by the Aboriginal packers met by explorer Seton-Karr in 1890. Seton-Karr encountered the party somewhere in the Klehini River basin, and the group pointed out the pass they had travelled through, reporting that they had travelled seven days overland from Dry Bay via this inland Tatshenshini River route (Seton-Karr 1891).

On paper at least, the Klehini/Rainy Hollow route over to the Tatshenshini would appear to be the most

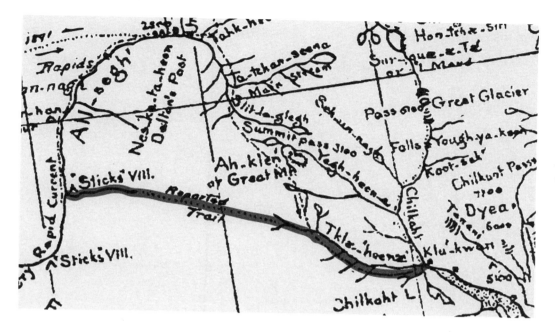

Figure 2.
A section of the map accompanying Davidson's (1901) publication of the Kohklux map, showing the reported Aboriginal trail (highlighted in grey) from Klukwan village to the "Sticks" Village.

direct path from the lower Chilkat River valley to the mouth of the O'Connor and the lower reaches of the Tatshenshini River. When viewed from the air, though, the portion of this route that lies within the upper O'Connor River basin appears to be difficult ground, with many deep canyons and steep cliff faces (Greer, pers. observation, 1995). We know that this route over to the middle Tatshenshini basin was not used by the Long Ago Person Found because the mineral grains in his gut place his travel route from the Klehini River to the Mineral Lakes area.

In addition to the two aforementioned routes (or possible routes) from the Klehini to the Tatshenshini River, sources indicate that there was at least one other route leading to the middle reaches of the Tatshenshini River. This section of the river is roughly between where it crosses the BC/Yukon border on the west and where it turns from flowing southward to westward around the mouth of either the Tomahnous River or Towagh Creek, downstream from the O'Connor River. Davidson's 1901 version of the 1867 Kohklux map shows a "reported trail" running westward from the headwaters of the Klehini River overland to the "Sticks Village" at the confluence of the O'Connor and Tatshenshini rivers

(fig. 2). As noted in chapter 10 on villages, this annotation on the trail between the middle reaches of the Tatshenshini River and the Klehini does not appear on either of the original versions of the Kohklux map that were drawn by Chief Kohklux and his wives, so it presumably derives from another source—perhaps the previously mentioned Seton-Karr reference.

Beyond the addition that appears on the Davidson revision of the Kohklux map, there is no information on the exact routing of the way from the Klehini over to the middle Tatshenshini. When asked a number of years ago where this route might have gone, the late Champagne and Aishihik Elder John Adamson said he understood that it went over the Samuel Glacier, but he could offer no specific details on the actual path (Adamson 1993). This is the route we believe the Kwädąy Dän Ts'ìnchį individual followed. The dynamic nature of the terrain and the fact that this section of his journey was above the treeline, where a single track may not have been used, mean it is best to characterize this as a *travel route* rather than a trail.

The Scientific Data on His Last Journey

ANALYSIS OF THE POLLEN, plant remains and mineral grains contained on or within the Kwädąy Dän Ts'ìnchį individual's body and clothing provide critical information on the last days, and therefore the route, of his final trek (Dickson et al. 2004; Dickson and Mudie 2008; chapter 28, this volume). Evidence from his gut also provides information on the time frame within which this travel took place—a span of a few days at most (Dickson et al. 2004). His stomach contained beach asparagus (*Salicornia*), most likely from the coast southwest of the mouth of the Chilkat River (see chapter 31), and pollen and seeds on his clothing indicate he had been in a coastal plant community (see chapter 28). Gut samples contained silt grains from the water he drank, with a mineral composition consistent with water from Mineral Lakes, located approximately halfway between Klukwan village and the discovery site. The Environmental Scanning Electron Microscope (ESEM) study of the pollen and microscopic plant and animal remains from the stomach and intestines (chapter 31) supports the idea that the Kwädąy Dän Ts'ìnchį man travelled up the Chilkat River, via Mineral Lakes, to the place where he lost his life, in the headwaters of Fault Creek (fig. 3).

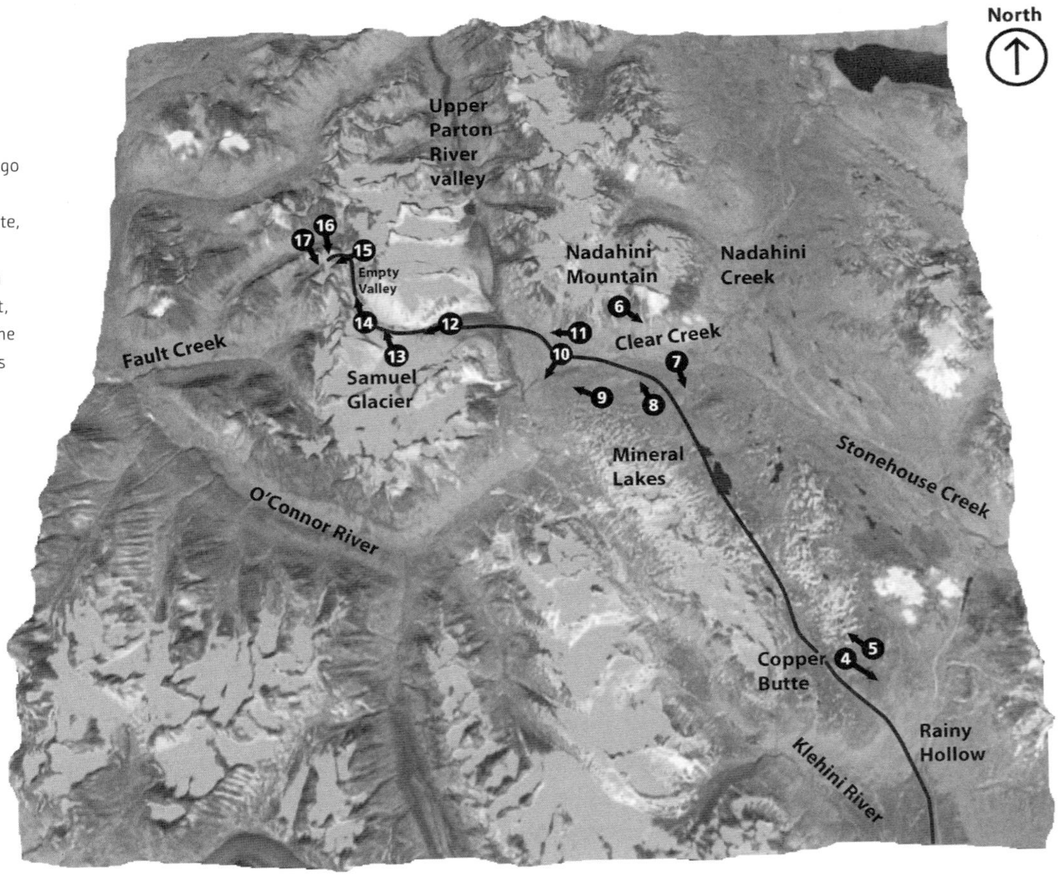

Figure 3. The route most likely taken by the Long Ago Person Found from Rainy Hollow to the discovery site, indicated by the line. The numbers correspond with figure numbers in the text, and the arrows indicate the direction the photographs were taken. Colour version on page 603.

Walking in the Kwäday Dän Ts'ìnchi Man's Footsteps: The Route

THE SOUTHERNMOST PORTION of the travel route that the Kwäday Dän Ts'ìnchi man followed after leaving the Haines/Klukwan area is well known, being largely followed by today's Haines Road. The starting point of his journey, however, has not been conclusively established. Despite the fact that fresh *Salicornia* (beach asparagus) pollen and fragments were found in his gut, as well as pollen on his clothing, the journey may not have started at salt water (where this plant grows). In early historic times boats travelled frequently up and down the Chilkat River and adjacent coastline, sometimes on a more than daily basis (Olson 1936). A similar transportation network may have also operated at the time the Long Ago Person Found lived, creating the possibility that fresh beach asparagus may have been carried upriver from the area where it grows to one of the villages in the lower Chilkat valley. The Kwäday Dän Ts'ìnchi man could have consumed this fresh plant at a village such as Klukwan, Kalwaltu or Yandestaki (fig. 1). Alternatively, the Long Ago Person Found may have travelled by boat himself, carrying beach asparagus from the seashore site where the plant grew to the location where his overland journey began. Together, these transport possibilities, the fact that all known village locations are situated upstream from modern sources of this plant, and the logical possibility that his last journey began at a village (although which one remains uncertain) mean that we identify the start of his last journey as being somewhere in the lower Chilkat River valley.

The specific routing of his travel from the lower Chilkat River to the Tatshenshini River drainage is uncertain, except for the gut mineral data showing that he travelled in the vicinity of Mineral Lakes. The most direct route from Mineral Lakes to where he lost his life takes a traveller over a major lobe of the Samuel Glacier, approaching the discovery site from the southeast.

To confirm the viability of the Samuel Glacier route option, archaeologist Darcy Mathews and writer Matt Simmons (hereafter referred to as the reconnaissance team) retraced what we believe is the route taken by the Kwäday Dän Ts'ìnchi man during his final day or two. The reconnaissance team covered a linear distance of approximately 31.5 kilometres, from Rainy Hollow on the Klehini River to a point about one kilometre from the discovery site (fig. 3).

The 2010 reconnaissance was a continuation of an earlier effort initiated by a team from Champagne and Aishihik First Nations. In 2001 Micheal Jim, Will Jones and the late Sarah Gaunt attempted to walk the route (chapter 4). Unfortunately, the Champagne and Aishihik party spent more time than anticipated accessing the lobe of the Samuel Glacier from the southeast and had to turn back after reaching it, as time and supplies ran out (M. Jim, pers. comm. to Greer, August 20, 2010).

The 2010 trek was conducted in early August, approximately the same time of year as the Kwäday Dän Ts'ìnchi man's journey (chapter 31). The reconnaissance team took five days to cover the distance from Rainy Hollow to a point near the discovery site. Much of that time was taken up with route-finding and inspecting snow and ice patches for perishable artifacts; it also included the return trip, walking out to the Haines Road. The Long Ago Person Found, likely knowledgeable about the trail and conditions, and being young and in excellent physical condition, would have covered this distance in considerably less time, perhaps travelling this section in as little as one day.

Lower Chilkat River Valley to Clear Creek via Copper Butte and Mineral Lakes

HISTORICALLY, a well-travelled foot trail from Kalwaltu and Klukwan villages (fig. 1) ran along the Chilkat River, then crossed the latter to follow the Klehini River (Davidson 1901; Krause 1956; Seton-Karr 1891); today's Haines Road follows this same route. The lower Chilkat and Klehini valleys are wide and densely forested with Western Hemlock, Sitka Spruce and thick, shrubby undergrowth. Riverbanks and islands are occupied by Black Cottonwoods, but gravel bars of the braided river system are open and sparsely vegetated, offering opportunities for easy and rapid transit (see fig. 4).

A few hundred metres above sea level, thick stands of Mountain Hemlock replace the lowland conifers. Terrain rises gradually to the northeast, toward the headwaters of the Klehini River at Rainy Hollow, where the ecosystem transitions to Mountain Hemlock parkland characterized by dense thickets of alder (see fig. 7, chapter 2). The historic trail suggests that despite the dense vegetation of the valley bottom, travel at low elevations was likely quite straightforward.

Rising up from the lower valley bottom area of the Klehini River at Rainy Hollow, the Kwädąy Dän Ts'ìnchį man ventured north-northwest toward Mineral Lakes (fig. 3). He had two options when leaving Rainy Hollow. The first was to follow either Wilson Creek or Inspector Creek (Klehini River tributaries) up onto the flat valley that runs north toward Mineral Lakes (fig. 4). The second was to leave the Klehini River at Rainy Hollow and travel directly toward Copper Butte by walking a relatively short distance uphill through slope-bottom shrubbery into the alpine tundra. While either option is plausible, the Wilson Creek/Inspector Creek route is slower due to thickets of alder and many saturated areas along the lower portion of this route.

Valley-bottom alder and willow shrub create a mass of generally horizontal stems that rise upward to chest level and higher. Pushing through these is a struggle, and tripping is a constant hazard. As well, herbaceous plants such as Cow Parsnip and Fireweed grow tall and dense in openings and are often full of biting insects. Creek courses in mountainous terrain have steep banks, boulders and highly uneven descents, requiring travellers to slowly pick their way upward. Furthermore, snow often persists long into the summer in the valley bottoms and adjacent chutes.

The alpine route straight to Copper Butte requires walking up a moderately steep but open area of alpine tundra. The distance uphill from Rainy Hollow before reaching alpine tundra is only about 1.5–2 kilometres, depending on the starting point. The alpine environment is a harshly beautiful, open and windy landscape of low, sparse tundra vegetation among patches of snow and rock. Open sightlines and limited ground cover provide not only opportunities for easy travel but also strategic locations for hunting animals such as Caribou. The constant breeze limits insects, and sources of drinking water abound.

Figure 4. Looking southeast from Copper Butte to Wilson Creek *(foreground)*, with the Klehini River valley in the background. Matt Simmons photograph. Colour version on page 604.

Figure 5. Panorama looking west *(left)* to Copper Butte and northwest *(right)* down the valley toward Mineral Lakes and Clear Creek valley in the distance. The Kwädąy Dän Ts'ìnchį man may have walked down the right side of this valley toward Mineral Lakes. Darcy Mathews photograph. Colour version on page 604.

Copper Butte is a large, conspicuous and isolated hill with steep sides and a small, relatively flat top (fig. 5). It is a distinctive landform and may have served as a navigation aid even to those who had only heard about but never seen it. The route by Copper Butte also affords the possibility of finding native copper, and although the reconnaissance team observed no native copper during their quick passage, copper ore is relatively common in the area.

From Copper Butte the Kwädąy Dän Ts'ìnchį man would have continued northwest, down the wide, flat valley toward Mineral Lakes, walking along the east side of the valley (fig. 5). Inspector Creek runs along the west side of the valley and is deeply incised in places, making walking slower and requiring more effort. It is a distance of eight kilometres from Copper Butte to the southernmost of the two Mineral Lakes and an additional five kilometres to the centre of Clear Creek Valley. The reconnaissance team was prevented from walking directly between Copper Butte and Mineral Lakes by a Grizzly Bear sow and cub, but looking at the intervening distance from either end, it is clear that contouring along the right side of the Mineral Lakes valley would require only a few small stream crossings and an elevation drop of 160 metres across relatively flat, well-drained, open tundra. In transit to Mineral Lakes few signs of big game were noted, but ground squirrels (or gophers, as they are locally known) were abundant.

In the vicinity of Mineral Lakes the Kwädąy Dän Ts'ìnchį man stopped to drink the water (fig. 6). Two kilometres beyond Mineral Lakes, he came to the Clear Creek (Stonehouse) valley and turned west toward the Samuel Glacier (fig. 7). With Nadahini Mountain looming to the north, terrain in the Clear Creek valley bottom is relatively flat and fed by many small tributaries (fig. 8). Although there is more alder in this valley than the one between Copper Butte and Mineral Lakes, walking is easy.

Within an hour after passing the headwaters of Clear Creek, the first glimpses of the Samuel Glacier come into view. The valley broadens and holds a large, shallow wetland bisected by many small streams. Within two kilometres of the Samuel Glacier, this wetland gives way to dry alpine tundra and stunning views of the main lobe of the ice flow (fig. 9).

Crossing the Samuel Glacier

FROM THE POINT AT WHICH one arrives at the Samuel Glacier, the route to where the Long Ago Person Found died heads in a general northwest direction. After crossing the glacier, there is a stretch of valley that is free of ice—colloquially referred to here as Empty Valley—followed to the west by a pass to the headwaters of Fault Creek's middle reach. The glacier at the head of

Figure 6. Mineral Lakes, looking southeast from the base of Nadahini Mountain. Clear Creek is in the foreground. The Kwädąy Dän Ts'ìnchi man would have walked north through this valley from Copper Butte, past these lakes. Darcy Mathews photograph.

Figure 7. Looking south-southeast from the Clear Creek valley (in the foreground) toward Mineral Lakes, over the route it is believed the Kwädąy Dän Ts'ìnchi person travelled. At about this point he would have turned west (toward the right side of the photograph) to head toward the Samuel Glacier. Darcy Mathews photograph.

Figure 8. View north over Clear Creek, looking toward Nadahini Mountain. When the Long Ago Person Found walked north from Mineral Lakes (behind the photographer, see also figs. 8 and 9) to this valley, he would have turned west *(left)* to head toward the Samuel Glacier. Darcy Mathews photograph.

Colour versions on page 605.

Figure 9. View west from the headwaters of Clear Creek to the lobe of the Samuel Glacier that was traversed to reach the discovery site. Darcy Mathews photograph. Colour versions on page 606.

this tributary is where the Kwädąy Dän Ts'ìnchį man's remains were found.

If the Tatshenshini River was his destination, then there was the possibility of heading south along the foot of the ice rather than heading north-northwest across the Samuel Glacier. Meltwater from the glacier feeds into the O'Connor River headwaters via a narrow valley that turns westward (figs. 3 and 10). This might seem on paper to be a logical route if one was heading to the confluence of the Tatshenshini and O'Connor rivers, but problems with this route can be discerned on the ground. The valley between the terminus of the Samuel Glacier lobe and the headwaters of the O'Connor River is extremely steep-sided. While passable today, Little Ice Age temperatures were colder and glaciers more extensive during the time of the Long Ago Person Found (see chapter 3). The lateral and terminal moraine limits of the Samuel Glacier are visible along the sides and bottoms of the recently deglaciated valleys. The

steep-sided, narrow valley of the O'Connor River headwaters may have been full of deeply crevassed ice falls and therefore a treacherous route during the time of the Long Ago Person Found's journey. Today the valley bottom of the upper O'Connor River is full of dense alder thickets, as it may have been two or three centuries ago, making travel difficult. The O'Connor headwaters route is therefore less desirable than that in the high alpine leading to Fault Creek—the route followed by the Kwädąy Dän Ts'ìnchį man. It is also possible that the Long Ago Person Found was attracted to the Samuel Glacier/Empty Valley/Fault Creek route for other reasons, such as sheep hunting. We propose, however, that he chose that route because it offered a more direct or easier way to his intended destination.

The Kwädąy Dän Ts'ìnchį man apparently headed north, walking across part of the Samuel Glacier into the Empty Valley, and then turned westward to the pass into the headwaters of Fault Creek's middle reach

Figure 10. Looking south to the headwaters of Clear Creek and the route likely taken by the Kwädąy Dän Ts'ìnchį person, shown with dashed line, between Mineral Lakes (to the left) and the Samuel Glacier (to the right). The headwaters of the O'Connor River at the toe of the Samuel Glacier are in the distance and beyond that the O'Connor River valley is visible. The valley to the left leads to Rainy Hollow and, as discussed in the text, is a known route between the Klehini and O'Connor/Tatshenshini rivers. The mineral grain data from the Kwädąy Dän Ts'ìnchį man indicate he did not travel from the Klehini drainage over to the O'Connor basin via the valley system shown in this image, even though this route may have been a variant of the Chilkat Trail. Darcy Mathews photograph. Colour version on page 606.

Figure 11. Looking west toward the lobe of the Samuel Glacier that the Kwädąy Dän Ts'ìnchį man travelled over to reach the place where he died. Following this glacier upward takes one to the first valley showing on the right. It is believed that he walked along the north margin (right side) to avoid crevasses and icefall on the inside bend of the glacier. Darcy Mathews photograph. Colour version on page 607.

(fig. 3). On this lobe of the Samuel Glacier, the Long Ago Person Found had to cross approximately five kilometres of ice (fig. 11), with its inherent dangers like crevasses. Accessing this lobe of the Samuel Glacier today is difficult. A narrow, steep-sided valley runs north-south along the edge of the ice (fig. 3). The valley's almost vertical sides consist of unstable glacial till that was deposited during the Little Ice Age and exposed during the last few decades as glacial ice receded. The reconnaissance team needed a day of careful route-finding to locate a safe means down this slope, through a series of interconnected and steep, ephemeral gullies draining the snowpack on the west face of Nadahini Mountain. Once at the bottom of the valley, the reconnaissance team crossed a deep and fast-flowing braided meltwater stream coming out of a northern snout of the Samuel Glacier, then walked through a field of dangerously unstable end and lateral moraine. This descent to the ice margin was the most challenging part of the entire trek. During the Little Ice Age, however, this section of the trip may have been easier. Glacial ice likely filled most of the valley of the upper O'Connor River, meaning that the Kwädąy Dän Ts'ìnchį person could have more or less stepped directly onto the ice from the flat terrain of the Clear Creek headwaters and proceeded up the Samuel Glacier without having to navigate the now deglaciated and therefore treacherous valley walls and bottom.

Stepping onto the glacier, it was immediately clear to the reconnaissance team that there were very few major crevasses and that the surface of the ice was covered with layers of frozen moraine, providing traction. The gradient was also very gentle, making the glacier surface a very accessible walkway (fig. 12). The maximum extent of Little Ice Age ice is discernable along the ridge that defines the northern margin of this lobe of the Samuel Glacier (fig. 12). Ongoing ablation of the ice has left very loose debris along the lower third of this ridge. Rockfall from this steep and unstable moraine presented significant danger to the reconnaissance team, a danger that the Kwädąy Dän Ts'ìnchį man may not have had to face. The reconnaissance team therefore walked well away from the valley margin to avoid the frequent rocks rolling down onto the ice. Judging from the extent of the Little Ice Age moraine (fig. 12), the slope gradient of the Samuel Glacier was likely the same during the Kwädąy Dän Ts'ìnchį man's time as it is today, with the ice being just a little higher and thicker when he travelled. It is therefore likely that the glacier's surface had few crevasses compared to other glaciers in the region. We are relatively certain that the Kwädąy Dän Ts'ìnchį man walked on the glacial ice, because even without the loose moraine, the valley margins are very steep and the adjacent ridge to the north is too sheer to ascend and walk along (fig. 12).

Not surprisingly, the Samuel Glacier is inhospitable, with constantly changing weather and

Maximum extent of Little Ice Age ice

Figure 12. Walking southwest along the north edge of the lobe of Samuel Glacier to reach the discovery site. Note the gentle gradient of the ice and lack of major crevasses. The dashed line denotes the maximum extent of Little Ice Age ice. Darcy Mathews photograph. Colour version on page 607.

Figure 13. Panorama taken on the Samuel Glacier, looking northwest *(left)* toward Empty Valley and the remnant of the hanging glacier above the valley. A large lateral moraine, centre, separates these two glaciers. At this point the Long Ago Person Found turned off of the Samuel Glacier toward the place where his body was found. Darcy Mathews photograph.

Figure 14. Looking northwest from Samuel Glacier up Empty Valley, toward the pass where the Kwädąy Dän Ts'inchį person was found. The arrow points to the area of the discovery site, which is on the opposite side of the pass and out of view. Darcy Mathews photograph. Colour version on page 608.

ice-cooled winds blowing down the valley from the main body of the glacier farther to the southwest. Walking this five-kilometre section of the trek on the ice was relatively simple, as crevasses were easily seen and bypassed or jumped over. Traversing this glacier while it is covered in snow, however, could be extremely dangerous—a peril that could confront travellers at any time of the year. During the Little Ice Age it was likely an exceptional summer when snow did not blanket the glacier's upper reaches. Under such conditions the 3.9-metre-long gaff pole or walking stick found in the site area (see chapter 23)

might have been very useful for probing the snow in search of crevasses.

Turning north into Empty Valley (figs. 2 and 13), the reconnaissance team had to cross a massive lateral moraine from the main Samuel Glacier, then negotiate stagnant ice from a hanging glacier on the west side of the valley (fig. 14). This melting ice created streams several metres deep and included large areas of slush, concealing much deeper holes and crevasses and making for dangerous walking. During the Long Ago Person Found's lifetime the glacier likely filled the entire mouth of this valley and merged with the main flow of the Samuel Glacier. Today the transition from the main flow of the Samuel Glacier to the valley that leads up to the Kwädąy Dän Ts'ìnchį site is the crux of the glacier crossing (fig. 13), but during the time of the Long Ago Person Found this transition mąy have been nothing more than a narrow band of moraine between Samuel Glacier and the hanging glacier. Ice at the north end of the hanging glacier feathers into the ice-cored moraine, making it easy to descend from the glacier and proceed uphill toward the mountain pass (fig. 14).

The steep-sided, U-shaped Empty Valley extends north from the Samuel Glacier to the Kwädąy Dän Ts'ìnchį discovery site, with a gradual elevation gain of 240 metres up its 3-kilometre length. The valley bottom now consists of moraine, meltwater streams and seasonal snow patches, but it was obviously filled with ice in times past (fig. 14). The moraine in the upper half of this valley appears more stabilized than Little Ice

Age moraine elsewhere on the route, making walking easier and safer. In 2005 a piece of cut birch bark was found near a large, distinctive glacial erratic located in the valley bottom (see chapter 23). A search around this same boulder by the reconnaissance team five years later did not identify any perishable material. From Empty Valley, the route to the discovery site turns westward, heading into the pass at the headwaters of Fault Creek.

The Place of His Passing

THE PLACE where the Long Ago Person Found perished is a narrow, ice-filled pass just beyond the highest point of the route between Lynn Canal and the Tatshenshini River. It is a frigid and windswept landscape of rock, ice and perennial snow. The remains were found just beyond the summit of the pass, melting out of ice below a small cirque that overlooks the site. This location

Figure 15. Looking west from the height of the pass down to the Kwädąy Dän Ts'ìnchį site. It is likely that any trail running through this pass would have continued on toward Fault Creek. Al Mackie photograph. Colour version on page 608.

Figure 16. Aerial view looking south to the route followed by the Long Ago Person Found (marked with a dashed line). He crossed the Samuel Glacier (in the distance), turned north and followed the Empty Valley over the pass *(centre right)*. The discovery site is just out of the frame to the right. Sheila Greer (CAFN) photograph. Colour version on page 609.

suggests he had just started the descent from the high alpine when he died (fig. 15). The ice that enclosed his body is situated near the headwaters of two small alpine tributaries of Fault Creek (figs. 3, 16 and 17).

Conclusions

WE SUGGEST that on his last journey, the Kwädąy Dän Ts'ìnchi man was travelling on a variant of the western routing of the Chilkat Trail—a routing that was taking him to some point in the middle reaches of the Tatshenshini River. Based on the scientific analysis of his remains, it is proposed that he travelled from the lower Chilkat River valley to the site of his death in two days. While travelling, he covered a distance of approximately 77.5 kilometres (assuming Klukwan village as his starting point) and gained about 1,600 metres in elevation, much of it a gradual rise. Once in the open alpine country, he encountered generally favourable and relatively level walking conditions. While it may seem difficult today for some to imagine walking 50 kilometres a day through mountainous terrain, the Kwädąy Dän Ts'ìnchi man was young and in good physical condition. He was likely a knowledgeable and experienced traveller, well prepared with food and equipment, and following a known route. The artifacts recovered in association with his remains suggest that he was travelling light, allowing for relatively rapid walking.

The reconnaissance team confirmed that the section of the travel route from the Mineral Lakes area to the place of his death saw the young man walking over a lobe of the Samuel Glacier. Despite the elevation gain and the inherent dangers, the route was found to be relatively fast, as there is no vegetation to slow progress in the alpine zone. Past the point that the reconnaissance team travelled, the route also affords, in the north Fault Creek tributary, opportunities for hunting Dall's Sheep.

Samuel Glacier

Empty Valley

Discovery Site

Figure 17. Aerial photograph looking south to the glacier where the remains were recovered. The lobe of the Samuel Glacier traversed by the reconnaissance team, as well as the valley empty of ice through which the Kwädąy Dän Ts'ìnchį man travelled, is shown on the left. Middle Fault Creek drains to the right. Sheila Greer (CAFN) photograph. Colour version on page 609.

While we may never know for certain why the Long Ago Person Found chose this specific route, the Dän, Tagish and Tlingit narratives of glacier travel (see chapter 29) indicate that glaciers were known as effective travel routes, well worth the risk they posed. The reconnaissance team's experience is consistent with Indigenous narratives of glacial travel, and we conclude that it is a logical and viable, albeit inherently dangerous, route. The Kwädąy Dän Ts'ìnchį man was living and moving knowledgeably within a landscape rich in the history and memory of his people. As researchers striving to understand details about the Long Ago Person Found man today, we worked with the knowledge of the traditional Dän, Tagish and Tlingit trails, coupled with the scientific information about

his movements gleaned from analysis of artifacts and of the Kwädąy Dän Ts'ìnchį man's actual remains. Experiencing the landscape directly provides insight into how this area might have been traversed in the past. The reconnaissance team first perceived this landscape as a remote and isolated place, known to us only through maps and aerial images. But we came to know much more of this world by moving through it. Our experience of dwelling even briefly where the Kwädąy Dän Ts'ìnchį person walked revealed a powerful landscape that we could traverse and, in some small way, share with him. In this, we are humbled by this place and the enduring presence of the Kwädąy Dän Ts'ìnchį man. The authors who have spent time at the discovery site and in the surrounding area (Mackie,

Hebda and Greer) similarly note that they were humbled by the strength, endurance and knowledge that the Kwädąy Dän Ts'ìnchį individual and other ancient travellers must have had.

We close with a reminder that ground reconnaissance of the Samuel Glacier route should not be attempted without the express permission of Champagne and Aishihik First Nations and BC Parks, who together co-manage Tatshenshini-Alsek Park. We also note that parts of the route discussed above are very dangerous. Despite the experience, knowledge, preparation and physical capability of the Kwädąy Dän Ts'ìnchį man, this environment claimed his life—a reminder to others that the landscape where the Long Ago Person Found died is a place to be respected.

ACKNOWLEDGEMENTS

We are grateful to Champagne and Aishihik First Nations for permitting the reconnaissance effort described here and for supporting all studies related to the Kwädąy Dän Ts'ìnchį discovery. We also thank Matt Simmons for arranging the logistics, participating in the field reconnaissance and kindly providing a photograph. Kathleen Matthews, University of Victoria library, ably assisted with digital mapping data. We also acknowledge the dedication and scientific expertise of project researchers James Dickson and Petra Mudie, whose investigations established the basic details of the travel route for the Long Ago Person Found's final journey. Our reconnaissance effort builds upon their research results.

6

CONNECTIONS

INTRODUCTION

Sheila Greer and Richard J. Hebda

Central to this book are the lessons learned from connections between the Kwäday Dän Ts'ìnchį man, the local landscape and the people living in the area. Previous chapters demonstrated his link to several places in the region at a time when relationships between the coastal and interior peoples were strong, with regular contact and exchange. This final portion of the volume connects the Long Ago Person Found to living people and contemporary Indigenous cultures, demonstrating perhaps the most valuable lessons to be gained from this discovery.

PART 6 BEGINS with a powerful chapter in which the Long Ago Person Found is directly connected to his living relatives through a study of community DNA. The study's results establish his clan relationship, which is a vital element of regional First Nations cultures and central to important obligations and protocols.

The remarkably preserved objects found at the discovery site also reconnect people to waning traditional practices that span centuries. Various workshops were undertaken to learn how to make objects like those found frozen in the ice. For instance, people learned how much knowledge was required to make gopher-skin robes and discovered that such robes were common household possessions as well as major

items of trade (the gophers themselves were also good to eat). Similarly, the wonderfully crafted spruce-root hat stimulated gatherings where people made other spruce-root objects, a practice carried on until recent times. Community members shared knowledge and experience in peeling and splitting roots, and they learned that, as with gophers, where and when one collected the raw material was vital to its quality and workability.

As one contributor describes, working together during the entire Kwäday Dän Ts'ìnchį project made and strengthened connections between cultures and ways of thinking. People from First Nations, governments and scholarly institutions recognized the necessity of achieving common goals, and they found new ways to understand and respect each other's needs and wants. The result was a unique collaboration, providing an example from which others can learn.

Champagne and Aishihik people have a special term for the bringing of people together to share learning: *Kets'ädän*. Community leader Diane Strand articulates how important the Kwädąy Dän Ts'ìnchį discovery has been in building connections—between the community, project participants, the neighbouring First Nations and tribes and the broader world outside.

The concept of *Kets'ädän* is a fitting theme to summarize this remarkable project and the still vital teachings of the young man who died on a BC glacier long ago.

WOLF ANCESTOR

Sheila Clark

Uncovered sleeping
a young man
in ancient repose

White blanket dishevelled
a lonely mountain bed
revealed

From the great beyond
you arouse my imagination

Whispering to me
in immortal language

Your poetry
In my DNA

In the expanse
of my heart
we've met

You and I

In the eyes of my mother
a glimpse

In the music of my son's laughter
a note

In the wingspan of my nephew's Eagle dance
a shadow

In the cadence of my grandfather's drum
a heartbeat

You

Never lost
Long Ago Person Found

Entombed in ice
the body abandoned
You left without leaving
Your life gifted to me

Dancing in this
Ceremonial Potlatch

I long for you

You are
the Ocean inside me
the Rivers that call

Tlingit brother
Your heart beats
In mine.

33

THE KWÄDĄY DÄN TS'ÌNCHĮ COMMUNITY DNA STUDY

The Search for Living Relatives

Sheila Greer, Karen Mooder and Diane Strand

Who was the young man whose remains were found on the northwestern BC glacier? Who were his people; who did he "belong to"? While other contributors explore these questions by investigating his travels or studying his garments and belongings, we approach these questions through the study of his genetic and family history.

WHEN THE REMAINS of the young man were discovered in 1999, questions of his identity and cultural connection became immediate concerns for the Champagne and Aishihik people and their neighbours. As southern Yukon First Nations and the tribes of southeastern Alaska all follow a matrilineal descent system, identifying the young man meant learning which side, or clan, he belonged to. That is, was he Crow/Raven or Wolf/Eagle? Knowing his clan was seen as the fundamental link to knowing who he was. If that detail could be established, there was hope of finding a story or song that referenced a lost person; it might even be possible to learn his name.

Here we report on the objectives, methods and results of the community DNA study (the search for living relatives). The study was initiated in 2000 and, after encountering various delays, was successfully concluded in the spring of 2008. Background information is first presented on the social context within which the Kwäday Dän Ts'ìnchį discovery was made and evaluated, as this information is crucial in providing an appropriate framework for the mitochondrial DNA (mtDNA) data as a means of identifying maternal relatives. It also helps contextualize the study results, illustrating that although 17 individuals were recognized as having the same mtDNA sequence as the Long Ago Person Found (therefore as belonging to the same maternal lineage as him), *the actual number of living relatives is most likely many times greater.*

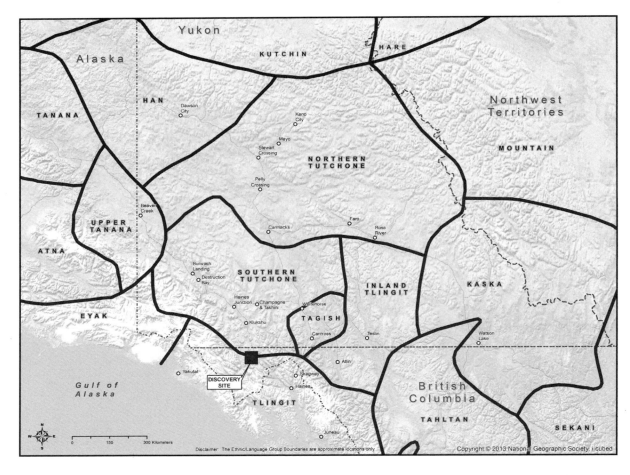

Figure 1. Distribution of Aboriginal cultural/language groups in the region and sample-collecting locales and communities mentioned. Boundaries indicated are not political and do not correspond with the traditional territories of recognized tribes and First Nations. Adapted from McClellan (1975).

We strongly emphasize that there are most probably also living individuals who would be related to the Long Ago Person Found in ways other than through the direct maternal line. Regrettably, our study did not investigate additional types of genetic connections between the Kwädąy Dän Ts'inchį individual and the contemporary Indigenous peoples of the southern Yukon, northwestern BC and southeastern Alaska.

Traditional Social Structure of the Indigenous Peoples of the Greater Study Area

THE CHAPTER reviewing the history of the Tatshenshini-Alsek area provides an introduction to the Indigenous cultures of the region and their respective histories in the Tatshenshini-Alsek country (see chapter 9). We begin here by briefly reviewing the traditional social structure of the contemporary Indigenous cultures in the greater region in order to explain the social context within which the community DNA study took place. The greater study area, from the

Yukon interior to the Pacific coast (fig. 1), straddles two different culture areas: the Northwest Coast (cf. Suttles 1990) and the Subarctic (cf. Helm 1981). Southeastern Alaska, at the northern end of the Northwest Coast culture area, is the homeland of the Tlingit, a stratified and complex society whose traditional livelihood focused on fishing, hunting sea mammals, harvesting shellfish and trading. The peoples and cultures of the northwestern interior—including the Dän (Tutchone), Tagish and Tahltan, whose languages belong to the Athapaskan language family—traditionally made their living by hunting and fishing. Athapaskan societies were not ranked to the same extent as the Tlingit, and their lifestyles were more nomadic, as they moved with the seasons to harvest resources at different locations in their territory. The Tlingit of southeastern Alaska, while known as traders and travellers, lived for most of the year in permanent villages.

While the Indigenous peoples of the two different culture areas had distinct economic patterns and population densities, these groups were—and still are—highly interconnected and interrelated. In the past few centuries at least, relationships based on trade and marriage have been strong between them (McClellan 1975). Because of this complex and interrelated history of coastal and interior peoples, it is difficult to search for neatly bounded social units, and even more so to place ethnic labels on groups of peoples. Rather, ethnic identity is, historically and today, a family if not individual matter that depends on many variables.

Despite variations in cultural practices and cultural appearance, a common social pattern prevails among the Indigenous cultures in the greater study area. They all (Tlingit and Athapaskan) recognize two basic divisions in their society: an individual is either a Crow/Raven or a Wolf/Eagle. Affiliation in one of the two groupings is inherited from one's mother, and it does not change over one's life (McClellan 1975; de Laguna 1975). Anthropologists refer to societies that recognize descent through the female line as matrilineal, and they use the term "moiety" (from the French word for half) to refer to this fundamental societal division. Figure 2 shows a generic genealogical chart illustrating various

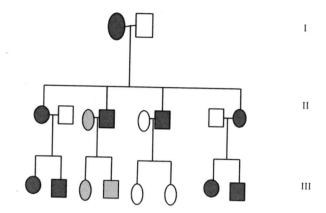

Figure 2. Schematic of mitochondrial DNA inheritance. Females are represented by circles and males by squares. The female in row I transmits her mtDNA to all her offspring as indicated by dark circles and squares in row 2. Her female offspring then transmit their mtDNA to their mother's offspring. Dark symbols represent multi-generational individuals belonging to the same matrilineage; light grey symbols represent a change in the mitochondrial DNA.

scenarios of descent through the matrilineal line over three generations. In the same way that mtDNA is passed on from a female to her offspring, so also are clan or moiety affiliations within matrilineal descent societies. Today First Nations members or citizens living in the interior (Yukon and BC) are apt to use the term "clan" to refer to this societal division—e.g., "I belong to the Wolf clan". Clan (or moiety) affiliation is so fundamental that in the Yukon at least, it has been described as "Indian law", put in place by the Creator. Anthropologist Catharine McClellan observed that in the mid 20th century the first question asked when meeting someone was: Are you a Crow or a Wolf? (McClellan 1975, 439; McClellan et al. 1987, 175).

McClellan and fellow anthropologist Frederica de Laguna, both recognized authorities on Tlingit and Athapaskan cultures, believed that the matrilineal moiety system shared by the Tlingit and the interior Athapaskan societies was ancient (de Laguna 1975; McClellan 1975). These two researchers also believed the matrilineal moiety system to have originated among peoples living in the interior and later spread to coastal southeastern Alaska. Linguists have similarly suggested that Tlingit and the various languages of the

Athapaskan family derive from a common ancestral language, albeit in the far distant past (Krauss 1980).

The Crow and Wolf terms are used by the Dän (Southern Tutchone), Northern Tutchone, Tagish, Tahltan and other Athapaskan peoples of the interior (McClellan 1975; McClellan et al. 1987). The Raven term is used by the Tlingit, whether residents of southeastern Alaska or in the interior (Yukon or northern BC), to refer to the societal division that is equivalent to the interior Crow group. The moiety opposite to the Raven is known as Wolf in the southern Tlingit area, whereas in the northern Tlingit area it is called Eagle (Campbell 1989; de Laguna 1975, 1990; Emmons 1991; Olson 1997). Table 1 shows the Indigenous-language terms for these two most basic social groupings.

The Tlingit social system features further complexity beyond this basic duality, recognizing divisions within each of the two halves or moieties. While anthropologists often use the term "sib" to denote these further divisions, community members most commonly use the term "clan". Across the entire Tlingit culture area some 38 different clans

Table 1. Indigenous language (Tlingit, Southern Tutchone) clan/moiety terms.

Clan/moiety (including pronunciation/ spelling variations)	Translation/meaning	Language	Source
Yêł Yéil	raven (the bird); also refers to Raven moiety	Tlingit	Leer et al. 2001 McClellan et al. 1987; Tlingit online dictionary
Ch'áak Ch'áak'	eagle (the bird); also refers to Eagle moiety	Tlingit	McClellan et al. 1987 Tlingit online dictionary
Ghùch	wolf (the animal); also refers to Wolf moiety	Tlingit	Leer et al. 2001
Ts'irki	raven the bird	Southern Tutchone	Cruikshank et al. 1990
Kajìt Käjèt Khäjèt Khänjèt Khanjet	Crow clan	Southern Tutchone	McClellan et al. 1987 STTC 1998 STTC 1998 STTC 1998 Tlen 1993
Ägäy	wolf (the animal)	Southern Tutchone	Tlen 1993
Agunda Aguna	Wolf clan	Southern Tutchone	McClellan et al. 1987; STTC 1998 STTC 1998

are recognized within the Raven moiety, and almost the same number within the Eagle/Wolf moiety (de Laguna 1990, Emmons in Hope and Thornton 2000). Tlingit clans also have associated crests. Not all clans, however, are represented in all Tlingit regional groupings—*kwans*, as they are termed in Tlingit. The clans represented in the various *kwans* located closest to the study area are listed in table 2. In the Tlingit system there may also be different houses within each clan—for example, Ch'aal' Hít (Willow House) within the Yanyeidí Wolf/Eagle moiety of the T'aaku Kwáan

(i.e., Taku River regional grouping of Tlingit) (Hope and Thornton 2000, 152). Traditional personal names are owned by the clans (or the two moieties, in the interior system) and are passed down only through clan/moiety lines. For the interior Athapaskan peoples, for instance, a female name belonging to the Wolf clan is only passed down to a Wolf woman, and a Crow male name only to a Crow man. In the Tlingit multi-clan system, personal names are similarly passed on to fellow clan members—for example, a Yanyeidí female name is only passed on to a Yanyeidí woman.

Table 2. Clans represented in region, listed by area or *kwan* (i.e., Tlingit regional group).

Raven/Crow	Eagle/Wolf
Laaxaayík Kwáan: Yakutat area (Near the Ice People)	
L'uknax.ádi; Kwaashk'i Kwáan	Kaagwaantaan; Lkuweidí; Teikweidí; Dagisdinaa
Gunaaxoo Kwáan: Dry Bay (Among the Athabascans Tribe)	
X'at'ka,aayí; Koosk'eidí; L'uknax.ádi; Lukaax.ádi	Dagisdinaa
Jilkaat Kwáan: Klukwan (Chilkat Tribe)	
Gaaaxteidí; Lukaax.ádi; Naach'ooneidí; Noowshaka.aayí	Kaagwaantaan; Dagisdinaa; Dakl'aweidí
Jilkoot Kwáan: Haines (Chilkoot Tribe)	
Lukaax.ádi	Shangukeidí; Kaagwaantaan
Xunaa Kwáan a.k.a. Káawu: Hoonah (Tribe or People From the Direction of the Northwind)	
T'akeintaan; Koosk'eidí; Gaanax.ádi	Wooshketaan; Chookaneidí; Kadakw.ádi; Kaagwaantaan
T'aaku Kwáan: Taku (Geese Flood Upriver Tribe)	
Gaanax.ádi; Ishkahittaan; Kookhittaan; Tooka.ádi	Yanyeidí; Tsaateeneidí; S'eet kweidí
Deisleen Kwáan: Teslin (Big Sinew Tribe)	
Ishkeetaan; Kookhittaan; Deisheetaan	Dakl'aweidí; Yanyeidí
Áa Tlein Kwáan: Atlin (Big Lake Tribe)	

Raven/Crow	Eagle/Wolf
Ishkeetaan; Kookhittaan; Deisheetaan	Yanyeidí; Dakl'aweidí
Aak'w Kwáan: Auke Bay (Small Lake Tribe)	
L'eeneidí; L'uknax.ádi; Gaanax.ádi	Wooshkeetaan
Xutsnoowú Kwáan: Angoon (Brown Bear Fort, a.k.a. Xudzidaa Kwáan – Burnt Wood Tribe	
Deisheetaan; Kak'weidi; L'eeneidí	Dakl'aweidí; Teikweidí; Wooshkeetaan
Carcross-Tagish	
Deisheetaan; Ganaxtedi; Kookhittaan; Ishkahittaan	Daklaweidi; Yan Yedi
Northern and Southern Tutchone and other interior peoples	
Crow	Wolf In the period 1940–70s, some Southern Tutchone were reporting their clan affiliation as Cankukedi, Dakl'aweidí, or Ketlimbet (McClellan in de Laguna 1975,85).
Sources: Hope and Thornton 2000, with the addition of data relevant to Carcross-Tagish people from Carcross-Tagish First Nation 2008, and to the Southern Tutchone from McClellan et al. 1987.	

For the Champagne and Aishihik people, as well as the citizens of the neighbouring First Nations and tribes, knowing which moiety or clan the Long Ago Person Found belonged to is a critical part of knowing who he was. This information was needed in order to hold any kind of funeral service or memorial potlatch for the Long Ago Person Found because individuals of the matrilineal moiety or clan opposite to that of the deceased help the bereaved family at the time of their loss—by providing the funeral feast, preparing the remains for burial (in earlier times, preparation for cremation), digging the grave and raising the headstone (McClellan 1975). At the deceased's headstone or memorial potlatch, the clan or moiety of the deceased thanks the opposite side for the support and assistance provided. These traditions and ceremonies are necessary to ensure the safe passage of the deceased's spirit to the other side or spirit world (McClellan 1964;

de Laguna 1975, 29), and they continue to be practised among all Indigenous peoples of the greater study area (cf. Central Council of the Tlingit and Haida Indian Tribes of Alaska 2008, Champagne and Aishihik First Nations Heritage Office 2007).

When mtDNA analysis was first proposed as a scientific means to identify the Long Ago Person Found's living relatives through the female line, it was embraced as the perfect fit to answer, *in Indigenous terms*, who he was.

What is Mitochondrial DNA?

MITOCHONDRIAL DNA is one of two types of DNA found within the cells of all plants and animals. Mitochondria are responsible for generating cellular energy and, as such, are crucial to the maintenance of normal physiology (Wallace 1994). The information encoded in mtDNA is passed unchanged from mothers to their offspring. This clonal character of mtDNA transmission makes it invaluable for tracing matrilineal relationships through time.

The region of mtDNA most frequently analysed in ancestry studies is referred to as the hypervariable region I (HVI; e.g., Handt et al. 1994). In contrast to the coding region of mtDNA, which contains 36 genes encoding proteins essential for life, the HVI contains no genes. Therefore, there has been little selective constraint (i.e., conservation) in this region, allowing mtDNA mutations to be randomly incorporated into the HVI, as well as the adjacent HVII and HVIII regions. Differences of two or more mutations observed between individuals effectively define a matrilineage. As more mutations accumulate in each matrilineage, they are subsequently observed to segregate into haplogroups. The number of mtDNA differences between any two haplogroups (and matrilineages within haplogroups) are directly proportional to the time that has elapsed since these haplogroups shared a common matrilineal ancestor. It is this characteristic of mtDNA matrilineages and haplogroups that allows us to estimate ancestral relationships between the Kwädąy Dän Ts'ìnchį man and contemporary peoples of the Tatshenshini-Alsek region.

Other Studies Linking Ancient DNA with Contemporary Peoples

ANCIENT MTDNA was first demonstrated to be an effective marker in tracing human evolution and ancestry over three decades ago (Pääbo et al. 1995). Since then, mtDNA has been used as a tool to link 7,000-year-old human remains in the United Kingdom with current-day inhabitants of the same region (i.e., Cheddar Man[1]). Analysis of the mtDNA of the 3,300-year-old iceman Ötzi, found in a northern Italian glacier (Handt et al. 1994), revealed that he belonged to a mtDNA haplogroup commonly found in modern peoples of central Europe.

Ancient DNA has also been used to both confirm and refute regional continuity on a population level. The varied population history of prehistoric Siberia and linkages to modern Siberian peoples have been demonstrated using ancient DNA (Mooder et al. 2006). Likewise, studies from the Four Corners region of the southwestern United States used ancient DNA to test hypotheses about the movement of ancient peoples (Kaestle and Smith 2001).

Challenges Faced by the Community DNA Study

THE STUDY TO IDENTIFY living relatives of the Long Ago Person Found encountered numerous delays and challenges, including a change in project researchers. The project was first proposed in 2000 by Dr. David Levin, who was then with the Department of Biology at the University of Victoria. Champagne and Aishihik First Nations agreed to the study with Dr. Levin as lead for the laboratory investigations, giving the First Nation responsibility for the community consultation and sample-collecting effort.[2] As described in the section on study methods, sample collection was complete by 2001, and the work of sample extraction and sequencing (mtDNA isolation and analysis) began at the University of Victoria. Because the lab work part of the study was never adequately funded, it progressed slowly over the next five years. By 2006 the lab advised that results had been obtained; unfortunately, data-handling problems were recognized, raising questions about the reliability of the results.[3]

Wanting to remove any doubt about the reliability of the data, and aware of its responsibility to the many individuals who had voluntarily participated in the study, Champagne and Aishihik First Nations realized that the most prudent course of action would be to have the lab work entirely re-done. Therefore, in 2007 the samples were returned from the University of Victoria and, under the direction of Dr. Mooder, forwarded to the Genome Quebec Innovation Centre at McGill University in Montreal for DNA extraction and sequencing. Dr. Mooder was tasked with interpreting the sequence data, including sorting the sequences obtained into genetic lineages.

The results we present here derive solely from the DNA extraction and sequencing work completed at the Genome Quebec Innovation Centre lab.

Study Methods

Recruitment

The target populations for recruitment were the citizens, members or beneficiaries of all First Nations and tribes that were known through community oral history to have some kind of connection, whether distant or direct, to the Tatshenshini-Alsek area. These include citizens

Figure 3. Map of Yukon First Nations traditional territories; based on Government of Yukon data.

or members of the Yakutat Tlingit Tribe, Chilkat Indian Tribe, Chilkoot Indian Association, Sealaska Heritage Inc., Kluane First Nation, White River First Nation, Carcross-Tagish First Nation, Kwanlin Dün First Nation, Ta'an Kwäch'än Council, Teslin Tlingit Council, Taku River Tlingit, Little Salmon-Carmacks First Nation, Selkirk First Nation, First Nation of Nacho Nyak Dun and Champagne and Aishihik First Nations. This same list represents the First Nations and tribes that were invited to send delegates to the regional meeting held in 2001 (figs. 1 and 3; see chapter 8).

All individuals who self-identified as Aboriginal were encouraged to participate in the study regardless of their personal family background specifics. In the early days of the study we had also hoped to use the social lineage data to strengthen the study sample quality by using existing genealogical data to identify known social matrilineages. The idea was that this information would then be used to recruit study participants, in order to ensure that at least one representative of each known social lineage was participating in the study. Unfortunately, funding limitations precluded us from taking such an approach.[4]

Sample collecting began in the spring of 2001, with a Champagne and Aishihik First Nations citizen providing the first sample. In May of that year, the study was introduced at the Regional Meeting of First Nations and Tribes, where it received a strong endorsement. Many of the delegates attending the meeting participated in the study at that time.

Following the regional meeting, the Champagne and Aishihik First Nations sample-collecting team of CAFN staff member Sarah Gaunt and Kluane Martin, a Kluane First Nation member and nurse, travelled to various communities in the region. Sample-collection sessions were held in Haines, Klukwan, Yakutat, Juneau (Sealaska Heritage offices), Carmacks and Dawson, at the local First Nation office or ANB (Alaska Native Brotherhood) facility (fig. 1). Sessions were also held at Champagne and Aishihik First Nations facilities in Haines Junction, Klukshu, Champagne, Takhini and Whitehorse. The sample-collecting events were publicized in newsletters and via notices produced

Table 3. Number of study participants and sequences obtained by sample collection locale.

Suffix code	Sample-collection locale	Participants	Sequences obtained	Tribe or First Nation reported/indicated
01	Champagne, Haines Junction, Klukshu, Klukwan, Takhini, Whitehorse	50	50	Champagne and Aishihik First Nations
02	Dawson, Haines Junction, Whitehorse	7	6	Teslin Tlingit Council
03	Dawson, Haines, Haines Junction, Whitehorse	5	5	Kluane First Nation
04	Carmacks, Haines Junction	11	11	Little Salmon-Carmacks First Nation
05	Dawson	1	1	Selkirk First Nation
06	Haines, Haines Junction, Juneau, Yakutat	36	33	Yakutat
07	Haines, Haines Junction, Juneau, Klukwan, Yakutat	53	53	Chilkoot Indian Association, Chilkat Indian Village, or tribal affiliation not reported/indicated
08	Dawson, Haines Junction, Klukwan, Whitehorse	6	6	Carcross-Tagish First Nation
09	Dawson, Haines Junction, Klukwan, Whitehorse	6	6	Kwanlin Dün First Nation, Champagne and Aishihik First Nations
10	Whitehorse	1	1	White River First Nation
11	not used	0	0	
12	Whitehorse	1	1	Tahltan (BC)
13	Carmacks, Dawson, Mayo	8	7	Nacho Nyak Dun, Ta'an Kwäch'än Council
14	Juneau	53	53	Yakutat, Chilkoot Indian Association, Chilkat Indian Village, or Alaskan but tribal affiliation not reported/indicated
15	Dawson	5	5	Ta'an Kwäch'än Council
16	Dawson	1	1	Dene-Chipewyan
17	Champagne, Klukwan	2	2	Hesquiaht, Haida
	Totals	**246**	**241**	

by the various First Nation and tribal organizations. Table 3 provides information on the number of samples collected at the different venues.

Sample-Collection Procedures

With the assistance of someone with nursing training, sampling was done by pricking each study volunteer's finger to produce a drop of blood, which was collected on Whatman FTA cards. The cards were then placed within small paper envelopes and labelled by the sample collector and data recorder with an assigned sequential sample reference number (e.g., 01, 02, 03). Sample reference numbers were also assigned a suffix code number (e.g., 02-01, 62-02), which identified the individual's home community and/or tribal or First Nation affiliation. If this information was lacking, the suffix code at least noted the collection locality.[5] As noted in table 3, the suffix code information provides an *approximation* of the geographic distribution of study participants.

Study participants were asked to provide their name and to sign a consent form. The form acknowledged that the blood sample being collected and the data generated from it would be used solely for the purpose intended—that is, determining relatedness to the Long Ago Person Found. It was also agreed that participants' names would not be publicly released.

Following the first large sample-collecting effort, the consent form was revised slightly, and participants were asked to provide some basic genealogical and clan affiliation data. Regrettably, sample collectors were not assertive about ensuring that all study participants provided information on the tribe or First Nation they were enrolled with or what clan/moiety they belonged to. Consequently, the assembled data set is incomplete in these regards.

All information (sample number, suffix code, participant's name, sample-collection date and collection locality) was duly noted in the handwritten sample reference log. The data key that linked individual study participants to their assigned study number was held solely by Champagne and Aishihik First Nations. This information was not provided to those members of

the study team responsible for processing samples in the lab and analysis of the sequences obtained.[6] Greer, as holder of the identification key (linking subject reference number to actual study participants), was responsible for comparing the biological lineage data with the social lineage data. At the time the samples were being collected, Strand, as CAFN Heritage Officer, was responsible for coordinating the genealogical research.

By the end of 2001 sample collection was completed, with 246 samples obtained.

Characteristics of the Study Sample

Budget limitations precluded a vigorous program to recruit study participants, which made it impossible for the study sample to be statistically representative of the social makeup of the region's Aboriginal population. Therefore, it is worth reviewing the geographic distribution and representativeness of the study sample. A total of 246 contemporary mtDNA samples were available for study, with samples obtained from members of all communities known to have some connection to the Tatshenshini-Alsek area. Table 3 shows the First Nations and tribal affiliation of the study participants, which breaks down as Alaskans 58 per cent (n=142); Yukoners 41 per cent (n=100); northern BC (that is, Tahltan) <1 per cent (n=1); and other First Nations 1 per cent (n=3).

For further context, the population of southeastern Alaska was reported as being *ca.* 73,000 in the year 2000, while Sealaska—the regional corporation representing the Tlingit and Haida of southeastern Alaska—reports present shareholder numbers as 17,000–19,000. The Yukon population is currently *ca.* 33,000, and 26 per cent (or 8,500) are of Aboriginal descent. The ratio of Alaskan to Yukon study participants is thus more than 3:1. Given that the total southeastern Alaska Indigenous population is slightly more than double that of the Yukon Indigenous population, it could be argued that Yukoners are underrepresented in the study sample.

Northern BC First Nations are not well represented in the study sample. For example, there are no participants from the community of Atlin (population

460), the home community of the Taku River Tlingit. The Atlin Tlingit are historically and genealogically connected to Carcross-Tagish and Teslin people (cf. McClellan 1975; Nyman and Leer 1993). In light of the clan affiliation of the matched relatives as discussed below, it is also regrettable that more Tahltan people did not participate in the study. One of the two clans identified as having living relatives of the Long Ago Person Found (Dak̲l'aweidí) is understood to have close historical association with the Tahltan homeland—that is, the interior country around the headwaters of the Stikine River.

Laboratory and Data Analysis Methods

Experimental design for the community DNA study was wholly informed by the region of the Kwädąy Dän Ts'ìnchį man's mtDNA HVI analyzed by Victoria Monsalve and colleagues (Monsalve et al. 2002). Upon their return from the University of Victoria, the available contemporary blood samples were re-inventoried, and 246 were sent to the Genome Quebec Innovation Centre for DNA isolation and analysis. DNA isolation was attempted for all 246 FTA samples using the Qiagen (Gentra) Autopure LS platform. DNA was successfully retrieved from 241 of 246 samples. The remaining five card samples were found to have insufficient blood to facilitate DNA recovery.

DNA samples were amplified via the polymerase chain reaction using approximately 10 ng of DNA and mtDNA HVI-specific primers F16020 (5'-TGTTCTTTCATGGGGAAGCAG-3') and R16517 (5'-CCTGAAGTAGGAACCAGATGTCG-3'). Following appropriate DNA purification, amplified DNA was prepared for dideoxy sequencing on the 3730xl Genetic Analyzer using BigDye 3.1 (Applied Biosystems) and each of the F16020 and R16517 primers. In the case of individuals possessing the 16189 T to C change, creating a long tract of cytosines, samples were sequenced twice for both the forward and reverse primers.

Sequencing data for 241 individuals were downloaded from a central server housed at the Genome Quebec Innovation Centre. DNA polymorphisms

relative to the Cambridge Reference Sequence (CRS; Andrews et al. 1999) were ascertained after DNA sequence alignment using the software program Sequencher (Gene Codes). All DNA positions observed to differ from the CRS were annotated by individual in an Excel database and subsequently compared to the DNA sequence previously obtained for the Kwädąy Dän Ts'ìnchį individual (Monsalve et al. 2002). Resulting sequences were also evaluated for mtDNA haplogroup membership using comparative phylogenetic data (e.g., Achilli et al. 2008). To better understand ancestral relationships between community matrilineages and that of the Kwädąy Dän Ts'ìnchį man, the software package PHYLIP (Felsenstein 1989) was used. The resulting matrilineal tree was visually evaluated to assess the relationship of the Kwädąy Dän Ts'ìnchį man and matching matrilineages to the rest of the samples.

Results

AFTER SEQUENCE ALIGNMENT and phylogenetic analysis, 17 of 241 samples were found to have an mtDNA sequence either identical to or closely matching (i.e., one difference) the mtDNA haplotype of the Kwädąy Dän Ts'ìnchį individual. Along with the characteristic haplogroup A2 substitutions (16111, 16223, 16290 and 16319), the Kwädąy Dän Ts'ìnchį haplotype is also defined by an uncommon T to C change at position 16189. The majority (80 per cent) of the remaining community samples fell into various A2 subclades, including A2, A2a, A2c and A2f, while a further 12 per cent possessed mtDNA haplogroups other than A2.

These results are not unexpected, based on the oral history of the region's Indigenous people (as discussed here and in chapter 9). Traditional stories suggest that relations between the coastal and interior peoples were particularly strong, and this is indeed reflected in the geographic distribution of the matches (eight Yukon, nine Alaskan).

While we recognize that this same social history data guided the study's recruitment approach, and that the pool of mtDNA samples obtained is not entirely representative of the demographic and social makeup of the greater study area, the results have validity and can confidently be used for their intended purpose—to identify the clan/moiety affiliation of the Kwädąy Dän Ts'ìnchį individual, as discussed below.

We note that this study was designed and undertaken solely to answer the question of whether or not contemporary Aboriginal peoples of the greater region of southern Yukon, northwestern BC and southeastern Alaska are related to the Kwädąy Dän Ts'ìnchį man through the matrilineal line. While the genetic and social data collected for the study could be used to explore broader kin relationships between contemporary peoples inhabiting the region, this was not the intent of the study and has therefore not been considered here.

Identifying Clan/Moiety by Linking the Genetic and Social Lineage Data

ONCE mtDNA MATCHES had been identified, the clan/moiety of the individuals identified was checked for patterns. As the information files were incomplete, some study participants identified as having matching mtDNA had to be asked to provide their clan/moiety affiliation after they were informed of their status as matches. However, the known clan/moiety affiliation of the matched individuals had not been publicly released at that time, nor were the names of the matches provided to the other matches.[7] The clan/moiety affiliation of the 17 matched individuals is reported in table 4 (page 505).

All individuals in the study sample having the same mtDNA matrilineage as the Kwädąy Dän Ts'ìnchį individual are on the Wolf/Eagle side. For those who follow the Tlingit system, either of two clans, Dakl'aweidí and Yanyeidí, were reported.[8] The history of these clans is discussed below. The clan affiliation of the individuals recognized as matches fits with the traditional stories about the history of these social units, and the results also demonstrate the historical depth and integrity of the clan/moiety system.

Two close family groupings were among the matches—a mother and her two daughters, who reside

in Juneau, and a mother and her son and daughter, who are Whitehorse residents. A few other matches knew they were related to each other or knew each other. None of the matches knew all of the others identified as being similarly related through their female line to the Kwädąy Dän Ts'ìnchį individual.

Negative Results and False Negative Data

IT IS NOTED that sequences were not obtained for five samples (table 3). For four of these, there was insufficient blood remaining on the sample cards, and the fifth sample card went missing during the change in labs.[9] Three of these five sample donors self-identified as being Crow/Raven, and another one as Eagle but belonging to the Kaagwantaan moiety. The clan/moiety of the matched individuals suggests that it is unlikely the sequences of these four individuals would have been matches. The fifth individual for whom a sequence was not obtained, a Yukoner, did not report a clan affiliation, so it is not known if this individual might have been a match that was missed.

Despite the known historical connection of the Yakutat Tlingit to the Tatshenshini-Alsek area, no matches from this community were identified. As the identified matches belong to two specific clans that are not historically associated with the Yaktutat or Dry Bay country (table 2), the lack of matches in this group of samples is not unexpected.

Beyond the Matches: the Pool of Maternal Relatives

WHILE 17 STUDY PARTICIPANTS were identified as being related to the Kwädąy Dän Ts'ìnchį man through the female line, the actual pool of living maternal relatives is significantly larger. This is because each of these individuals likely has other family members from the same matrilineage. The mtDNA sequences of these individuals, who can be referred to as "relatives of the matched maternal relatives", would be expected to similarly match that of the Kwädąy Dän Ts'ìnchį individual.

Figure 2, showing descent through the matrilineal line, can be used to demonstrate how the relatives of the matched relatives are recognized, and table 5 presents this same information in text form. For example, in the case of a female matched relative, her children would be relatives, her mother (if living) would be a relative, and her mother's sisters would be maternal relatives of the Long Ago Person Found. If this same matched female is a grandmother, the children of her daughters would also be maternal relatives. Her brother's children, however, would not be relatives through the maternal line, since they inherit their mtDNA from a different maternal lineage.

At present it is impossible to know just how many living maternal relatives there are, but the number likely runs into the hundreds, if not more. The identified maternal relatives have helped by mentioning people that they are related to through their mother's line, and it is expected that the list of known living maternal relatives will continue to grow over time, as more relatives of the matched maternal relatives are reported.

In addition to the family information provided by the 17 matches, tribal and First Nation enrollment records are expected to be an important source of genealogical information for identifying further maternal relatives. Champagne and Aishihik First Nations, for example, is able to place 5 of the matched 17 on one family genealogical chart, and in doing so, have identified more relatives through the female line. Public genealogical records are similarly recognized as important sources for identifying further relatives of the 17 matched maternal relatives.[10]

As the pool of living maternal relatives grows with the identification of further relatives of the matched maternal relatives, we can gain greater insight into the history of the two clans. One way is through consideration of the different versions of the clan histories and another is through study of the personal names held by the clan members, since personal names are passed down through clan lines.[11]

Dakl'aweidí and Yanyeidí History

OF THE 17 INDIVIDUALS with the same mtDNA sequence as the Kwädąy Dän Ts'ìnchį man, 14 reported that they belonged to either the Dakl'aweidí or Yanyeidí clan. It is therefore worth reviewing some information on the two clans to show how their shared history is illustrated by the mtDNA results.

Both clans belong to the northern Tlingit area from around Juneau/Douglas and north, and they are closely related. Traditional clan stories report that way back in time, the ancestors of both clans lived in the interior.[12] Then, at some point in the past, clan ancestors descended rivers such as the Taku or Stikine to the coast. These trips involved travelling underneath a glacier that blocked river travel at the time (Emmons in Hope and Thornton 2000; de Laguna 1975, 83).

The history of individual Tlingit clans is complex, with variations in members' understanding of their clan's past, depending on what community they are from. Some sources are quite specific in reporting that the Dakl'aweidí are descended from (i.e., an offshoot of) the Yanyeidí (de Laguna 1975, 67, 83; McClellan 2007, 223). Teslin Yanyeidí Elder Harry Morris understood that the clans split in the not-too-distant past, at a time when the group was living along the lower reaches of a river near the coast (Greer 2004a). Other stories and sources are more vague about the exact nature of the relationship. These differences make it somewhat difficult, if not impossible, to fit the stories together into one consistent whole (cf. McClellan 1975, 446).

Regardless, it is clear that the two clans share some history (McClellan 1975, 441–43). The DNA study results verify the historical relationship of the two clans, showing that the Kwädąy Dän Ts'ìnchį man's lineage predates the split.

The history of the two clans after the split, until they arrive at their historic or contemporary distribution, is similarly complex. The Yanyeidí have stayed associated with the Taku River basin but also spread to some of the northern island communities of the southeast. They are associated with the Alaskan communities of Yakutat, Hoonah and Sumdum and the northern BC and Yukon interior communities of Teslin, Atlin and Carcross (de Laguna 1990, 227; Nyman and Leer 1993; Hope and Thornton 2000, 148–59).[13] The Dak̲l'aweidí likewise spread out from their initial home area in Tahltan country (cf. de Laguna 1975, 76). The clan migration story recorded by McClellan from Carcross-Tagish Dak̲l'aweidí Elder Patsy Henderson notes that his people reached their present territory via a circuitous route (McClellan 2007, 223). From salt water the Dak̲l'aweidí expanded northward until they ended up at Klukwan, and from there they expanded inland to Shäwshe/Dalton Post, then eastward to the Carcross area (McClellan 1975, 446; 2007, 223). The Dak̲l'aweidí were and are represented in the communities of Chilkat (Klukwan), Hutsnuwu/Angoon and Tongass in the Alaskan southeast, and at Teslin, Atlin and Carcross in the interior (de Laguna 1990, 227; Carcross-Tagish First Nation 2008; Hope and Thornton 2000, 148–59). Emmons also observed that the Dak̲l'aweidí were once most numerous at Chilkat (Klukwan), where they "equalled or exceeded the other sibs (i.e., clans) in numbers" (Emmons as reported in McClellan 1975, 449).

Both clans have associated crests. Currently, a prominent Dak̲l'aweidí crest in southeastern Alaska and the Yukon is Killerwhale. The origin of this crest derives from a clan story about an ancestor who carved the original killer whale out of wood (de Laguna 1975, 72; Greer 2004b). The Ketlimbet variant of the Dak̲l'aweidí is also understood to have had the Killerwhale crest. The Tahltan Dak̲l'aweidí are reported to claim the Brown/Grizzly Bear and Eagle as crests, while the Tagish Dak̲l'aweidí claim the Wolf (de Laguna 1975, 72; McClellan 1975, 453). The Yanyeidí are primarily associated with the Wolf crest, and de Laguna notes that the Inland Tlingit Yanyeidí actually declare themselves to be the "real wolves" (1975, 80).

Today few Dän (Southern Tutchone) report affiliation with any particular clan, if one uses "clan" in the Tlingit sense. Most identify themselves as just Wolves, though one elder reported to Strand that there are two Wolf clans among the Champagne and Aishihik people. Still, when McClellan was conducting her research with the Dän in the mid 20th century, the situation was somewhat different. At that time, a number of Southern Tutchone Wolves at Champagne indicated that they were members of either the Cankukedi or the Dak̲l'aweidí clan (1975, 441). Based on information provided by Southern Tutchone informants, McClellan also reported that both of these Wolf/Eagle clans were formerly present at the old Dän villages in the Tatshenshini basin such as Shäwshe (also known as Dalton Post and by the Tlingit name Nesketahin) and Núghàyík (1975, 441, 442). She also noted that these Southern Tutchone used Ketlimbet as an alternative name for the Dak̲l'aweidí clan name (1975, 442).

This history of and relationship between the Dak̲l'aweidí and Yanyeidí clans shows that mtDNA studies have the potential to both confirm clan history as known through traditional stories and expand upon our understanding of this same history, illuminating details that may not have been passed down through oral history.

On Relatedness and Study Limitations

WHILE THIS STUDY focused solely on identifying relatives through the maternal line, mtDNA is only one part of the story of genetic relatedness. When a person is born, they inherit genetic material from their mother and father. This study did not investigate genetic inheritance through the paternal line (from fathers to daughters) or genes passed from mothers to sons; therefore, it did not identify people who would be genetically related to the Kwädąy Dän Ts'ìnchị man in other ways. This means many other people are likely just as closely related, perhaps even more closely related, to the Kwädąy Dän Ts'ìnchị individual as the 17 recognized maternal relatives or their broader group of maternal relatives that has been identified.

The study also did not attempt to measure the distance by which any of the identified maternal relatives are genetically related to the Long Ago Person Found—that is, the number of generations between a matched individual and the Kwädąy Dän Ts'ìnchị man. Variance is expected in this regard, with some of the positive matches being less distantly related than others. While we know that the Kwädąy Dän Ts'ìnchi man was Wolf/Eagle and that he has maternal relatives belonging to the Yanyeidí and Dakl'aweidí clans, we do not know what clan he belonged to. Geography and oral history give us clues but no firm conclusions. As previously noted, traditional stories indicate that individuals belonging to the Dakl'aweidí (and Ketlimbet) clans once lived in the Tatshenshini-Alsek basin. The historical importance of the Dakl'aweidí at the Chilkat village of Klukwan, whose people were known to have traded with and migrated into the Tatshenshini-Alsek area, has already been mentioned (see chapters 9 and 10). Thus, of the two Tlingit Wolf/Eagle clans with members identified as belonging to the same maternal lineage as the Long Ago Person Found, the Dakl'aweidí are most closely associated with the geographical area of the Kwädąy Dän Ts'ìnchi discovery. This hints, but does not confirm, that the Long Ago Person Found might have been Dakl'aweidí.

Conclusions

THE COMMUNITY DNA STUDY to identify living maternal relatives of the Kwädąy Dän Ts'ìnchi man experienced significant delays but eventually had a successful outcome. All of the 17 matched individuals are Wolf or Eagle clan/moiety members, suggesting that the Long Ago Person Found also belonged to the Wolf or Eagle clan/moiety. Thus, we have identified, in Indigenous terms, who the Long Ago Person Found was.

The study also provides additional insights. That all of the identified living maternal relatives belong to the same clan/moiety was not unanticipated by members and citizens of First Nations and tribes within the greater study region of southern Yukon, northwestern BC and southeastern Alaska. The results are seen as authenticating the integrity and long history of our culture's matrilineal descent system—a system in which clan/moiety affiliation is inherited from one's mother, just as mtDNA is. The study outcome has also highlighted the historical connection between the two Tlingit clans, Yanyeidí and Dakl'aweidí, a history documented in traditional stories and now confirmed through genetic studies.

Despite these positive outcomes, the study has raised as many questions as it has answered. We do not know if the Long Ago Person Found would have considered himself a member of a specific Tlingit clan such as Yanyeidí and Dakl'aweidí and, if so, what clan that would be. We also do not know who among the matched study participants are his genetically closest maternal family members. As well, a large pool of maternal relatives likely exists, perhaps numbering in the hundreds of individuals, and no information was gathered on his non-maternal relatives. Therefore, a large pool of individuals who would be genetically related to the Kwädąy Dän Ts'ìnchi man in other ways also likely exists throughout the study area.

The Kwädąy Dän Ts'ìnchi community DNA study reinforces the importance of evaluating genetic data within an appropriate context, whether it be archaeological, ethnographic or genealogical. The collaboration between the participant communities of the greater region surrounding the Tatshenshini-Alsek area demonstrates how it is possible for Indigenous communities to initiate and conduct genetic studies, effectively maintaining autonomy over the research questions and the process by which the research is carried out.

ACKNOWLEDGEMENTS

Champagne and Aishihik First Nations and the authors wish to acknowledge the contribution of all study participants. By stepping forward to donate a sample of their DNA, as well as information on their clan affiliation, they allowed an important question to be answered.

Champagne and Aishihik First Nations are grateful for the support received from neighbouring First Nations and tribes since the study was first initiated. Various of these neighbours assisted with the sample-collecting efforts by providing advertising and making their facilities available for this purpose. Supplementary contributions in support of the study were received by Champagne and Aishihik First Nations from a number of organizations in the first year of the study. These include donations from the Chilkoot Indian Tribe (Haines), facilitated by Lee Clayton, Chairman (or Chief) at the time; BC Parks (Ministry of Environment, Lands and Parks), facilitated by Peter Levy; and BC Archaeology Branch (Ministry of Small Business, Tourism and Culture), facilitated by Al Mackie.

Kluane Martin, a registered nurse and citizen of Kluane First Nation, was responsible for collection of the contemporary DNA samples. The late Sarah Gaunt, Champagne and Aishihik First Nations Heritage staff member, liaised with the neighbouring First Nations and tribes and organized the various sample-collection events.

The bulk of financial support for the community DNA study was provided by Champagne and Aishihik First Nations. The authors and the Champagne and Aishihik First Nations representatives on the Kwädąy Dän Ts'inchį management committee (past and present) are grateful for the support provided by Champagne and Aishihik First Nations'Chiefs and Council, from 1998 to present, under the leadership of Chief Bob Charlie, Chief James Allen, Chief Diane Strand and Chief Steve Smith.

Table 4. Clan moiety affiliation of the matched samples.

Sample ref. #	Clan/moiety affiliation	Source of clan/moiety information
040-01	Wolf	From individual (verbal)
198-01	Wolf	Consent form
061-02	Yanyeidí	Confirmed by family member following notification of match
062-02	Yanyeidí	Confirmed by family member following notification of match
063-02	Yanyeidí	Confirmed by individual following notification of match
181-02	Dakl'aweidí	Consent form
004-04	Wolf	Confirmed by family member following notification of match
013-08	Yanyeidí	From individual
209-07	Dakl'aweidí	Confirmed by family member following notification of match
076-14	Dakl'aweidí	From individual
082-14	Dakl'aweidí	Confirmed by family member following notification of match
085-14	Yanyeidí	Confirmed by individual following notification of match
106-14	Yanyeidí	Confirmed by family member following notification of match
110-14	Dakl'aweidí	Confirmed by family member following notification of match
117-14	Eagle	Consent form
123-14	Yanyeidí	Confirmed by individual following notification of match
124-14	Yanyeidí	Confirmed by individual following notification of match

Table 5. Determining which biological relatives of the matched study participants are also maternal relatives (adoptions excluded).

	Female matched relative	Male matched relative
his/her children	yes	no
his/her daughter's children	yes	no
his/her son's children	no	no
his/her mother	yes	yes
his/her siblings	yes	yes
his/her sister's children	yes	yes
his/her brother's children	no	no
his/her mother's sisters	yes	no
his/her mother's sister's children	yes	yes
his/her mother's brothers	yes	yes
his/her mother's brother's children	no	no
his/her father	no	no
his/her father's children	no	no
his/her father's brother	no	no
his/her father's sister	no	no
his/her father's brother's children	no	no
his/her father's sister's children	no	no

34

THE KWÄDĄY DÄN TS'ÌNCHĮ PROJECT

A Successful Collaboration

Alexander P. Mackie, Grant Hughes and Sheila Greer

My view of a successful conclusion to this project is when the science ... is married to our traditional knowledge, is married to our oral history and that we are able to identify the name of this long ago person that was found, as well as a bit of his life story.

—LAWRENCE JOE

COLLABORATIONS BETWEEN SCIENTISTS AND FIRST NATIONS are essential to the success of complex archaeological projects in Canada, especially those that involve human remains. The Kwäday Dän Ts'ìnchį project has been noted for its cooperative and productive approach to understanding the accidental death of an Aboriginal man hundreds of years ago (cf. Cruikshank 2007, 364; Moss 2011, 142–46; Thomas 2006, 235–43; Watkins 2000, 157–58). In 2002 the project partners received the Award of Merit from the British Columbia Museums Association in recognition of the partnerships created and the ongoing collaborative efforts. Such working partnerships can be difficult, even impossible, to achieve for a number of reasons, including a historical lack of trust by First Nations for archaeological projects and insufficient autonomy for the First Nations in decision-making. Also, there are often huge gaps between cultural sensitivities and scientific procedures.

Insufficient time, money and personnel to devote to relationship-building are other contributing factors. This is especially true given the urgency when human remains are discovered, which can leave little time to develop the necessary partnerships before decisions are made and actions taken.

In this chapter we discuss how the successful collaboration between the Champagne and Aishihik First Nations, BC Archaeology Branch, Royal BC Museum and scientific community came about. We hope that by examining our working relationship in detail, illustrating some of the initial differences and subsequent resolutions to challenges encountered, the project can contribute to future successful partnerships. We begin by summarizing pre-existing agreements and conditions that set the stage for this project to move forward in a cooperative manner. Then we recount the process of planning and implementing the recovery of the remains, as well as the subsequent negotiations to determine an appropriate level and period of study plus an overall management framework for the project. The resulting formalized co-management agreement, which guided all ensuing project efforts, allowed each of the parties to take on different roles and areas of responsibility within the agreed-upon framework of decision-making.

Reflections by individual Champagne and Aishihik First Nations (CAFN) citizens who were involved in the project can be found in chapters 7, 30 and 37; perspectives on the responses of neighbouring First Nations and tribes to the handling of the discovery are presented in chapter 8. This chapter provides insights into the partners' varying experiences with this collaboration. We learn that the parties jointly defined the project's goals and then achieved these objectives by working together—and, most importantly, without compromising each other's central values and priorities. Critical to our success was mutual respect and acceptance of each other's motivations for involvement. As noted in chapter 8, the First Nations' motivations were cultural and jurisdictional, rather than a matter of ownership. The government partners were highly aware of sensitivities around the find and the challenges the First Nations faced in engaging in a venture of this kind. While the Champagne and Aishihik First Nations' respect for their partners was something that grew over time, the relationships established in the process meant that the nation was committed to the success of the project and the production of this volume.

Background

THE DAY AFTER THE REMOVAL of the man's remains and belongings from the glacier, Champagne and Aishihik Chief Bob Charlie and BC Minister of Small Business, Tourism and Culture Ian Waddell stated in a joint press release: "Our governments will be working in partnership to ensure the preservation of this important find, and to develop a plan for the study of this person and their tools and clothing" (Champagne and Aishihik First Nations 1999).

Diane Strand, Heritage Resource Officer at the time of discovery and later elected Chief of Champagne and Aishihik First Nations, told a reporter during the final days of negotiating the Kwädąy Dän Ts'ìnchį Management Agreement:

> We strongly believe there's no reason to be butting heads against people when you can both be working cooperatively and moving forward.... We have a good relationship with archaeologists and we understand the importance of the scientific knowledge and how it can be used to gain a better understanding about who we are as a people. (Smith 1999)

There were many contributing factors behind these early signs that a cooperative approach was possible for such a sensitive find. Crucial to the partnership was a determination by the Champagne and Aishihik Elders Council, reached prior to discussions about recovery of the remains, that they were concerned about who this person was, wanting to know when he lived, how his tools and clothes were made and how he came to die. A history of collaboration between First Nations, the BC Archaeology Branch and the Royal BC Museum was one factor in the project's success.

Three important preconditions from the mid to late 1990s also contributed to CAFN's ability to enter into a collaborative relationship for this kind of project. First was the existence of the Tatshenshini-Alsek Park Management Agreement, a bilateral accord between the province of British Columbia and Champagne and Aishihik First Nations. The agreement, authorized by BC Order in Council 555/96, covers much of Champagne and Aishihik's traditional territory that lies within BC. Achieving the park management agreement led CAFN to suspend its treaty negotiations with the province, choosing instead to focus its efforts on cooperative management initiatives. As the Kwädąy Dän Ts'ìnchį discovery was located within this provincial park, it was subject to the park management agreement.

In a 2008 public talk in Victoria, CAFN's Lawrence Joe summarized aspects of the history and content of the park management agreement that he saw as relevant, indeed critical, to this project. We include some of his spoken comments in this chapter (further details can be found in chapter 7). He said:

> In 1993, after the area was declared as a park without any consultation, we entered into land claims negotiations.... We were not successful in concluding a treaty or an agreement in British Columbia. What we were successful in doing, as an alternative, was establishing a co-management agreement—a co-management agreement where we have shared decision-making for 80 per cent of our lands in British Columbia. (Joe 2008)

The Tatshenshini-Alsek Park Management Agreement (British Columbia 1996) includes clauses that are particularly pertinent to the Kwädąy Dän Ts'ìnchį project, as illustrated in figure 1. As Lawrence Joe pointed out:

> Champagne-Aishihik has exclusive responsibility for Aboriginal languages, for Aboriginal place names, naming of our sites and routes, for the interpretation of our culture and of our history. We have the authority for the conservation, protection and management of heritage site areas. We also have the ability, and have taken advantage of the opportunity, to take over the operations and maintenance of Tatshenshini-Alsek Park, and for a number of years now we have provided Champagne-Aishihik park rangers that are the only people on the ground in the park. It has worked out to be a very positive working relationship. (Joe 2008)

In addition to the 1996 park management agreement, a second important precondition occurred when the government of Canada, the government of Yukon and the Champagne and Aishihik First Nations finalized an agreement to address Champagne and Aishihik's claim to their lands that lay within Yukon Territory. The resulting modern-day treaty, Champagne and Aishihik First Nations Final Agreement (Government of Canada et al. 1992), gave CAFN

autonomy over a portion of their traditional lands that lie within the Yukon. The agreement also addresses a wide range of issues including eligibility and enrolment, personal taxation, First Nations' (collective) ownership of land, taxation of First Nations land, economic development and resource royalty sharing. Also part of the agreement are the management of lands and resources (fish and wildlife, forestry, water) within the First Nation's traditional territory, land use planning, the development assessment process and the creation of special management areas—e.g., parks and protected areas.

This land claim agreement and its related companion document, the Champagne and Aishihik First Nations Self-Government Agreement (Government of Canada et al. 1993), brought changes to this Indigenous government, moving it beyond the operations of a typical Indian band. The First Nation had to design and implement various governance programs to replace the previous administrative structures and operations arising from the federal *Indian Act*. By 1999 CAFN's transition to its new self-government structure was well underway, with the First Nation developing capacity in many areas. Staff members who had been involved in BC and Yukon treaty negotiations turned their attentions to developing a contemporary First Nations government, one that included a heritage program with experienced and knowledgeable people.

In their dealings with government (provincial, territorial and federal), Champagne and Aishihik had shifted from a pre-treaty concentration on redress of past wrongs to a post-treaty focus on present and future opportunities. There was a strong desire to show the world how a First Nation could successfully operate in Canadian society while using, retaining and respecting its traditional cultural values. The Kwädąy Dän Ts'ìnchį project was seen as an opportunity to show how science and tradition could be integrated in a way that would be meaningful and useful to both First Nations and scientific communities. As Lawrence Joe said:

> We all have an opportunity to change our attitudes, we have an opportunity to also share some of our

2.0 Objectives:

2.1(d) to provide for the planning, management and operation of the Park by the Parties [CAFN and BC Parks] in a manner which:

(ii) recognizes and protects the traditional and current use the Park by the Champagne and Aishihik First Nations and its citizens in the exercise of their traditional rights;

(iii) recognizes protects and preserves the rich history of the area comprising the Park, including the culture, history and traditions of the Champagne and Aishihik First Nations;

(v) conserves the natural resources of the Park for their intrinsic and scientific values and for compatible recreational opportunities;

(vi) integrates traditional and scientific knowledge in the management of the natural and cultural resources in the Park

5 .1 In accordance with their respective authorities, the Parties will establish a Park Management Board (the "Board") on or before June 1, 1996.

5.2 The Board will consist of two (2) representatives from British Columbia, one of whom will be the District Parks Manager or designate and two (2) representatives from the Champagne and Aishihik First Nations, one of whom will be the Director, Lands and Resources or designate.

5.4 The Board will be jointly chaired and consensus based.

9.2 The Champagne and Aishihik First Nations have *sole* authority over the following matters related to the Park [emphasis added]:

(a) the use of aboriginal languages;

(b) the provision of aboriginal place names;

Figure 1. Extracts from Tatshenshini-Alsek Park Management Agreement. OIC 555/96 can be accessed online through BC Laws' "Historical Orders in Council" page at http://www.bclaws.ca/civix/document/id/oic/arc_oic/0555_1996 or at the BC Legislative Library.

knowledge of this area.... We have an opportunity to gain a better understanding of self-governing First Nations and our role in Canada. (Joe 2008)

He also summed up the importance of self-government:

To us, self-government is our foundation. It provides us with the tools for programs, for legislation that's paramount to the territorial laws, that can be paramount to federal laws, and it really allows us to make decisions affecting our communities, our

lands, our government, our people and our rights. (Joe 2008)

The third important precondition for the First Nation was the Ice Patch Project, an effort that began with the 1997 discovery of well-preserved ancient hunting implements melting from alpine ice patches in the southern Yukon (Kuzyk et al. 1999; Farnell et al. 2004; Hare et al. 2004; Yukon 2011). These extraordinary archaeological discoveries led to the formation of a multidisciplinary collaboration between First

Nations, the Yukon Heritage Branch and the scientific community, with Champagne and Aishihik coordinating the involvement of First Nations.

In its earliest stages, the Kwäday Dän Ts'ìnchį project clearly benefitted from these established collaborations, and the project owes much of its subsequent success to the years of hard work done in the Yukon to develop meaningful partnerships that allowed participants to envision how a project like this one might be possible:

> Now, if the discovery had happened as little as four or five years previously, things might have had an entirely different outcome—because we would not have had the Tatshenshini-Alsek Park Management Agreement, we would not have had a final agreement in the Yukon, we would not have had the standing as a government, the access to resources and the access to staff. (Joe 2008)

Several important conditions at the BC Archaeology Branch also contributed to the project's success, including experience working with First Nations regarding human remains and the resources necessary to allow involvement in a complex and demanding project. The branch and its predecessors had been working directly with First Nations on archaeological projects for much of its existence. For 10 years, from 1973 onward, the Archaeology Sites Advisory Board (later the Provincial Heritage Advisory Board) included two First Nations representatives. The board had various responsibilities, including review of permit applications. Archaeologists considering excavations were required to obtain a Band Council Resolution as a condition of their permit. Following termination of the board, the branch required, through various processes, that archaeologists be in touch with local First Nations and, where possible, work with them on their projects. In 1994 new legislation was introduced and procedures changed so that each permit application was referred to local First Nations for review and comment; this has resulted in consulting archaeologists often working closely with the concerned Indigenous government(s).

When Aboriginal human remains are found in British Columbia, the BC Archaeology Branch becomes involved as the government agency mandated through the *Heritage Conservation Act* to deal with human remains in archaeological contexts. The branch's first priority is to work with the local First Nation(s) to determine the most appropriate course of action and to assist in carrying out those actions, where possible. At the time of the Kwäday Dän Ts'ìnchį discovery, the BC Archaeology Branch was formalizing a Found Human Remains policy, which was published on its website in September 1999 (British Columbia 1999). Pertinent extracts are included in figure 2.

In the 1990s as many as 40 accidental discoveries of Aboriginal skeletal remains were made in the province each year. At that time there were sufficient resources within the BC Archaeology Branch to help fund a physical anthropologist to assist with such discoveries as necessary. In the case of the Kwäday Dän Ts'ìnchį discovery, there were also resources to cover some fieldwork costs, as well as to keep Al Mackie, the BC Archaeology Branch's representative on the project, in the Yukon for the time necessary for recovery of the remains and the resulting negotiations to address how the project would be managed. Resources allowing Mackie to work full-time on the project during its first two years were also found.

In sum, at the time of the discovery, the BC Archaeology Branch had the expertise for working with First Nations in cases of found human remains as well as the resources to concentrate attention on the project at its busiest and most critical stage.

The Royal BC Museum has a long history of respecting the traditional cultural authority of First Nations, dating back to 1886 when the museum was established. This history has included many field archaeological studies with Aboriginal participation to better understand the traditional territory and practices of First Nations in the province. Conservation and preservation of totem poles and cultural objects has been undertaken in many locations, and loans of ceremonial objects by Aboriginal peoples to the museum for "safekeeping" have enabled the Royal BC Museum to

PROCEDURES

The following procedures will normally apply in cases where human remains are discovered fortuitously through various land altering activities such as house renovations, road construction or natural erosion; or during archaeological studies conducted under an *HCA* permit:

1. Fortuitous Discoveries

In cases where the branch has been notified that human remains have been discovered by chance, the following procedures should normally apply:

- the Coroner's Office and local policing authority should be notified as soon as possible.
- if the Coroner's Office determines the reported remains are not of forensic concern, the branch will attempt to facilitate disposition of the remains.
- if remains are determined to be of aboriginal ancestry, the branch will attempt to contact the relevant First Nation(s).
- generally, if remains are still interred and are under no immediate threat of further disturbance, they will not be excavated or removed.
- if the remains have been partially or completely removed, the branch will facilitate disposition.
- if removal of the remains is determined to be appropriate, they will be removed under authority of a permit issued pursuant to section 12 or 14, or an order under section 14 of the *HCA*, respecting the expressed wishes of the cultural group(s) represented to the extent this may be known or feasible.
- if circumstances warrant, the branch may arrange for a qualified physical anthropologist or an archaeologist with training in human osteology to provide an assessment of the reported remains in order to implement appropriate conservation measures.
- analysis should be limited to basic recording and in-field observations until consultation between the branch and appropriate cultural group(s) has been concluded.

Figure 2. Extracts from the BC Archaeology Branch 1999 Found Human Remains policy, which was accessed at http://www.for.gov. bc.ca/archaeology/ policies/found_ human_remains.htm.

engender the trust of many First Nations communities. Cultural traditions has also been supported through the building of the Mungo Martin Big House (Wawadiƚła), which serves as a location for cultural celebration at the museum, including, in 1953, the first potlatch to be held in Canada following the repeal of the anti-potlatch provisions of the *Indian Act*.

By the 1990s the Royal BC Museum had formalized its collaborative approach to working with First Nations, respecting and honouring their cultures, through the Aboriginal Materials Operating Policy. This policy follows the intent and spirit of the Task Force on Museums and First Peoples, which consisted of representatives from the Assembly of First Nations and the Canadian Museums Association. The museum, as an agent of the provincial Crown, is connected to provincial interests such as park management agreements and the

policies of the BC Archaeology Branch. Through history, policy and practice, the museum has had an ongoing interest and responsibility to collaborate on projects with First Nations.

Project Beginnings

IN CHAPTER 1 OF THIS VOLUME Bill Hanlon, Mike Roch and Warren Ward describe their discovery of the Kwädąy Dän Ts'ìnchį site on August 14, 1999, during a hunting trip in Tatshenshini-Alsek Park, and their return to Whitehorse to let people know of the find. The immediate reaction in the Yukon was to get cultural and archaeological experts up to the site to verify and assess the find. This visit occurred on August 17, 1999, the day after the hunters reported their discovery in Whitehorse. The assessment trip included several Champagne and Aishihik staff, as well as personnel representing BC Parks and Yukon Heritage Branch archaeologists. Delmar Washington, the pilot on this and all other flights to the site, is a Champagne and Aishihik citizen. Use of a single transportation provider throughout the project ensured greater security for the site location and increased the comfort of everyone involved.

The assessment visit confirmed the find, and a number of artifacts considered at risk were collected. Upon the group's return to Whitehorse, the BC Archaeology Branch was informed of the discovery, as were the Royal Canadian Mounted Police. At about the same time, Champagne and Aishihik leadership headed out on a long-planned retreat to a location with no road or phone access. The retreat's purpose was to discuss treaty issues with a neighbouring First Nation. Not surprisingly, the Kwädąy Dän Ts'ìnchį discovery was a topic for discussion as well. After the retreat, the discovery was also discussed at a meeting of First Nations elders and staff, where the elders expressed a desire that sufficient research be conducted to find out who the man was so that his remains could be buried by the appropriate people, with the required ceremony. At this same meeting, the elders also named the discovery.

The Kwädąy Dän Ts'ìnchį discovery was an unusual find that warranted a considered response, but such deliberations take time and mutual trust. In this case, there were some impediments to developing trust. There was no past relationship between the BC Archaeology Branch and Champagne and Aishihik First Nations to draw upon. Additionally, the CAFN decision-makers were out of phone contact, so they and the branch were unable to establish a dialogue or build a relationship during the initial planning stage. Both parties were thus working without key information regarding potential next steps and means for their implementation. Either party could easily have made decisions and taken actions that would have compromised understandings and future actions at this very early stage of the project.

Planning for Recovery of the Body

WHILE THE CAFN DECISION-MAKERS were at the retreat, the BC Archaeology Branch was fortunate to have the assistance and advice of Greg Hare and Dr. Ruth Gotthardt, Yukon Heritage Branch archaeologists, and Gord MacRae, the BC Parks supervisor with responsibility for Tatshenshini-Alsek Park. All three had been to the site on the assessment visit and had previous collaborative experience with Champagne and Aishihik. Their input and advice was critical in putting the project on a track to success.

Most of the early planning for dealing with the discovery on the province's behalf was done by the late Brian Apland, then director of the BC Archaeology Branch, in Victoria. Preparing for a recovery operation as a likely option, Apland sought out the expertise of University of Alberta forensic anthropologist Dr. Owen Beattie, who agreed to be available on short notice to advise the branch and CAFN on possible courses of action. At that time, Dr. Beattie was one of few people in the world with experience in frozen human remains from archaeological contexts. He was also available to assist in a recovery operation if that was decided upon, as were Gotthardt and Hare. Dr. Erik Blake, a Whitehorse-based glaciologist who was working on the Ice Patch Project with CAFN and the Yukon Heritage Branch, was also engaged to assist with the possible recovery.

With the understanding that there was no immediate threat to the stability of the remains, the branch continued with its initial scoping and planning while waiting to establish contact with CAFN, still on retreat. Then, on the morning of August 20, advisor Owen Beattie received the images taken during the August 17 site assessment visit. To Beattie, these images suggested that animals had been scavenging the human remains and that further serious damage was likely. It later turned out, thankfully, that this was not the case; what Beattie had thought were classic forensic signs of scavenging turned out to be the highly fragmented gopher-skin robe that had bunched and twisted from ice movement or wind.

With the Champagne and Aishihik retreat coming to an end the next day, a meeting seemed possible. Within a few hours of analysing the initial photos from the site, Al Mackie and Owen Beattie began their travel to Whitehorse on the chance that a rapid recovery operation might be attainable. Prior to departing for the north, the BC Archaeology Branch issued a Ministerial Order, a legal document that gave its representative (Mackie) the necessary authority to work on the site, should the parties make the decision to proceed with a recovery (fig. 3). Upon their arrival in Whitehorse, Beattie and Mackie were met by Yukon Heritage Branch staff, viewed the artifacts recovered and began discussions on preparations for work at the site.

Early on the morning of August 21, 1999, 10 people representing the BC Archaeology Branch, Champagne and Aishihik First Nations (councillors and staff) and the Yukon Heritage Branch, as well as Owen Beattie and Erik Blake, met to discuss the find. This meeting started over breakfast and continued at the Yukon Heritage Branch for most of the morning. The discussion ranged widely. Beattie presented information on frozen human remains and what can be learned from them. The CAFN representatives discussed their park co-management agreement, which was news to the BC government representative. They also spoke about cultural concerns and the need for an appropriate repository for the human remains and associated artifacts, if these were to be brought out from the site. Al Mackie talked about BC Archaeology Branch procedures and provided Lawrence Joe with a copy of the ministerial order (fig. 3). Given the site's remote location, logistics were also reviewed in detail.

To divert briefly from the narrative, the issuance of a ministerial order was cause for considerable concern to Champagne and Aishihik First Nations. Provincial representatives only became aware of these concerns years later, when we were preparing an early version of this chapter in 2008. For the branch, ministerial orders were routine parts of business during the 1990s, being issued frequently for cases of accidentally found human remains. It never occurred to branch staff that the document might be construed as anything more

than the enabling document it was intended to be, though consideration was given to the ramifications of issuing an order without consultation. For Champagne and Aishihik, however, receipt of a ministerial order was perceived as limiting their authority, their choices and their responsibilities; the order was taken as an indication that a decision had already been made to recover the human remains.

There is a long history of this kind of misunderstanding between First Nations and government officials when it comes to such pieces of paper. Because of the sensitivities that this history raises as we write about it here, and because of the nature of this misunderstanding, we now realize that the parties had arrived at a moment when the chance for collaboration could have fallen through, though perhaps none of us was fully conscious of this at that time. For CAFN at this point in the process it meant seeking the best outcome possible, one that would protect their interests and honour their cultural values to the greatest degree possible in a recovery scenario.

Lawrence Joe recounted this situation in his 2008 talk:

> Well, after, Al Mackie and Owen Beattie travelled to the Yukon, and we received word that they were arriving with a ministerial order to recover the body. We were a little bit concerned that they were overstepping their bounds and stepping on what we considered to be our authority, our responsibility, because, as I mentioned, we already had a park agreement that identified that. So we made sure they were fully aware of what our concerns were, and we have been able to establish a very positive relationship after what was initially a very testy start. (Joe 2008)

While the purpose of the order was not to require removal of the remains, there was indirect truth in CAFN's perceptions. Mackie had been instructed to make every effort to contact Champagne and Aishihik and to seek their support, and if he did not receive cooperation, he was to try and remove the remains

BRITISH COLUMBIA

Ministerial Order **1999-002**

ORDER OF THE MINISTER

PURSUANT to Section 14(4) of the *Heritage Conservation Act*, I hereby order that a heritage investigation be conducted for **recovery of human remains and associated materials reported to be melting from a glacier and potentially being scavenged by wildlife at archaeological site IkVf 1 located at the headwaters of Fault Creek near Samuel Glacier.**

The purpose of the heritage investigation is the expeditious assessment of reported "accidentally found human remains" and the implementation of appropriate conservation measures for materials exposed and disturbed by natural ablation of the glacier and apparent scavenging of thawed remains by wildlife.

The heritage investigation is to be conducted by **Alexander P. Mackie of Archaeology Branch, 5th Floor, 800 Johnson Street, Victoria, BC, V8W 9W3.**

This order is to remain in effect until **30 September 1999.**

Issued this **20th** day of **August, 1999.**

Minister of Small Business, Tourism and Culture

Per _____

Ministry of
Small Business,
Tourism and Culture

Archaeology
Branch

Mailing Address:
PO Box 9816 Stn Prov Govt
Victoria BC V8W 9W3

Location:
Fifth Floor
800 Johnson Street
Victoria

Figure 3. Kwäday̓ Dän Ts'ìnchį ministerial order, *Heritage Conservation Act.* On file, BC Archaeology Branch.

anyway. This was not made known to CAFN in 1999 and only came out nine years later as this chapter was in preparation. The branch's intentions came out of the position that it was not appropriate to leave human remains in the open where they could be scavenged by animals. It has to be noted, too, that logistical arrangements for carrying out a recovery effort would have been difficult or insurmountable without Champagne and Aishihik support and participation, as the branch did not have a precise location for the site at the time and obtaining this would have required the cooperation of BC Parks, which was focused on developing its own successful co-management regime with the First Nation in regard to the administration of Tatshenshini-Alsek Park.

In any case, the two parties with central interests and authority regarding the discovery—the province and Champagne and Aishihik First Nations—managed at this point to find common ground. Specifically, they shared a cross-cultural perspective that the remains should not be left on the mountain to be scavenged or otherwise interfered with. This led the parties to verbally agree to a recovery operation, one that would see all materials taken from the site brought to the Yukon, where they would be stored until decisions could be jointly made as to what could or should be done with them.

Although it was not mentioned at the time, removing the materials from BC effectively took them out of provincial jurisdiction and likely would have given Champagne and Aishihik a legal advantage in the unlikely event of a dispute about how to proceed after recovery.

With the recovery agreed upon, equipment began to be organized and supplies purchased. Most of the supplies and equipment for the recovery effort were provided by the Yukon Heritage Branch, Erik Blake and CAFN Councillor Ron Chambers.

Recovery 1999

ON AUGUST 22 it was clear enough to fly, so the recovery party shuttled equipment and personnel to the glacier. Eight individuals were present at the site that day, demonstrating the project's strongly collaborative approach even at this early stage, with Champagne and Aishihik, provincial and scientific representation. Present were

- Owen Beattie, with his vast experience in physical anthropology, forensics and frozen human remains;

- Erik Blake, glaciologist, with an intimate knowledge of ice and mountains and experienced with a laser theodolite used for mapping;

- Ron Chambers, member of the Tatshenshini-Alsek Park Management Board, councillor and Deputy Chief with Champagne and Aishihik First Nations, whose many talents include working on archaeology projects and guide outfitting in the mountains;

- Sarah Gaunt, Champagne and Aishihik heritage planner, accomplished photographer and Tatshenshini-Alsek Park Management Board member;

- Greg Hare, archaeologist for the Yukon government for many years, experienced with forensic cases in the territory and in the recovery of artifacts from ice;

- Al Mackie, a coastal archaeologist whose pertinent experience encompassed a consulting background that included working for and with First Nations, working on waterlogged sites with preserved organics and managing large-scale archaeological projects;

- Gord MacRae, Tatshenshini-Alsek Park supervisor who spent weeks or months of each year in the park and has a large and diverse knowledge of the area; and

- Delmar Washington, CAFN citizen, pilot and owner of Capital Helicopters.

Before the recovery began at the site on August 22, the crew assembled in a circle at the edge of the glacier, and Ron Chambers spoke quiet words of respect. For Mackie and others present, this brief ceremony was a powerful and moving moment that permeated the rest of the day, as team members focused on their respective tasks.

The morning was spent inspecting the site and surrounding areas, identifying items for collection and recovering the man's body from the ice. In the afternoon the human remains were lifted out of the ice, and soon after, with the recovered materials safely packed, the recovery team departed the mountain by helicopter. Out of the field, a late evening was spent planning the following day's work. Beattie had to return to Edmonton the next day, so he established access controls and proper storage regimes before departing. Hare and Gaunt both had commitments for the 23rd, but fortunately, others were available to participate in the recovery effort. Ty Heffner, an archaeology graduate student with physical anthropology skills, volunteered to assist, as did John Fingland from Champagne and Aishihik's Heritage program.

The second day of recovery fieldwork involved clearing snow so that artifacts and other materials spread over a wide area could be picked up. Near the end of the day, the group decided enough had been done and flew out while the weather held.

Back in Whitehorse, after being inundated by calls from the press, Champagne and Aishihik had arranged for a press conference the following morning, August 24. Communication with the media was also necessary because a BC government communications officer had by this point announced the discovery to the press, after failing to make contact with anyone in authority

at the First Nation. Champagne and Aishihik later required reassurance from the province that no slight was intended by the unilateral media release. At this point, both parties realized that the communications aspect of the project demanded attention. Working late that evening in tandem with branch representatives in Victoria, the province and the First Nation drafted a joint press release (fig. 4).

On August 24 the press conference was held in the Council Chambers of the Champagne and Aishihik First Nations administration building in Haines Junction. It was the project's first taste of an intensely interested press. CAFN Chief Bob Charlie, who has a broadcasting background, read the joint announcement, while Al Mackie, Gord MacRae, Sarah Gaunt, Diane Strand and others spoke with the press. Various TV and print media were present despite extremely short notice, including CBC TV, CBC Radio, CTV, *National Post*, *Whitehorse Star*, *New York Times*, *Los Angeles Times* and others.

Our experiences with the press at this point helped us understand how important and necessary good communications are to a project of this nature. For two or three days it was difficult to get on with the important business of developing a management framework for the discovery, as well as planning for next steps, because of the demands and distractions of the press. The group quickly established a routine where the morning was devoted to dealing with the media and the afternoon and evening to moving the project forward. In hindsight, there was value in having common problems to jointly overcome, and the late-night meetings helped forge mutual trust, a necessary ingredient for successful collaborations.

The afternoon of the first press release, Mackie met with CAFN Heritage program staff, CAFN Councillor Kathy Kushniruk and MacRae to rough out a management framework for the project. The concept of a Kwäday Dän Ts'ìnchi Management Group, operating under a formal management agreement, was proposed and accepted as a means to govern the project. Drafting the management agreement was not a simple process, however, as discussed later in this chapter.

CHAMPAGNE and AISHIHIK First Nations

NEWS RELEASE

FOR IMMEDIATE RELEASE
August 24, 1999

KWADAY DĀN SINCHÌ

(Long ago person found)

HAINES JUNCTION, YUKON - The Champagne and Aishihik First Nations and Archaeology Branch of the BC Ministry of Small Business, Tourism and Culture wish to announce the discovery of ancient human remains within a glacier. The remains were discovered a week ago by hunters travelling on foot. These people reported their discovery to the staff of the Yukon Heritage Branch's Beringia Centre in Whitehorse on Monday, August 16th. The remains were found on a glacier at high altitude within the Traditional Territory of Champagne and Aishihik First Nations, in the extreme northwest of the province of British Columbia.

The human remains are located within the Tatshenshini – Alsek Park, a protected area within the BC Parks system and is one of four adjacent parks forming a UNESCO World Heritage Site. This park was established in 1993, and is co-managed by BC Parks and Champagne and Aishihik First Nations. At present, there are no offices or facilities within the park, and the local operations centre for the park is here in Haines Junction.

The find consists of the remains of a human individual, and parts of his or her clothing and associated tools and equipment. The design of the clothing and tools indicate that the individual was aboriginal, and died in pre-contact times, before European people came into the area. We cannot at present specify what cultural group this person belonged to, or determine how long ago he or she died. The cause of the death is not known at present, but preliminary evidence indicates that this person died after falling into a glacier crevasse. An ancient foot trail runs through this area and across this glacier. Perhaps the individual, most likely a male, was travelling on this trail, or hunting in the area when the tragedy occurred.

A team of relevant specialists was quickly put together for the recovery of the human remains. This team included archaeologists, a forensic anthropologist, a glaciologist, an artifact conservator, and Champagne and Aishihik First Nations' and BC Parks representatives. Prior to the removal of the human remains,

Box 5309, Haines Junction, Yukon Y0B 1L0
Phone: (867) 634-2288   Fax: (867) 634-2108

Figure 4. First joint press release.

quiet words of respect were spoken by representatives of Champagne and Aishihik First Nations. Dr. Owen Beattie, University of Alberta, was present to give forensic advice and guidance during removal of the remains.

The collected remains and artifacts have been transferred to Whitehorse, where they are being appropriately stored and monitored by specialists. We have consulted with our Elders Council and members, to obtain their guidance on this find. The Elders have been very helpful in helping us identify the artifacts collected, and, in the past, we have worked with them to record place names and the history of this area. The Elders have indicated that we should use this situation, what appears to be an ancient tragedy, to learn more about this person; when he lived, and how his clothes and tools were made and how he died. This person will have much to tell us, to help us understand our past, and the history of our homeland. We wish to see these human remains treated with dignity and respect and to see the most positive outcome to this long ago event.

Our governments will be working in partnership to ensure the preservation of this important find, and to develop a plan for the study of this person and their tools and clothing.

On behalf of all involved, Chief Bob Charlie and BC's Minister Ian Waddell express sincere thanks to the men who reported the find, to staff of the Yukon Heritage Branch's Archaeology Program, who recognized the significance of the discovery, and brought it to our attention, and all those who assisted in the timely removal of the remains.

- 30 -

Contact:

Diane Strand
Heritage Resource Officer
Champagne and Aishihik First Nations
Haines Junction, Yukon
Ph: (867) 634-2331
Fax: (867) 634-2108

Sarah Gaunt
Champagne and Aishihik Representative
Tatshenshini-Alsek Park Management Board
Whitehorse, Yukon
Ph: (867) 667-7825
Fax: (867) 667-6202

Paige MacFarlane
Communications
Small Business, Tourism, and Culture
Province of British Columbia
Ph: (250) 953-4692, Fax: (250) 387-3798

Forensic Controls

FORENSIC CONTROLS for the access and handling of human remains were put in place under the direction of Owen Beattie. Described in detail in chapters 4 and 12 of this book, these forensic controls emphasized the need for special care and handling of the remains.

Forensic controls are in many ways complementary to the cultural values and protocols, common to different cultures, that specify how human remains should be handled. Cultural protocols ensure that a body is treated with dignity and respect as well as with consideration for the spiritual well-being of those involved with the situation—this would include the living and the dead. Awareness of Champagne and Aishihik cultural protocols was beneficial to the researchers and museum staff with responsibility for these matters and influenced the way in which the remains were stored, accessed and handled.

As recounted below, the parties eventually agreed that the body would be subject to an autopsy in order to learn more about this individual, including the cause of death. The Royal BC Museum was selected as the most suitable location for the autopsy to take place, and the remains were transferred to Victoria for this purpose. Except during those intervals when the remains were actively undergoing autopsy examinations, the man's body was safely and securely stored in the dedicated freezer at this institution (see chapter 11). Later, it was decided that the remains would be cremated before being laid to final rest. Immediately prior to their transfer to the Victoria crematorium, the remains were wrapped in the traditional manner, in preparation for this next transition.

Framing the Project

AFTER THE REMOVAL of the remains from Tatshenshini-Alsek Park to Whitehorse, Mackie remained in the Yukon for a period of 10 days acting as liaison between Champagne and Aishihik First Nations, the BC Archaeology Branch and the Royal BC Museum.

During this period, he received frequent support, via telephone, from his Victoria colleagues Brian Apland and Grant Hughes.

This period of negotiating next steps for the recovered remains focused on the following key issues.

1. Determining if the remains and artifacts could be studied.
2. If so, establishing what kinds of studies would be acceptable to the parties.
3. Identifying what types of facilities and expertise would be needed to house and conserve the human remains and artifacts during this possible period of study.
4. Identifying where such facilities existed and whether they would be willing and able to accept the responsibilities of housing the items.
5. Establishing the respective areas of responsibility for the province and Champagne and Aishihik First Nations.

Very early in the multi-party discussions the concept of a project management group, operating under a management agreement, was adopted as the management framework for the discovery. In order to move discussions on the above issues forward, without addressing them at too early a stage in the negotiations, the group focused on outlining the purpose and mission of the management group. This allowed us to communicate our respective goals and values, and find common ground.

The goals and concerns for the province were to ensure that the project's management agreement included conditions that would allow for the best scientific enquiry possible. BC also wished to avoid wording that spoke to the matter of the "ownership" of the discovery, as the concept was considered both problematic for human remains and inappropriate for legal reasons. We should clarify here that the province's *Heritage Conservation Act*, which regulates these kinds of discoveries, is silent on the matter of ownership.

Champagne and Aishihik's priorities in these negotiations arose from the First Nation's past experience in establishing new working relationships with federal and Territorial governments, as well as from its own cultural values and standards. According to Lawrence Joe, a participant in the negotiations:

> This young man that met his end high in the mountains on a glacier long ago was someone's father, their son, their brother, their uncle. We have a responsibility to the people that are in our territory. Not only a cultural responsibility, but we have an obligation to ensure that our First Nation values, that our First Nation beliefs, are respected in this process. It has been challenging. It has been a difficult position to put us in because … our culture places strong direction upon us to do things right, especially when we are dealing with a death. (Joe 2008)

CAFN also required mechanisms ensuring that respect for the man and his dignity were maintained at all times. Furthermore, they wanted their government involved in all decision-making. The First Nation was also concerned that the recovered materials would leave and never return to their traditional territory. It was important to them that they have full authority for the long-term disposition of the man's remains, as well as his possessions and the other artifacts recovered.

Kwädąy Dän Ts'ìnchį Management Agreement

THE GROUP TASKED with drafting the management agreement focused on immediate needs and discussed possible membership on the team, including a role for Owen Beattie as project scientific advisor. As negotiations progressed, other goals were identified, such as arrangements for transport of the human remains to the location where they could be safely and securely stored during the autopsy period—which, by this point, was identified and agreed upon as the Royal BC Museum. In transferring the remains, the parties sought transport that was affordable, direct and circumspect, avoiding a media circus at the point of delivery. A news blackout on the management

agreement negotiations and the transport arrangements was recognized as prudent. The parties also prepared video footage on the project, which was jointly released. This footage emphasized the central themes of treating the man with dignity and respect and maintaining the best scientific conditions possible.

The agreement-drafting process, which essentially framed the project, took a full week and involved three or four people working on Champagne and Aishihik's behalf, about the same number for the province, as well as input from many others. Drafts and counter drafts were circulated before a version was achieved that the parties each believed could work for them.

The resulting Kwädąy Dän Ts'ìnchį Management Agreement (fig. 5) outlines project goals and limits, as well as the roles and responsibilities of parties in regard to the discovery. The agreement was signed on August 31 by Brian Apland, on behalf of the province, and Lawrence Joe, representing Champagne and Aishihik. This was just two weeks and one day after the hunters reported their find in Whitehorse.

Loan Agreement between CAFN and Royal BC Museum

BY THIS POINT it had been agreed that studies on the human remains would proceed. The type of permitted studies had also been agreed upon; most significant was that they would be similar to an autopsy in scope. Still, further negotiations were required in order to address First Nations' concerns. A key consideration was to find a way for the human remains and artifacts to be held at the Royal BC Museum for the agreed-upon study period without being accessioned into museum collections. It was important to the First Nation that the materials being transferred to the institution ultimately stayed under First Nation control. The solution was to treat the incoming materials as a loan from the First Nation to the museum for the purposes of safekeeping and study. This arrangement was formalized with a loan agreement (fig. 6), which was signed on September 2, 1999.

KWADAY DÄN SINCHĮ

Agreement made this 31ª day of August 1999, between the British Columbia Archaeology Branch, Ministry of Small Business, Tourism and Culture, and the Champagne and Aishihik First Nations, respecting the management of human remains and associated artifacts from the Tatshenshini-Alsek Park, British Columbia.

1. This Agreement is without prejudice to the rights and interests of Champagne and Aishihik First Nations and the British Columbia Archaeology Branch, Ministry of Small Business, Tourism and Culture, hereafter referred to as the Parties.

2. The Parties wish to cooperate with one another in all decisions regarding the future management and study of remains and associated artifacts of an ancient person discovered in August of 1999 in the Tatshenshini-Alsek Park. These remains and artifacts have been recorded as Archaeological Site IkVf-001, and named Kwaday Dän Sinchį (Long Ago Person Found).

3. The Parties recognize the importance of Kwaday Dän Sinchį as an opportunity to learn about a past time in human use of the Tatshenshini area, and recognize that they are making a significant contribution to future generations through their mutual desire to protect and study these ancient remains.

4. The Parties agree that remains of Kwaday Dän Sinchį will be treated with respect and dignity throughout the scientific examinations.

5. The Parties agree that decisions regarding the final disposition of Kwaday Dän Sinchį, including associated artifacts, will be the responsibility of Champagne and Aishihik First Nations.

6. The Parties agree that when either has care, control or custody of Kwaday Dän Sinchį and associated artifacts, they will make best efforts to protect and treat the remains and artifacts in a respectful, dignified and scientifically sound manner.

7. The Parties have agreed to establish the Kwaday Dän Sinchį Management Group as described below to oversee the future management and study of Kwaday Dän Sinchį.

    a) The Management Group shall consist of three members selected by Champagne and Aishihik First Nations and three members selected by the BC Archaeology Branch. The Management group may agree to increase their membership.

    b) The Management Group shall be co-chaired and operate by consensus.

Figure 5. Kwädąy Dän Ts'ìnchį Management Agreement.

-2-

c) The Management Group will set out the relevant terms for the scientific study of the remains of Kwaday Dän Sinchí.

d) The Management Group may convene expert advisory panels to provide direction to the Group.

e) All public statements will be mutually endorsed by the co-chairs.

f) The Management Group will seek funding to cover expenses related to the transportation, storage and study of Kwaday Dän Sinchí.

g) Monthly reports, or as often as deemed necessary by the Management Group, will be made to the Chief and Council of Champagne and Aishihik First Nations and to the British Columbia Ministry of Small Business, Tourism and Culture.

8. In recognition that specialized care and treatment of the human remains will be required to provide for a period of time in which scientific studies can be conducted, the Parties agree that the remains will be released to the BC Archaeology Branch to arrange for such care and safekeeping at the Royal British Columbia Museum, or another agreed upon facility, for a period not less than 15 months.

9. The Parties agree that the remains will be returned for final disposition to Champagne and Aishihik First Nations by December 31, 2000 or at a later date, by agreement of the Management Group.

10. Any economic or scientific benefits or other considerations will be shared between the Parties in any matters negotiated by the Management Group.

11. To achieve consensus on disputed matters, the Parties will employ dispute resolution measures in a non-adversarial and informal manner. These measures may include mediation and, with the agreement of all members, other dispute resolution procedures which may assist the Management Group members to achieve consensus. The costs of the dispute resolution process will be borne equally by the Parties.

12. The Management Group may consider additional matters jointly agreed to by the Parties.

Director, BC Archaeology Branch	Director, CAFN Lands and Resources

LOAN AGREEMENT

Between: Champagne and Aishihik First Nations

And: Anthropological Collections Section
             Royal British Columbia Museum

Champagne and Aishihik First Nations agrees to lend the material listed below to the Anthropological Collections Section, Royal British Columbia Museum, from September 2, 1999 to December 31, 2000 for the purpose of safekeeping and scientific study under the direction of the Kwaday Dän Sinchí Management Group.

The material will be deposited with the same care it would receive if it were the property of the Royal British Columbia Museum, which assumes no responsibility in case of loss or damage or theft, fire or other events over which it has no control.

Description: Human Remains known as Kwaday Dän Sinchí from site IkVf-1, Tatshenshini-Alsek Park.
                 See attached list, **IkVf-1, Human Remains Collections Listing** prepared by Dr. R. Gotthardt, Yukon Heritage Branch, September 1, 1999.

_____   September 2, 1999
Director, Curatorial Services Branch   Date
Royal British Columbia Museum

_____   September 2/1999.
Lender   Date
Sarah Gaunt, on behalf of Champagne and
Aishihik First Nations

Figure 6. CAFN loan agreement with Royal BC Museum.

Figure 7. Loading the freezer into the chartered airplane. *(left to right)* Alkan employee, Harold Johnson, John Fingland (CAFN), Greg Hare (Yukon Heritage), Al Mackie (BC Archaeology) and Alkan employees. Sarah Gaunt (CAFN) photograph.

Transfer of the Remains

WITH THE MANAGEMENT AND LOAN AGREEMENTS in place, arrangements could be made for transporting the human remains as well as the most fragile of the man's belongings to the Royal BC Museum in Victoria (fig. 7). Champagne and Aishihik traditional practice dictates that a body should always be accompanied on such journeys—for instance, when remains are returned from a hospital to the home community for burial. In our case, Harold Johnson of Champagne was chosen by CAFN elders to accompany the remains to Victoria. By happy circumstance, one of the pilots chartered to fly the remains south was George Bahm, of Teslin Tlingit descent. Also flying south on the September 2 charter flight were Mackie and Yukon Heritage Branch conservator Diana Komejan. Those accompanying the remains thus ensured that both cultural and forensic needs of the project were respected at all times during the transfer.

Members of the Management Group

THE KWÄDĄY DÄN TS'ÌNCHĮ MANAGEMENT AGREEMENT established the project management group as the team to guide the management of the discovery. It was agreed that the management group would be co-chaired, with equal representation from CAFN and the province, and the museum was to be considered part of the province team. The management group was to operate by consensus.

There have been various changes to the makeup of the management group in the years since it was established, although some key personnel have retained a seat at the table throughout this period. The list of past and present members includes the following people:

- Co-chair: Lawrence Joe (CAFN), 1999–time of writing
- Co-chair: Brian Apland (BC Archaeology Branch), 1999–2001; replaced as co-chair by Grant Hughes (Royal BC Museum), 2001–09, who in turn was replaced by Kelly Sendall, 2009–time of writing

- Member (CAFN): Diane Strand, 1999–2006; Frances Oles, 2006–08
- Member (CAFN): Sarah Gaunt, 1999–2003; Sheila Greer, 2003–time of writing
- Member (BC Archaeology Branch): Al Mackie 1999–time of writing
- Member (Royal BC Museum): Grant Hughes, 1999–2001; replaced by Jim Cosgrove, 2001–07; Kelly Sendall, 2007–09
- Scientific Advisor: Owen Beattie, 1999–time of writing (included in meetings as needed, in person or via submissions)

Call for Proposals

THERE WERE MANY QUESTIONS surrounding the discovery, such as: What people or culture did the Kwädąy Dän Ts'ìnchį man belong to? When did he live? What was the cause of death? Lacking the resources to directly engage researchers who could seek the answers to these and other questions, the management group realized it was necessary to establish working relationships with reputable scientists who could conduct research without direct cost to the project. It was also important that any researchers involved shared interests similar to those of the management group, would respect the goals and operating framework of the project and could produce excellent work.

Most critically, the management agreement specified that the human remains were to be available for the autopsy investigations for a limited period—a maximum of 15 months—after which the remains were to be returned to the First Nation for final disposition. This requirement made it imperative that any studies to be done proceeded in a timely fashion. With this pressure at hand, Beattie led the effort in developing the call for proposals, asking qualified researchers to submit their ideas for possible research studies. In November 1999 the call went out to all Canadian universities as well as other select institutions in Canada and around the world.

There was a strong response, but a few crucial research areas were not among the submissions received. Accordingly, the management group decided to request proposals from researchers with the desired expertise, for instance some aspects of physical anthropology and the identification of wood species from artifacts. This process of identifying and acting upon recognized research gaps continued for several years, particularly when a line of enquiry raised additional questions. Methodological advances similarly meant that lines of investigation could be pursued that previously had not been feasible. For example, progress in the study of ancient DNA from animal remains allowed analyses to be completed in 2007 that were not successful at the beginning of the project. Researchers frequently teamed up with other experts to access specialized equipment or knowledge.

Overall, 36 proposals in 23 subject areas were received and reviewed by the management group. Studies of merit that addressed a research question or topic of identified interest were approved by consensus, one study per topic. Scientists with a broad range of expertise, located across Canada as well as in the United States, England, Scotland, Germany and Australia, eventually became involved in the project.

Researcher Agreement

WITH ASSISTANCE from Dr. Howard Brunt of the office of the Vice-President Research at the University of Victoria, the BC Archaeology Branch and the Royal BC Museum took the lead in developing a document that would spell out the working relationship between scientists and the project partners. The agreement template had to reflect the overall interests of the project, as articulated in the management agreement, and respect the interests of the scientists who were to be involved (fig. 8). The financial situation meant that the agreement had to make clear that participating scientists were responsible for securing the funding necessary to carry out their proposed studies.

THIS AGREEMENT pertains to research (the "Research") under the direction of Researcher Name at the Institution detailed in a proposal entitled: "Proposal Title" and dated Proposal Date (the "Proposal").

Whereas:

A.   The Archaeology Branch of the Ministry of Small Business, Tourism and Culture has an agreement with the Champagne and Aishihik First Nations, dated August 31, 1999, respecting the management of Kwaday Dan Ts'inchi and the Museum has an agreement with the Champagne and Aishihik First Nations, dated September 2, 1999, respecting the loan of human remains and artifacts for study and conservation; and

B.   The terms of this agreement are consistent with and have been guided by those agreements referred to in recital A above.

For good and valuable consideration, the Parties agree as follows:

1.   The Museum will provide the Researcher with access to certain human remains which were recovered from archaeological site IkVf-1, in Northwestern British Columbia and which are referred to as "Kwaday Dän Ts'inchí" between date and date, on the days and times which are convenient to the Museum.

2.   The Museum will permit the Researcher to remove samples of the human remains for the purpose of the Research. The Museum and the Researcher may agree to substitute equivalent samples to those described in the Proposal.

The Museum makes no representations regarding the quality of the samples or whether they will be useful for the purpose of the Research. The Museum reserves the right to distribute samples to other researchers and to use samples for its own purposes.

3.   The Researcher shall use the samples only for the purpose of the Research and after completion or termination of the Research shall return the sample material or sample residue, as applicable, to the Museum or dispose of it in accordance with the directions of the Museum.

4.   If the Researcher wishes to carry out work in addition to that set out in the Proposal, the Researcher will submit further proposals to the Museum for consideration.

5.   The Researcher will not make any commercial use of the sample materials or the results of the Research, without the prior written consent of the Museum.

6.   The Researcher will not, at any time, whether before, during or after the term of this agreement, make any claim of any type of ownership (including any claim to patent rights) to the samples or other genetic information found in the samples.

7.   The Researcher will make reasonable efforts to publish the Research.

Figure 8. Agreement wording for human remains researchers.

It was important that the researcher agreement not restrict an investigator's ability to publish scientific papers on the results of his/her studies, since publication is fundamental to a university-based researcher's career. At the same time, publication of culturally sensitive data was a concern for Champagne and Aishihik First Nations. Recognizing that there was limited chance of getting researchers to take on studies if editorial control was obligatory, that requirement was not specified in the research agreement. At the time, working arrangements that respected First Nations interests, such as those outlined in the OCAP (Ownership, Control, Access and Permission) provisions (see Schnarch, 2004), were being implemented in Canada but only in the First Nations health and medical research context. We note, though, that relationships between First Nations and researchers have evolved considerably in the time since our researcher agreement was crafted. A greater standard of mutual respect now exists, and at least some publication restrictions—such as editorial oversight—are becoming more acceptable, if not standard, in many academic research contexts in Canada involving First Nations.

With no formal requirement for researchers to submit their text to the management group for review prior to publication, efforts focused instead on developing positive relationships with the leading research project investigators. These relationships provided an opportunity to share the different points of view and understandings that the project grew out of, particularly the perspective of the First Nation. Nearly all researchers submitted their manuscripts to management team members Mackie and Greer (Sarah Gaunt before her) for input prior to publication and, with little discussion, incorporated the recommended changes. Editorial suggestions most commonly concerned the background setting sections, wording around cultural interpretations and text addressing the need to maintain the dignity of and respect for the deceased, rather than the scientific process and results being reported.

Ethical Conduct of Research

GIVEN THE IMPORTANCE of ethical considerations in dealing with a find of this nature, and recognizing that ethics approval would be required for all university-based researchers proposing studies related to the discovery, the management group accepted the policies and guidelines of Canada's national research bodies, the Tri-Council, as its operating standard. The Tri-Council's comprehensive guidelines, relatively new at the time, outline how research concerned with human remains and human subjects is to be conducted. The guidelines are obligatory for all Canadian university researchers.

For Canadian researchers proposing studies related to the discovery, the ethics review would be required by, and conducted within, their home institution. Projects proposed by researchers based outside of Canada presented a challenge, as the management group lacked the skills and expertise to review the proposed research for compliance with Tri-Council guidelines. The University of Victoria office of the Vice-President Research, under the direction of Dr. Howard Brunt, agreed to provide assistance to address this gap in expertise. An agreement was reached to have this office conduct, on behalf of the management group, the ethics review for the proposals originating outside of Canada, and a special committee was established at the University of Victoria to review the non-Canadian proposals. Committee members included Dr. Brunt, Dr. Roger Dixon, Dr. Michael Roth and Dr. Taiaiake Alfred, a Kanien'kehaka (Mohawk) citizen.

Loan Agreement for Samples Sent to Researchers

SAMPLES OBTAINED during the autopsy process required more detailed study than could be done during the actual autopsy sessions in order to yield the answers being sought. Samples originating from the artifacts and belongings also needed specialized equipment or processing that couldn't be done at the Royal BC Museum. In order to facilitate these further lines of inquiry, the museum entered into agreements to permit

Figure 9. Example of Royal BC Museum researcher loan agreement wording with researchers.

> **LOANS TO OTHER INSTITUTIONS**
>
> **RBCM LOAN NO.: KDS 001:2000**
>
> Loaned to: Dr. _____
> University of _____
>
> Date of loan: Aug. 13, 2001
> **Due date: Aug. 13, 2002**
>
> Authorized by: Grant W. Hughes, Director, Curatorial Services
>
> Four hairs from Kwaday Dan T'sinchi : KDS Sample # 24-35-1-1
>
> For DNA analysis. The borrower agrees to the following terms and conditions.
>
> 1. The Researcher shall use the samples only for the purpose of the Research and after completion or termination of the Research shall return the sample material or sample residue, as applicable, to the Museum or dispose of it in accordance with the directions of the Museum.
> 2. The transfer of materials from the borrower to a third party is not permitted unless expressly authorized in the loan agreement.
> 3. The Museum makes no representations regarding the quality of the samples or whether they will be useful for the purpose of the Research. The Museum reserves the right to distribute samples to other researchers and to use samples for its own purposes.
>
> Sample list attached: Yes   No X
>
> After samples have been received, **please sign the pink** copy and return it to NH Section, 675 Belleville Street, Victoria, BC, V8W 9W2.
>
> Borrower's Signature: _____ Date: _____

these samples and specimens to be loaned to researchers for more thorough investigation.

Like the researcher agreement, the resulting sample loan agreement (fig. 9) speaks to key principles: respectful treatment, continuous and transparent stewardship and adequate physical and records care. As well, the agreement underscores the goal of obtaining maximum information value from samples. The loan agreement indicates who has possession and control of, as well as responsibility for, particular samples. The type of research to be conducted was also specified, as was a due date for return of the sample(s) or sample residues.

It is the business and mandate of museums to maintain scientific, scholarly and museological professionalism. The researcher agreement and the sample loan agreement both reflect the Royal BC Museum's commitment to high standards of stewardship. These protocols for dealing with research

materials helped instil confidence among the project's First Nation partners that the museum was being sensitive to cultural values underlying, or in some way connected to, the materials in their care. The handling of samples collected during the community DNA study is discussed elsewhere (chapter 33).

Management Group Operation and Division of Responsibilities

BROAD PROJECT MANAGEMENT and longer-term goals were, and continue to be, discussed and addressed jointly through management group meetings. Decisions about the selection of researchers and the ethical guidelines under which the project would operate were made by consensus. Also included were decisions about what contextual studies were needed at the site, when communications to the public and fellow researchers were required, and what the content of these communications would be. Matters surrounding the integration and interpretation of various parts of the project were also agreed upon jointly. In some cases, specific members of the management group prepared background papers on particular topics in order to facilitate discussions.

Within the first year of operation, the management group recognized the prudence of subdividing areas of responsibility so that efficient progress could be made without constant need for approvals from the larger group (see table 1). An early decision was made to divide work along geographic lines: the Royal BC Museum would focus on recovered materials housed in Victoria, and CAFN on those which remained in the Yukon. The parties also gave each other permission to speak about the project to their local audience without needing management group approval. Communications with wider audiences, including presentations at national and international conferences or media announcements and similar presentations, were to be joint efforts.

Responsibilities: BC Archaeology Branch and the Royal BC Museum

THE BC ARCHAEOLOGY BRANCH and the Royal BC Museum worked together to fulfil the responsibilities that had fallen to the Victoria-centred part of the project. They led the development and management of research agreements for scientists working with the human remains, ensuring that all research proposals underwent ethics reviews.

Mackie was responsible, on behalf of the BC Archaeology Branch and for much of the project at the museum, for communications with CAFN on routine project business, usually directly to his CAFN counterpart (first Gaunt, later Greer). He also acted as the principal liaison with the researchers working on the human remains, and sometimes with museum researchers as well. Mackie, together with Nick Panter at the museum, sorted through and catalogued all the bags of material collected from the glacier (mostly fragments of the salmon and robe).

Table 1. Responsibilities assumed for various aspects of the Kwäd̲ay̲ Dän Ts'ìnchi project. RBCM refers to the Royal British Columbia Museum, CAFN to Champagne and Aishihik First Nations

Responsibility	BC Archaeology Branch	RBCM	CAFN	Scientific advisor
Broad policy and process formulation (via management group)	+	+	+	+
Decide ethical principles for research	+	+	+	+
Decide contextual studies required at site	+	+	+	
Communications strategies	+	+	+	
Broad logistical timing of various project operations	+	+	+	
Liaison with other First Nations			+	
Annual monitoring of site			+	

Responsibility	BC Archaeology Branch	RBCM	CAFN	Scientific advisor
Community DNA study			+	
Artifact replication workshops			+	
Decide and implement disposition of human remains and artifacts			+	
Cultural interpretation of all results			+	
Funeral arrangements			+	
Memorial arrangements			+	
Decisions regarding recovery of additional human remains			+	
Recovery of additional human remains	+	+	+	
Human biology studies				
Set up facility for safekeeping and conservation of human remains		+		+
Store human remains for safekeeping and study		+		
Prepare, distribute and track call for proposals to researchers	+	+		+
Review proposals, select researchers	+	+	+	+
Develop and arrange research agreements and ethics reviews	+	+		+
Liaise with researchers	+	+		
Manage research agreements, loans of materials to researchers, collections of samples, etc.		+		
Make autopsy and imaging arrangements		+		+
Review research manuscripts	+		+	+
Processing and storage of bags of materials collected from ice	+	+		
Hide and sinew artifact conservation, storage and study (robe and similar objects)		+		
Plant fibre artifact conservation and study (hat, sticks and similar)			+	
Finding and contracting with researchers for plant fibre artifacts			+	
Natural history research		+		

Responsibility	BC Archaeology Branch	RBCM	CAFN	Scientific advisor
Public outreach	+	+	+	
Website development	+	+	+	
Symposia/conference organization and presentations	+	+	+	
Temporary and short-term exhibits		+	+	
Media strategies, interviews and press conferences	+	+	+	
Book editorial committee	+	+	+	+
Book production		+		

The Royal BC Museum handled the conservation and study of the robe and other hide and sinew artifacts, as these had specialized conservation needs not readily accommodated in the Yukon. Considerable research was conducted around the gopher-skin robe, which broadened to involve many areas of expertise. Museum researchers also participated in studies to identify naturally occurring plants and animals found on the ice and surrounding areas and collaborated with researchers on other projects. The Victoria-based management group members had joint responsibility, with Beattie, for organizing research into the human remains. This included organizing the autopsy (conducted by Straathof and Beattie), collecting samples for research, arranging for medical imaging of the body at Victoria General Hospital, documenting all these processes, maintaining a catalogue of collected materials and dealing with researchers concerning their work.

The museum was responsible for the security of the artifacts and the human remains while they were in Victoria. A log was kept of all entries to the freezer in which the human remains were stored. Environmental conditions in the walk-in freezer, as well as in the chest freezer that had been brought from the Yukon (which contained the body and was kept inside the large walk-in freezer), were also tracked. The walk-in freezer was cleared of all contents and sterilized prior to delivery of the remains and the belongings. The walk-in freezer did not return to general use until after the human remains had been cremated and the larger pieces of the robe were completely dried and moved into storage outside the freezer. Chapter 11 (Cosgrove et al.) reports on the conservation regime established and implemented by Beattie and museum staff.

An important aspect of the Victoria work was to communicate progress to British Columbians, whose tax dollars were being spent on the project. To that end, the museum, sometimes in collaboration with the BC Archaeology Branch and always in consultation with the Champagne and Aishihik First Nations, undertook the following initiatives:

- Hosted a press conference 1999 and participated in ongoing joint responses to media requests

- Organized a public lecture presented by various researchers in November 2000

- Developed Robe Weekend, a two-day exhibit at the museum in February 2003, with researchers on hand to answer questions. The event was attended by all parties to the management agreement

- Installed a small exhibit at the Royal BC Museum in 2005 describing the discovery

- Participated in symposium planning during 2007–08 as part of the Northwest Anthropology Annual Meeting held in Victoria in 2008, and organized associated public lectures

- Developed a symposium website with abstracts of all the papers to be presented, and maintained it online for several years

- Hosted an online summary of the contents of this book for several years

In addition, branch staff, museum staff and Beattie participated in presentations about the project to CAFN communities in 2000 and at the symposium held in Haines Junction in 2008, both of which were organized by the First Nation (see chapter 8).

Responsibilities: Champagne and Aishihik First Nations

CHAMPAGNE AND AISHIHIK FIRST NATIONS have never claimed ownership of the Kwädąy Dän Ts'ìnchį discovery, rather, responsibility for it. This position resulted in CAFN consulting with neighbouring First Nations in Canada and Tribes in the United States to see how others felt about working with the BC government on a collaborative project. CAFN also consulted with their neighbours on how to proceed with disposition of the human remains and the man's belongings. Following these efforts, the First Nation made the decision about final disposition of the human remains (see chapter 8). In the summer of 2001 these remains were cremated and returned to the mountain where the Long Ago Person Found lost his life. Human remains discovered after the glacier melted were not removed from the mountain; instead they were buried with the cremated remains. The belongings and artifacts recovered from the site and surrounding area have been retained and currently are being properly cared for, though not in public storage.

As the formally recognized lead for all cultural matters, Champagne and Aishihik also organized all ceremonies held to honour the Long Ago Person Found, with CAFN staff helping move these events forward, in trust for the man's family. The First Nation government has also had responsibility for the cultural interpretation of the scientific discoveries.

As well, the First Nation has directed a number of specific studies and subprojects, including the community DNA study. They have been, and remain, the sole agency for the study and treatment of those artifacts that stayed in the Yukon. Champagne and Aishihik have coordinated the land-based research, in part because of their role as co-managers of the park. They took the lead on the cultural and historical research, including collating and gathering stories about travelling through glacier landscapes. The First Nation was responsible for collection and synthesis of ethnographic and historical information from the region, and for directing community-based studies focused on specific artifacts, such as the gopher robe. Inspired by the Kwädąy Dän Ts'ìnchį hat, Champagne and Aishihik helped their Alaskan neighbours in Klukwan and Haines with community-based workshops related to spruce-root weaving.

The First Nation government has been part of all project decisions, especially those concerned with the research and care of the human remains but those related to the belongings and artifacts as well. They have also been part of the development of all project communication efforts subsequent to the first release issued by the province.

Members of the First Nations' project team have made numerous presentations to various communities in the north and arranged for researchers to present their results to northern communities. In chapter 37 (on the concept of kets'ädän) one of the CAFN project participants explores some of these activities and the learning process from the First Nations' perspective.

Conclusion

THE KWÄDĄY DÄN TS'ÌNCHĮ PROJECT has been a successful collaboration for the parties involved. The First Nation's prior positive experiences in government-to-government and collaborative contexts, as well as in research situations focused on scientific investigations of First Nations culture and history, were essential to the project's favourable outcome. Another important factor was the existing Tatshenshini-Alsek Park Management Agreement, a document that clearly recognized the province and the First Nation as having shared responsibility for resource management in the protected area where the body was found, with the First Nation having sole authority over culture and heritage sites.

The Royal BC Museum's and the BC Archaeology Branch's past experiences working with Indigenous partners also contributed to the project's success. Both organizations agreed to work at a pace and in a manner that allowed the First Nation government to feel comfortable. The respect shown by all parties has made it possible to learn so much about the Long Ago Person Found without compromising either scientific standards or the cultural values fundamental to the First Nation.

Along with these pre-existing conditions, respect meant that the Indigenous government of Champagne and Aishihik could follow their tradition of stewardship and participate from a position of strength. They stepped forward to assume responsibility, on behalf of the larger Indigenous community, for this unusual find. The bravery of the First Nation in taking such a stand has to be acknowledged. This decision was made when the controversy over "Kennewick Man"—or the "The Ancient One", as he is known to the Confederated Tribes of the Umatilla Reservation—was at its height, with many tribes and First Nations having little interest in or respect for archaeological investigations into their past (Thomas 2000).

In addition to mutual respect, acceptance of each other's motivations for being interested and involved in the project has been critical to its success. All project participants were inherently committed to the project, and those involved used a flexible approach in order to ensure priorities could be acted upon and goals achieved. These considerations, including awareness of the sensitivities of the find, have helped the collaborating participants through some of their more difficult moments. As noted elsewhere (see, for example, chapter 37), the project has been a challenge for Champagne and Aishihik as a government, as well as for staff of the First Nations government tasked with dealing with the discovery.

Beyond these factors, it is apparent the parties as well as all team members who represented the parties were inherently committed to the project. Those involved used a flexible approach in order to ensure priorities could be acted upon and goals achieved. The key players worked well together, trusted each other and shared a common interest.

The Kwädąy Dän Ts'ìnchį project will continue while final details are wrapped up. But in many ways, this book represents the end—or at least the final stage—of a successful collaboration. We hope and trust that the knowledge gained, opportunities provided and relationships established will persist long past the need for an official management group.

> We all, especially First Nations people, are lucky to live in this special time in our history and the history of First Nation people in Canada, and particularly in the north. Our youth have tremendous opportunities for the future that we are growing … and I believe that this project, even though it has had some difficulties, is one of those wonderful opportunities that our youth will benefit from in the future. (Joe 2008)

35

OUR GOPHER ROBE (*SÄL TS'ÄT*) PROJECT

Bringing an Old Art Form Back to Life

Champagne and Aishihik First Nations, Frances Oles and Sheila Greer

Almost every older Indian woman in southern Yukon owns a "gopher skin" robe, and almost everybody has snared or trapped, then eaten with relish, hundreds of "gophers."
(McClellan 1975, 158)

IN THIS PAPER we report on community studies related to the robe or blanket made from the pelts of gophers, *Spermophilus* (now *Urocitellus*) *parryii*, also known as the Arctic Ground Squirrel—*säl*, as they are called in the Dákwanjè (Southern Tutchone) language. Since the Dákwanjè word *ts'ät* refers to both robe and blanket, we use the terms interchangeably and refer to the gopher robe/blanket as *säl ts'ät*. We begin by introducing gopher-pelt blankets or robes as a traditional art form, and provide background information from archival and community research. This gives us context for reporting on the community efforts to revitalize the art form. Note that we use the Dákwanjè language word for people, Dän, to refer to our people (Tlen 1993).

Champagne and Aishihik First Nations' *Säl Ts'ät*, or Gopher Robe/Blanket, Project was initiated in 2006. Our mandate was broad: to work with our elders to learn more about the process, everything from harvesting animals to the assembly of a robe/blanket. We wanted to ensure that another generation of our citizens was knowledgeable about this art form, and our intention was to help in

passing on this tradition within our community. We also wanted to highlight the importance of gopher as both a traditional food and a clothing source, thereby promoting the no-waste practice of our ancestors.

Inspiration for our project came from one undertaken by the Northwest Territories Gwich'in that involved the production of men's clothing made of Caribou hide (Thompson and Kritsch 2005). The multi-year Gwich'in effort saw women of various Mackenzie Delta area Gwich'in communities—including Inuvik, Tsiigehtchic, Aklavik and Fort McPherson—reproducing four different examples of traditional men's two-piece outfits based on specimens available in museum collections.

We also share a summary of what we learned about gophers' role in traditional Dän life, in order to help those who haven't had the opportunity to participate in community life as much as they might like. Thus, we describe our overall Gopher Robe Project as well as details on information gained about the manufacture and trade of *säl ts'ät* and other gopher-skin items.

Researchers who study Indigenous cultures have formulated the concept of "keystone species" (cf. Garibaldi and Turner 2004), referring to a critical economic resource that could be seen as the foundation of, or fundamental to, a traditional culture. It would be difficult to argue that a small mammal species like the gopher was a keystone species in traditional pre-contact Dän culture, as Caribou continue to be for our northern neighbours the Gwich'in of northern Yukon, western Northwest Territories and northeastern Alaska. Although gophers may not have supplied a huge proportion overall of Dän dietary needs in traditional times, we hope to show that these small mammals played a significant role in Dän life and that they can be perceived as iconic to the culture.

Champagne and Aishihik First Nations have been aware of the importance of gophers in Dän life and culture for a number of years. In 1996 CAFN's Heritage program held its first multi-day "gopher camp", which brought youth and elders together to help younger participants learn how to trap gophers and prepare the meat for food. A second gopher camp was held in 2009. In recognition of the gopher's importance in its culture, the First Nation instituted its "Gopher Buddies" program in 2006, providing a chance for youth to be "gophers" (i.e., helpers). Our Gopher Buddies youth help others by bussing tables and serving adults and Elders at First Nation events, such as our yearly summer general assembly. Also in 2006, the First Nation initiated its *Säl Ts'ät* Project, the focus of this paper.

Conservation and Study of the Kwädąy Dän Ts'ìnchį Robe

IN THE EARLY DAYS after the recovery of the Kwädąy Dän Ts'ìnchį remains in 1999, decisions were being made about the associated artifacts and belongings. Part of that involved determining which pieces could remain in the Yukon. Some of the belongings clearly required specialized conservation treatment and care, beyond what could be provided in Whitehorse, and were best treated at facilities located outside the territory.

Champagne and Aishihik First Nations wanted as many items as possible to remain in the north, however, so that they could be accessible to community members for study and learning, which was a priority for us. The gopher robe, *säl ts'ät*, was one of the items transferred to the Royal BC Museum in Victoria for study and stabilization (see chapter 26). We are pleased to report that our initial concern about the robe being conserved outside the north was misplaced; it did not prove to be a detriment to community learning about and from the Kwädąy Dän Ts'ìnchį robe. Information gained through robe conservation studies at the Royal BC Museum was eagerly shared by museum staff, and it triggered a process of remembering within the Champagne and Aishihik First Nations community. For example, author Frances Oles recalled gopher blankets from her childhood, when they were common in Dän households and either used as bed covers or stored away as family heirlooms. The robe's discovery is also partly responsible for Oles undertaking a process of experimentation with various tanning, sewing and ochre paint processes.

The Kwädąy Dän Ts'ìnchį robe thus became an inspiration to learn more about and to revitalize this traditional art form. We are grateful to the Royal BC Museum for its role in caring for the Kwädąy Dän Ts'ìnchį robe and bringing its story forward.

Gophers and Dän (Southern Tutchone) Culture

CONTEMPORARY TRAVELLERS to Dän country usually have their first encounter with a gopher (fig. 1) during the summer months along the Alaska Highway, when these little mammals can be seen scurrying across the road. Ubiquitous throughout the Yukon and northwestern BC, our region's ground squirrels, or gophers as they are locally known, are the largest of the ground squirrel family. Males reach up to two pounds in weight, females one and a half pounds (Forsyth 1985). The relatively dry interior country that is home to the Dän, Inland Tlingit, Tagish and Tahltan of southern Yukon and northwestern BC represents this species' southernmost range, which extends north to the Arctic coast. Ground squirrels are not found in the wetter northwest Pacific coastal environmental zone.

Explorer Edward Glave, who came to this country in 1890, was the first non-Aboriginal person to meet Dän in their homeland (see chapter 9). He created images of residents in the village known as Shäwshe, including a sketch of the man he identified as the village's "Second Chief" (Glave 1890–91; fig. 2). The sketch depicts a man sitting on what is likely a bale of furs and wearing a robe or blanket that we suspect might be a *säl ts'ät*. The following year, on his second trip into the Dän homelands, Glave travelled northwest from the Dezadeash Lake area toward Kluane Lake. On that expedition, he visited a late summer Dän camp located around the confluence of the Alsek and Kaskawulsh rivers and noted that "the women of the camp were catching *several hundred ground squirrels in a day*, the skins of which are patched in to robes; the meat is one of their favourite luxuries" (Glave 1892, 876, emphasis added).[1]

These words of Glave's, recorded at a key point in the history of the Dän homeland, are the oldest known documentary reference to Dän gopher harvesting.

ICK ARS, SECOND GUNENA CHIEF.

Figure 2. Sketch of Shäwshe chief wearing robe-type garment in 1890. Reproduced from Glave (1890–91).

Figure 1. A contemporary live gopher.

Later chroniclers of traditional Dän life, particularly anthropologist Catharine McClellan (1975), similarly referenced the significance of gophers to the Southern Tutchone, as well as to other southern Yukon and northwestern BC First Nations such as the Tagish and the Inland Tlingit of Atlin, Teslin and Carcross. The quote from McClellan at the beginning of this chapter strongly shows the importance of gophers as both a clothing and food source at the time of her research, in the middle of the 20th century.

McClellan also noted that when the pelts are in their prime in late August and September, a woman may spend two or three days setting out 100 or more gopher snares (1975, 158). This was in the 1940–60s. In addition to McClellan's observations, we found frequent references to gopher-harvesting activities in Champagne and Aishihik First Nations interview transcript files. It is clear that the practice was largely, though not entirely, dominated by women and, like berry picking, involved the participation of children. While a serious business, it was without a doubt viewed as a pleasant activity, a happy time, by those doing the work.

Gopher harvesting continues today, though the way it is practised varies somewhat from McClellan's mid 20th-century observations. Most contemporary gopher harvesting takes place at the end of the summer season, when the gophers have put on considerable fat for their winter hibernation. Gopher fat is highly nutritious; it also does not harden or solidify, as does the fat of large mammals like Moose, Caribou or sheep. Previously, the fat dripping from a roasting gopher would have been collected, but this is less commonly done today. Community members continue to recognize the health value of eating gopher that has been raised on a diet of nutritious foods from the land; gophers whose diet may be of questionable quality are not consumed.

When gophers come out of their hibernation burrows in the spring, they still have some fat stores. In times past their rich meat presented a pleasant dietary change at that time of year; it was a break from the traditional old-style winter diet, which emphasized dried meat or fish.

The meat of springtime-harvested gopher is plainer tasting in comparison to gopher harvested later in the season, when one can detect the flavours of sage or flowers eaten by the gopher during the summer months. Author Oles recalls that the spring gopher harvest was an important and highly anticipated family activity when she was a youth in the 1950s and '60s. Family members, children and adults worked together to catch the gophers, always on the lookout for a government official who might question their right to do so. The spring gopher harvest was also a fun time because it represented the first campfire and picnic of the season out on the land. Today spring gopher hunting is practised by fewer individuals. This is likely because of the general shift in land-use patterns, as our people have become more involved in wage labour. But it also may be because a wider range of dietary options is now available, making spring gopher less unique as a food source.

Still, the season's first gopher sightings are eagerly reported among community members. The animals' reappearance each year is taken as a sign that winter's tough days have once again passed. It also heralds good things to come. Community members can be heard speculating about how good those gophers are going to taste or, if they have managed to harvest some, they may be bragging about how tasty the gophers were.

McClellan (1975, 210) discusses traditional methods of cooking and preserving gopher meat and fat. We suspect that in times past gophers harvested in the spring were usually singed, which is a means of preparing the animal for food when saving the pelt is not a consideration. This cooking method involves first singeing the fur over the campfire. The charred hair is then scraped off, and the small animal is mounted on a stick and roasted over the fire. Today singeing is the preferred means of preparation, especially enjoyed by the younger generations.[2] The term "yummy" is often used to describe the taste of singed or roasted gopher.

We note that gophers also used to be dried in the fall for winter food; cooking dried gopher requires soaking it in water before they are "boiled up". Today, gophers are commonly boiled or made into a soup broth, or they are cooked over an open fire. Gophers still remain a favoured Dän food, particularly for elders (see Wein 1994, Wein and Freeman 1995), and gopher continues to be served as a traditional dish at potlatches or other types of community gatherings hosted by the Dän.

In summary, in the decades since the mid 20th century the practice of gopher harvesting continues, but there have been subtle shifts in where it occurs, in who does it and who they do it with. We believe that a number of factors, both cultural and natural, account for these shifts. Habitat change, as well as cyclic changes in the species abundance and distribution, is among the natural phenomena affecting the gopher harvest. As discussed below, there have also been changes in how gophers are caught and in the technology used.

Gopher-Pelt Items in Museum Collections

IN THE MIDDLE OF THE 20TH CENTURY Champagne and Aishihik First Nations sources told McClellan that "Gopher-skin robes have always been popular" (1975, 304). Gopher robes or blankets (*ts'ät*) make very warm bed covers. Champagne and Aishihik Elder Fred Brown reported that gopher blankets also made the perfect bedroll for travel because they are "lightweight and very warm", similar to today's down sleeping bag (Greer 2008). In times past the robes were reported to have been "tied at the throat and belted, so that the folds made large sleeves" (McClellan 1975, 304; fig. 3).

In addition to being made into blankets and robes, gopher pelts were also sewn into shirt-type parka garments (*yųk'e kwä'ūr*) for adults; children's snowsuits, consisting of a coat and pants; hats (*ts'at*) and capes. The fur is also often used on the leg part of mukluks (*keshän thu*), as trim on mittens (*mbàt*) and to make purses and small bags. Gopher pelt was also used to make fur clothing for dolls (Champagne and Aishihik First Nations oral history transcripts and collections files). We note that the man's knife carried by the Kwädąy Dän Ts'ìnchį individual was found inside its own gopher-fur sheath (see chapter 26).

While an exhaustive search of museum collections has not been done for items made of gopher (ground squirrel) pelts, a number of pieces of interest have been identified in institutions holding ethnographic collections from the northwest subarctic interior (Yukon, Alaska and northwestern BC). These items include the following:

- Woman's dress, in the National Museum of the American Indian, Smithsonian Institution's Alaska collections.[3] This beautiful and rare garment was collected in Yukon Territory in 1917. Like the Kwädąy Dän Ts'ìnchį robe, the lateral seam on this dress is fringed and similarly painted with ochre (see chapter 26).

- Child's winter shirt or parka with hood, from the Tahltan area, in the collections of the Canadian

Figure 3. Mrs. Sophie Watt (Southern Tutchone) wearing a *säl ts'ät* (gopher robe/blanket) in the traditional manner, gathered around the waist; taken on Kluane Lake, 1948. Catherine McClellan photograph, collection #J2160, Canadian Museum of History.

Museum of Civilization, Hull. This rare piece is depicted in Thompson (2007, 134).[4]

- Bag made of gopher pelts, in the collection of the American Museum of Natural History. This bag was collected at Klukwan and is catalogued as Tlingit in origin. We suggest that this bag may have been manufactured in the interior (northwestern BC or Yukon) and subsequently traded as a finished item to the coast, as similar bags are known to have been made in the interior.[5]

- Hat made of gopher pelts and other fur, in the collection of the American Museum of Natural History. This piece was collected in the area of Lake Laberge and the "Lewis" (i.e., Lewes or Yukon) River around the time of the Klondike gold rush of 1897–98 by the writer Tappan Adney.[6]

- Woman's cape made of black gopher pelts, in the collection of the McBride Museum, Whitehorse. This stunning garment was made for Kate Carmacks of Carcross, a member of the party whose find triggered the Klondike gold rush of 1897–98. Several local populations of gophers of this single-colour variety are known to exist in the southern Yukon, but the locations are well-kept secrets.[7]

- Adult parka and adult cape made of gopher pelts, in the Kluane Museum of Natural History, Burwash Landing, Yukon. These late 20th-century pieces were made by Champagne and Aishihik Elder Marge Jackson of Haines Junction.

- Woman's cape made of gopher pelts and Moose skin, in the Yukon Permanent Art Collection, Whitehorse. This piece was also made by Champagne and Aishihik First Nations Elder Marge Jackson.

- Man's jacket (short style) made by Mrs. Mildred Sparks of Haines, of gopher pelts traded from Champagne and Aishihik country, in the collection of the Sheldon Museum and Curatorial Center, Haines, Alaska.

- Champagne and Aishihik First Nations' Heritage collection includes hats, purses and dolls made or donated by community members, as well as a child's coat made of gopher pelts and some type of animal hide. This piece was purchased on eBay, obtained from a collector who bought it in Whitehorse and understood it to have been made in the 1940s. We have not yet been able to identify the maker of this coat.[8]

Blankets or robes nonetheless represent the bulk of the ethnographic objects in museum collections made of Arctic Ground Squirrel fur. This excludes pieces that are from the Inuit culture area. We identified the following gopher (ground squirrel) blankets and robes in museum holdings:

- A gopher robe from the Tahltan area in the ethnographic collections of the Canadian Museum of Civilization, illustrated in Thompson (2007, 66). This beautiful robe is made of 86 pelts.[9] Its inside seams are painted red, as with the Kwädąy Dän Ts'ìnchį robe, and it features the same vertical seam fringing (the fringing appears on the fur side). Gopher tails hang from the bottom of the robe; the tails had been left intact on the lowermost row of pelts.

- Two blankets/robes in Seattle's Burke Museum collection. See catalogue #1-1982, a robe collected in the Tlingit culture area made of 84 pelts; and catalogue #2311, collected in the Chilkat Tlingit area, which appears to be made of 72 pelts.[10]

- Four small blankets or capes made of gopher pelts in the collections of the American Museum of Natural History. Item E/1581, identified as a "blanket", is made of 72 pelts (12 skins across and 6 high). Items 16.1/2486, 16.1/2487 and 16.1/2488 are all identified as "capes". The first and third pieces in this list are made of 55 and 72 skins, respectively. The photo of item 16.1/2487 shows the cape on a roller, so it is not possible to count the number of gopher skins used.[11]

- A gopher robe in the collections of the Royal BC Museum, Victoria.

- One gopher robe in the Dawson City (Yukon) Museum.

- One gopher blanket made of 126 pelts at the George Johnston Museum in Teslin, Yukon.

Important Trade Items: Gopher Robes/Blankets

DÄN ORAL HISTORY SOURCES leave no doubt that *säl ts'ät* were important trade items in the 19th century. McClellan (1975, 507) mentions being told of one woman from Nu gha hit village on the Tatshenshini River (see chapter 10) who, in the late 1800s, produced as many as 15 robes a year. The robes were reportedly sold at Klukwan where, according to local sources, they commanded high prices—as much as a dog team.[13]

The late 19th-century Tlingit of southeastern Alaska, particularly the northern Tlingit of the Haines and Klukwan areas, highly valued ground squirrel fur (Emmons 1991, 136). A gopher robe/blanket appears in an 1890s photo of the inside of the Whale House at Klukwan.[14] In this image, taken by the team of Winter and Pond and reproduced in Wyatt (1989, 119), the gopher robe clearly has a place of honour in the Whale House. Gopher robes or blankets appear in numerous late 19th-century photos taken by southeast Alaskan professional photographers such as Winter and Pond and Case and Draper. The robes or blankets are shown being worn by women or being used as props for staged photographs (fig. 5; cf. Wyatt 1989, 48, 49, 64, 65).[15]

Items made of gopher skins are also mentioned in early period documentary sources from southeastern Alaska. For example, in 1877 C.E.S. Wood—the first non-Tlingit to enter Glacier Bay—prepared for a mountain goat hunting trip with the local residents. For this trip, Wood reported donning "belted shirts made of squirrel skins" (Wood 1882, 334).

It is our understanding that most southeastern Alaska gopher robes/blankets originated somewhere in the interior. They would have been manufactured by the Tutchone, Tagish, Inland Tlingit or Tahltan and then traded to the coast.

There is another possibility, that these robes may also have been made by a woman who was residing on the coast but originally from the interior, as pointed out by the late Elder Mrs. Kitty Smith (quoted in Sidney et al. 1979, 94). Gopher robes continue to be used as regalia by the southeast Alaska Tlingit today. One was worn, for example, by a Haines resident during the

Figure 4. Champagne and Aishihik First Nations citizen Lori Strand models heirloom *säl ts'ät*. Ukjese van Kampen photograph. Colour version on page 610.

- A gopher blanket made of *ca.* 100 pelts in the Joe family museum at Klukshu village (Yukon).

- One gopher blanket made of 84 pelts, on loan from a community member (fig. 4), in the Champagne and Aishihik First Nations Heritage collection.[12]

In addition to known museum holdings, various Champagne and Aishihik households also have family heirloom gopher robes in their possession.

Figure 5. Archival photo showing Juneau Tlingit women wearing gopher robes/blankets; robes would have been traded to southeastern Alaska from the interior. Winter and Pond photograph. Alaska Digital Archives #ASL-P87-0075.[24]

funeral services for the Long Ago Person Found, held at Klukshu in July 2001 (see chapter 8). One was also seen being worn at the Tlingit Celebration event in Juneau in 2008. A small robe produced by our Gopher Robe Project is now being used by the Dakwakada Dancers, the Champagne and Aishihik First Nations youth dance group.

Anthropologist Frederica de Laguna, who worked extensively with the Tlingit of the Dry Bay and Yakutat area in the middle of the 20th century, reported that products made of gopher skins, obtained in trade from the interior, were highly valued (1972, 350, 436). She recorded two gopher songs at Yakutat: "Ground Squirrel Song for Good Weather" and "*Teisle duy'a'ya'*—ground squirrel hunting song". While not among those songs that were published in *Under Mount Saint Elias* (de Laguna 1972), the recordings and transcriptions of these songs are in the collections of the Smithsonian Institution in Washington, DC. It is understood that these songs originated in the interior and were "traded to the coast".[16]

Based on our informal research and conversations with community members, we have identified the following Dän women as noted robe/blanket-makers: Mrs. Shorty, Maggie Jim, Annie Stick Kershaw, Lily Jackson, Annie Ned and Jessie Joe, all born sometime between 1880 and 1920, and K'ukewaá (mother of Maggie Jim), born in the mid-19th century. This list is by no means complete, and we believe there are other noted robe-makers who should be added. We are also certain that these women were preceded by many generations of robe/blanket-makers.

Our Gopher Robe Project

A PROJECT TO REVITALIZE the practice of making *säl ts'ät* took shape in 2005 when we realized how common gopher robes/blankets used to be in Dän households, yet how few community members were now sewing with the skins (as the practice of singeing and roasting gophers had become more common). We also recognized the urgency of revitalizing the tradition and practice of making *säl ts'ät* while it could still be salvaged.

A senior advisor for our project was Elder Marge Jackson, who was born in 1918 (Jackson 2006) and who passed away in 2013. Like most Dän female elders, Mrs. Jackson was an accomplished seamstress. When she was young, she regularly helped her mother, Maggie Jim, make robes or blankets from gopher skins; however, Mrs. Jackson had never personally made a robe/blanket from start to finish. Our project also benefited from the knowledge of elders Mary Long, Rosalie Washington, Audrey Brown, Frances Joe, Stella Boss and Jenny Moose, all of whom were familiar with tanning and either with sewing gopher skins or with the gopher blankets that were around their homes in times past. But none of these women had personally made a *säl ts'ät*.

Our project went through various stages, which took place over roughly eight months, with mixed age groups participating. The sessions were about learning and teaching, and were experiential in nature. The steps, as listed here, focus on the pelt and skin process but also include the meat and fat processing.

> Step 1, harvesting of gophers and skinning of pelts
> Step 2, first stage of tanning: removal of fat,
> soaking of skins, etc.
> Step 3, tanning
> Step 4, design, layout and sewing

Harvesting Gophers

ELDER MARGE JACKSON reported that the robes her mother made were constructed of 120 skins. The robe in the Champagne and Aishihik First Nations Heritage collection features 84 pelts. The Kwädąy Dän Ts'ìnchį robe is made of roughly 95 skins. McClellan (1975, 304) was told by her southern Yukon sources that gopher robes were usually "seven or eight skins across, and 12 or 14 skins down", which means they were constructed of between 84 and 112 pelts. The various museum examples, as noted above, feature 84, 72 or fewer pelts,

but the smaller ones are reported to be capes rather than blankets or full robes. We therefore conclude that between 72 and 120 skins are needed to make a *säl ts'ät*, depending on the size of the robe/blanket desired.

A general account of the traditional means of harvesting gophers in the southern Yukon is worth sharing here, as it describes snaring—formerly the most common means of catching these small mammals:

> Women used to set up long lines of gopher snares around their camps. To do this, a woman set out from camp with ten or twenty snares in her belt and a bundle of willow sticks to be used as spring poles for the snares. She also carried a stick with a hook notched in the end. The noose of the gopher snares was made from the springy midrib of an eagle feather. This was attached to a thong of moose or caribou babiche. At the top of the noose was a small wooden toggle pin.
>
> First the woman made a small hole about a metre from the inside of the gopher burrow. She stuck the base of the spring-pole into this hole. Next, she used her hooked stick to make another hole in the roof of the gopher tunnel near its opening. She set the noose in the tunnel and pulled the thong through the hole in the roof. Then she bent the spring-pole over and tied the end of the thong to it. When a ground squirrel tried to run out of the burrow, it was caught in the noose. The willow rod sprang up, tightening the noose around the gopher and holding it against the burrow roof. (McClellan et al. 1987, 128)

Gopher snares (*säl tth'émèl, säl tth'émèn*; fig. 6) are very light in weight.[17] This was an important concern in older times, when people walked everywhere. In the early decades after the Alaska Highway was put through southern Yukon in the 1940s—including the period when McClellan visited our area—it was common to see an older woman or couple accompanied by their pack dog, walking along the highway corridor as they snared or trapped gophers. Today our citizens have access to vehicular transportation, so weight is less of a consideration, and most find it more practical to

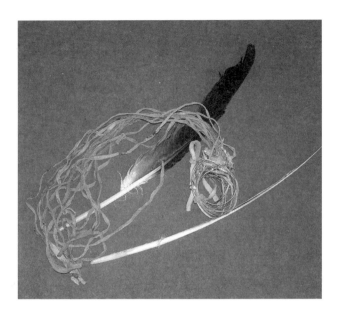

Figure 6. Traditional gopher snare made of sinew and eagle feathers.
Frances Oles (CAFN Heritage) photograph.

sticks extant within the Champagne and Aishihik community is discussed elsewhere in this volume (see chapter 23).

Further descriptions of gopher snaring can be found in Johnson and Raup (1964, 194) and McClellan (1975, 158–59). Sketches showing a snare, including close-up detail of the trigger mechanism and the snare set over the gopher hole, are included in McClellan. A photo of a gopher stick is presented in Johnson and Raup (1964, fig. 53).

McClellan was also told that gophers used to be taken with blunt or bunting arrows (1975, 159). Today people use the 22 gauge rifle. This type of harvesting happens in the spring, when the animals can be "called" and are taken for food only, as the pelts are of poorer quality. Shooting gopher is important hunting training for Dän youth, and making a "head shot" is a recognized achievement, with the dead gopher proudly turned over to an auntie or grandma. Youth may also be tasked with quickly scrambling to catch a wounded gopher before it retreats down its hole.

trap rather than to snare gophers, since traps are more easily set. Commercially available long, metal spring traps (size 0 or 1) are the preferred trap type used today.

Snares may still be used occasionally, but now they are made of brass wire rather than the traditional babiche and eagle feathers. Many Champagne and Aishihik families nonetheless retain their grandmothers' sets of gopher snares as family heirlooms, and perhaps her "gopher stick" (*tth'émèl ghät*)—the notched stick shaped like a large crochet hook with a sharpened point on the opposite end, used to set the old-style snares. The similarity between the carved and notched stick recovered from the Kwädąy Dän Ts'ìnchį site and the many family heirloom gopher

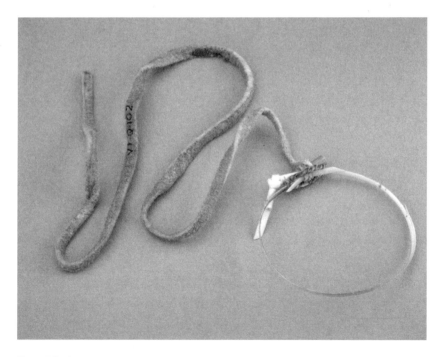

Figure 7. Gopher snare. Courtesy of Canadian Museum of History. VI-Q-102, S97-14310.

(Left) Figure 8.
Lorraine Allen *(right)* shows youth Tyrel Green *(left)* and Jonni-lynn Kushniruk how to set a gopher trap at Champagne and Aishihik First Nations language immersion camp.
Sheila Joe (CAFN Heritage) photograph.

(Below) Figure 9.
Skinning a gopher with a metal *tagwät*.
Sheila Joe (CAFN Heritage) photograph.

Various harvesting events occurred for our Gopher Robe Project. Some were formally part of the project, while additional trapping sessions were tacked onto other activities, such as language immersion camps held at Aishihik and Klukshu during the summer of 2007 (fig. 8). Teachers at these camps included Lorraine Allen, Martha Smith, Vivian Smith, Kathy Birckel, Frances Joe and Frances Wellar.

Our principal gopher-trapping camp took place in late August 2007, in the Cottonwood Trail area of Kluane National Park. Elders, community, staff and others interested in participating were invited. The camp was established around the treeline in the mountains north of Mush Lake, and the harvesting took place in the shrub and alpine tundra setting above this. The days were busy with trapping and skinning work (fig. 9), while stories were shared in the evenings. Teachers at this camp included Marge Jackson, Paddy Jim, Paul Birckel, Kathy Birckel, Chuck Hume and Mary Jane Jim.

A number of learning outcomes took place over the course of this camp, and by its end, all novices had gained experience in a variety of skills necessary for

successful gopher trapping. These included knowing the kinds of locations where traps should be set and how many traps to set in any one gopher hole area; when the animals were most active, which affects when you should set and check the traps; how to disguise the human scent on the trap; how to mark the location of the set traps so they could be found again; the quickest and most humane way to kill trapped gophers; how to skin the animal, and the care that had to be taken in doing so; how to process the meat of the skinned gopher; and the effects of the weather and other variables on all of the above. We learned the hard way, for example, that newly harvested skins have to be kept out of the rain, lest the fur "slip". We struggled with difficult parts of the process, such as skinning the head, and admired the skill of the experienced hands who were able to keep the fat on the animal rather than removing it with the pelt. It was challenging work, with the novices learning not only skills but patience—and having lots of fun. By the end of the camp, more than 60 gophers had been collected, which made a significant dent in the 100 plus needed to make a robe.

Another gopher harvesting event in August 2007 took place at *Tan ya*, in Kluane National Park, west of Klukshu village. During the early 20th century families based at Klukshu would take a break from their summer fishing to harvest gophers at *Tan ya*. Women and children would climb the long ridge up to this high plateau above the treeline, where they would camp for several days and snare gophers. In contrast, the 2007 visit to *Tan ya* was a day trip by helicopter and involved various community members, including youth and elders. The participating elders hadn't visited *Tan ya* since they were children, and they were very emotional about the opportunity to revisit a site they had frequented regularly in their youth. The weather, though, wasn't the most cooperative. It was raining, and although we heard gophers chirping in every direction, none were caught. Still, this provided an educational opportunity, as the elders were able to talk about how the weather affects gopher behaviour.

The gopher harvesting continued on weekends that year, during the remainder of the late summer to early fall period, but it ceased once the gophers began hibernation in early October. By the end of the season, we had assembled some 70 gopher skins for the robe project: those caught during the 2007 season plus skins donated by community members Mary Deguerre and Glen Kane.

The harvested skins were dried or frozen until we were ready to start the tanning process. Later, during the skin-processing phase of the project, we learned that not all of the skins we had gathered were suitable for our proposed robe manufacture. The pelts of younger or first-year gophers, born in the spring of the year they were harvested, are too thin for robe/blanket construction. Elder Marge Jackson advised that spring gopher pelts are "too dry, you can't soften them". When tanned, thin skins rustle like paper if rubbed, and spring skins also tear easily. But we learned that skins harvested in August may not be prime and suitable for robe production either. It turns out that we had held our gopher camp too early in the season. The date for our harvest camp had been chosen because we didn't want to make our elder teachers camp out in the mountains in September, when the weather is poorer. The result was that many of the skins we had gathered in August were not of suitable quality for robe/blanket construction.

We concluded that one of the most challenging parts of making a robe, in the contemporary context, is obtaining the required number of prime quality and matching pelts.

Processing and Tanning Skins

THE FIRST PART of the tanning process took place on an intermittent basis, mostly evenings and weekends, between October 2007 and January 2008. A one-day workshop in January focused specifically on the softening part of the tanning process.

The same year as our *Säl Ts'ät* Project, the Champagne and Aishihik Heritage program also held various moosehide-tanning workshops. The small gopher skins are easy to handle in comparison to an untanned moosehide, which is many, many times larger and heavier. A moosehide is also a major time commitment; many hours of work are needed to take the hide through one stage of the tanning process, and one cannot move on to the next stage until the previous stage is completed. In contrast, gophers can be worked on in smaller batches, a few pelts at once, as time permits. We can therefore say that the tanning stage is not the most difficult or challenging part of making a gopher robe/blanket.[18]

Figure 10. Gopher skins hanging on the line for winter freeze-drying during the tanning process. Frances Oles (CAFN Heritage) photograph.

To begin the tanning process, the gopher pelts were soaked in water, with pure soap added to soften and degrease the hide. This water has to be changed repeatedly. When asked what was used in the old days when pure soap wasn't available, our advisor Marge Jackson indicated that she didn't know if anything was. We suggest that perhaps nothing was added to the soaking water in times past, since the women skinning the gophers would have been more experienced and left less grease on the pelts.[19]

Mrs. Jackson cautioned us to work on batches of a few (e.g., five) hides at a time, in order to stay on top of the process. Between soakings, or when not being worked on, the partially tanned hides were hung outside, where they naturally freeze-dried in the cold winter temperatures (fig. 10). This freezing helps break down the collagen and cell walls of the pelt, which is a desired outcome of the tanning process.

Unwanted material—that is, excess fat and membranes—was removed from the skin's inner surface using small versions of the tool known as a *tagwät*, the tool used to skin the gophers. A *tagwät* (fig. 11) is a vertically oriented hide-working tool with a serrated distal end. In the ethnographic and archaeological literature, they are often referred to as "fleshers" or defleshers (see Clark 1981, 110). According to Elder Marge Jackson, the traditional form of this tool was made from a Caribou lower leg bone (i.e., metapodial) (Greer 2008b). We also experimented with wooden *tagwäts* made by Elder Paddy Jim, a recognized community expert on Dän traditional technology, during the August gopher harvest camp. We found that the wooden version of the *tagwät* worked and could do the job in a pinch. The bulk of the skinning and pelt-cleaning work, however, was done with metal *tagwäts* that had been made out of flattened spring metal from small leg-hold traps for the

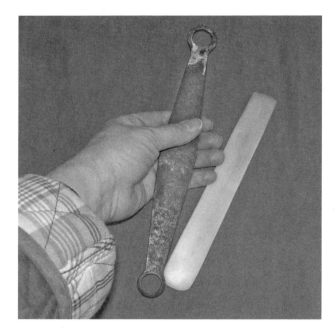

Figure 11. Tools. *Left,* metal *tagwät* (flesher) made from spring trap; *right,* traditional bone *tagwät*. Frances Oles (CAFN Heritage) photograph.

A *thechel* (figs. 12 and 13), the Dákwanjè name for the horizontally oriented hide-tanning tool, is the principal tool used during the tanning process. The *thechel* tool form is commonly found in southern Yukon archaeological sites, including sites in Champagne and Aishihik First Nations traditional territory.[20] *Thechels* from such sites are depicted in Workman (1978, fig. 22), who uses the term "tabular biface" to describe this tool form. In the Dän way, *thechels* are always hand-held rather than mounted on a pole or shaft, as reported for Yukon Tlingit communities (see fig. VIII d in McClellan 1975). To protect the hand, a cloth is often wrapped around the side of the tool.

Thechels were traditionally made of stone, from some type of flat, tabular rock. Once sheet metal became available from sources such as fuel drums, the Dän began making metal *thechels* as well. The Champagne and Aishihik Heritage program collection includes examples of both metal and stone *thechels* that were donated, made by community members or collected by citizens from family heritage sites (fig. 12).

1996 gopher camp. *Tagwäts* were also used occasionally as a scraping tool during the later part of the tanning process, when the pelt is softened.

Working with the wet hides led to considerable learning. For example, when novices remove the skin from the animal, they often take off too much fat, leaving it on the skin rather than on the carcass (where it is desirable). Extra effort is then required to get this fat off the skin and hair, which sometimes results in the skins being thinned too much. We also found it a challenge to work with older hides that had only been partially processed. On these older, dried hides the fat had oxidized and saturated the skin; despite all our efforts, they would not soften up like fresh hides.

Once the hides have been completely cleaned and dried, with all unwanted material removed from the inside surface, the true tanning part of the process takes place. This is begun by cutting the skins vertically down the belly, so that a flat surface is achieved. An X-acto knife or razor blade was found to work well for this process, as well as for cutting off the feet.

Figure 12. Small-sized metal *thechel* made specifically for working gopher skins. Frances Oles (CAFN Heritage) photograph.

Figure 13. *Thechels*, hide-scraping and -softening tools. *Upper left*, recently made stone example; *lower.left*, heritage stone *thechel*; *right*, heritage metal *thechel*. The latter two are from CAFN Heritage collection. Sheila Greer (CAFN Heritage) photograph.

Both stone and metal *thechels*, as well as bone *tagwäts*, were used in tanning our gopher skins, and all were found to work equally well. Some of our participants preferred working with the stone *thechel*; others, the metal type. A smaller *thechel* (fig. 13), made specifically for working with gopher pelts, was found to be especially useful. Using a *thechel* means the work proceeds with the worker seated, holding the tool in one hand and the pelt in the other. The hand holding the pelt is braced on one's knee, as the tool is pressed against and worked over the pelt's inside surface. The knee brace provides a solid surface but also gives a little as the pelt is worked with a firm hand—not as much force as required to soften a moosehide, though. The objective is to break down the pelt's cell structure through repeated stretching and working so that the skin remains soft when dry. One works at a skin repeatedly, and it gradually softens.

Our one-day tanning workshop, held in January 2008, attracted many visitors, including the high school students of the outdoor education class of our local school (St. Elias Community School, Haines Junction). With the students helping with the tanning work, the softening of the pelts was completed in short order.

Design, Assembly and Sewing of the *Säl Ts'ät*

THIS STAGE OF OUR ROBE PROJECT took place in a three-day workshop that was held in February 2008. Eighteen individuals participated, including Champagne and Aishihik Heritage program staff, elders and other community members, as well as Royal BC Museum Conservator Kjerstin Mackie.

McClellan (1975, 304) reports that the robe assembly process begins by trimming the pelts into rectangles, then sewing them together. While she is correct in noting that the pelts are trimmed, as we learned through the course of our workshop, the design and assembly process is somewhat more complicated than simply sewing fur rectangles together (fig. 14).

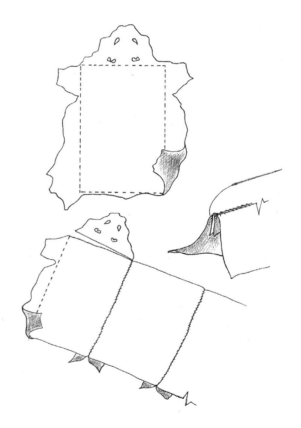

Figure 14. Sketches showing initial stages of *säl ts'ät* assembly. Shading indicates fur side. *Upper*, outlining the rectangle on the pelt. *Lower*, vertical row of pelts being assembled, stitching on the inside. Frances Oles illustration.

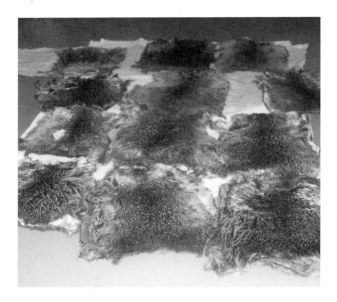

Figure 15. Sorting and colour-matching the pelts.
Frances Oles (CAFN Heritage) photograph.

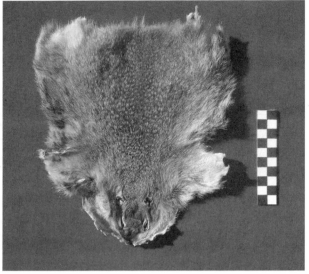

Figure 16. Immature gopher with late summer pelt; harvested at Upper Farm Field, Haines Junction area. Frances Oles (CAFN Heritage) photograph.

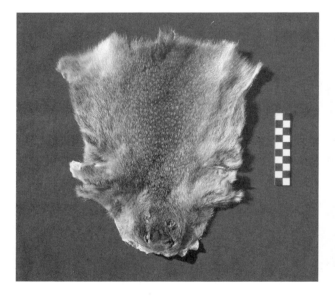

Figure 17. Mature, late summer pelt from Upper Farm Field, Haines Junction area: golden, with some rubbing near back of neck; not prime. Frances Oles (CAFN Heritage) photograph.

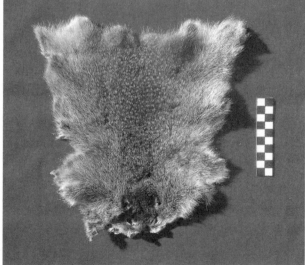

Figure 18. Mature, early fall pelt (described as "white" by one elder), from subalpine environment, Jarvis Creek basin. Note lack of yellow or reddish tones, but still not yet prime. Frances Oles (CAFN Heritage) photograph.

The robe assembly process begins by laying out the available skins so that they can be sorted by pelt colour and size, according to the robe/blanket designer's preference (fig. 15). Pelts can vary considerably depending on the animal's age when it was harvested, as well as when and where it was harvested (figs. 16, 17 and 18). The pelts get richer in colour as the season progresses, becoming golden in September, with mature males and females reaching an almost silver to white hue just before hibernation. This maturing sequence varies by location and is dependent on local conditions. Gophers harvested in mid September in one place may vary in colour and in colour pattern from those harvested at a different elevation or location during the same period.

Looking at existing robes, it is clear that considerable effort was taken by the designer to ensure that the pelts matched in size and colour. We are not exactly sure just how the *säl ts'ät* maker achieved this. It might have been by repeated harvesting in the same area over multiple years, so that the skins used on a given robe came from a specific local family or population of gophers. One community member suggests that the species were intentionally managed, as other wildlife species are known to have been, and this may have helped achieve matching pelts.[21] There may also have been some trading of pelts between blanket-makers.

Once the sorting of pelts is completed, the robe assembly process next involves trimming the pelts to ensure they are all the same dimension. Advisor Marge Jackson instructed us to use a template to mark out the desired rectangular shape on the inside of the pelt. The excess beyond the top and bottom edges of the rectangle outline is then trimmed off. We used an X-acto type of knife for this trimming, which results in the removal of the animal's four limbs. The excess on the sides or lateral edges of the rectangle is not removed at this stage. In fact, this lateral selvage is not removed, if at all, until two pelts are joined together along the side seam. As noted below, on many gopher-pelt items the side selvage was left on and then cut into fringing.

Sewing the *Säl Ts'ät*

THE PELTS ARE FIRST SEWN TOGETHER into horizontal rows, with vertical seams along the lateral edges of each pelt (figs. 19 and 20). These horizontal rows are then joined together to form the overall robe. The Kwäday Dän Ts'ìnchi robe appears to have been assembled in this manner (see chapter 26), as do all known robes/blankets in the Champagne and Aishihik community that we have inspected. Figures 14 and 20 provide schematic drawings of the robe/blanket assembly process.

A distinctive type of seam was used for joining the vertical seams on the historic gopher robes/blankets that we have examined. McClellan did not describe it in her recounting of the robe manufacturing process (1975, 304), nor was it readily apparent when we first examined contemporary robes. Mackie's analysis of the Kwäday Dän Ts'ìnchi robe (chapter 26) brought this technique to our attention, and once we were aware of it, we saw that it had been used on all *säl ts'ät* that we examined—including

Figure 19. February 2008 sewing workshop. *Left to right,* Alexia McKinnon, Mary Shadow, Marge Jackson and May Long. Note assembled horizontal row of pelts in front of Mrs. Long in foreground. Frances Oles (CAFN Heritage) photograph.

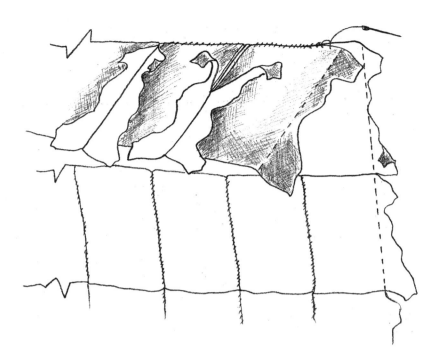

Figure 20. Sketch showing later stages of *säl ts'ät* assembly. Shading indicates fur side. Horizontal rows of pelts are assembled together by stitching on the fur side; the stitching is not visible on the inside. Selvage on vertical rows is then trimmed. Frances Oles illustration.

some heavily repaired ones, where the construction technique was initially difficult to establish.

In this technique, the pelts' lateral edges are sewn together using a simple over-cast stitch. The two pelts, as well as the selvage of the two pelts, are stitched together along the vertical seams so that a total of four layers of skin are being sewn together. Elder Marge Jackson advised that her mother employed the four-layer seam technique in sewing gopher pelts because it made the seam both stronger and more waterproof. Many of our community seamstresses also use the four-layer seam to attach the vamp to the moccasin sole when assembling moccasins (*ke* or *kenkàt ke*) (see Jackson 2006).

All of the older *säl ts'ät* that we examined—and of course, the Kwädąy Dän Ts'ìnchį robe—were sewn with homemade sinew (see chapter 26). True sinew, made from the tendons that run along the vertebral column of large cervidae such as Moose and Caribou, was commonly used in our area until the 1960s, when it was replaced by the more convenient, but less strong, wax/nylon artificial sinew. While many of our community seamstresses use Koban brand thread today to sew

items such as moccasins and vests, we used cotton thread to sew our blanket pieces together. The sewing was done with simple straight needles rather than the three-cornered steel needles that are now used to sew hide and leather, as three-cornered needles can tear the fragile gopher skins.

Traditionally, awls rather than needles were used to sew hides. The sewing proceeded by using the awl to pierce a hole in the skin, and then, while pulling the awl back between forefinger and middle finger, pushing the sinew thread softened by chewing through to the other side (McFadyen-Clark 1974, 96). Older community seamstresses reported that they switched from using the awl to the three-cornered needle to sew hides sometime between the 1950s and 1970s (Greer 2011).

"Got to be fine when you sew gopher skin, so it don't pull apart." (Mrs. Kitty Smith in Sidney et al. 1979, 94). Mrs. Bessie Allen also talks about the use of the awl in sewing gopher skins, mentioning an awl made from the hind leg bone of a lynx; this same source also refers to spruce root being used as thread, an alternative to sinew (2006, 19, 20). Despite our best efforts and having access to modern technology, our stitching was not as fine and

Figure 21. Our finished *säl ts'ät. Left to right*, elders Stella Boss, Eileen Miersch, Mary Shadow, Frances Joe and Rosalie Washington. Frances Oles (CAFN Heritage) photograph.

regular as most of the stitching on the Kwädąy Dän Ts'ìnchį robe (see chapter 26).

On the Kwädąy Dän Ts'ìnchį robe, as well as some of the museum examples of gopher-pelt garments, the selvage on the lateral seams was subsequently cut into fringing. The selvage on both the woman's dress in the Smithsonian Institution and the gopher robe from the Tahltan area in the Canadian Museum of Civilization was also treated in this manner. Elder Marge Jackson advised that this fringing is cut only after the robe/blanket or garment has been fully assembled, in order to avoid tears from cutting too close to the seam line.

The final assembly of a *säl ts'ät* involves sewing the strips of horizontal rectangles together, also using an over-cast stitch. This sewing is done on the hair side, so that when you look at the finished skin side, the stitching is not visible. Because the hairs are not parted at the seam line, this is likely a stronger seam as well.

While we were able to determine, over the course of our workshop, the *säl ts'ät* assembly sequence as described above (and as depicted in figures 14 and 20), it should be mentioned that we didn't actually follow this process when assembling our own *säl ts'ät*. Our workshop was, after all, a process of discovery for all, and work went on even in the absence of teacher Marge Jackson, resulting in some pelts being sewn together into vertical rather than horizontal strips. With various seamstresses involved in our project, we also noted differences in stitching tension and in the spacing of stitches. When compared to the regular stitching on the older robes/blankets that we have examined, the variance in our stitching suggested that traditionally, *säl ts'ät* would have likely been sewn by one individual or possibly a mother and daughter team.

In making our *säl ts'ät*, challenges were also experienced in the pelt-sorting process. After some initial sewing had been done, but still during the early stage where strips of pelts were being sorted, it became apparent that we had two different lots (different sizes and colours) of pelts being sewn in two different ways. At this point, a group decision was made to make not one but two garments. Thus our workshop ended up producing a smaller blanket made of 50 pelts (fig. 21) and a child-sized cape made of 39 pelts.

More recent *säl ts'ät*, including ours, are finished with calico or fabric lining, with felt edges attached along the borders. This treatment is similar to the treatment of edging on taxidermy fur rugs. We suggest that the taxidermy style of finishing may have inspired local craftspeople to begin finishing their gopher robes/blankets in this way. The older robes that we examined are not bordered on the sides or bottom but have a reinforcing hide (Moose/Caribou) at the top, where a tie attachment is sewn. The Kwädąy Dän Ts'ìnchį robe is also finished in this manner (see chapter 26). This hide strip is presumably to strengthen the top, which would be subject to more stresses and strains from wear or hanging. Adding this tie attachment to the top also effectively turns a blanket into a robe.

On some of the older pieces that have been studied, including the Kwädąy Dän Ts'ìnchį robe, inside seams have been treated with ochre. The Dákwänjè term for ochre is *si* (also spelt *tsi*), with various sources of the pigment known within Champagne and Aishihik First Nations traditional territory.[22] When asked about this treatment, Elder Marge Jackson indicated that in times past *si* was used to outline the rectangles, which were subsequently trimmed and sewn together to construct the *säl ts'ät*. Mrs. Jackson also said she understood that *si* was applied directly on prepared skins and pelts in dry form; she did not know if a medium was employed to bind the pigment. Author Oles found the pigment difficult to handle dry and therefore undertook some experimental work; an egg tempera mix was found to work quite well as a binding agent for *si*.

Outcome of Our *Säl Ts'ät* Project

OUR *SÄL TS'ÄT* PROJECT was a success, with numerous positive outcomes. There is now greater awareness and appreciation within our community of the beauty and unique qualities of this traditional Dän (Southern Tutchone) art form. We also have a better understanding of the role of gophers in traditional Dän culture, including how gopher robes/blankets were prized by other peoples but were also in everyday use by our families. The younger members of our community now have a better appreciation of the special characteristics of gopher fur—the warmth it provides and how lightweight it is, and therefore how perfectly suited it was to the traditional lifestyle of our ancestors.

Through a process of discovery and rediscovery, we were able to work out the basic production sequence of a gopher robe/blanket, beginning with the harvest of the gophers, the preparation and use of the meat and the tanning and processing of the pelts. We learned, for example, about the importance of matching hides for size and colour and about the unique four-layer seam used to make a stronger and more waterproof vertical seam.

Our community also learned about the effectiveness of traditional tools such as the *tagwäts* and *thechels* used to process gopher pelts. Existing heirloom gopher blankets held by Champagne and Aishihik families have also been brought out of storage closets so they can be admired by and shared with others in our community.

Individuals who participated directly in the project have become much more knowledgeable about gopher habits and behaviours, and they now have the confidence to do their own trapping and hide-processing. Those who are new to trapping and processing the skins admit they still have much to learn; refining one's trapping skills in order to harvest only the prime adults can take years to master. Still, some project participants have been sufficiently inspired that they want to make their own *säl ts'ät* and have begun the process of accumulating the pelts needed to do so, realizing that they are embarking on a multi-year process.[23]

The small blanket made during the project is now part of a rotating display at Da Kų, the Champagne and Aishihik First Nations Culture Centre in Haines Junction. It is part of an interpretive feature on blankets and other items made from gopher pelts. The gopher cape made by our project is now proudly worn by the youth performers of our local dance group, the Dakwakada Dancers. The cape is considered to be a special piece of regalia that all of our younger dancers get a chance to wear.

Our *Säl Ts'ät* Project has made us aware and appreciative of the skills, talents and culture of our ancestors. We should not have been surprised that the most difficult aspect of making a gopher robe/blanket is the accumulation of the necessary quantity of prime quality pelts—something that might best be characterized as a process rather than a simple task or skill. It takes time, patience and an intimate connection to the land, all of which were important parts of the lives of our ancestors. The Kwädąy Dän Ts'ìnchį robe took us down a path of rediscovery, learning and appreciation—a journey we are grateful for having been inspired to take.

ACKNOWLEDGEMENTS

Thanks to the many teachers who helped with this project. They include Kathy Birckel, May Long, Rosalie Washington, Audrey Brown, Eileen Miersch, Fred Brown, Paddy Jim, Paul Birckel, Chuck Hume, Mary Jane Jim, Lorraine Allen, Vivian Smith, Lena Tutin, Martha Smith and the late Marge Jackson, Moose Jackson, Mary Shadow, Frances Joe, Stella Boss, Jenny Moose, and Frances Wellar. They must be credited for sharing their knowledge and experience and for their patience and good humour. We acknowledge the many contributions of the late Catharine McClellan, an early documenter of traditional Dän culture, and of the late Kathy Kushniruk, a community member who made us aware of the importance of gophers in traditional Dän culture. We acknowledge, as well, those individuals who currently own or are family custodians of robes and allowed us to study the pieces under their care. Thanks to Kjerstin Mackie and the Royal BC Museum for sharing what was learned about the making of the *säl ts'ät* from their studies of the Kwädąy Dän Ts'ìnchį robe. We are grateful to Wayne Howell, archaeologist with Glacial Bay National Park, for leading us to the C.E.S. Wood quote and for the information on the songs recorded by de Laguna. Assistance with spelling Dákwanjè words came from Martha Smith and Lena Tutin. CAFN Heritage, Lands and Resources Department staff Paula Banks, Lawrence Joe, Sheila Joe-Quock, Richard Smith, Alexia McKinnon, Rose Kushniruk and Harry Smith assisted with the various workshops, harvest activities and editing.

Financial support for the *Säl Ts'ät* Project came from the Canada Council for the Arts, Government of Canada's International Polar Year program and Parks Canada's Kluane National Park Healing Broken Connections project. Like all studies directed by Champagne and Aishihik First Nations related to the Long Ago Person Found, the *Säl Ts'ät* Project has also been supported by Champagne and Aishihik First Nations government through both staff time and in-kind contributions. *Gunnalchish*.

36

SPRUCE-ROOT WEAVING

Hats, Baskets and Community Experiences

Sheila Greer

Like the gopher robe, the spruce hat found with Kwädąy Dän Ts'ìnchį man provided an opportunity for people to reconnect with the remarkable skills of their ancestors and elders. It is understood that prior to the hat's discovery, few in the Haines and Klukwan area were actively working with the material, and no one in the Yukon was doing so (though see Dangel 2005 for an inventory of weavers in the Sitka area). This chapter highlights community initiatives of learning and rediscovery that were inspired by the Kwädąy Dän Ts'ìnchį hat and describes how people were spurred to learn more about the use of spruce root, particularly for weaving. First, I provide cultural background on the art and practice of spruce-root weaving. This information comes from published sources and draws heavily on a book by the late Frances Paul of Sitka, who was a noted Tlingit weaver. Then I describe the post-discovery community weaving events that involved residents of local Alaskan and Yukon communities.

A Long Tradition

THE USE OF SPRUCE ROOT has a long tradition among the Indigenous people of southeastern Alaska and northwestern BC (Busby 2003) and Yukon (McClellan 1975). Coastal residents worked with the roots of Sitka Spruce (*Picea sitchensis*), and those in the interior with White Spruce (*Picea glauca*) and Black Spruce (*Picea mariana*). Spruce roots—called *seet x̲aat* in Tlingit and *khay ghät* in Dákwänje—were the raw material for twine or rope in olden days and had many uses. For example, they were preferred for lashing birch-bark baskets and securing fish traps because the roots do not stretch or deteriorate under wet conditions.

Beautiful baskets were also made from spruce roots. Although both the Tlingit and the Dän (Southern Tutchone) traditionally produced woven baskets (McClellan 1975, 280; Emmons 1991), there are no known examples of Dän baskets. It appears that the practice of basket-making disappeared quickly once metal pots and dishes became available to the interior peoples living in the Tatshenshini-Alsek country. The Tlingit of southeastern Alaska, however, continued to weave, and examples of the finest 19th- and 20th-century Tlingit woven spruce-root baskets and hats can be found in museums around the world, as well as in family and clan collections (e.g., Weber and Emmons 1986).

Thought to be one of the oldest and most important arts in the Tlingit traditional economy (Sheldon Museum, n.d.), spruce-root weaving became a high art form among the Tlingit ancestors. Emmons credited the northern Tlingit (i.e., the Tlingit of the Yakutat, Haines and Klukwan area) with producing "the most elaborately beautiful basketry of the entire coast, in perfection of weave, decorative embroidery, variety of geometric designs, and wealth of colour" (Emmons 1991, 213).

Traditionally, spruce-root basketry was solely a women's area of expertise. Women were responsible for all stages of the process, from the gathering and processing of the plant material to the detailed, time-consuming weaving:

To be a prolific weaver takes great commitment of time, staying power, and energy. I remember watching my grandmother weaving as I lay on my bed before falling asleep. When I woke, she was still sitting at her weaving, and wove all day, taking breaks only when they were absolutely necessary (Nora Dauenhauer in Paul 1982).

Baskets were made in different shapes and for a variety of purposes, including cooking, transporting water, serving food and collecting and storing berries (Dangel 2005; Emmons 1991; Paul 1982; Shotridge 1921). According to Mrs. Paul, five different weaving techniques were used, and the woven patterns told stories (Paul 1982; cf. Gunther 1990). In the Tlingit way, clan heirlooms (*At.óow*)—which could include property such as hats, blankets and rattles—are often named and have stories associated with them (Dauenhauer 2000). Mrs. Paul reports that only one basket, a piece made by Chilkat women and known as the "Mother Basket", ever achieved such status (1982, 69).

Spruce roots were woven into hats of varying styles. Mrs. Paul (1982, 38) notes that traditionally, three types of spruce-root hats were made. Simple, everyday hats, known in Tlingit as *zauk-kaht* (meaning "root hat"), were worn by both men and women. Figure 1 shows a late 19th-century Tlingit grandfather at Dyea wearing what appears to be a *zauk-kaht*. As well, larger hats featuring painted designs, known as *zauk-klen* (big hat) or *kum-doo-djee-kit-dee-zauk* (written-on or painted hat), were worn by the wealthy (Paul 1982).

The third type of hat made from woven spruce roots is the clan or crest hat, worn for ceremonial occasions (Paul 1982). Crest hats, also large and painted, feature the finest weaving and are topped with a series of stacked, cylindrical woven ornaments. The clan or crest hat is named *shah-dah-kookh* (Paul 1982; in Emmons 1991, 221 the name is spelled *shar-dar-khuke*). The clan hat name translates as "above the head thing like a celery top" and refers to the way that the hat makes its wearer "stand out" as the wild celery plant does. According to Emmons (1991, 221), the number of cylinders on a crest hat varied according to the rank of the wearer.

Figure 1. Nineteenth-century photo of Tlingit family at Dyea, with grandfather wearing an everyday hat made of woven spruce roots.
Courtesy of Alaska State Library.

Made of very finely woven spruce roots, the Kwädąy Dän Ts'ìnchį hat would appear, in form at least, to be an everyday one, a *zauk-kaht*. Such hats, according to Mrs. Paul, were completely daubed over with paint to help make them waterproof (1982, 46).

Community Learning

THE KWÄDĄY DÄN TS'ÌNCHĮ HAT triggered considerable interest, if not excitement, within the Klukwan- and Haines-area Tlingit community. Having grown up around spruce-root baskets and hats and having seen

their elders weaving when they were young, a number of community members were eager to learn more about this traditional art form and try their hand at the ancient art.

Klukwan Elder Ruth Kasko proposed that the Kwädąy Dän Ts'ìnchį hat be used as a teaching model for a University of Alaska Southeast accredited course in spruce-root weaving. Delores Churchill of Haida Gwaii, a master weaver and specialist in Haida and Tlingit spruce-root weaving, studied the Kwädąy Dän Ts'ìnchį hat (fig. 2) and was engaged to teach the approved course. The course was held in two workshop sessions, each of two weeks, which took place at Klukwan in September 2001 and September 2003.

In 2004 and 2005 plant fibre-weaving workshops of shorter duration, and not part of the accredited course, were held in Yukon. Mrs. Churchill provided instruction at the session in Haines Junction, while Frances Oles (Champagne and Aishihik First Nations) led the session in Teslin.

Both Sitka Spruce and White Spruce were used in all the workshops. The former was collected in the Klukwan area, and the latter around Champagne. Both were found to yield quality root suitable for twining. Information on where to obtain roots was provided by

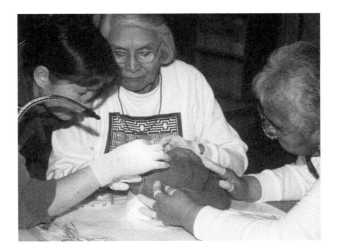

Figure 2. Frances Oles, Ruth Kasko and Delores Churchill counting the weaving rows on the Kwädąy Dän Ts'ìnchį hat, Whitehorse, 2000.
Sarah Gaunt (CAFN Heritage) photograph.

Figure 3. Marsha Hotch of Klukwan sorting roots at the 2001 workshop. Sarah Gaunt (CAFN Heritage) photograph.

less familiar with weaving but nonetheless knew how to prepare the roots.

Aware of the expertise required to weave spruce root, Mrs. Churchill taught the Tlingit twining technique using prepared yellow and red cedar bark. Cedar bark is more forgiving to work with than spruce root and therefore much easier for beginners' hands to manipulate. This approach was used at all of the workshops. After mastering cedar-bark weaving, participants then worked with spruce root.

Participants in the various workshops reported that they enjoyed being immersed in this art form, saying that it provided them with a connection to their ancestors. They appreciated the attention to detail and commitment to perfection demonstrated by the weavers of the past. Those who helped with the root-harvesting activities also took pride in gathering roots for their own use, from their home area.

Spruce-root weaving is a time-consuming process. It takes many hours of work and considerable personal commitment to produce a finished object, and years to master the technique. Some workshop participants were able to finish small baskets (fig. 4). Those involved in the Klukwan workshops started hats (fig. 5). In the years following the Klukwan sessions, individuals continued to weave on their own to complete their hats; some are still being worked on. In 2008 the hat

community elders, who advised that harvesting roots in areas of sandy rather than rocky soil is easiest, with riverbank and lakeshore environments reported as favoured places. Roots collected from these settings are also straighter and therefore easier to weave. Those participating in the root-harvesting activities were cautioned to approach the task with care, since taking too many roots can damage a tree.

Instruction was also provided in gathering and preparing the root. Preparing involves peeling the bark, then splitting and sorting the root into bundles ready for weaving. Having seen their elders weaving when they were young, many Klukwan workshop participants were familiar with the overall process (fig. 3). Elders who participated in the Haines Junction workshops were

Figure 4. Small baskets made by one of the participants in the weaving workshop held in Haines Junction in 2005. Frances Oles (CAFN Heritage) photograph.

Figure 5. Ruth Kasko of Klukwan in the early stages of weaving her hat. Sarah Gaunt (CAFN Heritage) photograph.

finished by Lorraine Kasko of Klukwan was given to the head of the Da<u>k</u>l'aweidí clan at Klukwan following the announcement of the community DNA study results (Marsha Hotch, pers. comm. 2010; see also chapter 33).

The Origin of Basketry

IT IS APPROPRIATE TO CLOSE this brief discussion on spruce-root weaving by sharing a traditional Tlingit story that recounts the origin of basketry. Told by noted Tlingit weaver Frances Paul, the story credits the women of the Alsek River country with making the first woven baskets.

It happened in those mysterious times when Raven still walked among men, exercising the cunning of his mind in bringing good to his creatures by ways strange and inexplicable to mankind. Already his greatest works had been accomplished. He had stolen the Sun, Moon and Stars from his grandfather, the great Raven-who-lived-above-the-Nass River, Nass-shah-kee-yalhl, and thus divided the night from the day. He had set the tides in order. He had filled the streams with fresh water and had scattered abroad the eggs of the salmon and trout so that the Tlingit might have food. But not yet had Raven disappeared into the unknown, taking with him the power of the spirit world, to mingle with mankind.

In those days a certain woman who lived in a cloud village had a beautiful daughter of marriageable age. She was greatly desired by all mortals, and many came seeking to mate with her. But there [sic] wooing was in vain. At last it was chanced that the eyes of the Sun rested with desire upon the maiden, and at the end of his day's travel across the sky, he took upon himself the form of a man and sought her for his wife.

Long years they lived together in the Sky-land, and many children came to them. But these children were of the Earth-world like their mother and not of the Spirit-world of their father, Ga-gahn. One day as the mother sat watching her children frolicking in the fields of the Sun-land, her mind filled with anxiety over their future, she plucked some roots and began idly to plait them together in the shape of a basket. Her husband, the Sun, had divined her fears and perplexities. So he took the basket which she had unknowingly made and increased its size until it was large enough to hold the mother and her eight children. In it they were lowered to their homeland, the Earth. Their great basket settled near Yakutat on the Alsek River, and that is the reason that the first baskets in southeastern Alaska were made by the Yakutat women (Paul 1982, 9).

ACKNOWLEDGEMENTS

The late Sarah Gaunt of Whitehorse did the initial planning and organization for the Klukwan workshops that were taught by Delores Churchill of Haida Gwaii. The late Ruth Kasko of Klukwan arranged for workshop participants to receive course credit with the University of Alaska Southeast. Champagne and Aishihik First Nations helped cover some of the tuition fees for the Klukwan workshop.

CAFN's Frances Oles organized the plant fibre–weaving workshops that took place in the Yukon. Oles, along with workshop participant Marsha Hotch of Klukwan, provided information on the activities and outcomes of the various teaching sessions.

———

37

KETS'ÄDÄN

Learning from the Discovery

Diane Strand (Xíxch'tláa)

The Kwädąy Dän Ts'ìnchį story is a fascinating one, and there is much learning shared about the Long Ago Person Found in this volume. We learn about the vast area where the Long Ago Person Found travelled, and of his relationship to the land. We learn about his last day and his last meal, even the pollen that was on his robe, as scientists bring forth details of one individual's life. We learn about cultural artifacts and practices and about our incredible ancestors who lived and survived in this most beautiful yet challenging land.

Dännch'e (How Are You?)

I HOPE EVERYONE READING THIS BOOK, scientists and non-scientists alike, will look beyond the individual to his society—think of his people, as we First Nations people tend to do. For many readers, much that they encounter in this volume will be new, whether it be the results of the scientific studies or discussion of the cultural context within which the discovery has been managed. Readers learn not just facts about where the Long Ago Person Found travelled and when he lived, but also about our cultures, including our values, traditions and the importance of relationships between different peoples. Of course, these lessons also tell about people's connection to the land, and this connection to place is an important part of the Kwädąy Dän Ts'ìnchį story.

From the very beginning, when the discovery first came to our attention, Champagne and Aishihik First Nations had claimed stewardship—not ownership—of the Long Ago Person Found. We saw it as our responsibility to take care of him because his remains were found in our traditional territory, and that is our way. When you are stewards of the land, you step forward to fulfil your duty, which for us meant ensuring that Indigenous values were respected. It also meant fulfilling obligations to others, because as Indigenous peoples, we do not stand alone. The book talks about all of the decisions regarding the find, how these decisions

Figure 1. Champagne and Aishihik Elder Paddy Jim, with his son Wayne Jim behind him, drumming during ceremony at the Kwädąy Dän Ts'ìnchị Symposium in Victoria, April 2008. Rose Kushniruk (CAFN) photograph.

were made in a regional and clan context, how we consulted with other tribes and First Nations, and how the ceremonies that were conducted reflect the traditions of Indigenous peoples from across the greater region.

The northerners who travelled to Victoria to attend the Kwädąy Dän Ts'ìnchị Symposium in April 2008 (figs. 1, 2 and 3; see also chapter 8) were not all from the same community. Various First Nations besides my own were able to participate, including Carcross-Tagish First Nation, Teslin Tlingit Council and Tahltan First Nation. While from different nations, we northerners are all related or connected in some way, either through marriage, birth or, more importantly, our clan system—a system which has been passed down from generation to generation. For First Nations people from Yukon, as well as southeastern Alaska and northern BC, clan affiliation is a key part of our personal identity. We recognize ourselves as belonging to either the Wolf/Eagle moiety/clan or the Crow/Raven moiety/clan. I am a Crow.

Who Are You? Ethnic Labels and Identity

IN ADDITION TO OUR CLAN AFFILIATION, you may also hear Indigenous people of the southern Yukon, northwestern BC and southeastern Alaska refer to themselves as Tlingit or Southern Tutchone (Dän, as we say in our own language), Tagish or Tahltan and so on. But identity, including ethnic labels that outsiders (such as anthropologists) applied to us, is a tricky business. I can illustrate this with my own story. Whether Tagish, Tutchone, Tlingit, Tahltan, and so on, the Indigenous people of the greater area around the Kwädąy Dän Ts'ìnchị discovery site continue to pass on their traditional Indigenous personal names, which we refer to as our "Indian names". When I was born, my grandmother, the late Annie Ned, a highly respected woman and recipient of the Order of Canada, gave me my Indian name: Xíxch'tláa. The name Xíxch'tláa is Tlingit and means "frog mother". It is an old Crow/Raven name that goes back countless generations. It was also my grandmother's name, and I feel very honoured and privileged to carry it.

Though my grandmother's first language was Dákwanjè (Southern Tutchone), and she self-identified as Dän, the name Xíxch'tláa is, as I mentioned, Tlingit. Now, you ask, why did my maternal grandmother have a Tlingit name, and why did she pass it on to me? You will understand this once you know my family history and genealogy. My great-grandfather (mother's mother's father), a Dän known as Hutchi Chief, had a Tlingit trading partner from the southeastern Alaska coast. One of my great-grandmothers (mother's father's mother) also came from the coast and married inland to a Dän. My grandmother was from the interior, her mother-in-law from the coast. There are many family genealogies like mine throughout our area of the north. I was given a Tlingit name, an old Tlingit name, because it was important to my grandmother that these connections between our peoples be maintained and recognized in a public manner—in this case, through a personal name. The connections between the Dän of the interior southern Yukon and northwestern BC and the Tlingit of southeastern Alaska, southern Yukon and northwestern

Figure 2. Some of the northern delegation attending the Kwädąy Dän Ts'ìnchį Symposium in Victoria, April 2008. Rose Kushniruk (CAFN) photograph.

Figure 3. Yukon residents, including some of the living maternal relatives, examining the fragments of the Kwädąy Dän Ts'ìnchį robe in the conservation lab at the Royal British Columbia Museum, April 2008. Rose Kushniruk (CAFN) photograph.

BC is centuries old, and enduring. My Indian name comes out of this context and is a reflection of the shared history of two peoples, Dän and Tlingit. Given such complex genealogies, how one labels oneself is often a matter of personal choice. While I consider myself a proud and strong Dän, Tlingit is part of my history, too.

Learning Journeys

DURING THE YEARS 2006–2010, I served as Champagne and Aishihik First Nations' Chief, and before that, I worked in the heritage field for many years. Through that work, I came to understand the fascination anthropologists and archaeologists have with studying First Nations people—studying our culture, who we are, where we come from, and the things that our people have made, like tools that have made our life easier. I have also seen, over the years, an evolution in how these researchers do their work. The first anthropologists that came to our traditional territory many years ago believed that they were documenting a dying culture. But this hasn't turned out to be the case. Today's anthropologists understand that ours is a living culture, one that is connected to its past but also very much here now and going to continue into the future. They recognize, too, that Indigenous peoples are not museum specimens or artifacts to be studied but people with differing values and concerns. The shift in academic perspective is evident throughout this volume. I thank all of the researchers who have moved into this new way of thinking, noting, too, that part of this shift in anthropology and archaeology is that some of our own people are now studying and working in these fields. I am very proud of that.

As you read about the scientific and cultural studies, you will learn about our journey of understanding the Long Ago Person Found. In many ways, scientists have completed their *kets'ädän*, their learning, but for our community and those of our neighbouring First Nations and tribes, further learning begins with the publication of this book. The findings presented in this volume help the Indigenous people of southern Yukon, northwestern BC and southeastern Alaska learn more about their common history and who they are as a collective people. It provides an opportunity to celebrate the values and traditions we share that have been passed on for generations.

More than a decade ago Champagne and Aishihik First Nations were faced with the challenge of trying to figure out what to do about the Kwädąy Dän Ts'ìnchį discovery and how to do it. Our elders had said to us: "If you are going to learn, then it'll be okay." I would like to acknowledge them for their wisdom and guidance, and commend the entire northern community for its patience. Our *kets'ädän* is just beginning.

The act of consulting with our neighbours on project matters (see chapter 8) has also brought many blessings. It has renewed and strengthened ties and relationships with friends and relatives across the southern Yukon, northwestern BC and southeastern Alaska. The connection that we have with our Alaskan Tlingit neighbours has been incredible, and we are most appreciative of their support and the circumstances that reminded us of our shared history and common values and traditions. Champagne and Aishihik First Nations has citizens that reside in Haines and Klukwan, Alaska, but also further afield—in Juneau, California and Idaho. We are very proud and happy for the new and renewed connections that have been made as a result of this project. The results of the community DNA study have similarly brought people together.

Challenges and Coping

I SPOKE BEFORE about my community's traditions, when a person passes on. We take great responsibility for caring and ensuring that the passage is done with dignity and respect. That is what we sought for the Long Ago Person Found, to bring science and tradition together. My community has experienced considerable trials and tribulations related to this discovery, and I am very proud of the growth that has occurred through this process, but I also have some reflections.

When the Long Ago Person Found was discovered, it made many people in my community look within themselves, questioning their own beliefs and spirituality. Should the remains have been removed from the glacier? Are we doing this right and following the proper protocol?

Doing it right, in the Dän way, means following the traditional teachings, our *Aduli* laws. *Aduli* can pose challenges for us, however, since Dän culture also gives highest respect to individual experiences, knowledge and ways of learning. This means we each may have different understandings of what is appropriate. Some claimed that our *Aduli* teachings dictated that things had to be done a certain way, while others believed we should be doing something different. Individuals had varying insights on what had been found and what it meant. At the same time, our culture is very strict about some things, especially protocols related to death, as the spirits of our deceased loved ones are never far from us. Traditionally, children never went to funerals and graveyards. The location of both death and the final resting place are very sacred places, imbued with spirits and power. Traditions change, though, with new ways of thinking, and children today have somewhat different experiences.

Still, our culture has protocols for dealing with the dead, as well as with spirits and power. When followed, these practices offer protection from the unknown. Elder Paddy Jim, who has been a key advisor to our Heritage program staff in handling the Kwädąy Dän Ts'ìnchį project, has stressed the importance of being prepared when dealing with such matters.

> There is no need to be scared of it, you know … That's why I say, it wouldn't affect us, if every time we going to talk about it, keep that ashes in the pocket. (Jim 2012)

Regardless of one's understanding, or the precautions that may have been taken, having to deal with the Kwädąy Dän Ts'ìnchį discovery made almost everyone involved think deeply about their spirituality, their beliefs regarding these most important matters

that are part of our collective human experience. Our citizens asked themselves: "What does *Aduli* really mean? Are we giving this person who died on the mountain the respect and dignity he deserves?" Those questions, and more, challenged many people (see chapter 8). I believe that we accepted that test with great honour, and at the end of the day, our values and traditions as a people were reinforced. Yes, this project has been a challenge for our government, but it has also been a blessing for our people, as everyone involved became more aware of our traditions and the values that they are based on.

We were not alone in facing challenges. Questions and uncertainties also confronted the researchers involved in the project. The scientists went through their own *kets'ädän* process. I want to acknowledge their growth, their openness to other ways of knowing and understanding, and I thank them for the respect shown for our cultural beliefs and protocol. I'll share one example. The Kwädąy Dän Ts'ìnchį individual wore a small leather pouch around his neck. Representatives of Champagne and Aishihik First Nations identified this item as his medicine pouch. Project researchers were curious, wanting to open the pouch so that the contents could be determined. Our traditions dictate, however, that such items are very personal and private. We said: "No. It belongs to him, and it won't be opened up", and it never was. The pouch was cremated along with the Long Ago Person Found's remains. The scientists were not in charge in this situation, and they had to "let go"—which may have been new for some of them.

There are many other examples of the growth and respect shown by all involved in the project. When the burial team prepared to return the remains of the Kwädąy Dän Ts'ìnchį individual to the glacier where he lost his life, we asked them to put rose bush thorns on their arms. They didn't ask why; they simply did it. Similarly, because depictions of the deceased are a concern for our culture, you will never see photos of the remains. Our instructions on such matters were totally respected by the researchers.

Shäw Nithän: You've Done Well

IN ENGLISH, we say "thanks" or "thank you"; in Tlingit, *kwanischis*; and in Dákwanjè, *shäw nithän* (*shäw nithän* translates as "you did good"). There is no word or phrase that specifically expresses thanks in the Dákwanjè language. Community members often use the Tlingit *kwanischis* or, alternatively, the phrase *Massi* or *Massi cho*. The latter originates in the Gwich'in cultural area and derives from the French word *merci*; *cho* means "big" in Southern Tutchone.

I have many thanks to give, both from myself and on behalf of my First Nation. I begin by thanking the hunters who made the original discovery—Bill Hanlon, Mike Roch and Warren Ward, teachers all, who had the foresight and understanding to do the right thing and report what they found. *Kwanischis.*

I also thank our project partners. This includes our official partners, the Royal British Columbia Museum and the BC Archaeology Branch, and our silent partner, Yukon Heritage. I thank all the scientists who have participated in the project, sharing their knowledge and expertise. I personally must give a special thank you to Champagne and Aishihik First Nations Council, in particular past Chiefs Bob Charlie and James Allen, and Councillor Gerald Brown. The support of our chiefs and councils over the years has been incredible, and I am grateful for their belief that we were going to have a positive outcome, as we did.

In my culture it is important to acknowledge the people who assisted or took care of deceased family members. This could mean helping in a physical sense, by being pall bearers or cooks at a potlatch, or providing emotional support for family. It could be helping with mental preparation, by ensuring that all the ceremonies and protocols were done in a proper manner. It could also be helping in a spiritual context, or it may be assisting without even realizing what their help really meant to those affected by the passing of a family member. Such would be the case of the Esquimalt and Songhees First Nations. While the Kwädąy Dän Ts'ìnchį man's remains were at the Royal British Columbia Museum for the autopsy (1999–2001), he was in their

traditional territory. It is important for the people from the north to say thank you for this, to acknowledge the people on whose territory the Long Ago Person Found was received.

We also thank the Esquimalt and Songhees First Nations for allowing us to gather together on their traditional territory for the Kwädąy Dän Ts'ìnchį Symposium, held in Victoria in April 2008. I thank the people that represented these nations at the symposium: Chief Thomas and Councillor George, respectively. I thank their ancestors and the people yet to come. I believe that having their traditional territory within the provincial capital has not only benefits and rewards but also challenges. This must be acknowledged, and I'd like to pay respect and homage for it, for all that these nations have endured for so many years. I'd also like to thank the Northwest Anthropological Association for allowing the Kwädąy Dän Ts'ìnchį Symposium to be part of their conference, and I would like to acknowledge all the northerners who attended the symposium to learn more.

We also thank those who travelled with the remains as they were transported, including the pilot, Delmar Washington, who brought the remains off the mountain; Harold Johnson, who escorted them to Victoria; and Sarah Gaunt, who returned the cremated remains to Yukon. The scientists who prepared the remains for travel, Owen Beattie, Greg Hare and Al Mackie, are also to be thanked.

Witnessing

WITNESSING, in particular witnessing ceremony, is an important part of our cultural tradition. Because ours was an oral rather than written culture, witnessing was the means by which our history and traditions were passed down from generation to generation. By reading this book, you are in a sense witnessing the Kwädąy Dän Ts'ìnchį story, just as the Honourable Steven Point, then Lieutenant Governor of British Columbia, witnessed the opening of the 2008 symposium, including the gifting that occurred at that time.

At the beginning of the symposium we also conducted a cultural ceremony. We did this in order to ensure things went right that weekend. While the context of this ceremony was novel, it was based on traditional practices and is worth elaborating on here. The ceremony began with drumming, as do all our ceremonies (fig. 1). All the Wolf/Eagle clan elders in attendance were then asked to come forward, with Wolf Elder Paddy Jim acting as spokesperson for the group, leading the group in prayer. Following the prayer by the Wolf elders, our *Nakhäni* (the "fixer", one who negotiates a solution between parties in disagreement) said a prayer. At the symposium the *Nakhäni* was Crow clan member Wayne Jim. The Wolf elders followed his prayer by singing a Southern Tutchone potlatch song. We refer to the song that the Wolf elders sang as the "Finishing Song". It is sung at a Memorial or Headstone potlatch (see chapter 8), and after this song is finished, your grieving is done, over. This song has never been put down on paper or recorded, nor should it ever be. Therefore, I can't provide the words to readers. But I can tell you that this song represents the closing of one chapter and the beginning of another. It represents a transition to better times, to happy things—just as the symposium represented the end of a chapter in the Kwädąy Dän Ts'ìnchį story, and the publication of this volume represents the beginning of yet another.

The morning the symposium began, the Wolf elders also sang a Southern Tutchone friendship song. This song's purpose was to ensure that we moved on to good things, which we did. The symposium was an incredible time of learning and sharing, and by the end of the second day, there was lots of dancing, which ended with a group photo (fig. 2).

You have read or witnessed some of the Kwädąy Dän Ts'ìnchį story. I encourage you to continue on, but also to remember that when you witness something, it is your responsibility to pass on what you know, or have seen, to those who are not able to attend and witness. By reading this book, you have become obliged to pass on to others what you have learned.

Sheila Greer, Richard J. Hebda, Alexander P. Mackie

Conclusions

Teachings Across Cultures and Times

IN 1999, WHEN THREE TEACHERS WERE HUNTING in Tatshenshini-Alsek Park and came across an unusual find, could they ever have imagined what their discovery would lead to or the teachings that would come from it? Could they have foreseen how it would bridge cultures and times, connecting the present with the past? They triggered a complex but powerful process of learning—or *kets'ädän*, as it would be said in Dákwanjè, the language of the Southern Tutchone people. The resulting story about the Kwäday Dän Ts'ìnchį man found on a remote glacier is more than a simple discovery of human remains and artifacts. It is a journey of learning that the hunters, and the many project participants, could never have imagined.

To begin with, we have discovered a great deal about a remarkable region. We have learned not only about one person but about a people—or peoples, depending on where and how one draws cultural boundaries. We learned about the vibrant Aboriginal history of this far northwestern corner of British Columbia, a unique landscape of rugged inland mountains accessible from the nearby Alaskan seacoast. We have come to know that the Long Ago Person Found lived at a time when crossing glaciers was not uncommon, and that these distinctive landscape features had their considerable dangers. Though daunting, the area now managed as a park by the province and the CAFN is rich in resources, including salmon and minerals. For many, it represents a true wilderness.

Various scientific investigations have established key facts about the Kwäday Dän Ts'ìnchi man and his possessions. Alive sometime between 1720 and 1850 AD, according to modern dating methods, he was a young man, essentially healthy, but also a carrier of a latent form of tuberculosis. He had eaten seafood much of his life and likely grew up close to salt water. He started his final journey in the late summer on or near the coast in the Haines-Klukwan area of Alaska and died only a few days later on the glacier. He had probably made other trips inland from the coast, and in his last year, he subsisted mostly on a diet of land meat.

For the local Tlingit living at or near salt water, as well as the Dän of the interior, the world was understood as the place that Raven (or Crow) "fixed". Storytelling was the traditional means of education, and one married a partner from the opposite clan/moiety. Clan identity is a central part of Indigenous cultures in the region, and using both contemporary scientific and traditional knowledge, identifying the Long Ago Person Found's clan was a key part of understanding who he was.

Studies have not been able to pinpoint a cause of death, but it appears to have been non-violent. We do not know what the purpose of his trip may have been, or his intended destination. He may have been heading to one of several villages known to have been located on the Tatshenshini River, west of the pass where he perished.

He had with him clothing of both coastal and inland traditions: a spruce-root hat to protect his head and an Arctic Ground Squirrel robe to shelter his body. Both hat and robe were normal items of apparel for his time. A number of wooden artifacts were also found near the body, some similar to tools still used in the area today.

During the time of the Long Ago Person Found, the first European trade metals were introduced to the peoples of the region, yet copper, a local metal, continued to be highly valued for trade and ornaments—like the copper bead found with the body. Aboriginal trade networks of the day (that is, of the 18th and 19th centuries) were transforming local cultures as coastal people began to move inland. By the time non-Indigenous outsiders first came into the Tatshenshini-Alsek country in the 1890s, they encountered a bicultural population where two languages, Tlingit and Dákwanjè, were spoken. Through his physical remains and his belongings, we now understand that the Kwädąy Dän Ts'ìnchi individual represented—literally and symbolically—the cultural exchange taking place between the Tlingit and the Dän in those centuries.

The project relied upon different ways of knowing. It included scientific studies involving scholars at many institutions, insights from old documentary sources and learning at the feet of knowledgeable First Nations elders—all equally important ways of "getting smart". But perhaps most importantly, we have learned about the crucial role that respect plays in projects that arise from, and occur within, a framework of many cultures. The project demonstrates that good research is about more than just fulfilling technical requirements, that how something is done is equally important as what gets done. This kind of research has to be undertaken in a shared and respectful manner; it has to honour all forms of knowledge and learning. We realized that just because a particular avenue of scientific investigation is technically possible, it does not have to be pursued. The social environment and how people feel about a research question is also an important factor.

The project has also taught patience and understanding. Those involved learned to be thoughtful and to listen carefully in order to hear and appreciate the views of others. It takes time to recognize how those who come from a different cultural background may see and understand the world. Similarly, it takes time to think about unfolding events and what they mean. It also takes time to process the findings made by others, to understand what those discoveries may mean for oneself and one's community.

More than 17 years have passed from the time of the original discovery in August 1999 to the production of this volume. As you have read or thumbed through the book, you likely have come to appreciate how complex this undertaking was, the range of studies involved and the number of discussions that took place. The project included everything from the modern technical science of DNA analysis to the consideration of personal experiences of travelling through glacial landscapes. These were not an empty 17 years, but involved long cycles of study, research and consultation, with some avenues of inquiry triggering further investigation. Time was taken to reflect on the results obtained and to consider their significance to other project participants.

As the final section of the volume shows, the project involved many kinds of connections—between individuals and cultures, of people to the land, between different landscapes and particularly the relationship between coastal and interior regions. The project connected the past and the present, allowing traditional skills to be admired and honoured in contemporary times, bringing the former alive in the latter. Certain traditional skills and practices were revived and passed on to future generations.

Through the involvement of the region's Indigenous peoples, the project became not just about the past and about science, but also about people alive today. The Champagne and Aishihik First Nations, the local Tlingit, Tagish and Dän, recognized the project's potential; they believed that something good could come out of an ancient tragedy. They gave the project its contemporary meanings.

We hope that in its full form this book and the work it represents might serve as a model for future collaborations involving First Nations, the scholarly community, governments and institutions. Perhaps it might serve as a broad example to society at large. For are we not all communities of people with interests and histories of our own, deserving of respect and understanding, while at the same time wanting to advance our own knowledge and skills in a world of rapidly advancing technology?

The Kwäday Dän Ts'ìnchi project has taught everyone involved an enormous amount about ourselves, regardless of our respective cultural backgrounds. The editors of this volume are grateful to the Long Ago Person Found for the journey on which he sent us. Clearly, his life and his story did not end on that windswept glacier in Tatshenshini-Alsek Park two centuries ago. They live on, giving us new understanding for the days in front of us. It has been an honour and a privilege to have been involved in this remarkable discovery.

Chapter 20. Identification of Sockeye Salmon from the Kwädąy Dän Ts'ìnchj Site
Freshwater and estuarine fish species inhabiting the discovery region

Lampetra ayresi	River Lamprey	*O. nerka*	Sockeye Salmon
L. richardsoni	Western Brook Lamprey	*O. tshawytscha*	Chinook Salmon
L. tridentata	Pacific Lamprey	*Salvelinus confluentus*	Bull Trout
Acipenser medirostris	Green Sturgeon	*S. malma*	Dolly Varden
A. transmontanus	White Sturgeon	*S. namaycush*	Lake Trout
Alosa sapidissima	Shad (introduced)	*Coregonus clupeaformis*	Lake Whitefish
Couesius plumbeus	Lake Chub	*Prosopium coulteri*	Pygmy Whitefish
Catostomus catostomus	Longnose Sucker	*P. cylindraceum*	Round Whitefish
Esox lucius	Northern Pike	*P. williamsoni*	Mountain Whitefish
Osmerus dentex	Rainbow Smelt	*Thymallus arcticus*	Arctic Grayling
Spirinchus thaleichthys	Longfin Smelt	*Lota lota*	Burbot
Thaleichthys pacificus	Eulachon	*Gasterosteus aculeatus*	Threespine Stickleback
Oncorhynchus clarki clarki	Coastal Cutthroat Trout	*Cottus aleuticus*	Coastrange Sculpin
O. gorbuscha	Pink Salmon	*C. asper*	Prickly Sculpin
O. keta	Chum Salmon	*C. cognatus*	Slimy Sculpin
O. kisutch	Coho Salmon	*Leptocottus armatus*	Pacific Staghorn Sculpin
O. mykiss	Rainbow Trout, Steelhead Salmon	*Platichthys stellatus*	Starry Flounder

Chapter 23. Wooden Artifacts from the Kwädąy Dän Ts'ìnchį Site and Surrounding Area

Kwädąy Dän Ts'ìnchį Belongings

Figure 2. Schematic drawing of selected artifacts collected at IkVf-1, to scale. The objects are illustrated with the distal end of the tool on the right and the proximal or base end on the left. The pieces inside the box were found in direct association with the Long Ago Person Found. Sheila Greer illustration.

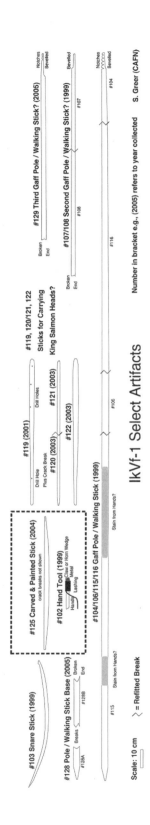

Chapter 25. Examination and Analysis of the Kwädąy Dän Ts'ìnchį Knife
Analytical Methods

X-radiography was carried out using either a Philips Mg 105Be X-ray tube (in 2006) or a Lorad LPX 160 X-ray tube (in 2008). In order to obtain optimal imaging of both the organic and metallic features, multiple X-radiographs were taken using tube voltages of either 35 or 55 kV and exposure times between 15 and 120 seconds. A tube current of 5 mA and a film to focal distance of 125 cm were employed.

Scanning electron microscopy/energy dispersive spectrometry was performed using an Hitachi S-3500N VP SEM integrated with an Oxford Inca X-act analytical silicon drift X-ray detector and an Inca Energy+ X-ray microanalysis system. The SEM was operated in high-vacuum mode using either a backscattered or secondary electron detector. Using this technique, elemental analysis of volumes down to a few cubic micrometres can be obtained for elements from boron (B) to uranium (U) at a level of approximately 0.1–1 per cent by weight.

For Fourier transform infrared spectroscopy, a Bruker Hyperion 2000 microscope interfaced to a Tensor 27 spectrometer was used. A portion of the sample was positioned on a diamond microsample cell and analysed in transmission mode by co-adding 200 scans and using a 4 cm^{-1} resolution.

For X-ray diffraction, patterns were obtained with a Bruker D8 Discover with GADDS (general area detector diffraction solution) equipped with a rotating anode and cobalt target. The patterns were measured at 45 kV and 75 mA, using a 0.5 mm collimator.

The scraping from the wood was analysed by gas chromatography/mass spectrometry using an Agilent 6890 gas chromatograph interfaced to an Agilent 5973 mass spectrometer. Prior to analysis, the samples were derivatised in Meth Prep II—m(trifluoromethyl) phenyltrimethylammonium hydroxide, TMTFTH, 0.2 N in methanol—from Alltech Associates. The main components in the resulting chromatogram were identified using published reference mass spectra.

Microscopic wood sections were examined by transmitted light microscopy in order to document the wood's anatomy in the transverse, tangential and radial planes. The results were compared with the anatomy of known woods to determine the type (genus) composing the knife's handle.

A microscopic sampling of collagen fibres from the lashing was immersed in distilled water and heated from 10°C to 95°C at 2°C per minute using a thermal microscope. Over this temperature range, the fibres' collagen molecules change in organization from ordered and semi-crystalline to more randomly oriented and amorphous. The visible consequence of this change, the shrinkage of the fibres, was recorded by capturing a digital image of the sample at intervals of 1.0°C over the full heating range. The resulting 85 images were combined to produce a time-lapse video of the fibres shrinking. Image processing and analysis extracted numerical data from the video, and these were graphed and analysed further to produce an onset shrinkage temperature (the temperature at the low end of the range over which shrinkage occurs) and a derivative peak temperature (the temperature at which the greatest rate of shrinkage occurred).

Chapter 1. Figure 1. Mike Roch (*left*) and Warren Ward beside the Tatshenshini River. Bill Hanlon photograph.

Chapter 1. Figure 3. First ice patch at the head of the west fork of the Fault Creek Glacier with Bill Hanlon (*left*) and Warren Ward at the base. Mike Roch photograph.

Chapter 1. Figure 4. Bill Hanlon looking southeast down the so-called Empty Valley from the ridge above the discovery site. The Samuel Glacier is visible in the background. Mike Roch photograph.

Chapter 1. Figure 6. Mike Roch (*left*) and Bill Hanlon with full packs, just west of Fault Creek. Warren Ward photograph.

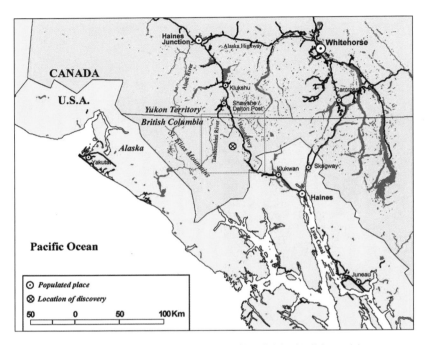

Chapter 2. Figure 1. The study region spans the borders of British Columbia, Yukon and the state of Alaska. The circled 'x' marks the study site and the red rectangular outline denotes the area represented in figure 2, below.

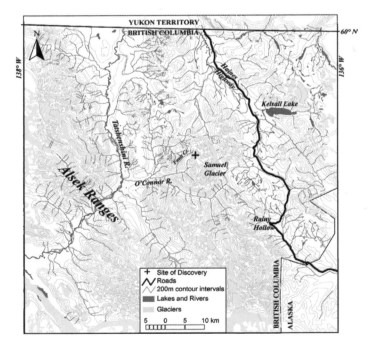

Chapter 2. Figure 2. The discovery site in Tatshenshini-Alsek Park, British Columbia.

Chapter 2. Figure 3. A rolling alpine landscape in the Chilkat Pass, Haines Highway, Alaska. Richard Hebda photograph.

Chapter 2. Figure 4. A braided channel flood plain on the Chilkat River, Haines Highway, Alaska. Richard Hebda photograph.

Chapter 2. Figure 7. Mountain Hemlock parkland in a sea of alder thickets. Richard Hebda photograph.

Chapter 2. Figure 9. Alpine tundra and rolling terrain looking west to the discovery site and Samuel Glacier. Richard Hebda photograph.

Chapter 2. Figure 10. Boreal forest north approach to Chilkat Pass Haines Highway. Richard Hebda photograph.

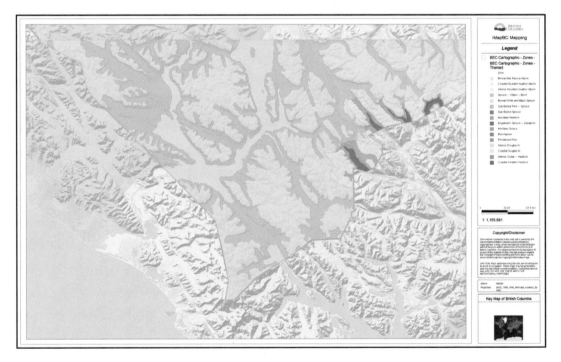

Chapter 2. Figure 11. Biogeoclimatic (BEC) zones, Haines Triangle.

Chapter 2. Figure 12. Alpine above spruce, willow and birch scrub at north approach to Chilkat Pass. Richard Hebda photograph.

Chapter 2. Figure 14. Alpine tundra in Chilkat Pass, Haines Highway, Alaska. Richard Hebda photograph.

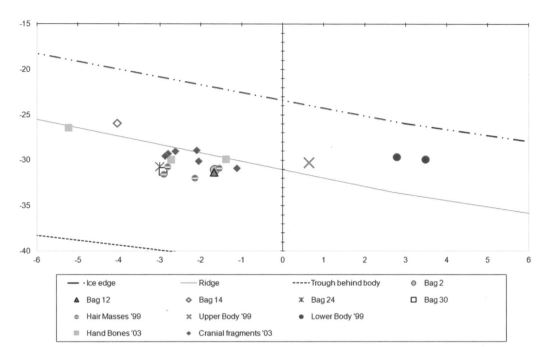

Chapter 4. Figure 8.
Horizontal distribution of surveyed human remains and principal artifact collections for 1999 and 2003. The scale is in metres relative to a reference point. North is to the top.

Chapter 4. Figure 12. Sarah Gaunt placing flowers on the burial cairn on behalf of Champagne and Aishihik women with Elder John Adamson, July 18, 2001. Al Mackie (BC Archaeology Branch) photograph.

Chapter 5. Figure 4. Near location 1, showing the banded rock (semiperlite), whitish-grey lichen *Stereocaulon* (Cauliflower Foam), dark yellowish-green moss *Dicranowiesia crispula* (Mountain Pincushion) and the pink flower of *Saxifraga oppositifolia* (Purple Mountain Saxifrage). July 21, 2001. Al Mackie photograph.

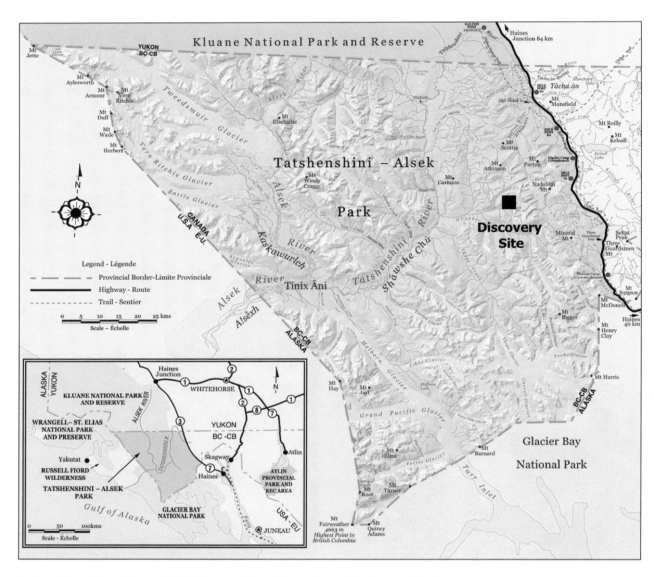

Chapter 7. Figure 1. Map of Tatshenshini-Alsek Park, showing traditional villages and the location of the discovery site. Based on a BC Parks map.

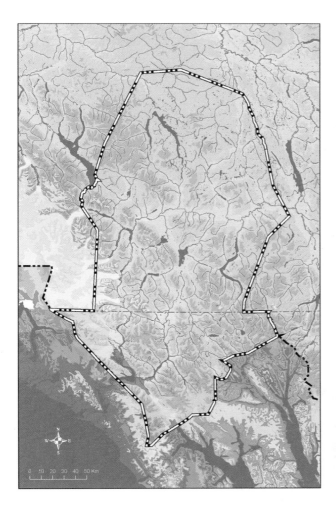

Chapter 7. Figure 2. Champagne and Aishihik First Nations traditional territory (outlined), which includes lands in southern Yukon and northwestern British Columbia.

Chapter 7. Figure 6. Lily Hume with salmon at the village of Klukshu, 1949. Catharine McClellan photograph. Courtesy of the Canadian Museum of History.

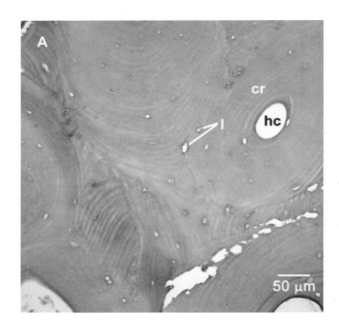

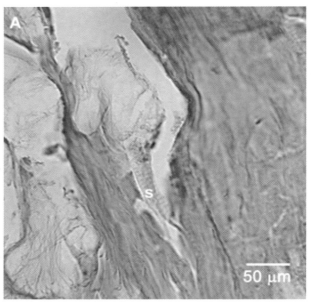

Chapter 18. Figure 1. Lumbar vertebral bone. *A,* Light micrograph of transverse section through a lumbar vertebral bone, Haematoxylin- and Eosin-stained: *hc,* Haversian canals; *cr,* concentric rings of lamellae; *l,* lacunae. Courtesy of John Wiley and Sons, Inc.

Chapter 18. Figure 2. Biceps branchii muscle. *A,* Light micrograph in longitudinal section showing good preservation with cross striations (*s*). Courtesy of John Wiley and Sons, Inc.

Chapter 20. Figure 1. One of several pieces of intact Sockeye Salmon flesh and skin (sample 105), as found in situ at the Kwädąy Dän Ts'ìnchį discovery site. Al Mackie photograph.

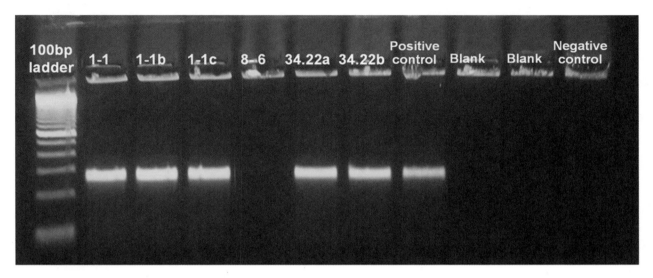

Chapter 20. Figure 7. This electrophoresis gel image shows some of the successfully amplified D-loop DNA samples (the bright band shows the presence of a 249 bp DNA fragment; some of the samples were amplified multiple times to replicate results). Two of the samples (1-1 and 34.22) yielded DNA, and one sample failed to amplify (8-6).

Chapter 20. Figure 8. Part of the amplified cytochrome b sequence aligned with reference sequences retrieved from GenBank: Rainbow Trout (NC_001717), Chum (AJ314561), Coho (AJ314563), Sockeye (AJ314568), Pink (AJ314562), Chinook (AJ314566) and Atlantic salmon (NC_001960). Dots indicate identical bases with Rainbow Trout. The ancient sequence recovered from the Kwädąy Dän Ts'ìnchį discovery samples is represented by a DNA consensus and matches identically with the Sockeye Salmon reference.

Chapter 21.
Figure 1.
The basketry hat partially reshaped. Scale in centimetres. Valery Monahan photograph.

Chapter 21.
Figure 4.
Inside view of the basketry hat showing attached headband. Loose ends of splices and inserted warps give the inside surface a slightly rough appearance. Scale in centimetres. Valery Monahan photograph.

Chapter 22. Figure 1. Range in wood condition. Wood on the left is lightweight, grey and fibrous. Left to right, condition improves. The fragment at the far right exhibits physical damage in the form of cracks, but the wood is dense, with crisp surface details. All scales are in centimetres. Government of Yukon.

Chapter 22. Figure 2. Wooden artifact IkVf-1:106, after drying. The left fragment was recovered from ice. It is stained, but surface details like wood grain are visible. The right fragment has weathered and lost surface details. Dark scratches show that weathering extends only a few millimetres at the surface. Scale in centimetres. Government of Yukon.

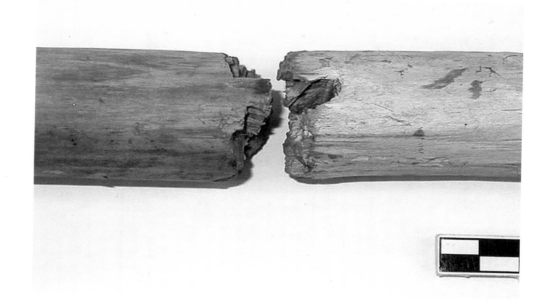

591

Chapter 22. Figure 4. Twined sinew has been used to repair a slit in the woven hat. Stitches are 0.4–0.7 centimetres apart. Government of Yukon.

Chapter 22. Figure 5.
Gentle cleaning of the hat's exterior reveals bright ochre in recesses of the weaving.
Government of Yukon.

Chapter 22. Figure 6. Hide tie strap. This section is unevenly cut with a clear pattern of hair follicles. Government of Yukon.

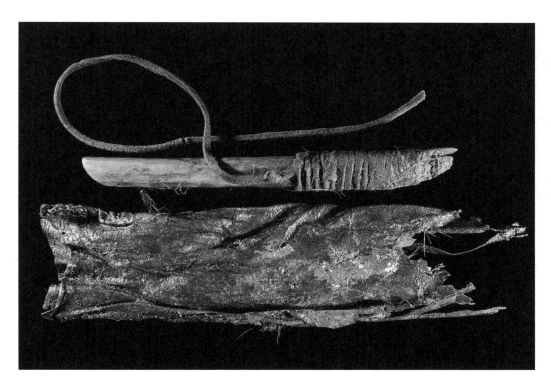

Chapter 23.
Figure 3. Photograph of IkVf-1:102, man's knife *(upper)* with its fur sheath *(lower).* The man's knife has been radiocarbon dated to 150 ± 50 c14 years and was found in association with human remains. Ruth Gotthardt (Yukon government) photograph.

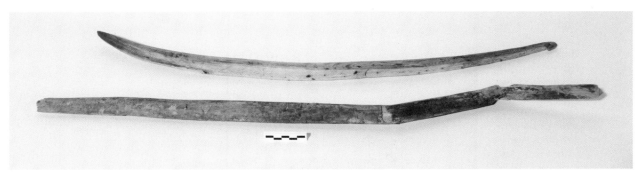

Chapter 23. Figure 4. *(top)* Photograph of IkVf-1:125, carved and painted stick. It has not been radiocarbon dated. Function is unknown; wood not identified. Found in association with human remains. *(bottom)* Photograph showing both IkVf-1:103 (fig. 5) and IkVf-1:125. Robin Armour (Yukon government) photograph.

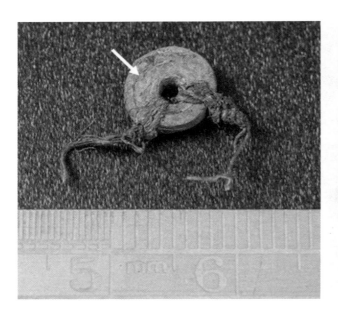

Chapter 24. Figure 1. Copper bead and thong fragments found with the Kwädąy Dän Ts'inchį man. *A*, Plan view showing thong attachment, spiral structure and millimetre scale. *B*, Edge view of bead showing flattened edge, same scale as 1*A*.

**Chapter 24.
Figure 3.** Major element composition of the Kwädąy Dän Ts'inchį bead. Vertical axis is a logarithmic scale of hits received by the detector. The horizontal axis is in seconds.

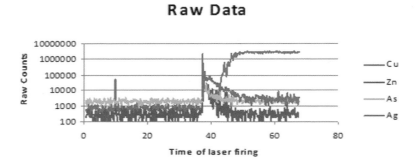

**Chapter 24.
Figure 4.** Minor element composition of the Kwädąy Dän Ts'inchį bead. Vertical axis is a logarithmic scale of hits received by the detector. The horizontal axis is in seconds.

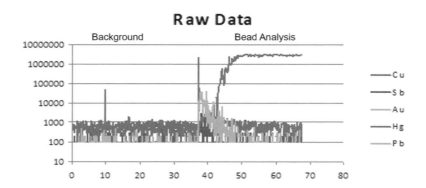

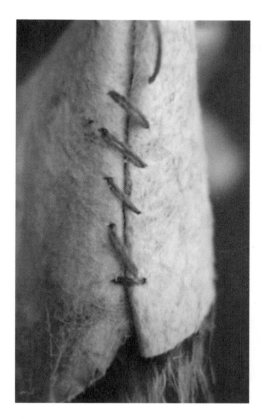

Chapter 26. Figure 2. Stitching on contemporary ground squirrel skins showing how two skins were stitched together in the Kwädąy Dän Ts'ìnchį robe.

Chapter 26. Figure 3. Kwädąy Dän Ts'ìnchį robe fragment from centre back area showing original stitching, patches and red ochre pigment visible from the skin side.

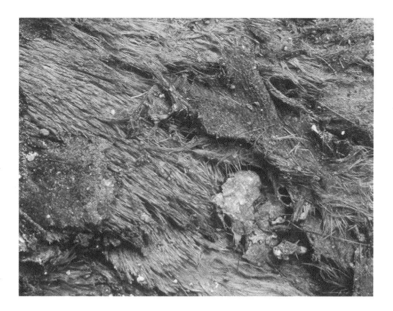

Chapter 26. Figure 4. Beaver-skin bag: skin side showing loss of skin; hair follicles are visible in situ.

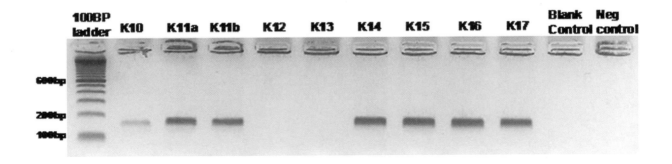

Chapter 27. Figure 2. Electrophoresis gel displaying some positively amplified samples using the universal 12S primer set designed to amplify fragments of approximately 120 bp. Note the lack of amplification in K12 and K13; these two samples consistently failed to amplify, despite repeated efforts with a variety of primers.

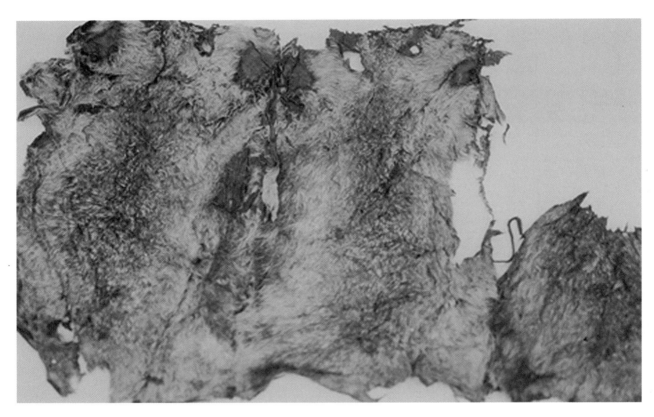

Chapter 28. Figure 1. Robe fragment showing general conditions of preservation and Arctic Ground Squirrel fur.

Chapter 28.
Figure 2. Robe mass before conservation.

Chapter 28.
Figure 3. Richard Hebda and Kjerstin Mackie at the Royal British Columbia Museum sampling the smooth inside surface of the robe using distilled water and a fine brush.

Chapter 28.
Figure 4. Image of a typical inside surface of robe pollen sample showing a large pollen grain of fireweed (*Epilobium* sp.) at upper right and charcoal (angular black objects) at centre left.

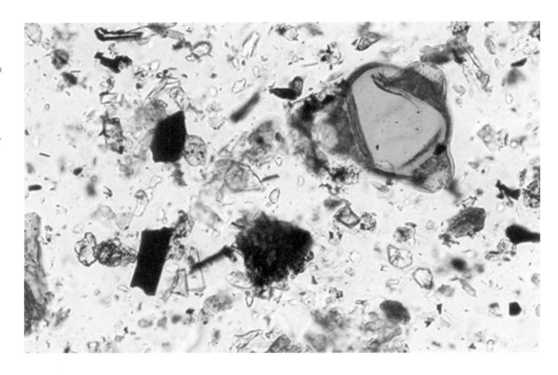

Chapter 28.
Figure 5. Pollen and spore assemblages from robe samples showing percentages of major types. *Artemisia* and *Salicornia* pollen is included in the "non-trees" value. Analyst Richard Hebda.

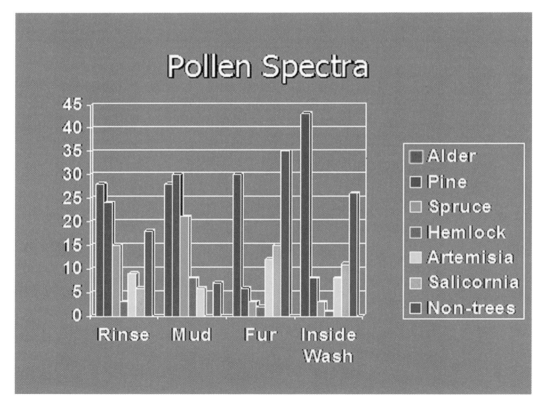

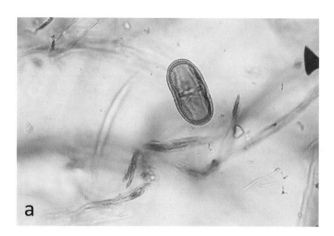
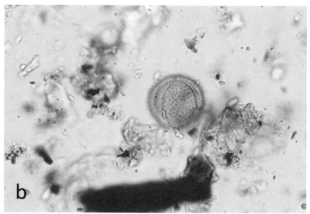

Chapter 28. Figure 6. Microscopic pollen of two relatively common non-arboreal (non-tree) pollen types recovered from robe samples. 6*A Heracleum maximum* (Cow Parsnip), 6*B Artemisia* (sagewort).

Chapter 28. Figure 8. The finely divided and layered branches of *Hylocomium splendens* (Step Moss), a lowland forest moss found in association with the robe. Jim Dickson photograph.

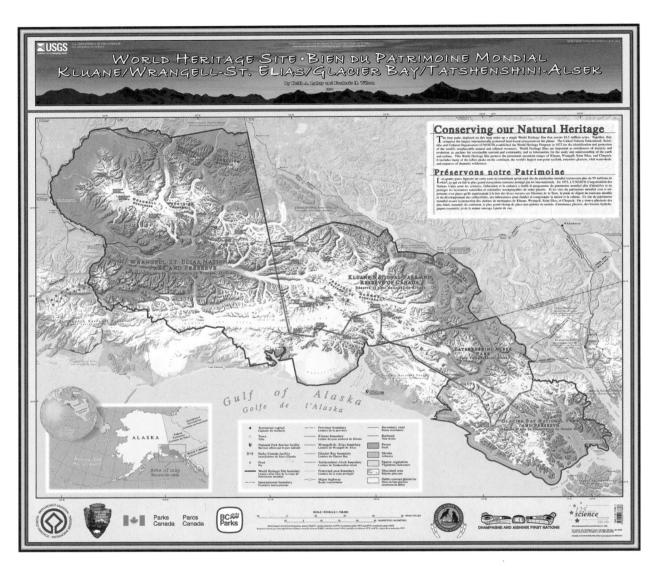

Chapter 29. Figure 2. The UNESCO World Heritage Site comprising Wrangell-St. Elias National Park, Glacier Bay National Park and Tongass National Forest (Alaska); Kluane National Park (Yukon); and Tatshenshini-Alsek Park (BC). White areas indicate glaciers. The Kwädąy Dän Ts'ìnchį discovery site is located in the lower right of the map. Courtesy of the USGS.[10]

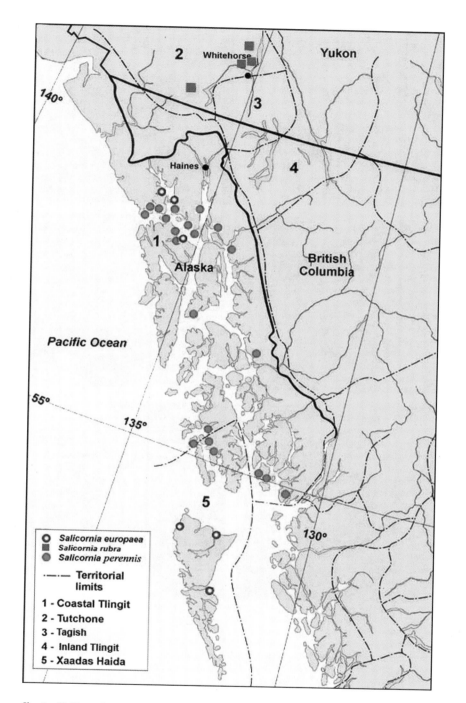

Chapter 31. Figure 2. Map showing distribution of *Salicornia* sites in Yukon Territory, southeastern Alaska and adjacent British Columbia. Dotted broken line shows the rough boundaries of Aboriginal territories according to the source map from the Vancouver Museum of Anthropology, 2002.

Chapter 31. Figure 7.

Inflorescences of *Salicornia* species and varieties.

A and *C. Salicornia pacifica* plants from the southernmost study area, showing the relatively long flowering spikes with more than eight rows of bracts. *A.* George Inlet; *C.* Esquimalt Lagoon.

B and *D. Salicornia perennis*, showing the shorter, more knobby flower spikes with fewer than eight rows of bracts, found on plants from the northern islands and mainland. *B.* Swanson Harbor; *D.* Pleasant Island.

E. Salicornia europaea, Hoonah salt marsh, showing the small annual coastal herb with pointed flowering spike tip and barely visible stamens *(inset)*.

F. Salicornia rubra, showing the distinctive colour, small size and almost sunken anthers of the annual saltwort (Red Samphire) from the Dawson Creek Trail near Whitehorse.

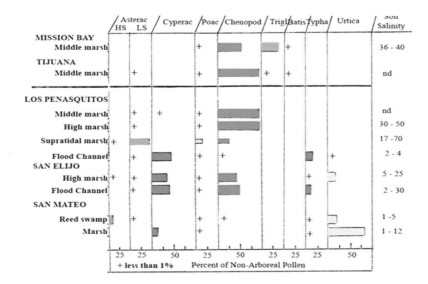

Chapter 31. Figure 8. Chenopod flowering time and marsh distribution.

A. Percentage of *Salicornia* pollen in air trap samples of *Salicornia perennis* marsh at Boundary Bay, southern BC (from Hebda 1977).

B. Relative abundances of herb pollen in two intertidal salt marshes (Mission Bay and Tijuana), in the salt marsh and adjacent flood plain freshwater marsh of Los Penasquitos Lagoon, and two more brackish water marshes at San Elijo and San Mateo lagoons, San Diego County, California (Scott et al. 2014). Asterac., Asteraceae; HS, high spine; LS, low spine; Cyperac., Cyperaceae; Poac., Poaceae; Chenopod., Chenopodiaceae; Trigl., *Triglochin*; *Batis*; *Typha*; *Urtica*.

+ less than 1%; percentages are based on total non-arboreal pollen.

Chapter 32. Figure 3.
The route most likely taken by the Long Ago Person Found from Rainy Hollow to the discovery site, indicated by the red line. The numbers correspond with figure numbers in the text, and the arrows indicate the direction the photographs were taken.

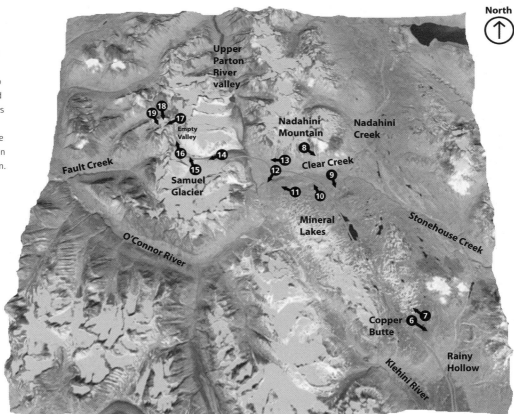

Chapter 32. Figure 4. Looking southeast from Copper Butte to Wilson Creek *(foreground)*, with the Klehini River valley in the background.
Matt Simmons photograph.

Chapter 32. Figure 5. Panorama looking west *(left)* to Copper Butte and northwest *(right)* down the valley toward Mineral Lakes and Clear Creek valley in the distance. The Kwädąy Dän Ts'ìnchį man may have walked down the right side of this valley toward Mineral Lakes. Darcy Mathews photograph.

Chapter 32.
Figure 6. Mineral Lakes, looking southeast from the base of Nadahini Mountain. Clear Creek is in the foreground. The Kwädąy Dän Ts'ínchį man would have walked north through this valley from Copper Butte, past these lakes. Darcy Mathews photograph.

Mineral Lakes, Copper Butte

Chapter 32.
Figure 7. Looking south-southeast from the Clear Creek valley (in the foreground) toward Mineral Lakes, over the route it is believed the Kwädąy Dän Ts'ínchį person travelled. At about this point he would have turned west (toward the right side of the photograph) to head toward the Samuel Glacier. Darcy Mathews photograph.

Chapter 32.
Figure 8. View north over Clear Creek, looking toward Nadahini Mountain. When the Long Ago Person Found walked north from Mineral Lakes (behind the photographer, see also figs. 8 and 9) to this valley, he would have turned west (*left*) to head toward the Samuel Glacier. Darcy Mathews photograph.

Chapter 32. Figure 9. View west from the headwaters of Clear Creek to the lobe of the Samuel Glacier that was traversed to reach the discovery site.
Darcy Mathews photograph.

Chapter 32. Figure 10. Looking south to the headwaters of Clear Creek and the route likely taken by the Kwädąy Dän Ts'ìnchį person, shown with dashed line, between Mineral Lakes (to the left) and the Samuel Glacier (to the right). The headwaters of the O'Connor River at the toe of the Samuel Glacier are in the distance and beyond that the O'Connor River valley is visible. The valley to the left leads to Rainy Hollow and, as discussed in the text, is a known route between the Klehini and O'Connor/Tatshenshini rivers. The mineral grain data from the Kwädąy Dän Ts'ìnchį man indicate he did not travel from the Klehini drainage over to the O'Connor basin via the valley system shown in this image, even though this route may have been a variant of the Chilkat Trail. Darcy Mathews photograph.

Chapter 32. Figure 11. Looking west toward the lobe of the Samuel Glacier that the Kwädąy Dän Ts'ìnchį man travelled over to reach the place where he died. Following this glacier upward takes one to the first valley showing on the right. It is believed that he walked along the north margin (right side) to avoid crevasses and icefall on the inside bend of the glacier. Darcy Mathews photograph.

Chapter 32. Figure 12. Walking southwest along the north edge of the lobe of Samuel Glacier to reach the discovery site. Note the gentle gradient of the ice and lack of major crevasses. The dashed line denotes the maximum extent of Little Ice Age ice. Darcy Mathews photograph.

Chapter 32. Figure 14. Looking northwest from Samuel Glacier up Empty Valley, toward the pass where the Kwädąy Dän Ts'ìnchį person was found. The arrow points to the area of the discovery site, which is on the opposite side of the pass and out of view. Darcy Mathews photograph.

Discovery site

Chapter 32. Figure 15. Looking west from the height of the pass down to the Kwädąy Dän Ts'ìnchį site. It is likely that any trail running through this pass would have continued on toward Fault Creek. Al Mackie photograph.

Chapter 32.
Figure 16. Aerial view looking south to the route followed by the Long Ago Person Found (marked with a dashed line). He crossed the Samuel Glacier (in the distance), turned north and followed the Empty Valley over the pass *(centre right)*. The discovery site is just out of the frame to the right. Sheila Greer (CAFN) photograph.

Chapter 32.
Figure 17. Aerial photograph looking south to the glacier where the remains were recovered. The lobe of the Samuel Glacier traversed by the reconnaissance team, as well as the valley empty of ice through which the Kwädąy Dän Ts'ìnchį man travelled, is shown on the left. Middle Fault Creek drains to the right. Sheila Greer (CAFN) photograph.

Chapter 35, Figure 4. Champagne and Aishihik
First Nations citizen Lori Strand models heirloom
säl ts'ät. Ukjese van Kampen photograph.

Notes

Chapter 8

1. See Carcross-Tagish First Nation and Greer (1995, 52ff.) for a discussion of the conflicts that occurred over packing on the Chilkoot/Dyea trail in the 1890s as a result of the liability issue; the latter determined which clans had the right to pack on the trail.

2. In the decades since the discovery was named, owing to the limitations of word-processing software, there has been some variance in how the Dákwänjè name for the discovery has been spelt. Newer software is better at reproducing the diacritics that indicate the different vowel tones (high, low, nasal), which are a key to proper pronunciation of the name.

3. When the Alaska Highway was built through the southern Yukon in the early 1940s, many First Nations graves were looted by the newcomers, who pilfered the personal items or "grave goods", as archaeologists would term them. The disrespect shown to the graves of their loved ones is still a matter of serious concern to Yukon First Nations.

4. The option of preserving or retaining the human remains was not put forward for discussion at the regional meeting as it was contrary to regional First Nations and tribal values.

5. For more information on the clan conference go to http://clanconference.org/. Accessed March 2010.

Chapter 9

1. For the comprehensive nature of their work, see Leechman 1946, 1949, 1950, 1962a, 1962b, n.d.; McClellan 1950, 1964, 1975a, 1975b, 1981a, 1981b, 1989, 2007a; McClellan et al. 1987; Cruikshank 1974, 1979, 1980, 1982, 1990, 1991, 2005; Cruikshank et al. 1990; O'Leary 1979, 1980, 1984, 1992a, 1992b; see also Ned 1984; Smith 1982.

2. Jukebox.uaf.edu/drybay/home/htm.

3. www.sheldonmuseum.org/.

4. Readers are directed to the websites hosted by various Indigenous governments and organizations: https://www.ytttribe.org/; www.chilkatindianvillage .org; http://www.chilkoot-nsn.gov/; http://www .skagwaytraditional.org/; www.cafn.ca.

5. The actual location of Stonehouse has yet to be pinpointed. Though de Laguna (in Emmons 1991, 83) doubts Emmons visited this place, an Emmons 1905 diary in the BC Archives, which includes a sketch map showing the location of Stonehouse as well as an account of his journey to it, suggests that he may have. The location Emmons gave for Stonehouse—"beyond the summit, on a level moss-covered plain, free from any obstructions and visible for all directions" (Emmons 1991, 83)—matches most closely information provided by CAFN Elder John Adamson (1993) and information provided on an unpublished map depicting the original routing of the Haines Road, on file at the Sheldon Museum in Haines. The Emmons diary indicates that he took photographs of Stonehouse, but these have not

yet been located; they do not seem to be part of the Emmons files in the Newcombe family collection at the BC Archives. In the 1990s, staff of the CAFN Heritage program tried unsuccessfully to relocate Stonehouse, as have others hiking in the area (Al von Finster comm. to Greer, April 5, 2008). The matter is somewhat complicated by the switching of creek names. On the Haines Road today, the Chilkat basin tributary indicated as Stonehouse Creek is not the real Stonehouse Creek, which is the watercourse now signed as "Clear Creek".

6. Judson Brown was also a Champagne and Aishihik First Nations citizen. His maternal grandmother, Susan (Kin-dah-ant), was a Tatshenshini basin resident (CAFN genealogical records), and Paddy Duncan, his grandmother's brother, was the Wolf clan leader at Shäwshe during the early part of the 20th century. Their father may have been the Chief named Tinneh Ts'arti, who lived at the settlement known as Núghàyík.

7. It is thought that the people living at Núghàyík at the time of Mrs. Smith's childhood trip were likely local Dän or Dän/Tlingit who would be fishing in the area, just as they were in the summer of 1890 when Glave and Dalton went down the river. The Nu gha people (or Nua Quas) had by this time abandoned the settlement.

8. Based on the buying and selling prices of marten fur, anthropologist Dominique Legros estimated that the Chilkat made a profit of 2000 per cent in these exchanges (Legros 1984). Marten were bought in the interior at the equivalent of $.50 (based on the cost to the Chilkat of an item traded for one marten skin) and sold on the coast for two to three dollars. While this rate of return is high by modern standards, it is not out of line for the period; the Hudson's Bay Company was making at least 1000 per cent profit in its fur trading efforts out of Port Simpson, located to the south on the BC coast (Davidson 1867–68). See also Carcross-Tagish First Nation and Greer 1995.

9. Dalton is understood to have married one or possibly two Indian ladies in the interior (cf. Kitty Smith in Cruikshank 1991, 92; McClellan 2007a, 51).

Chapter 10

1. Most of the information presented in these newspaper articles also appears in articles published by Glave the following year (Glave 1892a, 1892b). Glave's unpublished journal (Davey n.d.), housed at the University of Alaska Fairbanks Archives—the source from which the published account was produced—contains a few points of information omitted from the published accounts. Glave's journal contains hand-drawn maps related to the 1891 trip, but none from the 1890 expedition.

2. Champagne and Aishihik Elder Marge Jackson reported that her step-father, Big Jim Fred, obtained his first metal traps from Gach'alwa. She was also told that this man had a much younger wife and that he kept up his appearance for her (Jackson 2006; Greer 2005).

Chapter 23

1. There is some variation in how artifact fragments were inventoried and labelled over the years. Pieces that were not immediately recognized as being part of the same artifact were assigned separate numbers. This was the case for IkVf-1:104, IkVf-1:106, IkVf-1:115 and IkVf-1:116, which were refitted in 2001. In other cases, different fragments of the same artifact were considered subsets of one catalogue number, as was the case for IkVf-1:109, which represents four fragments of the same artifact.

2. Note that this catalogue does not include paleontological (i.e., natural history) specimens collected from the IkVf-1 site area, which are housed at Yukon Heritage, Paleontology, in Whitehorse. The latter includes a dog or wolf skull collected in 2005.

3. When perusing the online catalogues of museums holding items collected by Emmons, such as the AMNH or the Burke Museum, one should note that there are a few pieces reportedly obtained from Klukwan or Haines—and therefore classified as Tlingit in manufacture—that were more than likely made by residents of the interior. See, for example, AMNH collection items #E/2043, a bag; #19/192, a robe; and #E/1581, a blanket, all made of ground squirrel skins. In the collections of the Burke, see pieces such as items #1168, Man's Shirt, and #2284, Man's Garters, both

made of Caribou skin. In some cases, while indicated as having been collected at Klukwan or Haines, the interior point of origin was also noted in the original catalogue. See, for example, AMNH item #E/2315, a quiver with arrows, or Burke #1753, arrow. See www.anthro .amnh.org/anthro_coll.htm and www.washington.edu/ burkemuseum/collections/ethnology/collections/index .php, both accessed March 3, 2008.

4. See http:/vilda.alaska.edu/.

5. Plate 79, in de Laguna 1972, part 3, is an exception. In this image a gaff pole is identified leaning on a cache in a tent camp in the Yakutat area.

6. See www.alaskool.org/language/dictionaries. Accessed March 2008.

7. In 2000 a sample of the hide lashing from the man's knife was given to Dr. A. Cooper, then of Ancient Biomolecules Laboratory, Oxford University, for species identification. No results on this sample have been forthcoming, and Dr. Cooper is no longer with this institution. Correspondence on March 14, 2006, between the author and Dr. Cooper's successor at this lab, Dr. Beth Shapiro, suggested that the sample still remained at the lab, with the analysis work outstanding. Dr. Shapiro has also subsequently left this institution. We have requested that this sample be returned, but as of the date of writing, our request has not been fulfilled.

8. By the time of first publication (Beattie et al. 2000), the #104 and #116 fragments had been refitted together and labelled as the multi-notched stick, while the #106 and #115 fragments had been refitted together and labelled as the bear stick. While working with the various pieces in 2001, Yukon conservator Valery Monahan realized that the end of the #116 piece fit with the end of the #106 piece, making a more or less complete artifact close to 13 feet in length.

9. See www.washington.edu/burkemuseum/collections/ ethnology/index.php, consulted November 26, 2008.

10. Visit http://anthro.amnh.org/anthro_coll.shtml, select North American Ethnographic Collections, then select Northwest Coast under "region," then Tlingit under "culture". In the latter catalogue, these items are labelled as "carving knives" to distinguish them from other types of knives, such as fighting knives, potato knives, knives for splitting spruce roots, etc. Sources

consulted indicate that there are various Tlingit language terms for "knife", and it is not certain which would apply to this type.

11. See www.alaskool.org/language/dictionaries, accessed March 2008.

12. The only technology items related to marmot hunting that exist in museums are the carved pins or stakes that were used to secure the set snares (see Schwatka 1885). Many examples of these are found in museums; one example can be seen in the online ethnology collections of Seattle's Burke Museum. See http://www.washington .edu/burkemuseum/collections/ethnology/collections/ display.php?ID=26180, accessed November 26, 2008. It is understood that the carved heads on the sticks were thought to help lure the animal into the trap. These snare pins—or trap sticks, as they are also called—are known to have been used by the Tlingit. De Laguna (1972, 367) refers to the Southern Tutchone of the upper Alsek also using the snare sticks or pins, but hers is the only reference we are aware of for the Dän using of them.

13. Definition of alpenstock from Wikipedia, www.wikipedia .org/wiki/Alpenstock. Accessed March 2010.

14. See http://anthro.amnh.org/anthro_coll.shtml. Select North American Ethnographic Collections, then search under catalogue number 19/609. Accessed November 28, 2008.

15. In addition to being finely executed, another noteworthy detail about the notching on the IkVf-1:129 artifact is that the shape and distribution of its notches are similar to those on the halibut hook, another type of Tlingit fishing technology. Halibut hooks are large, wooden hooks that were suspended on fishing line to catch this species of ocean bottomfish. There are many examples of the exquisite and elaborately carved Tlingit halibut hooks in museum ethnographic collections (see, for example, the Anthropology collections of the American Museum of Natural History, www.anthro .amnh.org/anthro_coll.htm). Illustrations of halibut hooks can also be found in many published sources (e.g., de Laguna 1972, 389; Stewart 1977). Our concern, however, is not the elaborate carving but the basic design of this composite tool, which involves the joining together of two parts of the hook. This is achieved by carving flat surfaces on each of the two round-shaped wood pieces (i.e., sticks) that were to be joined, with a series of notches carved on each piece in the face

opposite to the flat surface. These notches allowed the lashing wound around the hook to firmly secure the two artifact pieces together. For an illustration of a Tlingit halibut hook that is missing its lashing, and thus where this basic construction design can be viewed more easily, see the photos of halibut hook #1477 in the collections of the Burke Museum at www.washington .edu/burkemuseum/collections/ethnology/collections/ index.php (accessed March 3, 2008). The pattern of carved notches on the IkVf-1:129 gaff pole specimen matches that on the Burke Museum example, as well as the halibut hooks in the collections of the Royal British Columbia Museum, which were examined by Greer.

Chapter 29

1. Champagne and Aishihik First Nations citizen and guide Ron Chambers at Yakutat, November 6, 2000; see Champagne and Aishihik First Nations 2000b, 14–15.

2. In this chapter we use the Southern Tutchone language term *Dän*, which means people, to refer to the Tutchone; the word *Tlingit* similarly means person or being.

3. The names of these various peoples refer to their Indigenous languages. Ahtna, Upper Tanana, Dákwanjè and Tagish languages all belong to the Athapaskan (Dene) family of languages. Tlingit and Eyak are separate and distinct languages, though linguistic research suggests that both have historical links to the Athapaskan language family, with Tlingit being more distant than Eyak (Krauss 1980).

4. A more in-depth analysis of this genre of First Nations oral literature can be found in *Do Glaciers Listen?* by Julie Cruikshank (2005).

5. The spelling used here, Qwantuk, is that provided by Wilfred Charlie's wife, Dawn Charlie, who transcribed the story as told by Wilfred. Anthropologist Dominique Legros, who worked extensively with Northern Tutchone Elder Tommy McGinty, notes that he recorded "Chief Kwanatak's stories" from McGinty (Legros 1999, 259). Though the Chief Kwanatak stories are not included in any of Legros' published works, it is thought that Kwanatak, as spelled by Legros, is the same person as Qwantuk.

6. In the video version, Adamson identifies the rescued trader as his great-, great-, great-grandfather and, later

in the same presentation, as his grandfather. In the audio version, he refers to this individual as his great-grandfather; the latter reference has been used here.

7. This story was widely known in the Champagne and Aishihik First Nations community, with other elders' telling of the events also documented (cf. Brown 1996; Banks 1999).

8. Most of Mrs. Sidney's descendants are citizens of the Carcross-Tagish First Nation, while Mrs. Ned's and Mrs. Smith's descendants belong primarily to the Champagne and Aishihik First Nations and/or to the Kwanlin Dün First Nation.

9. It is possible that the name of the Kwädąy Dän Ts'ìnchį individual might yet be discovered. Following the writing of this paper, a citizen reported at the 2009 Champagne and Aishihik First Nations general assembly that he had been told of a south Yukon resident who reportedly knows the name of the person who went missing in Tatshenshini-Alsek country so many years ago. This possible lead has not yet been investigated.

Chapter 31

1. Forensic palynology includes " ... the study of modern and fossil spores, pollen and other acid-resistant micro-plant remains in a legal context" (Mildenhall et al. 2006, 163).

Chapter 32

1. More than one routing from the upper Chilkat appears to have been used to access Kusawa Lake. The Frank Leslie expedition, led by a well-known guide from Klukwan, entered the Kusawa basin in May 1890 via the uppermost waters of the Takhini River, a Kusawa Lake tributary (Glave 1890; Schanz and Wells 1974). Moise and Moise (1986) retraced the Takhini variant of the eastern (Kusawa) Chilkat route in the 1980s. Another variant of the route from the upper Chilkat over to the Hendon River, which flows into the Kusawa near the head of the lake, is also understood to have been used (Jim 2009).

2. Today there is a watercourse along the Haines Road that is gazetted and signed as Stonehouse Creek, but CAFN elders knowledgeable about the history of this part of the traditional territory are emphatic that the

creek now labelled Clear Creek is the real Stonehouse Creek. Although CAFN Heritage staff attempted to locate the actual Stonehouse feature in the mid-1990s, as has Al von Finster (another author in this volume), it has yet to be rediscovered.

3. The Emmons diary indicates that it took him and his guides one day to travel from Klukwan to Porcupine. The second day they walked from Porcupine to Rainy Hollow. The third day on the trail they walked from Rainy Hollow to Stonehouse and back to Rainy Hollow. One of Emmons' guides on this 1905 trip was a man named "Paddy"; this was quite possibly Paddy Duncan, who would later be the Wolf clan leader at Shäwshe (Nesketahin) in the 1920s (see chapter 10 this volume).

Chapter 33

1. The Cheddar Man study is mentioned here because of the widespread publicity it received. It is noted, however, that this study is now largely discredited, because the authenticity of the sequence has been questioned.

2. Administratively, the community DNA study operated totally outside the purview of the Kwädąy Dän Ts'ínchį project's scientific review committee.

3. In 2004 Champagne and Aishihik First Nations (CAFN) engaged Dr. Mooder to review the work conducted and results obtained and to provide recommendations for improvement (Mooder 2005). Improved methods were implemented at UVic but when matches were· found and results forwarded to the CAFN they could not be matched to individuals who had participated in the study. Data-handling or "chain of custody" problems were recognized and Dr. Mooder was given responsibility of arranging for the lab work to be redone and for interpreting the new results.

4. The genealogical data for Champagne and Aishihik First Nations were reviewed, however, and 14 female lineages were recognized in the period between roughly 1850 and 1900. By chance rather than by intent, each of these lineages was found to have at least one member participating in the study.

5. The suffix codes were assigned by the sample collector, rather than based on information provided by the study participant, and are not necessarily an accurate representation of a participant's tribal or First Nation affiliation. Alaskan study participants more commonly reported clan affiliation rather than tribal affiliation on their consent forms. Consequently, data on the latter are weak.

6. In keeping with the agreement with the study participants, the blood samples used for DNA isolation were destroyed in Haines Junction, Yukon, by Greer in June 2009, with no sample remnants or residues remaining.

7. The consent form signed by the study participants specified that the names of the matches would not be released to the public. All study participants will be notified, in writing, of their status as matches, or not. As mailing addresses were not obtained for all participants, this process is ongoing at the time of writing.

8. The clan affiliation of one of the Alaskan matches is not known, though the individual self-identified as belonging to the Eagle moiety on the consent form.

9. Following their return from the University of Victoria, the samples were inventoried and this particular sample was identified as missing at that point. Efforts to locate it were unsuccessful. The individual who provided the sample has been notified.

10. One genealogical source at Yukon Archives in Whitehorse that is recognized as being particularly valuable for identifying relatives is the so-called Bonar Cooley genealogical chart. This document was researched and assembled in the 1970s by Bonar Cooley, of Teslin, and others. The chart shows the Alaskan and coastal Tlingit genealogical connections of Teslin Tlingit families. It also includes linked genealogical data on the Carcross-Tagish and the Atlin (Taku River Tlingit) people, and on some Southern Tutchone individuals. The original hand-lettered genealogical chart, more than 6 metres (20 feet) wide, was created to demonstrate eligibility to be beneficiaries (shareholders) of Sealaska, the southeastern Alaska regional Native corporation created through the Alaskan Native claims process.

11. The possibility of personal names as an important avenue of research was brought forward at the first meeting of the matched maternal relatives of the Long Ago Person Found, held in October 2008, when it became apparent that some personal names were shared by the Dakl'aweidí and Yanyeidí clans.

12. Outside researchers and anthropologists have interpreted these clan stories to suggest that the two clans are historically related to the Athapaskan peoples of the interior (Emmons in Hope and Thornton 2000; de Laguna 1975; McClellan 1975).

13. While de Laguna (1990, 227) reports that the Yanyeidí are historically connected to the Yakutat area, Hope and Thornton (2000, 149) do not. As the latter source is based on more contemporary research efforts, that information is used in table 2.

Chapter 35

1. This quote also appears in Glave 1892, Tempermental Explorer, Part 2.

2. Prior to our Gopher Robe Project it was becoming common to singe gophers, rather than to save the fur, even when the pelt was at its prime. We believe our project has helped to reverse this trend within our community.

3. This woman's dress can be viewed online at http://alaska.si.edu/record.asp?id=135. Accessed December 31, 2009.

4. This children's parka can be viewed online at http://collections.civilization.ca/public/pages/cmccpublic/emupublic/Display.php?irn=37687&QueryPage=%2Fpublic%2Fpages%2Fcmccpublic%2Femupublic%2FAdvQuery.php&lang=0. Accessed December 31, 2009.

5. The bag can be viewed online at http://anthro.amnh.org/anthro_coll.shtml. Search under catalogue item #E/2043, or chose North American Ethnographic Collections, then Northwest Coast region, then Tlingit culture and scroll through the pages of artifacts. Accessed December 31, 2009.

6. The hat can be viewed online at http://anthro.amnh.org/anthro_coll.shtml. Search under catalogue item #16/7951, or chose North American Ethnographic Collections, then Tutchone culture. Accessed December 31, 2009.

7. The Kate Carmacks gopher robe can be viewed online at http://www.macbridemuseum.com/collection/imageDisp.php?pageNum_Recordset1=41. Accessed December 31, 2009.

8. The CAFN Heritage program is extremely grateful to Mr. Kelly Wroot of Silver City, Yukon, for bringing this child's coat to our attention when it was for sale on eBay.

9. The robe consists of 7 rows of 12 pelts (total 84); however, in the two bottom corners two pelts are sewn together, covering an area equivalent to one pelt, making a total of 86 pelts.

10. For the Burke Museum in Seattle go to http://www.washington.edu/burkemuseum/collections/ethnology/collections. Accessed December 31, 2009.

11. These pieces can be viewed online at http://anthro.amnh.org/anthro_coll.shtml. Search under catalogue item (see below), or choose North American Ethnographic Collections, then Northwest Coast region, then Tlingit culture and scroll through the pages of artifacts. Accessed December 31, 2009.

12. This particular blanket was purchased by the late Kathy Kushniruk at an antique store in Haines, Alaska.

13. Richard Smith, personal communication to Greer, June 22, 2009; Richard heard this from his father, Harold Kane.

14. This photo of the interior of the Whale House can be viewed online in the Alaska digital archives at http://vilda.alaska.edu/cdm4/search.php. Search ASL-P87-0010. Accessed December 31, 2009.

15. These images can also be viewed online in the Alaska digital archives at http://vilda.alaska.edu/cdm4/search.php. Accessed December 31, 2009.

16. Wayne Howell, an anthropologist who has worked with the Yakutat community for many years, reported that the Yakutat people understand that these songs originated in the interior (personal communication to Greer, February 19, 2008). Both McClellan (1975) and de Laguna (1972) refer to songs as well as material goods being traded between the Alsek River area of the coast and the interior Champagne and Aishihik country. When representatives of Champagne and Aishihik First Nations visited Yakutat on a cultural exchange in 1996, their Yakutat hosts reported that they had been "keeping" various Southern Tutchone songs for their "interior friends" (Chief Diane Strand, personal communication to Greer, 2008).

17. Gopher snares can be viewed in the online collections of the Canadian Museum of History.

18. The overall process for tanning small mammal skins is similar to that of processing the hides or skins of larger mammals, and the Southern Tutchone way of tanning the latter is not unlike that reported by Northern Tutchone woman Mrs. Gertie Tom (see Tom 1981).

19. Urine is also known to have been used as a degreaser in times past.

20. Similar tools found in the Gwich'in culture area of northern Yukon have been referred to as chi-thos (Le Blanc 1983), which is a variant of the Gwich'in term for the tool.

21. Champagne and Aishihik citizen Gary Darbyshire reported that he had observed his grandmother, Mrs. Joe Kane, a Champagne resident, transporting gophers in a gunny sack from one location to a meadow where they were no longer available—i.e., restocking the species. (R. Kushniruk, personal communication to Greer, January 2010). A gopher restocking effort took place more recently (2008) at Duke Meadows, near Burwash Landing.

22. Initial studies have been conducted on the *si* (ochre) on hunting artifacts that have been recovered from the ice patch archaeological sites in southern Yukon, including Champagne and Aishihik First Nations traditional territory, by Helwig, Poulin and Monahan (2004). The ochre on the Kwädąy Dän Ts'ìnchį artifacts, including the Kwädąy Dän Ts'ìnchį robe, has not been sourced.

23. It is unfortunate that this interest comes just at a time when our local gopher population appears to be down in numbers. Local community members are not exactly sure why the population has dropped, but some have attributed it to the unusual weather experienced in recent years, such as the rain that fell in Haines Junction and area on Christmas Day 2005. Once frozen, the rain left a layer of ice over the burrow holes and, the following spring, many gophers died because they couldn't break out of their holes.

24. Available online at http://vilda.alaska.edu/cdm4/item_viewer.php?CISOROOT=/cdmg21&CISOPTR=661&REC=3. Accessed December 31, 2009.

References

Chapter 2

Banner, A., W. MacKenzie, S. Haeussler, S. Thomson, J. Pojar and R. Trowbridge. 1993. *A Field Guide to the Site Identification and Interpretation for the Prince Rupert Forest Region*. Land Management Handbook Number 26. Victoria: Research Branch, Ministry of Forests.

BC Parks. 2001. *Management Direction Statement for Tatshenshini-Alsek Park*. Victoria: Ministry of Water, Land and Air Protection, BC Parks Division. http://www.env.gov.bc.ca/bcparks/planning/mgmtplns/tatshenshini/tat_alsek.html.

Campbell, R.B., and C.J. Dodds. 1983. *Geology of Tatshenshini River Map Area (114P), British Columbia*. Calgary: Geological Survey of Canada.

Champagne and Aishihik First Nation. 2006. *Little Klukshu Sockeye Habitat Restoration Project Final Report*. Haines Junction, YT: CAFN.

Coney, P.J., D.L. Jones and J.W.H. Monger. 1980. "Cordilleran Suspect Terranes." *Nature* 288:329–33.

Cui, P., and Y. Erdmer, compilers. 2009. Geoscience Map 2009-1. Geological map of British Columbia at 1:1,500,000. Victoria: British Columbia Geological Survey, Ministry of Energy, Mines and Petroleum Resources. http://www.empr.gov.bc.ca/Mining/Geoscience/PublicationsCatalogue/Maps/GeoscienceMaps/Documents/GM2009-1_LowRes.pdf.

Cwynar, L.C. 1988. "Late Quaternary Vegetation History of Kettlehole Pond, Southwestern Yukon." *Canadian Journal of Forest Research* 18:1270–79.

———. 1990. "A Late Quaternary Vegetation History from Lily Lake, Chilkat Peninsula, Southeast Alaska." *Canadian Journal of Botany* 68:1106–12.

Hebda, R.J. 1995. "British Columbia Vegetation and Climate History with Focus on 6 ka BP." *Géographie Physique et Quaternaire* 49:55–79.

Hebda, R.J., and C. Whitlock. 1997. "Environmental History of the Coastal Temperate Rain Forest of Northwest North America." In *The Rain Forests of Home: Profile of a North American Bioregion*, edited by P.K. Schoonmaker, B. von Hagen and E.C. Wolf, 225–54. Covelo, CA: Island Press.

Holland, S.S. 1976. *Landforms of British Columbia: A Physiographic Outline*. Bulletin 48. Victoria: British Columbia Department of Mines and Petroleum Resources.

J.S. Peepre and Associates. 1992. "Tatshenshini-Alsek Region Wilderness Study." Unpublished report for the Tatshenshini-Alsek Wilderness Steering Committee.

MacDonald, G.M., and L.C. Cwynar. 1985. "A Fossil Pollen Based Reconstruction of the Late Quaternary History of Lodgepole Pine (*Pinus contorta* ssp. *latifolia*)." *Canadian Journal of Forest Research* 15:1039–44.

MacIntyre, D.G. 1984. "Geology of the Alsek-Tatshenshini Rivers Area (114P)." In *Geological Fieldwork, 1983: A Summary of Field Activities*, edited by A. Panteleyev and W.J. McMillan, 173–84. Victoria: British Columbia Geological Division.

Mathews, W.H. 1986. *Physiography of the Canadian Cordillera*. Geological Survey of Canada, Map 1701A, scale 1:5,000,000. Ottawa: Geological Survey of Canada.

McPhail, J.D. 2007. *The Freshwater Fishes of British Columbia*. Edmonton: University of Alberta Press.

Meidinger, D., and J. Pojar. 1991. *Ecosystems of British Columbia*. Ministry of Forests Special Report 6. Victoria: Ministry of Forests.

Mihalynuk, M.G., M.T. Smith, D.G. MacIntyre and M. Deschênes. 1993. "Tatshenshini Project, Northwestern British Columbia: Part B, Stratigraphic and Magmatic Setting of Mineral Occurrences." In *Geological Fieldwork 1992*, edited by B. Grant and J.M. Newell, 189–202. Victoria: British Columbia Ministry of Mines and Petroleum Resources.

Pojar, J. 1993. "Biodiversity Inventory of the Tatshenshini-Alsek Region." Unpublished notes.

Ritchie, J.C. and G.M. MacDonald. 1986. "The Patterns of Post-Glacial Spread of White Spruce." *Journal of Biogeography* 13:527–540.

Spear, R.W., and L.C. Cwynar. 1997. "Late Quaternary Vegetation History of White Pass, Northern British Columbia." *Arctic and Alpine Research* 29:45–52.

Chapter 3

Barrand, N.E., and M.J. Sharp. 2010. "Sustained Rapid Shrinkage of Yukon Glaciers Since the 1957–1958 International Geophysical Year." *Geophysical Research Letters* 37 (L07501). doi:10.1029/2009GL042030.

Beattie, Owen, Brian Apland, Erik W. Blake, James A. Cosgrove, Sarah Gaunt, Sheila Greer, Alexander P. Mackie, Kjerstin E. Mackie, Dan Straathof, Valerie Thorp and Peter M. Troffe. 2000. "The Kwäday Dän Ts'ìnchi Discovery from a Glacier in British Columbia." *Canadian Journal of Archaeology* 24 (1–2): 129–47.

Bolch, T., B. Menounos and R. Wheate. 2010. "Landsat-based Inventory of Glaciers in Western Canada, 1985–2005." *Remote Sensing of Environment* 114:127–37.

Boon, S., M. Sharp and P. Nienow. 2003. "Impact of an Extreme Melt Event on the Runoff and Hydrology of a High Arctic Glacier." *Hydrological Processes* 17:1051–72.

Borns Jr., H.W., and R.P. Goldthwait. 1966. "Late-Pleistocene Fuctuations of Kaskawulsh Glacier." *American Journal of Science* 269: 600–19.

Church, A., and J.J. Clague. 2009. "Recent Deglacierization of the Upper Wheaton River Watershed, Yukon." In *Yukon Exploration and Geology 2008*, edited by L.H. Weston, L.R. Blackburn and L.L. Lewis, 99–112. Whitehorse: Yukon Geological Survey.

Clague, J.J. 1989. "Quaternary Geology of the Canadian Cordillera." In *Quaternary Geology of Canada and Greenland*. Vol. 1, *Geology of Canada*, edited by J.O. Wheeler and A.R. Palmer. Geological Survey of Canada.

Clague, J.J., J. Kock and M. Geertsema. 2010. "Expansion of Outlet Glaciers of the Juneau Icefield in Northwest British Columbia during the Past Two Millennia." *The Holocene* 20 (3): 447–61.

Clarke, G.K.C., and E. W. Blake. 1991. "Geometric and Thermal Evolution of a SurgeType Glacier in its Quiescent State: Trapridge Glacier 1969–1989," *Journal of Glaciology* 37 (125): 158–69.

Cruikshank, J. 2001. "Glaciers and Climate Change: Perspectives from Oral Tradition." *Arctic* 54 (4): 377–393.

Cuffey, K.M., and W.S.B. Paterson. 2010. *The Physics of Glaciers*. Tarrytown, NY: Elsevier.

Dickson, James H., Michael P. Richards, Richard J. Hebda, Petra Mudie, Owen Beattie, Susan Ramsay, Nancy J. Turner, Bruce J. Leighton, John M. Webster, Niki R. Hobischak, Gail S. Anderson, Peter M. Troffe and Rebecca J. Wigen. 2004. "Kwäday Dän Ts'ìnchí, the First Ancient Body of a Man from a North American Glacier: Reconstructing his Last Days by Intestinal and Biomolecular Analyses." *Holocene* 14 (4): 481–86.

Dixon, E.J., M. Callanan, A. Hafner, P.G. Hore. 2014. "The Emergence of Glacial Archaeology." *Journal of Glacial Archaeology* 1 (1): 1–9.

Farnell, R., P.G. Hare, E. Blake, V. Bowyer, C. Schweger, S. Greer, R. Gotthardt. 2004. "Multidisciplinary Investigations of Alpine Ice Patches in Southwest Yukon, Canada: Paleoenvironmental and Paleobiological Investigations." *Arctic* 57 (3): 247–59.

Flint, R.F., and B.J. Skinner. 1974. *Physical Geology*. Hoboken, NJ: John Wily and Sons.

Fountain, A.G., and J.S. Walder. 1998. "Water Flow through Temperate Glaciers." *Reviews of Geophysics* 36 (3): 299–328.

Glen, J.W. 1958. "The Flow Law of Ice: A Discussion of the Assumptions Made in Glacier Theory, Their Experimental Foundations and Consequences." *International Association of Scientific Hydrology Bulletin* 47:171–83

Haeberli, W. 1994. Accelerated Glacier and Permafrost Changes in the Alps. In *Mountain Environments in Changing Climates*, edited by M. Beniston. New York: Routledge Publishing.

Hare, P.G., S. Greer, R. Gottardt, R. Farnell, V. Bowyer, C. Schweger and D. Strand. 2004. "Ethnographic and Archaeological Investigations of Alpine Ice Patches in Southwest Yukon, Canada." *Arctic* 57 (3): 260–72.

Holzhauser, H., and H.J. Zumbühl. 1999. "Glacier Fluctuations in the Western Swiss and French Alps in the 16th Century." *Climate Change* 43:223–37.

Klock, R., E. Hudon, D. Aihoshi and J. Mullock. 2001. *The Weather of the Yukon, Northwest Territories and Western Nunavut, Graphic Area Forecast 35*. Ottawa: Nav Canada.

Klock, R., and J. Mullock. 2001. *The Weather of British Columbia, Graphic Area Forecast 31*. Ottawa: Nav Canada.

Matthes, F.E. 1939. "Report of the Committee on Glaciers, April 1939." *Transactions of the American Geophysical Union* 20:518–23.

McClung, D.M. 2003. "Magnitude and Frequency of Avalanches in Relation to Terrain and Forest Cover." *Arctic, Antarctic and Alpine Research*, 35 (1): 82–90.

McClung, D.M., and P. Schaerer. 2006. *The Avalanche Handbook*. 3rd ed. Seattle: Mountaineers Books.

Mueller, D.R., and W.H. Pollard. 2004. "Gradient Analysis of Cryoconite Ecosystems from Two Polar Glaciers." *Polar Biology* 27 (2): 66–74.

Østrem, G., and M. Brugman. 1991. *Glacier Mass-Balance Measurements*. National Hydrology Research Institute (NHRI) Science Report 4. Norwegian Water Resources and Energy Administration and Environment Canada.

Reyes, A.V., B.H. Luckman, D.J. Smith, J.J. Clague and R.D. van Dorp. 2006. "Tree-Ring Dates for the Maximum Little Ice Age Advance of Kaskawulsh Glacier, St. Elias Mountains, Canada." *Arctic* 59 (1): 14–20.

Richards, M.P., S. Greer, L.T. Corr, O. Beattie, A. Mackie, R.P. Evershed, A. von Finster and J. Southon. 2007. "Radiocarbon Dating and Dietary Stable Isotope Analysis of Kwäday Dän Ts'ìnchí." *American Antiquity* 72 (4): 719–33.

Tompkins, H. 2006. "Historical Climatology of the Southern Yukon: Paleoclimatic Reconstruction Using Documentary Sources from 1842–1852." Master's thesis, Queen's University.

Torsnes, I., N. Rye and A. Nesje. 1993. "Modern and Little Ice Age Equilibrium-line Altitudes on Outlet Valley Glaciers from Jostedalsbreen, Western Norway: An Evaluation of Different Approaches to Their Calculation." *Arctic and Alpine Res.* 25 (2): 106–16.

Weertman, J. 1973. "Creep of Ice." In *Physics and Chemistry of Ice*, edited by E. Whalley, S.J. Jones and L.W. Gold. Ottawa: Royal Society of Canada.

Williams, L.D. 1978. "The Little Ice Age Glaciation Level on Baffin Island, Arctic Canada." *Palaeogeography, Palaeoclimatology, Palaeoecology* 25 (3): 199–207.

Chapter 4

Andrews, Thomas D., G. MacKay and L. Andrew. 2012. "Archaeological Investigations of Alpine Ice Patches in the Selwyn Mountains, Northwest Territories, Canada." In "The Archaeology and Paleology of Alpine Ice Patches: A Global Perspective," supplement 1, *Arctic* 65:1–21.

Farnell, Richard, P.G. Hare, E. Blake, V. Bowyer, C. Schweger, S. Greer and R. Gotthardt. 2004. "Multidisciplinary Investigations of Alpine Ice Patches in Southwest Yukon, Canada: Paleoenvironmental and Paleobiological Investigations." *Arctic* 57 (3): 247–58.

Hare, P. Gregory, S. Greer, R. Gotthardt, R. Farnell, V. Bowyer, C. Schweger and D. Strand. 2004. "Ethnographic and Archaeological Investigations of Alpine Ice Patches, Southwest Yukon, Canada." *Arctic* 57 (3): 260–72.

Hare, P. Gregory, with S. Greer (Champagne and Aishihik First Nations), H. Jones (Carcross-Tagish First Nation), R. Mombourquette (Kwanlin Dün First Nation), J. Fingland (Kluane First Nation), M. Nelson and J. Shorty (Ta'an Kwäch'än Council) and T. Evans (Teslin Tlingit Council). 2011. The Frozen Past – The Yukon Ice Patches. Whitehorse: Government of Yukon. Accessed October 31. www.tc.gov.yk.ca/publications/The_Frozen_Past_the_Yukon_Ice_Patches_2011.pdf.

Hare, P. Gregory, C.D. Thomas, T.N. Toper and R.M. Gotthardt. 2012. "The Archaeology of Yukon Ice Patches: New Artifacts, Observations and Insights," supplement 1, *Arctic* 65: 118–35.

Chapter 5

Antoniades, D., P.B. Hamilton, M.S.V. Douglas and J.P. Smol. 2008. *Diatoms of North America: The Freshwater Flora of Prince Patrick, Ellef Ringnes and Northern Ellesmere Islands from the Canadian Arctic Archipelago.* Iconographica Diatomologica 17. Ruggell, Lichtenstein: Gantner Verlag.

Asmund, B., and D.K. Hilliard. 1961. "Studies on Chrysophyceae from Some Ponds and Lakes in Alaska. I. *Mallomonas* species examined with electron microscope." *Hydrobiologia* 17:237–58.

Beattie, Owen, Brian Apland, Erik W. Blake, James A. Cosgrove, Sarah Gaunt, Sheila Greer, Alexander P. Mackie, Kjerstin E. Mackie, Dan Straathof, Valerie Thorp and Peter M. Troffe. 2000. "The Kwäday Dän Ts'ìnchi Discovery from a Glacier in British Columbia." *Canadian Journal of Archaeology* 24 (1–2): 129–47.

Bondy, J. 2006. "Tracking Past Sockeye Salmon Presence in Relation to Oral Tradition in a Remote Sub-Arctic Lake (Yukon Territory, Canada)". BSc (Hons) thesis, Queen's University.

Brodo, I.M., S.D. Sharnoff, S.D. and S. Sharnoff. 2001. *Lichens of North America.* New Haven, CT: Yale University Press.

Cody, W.J. 2000. *Flora of the Yukon Territory.* 2nd ed. Ottawa: NRC Research Press.

Croasdale, H.T. 1962. "Freshwater Algae of Alaska. III. Desmids from the Cape Thompson Area." *Transactions of the American Microscopical Society* 81:12–42.

———. 1965. "Desmids of Devon Island, N.W.T., Canada." *Transactions of the American Microscopical Society* 84:301–35.

Cronberg, G. 1986. "Chrysophycean Cysts and Scales in Lake Sediments: A Review." In *Chrysophytes: Aspects and Problems,* edited by J. Kristiansen and R.A. Andersen, 281–315. Cambridge: Cambridge University Press.

Danby, R.K., and D.S. Hik. 2007. "Evidence of Treeline Dynamics in Southwest Yukon from Aerial Photographs." *Arctic* 60:411–20.

Dickson, J.H. 1973. *Bryophytes of the Pleistocene.* Cambridge: Cambridge University Press.

———. 1986. "Bryophyte Analysis." In *Handbook of Holocene Palaeoecology and Palaeohydrology,* edited by B.E. Berglund, 627–43. Chichester: John Wiley.

Dickson, J.H., S. Bortenschlager, K. Oeggl, R. Porley, and A. McMullen. 1996. "Mosses and the Tyrolean Iceman's Southern Provenance." *Proceedings of the Royal Society of London B* 263:567–71.

Dickson, J.H. and P.J. Mudie. 2008. "The Life and Death of Kwäday Dän Ts'ìnchí, a 550-year-old Frozen Body from British Columbia: Clues from Remains of Plants and Animals." *Northern Review* 28:27–50.

Dickson, James H., Michael P. Richards, Richard J. Hebda, Petra Mudie, Owen Beattie, Susan Ramsay, Nancy J. Turner, Bruce J. Leighton, John M. Webster, Niki R. Hobischak, Gail S. Anderson, Peter M. Troffe and Rebecca J. Wigen. 2004. "Kwäday Dän Ts'ìnchí, the First Ancient Body of a Man from a North American Glacier: Reconstructing his Last Days by Intestinal and Biomolecular Analyses." *Holocene* 14 (4): 481–86.

Douglas, G.W., D. Meidinger and J.L. Penny. 2002. *Rare Plants of British Columbia.* Victoria: BC Government.

Douglas, G.W., D. Meidinger and J. Pojar, eds. 2000. *Illustrated Flora of British Columbia.* Vol. 5, *Dicotyledons and Pteridophytes.* Victoria: Ministry of Environment, Lands and Parks and Ministry of Forests.

———. 2002. *Illustrated Flora of British Columbia.* Vol. 8, *General Summary, Maps and Keys.* Victoria: Ministry of Environment, Lands and Parks and Ministry of Forests.

Douglas, G.W. and D.H. Vitt. 1976. "Moss-lichen Flora of the St. Elias-Kluane Ranges, Southwestern Yukon." *The Bryologist* 79:437–56.

Duff, K.E., and J. Smol, 1988a. "Chrysophycean Stomatocysts from the Postglacial Sediments of Tasikutaaq Lake, Baffin Island, N.W.T." *Canadian Journal of Botany* 67:1649–56.

———. 1988b. "Chrysophycean Stomatocysts from the Postglacial Sediments of a High Arctic Lake." *Canadian Journal of Botany* 66:1117–28.

———. 1992. "Chrysophyte Cysts in 36 Canadian High Arctic Ponds." *Nordic Journal of Botany* 12:471–99.

———. 1994. "Chrysophycean Cyst Flora from British Columbia (Canada) Lakes." *Nova Hedwigia* 58:353–89.

Duff, K.E., B.A. Zeeb and J.P. Smol. 1995. *Atlas of Chrysophycean Cysts.* Vol. 1. Dordrecht, Netherlands: Kluwer Academic Publishers.

Edwards, S.R. 2003. *English Names for British Bryophytes.* 3rd ed. Loughton, UK: British Bryological Society.

English, J., and M. Potapova. 2009. "*Aulacoseira pardata* sp. Nov., *A. Nivalis* comb. Nov., *A. nivaloides* comb. et stat. nov. and Their Occurrences in Western North America." *Proceedings of the Academy of Natural Sciences Philadelphia* 158:37–48.

Geiser, L.H., K.L. Dillman, C.C. Derr and M.C. Stensvold. 1994. *Lichens of Southeastern Alaska.* Petersburg, Alaska: USDA Forest Service.

Goward, T. 1994. *The Lichens of British Columbia.* Part 1, *Foliose and Squamulose Species.* Victoria: Ministry of Forests.

———. 1999. *The Lichens of British Columbia.* Part 2, *Fruticose Species.* Victoria: Ministry of Forests.

Gritten, H.M. 1977. "The Fine Structure of Some Chrysophycean Cysts." *Hydrobiologia* 53:239–52.

Hastings, R.I., and H.C. Greven. 2007. "Grimmia." In *Flora of North America North of Mexico,* vol. 27, *Bryophytes: Mosses Part II.* Oxford: Oxford University Press.

Hilliard, D.K. 1965. "Studies on Chrysophyceae from Some Ponds and Lakes in Alaska. V. Notes on the Taxonomy and Occurrence of Phytoplankton in an Alaskan Pond." *Hydrobiologia* 24:553–76.

Jones, H.G., and J.W. Pomeroy. 2000. "The Ecology of Snow and Snow-Covered Systems: Summary and Relevance to Wolf Creek, Yukon." In *Wolf Creek Research Basin: Hydrology, Ecology, Environment,* edited by Penny Trischuk, J.W. Pomeroy and R.J. Granger. Saskatoon: National Water Research Institute.

Klinkenberg, Brian, ed. 2013. *E-Flora BC: Electronic Atlas of Brtish Columbia.* Accessed November 29, 2016. http://ibis.geog.ubc.ca/biodiversity/eflora/.

Lauriol, B., C. Prévost and D. Lacelle. 2006. "The Distribution of Diatom Flora in Ice Caves of the Northern Yukon Territory, Canada: Relationship to Air Circulation and Freezing." *International Journal of Speleology* 35 (2): 83–92.

Lotter, A.F., Pienitz, R. and Schmidt, R. 1999. "Diatoms as Indicators of Environmental Change Near Arctic and Alpine Treeline. In *The Diatoms: Applications for the Environmental and Earth Sciences,* edited by E. Stoermer and J.P. Smol, 205–26. Cambridge: Cambridge University Press.

Meidinger, D., and J. Pojar. 1991. *Ecosystems of British Columbia.* Victoria: Ministry of Forests.

Miller, N.G. 1984. "Tertiary and Quaternary Fossils." In *New Manual of Bryology,* edited by R.M. Schuster, 1194–232. Nichinan, Japan: Hattori Botanical Laboratory.

Mudie, P.J., S. Greer, J. Brakel, J.H. Dickson, C. Schinkel, R. Peterson-Welsh, M. Stevens, N.J. Turner, M. Shadow and R. Washington. 2005. "Forensic Palynology and Ethnobotany of *Salicornia* Species (Chenopodiaceae) in Northwest Canada and Alaska." *Canadian Journal of Botany* 83:111–23.

Patrick, R., and C.W. Reimer. 1966. *Diatoms of the United States.* Vol. 1. Monograph 13. Philadelphia, PA: Academy of Natural Science.

Pick, F.R., C. Nalewajko and D.R.S. Lean. 1984. "The Origin of a Metalimnetic Peak." *Limnology and Oceanography* 29 (1): 125–34.

Pojar, J., and A.C. Stewart. 1991a. "Alpine Tundra Zone." In *Ecosystems of British Columbia* edited by D. Meidinger and J. Pojar, 263–74. Victoria: Ministry of Forests.

———. 1991b. "Spruce–Willow–Birch Zone." In *Ecosystems of British Columbia* edited by D. Meidinger and J. Pojar, 251–62. Victoria: Ministry of Forests.

Prescott, G.W. 1953. "Preliminary Notes on the Ecology of Freshwater Algae in the Arctic Slope, Alaska, with Descriptions of Some New Species." *The American Midland Naturalist* 50:463–73.

Pringle, H. 2002. "Out of the Ice: Who Was the Ancient Traveller Discovered in an Alpine Glacier?" *Canadian Geographic,* July/August 2002, 56–64.

Richards, M.P., S. Greer, L.T. Corr, O. Beattie, A. Mackie, R.P. Evershed, A. von Finster and J. Southon. 2007. "Radiocarbon Dating and Dietary Stable Isotope Analysis of Kwäday Dän Ts'ìnchí." *American Antiquity* 72 (4):719–33.

Round, F.E., R.M. Crawford and D.G. Mann. 1990. *The Diatoms: Biology and Morphology of the Genera.* Cambridge: Cambridge University Press.

Sandgren, C.D. 1983. "Survival Strategies of Chrysophycean Cysts: Reproduction and the Formation of Resistant Resting Cysts." In *Survival Strategies of the Algae,* edited by G. Fryxell, 23–48. Cambridge: Cambridge University Press.

Schofield, W.B. 1976. "Bryophytes of British Columbia III. Habitat and Distributional Information for Selected Mosses." *Syesis* 9:318–54.

———. 1992. *Some Common Mosses of British Columbia.* Victoria: Royal British Columbia Museum.

Selbie, D.T., B.P. Finney, D. Barto, L. Bunting, G. Chen, P.R. Leavitt, E.A. MacIsaac, D.E. Schindler, M.D. Shapley and I. Gregory-Eaves. 2009. "Ecological, Landscape and

Climatic Regulation of Sediment Geochemistry in North American Sockeye Salmon Nursery Lakes: Insights for Paleoecological Salmon Investigations. *Limnology and Oceanography* 54 (5): 1733–45.

Sheath, R.G., M.O. Morison, J.E. Korch, D. Kaczmarczyk, and K.M. Cole. 1986. "Distribution of Stream Macroalgae in South-Central Alaska." *Hydrobiologia* 135: 259–69.

Takeuchi, N. 2001. "The Altitudinal Distribution of Snow Algae on an Alaskan Glacier (Gulkana Glacier in the Alaska Range)." *Hydrological Processes* 15: 3447-59.

Takeuchi, N., S. Koshima and T. Segawa. 2003. "Effect of Cryoconite and Snow Algal Communities on Surface Albedo on Maritime Glaciers in South Alaska." *Bulletin of Glaciological Research* 20: 21–7.

Thomson, J.W. 1997. *American Arctic Lichens*, vol. 2, *The Microlichens*. Madison, WI: Wisconsin University Press.

Vitt, D.H., J.E. Marsh and R.B. Bovey. 1988. *Mosses, Lichens and Ferns of Northwest North America.* Seattle: University of Washington Press.

Weber, W.A. 2003. "The Middle Asian Element in the Southern Rocky Mountain Flora of Western United States: A Critical Biogeographical Review." *Journal of Biogeography* 30: 649–85.

Weber, W.A., and R.C. Wittman. 2007. *Bryophytes of Colorado Mosses, Liverworts and Hornworts.* Santa Fe: Pilgrims Progress.

Wehr, J.D., and R.G. Sheath. 2003. *Freshwater Algae of North America*. Amsterdam, Netherlands: Academic Press.

Wilkinson, A.N., B.A. Zeeb and J.P. Smol. 2001. *Atlas of Chrysophycean Cysts*, vol. 2. Dordrecht, Netherlands: Kluwer Academic Publishers.

Wilson, S.E., and K. Gajewski. 2002. "Surface Sediment Diatom Assemblages and Water Chemistry from 42 Subalpine Lakes in the Southwestern Yukon and Northern British Columbia." *Ecoscience* 9 (2): 256–70.

Zeeb, B.A., and J.P. Smol. 1993. "Postglacial Chrysophycean Cyst Record from Elk Lake, Minnesota." *Geological Society America* Special Paper 276: 239–49.

———. 2001. "Chrysophyte Scales and Cysts." In *Tracking Environmental Change Using Lake Sediments*. Vol. 3, *Terrestrial, Algal and Siliceous Indicators*, edited by J.P. Smol, H.J.B. Birks and W.M. Last. Dordrecht, Netherlands: Kluwer Academic Publishers.

Chapter 6

Arneborg, Jette, Jan Heinemeier, Niels Lynnerup, Henrik I. Nielsen, Niels Rud and Arny E. Sveinbjörnsdóttir. 1999. "Change of Diet of the Greenland Vikings Determined from Stable Carbon Isotope Analysis and [14]C Dating of Their Bones." *Radiocarbon* 41:157–68.

Barrett, James H., Roelf P. Beukens and Don R. Brothwell. 2000. "Radiocarbon Dating and Marine Reservoir Correction of Viking Age Christian Burials from Orkney." *Antiquity* 74: 537–42.

Beattie, Owen, Brian Apland, Erik W. Blake, James A. Cosgrove, Sarah Gaunt, Sheila Greer, Alexander P. Mackie, Kjerstin E. Mackie, Dan Straathof, Valerie Thorp and Peter M. Troffe. 2000. "The Kwäday Dän Ts'inchi Discovery from a Glacier in British Columbia." *Canadian Journal of Archaeology* 24 (1–2): 129–47.

Brown, T.A., D. Erle Nelson, John S. Vogel and John R. Southon. 1988. "Improved Collagen Extraction by Modified Longin Method." *Radiocarbon* 30:171–77.

DeNiro, Michael J. 1985. "Post-mortem Preservation and Alteration of In Vivo Bone Collagen Isotope Ratios in Relation to Paleodietary Reconstruction." *Nature* 317:806–9.

Greer, Sheila C. 2005. *IkVf-1 – The Kwaday Dän Ts'inchí Site Catalogue (Whitehorse Collections; Revised Spring 2005)*. On file, Champagne and Aishihik First Nations, Haines Junction, Yukon.

Reimer, P., M. Baillie, E. Bard, A. Bayliss, J. Beck, P. Blackwell, C. Bronk Ramsey, C. Buck, G. Burr, R. Edwards, M. Friedrich, P. Grootes, T. Guilderson, I. Hajdas, T. Heaton, A. Hogg, K. Hughen, K. Kaiser, B. Kromer, F. McCormac, S. Manning, R. Reimer, D. Richards, J. Southon, S. Talamo, C. Turney, J. van der Plicht and C. Weyhenmeyer. 2009. "INTCAL09 and MARINE09 Radiocarbon Age Calibration Curves, 0–50,000 Years cal BP." *Radiocarbon* 51:1111–50.

Richards, Michael P., S. Greer, L.T. Corr, O. Beattie, A. Mackie, R.P. Evershed, A. von Finster and J. Southon. 2007. "Radiocarbon Dating and Dietary Stable Isotope Analysis of Kwaday Dän Ts'inchí." *American Antiquity* 72 (4):719–33.

Richards, Michael P., and Robert E.M. Hedges. 1999. "Stable Isotope Evidence for Similarities in the Types of Marine Foods Used by Late Mesolithic Humans at Sites along the Atlantic Coast of Europe." *Journal of Archaeological Science* 26:717–22.

Richards, Michael P., and J. Alison Sheridan. 2000. "New AMS Dates on Human Bone from Mesolithic Oronsay." *Antiquity* 74:313–15.

Robinson, Stephen W., and Gail Thompson. 1981. "Radiocarbon Corrections for Marine Shell Dates with Application to Southern Pacific Northwest Coast Prehistory." *Syesis* 14:45–57.

Southon, John R., and Daryl Fedje. 2003. "A Post-glacial Record of ^{14}C Reservoir Ages from the British Columbia Coast." *Canadian Journal of Archaeology* 27:95–111.

Southon, John R., D. Erle Nelson and John S. Vogel. 1990. "A Record of Past Ocean-Atmospheric Radiocarbon Differences from the Northeast Pacific." *Paleooceanography* 5:197–206.

Stuiver, Minze, and Thomas Braziunas. 1993. "Modelling Atmospheric ^{14}C Influences and ^{14}C Ages of Marine Samples to 10,000 BC." *Radiocarbon* 35 (1): 137–90.

Stuiver, Minze, and Paula J. Reimer. 1993. "Extended ^{14}C Data Base and Revised CALIB 3.0 ^{14}C Age Calibration Program." *Radiocarbon* 35 (1):215–30.

Chapter 8

Beattie, Owen, Brian Apland, Erik W. Blake, James A. Cosgrove, Sarah Gaunt, Sheila Greer, Alexander P. Mackie, Kjerstin E. Mackie, Dan Straathof, Valerie Thorp and Peter M. Troffe. 2000. "The Kwäday Dän Ts'ìnchi Discovery from a Glacier in British Columbia." *Canadian Journal of Archaeology* 24 (1–2): 129–47.

CAFN (Champagne and Aishihik First Nations). 1999. Transcript of Champagne and Aishihik First Nations Heritage visit to Haines, Alaska, Chilkoot Indian Association, October 6. On file, Champagne and Aishihik First Nations Heritage program, Haines Junction.

———. 2000a. Notes from meeting with Chilkoot Indian Association, October 26. On file, Champagne and Aishihik First Nations Heritage program, Haines Junction.

———. 2000b. Transcript of meeting with Chilkat Indian Tribe, Klukwan, October 27. On file, Champagne and Aishihik First Nations Heritage program, Haines Junction.

———. 2000c. Transcript of meeting of Champagne and Aishihik First Nations representatives and Yakutat Tlingit Tribe, Yakutat, regarding the Kwäday Dän Ts'ìnchi discovery, November 6. On file, Champagne and Aishihik First Nations Heritage program, Haines Junction.

———. 2000d. Summary and transcript of meeting with Carcross-Tagish First Nation regarding the Kwäday Dän Ts'ìnchi discovery, December 11. On file, Champagne and Aishihik First Nations Heritage program, Haines Junction.

———. 2001a. Minutes from Kwäday Dän Ts'ìnchi Regional Meeting, Haines Junction, May 4 and 5. On file, Champagne and Aishihik First Nations Heritage program, Haines Junction.

———. 2001b. Notes from teleconference regarding 40-Day Party, September 19. On file, Champagne and Aishihik First Nations Heritage program, Haines Junction.

———. 2007. January draft of "Funeral and Memorial Potlatch: A Tradition Today." On file, Champagne and Aishihik First Nations Heritage program, Haines Junction.

Carcross-Tagish First Nation and Sheila Greer. 1995. *Skookum Stories on the Chilkoot/Dyea Trail*. Carcross, YT: Carcross-Tagish First Nation.

Central Council of the Tlingit and Haida Indian Tribes of Alaska. 2008. "Traditional Tlingit Ceremonies for the Dead." Accessed November 18. www.ccthita.org/memorialparties.htm.

Charlie, Bob. 1999. Press release on behalf of Champagne and Aishihik First Nations, August 24.

de Laguna, Frederica. 1990. "Tlingit." In *Handbook of North American Indians*, vol. 7, *Northwest Coast*, edited by Wayne Suttles, 203–28. Washington, DC: Smithsonian Institution.

Easterson, Mary. 1992. *Potlatch: The Southern Tutchone Way*. Burwash Landing, YT: Kluane First Nation.

Emmons, George Thornton. 1991. *The Tlingit Indians*. Edited by Frederica de Laguna. New York: American Museum of Natural History; Seattle: University of Washington Press.

Gaunt, Sarah (compiler). 2000. Notes from discussions with 19 individuals from Champagne and Aishihik and 18 individuals from neighbouring nations and tribes regarding disposition options for Kwäday Dän Ts'ìnchi. On file, Champagne and Aishihik First Nations Heritage program, Haines Junction.

Greer, Sheila. 2000. "Treatment of the Dead in the Tatshenshini-Alsek Area 500 Years Ago." Memo prepared for Sarah Gaunt, Diane Strand and Lawrence Joe. October. On file, Champagne and Aishihik First Nations Heritage program, Haines Junction.

———. 2006. Notes with Marge Jackson. On file, Champagne and Aishihik First Nations Heritage program, Haines Junction.

Hughes, Grant, Sheila Greer and Alexander Mackie. 2009. "Collaboration in an Exceptional Context: Kwäday Dän Ts'inchi—Long Ago Person Found." *Muse* 27 (2): 46–49.

Kan, Sergei. 1989. *Symbolic Immortality: The Tlingit Potlatch of the Nineteenth Century*. Washington, DC: Smithsonian Institution.

———. 2000. "The Forty-Day Party: A Tlingit Ritual with Russian Orthodox Roots." In *Celebration 2000, Restoring Balance Through Culture*, edited by Susan W. Fair and Rosita Worl. Juneau: Sealaska Heritage Foundation.

McClellan, Catharine. 1975. *My Old People Say. An Ethnographic Survey of Southern Yukon Territory*. 2 vols. Publications in Ethnology 6. Ottawa: National Museums of Canada.

McClellan, Catharine, L. Birckel, R. Blinghurst, J.A. Fall, C. McCarthy and J.R. Sheppard. 1987. *Part of the Land, Part of the Water: A History of Yukon Indians*. Vancouver: Douglas and MacIntyre.

Tlen, Daniel L. 1993. *Kluane Southern Tutchone Glossary (English to Southern Tutchone)*. Occasional Papers of the Northern Research Institute, monograph 1. Whitehorse: Northern Research Institute, Yukon College.

White, Lily (Chookaneidí), and Paul White (Kaagwaantaan). 2000. "Koo.éex': The Tlingit Memorial Party." In *Celebration 2000, Restoring Balance Through Culture*, edited by Susan W. Fair and Rosita Worl. Juneau: Sealaska Heritage Foundation.

Chapter 9

ABT (Alaska Boundary Tribunal). 1904. Proceedings of the Alaska Boundary Tribunal (1902-03). S. Doc. 162, 58th Congress, 2nd session, vol. 1, part 2: Appendix to the Case of the United States, 1–1550. In *Senate Documents*, vol 15. Washington, DC: Government Printing Office.

Adamson, John. 1993. Interview by Sarah Gaunt on Tatshenshini/Chilkat area, January 15. Transcript on file, Champagne and Aishihik First Nations Heritage program, Haines Junction.

———. 1997. Interview by Barb Joe and Sheila Greer. August 20. Transcript on file, Champagne and Aishihik First Nations Heritage program, Haines Junction.

Bailey, Jeff. 1998. "X-Ray Fluorescence Analysis of Twelve Obsidian Samples from Eight Archaeological Sites in the Southwestern Yukon Territory." Unpuplished report prepared for Sheila Greer. On file, Champagne and Aishihik First Nations Heritage program, Haines Junction.

Banks, Paula. 1996. Notes from interview with John Brown, September 20 (see Brown 1996), including information not recorded on tape. On file, Champagne and Aishihik First Nations Heritage program, Haines Junction.

Beardslee, Lester A. 1880. "Chilkat and Chilkoot." *Forest and Stream,* November 25.

Betts, Martha F. 1994. "The Subsistence Hooligan Fishery of the Chilkat and Chilkoot Rivers." Technical Paper 213. Juneau: Division of Subsistence, Alaska Department of Fish and Game. www.subsistence.adfg.state.ak.us/TechPap/tp213.pdf.

Blackstock, Michael. 2001. *Faces in the Forest: First Nations Art Created on Living Trees*. Montreal and Kingston: McGill-Queen's University Press.

Boyd, Robert. 1999. *The Coming of the Spirit of Pestilence: Introduced Infectious Diseases and Population Decline Among Northwest Coast Indians, 1774–1874*. Vancouver: University of British Columbia Press.

Brown, John. 1996. Interview by Paula Banks. September 20. Partial transcript on file, Champagne and Aishihik First Nations Heritage program, Haines Junction.

CAFN (Champagne and Aishihik First Nations). n.d. Greer notes on sites in BC territory. On file, Champagne and Aishihik First Nations Heritage program, Haines Junction.

———. 2001. Kwädạy Dän Ts'ìnchị Regional Meeting, Haines Junction, May 4 and 5, 2001. Transcript on file, Champagne and Aishihik First Nations Heritage program, Haines Junction.

———. 2010. *Dän Kéyi Kwändür, Stories from Our Country*. Whitehorse: Champagne and Aishihik First Nations.

CAFN (Champagne and Aishihik First Nations), Sarah Gaunt and Sheila Greer. 1993. "Shäwshe–Neskatahin Background History Study." Manuscript prepared for National Historic Parks and Sites Directorate, Canadian Parks Service.

———. 1995. "Shawshe Chu-Alsexh, Tatshenshini River Ethnohistory." Manuscript on file, Champagne and Aishihik First Nations Heritage program, Haines Junction.

CAFN (Champagne and Aishihik First Nations) and Sheila Greer. 1998. "Shawshe-Dalton Post History Research." Manuscript prepared for Champagne and Aishihik First Nations and Yukon Heritage Branch. On file, Champagne and Aishihik First Nations Heritage program, Haines Junction.

Clarke, C.H.D. [1940s]. Undated correspondence to Douglas Leechman, National Museum, Ottawa.

Cooley, Bonar. [1970s]. Genealogical chart on Tlingit and Haida ancestry. R-273 (94/104). Bonar Cooley Fonds. Yukon Archives.

CPCGN (Canadian Permanent Committee on Geographical Names), n.d. Geographical Names Files. Ottawa: Geographical Names Board of Canada Secretariat, Natural Resources Canada.

Cruikshank, Julia M. 1974. *Through the Eyes of Strangers: A Preliminary Survey of Land Use History in the Yukon During the Late Nineteenth Century.* Report to the Yukon Territorial Government and the Yukon Archives.

———. 1979. *Athapaskan Women: Lives and Legends.* Ottawa: National Museum of Man.

———. 1980. *Legend and Landscape: Convergence of Oral and Scientific Traditions with Special Reference to the Yukon Territory, Canada.* Diploma in Polar Studies thesis, Scott Polar Research Institute, Cambridge.

———. 1985. "The Gravel Magnet: Some Social Impacts of the Alaska Highway on Yukon Indians." In *The Alaska Highway: Papers of the 40th Anniversary Symposium*, edited by Kenneth Coates, 172–87. Vancouver: University of British Columbia Press.

———. 1990. "Getting the Words Right: Perspectives on Naming and Places in Athapaskan Oral History." *Arctic Anthropology* 271:52–65.

———. 1991. *Reading Voices.* Vancouver: Douglas and McIntyre.

———. 2005. *Do Glaciers Listen? Local Knowledge, Colonial Encounters and Social Imagination.* Vancouver: University of British Columbia Press.

Cruikshank, Julie, Angela Sidney, Kitty Smith and Annie Ned. 1990. *Life Lived Like a Story.* Vancouver: University of British Columbia Press.

CTFN (Carcross-Tagish First Nation) and Sheila Greer. 1995. *Skookum Stories on the Chilkoot/Dyea Trail.* Carcross, YT: Carcross-Tagish First Nation.

Dauenhauer, Nora Marks, and Richard Dauenhauer, eds. 1987. *Haa Shuká, Our Ancestors: Tlingit Oral Narratives.* Seattle: University of Washington Press.

———. 1990. *Haa Tuwunáagu Yís, for Healing Our Spirit: Tlingit Oratory.* Seattle: University of Washington Press.

———. 1994. *Haa Kusteeyi, Our Culture: Tlingit Life Stories.* Seattle: University of Washington Press.

Davidson, George. 1867–68. Cruise of the U.S. Revenue Steamer Lincoln in Russian America. Appendix K: Report, in House Executive Document 177, 40th Congress, 2nd Session, edited by Captain W.A. Howard.

———. 1901. "Explanation of an Indian Map of the Rivers, Lakes, Trails and Mountains from the Chilkaht to the Yukon Drawn by the Chilkaht Chief Kohklux in 1869." *Mazama* 2 (2): 75–92.

———. 1903. *The Alaska Boundary.* San Francisco: Alaska Packers Association.

Dawson, George. 1887. "Report of an Exploration in the Yukon District, N.W.T. and Adjacent Northern Portion of British Columbia." *Geological and Natural History Survey of Canada, Annual Report* (New Series) vol. 3, part 1, 1887–88.

de Laguna, Frederica. 1953. "Some Problems in the Relationship Between Tlingit Archaeology and Ethnology." In *Asia and North America: Transpacific Contacts,* edited by Marion W. Smith, 53–57. Society for American Archaeology Memoir 9.

———. 1958. "Geological Confirmation of Native Traditions, Yakutat, Alaska." *American Antiquity* 23:434.

———. 1972. *Under Mount Saint Elias: The History and Culture of the Yakutat Tlingit.* 3 parts. Smithsonian Contributions to Anthropology 7. Washington, DC: Smithsonian Institution.

———. 1975. "Matrilineal Kin Groups." In *Proceedings: Northern Athapaskan Conference 1971,* 2 vols., edited by A. MacFadyen Clark, 17–145. National Museum of Man Mercury Series, Canadian Ethnology Service Paper No. 27. Ottawa: National Museums of Canada.

———. 1989. "Lieutenant Emmons and the Alaska-Canada Boundary Controversy 1902–03." *Borderlands: A Conference on the Alaska-Yukon Border*, proceedings. Whitehorse, Yukon Historical and Museums Association. www.yukonalaska.com/yhma/public.

———. 1990. "Tlingit." In *Handbook of North American Indians,* vol. 7, *Northwest Coast,* edited by Wayne Suttles, 203–28. Washington, DC: Smithsonian Institution.

de Laguna, Frederica, F.A. Riddell, D.F. McGeein, K.S. Lane, J.A. Freed and C. Osborne, 1964. *Archaeology of the Yakutat Bay Area*. Bureau of American Ethnology Bulletin 192. Washington, DC: Government Printing Office.

Emmons, George T. [1905?]. "Trip up Chilkat River over Divide." Unpublished field notebook. Reel A01769, Box 60 File 13, PM#10. Newcombe Family Papers, Ms-1077. BC Archives. Victoria.

———. 1916. *The Whale House of the Chilkat*. American Museum of Natural History Anthropology Papers 19, part 1. New York: American Museum of Natural History.

———. 1991. *The Tlingit Indians*. Edited by Frederica de Laguna. New York: American Museum of Natural History.

Fair, Susan W., and Rosita Worl, eds. 2000. *Celebration 2000. Restoring Balance Through Culture*. Juneau: Sealaska Heritage Foundation.

French, Diana. 1993. "Archaeological Survey of the Tatshenshini River, Northwestern British Columbia." Manuscript on file, BC Heritage Branch, Victoria.

Gates, Michael. 2010. "Old Maps Help Chart Yukon History (Dalton Trail)." *Yukon News*, October 8.

Gates, Michael, and F. Roback. 1972. "An Archaeological Survey and Ethnohistoric Study of the Tatshenshini River Basin, Southwest Yukon." Manuscript on file, Champagne and Aishihik First Nations Heritage program, Haines Junction.

Gaunt, Sarah. 1990. Notes on Tatshenshini-Alsek country, including information from Elijah Smith. On file, Champagne and Aishihik First Nations Heritage program, Haines Junction.

———. 1991. Notes on Tatshenshini-Alsek country, including information from Elijah Smith. On file, Champagne and Aishihik First Nations Heritage program, Haines Junction.

———. 1994. Notes on Tatshenshini-Alsek country from Paul Birckel. On file, Champagne and Aishihik First Nations Heritage program, Haines Junction.

———. 1996. Notes on Tatshenhini-Alsek country from Paul Birckel. On file, Champagne and Aishihik First Nations Heritage program, Haines Junction.

Gibson, James R. 1992. *Otter Skins, Boston Ships, and China Good: The Maritime Fur Trade of the Northwest Coast, 1785–1841*. Seattle: University of Washington Press.

Glave, Edward J. 1890–91. "Our Alaska Expedition: Exploration of the Unknown Alseck River Region." *Frank Leslie's Illustrated Newspaper*, September 6, 1890; November 15, 1890; November 22, 1890; November 29, 1890; December 6, 1890; December 13, 1890; December 20, 1890; December 27, 1890; January 3, 1891; January 10, 1891.

———. 1892a. "The Tempermental Explorer in Alaska." Parts 1 and 2. *The West*.

———. 1892b. "Pioneer Packhorses in Alaska. I. The Advance. II. The Return." *Century Illustrated Monthly Magazine* 44–45:671–82.

Goldschmidt, Walter R., and T.H. Haas, 1998. *Haa aaní, Our Land: Tlingit and Haida Land Rights and Use*. Edited by T.F. Thornton. Juneau: Sealaska Heritage Foundation.

Government of BC. n.d. Archaeological Site Files, BC Heritage Branch, Victoria.

Green, Louis. 1982. *"The Boundary Hunters: Surveying of the 141st Meridian and the Alaskan Panhandle."* Vancouver: University of British Columbia Press.

Greer, Sheila. 1995. "Sheila Greer Notes BC Claim Research Oct. 1995." On file, Champagne and Aishihik First Nations Heritage program, Haines Junction.

———. 2004. "Sheila Greer Notes with Marge Jackson, Nov. 20, 2004." On file, Champagne and Aishihik First Nations Heritage program, Haines Junction.

———. 2005a. "Sheila Greer Notes with Marge Jackson, Moose Jackson, Harry Smith and Frances Oles, November 2005." On file, Champagne and Aishihik First Nations Heritage program, Haines Junction.

———. 2005b. "Sheila Greer Notes with John Adamson, July 23, 24, 2005, Klukshu General Assembly." On file, Champagne and Aishihik First Nations Heritage program, Haines Junction.

———. 2008. "Sheila Greer Interview Notes with Harryet Rappier in Juneau, June 2008." On file, Champagne and Aishihik First Nations Heritage program, Haines Junction.

Grinev, Andrei Val'terovich. 2005. *The Tlingit Indians in Russian America, 1741–1867*. Translated by R.L. Bland and K.G. Solovjova. Lincoln, NE: University of Nebraska Press.

Hinckley, Ted C. 1996. *The Canoe Rocks: Alaska's Tlingit and the Euroamerican Frontier, 1800–1912*. Lanham, MD: University Press of America,

Hope, Andrew, III. 2000. "On Migrations." In *Will the Time Ever Come: A Tlingit Source Book,* edited by Andrew Hope III and T.F. Thornton, 23–33. Fairbanks: Center for Cross-Cultural Studies, University of Alaska.

Hume, David. 1979. Neskatahin Village Preservation Project interview by B. O'Leary and Trudy Long, Sept. 23. Transcript on file, Champagne and Aishihik First Nations Heritage program, Haines Junction.

Isaac, Sandra. n.d. Search of federal and provincial government records for information on Aboriginal occupation of the Tatshenshini basin and the BC portion of Champagne and Aishihik First Nations traditional territory. Research materials on file, Champagne and Aishihik First Nations Heritage program, Haines Junction.

Jackson, Marge. 1991. Transcript of place name discussion with Marge Jackson, July 13, 1991, at her home in Klukshu with Julie Cruikshank and Sarah Gaunt. Manuscript on file, Champagne and Aishihik First Nations Heritage program, Haines Junction.

———. 1993. Interview by Sarah Gaunt, Sheila Greer and Ray Jackson. January 13. Transcript on file, Champagne and Aishihik First Nations Heritage program, Haines Junction.

Jackson, Marge, with assistance from Beth O'Leary. 2006. *My Country Is Alive, A Southern Tutchone Way of Life.* Published by author, Haines Junction.

Jackson, Peter. 1979. Neskatahin Village Preservation Project interview by B. O'Leary. Transcript on file, Champagne and Aishihik First Nations Heritage program, Haines Junction.

Jarvis, A.M. 1899. "Annual Report of Inspector A.M. Jarvis." *North-West Mounted Police Report 1898*, part 3, *Yukon Territory*, appendix H, 95–110.

Jim, Paddy. 1993. Interview by Sarah Gaunt and Sheila Greer. On file, Champagne and Aishihik First Nations Heritage program, Haines Junction.

———. 1995. Interview by Sarah Gaunt, Sheila Greer and Sean Sheardown. On file, Champagne and Aishihik First Nations Heritage program, Haines Junction.

Johnson, Linda. 1984. "The Day the Sun Was Sick." *Yukon Indian News,* Summer 1984: 11–13.

Kan, Sergei. 1987. "Orthodox Christianity and the Tlingit." *Arctic Anthropology* 24 (1): 32–55.

Kane, Jimmy, 1971. Interview by Ron Chambers, September 5. Transcript by Sarah Gaunt on file, Champagne and Aishihik First Nations Heritage program, Haines Junction.

———. 1978. Interview by Jack Schick and Brent Liddle (Kluane National Park wardens). Recording on file, Yukon Archives. Transcript by Sarah Gaunt on file, Champagne and Aishihik First Nations Heritage program, Haines Junction.

———. 1979. Neskatahin Village Preservation Project interview by Trudy Long and B. O'Leary. Transcript on file, Champagne and Aishihik First Nations Heritage program, Haines Junction.

Kane, Parten. 1990. Interview by Sarah Gaunt and Lawrence Joe. Manuscript on file, Champagne and Aishihik First Nations Heritage program, Haines Junction.

Krause, Arthur. 1882. "Die Expedition der Bremer geographischen Gesellschaft nach de Tschuktschen-Halbinsel und Alaska, 1881–1882, IV." *Deutsche Geographische Blatter* 5 (4): 308–25.

Krause, Aurel, and Arthur Krause. 1981. *Journey to the Tlingits by Aurel and Arthur Krause, 1881/82.* Translated by Margot Krause McCaffrey. Haines, Alaska: Centennial Commission.

———. 1993. *To the Chukchi Peninsula and to the Tlingit Indians 1881/1882: Journals and Letters by Aurel and Arthur Krause.* Translated by Margot Krause McCaffrey. The Rasmuson Library Historical Translation Series 8. Fairbanks, AK: University of Alaska Press.

———. 1993. *To the Chukchi Peninsula and to the Tlingit Indians 1881/1882: Journals and Letters by Aurel and Arthur Krause.* Translated by Margot Krause McCaffrey. The Rasmuson Library Historical Translation Series Volume VIII. Fairbanks, AK: University of Alaska Press.

Krauss, Michael. 1980. *Alaska Native Languages: Past, Present and Future.* Alaska Native Language Center Research Papers 4. Fairbanks: Alaska Native Language Center.

Larson, Brendan, and Lauri Larson. 1977. "The Proud Chilkat." Manuscript prepared for the Education Committee, Alaska Native Brotherhood, Alaska Native Sisterhood, Camp No. 5. Printed by Chilkat Press, Haines, AK. On file, Sheldon Museum, Haines.

Leechman, Douglas. 1946. "Prehistoric Migration Routes Through the Yukon." *Canadian Historical Review* 27:383–90.

———. 1949. *Indian Summer.* Toronto: Ryerson Press.

———. 1950. "Yukon Territory." *Canadian Geographical Journal* 40 (6): 240–67.

———. 1962a. "Copper and Cats." *Daily Colonist*, March 1962, 12.

———. 1962b. "Well-Guarded Short-Cut to the Klondike Goldfields, The Dalton Trail." *Daily Colonist*, April 1962, 1–3.

———. n.d. "The Yukon Territory," unpublished manuscript. Typescript on file, Douglas Leechman Records, Add. MSS. 1290. BC Archives. Victoria.

Legros, Dominique. 1984. "Commerce entre Tlingits et Athapaskans Tutchones au XIX Siècle." *Recherches Amérindiennes au Québec* 14 (2):11–24.

Malloy, Mary. 1998. *"Boston Men" on the Northwest Coast: The American Fur Trade, 1788–1844.* Kingston, ON: Limestone Press.

Mandy, Joseph T. 1933. Tatshenshini River Section. pp. A90A93. Rainy Hollow Section pp. A63A64. in Report of the Minister of Mines, 1933. Victoria: British Columbia Government.

Mandy, E. Madge. 1992. *Our Trails Led Northwest.* Surrey, BC: Heritage House.

McArthur, J.J. 1898. "Exploration of the Overland Route to the Yukon by way of the Chilkat Pass." *Canada Department of the Interior Annual Report 1898, Part 2, Dominion Lands Surveys Report*, 128–40.

McCandless, Rob. 1985. *Yukon Wildlife: A Social History.* Edmonton: University of Alberta Press.

McClellan, Catharine. 1950. "Culture Change and Native Trade in Southern Yukon Territory." PhD dissertation, University of California, Berkeley.

———. 1964. "Culture Contacts in the Early Historic Period in Northwestern North America." *Arctic Anthropology* 2 (2): 3–15.

———. 1975a. *My Old People Say: An Ethnographic Survey of Southern Yukon Territory.* 2 vols. National Museums of Canada Publications in Ethnology 6. Ottawa: National Museums of Canada. Republished 2001 as Canadian Ethnology Service Paper 137, Mercury Series.

———. 1975b. "Feuding and Warfare Among Northwestern Athapaskans." In *Proceedings, Northern Athapaskan Conference 1971*, 2 vols., edited by A. MacFadyen Clark, 181–258. National Museum of Man Mercury Series, Canadian Ethnology Service Paper No. 27, Ottawa: National Museums of Canada.

———. 1981a. Intercultural Relations and Cultural Change in the Cordillera. In *Handbook of North American Indians*, vol. 6, *The Subarctic*, edited by J. Helm, 387–401. Washington, DC: Smithsonian Institution.

———. 1981b. "Tutchone." In *Handbook of North American Indians*, vol. 6, *The Subarctic*, edited by J. Helm, 493–505. Washington, DC: Smithsonian Institution.

———. 1981c. "Tagish." In *Handbook of North American Indians*, vol. 6, *The Subarctic*, edited by J. Helm, 481–492. Washington, DC: Smithsonian Institution.

———. 1989. "Before Boundaries: People of Yukon/Alaska." *Borderlands: A Conference on the Alaska-Yukon Border*, proceedings. Whitehorse: Yukon Historical and Museums Association. www.yukonalaska.com/yhma/public.

———. 2007a. *My Old People's Stories: A Legacy for Yukon First Nations. Part I: Southern Tutchone Narrators.* Occasional Papers in Yukon History No. 5 (1), Hude Hudan Series. Whitehorse: Cultural Services Branch, Government of Yukon. www.tc.gov.yk.ca/pdf/mcclellan_opyh_5(1).pdf.

———. 2007b. *My Old People's Stories: A Legacy for Yukon First Nations. Part II: Tagish Narrators.* Occasional Papers in Yukon History No. 5 (2), Hude Hudan Series. Whitehorse: Cultural Services Branch, Government of Yukon. www.tc.gov.yk.ca/pdf/mcclellan_opyh_5(2).pdf.

———. 2007c. *My Old People's Stories: A Legacy for Yukon First Nations. Part III: Inland Tlingit Narrators.* Occasional Papers in Yukon History No. 5 (1), Hude Hudan Series. Whitehorse: Cultural Services Branch, Government of Yukon. www.tc.gov.yk.ca/pdf/mcclellan_opyh_5(3).pdf.

McClellan, Catharine, with L. Birckel, R. Bringhurst, J.A. Fall, C. McCarthy and J.R. Sheppard. 1987. *Part of the Land, Part of the Water: A History of the Yukon Indians.* Vancouver and Toronto: Douglas and McIntyre.

Ned, Annie. 1984. *Old People in Those Days, They Told Their Story All the Time.* Compiled by Julie Cruikshank. Whitehorse: Yukon Native Languages Project.

Newton, Richard G., and Madonna L. Moss. 2005. *Haa Atxaahi Haa Kusteeyix Sitee, Our Food Is Our Tlingit Way of Life. Excerpts from Oral Interviews.* Juneau: US Department of Agriculture, Forest Service Alaska Region R10-MR-30.

Oberg, Kalvero. 1973. *The Social Economy of the Tlingit Indians.* Vancouver: J.J. Douglas.

O'Leary, Beth Laura. 1979. "Report to Champagne-Aishihik Band Council on Neskataheen Village Project." Manuscript on file, Champagne and Aishihik First Nations Heritage program, Haines Junction.

———. 1980. "The Neskataheen Village Preservation and Survey Project." Draft report prepared for Champagne and Aishihik First Nations, on file, Champagne and Aishihik First Nations Heritage program, Haines Junction.

———. 1984. "Early Explorations of Nesketaheen and the Dalton Trail: An Ethnoarchaeological Perspective." In *Proceedings of the Fall Meeting, 1983, Yukon Historical and Museums Association*, 33–41. Whitehorse: Yukon Heritage Branch.

———. 1992a. *Salmon and Storage: Southern Tutchone Use of an Abundant Resource*. Occasional Papers in Archaeology No. 2. Whitehorse: Yukon Heritage Branch.

———. 1992b. Neskatahin Village Preservation and Survey Project (NVPP) 1979: Archaeological Site, Isolated Occurrence and Historic Trail inventory forms, prepared for Champagne and Aishihik Indian Band. On file, Champagne and Aishihik First Nations Heritage program, Haines Junction.

Olson, Ronald L. 1936. "Some Trading Customs of the Chilkat Tlingit." In *Essays in Honor of Alfred Louis Kroeber*, edited by R.H. Lowie, 211–14. Berkeley: University of California Press.

———. 1968. *Social Structure and Social Life of the Tlingit in Alaska*. Anthropological Records vol. 26. Berkeley: University of California Press.

Olson, Wallace M. 1997. *The Tlingit: An Introduction to Their Culture and History*. 3rd ed. Auke Bay, AK: Heritage Research.

Ritter, John (Yukon Native Language Centre). 2009. Email correspondence to S. Greer, July 15, 2009, concerning the translation of the Núghàyík place name. Manuscript on file, Champagne and Aishihik First Nations Heritage program, Haines Junction.

Sackett, Russell. 1979. *The Chilkat Tlingit: A General Overview*. Anthropology and Historic Preservation, Cooperative Park Studies Unit. Occ. Paper No. 23. Fairbanks: University of Alaska.

Seton-Karr, H.W. 1891a. "Explorations in Alaska and North-west British Columbia." *Proceedings of the Royal Geographical Society* 2:65–86.

———. 1891b. *Bear Hunting in the White Mountains, or Alaska and British Columbia Revisited*. London: Chapman and Hall.

Shotridge, Louis. 1920. *Ghost of Courageous Adventurer*. The Museum Journal 11 (1): 11–26.

Sidney, Angela. 1983. *Haa Shagoon (Our Family History)*. Whitehorse: Yukon Native Language Centre.

Sidney, Angela, Kitty Smith and Rachel Dawson. 1979. *My Stories Are My Wealth*. Whitehorse: Council for Yukon Indians.

Smith, Kitty. 1982. *Nindal Kwadindur, I'm Going to Tell You a Story*. Recorded by J. Cruikshank. Whitehorse: Council for Yukon Indians and Government of Yukon.

Swanton, John R. 1908. "Social Conditions, Beliefs and Linguistic Relationship of the Tlingit Indians." In *26th Annual Report, Bureau of American Ethnology*, 391–485. Washington, DC: Bureau of American Ethnology.

———. 1909. "Tlingit Myths and Text." *Bureau of American Ethnology Bulletin* 39: 154.

Tero, Richard B. 1973. "E.J. Glave and the Alsek River." *Alaska Journal* 3 (Summer).

Thompson, Arthur R. 1905. *Gold Seeking on the Dalton Trail*. Boston: Little, Brown and Co.

Thompson, Laurence C.M., and M. Dale Kinkade, 1990. "Languages." In *Handbook of North American Indians*, vol. 7, *Northwest Coast*, edited by W. Suttles. Washington, DC: Smithsonian Institution.

Thornton, Thomas F. 2007. *Being and Place Among the Tlingit*. Seattle: University of Washington Press in association with Sealaska Heritage.

Thornton, Thomas F., with contributions by D. McBride, S. Gupta, Carcross-Tagish First Nation, Chilkat Indian Village, Chilkoot Indian Association and Skagway Traditional Council. 2004. *Klondike Gold Rush National Historical Park Ethnographic Overview and Assessment*. US National Parks Service. Accessed October 16, 2010. www.nps.gov/history/history/online_books/klgo/ethnographic_overview.pdf.

Tyrrell, J.B. 1957. "Dalton Trail, from Haines, Alaska, to Carmacks, on Lewes River, and Exploration of Nisling River." *Geological Survey of Canada Memoir 284*, 3–11. Ottawa: Queen's Printer.

Van Stone, James. 1982. "Southern Tutchone Clothing and Tlingit Trade." *Arctic Anthropology* 19 (1): 51–61.

Waddington, Cal. 1975. "The Dalton Trail." CBC Radio. Cassette tape 181 Side A, Yukon Archives. Transcript prepared 1992–93 by Champagne and Aishihik First Nations; on file, Champagne and Aishihik First Nations Heritage program, Haines Junction.

Watson, Kenneth DePencier. 1948. *The Squaw Creek–Rainy Hollow Area Northern British Columbia*. British Columbia Department of Mines Bulletin 25. Victoria: British Columbia Department of Mines.

White-Fraser, G. 1901. "Report of George White-Fraser, D.L.S. Latitude Determinations on the Boundary between the Province of British Columbia and the Yukon Territory." *Annual Report of the Department of the Interior for the Year 1900-01*, appendix 22, 68–75.

Wood, Z.T. 1899. "Annual Report of Superintendent Z.T. Wood." *Northwest Mounted Police Report, 1898*, part 3, *Yukon Territory*, appendix A, 32–56.

Workman, Margaret, ed. 2000. *Kwädąy Kwändür, Traditional Southern Tutchone Stories, As Told by Marge Jackson, Mary Jacquot, Jessie Joe, Jimmy Copper Joe, Copper Lily Johnson and Jessie Jonathan*. Whitehorse: Yukon Native Language Centre.

Worl, Rosita. 1990. "History of Southeastern Alaska Since 1867." In *Handbook of North American Indians*, vol. 7, *Northwest Coast*, edited by W. Suttles, 149–59. Washington, DC: Smithsonian Institution.

Yukon Archives. 1985. *Dalton Trail. A Bibliography of Sources Available*. Offprint available at the Yukon Archives.

YHMA (Yukon Historical and Museums Association). 1994. Poster reproduction of the Koh-klux' Map, Chilkaht. [Alaska and Yukon]. Original by Chief Chilkaht Kohklux, 1852. Whitehorse: Yukon Historical and Museums Association.

Chapter 10

Adamson, John. 1993. Interview by Sarah Gaunt. January 15. Transcript on file, Champagne and Aishihik First Nations Heritage program, Haines Junction.

Banks, Paula, 1996. Notes from interview with John Brown, September 20 (see Brown 1996), including information not recorded on tape. On file, Champagne and Aishihik First Nations Heritage program, Haines Junction.

Brown, John. 1996. Interview by Paula Banks. September 20. Partial transcript on file, Champagne and Aishihik First Nations Heritage program, Haines Junction.

CAFN (Champagne and Aishihik First Nations). 2010. *Dän Kéyi Kwändür: Stories from Our Country*. Whitehorse: Champagne and Aishihik First Nations.

CAFN (Champagne and Aishihik First Nations) and Sheila Greer. 1998. "Shawshe-Dalton Post History Research." Manuscript prepared for Champagne and Aishihik First Nations and Yukon Heritage Branch on file, Champagne and Aishihik First Nations Heritage program, Haines Junction.

CAFN (Champagne and Aishihik First Nations), Sarah Gaunt and Sheila Greer. 1993. "Shäwshe-Neskatahin, Background History Study." Manuscript prepared for National Historic Parks and Sites Directorate and Canadian Parks Service.

———. 1995. "Shawshe Chu-Alsexh, Tatshenshini River Ethnohistory." On file, Champagne and Aishihik First Nations Heritage program, Haines Junction.

Canada (government of), Champagne and Aishihik First Nations and government of Yukon. 1993. *Champagne and Aishihik First Nations Final Agreement*. www.ainc-inac. gc.ca/al/ldc/ccl/fagr/ykn/chama/cham-eng.asp.

Cruikshank, Julie. 1991. *Reading Voices*. Toronto: Douglas and McIntyre.

Cruikshank, Julie, Angela Sidney, Kitty Smith and Annie Ned. 1990. *Life Lived Like a Story*. Vancouver: University of British Columbia Press.

Davidson, George. 1901. "Explanation of an Indian Map of the Rivers, Lakes, Trails and Mountains from the Chilkaht to the Yukon Drawn by the Chilkaht Chief, Kohklux, in 1869." *Mazama* 2 (2): 75–92.

Davey, W.J. n.d. "Glave Explorations in Alaska." Unpublished typed manuscript on microfiche, Box 1, Folder 5. University of Alaska Fairbanks.

de Laguna, Frederica. 1972. *Under Mount Saint Elias: The History and Culture of the Yakutat Tlingit*. 3 parts. Smithsonian Contributions to Anthropology 7. Washington, DC: Smithsonian Institution.

Dickson, James H., Michael P. Richards, Richard J. Hebda, Petra Mudie, Owen Beattie, Susan Ramsay, Nancy J. Turner, Bruce J. Leighton, John M. Webster, Niki R. Hobischak, Gail S. Anderson, Peter M. Troffe and Rebecca J. Wigen. 2004. "Kwäday Dän Ts'ìnchí, the First Ancient Body of a Man from a North American Glacier: Reconstructing his Last Days by Intestinal and Biomolecular Analyses." *Holocene* 14 (4): 481–86.

Emmons, George T. 1991. *The Tlingit Indians*. Edited by Frederica de Laguna. New York: American Museum of Natural History.

French, Diana. 1993. "Archaeological Survey of the Tatshenshini River, Northwestern British Columbia." On file, BC Heritage Branch, Victoria.

Gates, Michael. 1974. "A Preliminary Brief Concerning the State of an Historic Site in the Yukon." On file, Champagne and Aishihik First Nations Heritage program, Haines Junction.

———. 1976. "Loss of an Historic Site in the Yukon." *The Arctic Circular* 24:1–3.

Gates, Michael, and F. Roback. 1972. "An Archaeological Survey and Ethnohistoric Study of the Tatshenshini River Basin, Southwest Yukon." On file, Champagne and Aishihik First Nations Heritage program, Haines Junction.

Gaunt, Sarah. 2001. "Sarah Gaunt Notes June 16." On file, Champagne and Aishihik First Nations Heritage program, Haines Junction.

Glave, Edward J. 1890–91. "Our Alaska Expedition: Exploration of the Unknown Alseck River Region." *Frank Leslie's Illustrated Newspaper*, September 6, 1890; November 15, 1890; November 22, 1890; November 29, 1890; December 6, 1890; December 13, 1890; December 20, 1890; December 27, 1890; January 3, 1891; January 10, 1891.

———. 1892a. "The Tempermental Explorer in Alaska." Parts 1 and 2. *The West*.

———. 1892b. "Pioneer Packhorses in Alaska." *Century Illustrated Monthly Magazine* 44 (5): 671–82.

———. 2013. *Travels to the Alseck, Edward Glave's Reports from Southwest Yukon and Southeast Alaska, 1890-91.* Edited by Julie Cruikshank, Doug Hitch and John Ritter. Whitehorse: Yukon Native Language Centre.

Goodwin, Lance. Personal communication to S. Greer, May 31, 2011.

Gotthardt, Ruth. 1992. "Archaeological Assessment of a Parking Lot and a Proposed Borrow Source at Dalton Post/Sha'washe, Southwest Yukon." On file, Yukon Heritage Branch, Whitehorse.

Greer, Sheila. 1995. "Sheila Greer Notes, BC Claim Research, October." On file, Champagne and Aishihik First Nations Heritage program, Haines Junction.

———. 1996. "Sheila Greer Klukshu Notes with Marge Jackson, August." On file, Champagne and Aishihik First Nations Heritage program, Haines Junction.

———. 1997. "Sheila Greer Notes with Marge Jackson, August 25." On file, Champagne and Aishihik First Nations Heritage program, Haines Junction.

———. 1998a. "Sheila Greer Notes, July 24, Re: Kluskhu Area and addendum on Aishihik Site Collections (includes information from Frances Joe and Gordon Allison)." On file, Champagne and Aishihik First Nations Heritage program, Haines Junction.

———. 1998b. "Memo to Gordon Allison (Champagne and Aishihik First Nations Lands Officer), July 13 (regarding Old K. Kluskhu)." On file, Champagne and Aishihik First Nations Heritage program, Haines Junction.

———. 1998c. "Memo to Gordon Allison (Champagne and Aishihik First Nations Lands Officer) and Diane Strand (Champagne and Aishihik First Nations Heritage Officer), June 29 (regarding Old Kluskhu)." On file, Champagne and Aishihik First Nations Heritage program, Haines Junction.

———. 2005. "Sheila Greer Notes with Moose Jackson, Marge Jackson, Ron Chambers, Frances Oles and Linaya Workman, February, related to Kluane National Park Signs Project." On file, Champagne and Aishihik First Nations Heritage program, Haines Junction.

———. 2007. "Sheila Greer notes regarding planned trip to Nuquiak (Núghàyík), July." On file, Champagne and Aishihik First Nations Heritage program, Haines Junction.

———. 2008. "Sheila Greer Notes, May 27, Kluskhu Creek." On file, Champagne and Aishihik First Nations Heritage program, Haines Junction.

Jackson, Marge. 1979. Neskatahin Village Preservation Project interview by Walter Workman, Summer 1979. Transcript on file, Champagne and Aishihik First Nations Heritage program, Haines Junction.

———. 1993. Interview by Sarah Gaunt, Sheila Greer and Ray Jackson, January 13. Transcript on file, Champagne and Aishihik First Nations Heritage program, Haines Junction.

Jackson, Marge, with assistance from Beth O'Leary, 2006. *My Country Is Alive: A Southern Tutchone Way of Life.* Published by author, Haines Junction.

Jackson, Marge. 2009. *Songs and Stories from My Country* (CD). Released by author, Haines Junction.

Jim, Paddy, and Stella Jim. 1996. Interview by Sarah Gaunt, Sheila Greer and Sean Sheardown. August 16. On file, Champagne and Aishihik First Nations Heritage program, Haines Junction.

Joe, Lawrence, and CAFN (Champagne and Aishihik First Nations). 1990. "Fisheries Research, Greater Kluane Land Use Plan." On file, Champagne and Aishihik First Nations Heritage program, Haines Junction.

Johnson, Linda. 1984. "The Day the Sun Was Sick." *Yukon Indian News*, Summer 1984: 11–13.

Kane, Jimmy. 1978. Interview by Jack Schick and Brent Liddle (Kluane National Park wardens). Recording on file, Yukon Archives. Transcription by Sarah Gaunt on file, Champagne and Aishihik First Nations Heritage program, Haines Junction.

———. 1979. Neskatahin Village Preservation Project interview by Trudy Long, Summer 1979, Interview No. 2, tape 1, side B. Transcript on file, Champagne and Aishihik First Nations Heritage program, Haines Junction.

Kane, Parten. 1979. Neskatahin Village Preservation Project interview by Beth O'Leary, October 10, Interview No. 3, tape 2, sides A and B. Transcript on file, Champagne and Aishihik First Nations Heritage program, Haines Junction.

Krause, Aurel. 1956. *The Tlingit Indians*. Vancouver: Douglas and McIntyre.

Krause, Aurel, and Arthur Krause. 1981. *Journey to the Tlingits by Aurel and Arthur Krause, 1881/82*. Translated by Margot Krause McCaffrey. Haines, Alaska: Centennial Commission.

———. 1993. *To the Chukchi Peninsula and to the Tlingit Indians 1881/1882: Journals and Letters by Aurel and Arthur Krause*. Translated by Margot Krause McCaffrey. The Rasmuson Library Historical Translation Series 8. Fairbanks, AK: University of Alaska Press.

McArthur, J.J. 1898. "Exploration of the Overland Route to the Yukon by way of the Chilkat Pass." In *Canada Department of the Interior Annual Report 1898*, part 2, *Dominion Lands Surveys Report*, 128–40.

McClellan, Catharine. 1975. *My Old People Say: An Ethnographic Survey of Southern Yukon Territory*. 2 vols. National Museums of Canada Publications in Ethnology 6. Ottawa: National Museums of Canada. Republished 2001 as Canadian Ethnology Service Paper 137, Mercury Series.

———. 2007. *My Old People's Stories: A Legacy for Yukon First Nations*. 3 parts. Occasional Papers in Yukon History No. 5 (1), Hude Hudan Series. Whitehorse: Government of Yukon Cultural Services Branch.

Mudie, Petra J., Sheila C. Greer, Judith Brakel, James H. Dickson, Clara Schinkel, Ruth Peterson-Welsh, Margaret Stevens, Nancy J. Turner and Rosalie Washington. 2005. "Forensics Palynology and Ethnobotany of *Salicornia* Species (Chenopodiaceae) in Northwest Canada and Alaska." *Canadian Journal of Botany* 83:111–23.

O'Leary, Beth Laura. 1977. "Report on Field Studies of a Tutchone Salmon Fishery, Yukon Territory, Canada, in the Summer of 1977." Manuscript on file, Champagne and Aishihik First Nations Heritage program, Haines Junction.

———. 1979. "Report to Champagne-Aishihik Band Council on Neskataheen Village Project from Beth O'Leary." Manuscript on file, Champagne and Aishihik First Nations Heritage program, Haines Junction.

———. 1984. "Early Explorations of Nesketaheen and the Dalton Trail: An Ethnoarchaeological Perspective." In *Proceedings of the Fall Meeting, 1983, Yukon Historical and Museums Association,* 33–41. Whitehorse: Yukon Heritage Branch.

———. 1992a. *Salmon and Storage: Southern Tutchone Use of an Abundant Resource*. Occasional Papers in Archaeology No. 2. Whitehorse: Yukon Heritage Branch.

———. 1992b. "Neskatahin Village Preservation and Survey Project (NVPP)-1979, Archaeological Site, Isolated Occurrence and Historic Trail Inventory forms." Prepared for Champagne and Aishihik Indian Band. Manuscript on file, Champagne and Aishihik First Nations Heritage program, Haines Junction.

Olson, Ronald L. 1967. *Social Structures and Social Life of the Tlingit in Alaska*. Anthropological Records 26. Berkeley: University of California Press.

Richards, Michael P., Sheila Greer, Lorna T. Corr, Owen Beattie, Alexander Mackie, Richard P. Evershed, Al von Finster and John Southon. 2007. "Radiocarbon Dating and Dietary Stable Isotope Analysis of Kwäday Dän Ts'ìnchí." *American Antiquity* 72 (4): 713–34.

Rutherford, Douglas. 1992. "Archaeological Reconnaissance of the Proposed Dalton Post Gravel Pit Location." Manuscript on file, Yukon Heritage Branch, Whitehorse.

Seton-Karr, H.W. 1891. "Explorations in Alaska and North-west British Columbia." *Proceedings of the Royal Geographical Society and Monthly Record of Geography* 13 (2): 65–86.

Shadow, Mary, and Marge Jackson. 1997. Interview by Paula Bank and Liza Jacobs. January 13. On file, Champagne and Aishihik First Nations Heritage program, Haines Junction.

Smith, Jason J., Bill Waugh, Peter Etherington, Kathleen Jensen and Sean Stark. 2009. "Alsek River Sockeye Salmon Radiotelemetry Studies, 2001–2003." Prepared for Pacific Salmon Commission. On file, Champagne and Aishihik First Nations Renewable Resources, Haines Junction.

Watson, Kenneth DePencier. 1948. *The Squaw Creek–Rainy Hollow Area, Northern British Columbia*. British Columbia Department of Mines Bulletin 25. Victoria: British Columbia Department of Mines.

YHMA (Yukon Historical and Museums Association). 1995. Poster reproduction of the Koh-klux' Map, Chilkaht. [Alaska and Yukon]. Original by Chief Chilkaht Kohklux, 1852. Whitehorse: Yukon Historical and Museums Association.

Chapter 11

Biosafety in Microbiological and Biomedical Laboratories (BMBL). 4th ed. 1999. US Department of Health and Human Services, Centers for Disease Control and Prevention and National Institutes of Health. Washington, DC: US Government Printing Office.

Hardy, Bob. "ITS-90 Formulations for Vapor Pressure, Frostpoint Temperature, Dewpoint Temperature, and Enhancement Factors in the Range -100 to +100 C." In *Papers and Abstracts from the Third International Symposium on Humidity and Moisture, London, England, April 1998*, vol. 1, 214–22.

Gaber, O., K.-H. Künzel, Herbert Maurer and Werner Platzer. 1992. "Konservierung und Lagerung der Gletschermumie." In *Der Mann im Eis, Band 1*, edited by F. Höpfel, W. Platzer and K. Spindler, 92–99. Innsbruck, Austria: University of Innsbruck.

Chapter 12

Beattie, Owen, Brian Apland, Erik W. Blake, James A. Cosgrove, Sarah Gaunt, Sheila Greer, Alexander P. Mackie, Kjerstin E. Mackie, Dan Straathof, Valerie Thorp and Peter M. Troffe. 2000. "The Kwäday Dän Ts'inchi Discovery from a Glacier in British Columbia." *Canadian Journal of Archaeology* 24 (1–2): 129–47.

Brooks, S., and J.M. Suchey. 1990. "Skeletal Age Determination Based on the *Os Pubis*: A Comparison of the Acsadi-Nemeskeri and Suchey-Brooks Methods." *Human Evolution* 5:227–38.

Fleckinger, A., ed. 2003. *Die Gletschermumie aus der Kupferzeit 2*. Vienna: Folio Verlag; Bolzano: South Tyrol Museum of Archaeology.

Gaber, O., K-H. Künzel, H. Maurer and W. Platzer. 1992. "Konservierung und lagerung der gletschermumie." In *Der Mann im Eis, Band 1*, edited by F. Höpfel, W. Platzer and K. Spindler, 92–99. Innsbruck, Austria: Eigenverlag der Universität Innsbruck.

Iscan, M.Y., S.R. Loth and R.K. Wright. 1985. "Age Estimation from the Rib by Phase Analysis: White Males." *Journal of Forensic Sciences* 29:1094–1104.

Rollo, F., S. Luciani, A. Canapa and I. Marota. 2000. "Analysis of Bacterial DNA in Skin and Muscle of the Tyrolean Iceman Offers New Insight into the Mummification Process." *American Journal of Physical Anthropology* 111:211–19.

Samadelli, M., ed. 2006. "The Chalcolithic Mummy," vol. 3. Vienna: Folio Verlag; Bolzano: South Tyrol Museum of Archaeology.

Sudtiroler Archaeologiemuseum, ed. 1999. "Die Gletschermumie aus der Kupferzeit." Vienna: Folio Verlag; Bolzano: South Tyrol Museum of Archaeology.

Trotter, M. 1970. "Estimation of Stature from Intact Long Limb Bones." In *Personal Identification in Mass Disasters*, edited by T.D. Stewart, 71–83. Washington, DC: Smithsonian Institution.

Chapter 13

Arriaza, B.T., W.L. Salo, A.C. Aufderheide and T.A. Holcomb. 1995. "Pre-Columbian Tuberculosis in Northern Chile: Molecular and Skeletal Evidence." *American Journal of Physical Anthropology* 98:37–45.

Aspöck, H., H. Auer and O. Picher. 1996. "*Trichuris trichiura* Eggs in the Neolithic Glacier Mummy from the Alps." *Parasitology Today* 12:255–56.

Baron, H., S. Hummel and B. Herrmann. 1996. "*Mycobacterium tuberculosis* Complex DNA in Ancient Human Bones." *Journal of Archaeological Science* 23:667–71.

Baxarias, J., A. Garcia, J. González, A. Pérez-Pérez, B.G. Tudó, C.J. García-Bour, D.D. Campillo and E.D. Turbón. 1998. "A Rare Case of Tuberculous Gonarthropathy from the Middle Ages in Spain: An Ancient DNA Confirmation Study." *Journal of Paleopathology* 10:63–72.

Beattie, Owen, Brian Apland, Erik W. Blake, James A. Cosgrove, Sarah Gaunt, Sheila Greer, Alexander P. Mackie, Kjerstin E. Mackie, Dan Straathof, Valerie Thorp and Peter M. Troffe. 2000. "The Kwäday Dän Ts'ìnchi Discovery from a Glacier in British Columbia." *Canadian Journal of Archaeology* 24 (1–2): 129–47.

Braun, M., D.C. Cook and S. Pfeiffer. 1998. "DNA from *Mycobacterium tuberculosis* Complex Identified in North American, Pre-Columbian Human Skeletal Remains." *Journal of Archaeological Science* 25:271–77.

Cano, R.J., F. Tiefenbrunner, M. Ubaldi, C.D. Cueto, S. Luciani, T. Cox, P. Orkand, K.-H. Künzel and F. Rollo. 2000. "Sequence Analysis of Bacterial DNA in the Colon and Stomach of the Tyrolean Iceman." *American Journal of Physical Anthropology* 112:297–309.

Christensen, C.R., M. Jackson, J. Zhao, W. Vogl and M.V. Monsalve. 2010. "Mid-infrared Analysis of the Ancient Human Remains Found in a Glacier in British Columbia." *Canadian Light Sources Inc. Activity Report* 21:56–57.

Crubézy, E., B. Ludes, J.D. Poveda, J. Clayton, B. Crouau-Roy and D. Montagnon. 1998. "Identification of *Mycobacterium* DNA in an Egyptian Pott's Disease of 5400 Years Old." *Conférences: Académie polonaise des sciences, Centre scientifique à Paris* 321:941–51.

Donoghue, H.D., M. Spigelman, P. Grant and A. Klein. 2007. "Microbiological Investigation of a Korean Medieval Child Mummy Found in Yangju." In *Proceedings of the VI World Congress on Mummy Studies* [Abstracts], 214. Teguise, Spain.

Donoghue, H.D., M. Spigelman, J. Zias, A.M. Gernaey-Child, and D.E. Minnikin. 1998. "Mycobacterium tuberculosis Complex DNA in Calcified Pleura from Remains 1400 Years Old." *Letters in Applied Microbiology* 27:265–69.

Faerman, M., R. Jankauskas, A. Gorski, H. Bercovier and C.L. Greenblatt. 1997. "Prevalence of Human Tuberculosis in a Medieval Population of Lithuania Studied by Ancient DNA Analysis." *Ancient Biomolecules* 1:205–14.

Fletcher, H.A., H.D. Donoghue, J. Holton, I. Pap and M. Spigelman. 2003. "Widespread Occurrence of *Mycobacterium tuberculosis* DNA from 18th–19th Century Hungarians." *American Journal of Physical Anthropology* 120:144–52.

Haas, C.J., A. Zink, E. Molńar, U. Szeimies, U. Reischul, A. Marcsik, Y. Ardagna, O. Dutour, G. Pálfi and A.G. Nerlich. 2000. "Molecular Evidence for Different Stages of Tuberculosis in Ancient Bone Samples from Hungary." *American Journal of Physical Anthropology* 113:293–304.

Leles, D., A. Araújo, L.F. Ferreira and A.C.P. Vicente. 2008. "Molecular Paleoparasitological Diagnosis of *Ascaris lumbricoides* in Coprolites: Implications in Its Paleodistribution." In *Mummies and Science, World Mummies Research*, edited by P.A. Peña, C.R. Martín and A.R. Rogríguez, 351–55. Santa Cruz de Tenerife: Academia Canaria de la Historia.

Liu, C., H.M. Park, M.V. Monsalve and D.D.Y. Chen. 2010. "Free Fatty Acids Composition in Adipocere of the Kwäday Dän Ts'ìnchi Ancient Remains Found in a Glacier." *Journal of Forensic Sciences*. 55 (4): 1039–43.

Mays, S., E. Fysh and G.M. Taylor. 2002. "Investigation of the Link Between Visceral Surface Rib Lesions and Tuberculosis in a Medieval Skeletal Series from England Using Ancient DNA." *American Journal of Physical Anthropology* 119:27–36.

Mays, S., G.M. Taylor, A.J. Legge, D.B. Young and G. Turner-Walker. 2001. "Paleopathological and Biomolecular Study of Tuberculosis in a Medieval Skeletal Collection from England." *American Journal of Physical Anthropology* 114:298–311.

Mims, C., H.M. Dockrell, R.V. Goering, I. Roitt, D. Wakeline and M. Zuckerman. 2004. *Medical Microbiology*, 3rd ed. Spain: Elsevier.

Minnikin, D.E., G.S. Besra, O.Y-C. Lee, A. Papaemmanouil, A.M. Gernaey, M. Spigelman, H.D. Donoghue and C.L. Greenblatt. 2007. "The Disease Status of the Granville Mummy: Lipid Biomarkers Support DNA Evidence for Tuberculosis." In *Proceedings of the VI World Congress on Mummy Studies* [Abstracts], 169. Teguise, Spain.

Monsalve, M.V., E. Humphrey, D.C. Walker, C. Cheung, W. Vogl and M. Nimmo. 2008a. "Brief Communication: State of Preservation of Tissues from Ancient Human Remains Found in a Glacier in Canada." *American Journal of Physical Anthropology* 137:348–55.

Monsalve, M.V., E. Humphrey, D. C. Walker, C. Cheung, T. Swanston, H. Deneer, P. Hazelton, G. Oda, E. Kahila, G. Bar-Gal and M. Spigelman. 2008b. "Report of a Multidisciplinary Workup of the Kwäday Dän Ts'ìnchi Human Remains from Canada: Microorganisms analyses." In *Mummies and Science. World Mummies Research*, edited by P.A. Peña, C.R. Martín and A.R. Rogríguez, 409–16. Santa Cruz de Tenerife: Academia Canaria de la Historia.

Morris, J.A., L.M. Harrison and S.M. Partridge. 2006. "Postmortem Bacteriology: A Re-evaluation." *Journal of Clinical Pathology* 59:1–9.

Nerlich, A.G., C.J. Haas, A. Zink, U. Szeimies and H.G. Hagedorn. 1997. "Molecular evidence for Tuberculosis in an Ancient Egyptian Mummy." *Lancet* 350:1404.

Rogers, H.J., and H.R. Perkins. 1968. *Cell Walls and Membranes*. London: Spon.

Rollo, F., S. Luciani, A. Canapa and I. Marota. 2000. "Analysis of Bacterial DNA in Skin and Muscle of the Tyrolean Iceman Offers New Insight into the Mummification Process." *American Journal of Physical Anthropology* 111:211–19.

Salo, W.L., A.C. Aufderheide, J. Buikstra and T.A. Holcomb. 1994. "Identification of *Mycobacterium tuberculosis* DNA in a Pre-Columbian Peruvian Mummy." *Proceedings of the National Academy of Sciences of the United States of America* 91 (6): 2091–94.

Seltmann, G., and O. Holst. 2002. *The Bacterial Cell Wall*, 1st ed. Berlin: Springer-Verlag.

Spigelman, M., and E. Lemma. 1993. "The Use of Polymerase Chain Reaction (PCR) to Detect *Mycobacterium tuberculosis* in Ancient Skeletons." *International Journal of Osteoarchaeology* 3:137–43.

Straathof, D., and O. Beattie. 2008. "A Review, Discussion and Interpretation of the Human Remains from the Kwäday Dän Ts'ìnchi Discovery." In *Program of the 61st Annual Northwest Anthropological Conference* [Abstracts], 90–91. Victoria.

Swanston, T., H. Deneer and E. Walker. 2008. "The Discovery of *Mycobacterium tuberculosis* Complex DNA in Tissues Associated with Kwäday Dän Ts'ìnchi." In *Program of the 61st Annual Northwest Anthropological Conference* [Abstracts], 91–92. Victoria.

Taylor, G.M., M. Crossey, J. Saldanha and T. Waldron. 1996. "DNA from Mycobacterium tuberculosis Identified in Mediaeval Human Skeletal Remains Using Polymerase Chain Reaction." *Journal of Archaeological Science* 23:789–98.

Taylor, G.M., M. Goyal, A.J. Legge, A.J. Shaw and D. Young. 1999. "Genotypic Analysis of *Mycobacterium tuberculosis* from Medieval Human Remains." *Microbiology* 145:899–904.

Zink, A., C.J. Haas, H. Hagedorn, U. Reischl, U. Szeimies and A. Nerlich. 2001. "Molecular Analysis of Skeletal Tuberculosis in an Ancient Egyptian Population." *Journal of Medical Microbiology* 50:355–66.

Chapter 14

Anderson, G.S. 2005. "Forensic Entomology." *Minerva Medicolegale* 125:45–60.

Anderson, K., and O. Halvorsen. 1978. "Egg Size and Form as Taxonomic Criteria in *Diphyllobothrium* (Cestoda, Pseudophyllidea)." *Parasitology* 76:229–40.

Aspöck, H., H. Auer and O. Picher. 1996. "*Trichuris trichura* Eggs in the Neolithic Glacier Mummy from the Alps." *Parasitology Today* 12:255–56.

Bathurst, R.R. 2005. "Archaeological Evidence of Intestinal Parasites from Coastal Shell Middens." *Journal of Archaeological Science* 32:115–23.

Baumgartner, D., and B. Greenberg. 1985. "Distribution and Medical Ecology of the Blow Flies (Diptera: Calliphordiae) of Peru." *Annals of the Entomological Society of America* 78:565–87.

Beattie, Owen, Brian Apland, Erik W. Blake, James A. Cosgrove, Sarah Gaunt, Sheila Greer, Alexander P. Mackie, Kjerstin E. Mackie, Dan Straathof, Valerie Thorp and Peter M. Troffe. 2000. "The Kwäday Dän Ts'ìnchi Discovery from a Glacier in British Columbia." *Canadian Journal of Archaeology* 24 (1–2): 129–47.

Beaver, P.C., R.C. Jung and E.W. Cupp. 1984. *Clinical Parasitology*. Philadelphia: Lea and Febiger.

Borror, D. 1989. *An Introduction to the Study of Insects*. 6th ed. Philadelphia: Saunders College Publishing.

Byrd, J.H., and J.L. Castner. 2001. *Forensic Entomology: The Utility of Arthropods in Legal Investigations*. Boca Raton, FL: CRC Press.

de Jong, G., and J. Chadwick. 1999. "Decomposition and Arthropod Succession on Exposed Rabbit Carrion During Summer at High Altitudes in Colorado, USA." *Journal of Medical Entomology* 36 (6): 833–45.

Dickson, James H., Michael P. Richards, Richard J. Hebda, Petra Mudie, Owen Beattie, Susan Ramsay, Nancy J. Turner, Bruce J. Leighton, John M. Webster, Niki R. Hobischak, Gail S. Anderson, Peter M. Troffe and Rebecca J. Wigen. 2004. "Kwäday Dän Ts'ìnchí, the First Ancient Body of a Man from a North American Glacier: Reconstructing his Last Days by Intestinal and Biomolecular Analyses." *Holocene* 14 (4): 481–86.

Garcia, L.S. 2007. *Diagnostic Medical Parasitology.* Washington, DC: ASM Press.

Hansen, J.P.H., J. Meldgaard and J. Nordqvist, eds. 1991. *The Greenland Mummies.* London: British Museum Publications.

Hilliard, D.K. 1972. "Studies of the Helminth Fauna of Alaska. LI. Observations on the Eggshell Formation in Some Diphyllobothriid Cestodes." *Canadian Journal of Zoology* 50:585–92.

Mackie, A.P. n.d. Personal Communication. Victoria: Royal British Columbia Museum.

Martinez, E., P. Duque and M. Wolff. 2007. "Succession Pattern of Carrion-Feeding Insects in Paramo, Colombia." *Forensic Science International* 166:182–89.

Mathewes, R. n.d. Personal Communication. Department of Biological Sciences. Simon Fraser University, Burnaby, BC.

McDonald, T.E., and L. Margolis. 1995. *Synopsis of the Parasites of Fishes of Canada: Supplement (1978–1993).* Ottawa: National Research Council of Canada.

Meyer, M.C. 1966. "Evaluation of Criteria for the Recognition of *Diphyllobothrium* Species." *Transactions of the American Microscopical Society* 85:89–99.

Mudie, P.J., S. Greer, J. Brakei, J.H. Dickson, C. Schinkel, R. Peterson-Welsh, M. Stevens, N.J. Turner, M. Shadow and R. Washington. 2005. "Forensic Palynology and Ethnobotany of *Salicornia* Species (Chenopodiaceae) in Northwest Canada and Alaska." *Canadian Journal of Botany* 83:111–23.

Rausch, R.L., and D.K. Hilliard. 1970. "Studies on the Helminth Fauna of Alaska. XLIX. The Occurrence of *Diphyllobothrium latum* (Linnaeus, 1758) (Cestoda: Diphyllobothriidae) in Alaska, with Notes on Other Species." *Canadian Journal of Zoology* 48:1201–19.

Reinhard, K., and O. Urban. 2003. "Diagnosing Ancient Diphyllobothriasis from Chinchorro Mummies." *Memórias do Instituto Oswaldo Cruz* 98 (suppl. 1):191–93.

Schmidt, G.D., and L.S. Roberts. 1989. *Foundations of Parasitology.* St. Louis, MO: Times Mirror/Mosby College Publishing.

Smith, R.J., and B. Heese. 1995. "Carcass Selection in a High Altitude Population of the Burying Beetle, *Nicrophorus investigator* (Silphidae)." *Southwestern Naturalist* 40 (1): 50–55.

VanLaerhoven, S.L., and G.S. Anderson. 1999. "Insect Succession on Buried Carrion in Two Biogeoclimatic Zones of British Columbia." *Journal of Forensic Science* 44 (1): 32–43.

Chapter 15

Allison, M.J., D. Mendoza and A. Pezzia. 1973. "Documentation of a Case of Tuberculosis in Pre-Columbian America." *American Review of Respiratory Disease* 107 (6): 985–91.

Allison, M.J., T. Bergman and E. Gerszten. 1999. "Further Studies on Fecal Parasites in Antiquity." *American Journal of Clinical Pathology* 112 (5): 605–9.

Atherton, J.C. 2006. "The Pathogenesis of *Helicobacter pylori*-Induced Gastro-Duodenal Diseases." *Annual Review of Pathology* 1:63–96.

Atherton, J.C., P. Cao, R.M. Peek, M.K.R. Tummuru, M.J. Blaser and T.L. Cover. 1995. "Mosaicism in Vacuolating Cytotoxin Alleles of *Helicobacter pylori*." *The Journal of Biological Chemistry* 270:17771–77.

Aufderheide, A.C., and C. Rodriguez-Martin. 1998. *The Cambridge Encyclopedia of Human Paleopathology.* Cambridge: Cambridge University Press.

Aviles-Jimenez, F., D.P. Letley, N. Salama Gonzalez-Valencia, J. Torres and J.C. Atherton. 2004. "Evolution of the *Helicobacter pylori* Vacuolating Cytotoxin in a Human Stomach." *Journal of Bacteriology* 186 (15): 5182–85.

Bishai, W. 2000. "Lipid Lunch for Persistent Pathogen." *Nature* 406:683–85.

Boyd, R. 1999. *The Coming of the Spirit of Pestilence.* Vancouver: UBC Press.

Braun, M., D. Collins Cook and S. Pfeiffer. 1998. "DNA from *Mycobacterium tuberculosis* Complex Identified in North American, Pre-Columbian Human Skeletal Remains." *Journal of Archaeological Science* 25 (3): 271–77.

Buikstra, J.E., and S. Williams. 1991. "Tuberculosis in the Americas: Current Perspectives." In *Human Paleopathology: Current Syntheses and Future Options,* edited by Donald J. Ortner and Arthur C. Aufderheide, 161–72. Washington, DC: Smithsonian Institution Press.

Carroll, I.M., A.A. Khan and N. Ahmed. 2004. "Revisiting the Pestilence of *Helicobacter pylori*: Insights into Geographical Genomics and Pathogen Evolution." *Infection, Genetics and Evolution: Journal of Molecular Epidemiology and Evolutionary Genetics in Infectious Diseases* 4 (2): 81–90.

Donoghue, H.D., M. Spigelman, C.L. Greenblatt, G. Lev-Maor, G.K. Bar-Gal, C. Matheson, K. Vernon, A.G. Nerlich and A.R. Zink. 2004. "Tuberculosis: From Prehistory to Robert Koch, as Revealed by Ancient DNA." *The Lancet Infectious Diseases* 4 (9): 584–92.

Eisenach, K.D., M.D. Cave, J.H. Bates and J.T. Crawford. 1990. "Polymerase Chain Reaction Amplification of a Repetitive DNA Sequence Specific for *Mycobacterium tuberculosis.*" *The Journal of Infectious Diseases* 161 (5): 977–81.

Falush, D., T. Wirth, B. Linz, J.K. Pritchard, M. Stephens, M. Kidd and M.J. Blaser. 2003. "Traces of Human Migrations in *Helicobacter pylori* Populations." *Science* 299 (5612): 1582–85.

Felsenstein, J. 1985. "Confidence Limits on Phylogenies: An Approach Using the Bootstrap." *Evolution* 39:783–91.

Fortuine, R. 2005. *"Must We All Die?" Alaska's Enduring Struggle with Tuberculosis.* Fairbanks: University of Alaska Press.

Ghose, C., G.I. Perez-Perez, M.G. Dominguez-Bello, D.T. Pride, C.M. Bravi and M.J. Blaser. 2002. "East Asian Genotypes of *Helicobacter pylori* Strains in Amerindians Provide Evidence for Its Ancient Human Carriage." *Proceedings of the National Academy of Sciences of the United States of America* 99 (23): 15107–11.

Higuchi, R., B. Bowman, M. Freiberger, O.A. Ryder and A.C. Wilson. 1984. "DNA Sequences from the Quagga, an Extinct Member of the Horse Family." *Nature* 312 (5991): 282–84.

Kaestle, F.A. 2002. "Ancient DNA in Anthropology: Methods, Applications, and Ethics." *Yearbook of Physical Anthropology* 45:92–130.

Kersulyte, D., A.K. Mukhopadhyay, B. Velapatino, W. Su, Z. Pan, C. Garcia and V. Hernandez. 2000. "Differences in Genotypes of *Helicobacter pylori* from Different Human Populations." *Journal of Bacteriology* 182 (11): 3210–18.

Letley, D.P., A. Lastovica, J.A. Louw, C.J. Hawkey and J.C. Atherton. 1999. "Allelic Diversity of the *Helicobacter pylori* Vacuolating Cytotoxin Gene in South Africa: Rarity of the vacA s1a Genotype and Natural Occurrence of an s2/m1 Allele." *Journal of Clinical Microbiology* 37 (4): 1203–05.

Lindahl, T. 1993. "Instability and Decay of the Primary Structure of DNA." *Nature* 362:709–15.

Marshall, B.J., and J.R. Warren. 1984. "Unidentified Curved Bacilli in the Stomach of Patients with Gastritis and Peptic Ulceration." *Lancet* 1 (8390): 1311–15.

Morales-Espinosa, R., G. Castillo-Rojas, G. Gonzalez-Valencia, S. Ponce de Leon, A. Cravioto, J.C. Atherton and Y. Lopez-Vidal. 1999. "Colonization of Mexican Patients by Multiple *Helicobacter pylori* Strains with Different vacA and cagA Genotypes." *Journal of Clinical Microbiology* 37 (9): 3001–04.

Morse, D. 1961. "Prehistoric Tuberculosis in America." *American Review of Respiratory Diseases* 83 (4): 489–504.

Nicholas, K.B., H.B.J. Nicholas and D.W.I. Deerfield. 1997. "Genedoc: Analysis and Visualization of Genetic Variation." *EMBNEW.NEWS* 4, 14. http://www.psc. edu/ biomed/dissem/genedoc/index.html.

Ortner, D.J. 2003. *Identification of Pathological Conditions in Human Skeletal Remains.* San Diego: Academic Press.

Pääbo, S., H. Poinar, D. Serre, V. Jaenicke-Despres, J. Hebler, N. Rohland, M. Kuch, J. Krause, L. Vigilant and M. Hofreiter. 2004. "Genetic Analyses from Ancient DNA." *Annual Review of Genetics* 38:645–79.

Pan, Z., D.E. Berg, R.W.M. van der Hulst, W. Su, A. Raudonikiene, S. Xiao, J. Dankert, G.N.J. Tytgat and A. van der Ende. 1998. "Prevalence of Vacuolating Cytotoxin Production and Distribution of Distinct vacA Alleles in *Helicobacter pylori* from China." *The Journal of Infectious Diseases* 178:220–26.

Richards, M.P., S. Greer, L.T. Corr, O. Beattie, A. Mackie, R.P. Evershed, A. von Finster and J. Southon. 2007. "Radiocarbon Dating and Dietary Stable Isotope Analysis of Kwäday Dän Ts'ìnchi." *American Antiquity* 72 (4): 719–33.

Roberts, C., and J.E. Buikstra. 2003. *The Bioarchaeology of Tuberculosis.* Gainesville: University Press of Florida.

Roberts, C., and S. Ingham. 2008. "Using Ancient DNA Analysis in Paleopathology: a Critical Analysis of Published Papers, with Recommendations for Future Work." *International Journal of Osteoarchaeology* 18 (6): 600–13.

Rothschild, B.M., L.D. Martin, G. Lev, H. Bercovier, G.K. Bar-Gal, C. Greenblatt, H. Donoghue, M. Spigelman and D. Brittain. 2001. "*Mycobacterium tuberculosis* Complex DNA from an Extinct Bison Dated 17,000 Years Before the Present." *Clinical Infectious Diseases* 33 (3): 305–11.

Sacchettini, J.C., E.J. Rubin and J.S. Freundlich. 2008. "Drugs Versus Bugs: In Pursuit of the Persistent Predator *Mycobacterium tuberculosis*." *Nature Reviews Microbiology* 6 (1): 41–52.

Salo, W.L., A.C. Aufderheide, J. Buikstra and T.A. Holcomb. 1994. "Identification of *Mycobacterium tuberculosis* DNA in a Pre-Columbian Peruvian Mummy." *Proceedings of the National Academy of Sciences of the United States of America* 91 (6): 2091–94.

Schulting, R.J., and A.D. McMillan. 1995. "A Probable Case of Tuberculosis from a Burial Cave in Barkley Sound, Vancouver Island." *Canadian Journal of Archaeology* 19: 149–53.

Spigelman, M., and E. Lemma. 1993. "The Use of the Polymerase Chain Reaction (PCR) to Detect *Mycobacterium tuberculosis* in Ancient Skeletons." *International Journal of Osteoarchaeology* 3 (2): 137–43.

Swanston, T. 2010. "Past Human Health and Migration: The Analysis of Microbial DNA Associated with Human Remains Recovered from a Glacier in Canada." Unpublished PhD dissertation, University of Saskatchewan.

Tamura, K., J. Dudley, M. Nei and S. Kumar. 2007. "MEGA4: Molecular Evolutionary Genetics Analysis (MEGA) Software Version 4.0." *Molecular Biology and Evolution* 24:1596–99.

Thomson, J.D., T.J. Gibson, F. Plewniak, F. Jeanmougin and D.G. Higgins. 1997. "The CLUSTAL_X Windows Interface: Flexible Strategies for Multiple Sequence Alignment Aided by Quality Analysis Tools." *Nucleic Acids Research* 25:4876–82.

van Doorn, L.J., C. Figueiredo, R. Rossau, G. Jannes, M. van Asbroeck, J.C. Sousa, F. Carneiro and W.G.V. Quint. 1998. "Typing of *Helicobacter pylori* vacA Gene and Detection of cagA Gene by PCR and Reverse Hybridization." *Journal of Clinical Microbiology* 36 (5): 1271–76.

World Health Organization. 2008. http://www.who.int/tb/en.

Yamaoka Y., T. Kodama, M. Kita, J. Imanishi, K. Kashima and D.Y. Graham. 1998. "Relationship of vacA Genotypes of *Helicobacter pylori* to cagA Status, Cytotoxin Production, and Clinical Outcome." *Helicobacter* 3:241–53.

Yamaoka, Y., E. Orito, M. Mizokami, O. Gutierrez, N. Saitou, T. Kodama, M. Osato, J. Kim, F. Ramirez, V. Mahachai and D. Graham. 2002. "*Helicobacter pylori* in North and South America before Columbus." *FEBS Letters* 517:180–84.

Yamazaki, S., A. Yamakawa, T. Okuda, M. Ohtani, H. Suto, Y. Ito, Y. Yamazaki, Y. Keida, H. Higashi, M. Hatakeyama and T. Azuma. 2005. "Distinct Diversity of vacA, cagA, and cagE Genes of *Helicobacter pylori* Associated with Peptic Ulcers in Japan." *Journal of Clinical Microbiology* 43 (8): 3906–16.

Yang, D.Y., B. Eng, J.S. Waye, J.C. Dudar and S.R. Saunders. 1998. "Technical Note: Improved DNA Extraction from Ancient Bones Using Silica-Based Spin Columns." *American Journal of Physical Anthropology* 105 (4): 539–43.

Zimmerman, M.R. 1998. "Alaskan and Aleutian Mummies." In *Mummies, Disease and Ancient Cultures*, 2nd ed., edited by Aidan Cockburn, Eve Cockburn and Theodore A. Reyman. 138–53. New York: Cambridge University Press.

Zimmerman, M.R., and G.S. Smith. 1975. "A Probable Case of Accidental Inhumation of 1600 Years Ago." *Bulletin of the New York Academy of Medicine* 51:828–37.

Zink A., and A. Nerlich. 2004. "Molecular Strain Identification of the *Mycobacterium tuberculosis* Complex in Archival Tissue Samples." *Journal of Clinical Pathology* 57:1185–92.

Chapter 16

Alves-Silva, J., S.M. da Silva, P.E.M. Guimaraes, A.C.S. Ferreira, H-J. Bandelt, S.D. Pena, and P.V. Ferreira. 2000. "The Ancestry of Brazilian mtDNA Lineages." *American Journal of Human Genetics* 67:444–61.

Ambrose, S.H. 1990. "Preparation and Characterization of Bone and Tooth Collagen for Stable Carbon and Nitrogen Isotope Analysis." *Journal of Archaeological Science* 17:431–51.

Anderson, S., A.T. Bankier, B.G. Barrell, M.H.L. de Bruijn, A.R. Coulson, J. Drouin, I.C. Eperon, D.P. Nierlich, B.A. Roe, F. Sanger, H. Schreier, A.J.H. Smith, R. Staden and I.G. Young. 1981. "Sequence and Organization of the Human Mitochondrial Genome." *Nature* 290:457–74.

Boles, T.C., C.C. Snow and E. Stover. 1995. "Forensic DNA Testing on Skeletal Remains from Mass Graves: A Pilot Project in Guatemala." *Journal of Forensic Sciences* 40:349–55.

Forster, P., R. Harding, A. Torroni and H-J. Bandel. 1996. "Origin and Evolution of Native American mtDNA Variation: A Reappraisal." *American Journal of Human Genetics* 59:935–45.

Gilbert, M.T.P., D.L. Jenkins, A. Gotherstrom, N. Naveran, J.J. Sanchez, M. Hofreiter, P.F. Thomsen, J. Binladen, T.F.G. Higham, R.M. Yohe II, R. Parr, L.S. Cummings and E. Willerslev. 2008. "DNA from Pre-Clovis Human Coprolites in Oregon, North America." *Science* 320:786–89.

Guhl, F., C. Jaramillo, G.A. Vallejo, R. Yockteng and F. Cardenas-Arroyo. 1999. "Isolation of *Trypanosoma cruzi* in 4000-year-old Mummified Human Tissue from Northern Chile." *American Journal of Physical Anthropology* 108:401–07.

Hagelberg, E., and J.B. Clegg. 1991. "Isolation and Characterization of DNA from Archaeological Bone." *Proceedings of the Royal Society of London, Biological Sciences.* 244:45–50.

Handt, O., M. Krings, R.H. Ward and S. Pääbo. 1996. The Retrieval of Ancient Human DNA Sequences. *American Journal of Human Genetics* 59:368–76.

Higuchi, R., C.H. von Beroldigen, G.F. Sensabaugh, and H.A. Erlich. 1988. "DNA Typing from Single Hairs." *Nature* 332: 543–46.

Innis, M.A., and D.H. Gelfand. 1990. "Optimization of PCRs." In *PCR Protocols. A Guide to Methods and Applications,* edited by M.A. Innis, D.H. Gelfand, J.J. Sninsky and T.J. White. San Diego, CA: Academic Press.

Kolman, C.J., and E. Bermingham. 1997. "Mitochondrial and Nuclear DNA Diversity in the Chocó and Chibcha Amerinds of Panamá." *Genetics* 147:1289–1302.

Lorenz, J., and G.D. Smith. 1996. "Distribution of Four Founding mtDNA Haplogroups among Native North Americans." *American Journal of Physical Anthropology* 101 (3): 307–23.

Malhi, R.S., B.M. Kemp, J.A. Eshleman, J. Cybulski, D.G. Smith, S. Cousins and H. Harry. 2007. "Mitochondrial Haplogroup M Discovered in Prehistoric North Americans." *Journal of Archaeological Science* 34:642–48.

Monsalve, M.V., and A.C. Stone. 2005. "mtDNA Lineage Analysis: Genetic Affinities of the Kwäday Dän Ts'inchi Remains with Other Native Americans." In *Biomolecular Archaeology: Genetic Approaches to the Past,* edited by D.M. Reed, 9–21. Occasional Paper No. 2 (2). Carbondale: Southern Illinois University Press.

Monsalve, M.V., A.C. Stone, C.M. Lewis, A. Rempel, M. Richards, D. Straathof and D.V. Devine. 2002. "Molecular Analysis of the Kwäday Dän Ts'inchi Ancient Remains Found in a Glacier in Canada." *American Journal of Physical Anthropology* 119:288–91.

Monsalve, M.V., A.C. Stone, C.M. Lewis, A. Rempel, M. Richards, D. Straathof and D.V. Devine. 2003. "mtDNA Analysis of Human Ancient Remains Found in a Glacier in Canada in 1999." In *Mummies in a New Millennium,* edited by N. Llynnerup, C. Andreasen and J. Berglund, 128–30. Proceedings of the 4th World Congress on Mummy Studies, 2001. Nuuk, Greenland: Greenland National Museum and Archive and Danish Polar Center.

Nei, M., and W.H. Li. 1979. "Mathematical Model for Studying Genetic Variation in Terms of Restriction Endonucleases. *Proceedings of the National Academy of Sciences of the United States of America* 76:5268–73.

Schneider, Stefan, David Roessli and Laurent Excoffier, eds. 2000. ARLEQUIN ver. 2.000: A Software for Population Genetic Analysis. Geneva, Switzerland: Genetics and Biometry Laboratory, University of Geneva.

Shields, G.F., A.M. Schmiechen, B.L. Frazier, A. Redd, M.I. Voevoda, J.K. Reed and R.H. Ward. 1993. "mtDNA Sequences Suggest a Recent Evolutionary Divergence for Beringian and Northern North American Populations." *American Journal of Human Genetics* 53:549–62.

Stone, A.C., and M. Stoneking. 1998. "mtDNA Analysis of a Prehistoric Oneota Population: Implications for the Peopling of the New World." *American Journal of Human Genetics* 62:1153–70.

Strand, D., K. Mooder and S. Greer. 2008. "The Kwäday Dän Ts'inchi Community DNA Study – The Search For Living Relatives." Presented at the 61st Annual Northwest Anthropological Conference, Victoria, April 2008.

Torroni, A., T.G. Schurr, C.C. Yang, E.J. Szathmary, R.C. Williams, M.S. Schanfield, G.A. Troup, W.C. Knowler, D.N. Lawrence, K.M. Weiss and D.C. Wallace. 1992. "Native American Mitochondrial DNA Analysis Indicates That the Amerind and the Nadene Populations Were Founded by Two Independent Migrations." *Genetics* 130: 153–62.

Torroni, A., T.G. Schurr, M.F. Cabell and M.D. Brown. 1993. "Asian Affinities and Continental Radiation of the Four Founding Native American mtDNAs." *American Journal of Human Genetics* 53:563–90.

Ward, R.H., B.L. Frazier, K. Dew-Jager and S. Pääbo. 1991. "Extensive Mitochondrial Diversity within a Single Amerindian Tribe." *Proceedings of the National Academy of Sciences of the United States of America* 88:8720–24.

Ward, R.H., A. Redd, D. Valencia, B. Frazier and S. Pääbo. 1993. "Genetic and Linguistic Differentiation in the Americas." *Proceedings of the National Academy of Sciences of the United States of America* 90:10663–67.

Yang, D.Y., B. Eng, J.S. Waye, J.C. Dudar and S.R. Saunders. 1998. "Technical Note: Improved DNA Extraction from Ancient Bones Using Silica-based Spin Columns." *American Journal of Physical Anthropology* 105:539–43.

Chapter 17

Kempson, I.M., W.M. Skinner and K.P. Kirkbride. 2007. "The Occurrence and Incorporation of Cu and Zn in Hair and Their Potential Role as Bio-indicators. A Review." *Journal of Toxicology and Environmental Health, part B* 10:611–22.

Kempson, I.M., W.M. Skinner and R.R. Martin. 2010. "Changes in the Metal Content of Human Hair During Diagenesis from 500 Years Exposure to Glacial and Aqueous Environments." *Archaeometry* 52:450–66.

Mackie, Kjerstin. 2005. "Long Ago Person Found: An Ancient Robe Tells a New Story." In *Recovering the Past: The Conservation of Archaeological and Ethnographic Textiles*, North American Conservation Conference preprints, compiled by Emilia Cortes and Suzanne Thomassen-Kruse, 35–46.

Macko, S.A., M.H. Engel, V. Andrusevich, G. Lubec, T.C. O'Connell and R.E.M. Hedges. 1999. "Documenting the Diet in Ancient Human Populations through Stable Isotope Analysis of Hair." *Philosophical Transactions of the Royal Society of London B* 354:65–76.

O'Connell, T.C., and R.E.M. Hedges. 1999. "Isotopic Comparison of Hair and Bone: Archaeological Analysis." *Journal of Archaeological Science* 26:661–65.

Potsch, L., and M.R. Moeller. 1996. "On Pathways for Small Molecules into and out of Human Hair Fibers." *Journal of Forensic Science* 41 (1): 121–25.

Radosevich, S.C. 1993. "The Six Deadly Sins of Trace Element Analysis: A Case of Wishful Thinking in Science." In *Investigations of Ancient Human Tissue: Analysis in Anthropology*, edited by M.K. Sandford, 269–332. Langhorne, PA: Gordon and Breach Science Publishers.

Rowe, W.F. 1997. "Biodegradation of Hairs and Fibers." In *Forensic Taphonomy: The Postmortem Fate of Human Remains*, edited by W.D. Haglund and M.H. Sorg, 337–51. Boca Raton, FL: CRC Press.

White, C.D. 1993. "Isotopic Determination of Seasonality in Diet and Death from Nubian Mummy Hair." *Journal of Archaeological Science* 20:657–66.

Wierzchos, J., C. Ascado, L.G. Sancho and A. Green. 2003. "Iron-rich Diagenetic Minerals are Biomarkers of Bacterial Activity in Antarctic Rocks." *Geomicrobiology Journal* 20:15–24.

Chapter 18

Ammitzbøll, T., R. Møller, G. Møller, T. Kobayasi, H. Hino, G. Asboe-Hansen and J.P. Hart Hansen. 1989. "Collagen and Glycosaminoglycans in Mummified Skin." In *Meddr Grønland, Man and Society 12*, edited by J.P. Hart Hansen and H.C. Gulløv, 93–99. Copenhagen: Commission for Scientific Research in Greenland.

Beattie, Owen, Brian Apland, Erik W. Blake, James A. Cosgrove, Sarah Gaunt, Sheila Greer, Alexander P. Mackie, Kjerstin E. Mackie, Dan Straathof, Valerie Thorp and Peter M. Troffe. 2000. "The Kwäday Dän Ts'inchi Discovery from a Glacier in British Columbia." *Canadian Journal of Archaeology* 24 (1–2): 129–47.

Bereuter, T.L., E. Lorbeer, C. Reiter, H. Seidler and H. Unterdorfer. 1996. "Post Mortem Alteration of Human Lipids, Part 1: Evaluation of Adipocere Formation and Mummification by Desiccation." In *Human Mummies: A Global Survey of Their Status and the Techniques of Conservation*, edited by K. Spindler, H. Wilfing, E. Rastbichler-Zissernig, D. zur Hedden and H. Nothdurfter, 265–73. Vienna: Springer.

Christensen, C.R., M. Jackson, J. Zhao, W. Vogl and M.V. Monsalve. 2009. "Mid-Infrared Analysis of the Ancient Human Remains Found in a Glacier in British Columbia." *Canadian Light Sources Activity Report* 21:58–59.

Colson, I.B., J.F. Bailey, M. Verautere, B.C. Sykes and R.E.M. Hedges. 1997. "The Preservation of Ancient DNA and Bone Diagenesis." *Ancient Biomolecules* 1:109–17.

Handt, O., M. Richards, M. Trommsdorff, C. Kilger, J. Simanainen, O. Georgiev, K. Bauer, A. Stone, R. Hedges and W. Schaffner. 1994. "Molecular Genetic Analyses of the Tyrolean Ice Man." *Science* 264:1775–78.

Hart Hansen, J.P. 1989. "The Mummies from Qilakitsoq: Paleopathological Aspects." In *The Mummies from Qilakitsoq*, edited by J.P. Hart Hansen and H.C. Gulløv, 69–82. Copenhagen: Commission for Scientific Research in Greenland.

Hart Hansen, J.P., and H.C. Gulløv, eds. 1989. *The Mummies from Qilakitsoq: Eskimos in the 15th Century.* Meddelelser om Grønland, Man and Society 12. Copenhagen: Commission for Scientific Research in Greenland.

Hart Hansen, J.P., and J. Nordqvist. 1996. "The Mummy Find from Qilakitsoq in Northwest Greenland." In *Human Mummies: A Global Survey of Their Status and the Techniques of Conservation*, edited by K. Spindler, H. Wilfing, E. Rastbichler-Zissernig, D. zur Hedden and H. Nothdurfter, 107–21. Vienna: Springer.

Haynes, S., J.B. Searle, A. Bretman and K.M. Dobney. 2002. "Bone Preservation and Ancient DNA: The Application of Screening Methods for Predicting DNA Survival." *Journal of Archaeological Science* 29:585–92.

Hedges, Robert E.M., Andrew R. Millard and A.W.G. Pike. 1995. "Measurements and Relationships of Diagenetic Alteration of Bone from Three Archaeological Sites." *Journal of Archaeological Science* 22 (2): 201–09.

Hess, M.W., G. Klima, K. Phaller, K.-H. Künzel and O. Gaber. 1998. "Histological Investigations of the Tyrolean Ice Man." *American Journal of Physical Anthropology* 106:521–32.

Kobayasi, T., T. Ammitzbøll and G. Asboe-Hansen. 1989. "Electron Microscopy of the Skin of a Greenlandic Mummy." In *The Mummies from Qilakitsoq*, edited by J.P. Hart Hansen and H.C. Gulløv, 100–05. Copenhagen: Commission for Scientific Research in Greenland.

Liu, C., H.M. Park, M.V. Monsalve and D.D.Y. Chen. 2010. "Qualification of Free Fatty Acids in Adipocere of the Kwäday Dän Ts'ínchi Ancient Remains Found in a Glacier in Canada." *Journal of Forensic Sciences* 55:1039–43.

Monsalve, M.V., and A.C. Stone. 2005. "mtDNA Lineage Analysis: Genetic Affinities of the Kwäday Dän Ts'inchi Remains with Other Native Americans." In *Biomolecular Archaeology: Genetic Approaches to the Past,* edited by D.M. Reed, 9–21. Carbondale: Southern Illinois University Press.

Monsalve, M.V., A.C. Stone, C.M. Lewis, A. Rempel, M. Richards, D. Straathof and D.V. Devine. 2002. Molecular analysis of the Kwäday Dän Ts'inchi Ancient Remains Found in a Glacier in Canada." *American Journal of Physical Anthropology* 119:288–91.

Monsalve, M.V., E. Humphrey, D G. Walker, C. Cheung, W. Vogl and M. Nimmo. 2008. "Brief Communication: State of Preservation of Tissues from Ancient Human Remains Found in a Glacier in Canada." *American Journal of Physical Anthropology* 137:348–55.

Strand, D., K. Mooder and S. Greer. 2008. "The Kwäday Dän Ts'inchi Community DNA Study – The Search for Living Relatives." Presented at the 61st Annual Northwest Anthropological Conference, Victoria, April 2008.

Thuesen, I., and J. Engberg. 1990. "Recovery and Analysis of Human Genetic Material from Mummified Tissue and Bone." *Journal of Archaeological Science* 17:679–89.

Wilson, A.S., T. Taylor, M.C. Ceruti, J.A. Chavez, J. Reinhard, V. Grimes, W. Meier-Augenstein, L. Cartmell, B. Stern, M.P. Richards, M. Worobey, I. Barnes and M.T.P. Gilbert. 2007. "Stable Isotope and DNA Evidence for Ritual Sequences in Inca Child Sacrifice." *Proceedings of the National Academy of Sciences of the United States of America* 104 (42): 16456–61.

Zimmerman, M.R. 1985. "Paleopathology in Alaskan Mummies." *American Science* 73:20–25.

Zimmerman, M.R., and A.C. Aufderheide. 1984. "The Frozen Family of Utqiagvik: The Autopsy Findings." *Arctic Anthropology* 21:53–63

Zimmerman, M.R., and G.S. Smith. 1975. "A Probable Case of Accidental Inhumation of 1600 Years Ago." *Bulletin of the New York Academy of Medicine* 51:828–37.

Zimmerman, M.R., and R.H. Tedford. 1976. "Histologic Structures Preserved for 21,300 years." *Science* 194: 183–84.

Chapter 19

Ambrose, Stanley H., and Lynette Norr. 1993. "Experimental Evidence for the Relationship of the Carbon Isotope Ratios of Whole Diet and Dietary Protein to Those of Bone Collagen and Carbonate." In *Prehistoric Human Bone: Archaeology at the Molecular Level*, edited by Karl Lambert and Gisella Grupe, 1–37. New York: Springer-Verlag.

Beattie, Owen, Brian Apland, Erik W. Blake, James A. Cosgrove, Sarah Gaunt, Sheila Greer, Alexander P. Mackie, Kjerstin E. Mackie, Dan Straathof, Valerie Thorp and Peter M. Troffe. 2000. "The Kwäday Dän Ts'inchi Discovery from a Glacier in British Columbia." *Canadian Journal of Archaeology* 24 (1–2): 129–47.

Brown, T.A., D. Erle Nelson, John S. Vogel and John R. Southon. 1988. "Improved Collagen Extraction by Modified Longin Method." *Radiocarbon* 30:171–77.

Chisholm, Brian S., D. Erle Nelson and Henry P. Schwarcz. 1982. "Stable Carbon Ratios as a Measure of Marine versus Terrestrial Protein in Ancient Diets." *Science* 216:1131–32.

———. "Marine and Terrestrial Protein in Prehistoric Diets on the British Columbia Coast." *Current Anthropology* 24:396–98.

Corr, Lorna T., Judith C. Sealy, Mark C. Horton and Richard P. Evershed. 2005. A Novel Marine Dietary Indicator Utilising Compound-Specific Bone Collagen Amino Acid ^{13}C Values of Ancient Humans." *Journal of Archaeological Science* 32 (3): 321–30.

Cruikshank, Julie. 2005. *Do Glaciers Listen? Local Knowledge, Colonial Encounters and Social Imagination.* Vancouver: University of British Columbia Press.

Davidson, George. 1901. "Explanation of an Indian Map of the Rivers, Lakes, Trails and Mountains from the Chilkaht to the Yukon Drawn by the Chilkaht Chief, Kohklux, in 1869." *Mazama* 2 (2): 75–92.

de Laguna, Frederica. 1972. *Under Mount Saint Elias: The History and Culture of the Yakutat Tlingit.* 3 parts. Smithsonian Contributions to Anthropology 7. Washington, DC: Smithsonian Institution.

DeNiro, Michael J. 1985. "Post-Mortem Preservation and Alteration of In Vivo Bone Collagen Isotope Ratios in Relation to Paleodietary Reconstruction." *Nature* 317:806–09.

Dickson, James H., Michael P. Richards, Richard J. Hebda, Petra Mudie, Owen Beattie, Susan Ramsay, Nancy J. Turner, Bruce J. Leighton, John M. Webster, Niki R. Hobischak, Gail S. Anderson, Peter M. Troffe and Rebecca J. Wigen. 2004. "Kwäday Dän Ts'inchí, the First Ancient Body of a Man from a North American Glacier: Reconstructing his Last Days by Intestinal and Biomolecular Analyses." *Holocene* 14 (4): 481–86.

Emmons, George T. 1991. *The Tlingit Indians.* Edited by Frederica de Laguna. Seattle: University of Washington Press.

Finney, Bruce P., Irene Gregory-Eaves, Jon Sweetman, Marianne S.V. Douglas and John P. Smol. 2000. "Impacts of Climatic Change and Fishing on Pacific Salmon Abundance over the Past 300 Years." *Science* 290: 795–99.

Glave, Edward J. 1890–91. "The Alaska Expedition." *Frank Leslie's Illustrated Newspaper* 70: 485, 572; 71: 84, 86, 87, 266, 286, 287, 310, 328, 332, 352, 374–76, 396, 397.

Greer, Sheila C., 2005. *IkVf-1 – The Kwädąy Dän Ts'inchj Site Catalogue (Whitehorse Collections).* Revised Spring 2005. On file, Champagne and Aishihik First Nations, Haines Junction.

Greer, Sheila C., and CTFN (Carcross-Tagish First Nation). 1995. *Skookum Stories on the Chilkoot/Dyea Trail.* Carcross, YT: Carcross-Tagish First Nation.

Howland, Mark R., Lorna T. Corr, Susan M.M. Young, Vicki Jones, Susan Jim, Nicholas J. van der Merwe, Alva D. Mitchell and Richard P. Evershed. 2003. "Expression of the Dietary Isotope Signal in the Compound-Specific ^{13}C Values of Pig Bone Lipids and Amino Acids." *International Journal of Osteoarchaeology* 13 (1-2): 54–65.

Jim, Susan. 2000. "The Development of Bone Cholesterol ^{13}C Values as a New Source of Palaeodietary Information: Models of Its Use in Conjunction with Bone Collagen and Apatite ^{13}C Values." PhD dissertation, University of Bristol.

Krause, Aurel. 1956. *The Tlingit Indians.* Translated by Erna Gunther. Vancouver: Douglas and McIntyre.

Lee-Thorp, J.A. 2008. "On Isotopes and Old Bones." *Archaeometry* 50:925–50.

Legros, Dominique. 1984. "Commerce entre Tlingits et Athapaskans Tutchones au XIXe siècle." *Recherches Amérindiennes au Québec* 14 (2): 11–24.

Lovell, Nancy C., Brian S. Chisholm, D. Erle Nelson and Henry P. Schwarcz. 1986. "Prehistoric Salmon Consumption in Interior British Columbia." *Canadian Journal of Archaeology* 10:99–106.

McClellan, Catharine. 1964. "Culture Contacts in the Early History Period in Northwestern North America." *Arctic Anthropology* 2 (2): 3–15.

———. (1975) 2001. *My Old People Say: An Ethnographic Survey of Southern Yukon Territory.* Hull, QC: Canadian Museum of Civilization.

Mudie, Petra J., Sheila Greer, Judith Brakel, James H. Dickson, Clara Schinkel, Ruth Peterson-Welsh, Margaret Stevens, Nancy J. Turner, Mary Shadow and Rosalie Washington. 2005. "Forensic Palynology and Ethnobotany of *Salicornia* Species (Chenopodiaceae) in Northwest Canada and Alaska." *Canadian Journal of Botany* 83:111–23.

Newton, Richard G., and Madonna L. Moss. 2005. *Haa Atxaahi Haa Kusteeyix Sitee, Our Food Is Our Tlingit Way of Life.* Forest Service Alaska Region Report #R10-MR-30. Juneau: US Department of Agriculture.

O'Connell, Tamsin C., and Robert E.M. Hedges. 1999. "Investigations into the Effect of Diet on Modern Human Hair Isotopic Values." *American Journal of Physical Anthropology* 108:409–25.

O'Connell, Tamsin, Robert E.M. Hedges, Matthew A. Healey and Andrew H.R.W. Simpson. 2001. "Isotopic Comparison of Hair, Nail and Bone: Modern Analyses." *Journal of Archaeological Science* 28:1247–55.

O'Leary, Beth L. 1992. *Salmon and Storage: Southern Tutchone Use of an Abundant Resource.* Occasional Papers in Archaeology No. 2. Whitehorse: Yukon Heritage Branch.

Peterson, Bruce J., Robert W. Howarth and Robert H. Garritt. 1985. "Multiple Stable Isotopes Used to Trace the Flow of Organic Matter in Estuarine Food Webs." *Science* 227:1361–63.

Peterson, Bruce J., and Brian Fry. 1987. "Stable Isotopes in Ecosystem Studies." *Annual Reviews of Ecological Systems,* 18:293–320.

Richards, M.P., S. Greer, L.T. Corr, O. Beattie, A. Mackie, R.P. Evershed, A. von Finster and J. Southon. 2007. "Radiocarbon Dating and Dietary Stable Isotope Analysis of Kwäday Dän Ts'ìnchj." *American Antiquity* 72 (4): 713–734.

Richards, Michael P., and Robert E.M. Hedges. 1999. "Stable Isotope Evidence for Similarities in the Types of Marine Foods Used by Late Mesolithic Humans at Sites along the Atlantic Coast of Europe." *Journal of Archaeological Science* 26:717–22.

Sealy, Judith. 2001. "Body Tissue Chemistry and Palaeodiet." In *Handbook of Archaeological Sciences,* edited by D.R. Brothwell and A.M. Pollard, 269–79. Chichester: John Wiley and Sons.

Seton-Karr, Heywood W. 1891. "Explorations in Alaska and North-west British Columbia." *Proceedings of the Royal Geographical Society* 2: 65–86.

Stott, Andrew W., Emma Davies, Richard P. Evershed and Noreen Tuross. 1997. "Monitoring the Routing of Dietary and Biosynthesised Lipids Through Compound-Specific Stable Isotope (^{13}C) Measurements at Natural Abundance." *Naturwissenschaften* 84 (2): 82–86.

Stott, Andrew W., Richard P. Evershed, Susan Jim, Vicki Jones, Juliet Rogers, Noreen Tuross and Stanley Ambrose. 1999. "Cholesterol as a New Source of Palaeodietary Information: Experimental Approaches and Archaeological Applications." *Journal of Archaeological Science* 26:705–16.

Swanton, John R. 1908. "Social Conditions, Beliefs and Linguistic Relationship of the Tlingit Indians." *Bureau of American Ethnology Annual Report* 26:391–485.

YHMA (Yukon Historical and Museums Association). 1994. *The Kohklux Map.* Whitehorse, Yukon. Poster reproduction of Koh--klux' Map, Chilkaht (Alaska and Yukon). Original by Chief Chilkaht Kohklux, 1852. Whitehorse: Yukon Historical and Museums Association.

Chapter 20

Billington, N., and P.D.N. Hebert. 1991. "Mitochondrial DNA Diversity in Fishes and Its Implications for Introductions." *Canadian Journal of Fisheries and Aquatic Sciences* 48:80–94.

Bilton, H.T., D.W. Jenkinson and M.P. Shepard. 1964. "A Key to Five Species of Pacific Salmon (Genus *Oncorhynchus*) Based on Scales Characters." *Journal of the Fisheries Research Board of Canada* 21 (5): 1267–88.

Bilton, H.T., and S.A.M. Ludwig. 1966. "Times of Annulus Formation on Scales of Sockeye, Pink and Chum Salmon in the Gulf of Alaska." *Journal of the Fisheries Research Board of Canada* 23 (9): 1403–10.

Brinkhuizen, D.C. 1997. "Some Remarks on Seasonal Dating of Fish Remains by Means of Growth Ring Analysis." In *Fish Remains and Humankind,* edited by A.K.G. Jones and R.A. Nicholson. Internet Archaeology 3. http://intarch.ac.uk/journal/issue3/ brink_toc.html.

Casteel, R.W. 1972. "A Key, Based on Scales, to the Families of Native California Freshwater Fishes." *Proceedings of the California Academy of Sciences* 39 (7): 75–86.

———. 1973. "The Scales of the Native Freshwater Fish Families of Washington." *Northwest Science* 47 (4): 230–38.

———. 1976. *Fish Remains in Archaeology and Paleo-environmental Studies.* London: Academic Press.

Deelder, C.L., and J.J. Willemse. 1973. "Age Determination in Freshwater Teleosts, Based on Annular Structures in Fin-Rays." *Aquaculture* 1: 365–71.

Fillatre, E.K., P. Etherton and D.D. Heath. 2003. "Bimodal Run Distribution in a Northern Population of Sockeye Salmon (*Oncorhynchus nerka*): Life History and Genetic Analysis on a Temporal Scale." *Molecular Ecology* 12 (7): 1793–1805.

Follett, W.I. 1967. "Fish Remains from Coprolites and Midden Deposits at Lovelock Cave, Churchill County, Nevada." University of California Archaeological Survey Reports 70: 94–115.

Hogman, W.J. 1968. "Annulus Formation on Scales of Four Species of Coregonids Reared Under Artificial Conditions." *Journal of the Fisheries Research Board of Canada* 25 (10): 2111–22.

Koo, T.S.Y. 1962. "Differential Scale Characters among Species of Pacific Salmon." In *Studies of the Alaska Red Salmon*, edited by T.S.Y. Koo, 127–35. Seattle: University of Washington Press.

McAllister, D.E. 1962. "Fish Remains from Ontario Indian Sites 700 to 2500 Years Old." *National Museum of Canada Natural History Papers* 17. Ottawa: National Museums of Canada.

Nielsen, J.L., C. Gan and W.K. Thomas. 1994. "Differences in Genetic Diversity for Mitochondrial-DNA between Hatchery and Wild Populations of *Oncorhynchus*." *Canadian Journal of Fisheries and Aquatic Sciences* 51:290–97.

Pearson, R.E. 1966. "Number of Circuli and Time of Annulus Formation on Scales of Pink Salmon (*Oncorhynchus gorbuscha*)." *Journal of the Fisheries Research Board of Canada* 23 (5): 747–56.

Struever, S., and P. Thurmer. 1972. "American Culture, 12 Layers Deep, Dug Up in Illinois." *Smithsonian* 3 (5): 26–33.

Watson, J.E. 1964. "Determining the Age of Young Herring from Their Otoliths." *Transactions of the American Fisheries Society* 93 (1): 11–20.

Waugh, B., P. Etherton, S. Stark and K. Jensen. 2004. "Abundance of the Sockeye Salmon Escapement in the Alsek River Drainage." Pacific Salmon Commission Technical Report No. 15.

Whaley, R.A. 1991. "An Improved Technique for Cleaning Fish Scales." *North American Journal of Fisheries Management* 11: 234–36.

Yang, D.Y., B. Eng, J.S. Waye, J.C. Dudar and S.R. Saunders. 1998. "Improved DNA Extraction from Ancient Bones Using Silica-Based Spin Columns." *American Journal of Physical Anthropology* 105:539–43.

Yang, D.Y., A. Cannon and S.R. Saunders. 2004. "DNA Species Identification of Archaeological Salmon Bone from the Pacific Northwest Coast of North America." *Journal of Archaeological Science* 31 (5): 619–31.

Chapter 21

Adovasio, J.M. 1977. *Basketry Technology: A Guide to Identification and Analysis*. Chicago: Aldine.

Beattie, Owen, Brian Apland, Erik W. Blake, James A. Cosgrove, Sarah Gaunt, Sheila Greer, Alexander P. Mackie, Kjerstin E. Mackie, Dan Straathof, Valerie Thorp and Peter M. Troffe. 2000. "The Kwäday Dän Ts'inchi Discovery from a Glacier in British Columbia." *Canadian Journal of Archaeology* 24 (1–2): 129–47.

Bernick, Kathryn. 1983. *A Site Catchment Analysis of the Little Qualicum River Site, DiSc 1: A Wet Site on the East Coast of Vancouver Island, BC*. National Museum of Man Mercury Series, Archaeological Survey of Canada Paper 118. Ottawa: National Museums of Canada.

——. 1987. "The Potential of Basketry for Reconstructing Cultural Diversity on the Northwest Coast." In *Ethnicity and Culture*, edited by Reginald Auger, Margaret F. Glass, Scott MacEachern and Peter H. McCartney, 251–57. Proceedings of the 18th Annual Chacmool Conference, Archaeological Association, University of Calgary.

——. 1998a. *Basketry and Cordage from Hesquiat Harbour, British Columbia*. Victoria: Royal British Columbia Museum.

——. 1998b. "Stylistic Characteristics of Basketry from Coast Salish Area Wet Sites." In *Hidden Dimensions: The Cultural Significance of Wetland Archaeology*, edited by Kathryn Bernick, 139–56. Vancouver: UBC Press.

——. 1999. "Lanaak (49XPA78): A Wet Site on Baranof Island, Southeastern Alaska." Permit 99-10. Unpublished report submitted to the Sitka Tribe of Alaska and the Alaska Office of History and Archaeology, Anchorage.

——. 2000. "Katete River Basketry Hat." Unpublished report. Submitted to Alan Hoover, Anthropological Collections, Royal BC Museum, Victoria.

——. 2001a. "Basketry Hat: A Technological and Stylistic Description." Unpublished report submitted to Champagne and Aishihik First Nations, Whitehorse, YT.

———. 2001b. "A Report on a Waterlogged Discovery." *Discovery* 29 (3): 5.

———. 2003. "A Stitch in Time: Recovering the Antiquity of a Coast Salish Basket Type." In *Emerging from the Mist: Studies in Northwest Culture History*, edited by R.G. Matson, Gary Coupland and Quentin Mackie, 230–43. Vancouver: UBC Press.

Croes, Dale R. 1977. "Basketry from the Ozette Village Archaeological Site: A Technological, Functional, and Comparative Study." PhD dissertation, Washington State University. University Microfilms, Ann Arbor.

———. 1987. "Locarno Beach at Hoko River, Olympic Peninsula, Washington: Wakashan, Salishan, Chimakuan or Who?" In *Ethnicity and Culture*, edited by Reginald Auger, Margaret F. Glass, Scott MacEachern and Peter H. McCartney, 259–83. Proceedings of the 18th Annual Chacmool Conference, Archaeological Association, University of Calgary.

Devine, Sue E. 1980. "Nootka Basketry Hats – Two Special Types." *American Indian Basketry* 1 (3): 26–31.

———. 1981. "Kwakiutl Spruce Root Hats." *American Indian Basketry* 1 (4): 24–27.

———. 1982. "Spruce Root Hats of the Tlingit, Haida and Tsimsian." *American Indian Basketry* 2 (2): 20–25.

Emmons, George T. 1993. *The Basketry of the Tlingit*. Reprint, bound with *The Chilkat Blanket*. Originally published 1903, Memoirs of the American Museum of Natural History, vol. 3, part 2. Sitka, AK: Sheldon Jackson Museum, Alaska State Museums.

Fifield, Terence E., and David E. Putnam. 1995. "Thorne River Basket: Description, Context, and Opportunity." Paper presented at the conference "Hidden Dimensions: The Cultural Significance of Wetland Archaeology," April 27–30, 1995, Vancouver.

Hodge, F.W. 1929. "A Nootka Basketry Hat." *Indian Notes* 6:254–58.

Laforet, Andrea. 1990. "Regional and Personal Style in Northwest Coast Basketry." In *The Art of Native American Basketry: A Living Legacy*, edited by Frank W. Porter III, 281–98. Westport, CT: Greenwood Press.

———. 2000. "Ellen Curley's Hat." In *Nuu-chah-nulth Voices, Histories, Objects & Journeys*, edited by Alan L. Hoover, 330–38. Victoria: Royal British Columbia Museum.

Paul, Frances. 1981. *Spruce Root Basketry of the Alaska Tlingit*. Reprint. Sitka, AK: Sheldon Jackson Museum.

Suttles, Wayne. 1990. *Handbook of North American Indians*. Vol. 7, *Northwest Coast*. Washington, DC: Smithsonian Institution.

Watkins, Frances E. 1939. "Potlatches and a Haida Potlatch Hat." *Masterkey* 12 (1): 11–17.

Young, Gregory. 2000. "Wood Identification of Artifacts Recovered with Kwaday Dan Ts'inchi." Canadian Conservation Institute Conservation Processes and Materials Research Report No. 76622. Unpublished report for Yukon Heritage Branch.

———. 2001. "Correction to Report: Wood Identification of Artifacts Recovered with the Kwaday Dan Sinchi." Canadian Conservation Institute Processes and Materials Research Report No. 76622. Unpublished report for Champagne and Aishihik First Nations and Yukon Heritage Branch.

Chapter 22

Beattie, Owen, Brian Apland, Erik W. Blade, James A. Cosgrove, Sarah Gaunt, Sheila Greer, Alexander P. Mackie, Kjerstin E. Mackie, Dan Straathof, Valerie Thorp and Peter M. Troffe. 2000. "The Kwäday Dan Ts'ìnchì Discovery from a Glacier in British Columbia." *Canadian Journal of Archaeology* 24 (1–2): 129–47.

Bernick, Kathryn. 2001. "Basketry Hat: A Technological and Stylistic Description." Unpublished report for Champagne and Aishihik First Nations Heritage program, Whitehorse, YT.

Cronyn, J.M. 1990. *The Elements of Archaeological Conservation*. London: Routledge.

Feist, William C. 1990. "Outdoor Wood Weathering and Protection." In *Archaeological Wood Properties, Chemistry and Preservation*, edited by Roger M. Rowell and R. James Barbour. Advances in Chemistry Series 225, papers from the Cellulose, Paper and Textile Division at the 196th National Meeting of the American Chemical Society, Los Angeles, CA, September 25–30, 1988. Washington, DC: American Chemical Society.

Florian, Mary-Lou E. 2005. "Identification of the Woods of Some of the Champagne and Aishihik First Nation Artifacts from the IkVf Site." Unpublished report for Champagne and Aishihik First Nations Heritage program.

Greer, Sheila C. *The Kwaday Dan Ts'inchi Site (IkVf-1) Catalogue – Whitehorse Collections*. With input from Valery

Monahan. Revised April 2006. On file, Champagne and Aishihik First Nations, Haines Junction, Yukon.

Grosjean, Martin, Peter J. Suter, Mathias Trachsel and Heinz Wanner. 2007. "Ice-borne Prehistoric Finds in Swiss Alps Reflect Holocene Glacier Fluctuations." *Journal of Quaternary Science* 22 (3): 203–07. doi: 10.1002/jqs.1111.

Hamilton, Donny L. 1998. "Wood Conservation." *Methods of Conserving Underwater Archaeological Material Culture.* Conservation Files: ANTH 605, Conservation of Cultural Resources I. Nautical Archaeology Program, Texas A&M University.

Komejan, Diana, Valery Monahan and Cathy Ritchie. 1999–2008. The Government of Yukon Conservation Program records of the Kwäday Dän Ts'ìnchi Collection.

Mager, Mary. 2001. "Electron Microscopic Analysis of a Pigment Sample from the KDT Hat." Unpublished report for Champagne and Aishihik First Nations Heritage program.

Monahan, Valery. 2004. "The Yukon 'Ice Patch' Collection: Conserving a Unique Archaeological Resource." Paper presented at the CAC/ACCR Conference in Quebec City May 26–30, 2004.

Scott, Rosalie, and Tara Grant, eds. 2007. *Conservation Manual for Northern Archaeologists.* Revised 3rd edition. Yellowknife, NT: Department of Education, Culture and Employment. Accessed June 2008. http://pwnhc.ca/programs/downloads/conservation_mannual.pdf.

Young, Gregory. 2000. "Wood Identification of Artifacts Recovered with Kwaday Dan Ts'inchi." Canadian Conservation Institute Conservation Processes and Materials Research Report No. 76622. Unpublished report for Yukon Heritage Branch.

———. 2001. "Correction to Report: Wood Identification of Artifacts Recovered with the Kwaday Dan Sinchi." Canadian Conservation Institute Processes and Materials Research Report No. 76622. Unpublished report for Champagne and Aishihik First Nations and Yukon Heritage Branch.

———. 2005. "Preservation Assessment of the Hat Band of Kwaday Dan Ts'inchi." Borden no. IkVf-1:113, *Canadian Conservation Institute Conservation Research Report No. 92545* for Yukon Heritage Branch on behalf of Champagne Aishihik First Nations Heritage program.

Chapter 23

Adamson, John. 1993. Interview by Sarah Gaunt on Tatshenshini/Chilkat area, January 15. Transcript on file, Champagne and Aishihik First Nations, Haines Junction.

Beattie, Owen, Brian Apland, Erik W. Blake, James A. Cosgrove, Sarah Gaunt, Sheila Greer, Alexander P. Mackie, Kjerstin E. Mackie, Dan Straathof, Valerie Thorp and Peter M. Troffe. 2000. "The Kwäday Dän Ts'ìnchi Discovery from a Glacier in British Columbia." *Canadian Journal of Archaeology* 24 (1–2): 129–47.

Borden, Charles E. 1952. "A Uniform Site Designation Scheme for Canada." *Anthropology in British Columbia* 3:44–48.

Brown, Steven C., ed. 2000. *Spirits of the Water: Native Art Collected on Expeditions to Alaska and British Columbia, 1774–1910.* Seattle: University of Washington Press.

Carcross-Tagish First Nation, Champagne and Aishihik First Nations, Kluane First Nation, Kwanlin Dün First Nation. 2002. *Ice Patch.* Newsletter published by Carcross-Tagish First Nation, Champagne and Aishihik First Nations, Kluane First Nation, Kwanlin Dün First Nation. On file, Champagne and Aishihik First Nations, Haines Junction, YT.

Carcross-Tagish First Nation, Champagne and Aishihik First Nations, Kluane First Nation, Kwanlin Dün First Nation, Ta'an Kwach'än Council, Teslin Tlingit Council. 2005. *Ice Patch*, Issue 2. Newsletter published by Carcross-Tagish First Nation, Champagne and Aishihik First Nations, Kluane First Nation, Kwanlin Dün First Nation, Ta'an Kwach'än Council, Teslin Tlingit Council. On file, Champagne and Aishihik First Nations. www.taan.ca/assets/files/Heritage/Icepatchnewsletter2005_lowres.pdf.

Champagne and Aishihik First Nations. 2000a. Transcript of meeting with representatives of Chilkat Indian Tribe with Champagne and Aishihik First Nations representatives, in regard to the Kwäday Dän Ts'ìnchi discovery, October 27. On file, Champagne and Aishihik First Nations Heritage program, Haines Junction, YT.

———. 2000a. Partial transcript of video recording of December 1 interview session with Bill Hanlon, Mike Roch and Warren Ward regarding Kwäday Dän Ts'ìnchi. On file, Champagne and Aishihik First Nations, Haines Junction, YT.

Clark, Annette M. 1974. *The Athapaskans: Strangers of the North.* Ottawa: National Museums of Canada.

Clark, D.W. 1990. "Prehistory of the Western Subarctic." In *Handbook of North American Indians*, vol. 6, *Subarctic*, edited by June Helm, 107–29. Washington, DC: Smithsonian Institution.

Cole, Douglas, and David Darling. 1990. "History of the Early Period." In *Handbook of North American Indians*, vol. 7, *Northwest Coast*, edited by Wayne Suttles, 119–34. Washington, DC: Smithsonian Institution.

Davis, Stanley. 1990. "The Prehistory of Southeastern Alaska." In *Handbook of North American Indians*, vol. 7, *Northwest Coast*, edited by Wayne Suttles, 197–202. Washington, DC: Smithsonian Institution.

———. 1996. "The Archaeology of the Yakutat Foreland: A Socioecological View." PhD dissertation, Texas A&M University.

de Laguna, Frederica. 1960. *The Story of Tlingit Community. A Problem in the Relationship between Archaeological, Ethnological, and Historical Methods*. Bureau of American Ethnology Bulletin 172. Washington, DC: Government Printing Office.

———. 1972. *Under Mount Saint Elias: The History and Culture of the Yakutat Tlingit*. 3 parts. Smithsonian Contributions to Anthropology 7. Washington, DC: Smithsonian Institution.

———. 1990. "Tlingit." In *Handbook of North American Indians*, vol. 7, *Northwest Coast*, edited by Wayne Suttles, 203–28. Washington, DC: Smithsonian Institution.

de Laguna, Frederica, and F.A. Riddell, D.F. McGeein, K.S. Lane and J.A. Freed. 1964. *Archaeology of the Yakutat Bay Area, Alaska*. Bureau of American Ethnology Bulletin 192. Washington, DC: Government Printing Office.

Emmons, George T. 1991. *The Tlingit Indians*. Edited by Frederica de Laguna. New York: American Museum of Natural History.

Fienup-Riordan, Ann. 2005. *Yup'ik Elders at the Ethnologisches Museum Berlin: Fieldwork Turned on Its Head*. Seattle: University of Washington Press.

Florian, Mary-Lou E. 2005. "Identification of the Woods of Some of the Champagne and Aishihik First Nations Artifacts from the IkVf Site." Manuscript prepared for Champagne and Aishihik First Nations Heritage program. On file, Champagne and Aishihik First Nations, Haines Junction, YT.

Greer, Sheila. 2005a. "Errata/Clarification Note to Report by Mary-Lou E. Florian, titled 'Identification of the woods of some of the Champagne and Aishihik First Nations artifacts from the IkVf Site'," September 16. On file, Champagne and Aishihik First Nations, Haines Junction, YT.

———. 2005b. The Kwaday Dan Ts'inchi Site (IkVf-1) Catalogue – Whitehorse Collections. With input from Valery Monahan. Revised April 2006. On file, Champagne and Aishihik First Nations, Haines Junction, YT.

———. 2005c. "Sheila Greer Notes with Marge Jackson, Moose Jackson, Harry Smith and Frances Oles. 2005." On file, Champagne and Aishihik First Nations. Haines Junction, YT.

———. 2005d. "Sheila Greer Notes and Transcript with John Adamson." Recorded at Klukshu General Assembly, July 23 and 24. Transcript on file, Champagne and Aishihik First Nations, Haines Junction, YT.

———. 2006a. "Sheila Greer Notes with John Adamson May 11, 2006." Manuscript on file, Champagne and Aishihik First Nations, Haines Junction, YT.

———. 2006b. "Sheila Greer 2006 Notes with Paddy Jim and Marge Jackson." Manuscript on file, Champagne and Aishihik First Nations Heritage program, Haines Junction, YT.

———. 2008. "Sheila Greer Southern Tutchone Vocabulary Work with Marge Jackson, Frances Joe and Hayden Woodruff; Spellings by Vivian Smith, Winter 2008." Manuscript on file, Champagne and Aishihik First Nations Heritage program, Haines Junction, YT.

Hare, P. Gregory, S. Greer, R. Gotthardt, R. Farnell, V. Bowyer, C. Schweger and Diane Strand. 2004. "Ethnographic and Archaeological Investigations of Alpine Ice Patches, Southwest Yukon, Canada." *Arctic* 57 (3): 260–72.

Hare, P. Gregory, C.D. Thomas, T.N. Toper and R.M. Gotthardt. 2012. "The Archaeology of Yukon Ice Patches: New Artifacts, Observations and Insights," supplement 1, *Arctic* 65: 118–135.

Harp, Elmer, Jr. 2005. *North to The Yukon Territory Via the Alcan Highway in 1948: Field Notes from The Andover-Harvard Expedition*. Occasional Papers in Anthropology No. 14. Hudé Hudän Series. Whitehorse: Yukon Archaeology Programme.

Helm, June, ed. 1981. *Handbook of North American Indians*, vol. 6, *Subarctic*. Washington, DC: Smithsonian Institution.

Helwig, Kate, Jane Sirois, Gregory Young, Jennifer Poulin and Jeremy Powell. 2008. "Scientific Examination of the Kwaday Dan Ts'inchi Hand Tool." Canadian Conservation Institute Report ARL 4444. Manuscript prepared for Champagne and Aishihik First Nations Heritage program, Haines Junction, YT.

Helwig, Kate, Jennifer Poulin and Valery Monahan. 2004. "The Characterization of Paint and Adhesive Residues on Hunting Tools from Southern Yukon Ice Patches." Paper presented at the Alaska Anthropological Association Annual Meeting, Whitehorse, YT.

Johnson, Frederick, and Hugh M. Raup. 1964. "Investigations in Southwest Yukon: Geobotanical and Archaeological Reconnaissance." In *Investigations in Southwest Yukon*, Numbers 1 and 2, 1–164. Papers of the Peabody Foundation for Archaeology 6. Andover, Massachusetts: Phillips Academy.

Jonaitis, Aldona. 1988. *From the Land of the Totem Poles: The Northwest Coast Indian Art Collection at the American Museum of Natural History*. New York: American Museum of Natural History.

Krause, Aurel. 1956. *The Tlingit Indians*. Translated by Erna Gunther. Vancouver: Douglas and McIntyre.

Krause, Aurel, and Arthur Krause. 1981. *Journey to the Tlingits by Aurel and Arthur Krause, 1881/82*. Translated by Margot Krause McCaffrey. Haines, Alaska: Centennial Commission.

———. 1993. *To the Chukchi Peninsula and to the Tlingit Indians 1881/1882: Journals and Letters by Aurel and Arthur Krause*. Translated by Margot Krause McCaffrey. The Rasmuson Library Historical Translation Series 8. Fairbanks, AK: University of Alaska Press.

Langdon, Steve. 1986. "Traditional Tlingit Stone Fishing Technologies." *Alaska Native News* 4 (3): 21–26.

Leer, Jeff, Doug Hitch and John Ritter. 2001. *Interior Tlingit Noun Dictionary: The Dialects Spoken by Tlingit Elders of Carcross and Teslin, Yukon, and Atlin, British Columbia*. Whitehorse: Yukon Native Language Centre.

Lohse, E.S. and Frances Sundt. 1990. "History of Research: Museum Collections." In *Handbook of North American Indians*, vol. 7, *Northwest Coast*, edited by Wayne Suttles, 88–97. Washington, DC: Smithsonian Institution.

Low, Jean. 1977. "George Thornton Emmons." *Alaska Journal* 7 (1): 2–11.

Mackie, Alexander P. 2000. Email correspondence to Sarah Gaunt and Sheila Greer, regarding February 4 interview with Warren Ward on provenience of artifacts collected from site. On file, Champagne and Aishihik First Nations, Haines Junction, Yukon.

McClellan, Catharine. (1975) 2001. *My Old People Say: An Ethnographic Survey of Southern Yukon Territory*. 2 vols. Publications in Ethnology 6. Ottawa: National Museums of Canada.

———. 1981. "Inland Tlingit." In *Handbook of North American Indians*, vol. 6 *Subarctic*, edited by June Helm, 469–80. Washington, DC: Smithsonian Institution.

———. 2007. *My Old People's Stories: A Legacy for Yukon First Nations*. Part 1, *Southern Tutchone Narrators*. Occasional Papers in Yukon History 5 (1). Hudé Hudän Series. Whitehorse: Yukon Cultural Services Branch.

McClellan, Catharine, and Glenda Denniston. 1981. "Environment and Culture in the Cordillera." In *Handbook of North American Indians*, vol. 6, *Subarctic*, edited by June Helm, 372–87. Washington, DC: Smithsonian Institution.

McClellan, Catharine, L. Birckel, R. Bringhurst, J.A. Fall, C. McCarthy and J.R. Sheppard. 1987. Part of the Land, Part of the Water. A History of Yukon Indians. Vancouver: Douglas and McIntyre.

Nelson, Richard K. 1983a. *Make Prayers to the Raven: A Koyukon View of the Northern Forest*. Chicago: University of Chicago Press.

———. 1983b. *The Athabaskans: People of the Boreal Forest*. Alaska Historical Commission Studies in History No. 27. Fairbanks: University of Alaska Museum.

Oberg, Kalervo. 1973. *The Social Economy of the Tlingit Indians*. American Ethnological Society Monograph 55. Seattle: University of Washington Press.

O'Brien, Thomas A. 1997. "Athabaskan Implements from the Skin House Days as Related by Reverend David Salmon." Master's thesis, University of Alaska Fairbanks.

O'Leary, Laura Beth. 1992. *Salmon and Storage: Southern Tutchone Use of an Abundant Resource*. Occasional Papers in Archaeology 2. Whitehorse: Yukon Heritage Branch.

Olson, Ronald. 1936. "Some Trading Customs of the Chilkat Tlingit." In *Essays in Anthropology Presented to A.L. Kroeber*, edited by Robert L. Lowie, 211–14. Berkeley: University of California Press.

Osgood, Cornelius. 1936. *Contributions to the Ethnography of the Kutchin*. Publications in Anthropology, 14. New Haven: Yale University.

———. 1937. *The Ethnography of the Tanaina*. Publications in Anthropology, 16. New Haven, CT: Yale University.

———. 1940. *Ingalik Material Culture*. Publications in Anthropology, 22. New Haven, CT: Yale University.

———. 1971. *The Han Indians*. Publications in Anthropology 74. New Haven, CT: Yale University.

Sackett, Russell. 1979. *The Chilkat Tlingt: A General Overview*. Cooperative Park Studies Unit Occasional Paper 23. Fairbanks: University of Alaska Fairbanks.

Scherer, J.A. 1981. "Repository Sources for Subarctic Photographs." *Arctic Anthropology* 18 (2): 59–65.

Schwatka, Frederick. (1885) 1983. *Along Alaska's Great River*. Rahway, NJ: W.L. Mershon and Co. Reprint, Anchorage, AK: Northwest Publishing Company.

Shotridge, Louis. 1920. "Ghost of Courageous Adventurer." *Museum Journal* 11 (1): 11–26.

Southern Tutchone Tribal Council. 1999. *Dän k'è Kwänjè Literacy Workshops*. Haines Junction: Southern Tutchone Tribal Council.

Stewart, Hilary. 1977. *Indian Fishing: Early Methods on the Northwest Coast*. Vancouver and Toronto: Douglas and McIntyre.

Suttles, Wayne, ed. 1990. In *Handbook of North American Indians*, vol. 7, *Northwest Coast*. Washington, DC: Smithsonian Institution.

Thompson, Judy. 2008. *Recording Their Story: James Teit and the Tahltan*. Gatineau: Canadian Museum of Civilization.

Thornton, Thomas F. 2004. *Klondike Gold Rush National Historical Park, Ethnographic Overview and Assessment, Final Report*. Anchorage, AK: US National Parks Service. www.nps.gov/history/history/online_books/klgo/ethnographic_overview.pdf.

Tlen, Daniel L. 1993. *Kluane Southern Tutchone Glossary*. Occasional Papers of the Northern Research Institute Monograph 1. Whitehorse: Northern Research Institute.

VanStone, James. 1974. *Athapaskan Adaptations: Hunters and Fishermen of the Subarctic Forests*. Chicago: Aldine.

VanStone, James, and William E. Simeone. 1986. *And He Was Beautiful: Contemporary Athapaskan Material Cultural in the Collections of the Field Museum of Natural History*. Fieldiana Anthropology, n.s., 10. Chicago: Field Museum of Natural History.

Wood, C.E.S. 1882. "Among the Thlinkits in Alaska." *Century Magazine* 24 (3): 323–39.

Wyatt, Victoria. 1989. *Images from the Inside Passage: An Alaskan Portrait by Winter & Pond*. Seattle: University of Washington Press.

Young, Gregory. 2001. "Correction to Report: Wood Identifications of Artifacts Recovered with Kwaday Dan Sinchi. Report No. 76622." Ottawa: Canadian Conservation Institute.

Chapter 24

Abercrombie, William R. 1900. "A Supplementary Expedition into the Copper River Valley, 1884." In *Compilation of Narratives of Explorations in Alaska*. Washington, DC: Government Printing Office.

Acheson, Steven. 2003. "The Thin Edge: Evidence for Precontact Use and Working of Metal on the Northwest Coast." In *Emerging from the Mist: Studies in Northwest Coast Culture History*, edited by R.G. Matson, Gary Coupland and Quentin Mackie, 213–29. Vancouver: University of British Columbia Press.

Allen, Henry T. 1887. *Report of an Expedition to the Copper, Tanana, and Koyukuk Rivers, in the Territory of Alaska, in the Year 1885*. Washington, DC: Government Printing Office.

Beattie, Owen, Brian Apland, Erik W. Blake, James A. Cosgrove, Sarah Gaunt, Sheila Greer, Alexander P. Mackie, Kjerstin E. Mackie, Dan Straathof, Valerie Thorp and Peter M. Troffe. 2000. "The Kwäday Dän Ts'inchi Discovery from a Glacier in British Columbia." *Canadian Journal of Archaeology* 24 (1–2): 129–47.

Boas, Franz. 2002. *Indian Myths and Legends from the North Pacific Coast of America: A Translation of Franz Boas 1895 Edition of Indianische Sagen von der Nord-Pacifischen Kuste Amerikas*. Translated by Deitrich Bertz. Vancouver: Talon Books.

Bostock, Hugh S. 1957. *Yukon Territory Selected Field Reports of the Geological Survey of Canada 1898 to 1933*. Ottawa: Department of Mines and Technical Surveys.

Broderick, Thomas M. 1929. "Zoning in Michigan Copper Deposits and Its Significance." *Economic Geology* 24:149–62.

Brooks, Alfred H. 1900. *A Reconnaissance from Pyramid Harbor to Eagle City, Alaska, Including a Description of the Copper Deposits of the Upper White and Tanana Rivers.* Washington, DC: Government Printing Office.

Cooper, H. Kory. 2006. "Copper and Social Complexity: Frederica de Laguna's Contribution to Our Understanding of the Role of Metals in Native Alaskan Society." *Arctic Anthropology* 43:148–63.

———. 2007. "The Anthropology of Native Copper Technology and Social Complexity in Alaska and the Yukon Territory: An Analysis Using Archaeology, Archaeometry, and Ethnohistory." PhD dissertation, University of Alberta.

Cooper, H. Kory, M. John M. Duke, Antonio Simonetti and GuangCheng Chen. 2008. "Trace Element and Pb Isotope Provenance Analyses of Native Copper in Northwestern North America: Results of a Recent Pilot Study Using INAA, ICP-MS, and LA-MC-ICP-MS." *Journal of Archaeological Science* 35 (9): 1732–47.

de Laguna, Frederica. 1956. *Chugach Prehistory: The Archaeology of Prince William Sound, Alaska.* Seattle: University of Washington Press.

———. 1960. *The Story of a Tlingit Community: A Problem in the Relationship between Archaeological, Ethnological and Historical Methods.* Bureau of American Ethnology Bulletin 17. Washington, DC: Government Printing Office.

———. 1972. *Under Mount Saint Elias: The History and Culture of the Yakutat Tlingit.* Washington, DC: Smithsonian Institution.

de Laguna, Frederica, and Catharine McClellan. 1981. "Ahtna." In *Handbook of North American Indians,* vol. 6, *Subarctic,* edited by June Helm, 641–63. Washington, DC: Smithsonian Institution.

de Laguna, Frederica, Francis A. Riddell, Donald F. McGeein, Kenneth S. Lane and J. Arthur Freed. 1964. *Archaeology of the Yakutat Bay Area, Alaska.* Bureau of American Ethnology Bulletin 192. Washington, DC: Government Printing Office.

Emmons, George T. 1991. *The Tlingit Indians.* Seattle: University of Washington Press.

Franklin, Ursula M. 1982. "Folding: A Prehistoric Way of Working Native Copper in the North American Arctic." *Museum Applied Science Center for Archaeology* 2:48–52.

Franklin, Ursula M., Ellen Badone, Ruth Gotthardt and Brian Yorga. 1981. *An Examination of Prehistoric Copper Technology and Copper Sources in Western Arctic and Subarctic North America.* Ottawa: National Museums of Canada.

Glave, Edward J. 1892. "Pioneer Packhorses in Alaska." *Century Magazine* 44:869–81.

Hayes, Charles W. 1892. "An Expedition through the Yukon District." *National Geographic* 4:117–62.

Kari, James. 2005. *Copper River Place Names: Report on Culturally Important Places to Alaska Native Tribes in South Central Alaska.* Fairbanks, AK: Bureau of Land Management.

Keddie, Grant. 1990. "The Question of Asiatic Objects on the North Pacific Coast of America." *Contributions to Human History* 3:1–26.

Keithahn, Edward L. 1964. "The Origin of the 'Chief's Copper' or Tinneh." *University of Alaska Anthropological Papers* 12:59–78.

Legros, Dominique. 1984. "Commerce entre Tlingits et Athapaskans Tutchones au XIX siecle." *Recherche Amerindiennes au Québec* 14:11–24.

McCartney, Allen P. 1988. "Later Prehistoric Metal Use in the New World Arctic." In *The Late Prehistoric Development of Alaska's Native People,* edited by Robert D. Shaw, Roger K. Harritt and Don E. Dumond, 57–94. Anchorage: Aurora.

McClellan, Catharine. 1964. "Culture Contacts in the Early Historic Period in Northwestern North America." *Arctic Anthropology* 2:3–15.

———. 1975. *My Old People Say: An Ethnographic Survey of Southern Yukon Territory.* Ottawa: National Museums of Canada.

———. 1981. "Tutchone". In *Handbook of North American Indians,* vol. 6, *Subarctic,* edited by June Helm, 493–513. Washington, DC: Smithsonian Institution.

———. 1987. *Part of the Land, Part of the Water: A History of the Yukon Indians.* Vancouver: Douglas and McIntyre.

Moffit, Fred H., and Adolph Knopf. 1910. *Mineral Resources of the Kotsina and Chitina Valleys, Copper River Region.* Washington, DC: Government Printing Office.

Moss, Madonna L., Jon M. Erlandson and Robert Stuckenrath. 1989. "The Antiquity of Tlingit Settlement on Admiralty Island." *American Antiquity* 54:534–43.

Orth, Donald J. 1967. *Dictionary of Alaska Place Names.* Washington, DC: Government Printing Office.

Patterson, Clair C. 1971. "Native Copper, Silver, and Gold Accessible to Early Metallurgists." *American Antiquity* 36:286–321.

Peter, J.M. 1989. "The Windy Craggy Copper-Cobalt-Gold Massive Sulphide Deposit, Northwestern British Columbia." *Geological Field Work 1988*, Paper 1989-1, 455-466. Victoria: Ministry of Energy, Mines and Petroleum Resources.

Powell, Addison M. 1909. *Trailing and Camping in Alaska.* Vancouver: Douglas and McIntyre.

Rapp, George Jr., James Allert, Vanda Vitali, Zhichun Jing and Eller Henrikson. 2000. "Determining Geologic Sources of Artifact Copper: Source Characterization Using Trace Element Patterns." Lanham, MD: University Press of America.

Richards, Michael P., Sheila Greer, Lorna T. Corr, Owen Beattie, Alexander Mackie, Richard P. Evershed, Al von Finster and John Southon. 2007. "Radiocarbon Dating and Dietary Stable Isotope Analysis of *Kwädąy Dän Ts'ínchị.*" *American Antiquity* 72 (4): 719–33.

Rogers, Edward S. 1965. *An Athapaskan Type of Knife.* Anthropological Papers 9. Ottawa: National Museums of Canada.

Sanborn, Michael, and Kevin Telmer. 2003. "The Spatial Resolution of LA-ICP-MS Line Scans across Heterogeneous Materials such as Fish Otoliths and Zoned Minerals." *Journal of Analytical Atomic Spectrometry* 18:1231–37.

Schwatka, Frederick. 1996. *Schwatka's Last Search: The New York Ledger Expedition through Unknown Alaska and British America: Including the Journal of Charles Willard Hayes, 1891.* Fairbanks: University of Alaska Press.

Shinkwin, Anne D. 1979. *Dakah De'nin's Village and the Dixthada Site: A Contribution to Northern Athapaskan Prehistory.* Ottawa: National Museums of Canada.

Swanton, John R. 1909. *Tlingit Myths and Texts.* Washington, DC: Government Printing Office.

Wayman, Michael L. 1989. "Native Copper: Humanity's Introduction to Metallurgy." In *All That Glitters: Readings in Historical Metallurgy,* edited by Michael L. Wayman, 3–6. Montreal: The Metallurgical Society of the Canadian Institute of Mining and Metallurgy.

Wayman, Michael L., Robert R. Smith, Clifford G. Hickey and M. John M. Duke. 1985. "The Analysis of Copper Artifacts of the Copper Inuit." *Journal of Archaeological Science* 12: 367–75.

Wayman, Michael L., Jonathan C.H. King and Paul T. Craddock. 1992. *Aspects of Early North American Metallurgy.* London: British Museum.

Workman, William B. 1976. "Archaeological Investigations at GUL-077: A Prehistoric Site Near Gulkana, Alaska." Unpublished report. Anchorage: Alaska Methodist University.

Chapter 25

Beattie, O., B. Apland, E.W. Blake, J.A. Cosgrove, S. Gaunt, S. Greer, A.P. Mackie, K.E. Mackie, D. Straathof, V. Thorp, and P. Troffe. 2000. "The Kwaday Dan Ts'inchi Discovery from a Glacier in British Columbia." *Canadian Journal of Archaeology* 24 (1–2): 129–47.

Buchwald, V.F. and R.S. Clarke, Jr. 1989. "Corrosion of Fe-Ni Alloys by Cl-containing Akagenéite (β-FeOOH): The Antarctic Meteorite Case." *American Mineralogist* 74: 656–67.

Burke Museum. 2008. Online ethnology collection accessed June 2008. http://www.washington.edu/burkemuseum/collections/ethnology/index.php.

Callister, W.D. 2003. *Materials Science and Engineering: An Introduction*, 6 ed. New York: John Wiley and Sons.

Chadefaux, C., C. Vignaud, M. Menu and I. Reiche. 2008. "Multianalytical Study of Palaeolithic Reindeer Antler. Discovery of Antler Traces in Lascaux Pigments by TEM." *Archaeometry* 50 (3): 516–34.

Egan, B. 1997. *The Ecology of the Mountain Hemlock Zone.* Victoria: Ministry of Forests. Accessed June 2008. http://www.for.gov.bc.ca/hfd/pubs/docs/Bro/bro51.pdf.

———. 1999. *The Ecology of the Coastal Western Hemlock Zone.* Victoria: Ministry of Forests. http://www.for.gov.bc.ca/hfd/pubs/docs/Bro/bro31.pdf.

Emmons, George Thornton. 1991. *The Tlingit Indians.* Edited by Frederica de Laguna. Seattle: University of Washington Press.

Keddie, G. 1990. *The Question of Asiatic Objects on the North Pacific Coast of America: Historic or Prehistoric?* Contributions to Human History 3. Victoria: Royal British Columbia Museum.

——. 2006. *The Early Introduction of Iron among First Nations of British Columbia*. Accessed December 2015. http://royalbcmuseum.bc.ca/staffprofiles/files/2013/08/An-Early-Introduction-to-Iron-in-B.C.-Grant-Keddie.pdf.

O'Connor, T.P. 1987. "On the Structure, Chemistry and Decay of Bone, Antler and Ivory." In *Archaeological Bone, Antler and Ivory*, Occasional Papers 5, 6–8. London: United Kingdom Institute for Conservation of Historic and Artistic Works.

Richards, M.P., S. Greer, L.T. Corr, O. Beattie, A. Mackie, R.P. Evershed, A. von Finster and J. Southon. 2007. "Radiocarbon Dating and Dietary Stable Isotope Analysis of Kwäday Dän Ts'ìnchj." *American Antiquity* 72 (4): 719-33.

Scott, D.A., and G. Eggert. 2007. "The Vicissitudes of Vivianite as a Pigment and Corrosion Product." *Reviews in Conservation* 8:3–14.

Selwyn, Lyndsie. 2004. *Metals and Corrosion: A Handbook for the Conservation Professional*. Ottawa: Ministry of Public Works and Government Services Canada.

Wainwright, I. 1990. "Examination of Paintings by Physical and Chemical Methods." In *Shared Responsibility: Proceedings of a Seminar for Curators and Conservators, National Gallery of Canada, Ottawa, Canada, 26, 27 and 28 October 1989.* 79–102. Ottawa: National Gallery of Canada.

Wayman, Michael L. 1989. "On the Early Use of Iron in the Arctic." In *All That Glitters: Readings in Historical Metallurgy*, edited by Michael L. Wayman, 94–100. Montreal: The Metallurgical Society of the Canadian Institute of Mining and Metallurgy.

Chapter 26

Boas, Franz. 1909. "The Kwakiutl of Vancouver Island." *Memoirs of the AMNH* 8, part 2, 301–522. New York: G.E. Stechert.

Grew, Francis, and Margrethe de Neergaard. 1988. *Shoes and Pattens – Medieval Finds from Excavations in London*. London: Museum of London.

McClellan, Catharine. 2001. *My Old People Say: An Ethnographic Survey of Southern Yukon Territory,* part 1. Mercury Series Canadian Ethnology Service Paper 137. Ottawa: Canadian Museum of Civilization.

Wyatt, Victoria. 1989. *Images from the Inside Passage: An Alaskan Portrait by Winter and Pond*. Seattle: University of Washington Press.

Chapter 27

Beattie O., B. Apland, E.W. Blake, J.A. Cosgrove, S. Gaunt, S. Greer, A.P. Mackie, K.E. Mackie, D. Straathof, V. Thorp, P. Troffe. 2000. "The Kwäday Dän Ts'inchi Discovery from a Glacier in British Columbia." *Canadian Journal of Archaeology* 24 (1–2): 129–47.

Burger J., S. Hummel, B. Herrmann and W. Henke. 1999. "DNA Preservation: A Microsatellite-DNA Study on Ancient Skeletal Remains." *Electrophoresis* 20:1722–28.

Burger J., I. Pfeiffer, S. Hummel, R. Fuchs, B. Brenig and B. Herrmann. 2001. "Mitochondrial and Nuclear DNA from (Pre)historic Hide-Derived Materials." *Ancient Biomolecules* 3 (3): 227–38.

de Laguna, F. 1972. *Under Mount Saint Elias: The History and Culture of the Yakutat Tlingit*. Washington, DC: Smithsonian Institution Press.

Dickson, James H., Michael P. Richards, Richard J. Hebda, Petra Mudie, Owen Beattie, Susan Ramsay, Nancy J. Turner, Bruce J. Leighton, John M. Webster, Niki R. Hobischak, Gail S. Anderson, Peter M. Troffe and Rebecca J. Wigen. 2004. "Kwäday Dän Ts'ìnchí, the First Ancient Body of a Man from a North American Glacier: Reconstructing his Last Days by Intestinal and Biomolecular Analyses." *Holocene* 14 (4): 481–86.

Eddingsaas, A.A., B.K. Jacobsen, E.P. Lessa and J.A. Cook. 2004. "Evolutionary History of the Arctic Ground Squirrel (*Spermophilus parryii*) in Nearctic Beringia." *Journal of Mammalogy* 85 (4): 601–10.

Emmons, G.T. 1991. *The Tlingit Indians*. Edited by Frederica de Laguna and with a biography by Jean Low. Seattle: University of Washington Press.

Gilbert, M.T.P., L.P. Tomsho, S. Rendulic, M. Packard, D.I. Drautz, A. Sher, A. Tikhonov, L. Dalen, T. Kuznetsova, P. Kosintsev, P.F. Campos, T. Higham, M.J. Collins, A.S. Wilson, F. Shidlovskiy, B. Buigues, P.G.P. Ericson, M. Germonpre, A. Gotherstrom, P. Iacumin, V. Nikolaev, M. Nowak-Kemp, E. Willerslev, J.R. Knight, G.P. Irzyk, C.S. Perbost, K.M. Fredrikson, T.T. Harkins, S. Sheridan, W. Miller and S.C. Schuster. 2007. "Whole-Genome Shotgun Sequencing of Mitochondria from Ancient Hair Shafts." *Science* 317 (5846): 1927–30.

Gilbert, M.T.P., A.S. Wilson, M. Bunce, A.J. Hansen, E. Willerslev, B. Shapiro, T.F.G. Higham, M.P. Richards, T.C. O'Connell, D.J. Tobin, R.C. Janaway and A. Cooper. 2004. "Ancient Mitochondrial DNA from Hair." *Current Biology* 14 (12): R464.

Hall, T.A. 1999. "BioEdit: A User-Friendly Biological Sequence Alignment Editor and Analysis Program for Windows 95/98/NT," *Nucleic Acids Symposium Series* 41:95–98.

Haynes, S., J.B. Searle, A. Bertman and K.M. Dobney. 2002. "Bone Preservation and Ancient DNA: The Application of Screening Methods for Predicting DNA Survival." *Journal of Archaeological Science* 29 (6): 585–92.

Higuchi, R., C.H. von Beroldingen, G.F. Sensabaugh and H.A. Erlich. 1988. "DNA Typing from Single Hairs." *Nature* 332 (7): 543–46.

Lassen, C., S. Hummel, B. Herrmann. 1994. "Comparison of DNA Extraction and Amplification from Ancient Human Bone and Mummified Soft Tissue." *International Journal of Legal Medicine* 107 (3): 152–55.

Leonard, J.A., O. Shanks, M. Hofreiter, E. Kreuz, L. Hodges, W. Ream, R.K. Wayne and R.C. Fleischer. 2007. "Animal DNA in PCR Reagents Plagues Ancient DNA Research." *Journal of Archaeological Science* 34 (9): 1361–66.

Lindahl, T. 1993. "Instability and Decay of the Primary Structure of DNA." *Nature* 362 (6422): 709–15.

Lydolph, M.C., J. Jacobsen, P. Arctander, M.T.P. Gilbert, D.A. Gilichinsky, A.J. Hansen, E. Willerslev and L. Lange. 2005. "Beringian Paleoecology Inferred from Permafrost-Preserved Fungal DNA." *Applied and Environmental Microbiology* 71 (2): 1012–17.

Mackie, K.E. 2004. "Conservation and Analysis of a Fur Garment from a Glacier." *Proceedings of Congreso Internacional Patrimonio Cultural* 20, Marzo, 2004.

———. 2005. "Long Ago Person Found: An Ancient Robe Tells a New Story. In *Recovering the Past: The Conservation of Archaeological and Ethnographic Textiles, North American Textile Conservation Conference,* edited by E. Cortes and S. Thomassen-Krauss, 35–46.

———. 2008. "Conservation of a Frozen First Nations Robe." In *Preserving Aboriginal Heritage: Technical and Traditional Approaches,* edited by C. Dignard, K. Helwig, J. Mason, K. Nanowin and T. Stone, 115–19. Ottawa: Canadian Conservation Institute.

Martin, R.R. 2001. "Advanced Analysis of Hair from Kwaday Dan Ts'inchi." *Canadian Journal of Analytical Sciences and Spectroscopy* 46 (3): 108–10.

Monsalve, M.V., A.C. Stone, C.M. Lewis, A. Rempel, M. Richards, D. Straathof and D.V. Devine. 2002. "Brief Communication: Molecular Analysis of the Kwäday Dän Ts'inchi Ancient Remains Found in a Glacier in Canada." *American Journal of Physical Anthropology* 119 (3): 288–91.

Nicholls, A., E. Matisoo-Smith, M.S. Allen. 2003. "A Novel Application of Molecular Techniques to Pacific Archaeofish Remains." *Archaeometry* 45:133–47.

Nozawa, H., T. Yamamoto, R. Uchihi, T. Yoshimoto, K. Tamaki, S. Hayashi, T. Ozawa and Y. Katsumata. 1999. "Purification of Nuclear DNA from Single Hair Shafts for DNA Analysis in Forensic Sciences." *Legal Medicine* 1 (2): 61–67.

O'Rourke, D.H., S.W. Carlyle and R.L. Parr. 1996. "Ancient DNA: Methods, Progress and Perspectives." *American Journal of Human Biology* 8 (5): 557–71.

Pääbo, S., R.G. Higuchi and A.C. Wilson. 1989. "Ancient DNA and the Polymerase Chain-Reaction: The Emerging Field of Molecular Archaeology." *Journal of Biological Chemistry* 264 (17):9709–12.

Poinar, H.N. 2003. "The Top 10 List: Criteria of Authenticity for DNA from Ancient and Forensic Samples." *International Congress Series* 1239:575–79.

Richards, Michael P., Sheila Greer, Lorna T. Corr, Owen Beattie, Alexander Mackie, Richard P. Evershed, Al von Finster and John Southon. 2007. "Radiocarbon Dating and Dietary Stable Isotope Analysis of Kwäday Dän Ts'ìnchí." *American Antiquity* 72 (4): 719–33.

Teasdale, M.D., N.L. van Doorn, S. Fiddyment, C.C. Webb, T. O'Connor, M. Hofreiter, M.J. Collins and D.G. Bradley. 2015. "Paging Through History: Parchment as a Reservoir of Ancient DNA for Next Generation Sequencing." *Philosophical Proceedings of the Royal Society B Biological Sciences* 370, 20130379.

Thompson, J.D., D.G. Higgins and T.J. Gibson. 1994. "CLUSTAL W: Improving the Sensitivity of Progressive Multiple Sequence Alignment through Sequence Weighting, Position-Specific Gap Penalties and Weight Matrix Choice." *Nucleic Acids Research* 22:4673–80.

Vuissoz, A., M. Worobey, N. Odegaard, M. Bunce, C.A. Machado, N. Lynnerup, E.E. Peacock and M.T.P. Gilbert. 2007. "The Survival of PCR-Amplifiable DNA in Cow Leather." *Journal of Archaeological Science* 34 (5): 823–29.

Willerslev, E., A.J. Hansen and H.N. Poinar. 2004. "Isolation of Nucleic Acids and Cultures from Fossil Ice and Permafrost." *Trends in Ecology and Evolution* 19 (3): 141–47.

Wilson, I.G. 1997. "Inhibition and Facilitation of Nucleic Acid Amplification." *Applied and Environmental Microbiology* 63 (10): 3741–51.

Yang, D.Y., A. Cannon and S.R. Saunders. 2004. "DNA Species Identification of Archaeological Salmon Bone from the Pacific Northwest Coast of North America." *Journal of Archaeological Science* 31 (5): 619– 31.

Yang, D.Y., B. Eng, J.S. Waye, J.C. Dudar and S.R. Saunders. 1998. "Improved DNA Extraction from Ancient Bones Using Silica-Based Spin Columns." *American Journal of Physical Anthropology* 105: 539–43.

Yang, D.Y., J.R. Woiderski and J.C. Driver. 2005. "DNA Analysis of Archaeological Rabbit Remains from the American Southwest." *Journal of Archaeological Science* 32 (4): 567–78.

Chapter 28

Bagnell, C.R. 1975. "Species Distinction among Pollen Grains of *Abies*, *Picea*, and *Pinus* in the Rocky Mountain Area (A Scanning Electron Microscope Study)." *Review of Palaeobotany and Palynology* 19:203–20.

Beattie, Owen, Brian Apland, Erik W. Blake, James A. Cosgrove, Sarah Gaunt, Sheila Greer, Alexander P. Mackie, Kjerstin E. Mackie, Dan Straathof, Valerie Thorp and Peter M. Troffe. 2000. "The Kwäday Dän Ts'inchi Discovery from a Glacier in British Columbia." *Canadian Journal of Archaeology* 24 (1–2): 129–47.

Bryant, V.M., and G.D. Jones. 2006. "Forensic Palynology: Current Status of a Rarely Used Technique in the United States of America." *Forensic Science International* 163: 183–97.

Bryant, V.M., and D.C. Mildenhall. 1998. "Forensic Palynology: A New Way to Catch Crooks." In *New Developments in Palynomorph Sampling, Extraction, and Analysis, American Association of Stratigraphic Palynologists Foundation*, edited by V.M. Bryant and J.W. Wrenn, 145–55. Contributions Series Number 33.

Campeau, S., R. Pienitz and A. Hequette. 1999. *Diatoms from the Beaufort Sea Coast, Southern Arctic Ocean (Canada)*. Bibliotheca Diatomologica, vol. 42, Stuttgart: J. Cramer.

Coyle, H.M. 2005. *Forensic Botany: Principles and Applications to Criminal Casework*. Boca Raton, FL: CRC Press.

Cwynar, L.C. 1990. "A Late Quaternary Vegetation History from Lily Lake, Chilkat Peninsula, Southeast Alaska." *Canadian Journal of Botany* 68:1106–112.

Dickson, J.H. 2000. "Bryology and the Iceman: Chorology, Ecology and Ethnobotany of the Mosses *Neckera complanata* Hedw. and *Neckera crispa* Hedw." In *The Man in the Ice*, vol. 4, edited by S. Bortenschlager and K. Oeggl, 77–88. Vienna: Springer.

———. 2003. "Low to Moderate Altitude Bryophytes at the Iceman Site and Their Significance." *Schriften des Südtiroler Archäologiemuseums*, 3:27–34.

Dickson, J.H., S. Bortenschlager, K. Oeggl, R. Porley and A. McMullen. 1996. "Mosses and the Tyrolean Iceman's Southern Provenance." *Proceedings of the Royal Society of London B* 263:567–71.

Dickson, J.H., and P.J. Mudie. 2008. "The Life and Death of Kwaday Dan Ts'inchi, an Ancient Frozen Body from British Columbia: Clues from Remains of Plants and Animals." *The Northern Review* 28:27–50.

Dickson, J.H., K. Oeggl and L.L. Handley. 2005. "The Iceman Reconsidered." *Scientific American Special Archaeology* 15:4–13.

Dickson, James H., Michael P. Richards, Richard J. Hebda, Petra Mudie, Owen Beattie, Susan Ramsay, Nancy J. Turner, Bruce J. Leighton, John M. Webster, Niki R. Hobischak, Gail S. Anderson, Peter M. Troffe and Rebecca J. Wigen. 2004. "Kwäday Dän Ts'ìnchí, the First Ancient Body of a Man from a North American Glacier: Reconstructing his Last Days by Intestinal and Biomolecular Analyses." *Holocene* 14 (4): 481–86.

Faegri, K., and J. Iversen. 1989. *Textbook of Pollen Analysis*. 4th ed. London: John Wiley.

Groenman-van Waateringe, W. 1993. "Analyses of Hides and Skins from Hauslabjoch." In *Die Gletschermummie vom Ende der Steinzeit aus den Ötztaler Alpen*, edited by M. Egg, R. Goedecker, W. Groenman-van Waateringe and K. Spindler, 114–28. *Jahrbuch des Römisch-Germanischen Museums* 39.

———. 1998. "Pollen in Animal Coats and Bird Feathers." *Review of Palaeobotany and Palynology* 103:11–16.

Hebda, R.J. 1995. "British Columbia Vegetation and Climate History with Focus on 6 KA BP." *Geographie Physique et Quaternaire*, 49:55–79.

Heiss, A.G., and K. Oeggl. 2009. "The Plant Macro-Remains from the Iceman Site (Tisenjoch, Italian-Austrian border, Eastern Alps): New Results on the Glacier Mummy's Environment." *Vegetation History and Archaeobotany* 18:23–35.

Hurlimann, J., P. Feer, F. Elber, K. Niederberger, R. Dirnhofer, D. Wyler. 2000. "Diatom Detection in the Diagnosis of Death by Drowning." *International Journal of Legal Medicine* 114:6–14.

Lotter, A.F., R. Pienitz and R. Schmidt. 1999. "Diatoms as Indicators of Environmental Change Near Arctic and Alpine Treeline." In *The Diatoms: Applications for the Environmental and Earth Sciences*, edited by E. Stoermer and J.P. Smol, 205–26. Cambridge: Cambridge University Press.

Mackie, K. 2005. "Long Ago Person Found: An Ancient Robe Tells a New Story." In *Recovering the Past: The Conservation of Archaeological and Ethnographic Textiles, North American Textile Conservation Conference,* edited by E. Cortes and S. Thomassen-Krauss, 35–46.

Mathewes, R.W. 2006. "Forensic Palynology in Canada: An Overview with Emphasis on Archaeology and Anthropology." *Forensic Science International* 163: 198–203.

Mudie, P.J., S. Greer, J. Brake, J.H. Dickson, C. Schinkel, R. Peterson-Welsh, M. Stevens, N.J. Turner, M. Shadow and R. Washington. 2005. "Forensic Palynology and Ethnobotany of *Salicornia* Species (Chenopodiaceae) in Northwest Canada and Alaska." *Canadian Journal of Botany* 83:1–13.

Oeggl, K. 2009. "The Significance of the Tyrolean Iceman for the Archaeobotany of Central Europe." *Vegetation History and Archaeobotany* 18:1–11.

Chapter 29

Adamson, John. 2006. Interview by Sheila Greer, May 11, related to the Kwädąy Dän Ts'ìnchį discovery. Transcript on file, Champagne and Aishihik First Nations, Haines Junction, YT.

Allen, Bessie (Äshènją), and Lorraine Allen (recorder and transcriber). 2006. *Family and Traditional Southern Tutchone Stories*. Whitehorse: Yukon Native Language Centre.

Banks, Paula. 1999. "Paula Banks Notes with Jimmy G. Smith at Kusawa Lake, August 19, 1999." Notebook on file, Champagne and Aishihik First Nations, Whitehorse.

Brown, John. 1996. Interview by Paula Banks, September 20. Transcript on file, Champagne and Aishihik First Nations, Haines Junction.

Champagne and Aishihik First Nations. 1999. "Transcript of Champagne and Aishihik First Nations Heritage visit to Haines Alaska, Chilkoot Indian Association, October 6, Evening Meeting." On file, Champagne and Aishihik First Nations, Haines Junction.

———. 2000a. "Transcript of Meeting with Chilkat Indian Tribe, Klukwan, October 27, 2000." On file, Champagne and Aishihik First Nations , Haines Junction.

———. 2000b. "Transcript of Meeting of Champagne and Aishihik First Nations Representatives and Yakutat Tlingit Tribe, Yakutat, Regarding the Kwaday Dan Sinchi Discovery, November 6, 2000." On file, Champagne and Aishihik First Nations, Haines Junction.

———. 2000c. "Summary and Transcript of Meeting with Carcross-Tagish First Nation Regarding the Kwaday Dan Sinchi Discovery, December 11, 2000." On file, Champagne and Aishihik First Nations, Haines Junction.

Charlie, Wilfred. 1999. "Qwantuk's Story." English-language recording, September 22, Carmacks. Transcript prepared by Dawn Charlie. On file, Champagne and Aishihik First Nations, Haines Junction.

Cruikshank, Julie. 2005. *Do Glaciers Listen? Local Knowledge, Colonial Encounters, and Social Imagination*. Vancouver: University of British Columbia Press.

Cruikshank, Julie, Angela Sidney, Kitty Smith and Annie Ned. 1990. *Life Lived Like a Story*. Vancouver: University of British Columbia Press.

Davidson, George. 1901. "Explanation of an Indian Map of the Rivers, Lakes, Trails and Mountains from the Chilkaht to the Yukon drawn by the Chilkaht Chief, Kohklux, in 1869." *Mazama* 2 (2): 75–92.

de Laguna, Frederica. 1972. *Under Mount Saint Elias: The History and Culture of the Yakutat Tlingit*. 3 parts. Smithsonian Contributions to Anthropology 7. Washington, DC: Smithsonian Institution.

———. 1975. "Matrilineal Kin Groups in Northwestern North America." In *Proceedings: Northern Athapaskan Conference, 1971*, vol. 2, edited by A. McFadyen Clark, 17–45. Canadian Ethnology Service, Paper No. 27. National Museum of Man Mercury Series. Ottawa: National Museums of Canada.

———. 1990. "Tlingit." In *Handbook of North American Indians*, vol. 7, *Northwest Coast*, edited by W. Suttles, 203–29. Washington, DC: Smithsonian Institution.

Emmons, George T. 1991. *The Tlingit Indians*. Edited by Frederica de Laguna. New York: American Museum of Natural History.

Greer, Sheila. 2005. "Sheila Greer Notes with John Adamson, July 24 and 25, Klukshu General Assembly." On file, Champagne and Aishihik First Nations, Haines Junction.

———. 2006. "Sheila Greer Notes with John Adamson May 11, 2006, Whitehorse." On file, Champagne and Aishihik First Nations, Haines Junction.

Greer, Sheila C., and CTFN (Carcross-Tagish First Nation). 1995. *Skookum Stories on the Chilkoot/Dyea Trail*. Carcross, YT: Carcross-Tagish First Nation.

Hope, Andrew, III, and Thomas F. Thornton, eds. 2000. *Will the Time Ever Come? A Tlingit Source Book*. Fairbanks: University of Alaska Fairbanks.

Ingelson, Allan, Michael Mahony and Robert Scace. 2001. *Chilkoot, An Adventure in Ecotourism*. Fairbanks: University of Alaska Press.

Jackson, Moose, and John Brown. 1997. Interview by Paula Banks and Liza Jacobs, January 28. Transcript on file, Champagne and Aishihik First Nations, Haines Junction.

James Gilbert Smith, Indian Name = "Gooch Sháa," May 30, 1919 to June 6, 2005. 2005. Funeral pamphlet on file, Champagne and Aishihik First Nations, Haines Junction, YT.

Kakúwät, Wilfred Albert Charlie, July 24th, 1937 to June 13th, 2005. 2005. Funeral pamphlet on file, Champagne and Aishihik First Nations, Haines Junction, YT.

Kha`Guxh- (Tlingit) Ká Go (S. Tutchone) John Wesley Adamson, January 27, 1917–June 18, 2008. 2008. Funeral pamphlet on file, Champagne and Aishihik First Nations, Haines Junction, YT.

Krause, Aurel. 1956. *The Tlingit Indians*. Vancouver: Douglas and McIntyre.

Krause, Aurel, and Arthur Krause. 1981. *Journey to the Tlingits by Aurel and Arthur Krause, 1881/82*. Translated by Margot Krause McCaffrey. Haines, Alaska: Centennial Commission.

———. 1993. *To the Chukchi Peninsula and to the Tlingit Indians 1881/1882: Journals and Letters by Aurel and Arthur Krause*. Translated by Margot Krause McCaffrey. The Rasmuson Library Historical Translation Series 8. Fairbanks, AK: University of Alaska Press.

Krauss, Michael E. 1980. *Alaskan Native Languages: Past, Present and Future*. Alaska Native Language Center Research Papers No. 4, Fairbanks, AK: University of Alaska Fairbanks Alaska Native Language Centre.

Leer, Jeff, Doug Hitch and John Ritter, compilers. 2001. *Interior Tlingit Noun Dictionary: The Dialects Spoken by Tlingit Elders of Carcross and Teslin, Yukon, and Atlin, British Columbia*. Whitehorse: Yukon Native Language Centre.

Legros, Dominique. 1984. "Commerce Entre Tlingits et Athapaskan Tutchones au XIX siècle." *Recherches Amérindienne au Québec* 14 (2): 11–24.

———. 1999. *Tommy McGinty's Northern Tutchone Story of Crow: A First Nation Elder Recounts the Creation of the World*. Canadian Ethnology Service Paper 133. Hull, QC: Canadian Museum of Civilization.

———. 2007. *Oral History as History: Tutchone Athapaskan in the Period 1840–1902*. 2 vols. Occasional Papers in Yukon History No. 3, Hudé Hudän Series. Whitehorse: Yukon Cultural Services Branch.

McClellan, Catharine. 1950. "Culture Change and Native Trade in Southern Yukon Territory." PhD dissertation, University of California, Berkeley.

———. 1964. "Culture Contacts in the Early Historic Period in Northwestern North America." *Arctic Anthropology* 2 (2): 3–15.

———. 1975. *My Old People Say: An Ethnographic Survey of Southern Yukon Territory*. Publications in Ethnology 6. 2 vols. Ottawa: National Museums of Canada. Republished in 2001 as Mercury Series Canadian Ethnology Service Paper 137, Canadian Museum of Civilization.

———. 1981. Intercultural Relations and Cultural Change in the Cordillera. In *Handbook of North American Indians*, vol. 6, *Subarctic*, edited by June Helm. Washington, DC: Smithsonian Institution.

———. 2007. *My Old People's Stories: A Legacy for Yukon First Nations*. Occasional Papers in Yukon History No. 5 (1), Hudé Hudän Series. Whitehorse: Yukon Cultural Services Branch.

McClellan, Catharine., L. Birckel, R. Bringhurst, J.A. Fall, C. McCarthy and J.R. Sheppard, 1987. *Part of the Land, Part of the Water. A History of Yukon Indians*. Vancouver: Douglas and McIntyre.

Ned, Annie. 1984. *Old People in Those Days, They Told Their Story All the Time*. Compiled by Julie Cruikshank. Whitehorse: Yukon Native Languages Project.

Nyman, Elizabeth, and Jeff Leer. 1993. *Gágiwduł.àt: Brought Forth to Reconfirm. The Legacy of a Taku River Tlingit Clan*. Whitehorse: Yukon Native Language Centre; Fairbanks: Alaska Native Language Center.

Shadow, Mary, and Marge Jackson. 1997. Interview by Paula Banks and Liza Jacobs, January 13. Transcript on file, Champagne and Aishihik First Nations Heritage program, Haines Junction.

Shotridge, Louis. 1920. "Ghost of Courageous Adventurer." *The Museum Journal* 11 (1): 11–26.

Sidney, Angela, 1982. *Tagish Tlaagu. Tagish Stories*. Whitehorse: Council for Yukon Indians and the Government of Yukon.

Sidney, Angela, Kitty Smith and Rachel Dawson. 1979. *My Stories Are My Wealth*. Whitehorse: Council for Yukon Indians.

Smith, Donna, producer. 2001. *Haa Shagoon: John Adamson Story*. Dákwanjè with English-language subtitles. Whitehorse: Northern Native Broadcasting. Videocassette (VHS), 24 min.

Smith, Jimmy G. 2000. Interview by John Fingland, March 3. Manuscript on file, Champagne and Aishihik First Nations, Haines Junction.

Smith, Kitty. 1982. *Nindal Kwadindur, I'm Going to Tell You a Story*. Recorded by J. Cruikshank. Whitehorse: Council for Yukon Indians and Government of Yukon.

Swanton, John. R. 1909. "Tlingit Myths and Text." *Bureau of American Ethnology Bulletin* 39, 154–65. Washington, DC: Smithsonian Institution.

Wilson, Clifford. 1970. *Campbell of the Yukon*. Toronto: Macmillan of Canada.

Workman, Margaret, compiler and translator. 2000. *Kwädąay Kwändür, Traditional Southern Tutchone Stories, As Told by Marge Jackson, Mary Jacquot, Jessie Joe, Jimmy Copper Joe, Copper Lily Johnson and Jessie Jonathan*. Whitehorse: Yukon Native Language Centre.

Wright, Allen. 1976. *Prelude to Bonanza*. Sidney, BC: Gray's Publishing.

YHMA (Yukon Historical and Museums Association). 1994. Poster. Reproduction of the Koh-klux' Map, Chilkaht. [Alaska and Yukon]. Original by Chief Chilkaht Kohklux, 1852.

Chapter 31

Alaska Department of Fish and Game. 2016. "Commercial Sea Urchin Dive Fisheries." Accessed December 1. http://www.adfg.alaska.gov/index.cfm?adfg= CommercialByFisheryDive.seaurchin.

Ball, Peter W. 2003a. "Salicornia." In *Flora of North America: North America*, vol. 4, *Magnoliophyta: Caryophyllidae*, part 1, edited by the Flora of North America Editorial Committee, 382–84. Oxford: Oxford University Press.

———. 2003b. "Sarcocornia." In *Flora of North America: North America*, vol. 4, *Magnoliophyta: Caryophyllidae*, part 1, edited by the Flora of North America Editorial Committee, 384–87. Oxford: Oxford University Press.

Chaves, S.A.M., and K.J. Reinhard. 2006. "Critical Analysis of Coprolite Evidence of Medicinal Plant Use, Piauí, Brazil." *Palaeogeography, Palaeoecology, Palaeoclimatology* 237:110–18.

Choudhary, A., D. Bodade and W. Pranita. 2015. "Drug Screening and Activity of Polyphenolic Compound against *Myobacterium tuberculosis* Using Algal Extracts." *Biochemical Physiology* 4 (2):1000158. Open Access publication. http://dx.doi.org/10.4172/2168-9652.1000158.

CSMCRI (Central Salt and Marine Chemicals Research Institute). 2015. "Discipline of Wasteland Research." Accessed December: http://csmcri.org/Pages/Research/Discipline_Wasteland_Research.php.

Cogo, R., and N. Cogo. 1974. *Haida Food from Land and Sea*. Written in collaboration with T.L. Pulu. Anchorage: University of Alaska Materials Development Center.

Crow, J.H., and J.D. Koppen. 1977. *The Salt Marshes of China Poot Bay, Alaska*. Vol. 10, *Environmental Studies of Kachemak Bay and Lower Cook Inlet*. Anchorage: Alaska Department of Fish and Game.

de Laguna, F. 1972. *Under Mount Saint Elias: The History and Culture of the Yakutat Tlingit*. Smithsonian Contributions to Anthropology 7. Washington, DC: Smithsonian Institution.

Dickson, J.H., and P.J. Mudie. 2008. "The Life and Death of Kwäday Dän Ts'ìnchi, an Ancient Frozen Body from British Columbia: Clues from Remains of Plants and Animals." *The Northern Review* 28:27–50.

Dickson, J.H., W. Hofbauer, R. Porley, A. Schmidle, W. Kofler and K. Oeggl. 2009. "Six Mosses from the Tyrolean Iceman's Alimentary Tract and Their Significance for His Ethnobotany and Events of the Last Days." *Vegetation History and Archaeobotany* 18 (1): 12–22.

Dickson, J.H., K. Oeggl, and L.L. Handley. 2003. "The Iceman Reconsidered." *Scientific American* 288:70–79.

Dickson, James H., Michael P. Richards, Richard J. Hebda, Petra Mudie, Owen Beattie, Susan Ramsay, Nancy J. Turner, Bruce J. Leighton, John M. Webster, Niki R. Hobischak, Gail S. Anderson, Peter M. Troffe and Rebecca J. Wigen. 2004. "Kwäday Dän Ts'inchí, the First Ancient Body of a Man from a North American Glacier: Reconstructing his Last Days by Intestinal and Biomolecular Analyses." *Holocene* 14 (4): 481–86.

Dmytryshyn, B., and E.A.P. Crownhart-Vaughan. 1976. *Colonial Russian America Kyrill T. Khlebnikov's Reports 1817–1832.* Portman: Oregon Historical Society.

Earlandson, J.M., and M.L. Moss. 2001. "Shellfish Feeders, Carrion Eaters and the Archaeology of Aquatic Adaptations." *American Antiquity* 66 (3): 413–32.

Emmons, G.T. 1991. *The Tlingit Indians.* Seattle: University of Washington Press.

Garza, D. 2007. "Surviving on the Foods and Water from Alaska's Southern Shores." *University of Alaska Marine Advisory Bulletin* 38. 2nd ed. http://seagrant.uaf.edu/bookstore/index.html.

Hall, J.K. 1995. *Native Plants of Southeast Alaska.* Juneau, AK: Windy Ridge Publishers.

Hebda, R.J. 1977. "The Paleoecology of a Raised Bog and Associated Deltaic Sediments of the Fraser River Delta." PhD thesis, University of British Columbia.

Heiss, A.G., and K. Oeggl. 2009. "The Plant Macro-Remains from the Iceman Site (Tisenjoch, Italian-Austrian Border, Eastern Alps): New Results on the Glacier Mummy's Environment." *Vegetation History and Archaeobotany* 18:23–35.

Holmberg, H.J. 1985. *Holmberg's Ethnographic Sketches.* Translated by Marvin W. Falk. Edited by Fritz Jaensch. Fairbanks, AK: Limestone Press.

Hutchinson, I. 1988. "The Biogeography of the Coastal Wetlands of the Puget Trough: Deltaic Form, Environment and Marsh Community Structure." *Journal of Biogeography* 15:729–45.

Krause, A. (1885) 1956. *The Tlingit Indians: Results of a Trip to the Northwest Coast of America and the Bering Strait.* Translated by Erna Gunther. Seattle: University of Washington Press.

Kumagai, Y., and A. Kubo. 1997. Calcium containing composition from sea urchin with high oral bioavailablity. US Patent EP0634105 B1, filed July 15, 1994. https://www.google.com/patents/EP0634105B1.

Lawrence, J.M. 2006. *The Biology and Ecology of Edible Sea Urchins.* Amsterdam: Elsevier Science.

Mildenhall, D.C., P.E.J. Wiltshire and V.J. Bryant. 2006. "Forensic Palynology: Why Do It and How It Works." *Forensic Science International* 163:163–72.

Moss, M.L. 1993. "Shellfish, Gender and Status on the Northwest Coast: Reconciling Archaeological, Ethnographic and Ethnohistorical Records of the Tlingit." *American Anthropologist* 95:631–52.

Mudie, P.J., and J.H. Dickson. 2004. "Report on Field Work in Alaska, July 29–August 31, 2004." Unpublished report on file, Sealaska, Juneau.

Mudie, P.J., S. Greer, J. Brake, J.H. Dickson, C. Schinkel, R. Peterson-Welsh, M. Stevens, N.J. Turner, M. Shadow and R. Washington. 2005. "Forensic Palynology and Ethnobotany of *Salicornia* Species (Chenopodiaceae) in Northwest Canada and Alaska." *Canadian Journal of Botany* 83:111–23.

Müller, W., H. Fricke, A.N. Halliday, T. McCulloch and J-A. Wartho. 2003. "Origin and Migration of the Alpine Iceman." *Science* 302:862–66.

Oberg, K. 1973. *The Social Economy of the Tlingit Indians.* American Ethnological Society, Monograph 55. Seattle: University of Washington Press.

Oeggl, K. 2009. "The Significance of the Tyrolean Iceman for the Archaeobotany of Central Europe." *Vegetation History and Archaeobotany* 18:1–11.

Oeggl, K., W. Kofler, A. Schmidl, J.H. Dickson, E. Egarter-Vigl and O. Gaber. 2007. "The Reconstruction of the Last Itinerary of 'Otzi', the Neolithic Iceman, by Pollen Analyses from Sequentially Sampled Gut Extracts." *Quarternary Science Reviews* 26 (7–8): 853–61.

Olson, R. 1936. "Fieldnotes – Klawock, Alaska." Bancroft Library, University of California, Berkeley.

Olson, W.M., and J.F. Thilenius. 1993. *The Alaska Travel Journal of Archibald Menzies 1793–1794.* Fairbanks: University of Alaska Press.

Parker, C. 2000. "Vascular Plant Inventory of Selected Sites Haines and Vicinity, S.E. Alaska, Summer 2000." University of Alaska Herbarium Report, Fairbanks.

Post, A., and G.P. Streveler. 1976. "Tilted Forest: Glaciological-Geologic Implications of Vegetated Neoglacial Ice at Lituya Bay, Alaska." *Quarterly Review of Biology* 6:111–17.

Radwan, H.M, N.M. Nazif and L.M. Abou-Setta. 2007. "Phytochemical Investigation of *Salicornia fruticosa* (L.) and Their Biological Activity." *Research Journal of Medicine and Medical Sciences* 2 (2): 72–78.

Scott, D.B., J. Frail-Gauthier and P.J. Mudie. 2014. "Coastal Wetlands of the World: Geology, Ecology, Distribution and Applications." New York: Cambridge University Press.

Swanton, J.R. 1909. *Tlingit Myths and Texts*. Washington, DC: Government Printing Office.

Thilenius, J.F. 1995. "Phytosociology and Succession on Earthquake-Uplifted Coastal Wetlands, Copper River Delta, Alaska." *General Technical Reports* PNW-GTR-346.

Thornton, T.F. 2004. "Klondike Gold Rush National Historic Park Ethnographic Overview and Assessment. Klondike Gold Rush National Historic Park Final Report." Whitehorse: Lost Moose Publishing; and National Parks Service, Alaska Regional Office, NPD-111.

Turner, N.J. 2003. "The Ethnobotany of Edible Seaweed (*Porphyra abbottae* and Related Species; Rhodophya: Bangiales) and Its Uses by First Nations on the Pacific Coast of British Columbia." *Canadian Journal of Botany* 81:283–93.

US EPA (United States Environmental Protection Agency) Region 10. 1995. Factsheet on proposed reissuance of a national pollutant discharge system (NPDES) permit to discharge pollutants pursuant to the provisions of the Clean Water Act (CWA), City of Haines, Alaska. http://yosemite.epa.gov/R10/WATER.NSF.

Chapter 32

Adamson, John. 1993. Interview by Sarah Gaunt. January 15. Transcript on file, Champagne and Aishihik First Nations Heritage program, Haines Junction, YT.

Cruikshank, Julie, 2005. *Do Glaciers Listen? Local Knowledge, Colonial Encounters, and Social Imagination.* Vancouver: UBC Press.

Davidson, George. 1901. "Explanation of an Indian Map of the Rivers, Lakes, Trails and Mountains from the Chilkaht to the Yukon Drawn by the Chilkaht Chief, Kohklux, in 1869." *Mazama* 2 (2): 75–92.

Deguerre, Mary. 2006. "Sheila Greer Notes with Mary De Guerre October 1 and 2, 2006, When Travelling Between Haines Junction and Haines." On file, Champagne and Aishihik First Nations Heritage program, Haines Junction, YT.

de Laguna, Frederica. 1975. *The Archaeology of Cook Inlet, Alaska.* Philadelphia: University of Pennsylvania Press. Reprinted with a new foreword by Karen W. Workman and William B. Workman and a new preface by Frederica de Laguna. Fairbanks: Alaska Historical Society.

Dickson, James H., Michael P. Richards, Richard J. Hebda, Petra Mudie, Owen Beattie, Susan Ramsay, Nancy J. Turner, Bruce J. Leighton, John M. Webster, Niki R. Hobischak, Gail S. Anderson, Peter M. Troffe and Rebecca J. Wigen. 2004. "Kwäday Dän Ts'ínchí, the First Ancient Body of a Man from a North American Glacier: Reconstructing his Last Days by Intestinal and Biomolecular Analyses." *Holocene* 14 (4): 481–86.

Dickson, J.H. and P. Mudie. 2008. "In the Footsteps of Kwäday Dän Ts'ínchí: A Botanical Journey." Presentation at the Kwäday Dän Ts'ínchí Symposium, Northwest Anthropological Conference, Victoria.

Emmons, George T. [1905]. "Trip Up Chilkat River and Over Divide 1905." British Columbia Provincial Archives, Manuscript 077, Newcombe Family Papers volume 60, folder 13. Available on microfilm. BC Archives, Victoria.

Gates, Michael. 2010. *History Hunting in the Yukon.* Madeira Park, BC: Harbour.

Glave, E.J. 1890. "The Alaska Expedition." *Frank Leslie's Illustrated Newspaper*, vol. 70: 485, 572; vol. 71: 84, 86, 87, 266, 286, 287, 310, 328, 332, 252, 374–76, 396, 397.

Greer, Sheila. 1995. "Sheila Greer Notes, BC Claim Research, October." On file, Champagne and Aishihik First Nations Heritage program, Haines Junction, YT.

Greer, Sheila, and CTFN (Carcross-Tagish First Nation). 1995. *Skookum Stories on the Chilkoot/Dyea Trail.* Carcross, YT: Carcross-Tagish First Nation.

Jim, Paddy. 1993. Interview by Sarah Gaunt and Sheila Greer at CAFN Whitehorse office. Transcript on file, Champagne and Aishihik First Nations, Haines Junction, Yukon.

———. 2009. Interview by Sheila Greer regarding the Kusawa area, June 25. Audio recording on file, Champagne and Aishihik First Nations Heritage program, Haines Junction, YT.

Kane, Jimmy. 1978. Interview by Jack Schick and Brent Liddle. Recording on file, Yukon Archives. Transcription by Sarah Gaunt on file, Champagne and Aishihik First Nations Heritage program, Haines Junction, YT.

Krause, Aurel. 1956. *The Tlingit Indians: Results of a Trip to the Northwest Coast of America and the Bering Strait*. Translated by Erna Gunther. Seattle: University of Washington Press.

Larson, Brendan, and Lauri Larson, 1977. *The Proud Chilkat*. Prepared for the Education Committee, Alaska Native Brotherhoood, Alaska Native Sisterhood, Camp No. 5. Haines, AK: Chilkat Press.

McClellan, Catharine, 1975. *My Old People Say: An Ethnographic Survey of Southern Yukon Territory*, part 1. National Museum of Man, Publications in Ethnology No. 6 (1). Ottawa: National Museums of Canada.

———. 2007. *My Old People's Stories: A Legacy for Yukon First Nations*. Occasional Papers in Yukon History, Hudé Hudän services. Whitehorse: Yukon Heritage.

Moise, Ben, and Cathy M. Moise. 1986. Retracing the Chilkat. *Canadian Geographic* February–March, 52–61.

Mudie, P.J., S. Greer, J. Brake, J.H. Dickson, C. Schinkel, R. Peterson-Welsh, M. Stevens, N.J. Turner, M. Shadow and R. Washington. 2005. "Forensic Palynology and Ethnobotany of *Salicornia* Species (Chenopodiaceae) in Northwest Canada and Alaska." *Canadian Journal of Botany* 83:1–13.

Neufeld, David, and Frank Norris. 1996. *Chilkoot Trail, Heritage Route to the Klondike*. Whitehorse: Lost Moose.

Olson, Ronald. 1936. "Some Trading Customs of the Chilkat Tlingit." In *Essays in Honor of Alfred Louis Kroeber*, edited by R.H. Lowie, 211–14. Berkeley: University of California Press, Berkeley.

Sackett, Russell. 1979. *The Chilkat Tlingit: A General Overview*. Anthropology and Historic Preservation, Cooperative Studies Park Unit Occasional Paper No. 23. Fairbanks: University of Alaska Fairbanks.

Schanz, A.B., and E.H. Wells. 1974. "From Klukwan to the Yukon, by A.B. Schanz and E.H. Wells of the *Frank Leslie's Illustrated Newspaper* Expedition to Alaska. An Edited Account by R. Sheman." *Alaska Journal* 4 (3): 169–80.

Scidmore, Eliza. (1898) 1973. "The Northwest Passes to the Yukon." *Alaska Journal* 3 (Summer 1973): 145. Reprint of *National Geographic*, April 1898.

Seton-Karr, H.W. 1891. *Bear-Hunting in the White Mountains: or, Alaska and British Columbia Revisited*. London: Chapman and Hall.

Sheldon Museum, n.d. Untitled files related to the history of the Haines Road. On file, Sheldon Museum, Haines, Alaska.

Thornton, Thomas F., with contributions by D. McBride, S. Gupta, Carcross-Tagish First Nation, Chilkat Indian Village, Chilkoot Indian Association and Skagway Traditional Council. 2004. *Klondike Gold Rush National Historical Park Ethnographic Overview and Assessment*. US National Parks Service. www.nps.gov/history/online_books/klgo/ethnograhic_overview.pdf.

Yukon Archives. 1985. *Dalton Trail: A Bibliography of Sources Available*. Offprint available at Yukon Archives, Whitehorse.

Chapter 33

Achilli, A., U.A. Perego, C.M. Bravi, M.D. Coble, Q.P. Kong, S.R. Woodward, A. Salas, A. Torroni, and H.J. Bandelt. 2008. "The Phylogeny of the Four Pan-American mtDNA Haplogroups: Implications for Evolutionary and Disease studies." *PLoS ONE* 3 (3): e1764.

Alaskool (Institute for Social and Economic Research, University of Alaska Anchorage). 2008. *Online Tlingit Dictionary*. Accessed December 28. www.alaskool.org/LANGUAGE/dictionaries/akn/tlingit/information/Index_TND.html.

Andrews, R.M., I. Kubacka, P.F. Chinnery, R.N. Lightowlers, D.M. Turnbull, and N. Howell. 1999. "Reanalysis and Revision of the Cambridge Reference Sequence for Human Mitochondrial DNA." *Nature Genetics* 23 (2): 147.

Campbell, C.R. 1989. "A Study of Matrilineal Descent from the Perspective of the Tlingit NexA'adi Eagles." *Arctic* 42 (2): 119–27.

Carcross-Tagish First Nation (CTFN). 2008. Website of the Carcross-Tagish First Nation. Accessed December 28. www.ctfn.ca.

Central Council of the Tlingit and Haida Indian Tribes of Alaska. 2008. Traditional Tlingit Ceremonies for the Dead. Accessed November 18. www.ccthita.org/memorialparties.htm.

Champagne and Aishihik First Nations Heritage Office. 2007. "Funeral and Memorial Potlatch: A Tradition Today." Draft, January 2007. Manuscript on file, Champagne and Aishihik First Nations Heritage program, Haines Junction, YT.

Cruikshank, Julie, Angela Sidney, Kitty Smith and Annie Ned. 1990. *Life Lived Like a Story*. Vancouver: University of British Columbia Press.

de Laguna, Frederica. 1972. *Under Mount Saint Elias: The History and Culture of the Yakutat Tlingit*. 3 parts. Smithsonian Contributions to Anthropology 7. Washington, DC: Smithsonian Institution.

———. 1975. "Matrilineal Kin Groups in Northwestern North America." In *Proceedings: Northern Athapaskan Conference, 1971*, vol. 2, edited by A. McFadyen Clark, 17–145. Canadian Ethnology Service Paper 27. National Museum of Man Mercury Series. Ottawa: National Museums of Canada.

———. 1990. "Tlingit." In *Handbook of North American Indians*, vol. 7, *Northwest Coast*, edited by W. Suttles, 203–29. Washington, DC: Smithsonian Institution.

Emmons, George Thornton. 1991. *The Tlingit Indians*. Edited by Frederica de Laguna. New York: American Museum of Natural History.

———. 2000. Excerpts from "The History of the Tlingit Tribes and Clans." In *Will the Time Ever Come? A Tlingit Source Book*, part 3, appendix A, edited by Andrew Hope III and Thomas F. Thornton, 131–47. Fairbanks: University of Alaska Fairbanks, Alaska Native Knowledge Network, Center for Cross-Cultural Studies.

Felsenstein, J. 1989. "PHYLIP – Phylogeny Inference Package (Version 3.2)." *Cladistics* 5:164–66.

Greer, Sheila. 2004. "Sheila Greer Notes with Harry Morris and Juanita Sydney (Kremer)." On file, Teslin Tlingit Council Heritage program, Teslin, YT.

Greer, Sheila. 2004a. "Sheila Greer Notes with John Adamson." On file, Champagne and Aishihik First Nations Heritage program, Haines Junction, YT.

Handt, O., M. Richards, M. Trommsdorff, C. Kilger, J. Simanainen, O. Georgiev, K. Bauer, A. Stone, R. Hedges, W. Schaffner. 1994. "Molecular Genetic Analysis of the Tyrolean Ice Man." *Science* 264 (5166): 1775–78.

Helm, June, ed. 1981. *Handbook of North American Indians*. Vol. 6, *Subarctic*. Washington, DC: Smithsonian Institution.

Hope, Andrew, III. 2000. "Tlingit Tribes and Clan Houses." In *Will the Time Ever Come? A Tlingit Source Book*, part 3, appendix A, edited by Andrew Hope III and Thomas F. Thornton, 148–59. Fairbanks: University of Alaska Fairbanks, Alaska Native Knowledge Network, Center for Cross-Cultural Studies.

Ingelson, Allan, Michael Mahony and Robert Scace. 2001. *Chilkoot: An Adventure in Ecotourism*. Calgary: University of Calgary Press.

Kaestle, F.A., and D.G. Smith. 2001. "Ancient Mitochondrial DNA Evidence for Prehistoric Population Movement: the Numic Expansion." *American Journal of Physical Anthropology* 115 (1): 1–12.

Krauss, Michael E. 1980. *Alaskan Native Languages: Past, Present and Future*. Alaska Native Language Center Research Papers No. 4. Fairbanks: University of Alaska Fairbanks. www.eric.ed.gov/ERICDocs/data/ericdocs2sql/content_storage_01/0000019b/80/14/ab/53.pdf. Accessed January 4, 2009.

Leer, Jeff, Doug Hitch and John Ritter, compilers. 2001. *Interior Tlingit Noun Dictionary: The Dialects Spoken by Tlingit Elders of Carcross and Teslin, Yukon, and Atlin, British Columbia*. Whitehorse: Yukon Native Language Centre.

McClellan, Catharine. 1964. "Culture Contacts in the Early Historic Period in Northwestern North America." *Arctic Anthropology* 2 (2): 3–15.

———. (1975). 2001. *My Old People Say. An Ethnographic Survey of Southern Yukon Territory*. 2 vols. National Museums of Canada Publications in Ethnology 6. Ottawa: National Museums of Canada.

———. 2007. *My Old People's Stories: A Legacy for Yukon First Nations*. Occasional Papers in Yukon History No. 5 (1), Hudé Hudän Series. Whitehorse: Yukon Cultural Services Branch.

McClellan, Catharine, L. Birckel, R. Bringhurst, J.A. Fall, C. McCarthy and J.R. Sheppard, 1987. *Part of the Land, Part of the Water: A History of Yukon Indians*. Vancouver: Douglas and McIntyre.

Monsalve, M.V., A.C. Stone, C.M. Lewis, A. Rempel, M. Richards, D. Straathof and D.V. Devine. 2002. "Brief Communication: Molecular Analysis of the Kwäday Dän Ts'inchi Ancient Remains Found in a Glacier in Canada." *American Journal of Physical Anthropology* 119 (3): 288–91.

Mooder, Karen. 2005. "Review of DNA Analysis of Kwaday Dän Ts'inchi: A Search for Living Relatives." Report prepared for Champagne and Aishihik First Nations.

On file, Champagne and Aishihik First Nations Heritage program, Haines Junction, YT.

Mooder, Karen P., T.G. Schurr, F.J. Bamforth, V.I. Bazaliiski, and N.A. Savel'ev. 2006. "Population Affinities of Neolithic Siberians: A Snapshot from Prehistoric Lake Baikal." *American Journal of Physical Anthropology* 129 (3): 349–61.

Nyman, Elizabeth, and Jeff Leer. 1993. *Gágiwduł.àt: Brought Forth to Reconfirm. The Legacy of a Taku River Tlingit Clan.* Whitehorse: Yukon Native Language Centre; Fairbanks: Alaska Native Language Center.

Olson, Wallace. 1997. *The Tlingit: An Introduction to Their Culture and History.* Auk Bay, AK: Heritage Research.

STTC (Southern Tutchone Tribal Council). 1998. *Dän K'e Kwänje Literacy Workshops.* Haines Junction, YT: Southern Tutchone Tribal Council.

Suttles, Wayne, ed. 1990. *Handbook of North American Indians,* vol. 7, *Northwest Coast.* Washington, DC: Smithsonian Institution.

Swanton, John R. 1909. "Tlingit Myths and Texts." *Bureau of American Ethnology Bulletin* 39, 154–65. Washington, DC: Smithsonian Institution.

Tlen, Daniel. 1993. *Kluane Southern Tutchone Glossary (English to Southern Tutchone).* Occasional Papers of the Northern Research Institute Monograph 1. Whitehorse: Northern Research Institute.

Wallace, Douglas C. 1994. "Mitochondrial DNA Sequence Variation in Human Evolution and Disease." *Proceedings of the National Academy of Sciences of the United States of America* 91:8739–46.

Chapter 34

British Columbia government. 1996. Tatshenshini-Alsek Park Management Agreement. http://www.elp.gov .bc.ca/bcparks/planning/mgmtplns/tatshenshini/ appendices.pdf.

British Columbia government. 1999. Found Human Remains policy of the Ministry of Forests, Lands and Natural Resource Operations. https://www.for.gov.bc.ca/ archaeology/policies/found_human_remains.htm.

Canada, government of Yukon and Champagne and Aishihik First Nations, 1992. *Champagne and Aishihik First Nations Final Agreement.* https://www.aadnc-aandc.gc.ca/eng/12 94331836730/1294331953744.

Canada, government of Yukon and Champagne and Aishihik First Nations, 1993. *Champagne and Aishihik First Nations Self-Government Agreement.* http://www.aadnc-aandc. gc.ca/eng/1100100030683.

Champagne and Aishihik First Nations. 1999. "Kwaday Dän Sinchì." News release, August 24. Haines Junction, YT.

Cruikshank, Julie. 2007. "Melting Glaciers and Emerging Histories in the Saint Elias Mountains." In *Indigenous Experience Today*, edited by Marison de la Cadena and Orin Starn, 355–78. Oxford: Werner Grenn Foundation for Anthropological Research.

Farnell, R., G. Hare, E. Blake, V. Bowyer, C. Schweger, S. Greer and R. Gotthardt. 2004. "Multidisciplinary Investigations of Alpine Ice Patches in Southwest Yukon, Canada." *Arctic* 57 (3): 247–59.

Hare, Greg, with contributions from Sheila Greer (Champagne and Aishihik First Nations), Heather Jones (Carcross-Tagish First Nation), Rae Mombourquette (Kwanlin Dün First Nation), John Fingland (Kluane First Nation), Mark Nelson and Jason Shorty (Ta'an Kwäch'än Council) and Tip Evans (Teslin Tlingit Council). 2011. *The Frozen Past. The Yukon Ice Patches.* Whitehorse: Yukon Heritage Branch. http://www.tc.gov.yk.ca/publications/The_ Frozen_Past_the_Yukon_Ice_Patches_2011.pdf.

Hare, P. Gregory, S. Greer, R. Gotthardt, R. Farnell, V. Bowyer, C. Schweger and Diane Strand. 2004. "Ethnographic and Archaeological Investigations of Alpine Ice Patches, Southwest Yukon, Canada." *Arctic* 57 (3): 260–72.

Kuzyk, Gerald W., Donald E. Russell, Richard E. Farnell, Ruth M. Gotthardt, P. Gregory Hare and Erik Blake. 1999. "In Pursuit of Prehistoric Caribou on Thandlat, Southern Yukon." *Arctic* 52 (2): 214–19.

Joe, Lawrence. 2008. Speech presented at Kwäday Dän Ts'ìnchį public lecture, University of Victoria, Victoria, BC, April 27, 2008.

Moss, Madonna. 2011. *Northwest Coast: Archaeology as Deep History.* Washington, DC: Society for American Archaeology.

Schnarch, Brian. n.d. *Ownership, Control, Access, and Possession (OCAP) or Self-Determination Applied to Research: A Critical Analysis of Contemporary First Nations Research and Some Options for First Nations Communities. Journal of Aboriginal Health, 1-1:80-95.* Accessed August 31, 2017 at http://www.naho.ca/jah/ english/jah01_01/journal_p80-95.pdf.

Smith, Stephen. 1999. "Archaeologists, Natives Open 'Incredible Windows'." *National Post*, August 30.

Thomas, David Hurst. 2000. *Skull Wars: Kennewick Man, Archaeology and the Battle for Native American Identity*. New York: Basic Books.

———. 2006. "Finders Keepers in Deep American History: Some Lessons in Dispute Resolution." In *Imperialism, Art and Restitution*, edited by John Henry Merryman, 218–53. Cambridge: Cambridge University Press.

Watkins, Joe. 2000. *Indigenous Archaeology: American Indian Values and Scientific Practice*. Oxford: Rowman and Littlefield.

Chapter 35

Allen, Bessie Äshènjà. 2006. *Family and Traditional Southern Tutchone Stories*. Recorded and transcribed by Lorraine Allen. Whitehorse: Yukon Native Language Centre.

Clark, Donald W. 1981. "Prehistory of the Western Subarctic." In *Handbook of North American Indians*, vol. 6, *Subarctic*, edited by June Helm, 107–29. Washington, DC: Smithsonian Institution.

de Laguna, Frederica. 1972. *Under Mount Saint Elias: The History and Culture of the Yakutat Tlingit*. Smithsonian Contributions to Anthropology 7. Washington, DC: Smithsonian Institution.

Emmons, George T. 1991. *The Tlingit Indians*, edited by Frederica de Laguna. New York: American Museum of Natural History.

Forsyth, Adrian. 1985. *Mammals of the Canadian Wild*. Camden East, ON: Camden House.

Garibaldi, Ann, and Nancy Turner. 2004. "Cultural Keystone Species: Implications for Conservation and Restoration." *Ecology and Society* 9 (3): 1–18. http://www.ecologyandsociety.org/vol9/iss3/art1.

Glave, Edward J. 1890–91. "Our Alaska Expedition: Exploration of the Unknown Alseck River Region." *Frank Leslie's Illustrated Newspaper*, September 6: 86–87; November 15, 1890: 262; November 22: 286–87; November 29: 310; December 6: 332; December 13: 352; December 20: 376; December 27: 396–97; January 3, 1891: 414; January 10: 438.

———. 1892. "Pioneer Packhorses in Alaska. I–The Advance. II–The Return." *Century Magazine* 44 (5): 671–82.

Greer, Sheila. 2008a. "Sheila Greer Notes Gopher Robe/Blanket Project File." On file, Champagne and Aishihik First Nations Heritage program, Haines Junction, YT.

———. 2008b. "Southern Tutchone Vocabulary Work with Marge Jackson, Frances Joe and Hayden Woodruff; Spellings by Vivian Smith." Winter 2008. Manuscript on file, Champagne and Aishihik First Nations, Haines Junction, YT.

———. 2011. "Sheila Greer Notes with Marge Jackson and Other Elders, Including Information from Sadie Brown via Paula Banks." On file, Champagne and Aishihik First Nations, Haines Junction, YT.

Helwig, Kate, Jennifer Poulin and Valery Monahan. 2004. "The Characterization of Paint and Adhesive Residues on Hunting Tools from Southern Yukon Ice Patches." Paper presented at the Alaska Anthropological Association Annual Meeting, Whitehorse.

Jackson, Marge, with assistance from Beth O'Leary. 2006. *My Country Is Alive: A Southern Tutchone Way of Life*. Published by author, Haines Junction, Yukon.

Johnson, Frederick, and Hugh M. Raup. 1964. *Investigations in Southwest Yukon: Geobotanical and Archaeological Reconnaissance*. Papers of the Robert S. Peabody Foundation for Archaeology 6 (1): 11–97. Andover, Massachusetts.

Le Blanc, Raymond L. 1983. *The Rat Indian Creek Site and the Late Prehistoric Period in the Interior Northern Yukon*. National Museum of Man Mercury Series, Archaeological Survey of Canada Paper 120. Ottawa: National Museums of Canada.

McClellan, Catharine. 1975. *My Old People Say: An Ethnographic Survey of Southern Yukon Territory*. Publications in Ethnology 6. Ottawa: National Museums of Canada. Republished as Mercury Series, Canadian Ethnology Service Paper 137, Canadian Museum of Civilization, 2001.

McClellan, Catharine, Lucie Birckel, Robert Bringhurst, James A. Fall, Carol McCarthy and Janice R. Sheppard. 1987. *Part of the Land, Part of the Water: A History of the Yukon Indians*. Vancouver: Douglas and McIntyre.

McClellan, Catharine, and Glenda Denniston. 1981. "Environment and Culture in the Cordillera." In *Handbook of North American Indians*, vol. 6, *Subarctic*, edited by June Helm, 372–86. Washington, DC: Smithsonian Institution.

McFadyen- Clark, Annette. 1974. "The Athapaskans: Strangers of the North." In *The Athapaskans: Strangers of the North: An International Travelling Exhibition from the Collection of the National Museum of Man, Canada and the Royal Scottish Museum*. Ottawa: National Museums of Canada.

Sidney, Angela, Kitty Smith and Rachel Dawson. 1979. *My Stories Are My Wealth*. With Julie Cruikshank. Whitehorse: Council for Yukon Indians.

Thompson, Judy. 2007. *Recording Their Story: James Teit and the Tahltan*. Vancouver: Douglas and McIntyre.

Thompson, J., and I. Kritsch. 2005. *Yeenoo Dai'K'è'tr'ijilkai' Ganagwaandaii, Long Ago Sewing We Will Remember*. Mercury Series, Canadian Ethnology Service Paper 143. Gatineau, QC: Canadian Museum of Civilization.

Thompson, Judy. 2007. *Recording Their Story: James Teit and the Tahltan*. Vancouver: Douglas & McIntyre.

Tlen, Daniel. 1993. *Kluane Southern Tutchone Glossary (English to Southern Tutchone)*. Occasional Papers of the Northern Research Institute, Monograph 1. Whitehorse: Northern Research Institute.

Tom, Gertie. 1981. *How to Tan Hides in the Native Way*. Whitehorse: Yukon Native Language Centre.

Wein, Eleanor E. 1994. *Yukon First Nations Food and Nutrition Study*. Report to the Champagne and Aishihik First Nations, the Teslin Tlingit Council, the Vuntut Gwich'in First Nation, the Yukon Department of Health and the National Institute of Nutrition. Edmonton: Canadian Circumpolar Institute, University of Alberta.

Wein, Eleanor E., and M.R. Freeman. 1995. "Frequency of Traditional Food Use by Three Yukon First Nations." *Arctic* 48 (2): 161–71.

Workman, William B. 1978. *Prehistory of the Aishihik-Kluane Area, Southwest Yukon Territory*. National Museum of Man Mercury Series 74. Ottawa: Archaeological Survey of Canada.

Wood, C.E.S. 1882. "Among the Thlinkets in Alaska." *Century Magazine* 24 (3): 323–39.

Wyatt, Victoria. 1989. *Images from the Inside Passage: An Alaskan Portrait by Winter and Pond*. Seattle: University of Washington Press.

Chapter 36

Busby, Shannon. 2003. *Spruce Root Basketry of the Haida and Tlingit*. Seattle: University of Washington Press.

Dangel, H.D. 2005. *A Celebration of Weavers: Catalog of Weavers and Baskets of the Doris Borhauer Collection, Sitka, Alaska*. Sitka, AK: Donning Company Publishers.

Dauenhauer, Nora. 1982. "Basketry Terms in New Orthography, with Notes." Appendix to *Spruce Root Basketry of the Alaska Tlingit*, by Frances Paul. Reprinted. Sitka, AK: Sheldon Jackson Museum.

———. 2000. "Tlingit At.óow." In *Celebration 2000, Restoring Balance Through Culture*, edited by Susan Fair and Rosita Worl, 101–06. Juneau, AK: Sealaska Heritage Foundation.

Emmons, George T. 1991. *The Tlingit Indians*. Edited by Frederica de Laguna. New York: American Museum of Natural History.

Gunther, Erna. 1990. *Design Units on Tlingit Basket*. Sitka, Alaska: Sheldon Jackson Museum.

McClellan, Catharine. 1975. *My Old People Say: An Ethnographic Survey of Southern Yukon Territory*. Publications in Ethnology 6. Ottawa: National Museums of Canada. Republished in 2001 as Mercury Series, Canadian Ethnology Service Paper 137. Ottawa: Canadian Museum of Civilization.

Paul, Frances. (1944) 1982. *Spruce Root Basketry of the Alaska Tlingit*. Education Division, US Indian Service, Department of the Interior. Reprinted Sitka, AK: Sheldon Jackson Museum.

Sheldon Museum. n.d. *The Spruce Root Basket*. Haines, AK: Sheldon Museum and Cultural Center.

Shotridge, Louis. 1921. "The Tlingit Woman's Root Basket." *University of Pennsylvania Museum Journal* 12 (3): 162–6. Reprinted 1984 by Sheldon Jackson Museum.

Weber, Ronald L., and G.T. Emmons. 1986. "Emmons's Notes on Field Museum's Collection of Northwest Coast Basketry: Edited with an Ethnoarchaeological Analysis." *Fieldiana*, n.s. 9. Chicago: Field Museum of Natural History.

Contributors

EDITORS

Richard J. Hebda, PhD, is a botanist who studies the vegetation and climate history of British Columbia, the ethnobotany of First Nations in BC, climate change and its impacts, ecology and origins of Garry Oak and alpine ecosystems, and the botany of grasses. He has been a curator at the Royal British Columbia Museum for more than 37 years and an adjunct faculty member at the University of Victoria for more than 33 years. Richard has served as BC's expert advisor on Burns Bog and as science advisor in paleontology. In 2013 he received the Queen's Diamond Jubilee medal for his services to paleontology and in 2015 the Canada-wide Bruce Naylor Award for curatorship in natural history.

Sheila Greer, Nthe aghajêl, is an adopted member of the Käjèt (Crow) clan. Her degrees in anthropology and archaeology lead to a consulting career, with CAFN and other Yukon First Nations being her principal clients. More recently, she became CAFN staff. In the role of Heritage Manager, Sheila assists the First Nation in fulfilling its responsibilities as steward of and voice for the many manifestations of its heritage and history.

Alexander P. Mackie has worked as an archaeologist on the west coast for 40 years. He spent 20 years with the BC Archaeology Branch, including 14 years as a member of the Kwäday Dän Ts'ìnchį Management Group with responsibility for liaison between the government of BC, the Champagne and Aishihik First Nations and the scientific research team. In 2013 Al returned to the private sector as a consultant and researcher.

AUTHORS

John Adamson Champagne and Aishihik First Nations

Gail S. Anderson Centre for Forensic Research, School of Criminology, Simon Fraser University

Owen Beattie Department of Anthropology, University of Alberta

Kathryn Bernick Research Associate, Royal BC Museum

Erik W. Blake Icefield Instruments Inc.; Yukon College, Yukon Cold Climate Innovation Centre

Ron Chambers Champagne and Aishihik First Nations

Wilfred Charlie Little Salmon-Carmacks First Nation

Claudia Cheung Department of Pathology and Laboratory Medicine, University of British Columbia

Sheila Clark Teslin Tlinglit Council

H. Kory Cooper Department of Anthropology, Purdue University

Brian Coppins Royal Botanic Garden, Edinburgh, UK

Jim Cosgrove Royal BC Museum

L. T. Corr Organic Geochemistry Unit, Biogeochemistry Research Centre, School of Chemistry, University of Bristol

Harry Deneer Department of Pathology and Laboratory Medicine, University of Saskatchewan

James Dickson School of Biology, University of Glasgow

R.P. Evershed Organic Geochemistry Unit, Biogeochemistry Research Centre, School of Chemistry, University of Bristol

John Fingland Champagne and Aishihik First Nations

Al von Finster Fisheries and Oceans Canada

Sarah Gaunt Champagne and Aishihik First Nations

Tara Grant Canadian Conservation Institute

Monique Haakensen Contango Strategies Ltd.

William H. Hanlon

Roxanne Hastings Royal Alberta Museum

Paul Hazelton Department of Medical Microbiology, College of Medicine, University of Manitoba

Kate Helwig Canadian Conservation Institute

Niki Hobischak Centre for Forensic Research, School of Criminology, Simon Fraser University

Grant Hughes Director of Curatorial Services, Royal BC Museum

Elaine Humphrey Bio-imaging Facility and Microscopy, University of British Columbia

Moose Jackson Champagne and Aishihik First Nations

Lawrence Joe Champagne and Aishihik First Nations

Sheila Joe Champagne and Aishihik First Nations

Ivan M. Kempson Future Industries Institute, University of South Australia

Bruce J. Leighton Department of Biological Sciences, Simon Fraser University

Kjerstin Mackie Textiles Conservator, Royal BC Museum

Kendrick L. Marr Curator of Botany, Royal BC Museum

Ronald R. Martin University of Western Ontario, Department of Chemistry

Darcy Mathews Department of Anthropology, University of Victoria

Karen Mooder Vancouver

Valery Monahan Yukon Department of Tourism and Culture

Maria Victoria Monsalve Department of Pathology and Laboratory Medicine, University of British Columbia

Petra J. Mudie Geological Survey Canada (Atlantic)

Mike Nimmo Department of Pathology and Laboratory Medicine, University of British Columbia

Frances Oles Champagne and Aishihik First Nations

Nicholas Panter Royal BC Museum

Michael Petrik Centre for Forensic Research, School of Criminology, Simon Fraser University; Salus University

Jennifer Poulin Canadian Conservation Institute

Michael P. Richards Department of Archaeology, Simon Fraser University; Department of Human Evolution, Max Planck Institute for Evolutionary Anthropology

Kelly Sendall Royal BC Museum

Jane Sirois Canadian Conservation Institute

Jimmy G. Smith Champagne and Aishihik First Nations

J. Southon Earth System Science Department, University of California Irvine

Camilla F. Speller BioArCh, Department of Archaeology, University of York, UK; Ancient DNA Laboratory, Department of Archaeology, Simon Fraser University

Dan Straathof Department of Laboratory Medicine and Pathology, Royal Columbian Hospital

Diane Strand Champagne and Aishihik First Nations

Treena Swanston Department of Archaeology and Anthropology, University of Saskatchewan

Kevin Telmer School of Earth and Ocean Sciences, University of Victoria

Frank Thomas Natural Resources Canada; Geological Survey Canada (Atlantic)

Peter M. Troffe InStream Fisheries Research Inc.

Wayne Vogl Department of Cellular and Physiological Sciences, University of British Columbia

David C. Walker Department of Pathology and Laboratory Medicine, University of British Columbia

Ernest Walker Department of Archaeology and Anthropology, University of Saskatchewan

Michael Wayman University of Alberta

John M. Webster Department of Biological Sciences, Simon Fraser University

Dongya Y. Yang Ancient DNA Laboratory, Department of Archaeology, Simon Fraser University

Gregory Young Canadian Conservation Institute

Jacksy Zhao Pathology and Laboratory Medicine, University of British Columbia

Index

Captions and tables indicated by page numbers in italics